Evolution

Theodosius Dobzhansky (1900–1975)

Evolution

THEODOSIUS DOBZHANSKY
FRANCISCO J. AYALA
G. LEDYARD STEBBINS
JAMES W. VALENTINE

University of California, Davis

W. H. FREEMAN AND COMPANY
San Francisco

Library of Congress Cataloging in Publication Data

Main entry under title:

Evolution.

Bibliography: p.
Includes index.
1. Evolution. I. Dobzhansky, Theodosius
Grigorievich, 1900–1975.
QH366.2.E853 575 77–23284
ISBN 0–7167–0572–9

Printed in the United States of America

9 8 7 6 5 4 3 2 1

Nothing in biology makes sense except in the light of evolution.

Th. Dobzhansky, 1973

Contents

Preface xiii

1 THE NATURE OF EVOLUTION 1
The Complexity of Interactions Between Population and Environment 2
The Reaction of Organisms to Environmental Complexity 3
Natural Selection: The Link Between Environmental Change and Organic Evolution 4
Subspecific and Transspecific Aspects of Evolution 5
The Role of Chance in Evolution 6
Definitions of Evolution 7
A Brief History of Evolutionary Theory 9

2 THE GENETIC STRUCTURE OF POPULATIONS 20
The Hereditary Materials 20
The Genetic Code 24
Genotype and Phenotype 28
Populations and Gene Pools 30
Genetic Variation and Evolution 31
Two Models of Population Structure 34
Allelic Variation at Individual Gene Loci 36
Genetic Variation and Artificial Selection 38
Concealed Genetic Variation Affecting Fitness 39
Molecular Techniques to Quantify Genetic Variation 46
Quantification of Genetic Variation in Populations 52

3 THE ORIGIN OF HEREDITARY VARIATION 57
Classification of Mutations 57
Historical Overview 58
Gene Mutations 59
Effects of Mutation 62
Mutation and Adaptation 64
Rates of Mutation 68
Rates of Mutation and Rates of Evolution 71
Evolution of Genome Size 72
Chromosomal Deletions and Duplications 75
Evolution of Duplicated Genes 78
Genes in Multiple Copies 83
Highly Repetitive DNA 85
Inversions and Translocations 87
Chromosomal Fusions and Fissions 92

4 NATURAL SELECTION 95
The Idea of Natural Selection 96
The Gene Pool and the Hardy–Weinberg Equilibrium 99
Darwinian Fitness or Selective Value 100
Normalizing Natural Selection 102
Genetic Burdens or Loads 105
Polymorphism and Balancing Natural Selection 107
Diversifying Natural Selection 116
Sexual Selection 118
Directional Selection 120
Natural Selection and Adaptedness 124
Group and Kin Selection 125

5 POPULATIONS, RACES, SUBSPECIES 128
Clones, Pure Lines, Populations 130
Microgeographic Races 132
Geographic Races or Subspecies of Animals and Plants 134
Human Racial Variation 138
Race Classification 144
Race Differences and Natural Selection 146
Adaptively Neutral Traits 153
Random Genetic Drift 157
Adaptively Neutral Race Differences 160
Haldane's Dilemma 163

6 SPECIES AND THEIR ORIGINS 165
Why Should There Be Species? 166
Discontinuity of Organic Variation 168
Reproductive Isolation 170

Examples of Prezygotic Isolating Mechanisms 171
Hybrid Inviability, Sterility, and Breakdown 177
Evolutionary Origin of Reproductive Isolation 179
Sibling Species and Semispecies 182
Speciation on Oceanic Islands 186
Interspecific Hybrids and Chromosome Comparisons 188
Species Differences in Allozymes 192

7 PATTERNS OF SPECIATION 195
Ecogeographic Factors Affecting the Origin of Species 196
Differences Between Speciation in Animals and Plants 202
Pollination Biology and Speciation in Plants 204
Postzygotic Isolating Mechanisms in Animals and Plants 210
Hybridization as an Evolutionary Catalyst 219
The Occurrence of Semispecies in Plants 221
The Evolutionary Significance of Polyploidy 224
The Relationship Between Polyploidy and Apomixis 229
The Species Problem in Primitively Asexual Organisms 230

8 TRANSSPECIFIC EVOLUTION 233
The Taxonomic Hierarchy 233
Nonadaptive Models 242
Adaptive Models 245
Genetic Patterns 254

9 PHYLOGENIES AND MACROMOLECULES 262
Homology and Analogy 263
Evaluation of Homologies 266
Chromosome Phylogenies 271
Methods of "Hybridizing" DNA 276
DNA Phylogenies 281
Electrophoretic Measures of Genetic Differentiation 284
Electrophoretic Phylogenies 285
Immunological Techniques 290
Amino Acid Sequences of Proteins 293
Protein Phylogenies 299
The Neutrality Theory of Protein Evolution 303
The Molecular Clock of Evolution 308

10 THE GEOLOGICAL RECORD 314
The Geological Time Scale 314
The Record of Ancient Environments 318
The Record of Ancient Life 323
The Fossil Record of Evolutionary Rates 327
Regulation of Diversity 337

Patterns of Extinction 342
Patterns of Diversification 346

11 COSMIC EVOLUTION AND THE ORIGIN OF LIFE 349
The Primitive Earth 351
Prebiotic Organic Compounds 354
Concentration of Prebiotic Compounds 356
The Origin of Natural Selection 358
The Nature of Protocells 360
Extraterrestrial Life 364

12 EVOLUTION OF PROKARYOTES AND UNICELLULAR EUKARYOTES 369
The Kingdoms of Organisms 369
Adaptive Radiation of Prokaryotes 377
Evolution in Viruses 380
The Origin of Eukaryotes 383
The Origin of Mitosis and Sex 389
The Origins of Multicellular Organisms 394

13 THE EVOLUTIONARY HISTORY OF METAZOA 397
Metazoan Grades and Major Body Plans 397
Evolution of the Tissue Grade 399
Evolution of the Triploblastic Grades 406
Evolution of the Coelom 408
Evolution of Early Metazoan Radiations 413
Evolution of the Chordates 417
Phanerozoic Marine Diversification and Extinction 421
The Terrestrial Invasions 429
Reptiles, Dinosaurs, and Mammals 432
Primates 436

14 EVOLUTION OF MANKIND 438
Structural Similarities and Differences 440
Chemical Similarities and Differences 443
Man's Early Ancestors and Relatives 445
Culture—The Human Domain 450
Mind, Self-Awareness, Death-Awareness 453
Ethics and Values 454
Does Mankind Continue to Evolve Genetically? 457
Euphenics and Eugenics 459

15 THE FUTURE OF EVOLUTION 464
The Evolutionary Future of Mankind 466
Artificial Control of Organic Evolution 468

The Guidance of Cultural Evolution 470
Toward a Cultural and Ecological Equilibrium 472

16 PHILOSOPHICAL ISSUES 474
Empirical Science 475
The Scientific Method 476
The Criterion of Demarcation 479
Mendel and the Scientific Method 482
Darwin and the Scientific Method 484
Mechanism and Vitalism 487
Reductionism versus Compositionism 490
The Reduction of Theories 491
Darwin's Conceptual Revolution 495
Teleological Explanations 497
Natural and Artificial Teleology 500
The Theory of Natural Selection 504
The Concept of Progress 507
The Expansion of Life 511
Particular Forms of Evolutionary Progress 513

Literature Cited 517
Index 545

Preface

The modern theory of evolution emerged in the 1930s and 1940s as a synthesis of genetic knowledge and the Darwinian concept of natural selection. Increasingly this theory of evolution has pervaded all biological disciplines and has contributed to their development and enrichment. The theory of evolution is, indeed, the most encompassing biological theory. At the same time, however, the modern theory of evolution has been expanded by the contributions from other biological disciplines, such as zoology, botany, anthropology, and paleontology; physiology, microbiology, and biochemistry; experimental and mathematical population biology; ecology and systematics; genetics and developmental biology. The modern paradigm of evolution is often called the synthetic theory precisely because it integrates the contributions of so many fields of knowledge.

The last 25 years have witnessed the revolutionary impact of molecular biology upon genetics and developmental biology, two fields of fundamental importance for evolutionary studies. As a result, evolutionists have acquired powerful new tools and concepts for investigating evolutionary processes. Even though molecular biology is still in an early stage of growth, many unexpected and revealing discoveries have already emerged. Three examples may be cited: the surprisingly large amounts of genetic variability harbored in living populations; the mechanisms of gene regulation; and the techniques to obtain quantitatively precise measures of rates of genetic evolution. Further molecular studies will certainly modify the paradigm of evolution as we now understand it, although the extent of change remains to be determined.

The synthetic theory of evolution continues to be expanded, modified, and enriched by studies in other biological disciplines besides molecular biology. Of special recent significance are the increased understanding of the ecological dimension of population biology, which has demonstrated ever more clearly the nature of the selective forces directing evolutionary change, and the revitalization of the earth sciences, which has led to new paleobiological models for understanding evolutionary processes stretching over vast spans of geological time.

The constant expansion of the evolutionary paradigm has reached the point that no one author can feel competent to deal authoritatively with all its aspects. For this reason this book was written by four authors, each with markedly different but complementary interests and expertise. Although this has broadened the range of topics that could be expertly treated, we have by no means exhausted the gamut of evolutionary issues. Perhaps this is the last time that a reasonably comprehensive account of the theory of evolution can be encompassed in a single volume.

While our collaboration has increased the scope of the book, it has also produced in one book four points of view that do not always coincide on unsettled issues. We regard this diversity of viewpoints as a strength—it signals where additional research is needed and suggests alternative approaches to the field in general. Each author wrote four chapters, for which he alone is ultimately responsible, although the scope and content of each chapter emerged from consultation by the authors with one another. Th. Dobzhansky has written Chapters 4, 5, 6, and 14; F. J. Ayala Chapters 2, 3, 9, and 16; G. L. Stebbins Chapters 1, 7, 12, and 15; and J. W. Valentine Chapters 8, 10, 11, and 13. Professor Dobzhansky died while the book was in preparation. The other three authors dedicate this book to him in tribute to his preeminence and in friendship. We have therefore listed him as the first author, while the other three authors are listed simply in alphabetical order.

We have written this volume to serve as a textbook in evolution courses, particularly for advanced undergraduates and for graduate students. The book is also intended for a more general reader who wants to know about the spectacular advances that have taken place in evolutionary studies during the last few years. Finally, we hope that this book will be useful to our colleagues, biologists as well as other scientists, as a summary of the current theory of evolution.

Francisco J. Ayala
G. Ledyard Stebbins
James W. Valentine

Davis, California
December 1976

Evolution

1

The Nature of Evolution

The concept of evolution, which is now basic to the life sciences, has provided new and in some ways revolutionary answers to questions men have been asking for centuries. The two most important of these are: "Why am I here, what is the purpose of human existence?", and "What is the nature of the world of life that surrounds us?" Evolution tells us that we are here because of a long series of past events that do not differ in kind from the events and processes that produced the millions of different organisms that surround us. The most important processes are: (1) interactions between organisms and their environment that are highly diverse both historically and geographically; (2) the continuity of heredity and cultural tradition; and (3) the occasional disturbance of these regularities by chance.

The effects of this revolution in thinking are just beginning to be realized. If man has arrived at his present state as a result of natural processes rather than a supernatural will, he can learn to control these processes. He can improve his condition by analyses of natural phenomena, syntheses of ideas, and action to realize well-defined objectives. The limitations we must overcome are chiefly our own difficulty in understanding the vast complexity of the world that surrounds us, as well as the weight of tradition, the conservatism of cultural heredity, and the egotism of individuals which restricts our ability to act on the basis of principles that underlie natural processes. In order to solve our present problems, therefore, we must first learn as much as we can about the nature and origin of living organisms, including our own complex societies; and second, we must devise sound and reliable methods of translating this knowledge into constructive action. The purpose of this book is to present a summary of our current knowledge about the origin and evolution of the world of life.

THE COMPLEXITY OF INTERACTIONS BETWEEN POPULATION AND ENVIRONMENT

In a world that may contain four to five million different kinds of organisms (Dobzhansky, 1970), exploiting in various ways a large number of habitats, the variety of possible interactions between populations and their environments is enormous. Equally vast is the scope of these interactions. At the bottom of the ocean fish depend for their existence on organisms that fall to these depths after having completed their life on the surface of the water, 6,000 meters above. On the slopes of Mount Everest, 6,500 meters above sea level, a few seed plants (*Delphinium*) grow in the shelter of lichen-covered rocks, providing food for a few insects, which in turn support a small number of spiders. In deserts such as those along the coast of Peru and Chile and the Sahara of Africa, plant seeds and highly resistant eggs of small animals can endure rainless periods lasting for many years, and then suddenly, when the rains finally arrive, produce a flush of growth and activity (Walter, 1971).

The complexity of ecosystems is based upon thousands of different ways of exploiting the same habitat. Consider, for instance, a forest of redwood or *Sequoia* in northwestern California. Organisms there range in size from trees 110 meters tall to bacteria only 0.000001 meters long. The lush undergrowth supports (or did before human interference) many kinds of animals that feed upon each other according to several food chains. One chain starts with shrubs that are browsed by deer or elk, which in turn are the prey of pumas. Another starts with the sap exuded from plants, which supports yeasts, which in turn are eaten by flies such as *Drosophila;* these provide food for bats or passerine birds, which in turn are the prey of hawks. Still another food chain starts with decaying leaves and their associated microorganisms, on which subterranean insects and other arthropods subsist; these form the prey of amphibians, which are preyed upon by snakes, which often become the food of owls. In tropical rain forests food chains are far more complex. Complex food chains exist in other rich biota, such as those of tide pools along the seashore, the plankton-rich surface of the ocean, and even habitats within individual organisms, such as the interior of the bovine rumen (Hungate, 1966).

This hierarchy of population–environment interactions is by no means constant, but has varied greatly over both short and long periods of time. In temperate woodlands and fields the seasonal succession of biotas is one of the most obvious facts of life. Larger, long-lived organisms react to the climatic cycle either by altering their phenotypes, becoming dormant during unfavorable seasons, or by regular migrations. On the other hand, populations of some short-lived species having rapid reproductive cycles, such as rodents (Gershenson, 1945) and flies (Dobzhansky, 1951), alter their genetic composition in response to these changes. During recorded human history greater and more permanent changes of the environment have taken place, accompanied by corresponding changes in populations. Examples are well documented of alterations in the genetic composition of populations in response to these changes, the best known of these being industrial melanism (Ford, 1971). During the entire period of the earth's history since the major phyla of organisms appeared, all of the regions of the earth have undergone complete alterations of their biota, and many of them have experienced several "ecological revolutions."

THE REACTION OF ORGANISMS TO ENVIRONMENTAL COMPLEXITY

The way in which populations of organisms have reacted and are reacting to environmental changes depends upon the genetic characteristics of individuals as well as the amount and kind of genetic variability in the population. With respect to individuals, the most important characteristics are the continuity of heredity, which depends upon the self-replication of chromosomal DNA; the capacity for change by mutation; and the harmonious integration of the genotype. With respect to populations, the most important characteristics are their unlimited capacity for increase, which requires the destruction of individuals in order to keep population numbers constant; their great store of genetic variability, which is discussed in later chapters; and the limitations of this variability in any particular population. The latter characteristic determines whether, in response to a particular environmental change, a population will adapt to the new environment or become extinct. As Simpson (1953) has pointed out, species become extinct not because some unexplained "evolutionary urge" has forced nonadaptive phenotypes upon them, or because they have lost essential genetic variability. If adaptive change or evolution takes place, the method of adaptation will depend on the kind of genetic variation available.

Given a new ecological niche to which it can become adapted, a population can achieve this adaptation in any one of several different ways. Although in many instances the particular adaptation that is adopted may depend entirely upon the chance appearance of certain favorable mutations or gene combinations, generally the direction taken depends chiefly upon preexisting capacities already present in the gene pools. If, for instance, a plant community becomes exposed to increasing aridity, those populations that already have somewhat succulent leaves or stems are likely to become more succulent; those whose leaf surfaces are protected from evaporation by thick, hard cuticles may evolve thicker, more complex cuticles; and those that already evade dry spells by losing their leaves and becoming dormant may evolve longer periods of dormancy and more rapid growth during the favorable season. If a community of small animals becomes exposed to a more efficient predator, those that have escaped previous predators by speed and agility may become faster and more agile; those that escape by burrowing in the ground may become more strongly fossorial; and those that because of a bad odor or other noxious properties are avoided by many predators may become even more unpleasant and noxious.

These divergent adaptations to the same environmental change can be characterized by a principle first expressed by the botanist W. F. Ganong (1901): adaptive modification along the lines of least resistance. This can sometimes lead to unexpected changes. For instance, the ability to digest cellulose has been acquired both in insects, such as cockroaches and termites (Moore, 1969), and ruminant mammals (Hungate, 1966) not by the evolution of new enzymes via gene mutation, but by acquisition of an internal biota of lignin-digesting microorganisms.

NATURAL SELECTION: THE LINK BETWEEN ENVIRONMENTAL CHANGE AND ORGANIC EVOLUTION

As Darwin first recognized, the link between variability or constancy of the environment and evolutionary change or stability is provided by natural selection (more recent evidence for this principle is presented later in this book). This process is the inevitable outcome of seven characteristics of populations: genetic variability, genetic recombination, hereditary continuity, the capacity for mutation, excess reproductive capacity, the integration of the genotype, and the limitations of the gene pool of a population. Given genetic variability and cross fertilization, the potential number of genetically different kinds of zygotes is enormous; a new generation can possess only a sample of these possible combinations. The interaction between organisms and their environment, which may or may not involve a "struggle for existence" between different organisms, will inevitably affect the genetic composition of this sample. The most fit individuals, in terms of reproductive capacity, contribute the largest proportion of genes or alleles to the next generation. The direction in which natural selection guides the population depends upon both the nature of environmental change or stability and the content of the gene pool of the population.

The diversity of population–environment interactions is responsible for the fact that natural selection can either promote constancy (normalizing selection), direct continuous change (directional selection), or promote diversification (diversifying selection), depending upon whether environments change and if so, upon the nature of the changes. (These three kinds of natural selection are discussed in Chapter 4.) If population–environment interactions remain constant through time, normalizing selection prevails, and evolutionary change is arrested. If a particular sequence of population–environment interactions changes constantly in one direction, the result is directional selection. This result is most likely to take place when specific interactions have been established between two different kinds of organisms, such as predator–prey, grazing animal–forage, or pollinator–flower. Thus, as Darwin pointed out (1872), if wolves prey chiefly upon rabbits or hares, the fleetest of these will most often escape, and so leave the largest number of progeny, while the fleetest wolves are likely to be the best nourished, and hence the most fecund. Continuous, directed evolution is produced by repeated feedback interactions between predator and prey. Similarly, during the evolution of horses on the plains of North America, those grasses that evolved the hardest leaves were the least likely to be damaged by grazing, while the horses having the hardest teeth (with the most complex patterns of enamel) were best equipped to feed on the largest number of grasses. Such instances of co-evolution stimulated by feedback interactions account for a large proportion of the continuous evolutionary trends paleontologists have observed in vertebrates (Simpson, 1953). They are likely to continue until the physiological limits to selection have been reached.

If a previously homogeneous habitat becomes diversified, the interactions between adjoining populations and their respective environments can diverge from each other, and the process of adaptive radiation can begin. This divergence would not become permanent, however, unless reproductive barriers, discussed in Chapters 6 and 7, emerged to keep the diverging lines genetically separate.

SUBSPECIFIC AND TRANSSPECIFIC ASPECTS OF EVOLUTION

So far we have summarized the *synthetic theory of organic evolution,* which will be elaborated in the next six chapters. The reader should already realize from this summary that three main processes—mutation, genetic recombination, and natural selection—are universal and inevitable consequences of the nature of individual organisms, of the genetic structure of populations, and of the diversity of population–environment interactions. Moreover, these processes are essential for evolutionary change and for divergence.

Few biologists now doubt that these three processes, acting concurrently, can be responsible for the origin of diverse races and species. In evolutionary language, evolution at the subspecific level by means of recognized processes is now regarded by most biologists as an experimentally demonstrated fact. On the other hand, some doubt the ability of these processes, by themselves, to give rise to genera, families, or any groups of organisms having "new" characteristics. "Transspecific evolution" is generally recognized as a reality, but its explanation on the basis of recognized processes is still in the stage of partly demonstrated theory or, in the case of the origin and early differentiation of the first cellular organisms, a working hypothesis.

One can easily understand why this should be so. As is amply documented in this and other books (Dobzhansky, 1970; Mayr, 1963, 1970; Grant, 1971), problems concerning the origin of races and species can be attacked and solved by means of carefully controlled quantitative experiments. The origin of taxa of higher categories, however, necessarily takes place over very long periods of time; these taxa are genetically isolated from each other, so their differences cannot be analyzed by means of controlled hybridizations. Nevertheless, as is explained in Chapter 8, the transition from the subspecific to the transspecific level has in many instances been shown to be gradual. Furthermore, no particular morphological or physiological differences between populations are peculiar to genera, families, or other higher categories. Differences that in some groups appear to be solely the products of evolution at the transspecific level can in other groups be recognized as separating closely related species or even different races of the same species.

Subspecific evolution and transspecific evolution are not different phenomena or processes: they are two aspects of the same series of processes and phenomena. The principal difference between them is the way in which the problems concerning them are attacked. Evolutionists investigate the subspecific aspect of evolution chiefly by studying populations or groups of closely related populations at a single time level, or at most over a small number of generations. They extrapolate to the transspecific level by comparing individuals or populations that have evolved over time spans measured in terms of millions of generations or years. Findings at the subspecific level can be applied to the transspecific level if one grants that the processes that promote evolutionary change in modern populations have been active throughout the millions of years during which organisms have existed. Extrapolation from the present to the past is therefore possible. Like all extrapolation, however, it can succeed only if it is based upon enough information of the right kind. The required information must be concerned not only with population–environment interactions and evolutionary processes in modern organisms, but also with the nature and probable environment of extinct populations, represented only by fossil forms.

THE ROLE OF CHANCE IN EVOLUTION

So far we have emphasized the deterministic factors and processes that affect evolution. Nevertheless, the effects of chance cannot be neglected. It operates at many levels. At the level of the gene pool, mutations produce variations that are random with respect to the environment, and thus cannot by themselves direct evolution toward new adaptations. Furthermore, evidence reviewed in Chapter 3 indicates that mutations having conspicuous effects on the phenotype usually reduce fitness, and tend to be rejected by natural selection. Adaptive mutations are usually those that have such small effects that they are of little consequence, but contribute to adaptiveness in combination with many other genes. The variability that makes possible adaptive adjustments of populations to their environment results from the fact that overall rates of mutations (the sum of those produced by all of the genes an organism possesses) are high. In addition, this initial random variability is increased to an enormous degree by genetic recombination. Each unitary random variation is therefore of little consequence, and may be compared to random movements of molecules within a gas or liquid. Directional movements of air or water can be produced only by forces that act at a much broader level than the movements of individual molecules, e.g., differences in air pressure, which produce wind, or differences in slope, which produce stream currents. In an analogous fashion, the directional force of evolution, natural selection, acts on the basis of conditions existing at the broad level of the environment as it affects populations. It can be effective only to the extent to which individual genes, whether new mutants or preexisting alleles, can combine to produce adaptive combinations capable of responding to environmental challenges.

With respect to population differentiation, the effects of chance have considerable influence when populations are small and there is little or no counteracting pressure of selection. The significance of chance or stochastic events in evolution is discussed in Chapters 5 and 9.

At the transspecific level there are good reasons for believing that chance has played a role in determining which of several possible evolutionary directions a population will take. For instance, the early Tertiary mammal *Hyracotherium* ("Eohippus") may have been the common ancestor of horses, rhinoceroses, and tapirs (Simpson, 1953; Colbert, 1969). Apparently, some time during the period of its existence diversifying selection took place, and caused separate populations of this animal to start evolving in very different directions. Although this divergence may have been triggered entirely by different environmental changes taking place in different parts of the range of the initial species or genus, the possibility cannot be ignored that, by chance, the composition of the gene pool was somewhat different in separate populations, so that in any one of them modification along the lines of least genetic resistance led in the direction of horses, in another toward rhinoceroses, and in a third toward tapirs.

We therefore recognize the importance of chance in evolution. We regard it as one of several factors, one that always produces a certain amount of "evolutionary noise," i.e., random deviation from the norm of evolutionary change, and under certain special conditions can have profound effects upon particular kinds of change.

DEFINITIONS OF EVOLUTION

On the basis of this preliminary summary of the modern synthetic theory of organic evolution, definitions of the concept of evolution as a whole can be formulated. To Darwin, evolution was synonymous with the origin of species. Organic evolution begins with the differentiation of populations within species, and continues above the species level to form genera, families, and other higher categories. Given an adequate fossil record, one can follow transspecific evolution through time by recognizing evolutionary lineages. The multiplication of species depends on the development of reproductive isolation, without which separate evolutionary lines cannot diverge from each other through exploitation of similar habitats in different ways. As Mayr (1942, 1963) has pointed out, the title of Darwin's epoch-making book is mistaken; although he entitled it *Origin of Species*, Darwin wrote about organic evolution as a whole. Similarly, definitions of evolution that emphasize the transspecific aspect* are also misleading, since they draw attention away from those phases of subspecific evolution that can most easily be studied experimentally and quantitatively. (Equally inadequate, but for the opposite reason

*For example, Petit Larousse: "Series of successive transformations, in particular those which living beings have undergone during geological times."

that they neglect the transspecific aspects of evolution, are definitions such as that of Wright (1942): "Evolution is the statistical transformation of populations.")

Clearly, a definition of organic evolution is needed that includes all of its aspects and at the same time distinguishes between evolution and mere change. The concept of "transformation" is valid if it includes partial transformations, such as the acquisition of DDT-resistance in races of flies. Transformation is a better concept than "change" since it implies that several different changes have occurred that make it impossible for the transformed population to revert completely to its prior condition. Evolution can thus be distinguished from cyclic changes in populations that are repeated year by year. Even though such changes may involve alterations in the genetic composition of the population, and so may become helpful aids for analyzing evolutionary processes, they cannot themselves be regarded as constituting evolution. Transformation, however, does not imply irreversibility in every single characteristic the transformed population has acquired. As Simpson (1953) has clearly pointed out with ample documentation, such individual characters as size, color, or adaptation to a generalized habitat, e.g., water, can always be reversed and frequently are. Irreversibility applies only to a complex ensemble of characteristics, controlled by many separate genes, and which accompany a marked change in the population–environment interaction.

A valid definition of evolution must not include the notion that evolution is always progressive, leading inevitably from simpler to more complex forms of life. To be sure, some of the most important evolutionary events have been increases in structural complexity, e.g., the evolution of the eukaryote cell and cellular differentiation in multicellular organisms, as well as that of the seed and the flower, the complex sense organs of vertebrates, the brain and spinal column, warm blood, and animal societies. However, such events are rare compared to the most ubiquitous evolutionary events, namely, adaptive radiations at one particular level of complexity. Furthermore, both older and modern theories recognize even degenerations of structure as evolution, provided they are products of adaptive alterations in population–environment interactions. The origin of blind races of animals in caves or such degenerate parasites as fleas and lice are just as much a part of evolution as the origin of an orchid flower or the human brain. (The concept of progress in evolution is discussed in Chapter 16.)

The following definition is an attempt to satisfy all these requirements.

Organic evolution is a series of partial or complete and irreversible transformations of the genetic composition of populations, based principally upon altered interactions with their environment. It consists chiefly of adaptive radiations into new environments, adjustments to environmental changes that take place in a particular habitat, and the origin of new ways for exploiting existing habitats. These adaptive changes occasionally give rise to greater complexity of developmental pattern, of physiological reactions, and of interactions between populations and their environment.

What is the relationship between organic and other kinds of evolution? During the century and more since Darwinism came into being, the concept of evolution has been applied not only to the living world but to the nonbiological as well. Thus, we speak of the evolution of the entire universe, the solar system, and the physical earth, apart from the organisms that inhabit it. As we shall show in Chapter 11, the origin of life is best explained as the outcome of pre-cellular chemical evolution, which took place over millions of years. We also speak of social or cultural evolution. As is explained in Chapter 14, the present state of mankind on the earth is the outcome of three kinds of evolution: chemical, organic, and social or cultural evolution. The present volume, although it is concerned principally with organic evolution, deals sufficiently with chemical and cultural evolution to place organic evolution in its proper theoretical perspective.*

A BRIEF HISTORY OF EVOLUTIONARY THEORY

The idea that different kinds of organisms can be transformed one into the other has existed in the minds of some men ever since the dawn of culture. It was expressed specifically by certain Greek philosophers, such as Anaximander and Empedocles. The transformation of human beings into animals and of dragons into human form is a familiar theme of mythology in many cultures. The belief that animals could arise from inorganic matter was widespread during the Middle Ages, and was regarded by theologians, such as Thomas Aquinas, as entirely compatible with Christian philosophy. Indeed, until the time of Redi and Spallanzani most people believed that decaying meat could be transformed into flies. The idea of the permanence of species as special creations entered the Judeo-Christian culture through myths such as those related in the book of Genesis. It became a basic tenet of natural history in the seventeenth and eighteenth centuries, partly as a result of such comprehensive compilations of natural history as those of Linnaeus.

Even during the eighteenth century, however, some observant naturalists began to doubt that species are unchanging separate creations. Buffon, for instance, was sometimes a champion of fixity, and in other instances favored transformism. His transformism, however, never approached the status of a coherent theory of evolution, but rather was more an expression of the "degradation" of organisms from an original, ideal type. Other eighteenth-century philosophers favored a kind of transformism that included elements of modern evolutionary theory. Maupertuis, in his *Vénus Physique* (1746), described albino negroes and suggested that their origin could be in chance alterations of "parts of the seminal fluid."

The first complete theory of evolution was that of Lamarck (1809). It contained two elements. The first and more familiar, but often misinterpreted

*For a more detailed holistic treatment of evolution see Lewontin, 1968.

(see Boesiger, 1974), is that organisms are capable of changing their form, proportions, color, agility, and industry in response to specific changes in the environment. This capability was regarded as an essential property of living organisms, and altered conditions were therefore believed to be hereditary. The other element of Lamarckian evolution, upon which he based a significant revision of the classification of invertebrate animals, was his belief in a discontinuous progression from simpler to more complex kinds of organisms. In his words, nature "has given to the acts of organization the faculty of making the organization itself more and more complex, by increasing the energy of movement of the fluids and therefore that of organic movement. . . ." Apparently, he believed that a tendency toward greater complexity is an inherent property of life itself.

Lamarck was the first modern naturalist to discard the concept of fixed species and instead view species as variable populations. He was also the first to state explicitly that complex organisms have evolved from simpler ones; thus, he regarded the simplest organisms known to him as the original organisms, calling them "Infusoires." Nevertheless, Lamarck is only in a limited sense the founder of evolutionary theory. Neither his concept of the inheritance of acquired adaptations nor that of an inherent tendency toward complexity can now be regarded as valid. Moreover, his theories are only scantily and vaguely documented by actual examples. One searches in vain through the *Philosophie Zoologique* for the kind of specific factual evidence that is the main body of *Origin of Species.* Darwin rejected Lamarck not because he disagreed with the theory of the inheritance of acquired adaptations, but because he found the *Philosophie Zoologique* and other writings by Lamarck devoid of the specific

Jean-Baptiste de Lamarck, the first scientist to formulate a complete theory of biological evolution. [Courtesy of the Museum of Comparative Zoology, and the Grey Herbarium, Harvard University.]

Charles Darwin, circa 1854, founder of the modern theory of biological evolution. [Courtesy of Professor G. Evelyn Hutchison.]

factual data that he needed for his own synthesis. Lamarck was by no means a synthesizer of knowledge. His works are highly subjective speculations about facts he observed directly. References to the writings of other naturalists are given only when he expresses disagreement with them. Because of their subjectivity, his theories could not be tested, as could Darwin's, by additional observations and experiments.

Charles Darwin was indeed the founder of the modern theory of evolution. The greatness of Darwin as an evolutionist was not only his recognition and development of natural selection as a unifying concept of evolutionary change and stability, but his preeminence as a synthesist. Inherent in his writings are two essential elements: a detailed and profound analysis of individual phenomena by means of carefully chosen observations and designed experiments, followed by a complete synthesis of all available information.

Darwin was a great synthesist; he developed an ability to see relationships between apparently unrelated facts even before conceiving the theory of evolution itself. Many of the phenomena that went into his final synthesis were observed during his voyage around the world on H. M. S. *Beagle*. His awe and wonder at the richness and diversity of life in the Amazonian forests brought forth some of the finest prose he wrote. The rocky wastes of Patagonia, with their incredibly rich beds of mammalian fossils, made him directly aware of the life of the past. Unusual and instructive patterns of the geographic distribution of animals and plants became evident to him at every stage of his voyage, but their significance became particularly clear after his visit to the Galápagos Islands. There he was again made aware of the significance of changed conditions of life for the differentiation of populations (through comparing populations of "Darwin's finches") and the importance of geographic isolation for the origin of species (through his observations of the tortoises). His observations of the Fuegian Indians, particularly his comparisons between the native communities and his shipmate Jeremy Button, who had been born in Fuegia but brought up in England, made him aware of a different element, cultural influence, which has played a dominant role in the evolution of human societies.

The remaining elements that went into his synthesis—the Malthusian doctrine of population increase, artificial selection as practised by English animal breeders, and the significant facts of comparative anatomy and embryology—were all acquired in a few years after the *Beagle* voyage, when Darwin was in his early thirties. One of the most remarkable features of Darwin's career is that he never published his "Essay of 1844," which contains all of the essential elements of *Origin of Species* (1859). For a quarter of a century he devoted his life to a prolonged, intensive and single-minded effort to obtain new evidence in favor of this theory. This single-minded attention to his goal was as much responsible as his superior intelligence for his success in giving the world the concept of evolution.

An additional, highly important factor that contributed to Darwin's success was the intellectual climate of the first half of the nineteenth century, and Darwin's ability to take advantage of it. The description of new species of animals and plants, stimulated by the introduction of Linnaeus' system in the late eighteenth century, was still in full swing in the nineteenth century, and was further advanced as a result of careful explorations in all continents. The geologists William Smith and Sir Charles Lyell, the latter a friend and mentor of Darwin, showed that successions of fossils give evidence of continuous change throughout geological ages. The new understanding of the geographical distribution of plants, developed by Darwin's friend Joseph Hooker and his American correspondent Asa Gray, formed another important element in Darwin's synthesis. The fact that *Origin of Species* appeared at the right time to be appreciated, while earlier works did not, is evident from the fact that Alfred Russell Wallace, working independently and in almost complete isolation in the East Indies, developed a theory of evolution much like that of Darwin, although his synthesis was much less complete. Wallace's letter to

Testudo microphyes, Isabela I.

Testudo abingdonii, Pinta I.

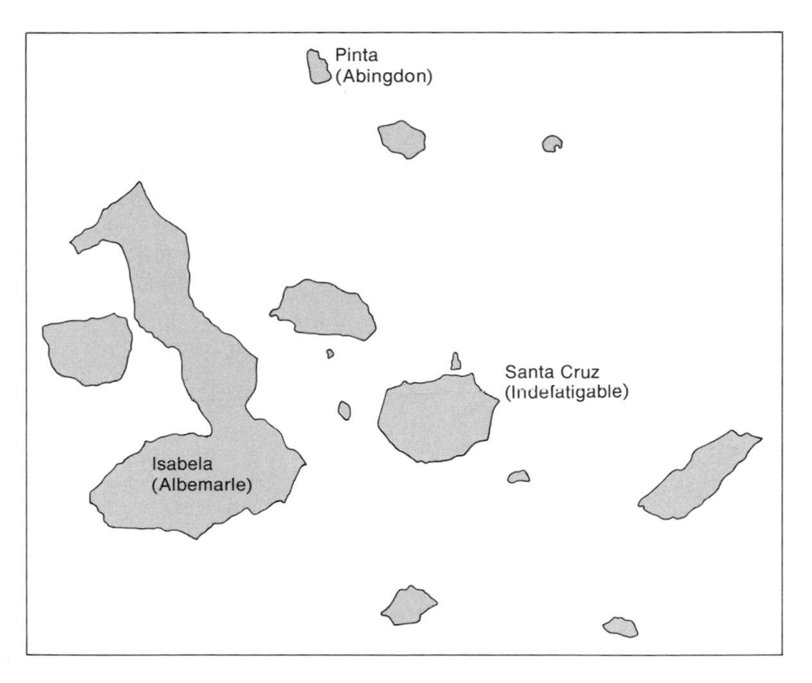

Testudo ephippium, Santa Cruz I.

FIGURE 1-1.
The Galápagos Islands, with drawings of three tortoises found on different islands.

Darwin, followed by the presentation of their theories at a meeting of the Linnean Society in 1858, at which neither author was present, was an event with dramatic consequences in the history of evolution since it led directly to the publication of *Origin of Species*. Perhaps more than any other single event, Darwin's reaction to Wallace's letter demonstrated his ability to accept and acknowledge the contributions of other scientists.

Several critics have attempted to belittle Darwin by maintaining that his experiments were poorly defined, not quantitative, without adequate controls, and thus unworthy of recognition by modern scientists. There is no doubt that Darwin's standards of experimentation were not up to those of modern scientists. Nevertheless, the standards of the experiments described in his books* must not be compared to modern standards, but should be viewed in the light of the biology and natural history of the 1830s and 1840s, when Darwin received his training and began his work. At that time, precise experimentation, though already well developed in physics and chemistry, was little known in biology, even in physiology, where it was most needed. The fathers of precise experimentation in biology, Claude Bernard (born 1813) and Louis Pasteur (born 1822), were both younger than Darwin. Bernard's *Introduction à l'étude de la médecine expérimentale* did not appear until 1865, six years after the first edition of *Origin of Species*. Among Darwin's contemporaries, Gregor Mendel used precise experimentation. However, Mendel's experiments were an isolated phenomenon. Their extraordinary nature may have contributed as much as Mendel's isolation from other scientists to preventing their recognition while he was still alive. Although Darwin cannot be regarded as a first-class experimenter by modern standards, one cannot deny that experimentation formed an integral part of his scientific method.

The importance of Darwin's contribution is reflected in the immediate acceptance of his theories by many contemporary biologists and in the vast amount of research on evolutionary problems that took place during the second half of the nineteenth century. Although this research did much to advance the theory of evolution, nevertheless it was limited in scope. It consisted chiefly in searching for additional examples of natural selection, in applying the techniques of comparative anatomy and embryology (particularly the concepts of homology, recapitulation, and embryonic similarity), in the discovery and description of many fossil forms, and in building largely speculative phylogenetic trees of various plants and animals. The difficulties of the theory of natural selection, which Darwin clearly recognized and set forth in the final edition of *Origin of Species* (1872), were not being cleared up.

The principal reason for this deficiency was the failure of biologists to recognize Mendel's laws of inheritance until they were rediscovered in 1900. As long as no coherent theory of heredity existed, the basis of natural selection could not be understood. Darwin's theory of pangenesis was an unfortunate

***Insectivorous Plants*, 1875; *Effects of Cross- and Self-Fertilisation in the Vegetable Kingdom*, 1877; *Climbing Plants*, 1875; and *The Formation of Vegetable Mould through the Action of Worms*, 1881.

anomaly. It was almost his only venture into the field of pure speculation. He admits in his autobiography that "an unverified hypothesis is of little or no value," but adds, "if anyone should hereafter be led to make observations by which some such hypothesis could be established, I shall have done good service." Unfortunately, such observations could never be made. One might speculate whether Darwin would have formulated his theory of pangenesis if he had been aware of Mendel's experiments.

Darwin's ignorance of the laws of heredity had two important effects. In the first place, it prevented him from ever resolving completely the relationship between natural selection and "the effects of use and disuse"; he never completely discarded the Lamarckian theory of the inheritance of acquired adaptations. Secondly, he was unable to answer the criticism of Fleeming Jenkin, who maintained that if hereditary traits were transmitted by particles contained in miscible fluids, the initial effects of natural selection would be nullified by crossing between selected individuals and the rest of the population. Only the fact of permanent, particulate, self-replicating genes could successfully answer this argument.

This being the case, the history of evolutionary theory during the three decades that followed the rediscovery of Mendel's laws is one of the most extraordinary paradoxes in the history of science. Far from lending strength to Darwin's theory of natural selection, the first decades of Mendelian genetics were largely responsible for a temporary decline in Darwin's reputation among biologists. Anti-selectionist works, such as A. F. Shull's textbook on evolution (1936) and *The Variations of Animals in Nature* by G. C. Robson and O. W. Richards (1936), became standard reading for many undergraduate and graduate students in biology, and many professors told their students that "Darwinism is dead," by which they meant that natural selection could not be regarded as a major agent of evolutionary change. The then current dogma on the mechanisms of evolution was well expressed by T. H. Morgan in *The Scientific Basis of Evolution* (1932). Morgan presented three possible explanations: the inheritance of acquired adaptations, which he rejected; natural selection, which he regarded only as a way of purifying the germ plasm from harmful mutations; and mutation, which he regarded as the only significant factor capable of bringing about evolutionary change, including the origin of more complex organisms. A common belief among biologists of this period was that some as yet unknown processes would have to be discovered before evolution could be fully understood.

Several factors contributed to this situation. First, neither the rediscoverers of Mendel's laws nor their immediate followers, such as Bateson, were naturalists. They knew little or nothing about variation patterns in populations of animals or plants in nature, or about interactions between organisms and their environment. Consequently, they viewed heredity and variation in terms of cultivated plants and domestic animals, garden cultures of introduced weeds, such as *Oenothera* in Europe, and laboratory cultures of *Drosophila*. The majority of naturalists during this period belittled mutation as a source of evolutionary change because they could not recognize in the laboratory and garden mutations described by the geneticists anything resembling the kinds of variation that appeared to be significant in natural populations. Second, probably because of their experience with individual progenies in the laboratory and garden, most of the early geneticists were essentially typologists. To de Vries, Bateson, and Morgan the mutant individual was more important than the population to which it belonged. They never developed the concept of evolution as a result of changes in gene frequency, a notion that dominates the thinking of modern population geneticists. Finally, being missionaries for a new theory, they naturally belittled any observations or phenomena that might appear to detract from its importance. To de Vries, Johannsen, and Bateson, Darwin's explanation that slight variations between individuals are due chiefly or entirely to phenotypic modification of a constant genotype appeared to be most compatible with the concept of Mendelism as a universal theory of particulate heredity. Even when the experiments of Nilsson-Ehle demonstrated that multiple-factor inheritance could explain quantitative inheritance on a particulate basis, other geneticists failed to explore this possibility, and many of them completely ignored it. This explains the curious fact that in the 1930s certain British geneticists redescribed multiple-factor inheritance under the name "polygenic inheritance," as if it were a new phenomenon.

The restoration of Darwinian natural selection as the principal guiding factor in evolution began with the birth of population genetics in the 1920s, based upon the work of S. S. Chetverikov, R. A. Fisher, J. B. S. Haldane, and S. Wright. Of these, the most complete application of genetics to the study of variation in populations was that of Chetverikov (1926), which until recently was not available to western scientists unfamiliar with the Russian language. Fisher's book *The Genetical Theory of Natural Selection* (1930) was the first systematic attempt in the English language to harmonize Darwin's observations on natural variation with Mendelian particulate genetics. He expresses clearly the concept of the harmonious integration of the genotype and the consequent higher probability that mutations having small phenotypic effects, rather than the familiar laboratory mutations having large effects, would be incorporated into the genotype and so contribute to evolution. The reception of Fisher's book is a clear indication of the climate of its time. One searches in vain through the issues of *Science* for 1930 and 1931 for a review of it. Apparently, the editors did not consider it important enough to be worth reviewing. The review in *Nature* by Bateson's disciple, R. F. Punnett (1930), contains several illuminating sentences. "Throughout the book one gets the impression that Dr. Fisher views the evolutionary process as a very gradual, almost impalpable one in spite of the discontinuous basis upon which it works. Perhaps this is because he regards a given population as an entity with its own peculiar properties as such, whereas for the geneticist it is a collection of individuals". And in relation to industrial melanism: "Surely in such cases the mutation can be said to have determined the *direction* [sic] of evolutionary change."

Haldane understood to a certain degree the genetics of populations, and made important contributions to the synthesis. But his book *The Causes of Evolution* (1932) is more in the style of previous books on the subject than the synthetic approach developed later by other evolutionists. The paper of Wright (1931) quantitatively treated changes in the genetic composition of populations that result from mutation and natural selection, as well as an additional factor, the effect of chance in small populations. In this paper Wright also developed the concept of adaptive "peaks" and "valleys." The paper, however, was written largely in mathematical formulas that appeared complex and abstruse to the great majority of evolutionists. Although the geneticists who attended the VII International Congress of Genetics, in 1932 regarded Wright's work with great respect and realized its importance, most of them did not understand its content any better than they would have understood a presentation of Chetverikov's work in Russian.

The modern synthetic theory as a generally accepted way of approaching problems of evolution was born in 1937 with the publication of Dobzhansky's *Genetics and the Origin of Species*. In this book the previous investigations on population genetics were reviewed in clear, understandable English, and integrated with the facts of the chromosome theory of heredity worked out by the Morgan school as well as with observations on variation in natural populations. Its wide acceptance set the stage for books of a similar kind that followed: Huxley (1942), Mayr (1942), Simpson (1944, 1953), and Stebbins (1950), among others. Since the publication of these books the synthetic theory has been widely recognized as the most plausible explanation of the causes of evolution, and has developed largely because of research stimulated by their publication.

The birth of molecular genetics in the 1950s affected the development of the modern synthetic theory to a considerable degree, but did not alter its fundamental nature. One effect was to dispose finally of any theoretical basis for the inheritance of acquired adaptations or for internally direct "orthogenetic" evolutionary trends. The last resort of biologists who favored Lamarckian explanations was the bacteria, to which Weismann's concept of the segregation of germ plasm and soma does not apply. The direct acquisition by bacteria of hereditary resistance to drugs and bacteriophages was postulated by a number of workers. This claim was seriously challenged by Luria and Delbrück (1943), who approached the problem statistically, and was definitely shown to be incorrect by Lederberg and Lederberg (1952) through the use of the replicate plating technique. The notion of the inheritance of acquired characteristics was finally demolished with the recognition of the molecular segregation of the germ line, in the form of DNA molecules that cannot alter their structure in an adaptive fashion in response to specific stimuli from the environment. Molecular genetics has also shown that morphological characteristics and physiological reactions are produced in a highly indirect manner, through the concerted action of scores or hundreds of genes that code either for the enzymes responsible for biosynthetic pathways of molecular synthesis, or for complex systems of regulators of gene action. The nature of this complexity precludes any possibility that successive mutations can automatically cause a characteristic to become altered progressively in a single direction.

Molecular genetics has provided the evolutionist with valuable new tools. The most widely used of these are allozymes as markers for genetic variation in populations, the comparative sequencing of proteins as a method of analyzing phylogenetic relationships, and the comparison of nucleotide sequences in DNA, by the annealing or "hybridization" method. These methods are discussed later in this book.

The differences between the modern synthetic theory and that of natural selection as set forth by Darwin reflect advances in scientific knowledge and method since Darwin's day. The most important is the addition of the Mendelian laws of particulate heredity. The knowledge that heredity is determined by specific genes located on chromosomes removes at once the objection of Fleeming Jenkin, since the swamping effect he pointed out can be overcome in various ways. It also gives the evolutionist a mechanism by which he can measure precisely the intensity of selection. Finally, the complexity of the action of natural selection has been established through experiments, described in Chapter 4, which demonstrate that the phenotypic characters upon which selection exerts its primary action often result from epistatic interactions between many genes.

Hardly less important have been the addition of two valuable methodological tools: the mathematics of population dynamics and the use of precise quantitative experiments. Both techniques are based upon the assumption that much can be learned through constructing models and extrapolating from them.

In addition to expanding the factual basis upon which the synthetic theory is founded, modern trends in scientific knowledge have shown that many basic problems can be understood only by assuming that all biological organization, down to the level of molecules, has evolved as a result of natural selection acting upon genetic variation. Enzyme molecules, equipped with active sites, binding sites, allosteric sites, and sequences of residues responsible for precise tertiary and quaternary structures, achieve this structural and functional complexity only via a long history of adaptive evolution. The same can be said of selectively permeable membranes, cells specialized for various functions, and many other features of biological organization. The enormous amount of research expended on discovering the mechanism of action of various hormones and growth substances in both plants and animals has shown that simple relationships between hormone and substrate or hormone and function do not exist, and that mere knowledge of their chemical structure contributes little toward solving the basic problems of origin and function. Multiple functions in a single hormone are the rule rather than the exception. This phenomenon is best explained by assuming that once a particular hormone acquired the ability to perform an important function in the ancestral stock of an evolutionary line, evolution by natural selection built up other features of structure and organization that took advantage of the hormone molecule already present, enabling it to perform other functions.

In short, the recent history of the life sciences has increasingly demonstrated that "Nothing in biology makes sense except in the light of evolution" (Dobzhansky, 1973a).

2

The Genetic Structure of Populations

THE HEREDITARY MATERIALS

Biological evolution consists of changes in the genetic constitution of populations. In most organisms genetic information is encoded in the chemical substance deoxyribonucleic acid (DNA); in some viruses it is encoded in a related chemical species, ribonucleic acid (RNA).

The nucleic acids were first described by F. Miescher in 1874. Avery, MacLeod, and McCarthy (1944) obtained results suggesting that DNA was the carrier of hereditary information. They obtained highly purified DNA from a virulent strain of pneumococcus *(Diplococcus pneumoniae)*. Nonvirulent pneumococci were incubated in the presence of this purified DNA, and some virulent pneumococci were recovered. A genetic *transformation* had occurred; DNA was described as a transforming agent. Mirsky and Ris (1949) found that all somatic cells of an organism contain, as a rule, the same amount of DNA, while gametic cells contain half as much as somatic cells. This is what would be expected of the genetic material. Proteins were at that time favored by many biologists as the most likely substances encoding hereditary information. Mirsky and Ris found different amounts of them in different cells of a given organism.

More direct evidence of DNA as the hereditary material was obtained by Hershey and Chase (1952). They demonstrated that when the bacteriophage virus *T2* infects the bacterium *Escherichia coli,* only the DNA of the phage enters the bacteria and brings about its own replication. The protein coat that envelopes the virus DNA neither enters the bacteria nor participates in the replication process. Gierer and Schramm (1956) and Fraenkel-Conrat and Singer (1957) showed that RNA is the hereditary material in certain viruses.

Watson and Crick proposed in 1953 the double-helix model of DNA. This model was consistent with all the information then available about the composition and structure of DNA. Moreover, as we shall show; the double-helix model provided a plausible explanation of the basic properties of the hereditary material—carrier of information and binary replication. The DNA model proposed by Watson and Crick has now been convincingly confirmed almost down to its finest details.

DNA and RNA are long chains (polymers) composed of four different units called nucleotides. Each nucleotide consists of a nitrogen-containing base linked with a five-carbon sugar (pentose) and a phosphate group. The pentose sugar is deoxyribose in DNA and ribose in RNA. Each phosphate group is covalently bound to the 5′-carbon of the sugar, and establishes also a covalent bond with the 3′-carbon of the sugar of a second nucleotide. The polynucleotide is held together by these 5′-3′ covalent diester bonds. Polynucleotide chains are structurally polarized since a 3′-hydroxyl (3′-OH) end and a 5′-phosphate (5′-P) end can be identified. The nitrogen bases in the nucleotides are of two classes: purines and pyrimidines. The two most commonly occurring pyrimidines in DNA are cytosine (C) and thymine (T); the two most common purines are adenine (A) and guanine (G). RNA generally contains uracil (U) instead of thymine.

The structure of single polynucleotide chains was already known in 1953. Watson and Crick proposed that DNA is organized as a double-helical molecule made up of two complementary polynucleotide chains, with the phosphate–sugar backbone on the outside and the nitrogen bases inside. The paired chains are held together by hydrogen bonds between bases in different chains, in such a way that A always pairs with T, and G always pairs with C. It follows that DNA has as many A as T bases, and as many G as C bases; that is, the ratios A:T and G:C equal unity. By contrast, the ratio (A + T):(G + C) varies from organism to organism. These properties of ratios between the bases had already been demonstrated by Chargaff (1951). The molecular structure of DNA double-helix is schematically represented in Figure 2-1.

The hereditary material of all organisms, except some viruses, is DNA organized in a double-helix. Five kinds of viruses are presently known to contain single stranded DNA. Most RNA-containing viruses have single-stranded RNA, although some have duplex RNA molecules. However, single-stranded RNA molecules often contain regions where the RNA polymer folds back on itself to form short double-helical segments. The structure of double helical RNA is probably similar to that of DNA, except that A forms hydrogen bonds with U rather than with T.

The genetic information is encoded in the sequence of the nitrogen bases in the nucleic acids. The nitrogen bases may be considered the letters of a genetic alphabet. A specific sequence of letters in the English alphabet can make up any word of the English language; a sequence of words conveys information. In an analogous fashion, one may think of genes as genetic "sentences." The genetic endowment of the individual may then be thought of as a "book"

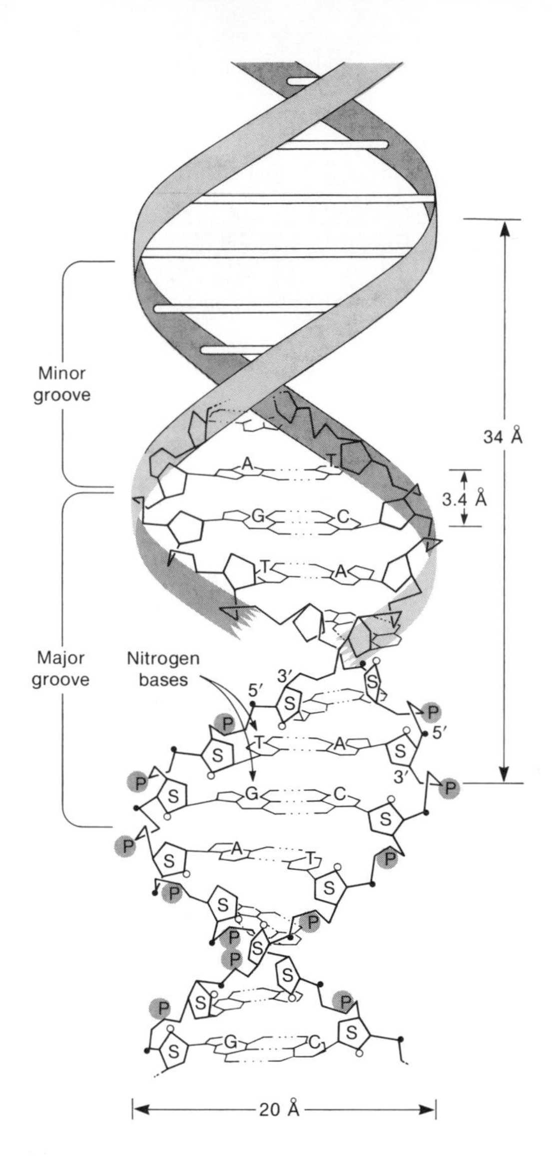

FIGURE 2-1.
The double-stranded helical configuration of the DNA molecule. The molecule consists of two polynucleotide chains. The outward backbone of the molecule is made of alternating deoxyribose sugars (S) and phosphate groups (P). The phosphate group of a given nucleotide is covalently bound to the 5′-carbon of the sugar and establishes an additional covalent bond with the 3′-carbon of the sugar of a second nucleotide. Nitrogen bases connected to the sugars project towards the center of the molecule. The two nucleotide chains are held together by hydrogen bonds between complementary purine-pyrimidine bases. Adenine (A) and thymine (T) form two hydrogen bonds; cytosine (C) and guanine (G) form three. The base pairs are stacked flat, one above the other, at intervals of 3.4 Å (one angstrom, Å $= 10^{-7}$ mm), and each pair is rotated 36°. Thus, each chain makes a complete rotation every 34 Å, with 10 base pairs per complete rotation. In cross section the two chains are separated by 120° in one direction and 240° in the other around the circumference of an imaginary cylinder containing the double-stranded helix, the diameter of which is 20 Å. As a consequence, there are two grooves along the sides of the molecule, one wider than the other. The two complementary chains run in opposite directions.

made up of genetic sentences. In the case of duplex DNA chains the information is encoded in the nucleotide sequence along both complementary chains. The bases are stacked one on top of the next without any stereochemical limitations on the sequence of bases along the double helix. This freedom contrasts with the strict determination of the bases paired between the two complementary chains.

The number of potentially different sequences of four kinds of bases in a chain with n nucleotides is 4^n. This becomes a staggeringly large number whenever the length of the chain is in the hundreds, as is generally the case for individual genes. The basic units of information, however, are not the individual bases, but discrete groups of three consecutive bases. Because of the redundancy of the genetic code (see below), there are only 21 different units of information among the triplets (61 triplets code for 20 amino acids, the other three triplets are termination signals). A polynucleotide chain with 600 nucleotides has 200 groups of three bases. The number of potentially different messages contained in chains of that length is $21^{200} = 10^{264}$, a number much greater than the number of atoms in the known universe. There is practically no limit to the number of different messages that can be encoded in long DNA chains.

The four bases, A, T, C, G, are the only ones generally found in DNA, although other chemically slightly different bases are occasionally found, particularly in some bacteria and bacteriophages. The near universality of the four bases suggests that the present structure of DNA may have evolved only once. It indicates, moreover, that the evolution of organisms has taken place by formation of ever new sequences, not by replacement of the bases. It is possible that among the primordial forms of life there may have been a greater diversity of bases in DNA, and that the four now existing replaced the others by natural selection (Dobzhansky, 1970).

The double helix suggests a mechanism for the precise replication of genes. The sequence along one of the strands unambiguously specifies the sequence along the complementary strand, owing to the strict determination of the base-pairing between the two DNA chains. According to Watson and Crick (1953), if the two strands in a given molecule were to separate by an unwinding process and become exposed to a pool of nucleotides, each strand might serve as a template for a complementary strand (Figure 2-2). Two new double-helices would be formed, which, because of the rules of base-pairing, would be identical to each other and to the parental double helix. The "daughter" double-helices would each have a complete chain from the parental double-helix and a newly made up chain—this has been called a "semiconservative" mode of replication.

This semiconservative model of DNA replication has been substantiated by ingenious experiments performed in *Escherichia coli* by Meselson and Stahl (1958). Cairns (1963) obtained microphotographs of *E. coli* chromosomes showing the replicating fork of the DNA as postulated by the model. The replication of DNA is also semiconservative in higher organisms, as first shown in the broad bean *Vicia faba* by Taylor and his colleagues (1957; see also Taylor, 1969).

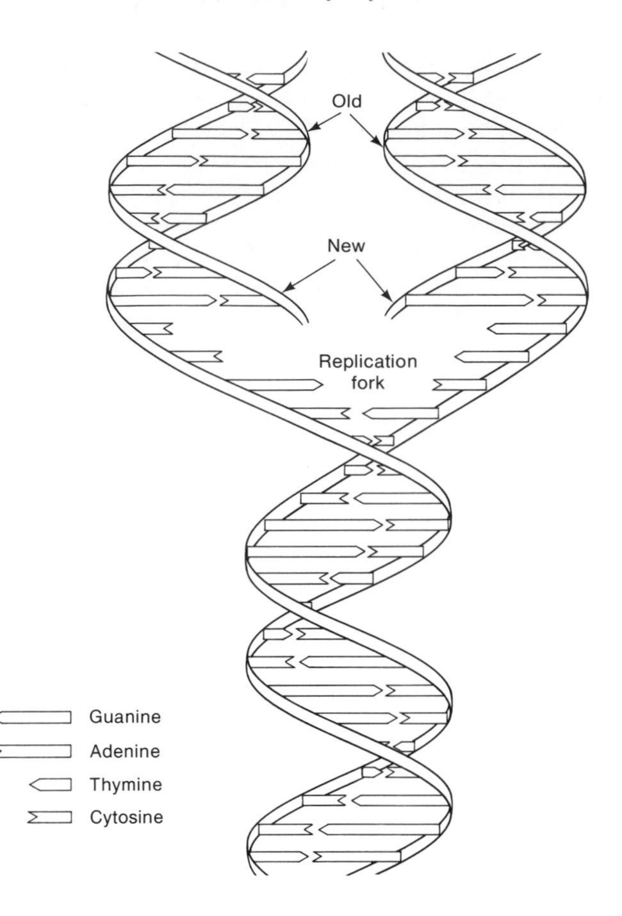

FIGURE 2-2.
Model of the replication of DNA. The two strands of the double helix separate by an unwinding process. Each strand serves as a template for the replication of a complementary strand, owing to the specificity of the base-pairing: adenine always pairs with thymine, guanine pairs with cytosine. The replication process results in two "daughter" double helices, identical to each other and to the parental double helix. Each daughter double helix has a complete chain from the parental molecule and a newly synthesized chain. The region in which the two parental strands are separating, and where the daughter strands are synthesized, is called the *replication fork*.

THE GENETIC CODE

The genetic information contained in DNA controls the development and metabolism of an organism through the processes of *transcription* and *translation.* Two classes of genes are sometimes distinguished, structural and regulatory. *Structural* genes are those whose RNA products are "translated" into proteins (polypeptides). *Regulatory* genes are those that control the activity of other genes. Structural and regulatory genes are not two mutually exclusive classes, nor do they include all kinds of genes. Some genes are both structural and regulatory, such as the genes coding for "effector" molecules that regulate the transcription of other genes by binding to their operators or promoter sites. And there are genes that, in a strict sense, are neither structural nor regulatory, such as the genes coding for ribosomal RNA (rRNA) and transfer RNA (tRNA).

Structural genes specify the sequence of amino acids in proteins. The processes of transcription and translation mediate the transmission of information from genes to proteins. These processes are represented in Figure 2-3. The process of transcription takes place in the nucleus (except for extranuclear genes, such as those in chloroplasts and mitochrondria). The base sequence of a DNA segment is transcribed into a complementary RNA molecule. Certain sites in the DNA, called promoters, contain the information as to where the transcription of a particular sequence should begin. Promoters are thought to be adjacent to the sequences to be transcribed, but the promoters themselves are not transcribed.

The transcription of the base sequence of the DNA into a complementary RNA sequence is essentially similar to the replication of DNA, except that ribonucleotides are matched with the DNA template, and the base uracil rather than thymine pairs with adenine. The process is mediated by an RNA polymerase enzyme. Only one of the two DNA strands is transcribed (Marmur *et al.*, 1963), but the strand transcribed need not be the same for all genes, at least in bacteriophages such as *T4* and λ.

The process of translation takes place in the cytoplasm and is mediated by the ribosomes, transfer RNA molecules, and several enzymes. The RNA transcribed from structural genes is called *messenger* RNA (mRNA). The mRNA's carry the information for the specific sequence of amino acids in polypeptides; one or more polypeptides make up a protein. Polypeptides are chains of many amino acids linked by peptide bonds joining the carboxyl (—COOH) terminal group of one amino acid and the amino ($—NH_2$) terminal group of the next amino acid. Although more than 200 amino acids are known, only 20 are common constituents of proteins.

Messenger RNA synthesized in the nucleus moves to the cytoplasm, where it becomes associated with groups of ribosomes. There the base sequence in the mRNA's is "read" with the aid of tRNA's that are specific for each kind of amino acid. The mRNA is translated in a step-wise fashion, starting at the 5′-carbon end and proceeding towards the 3′-carbon end. The polypeptides are synthesized proceeding from the amino end of the first amino acid to the carboxyl terminus of the last amino acid. The information is contained in mRNA in discrete sequences of three nucleotides, called *codons*. Each codon corresponds to a complementary sequence of three nucleotides, called an *anticodon,* at a particular site of an appropriate tRNA. Each kind of tRNA molecule associates with a specific amino acid. As succeeding codons of mRNA pass through the ribosomes, tRNA's are sequentially involved in codon–anticodon pairings. The corresponding amino acids are thus brought into position,

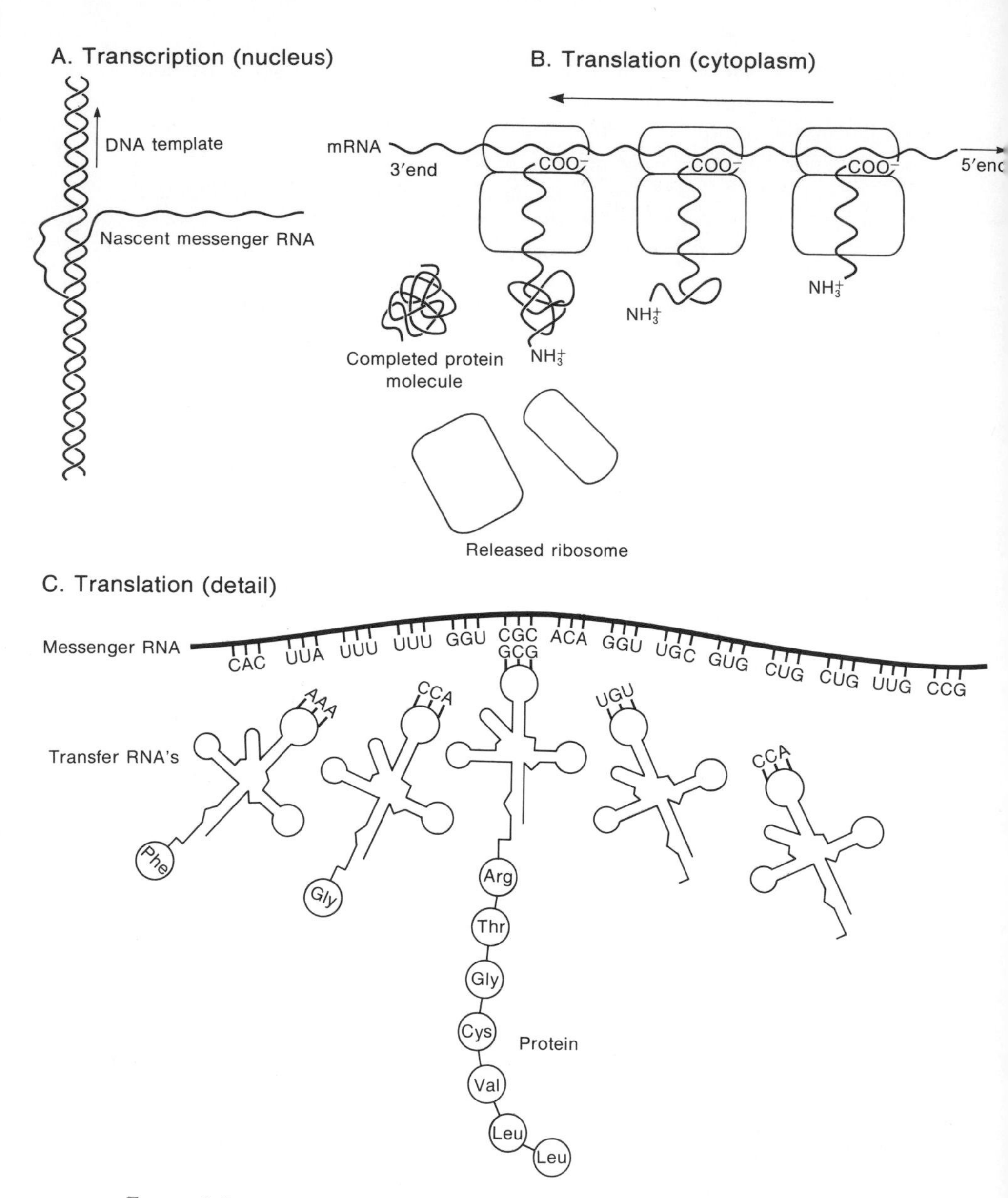

FIGURE 2-3.
Schematic representation of the processes of transcription and translation. **A.** Transcription. One strand of the DNA helix serves as a template for the synthesis of a complementary chain of messenger RNA. **B.** Translation. In eukaryotes the messenger RNA synthesized in the nucleus moves to the cytoplasm, where several ribosomes attach to it. Each ribosome synthesizes a polypeptide as it proceeds several nucleotides behind the preceding ribosome. Each codon in the messenger RNA is recognized by a complementary anticodon on a transfer RNA molecule carrying a particular amino acid. **C.** Detail of the translation process. The configuration of the transfer RNA molecules is an outline of the known structure of alanine transfer RNA.

peptide bonds form between adjacent amino acids, and a polypeptide is synthesized.

Amino acids combine with tRNA's by a reaction mediated by a set of activating enzymes called aminoacyl-tRNA ligases. In each organism there are at least, and probably no more than, 20 different ligases, one specific for each of the 20 amino acids. About 60 varieties of tRNA's exist in higher organisms, and at least 30 to 40 in *Escherichia coli.* Because only 20 amino acids are normally involved in protein synthesis, several types of tRNA may associate with the same amino acid. For example, five species of tRNA's in *Escherichia coli* associate with leucine, two of which exist in greater abundance than the other three.

The first amino acid incorporated into a polypeptide is always methionine, which is specified by the codon AUG in mRNA. The completed polypeptide may or may not have methionine as the initial amino acid, since the initial methionine is often cleaved off before protein synthesis ends. Two types of tRNA's associate with methionine: one recognizes initial AUG codons, the other recognizes all other AUG codons in mRNA sequences. The structures of these two species of tRNA are different in *Escherichia coli,* and probably also in other organisms, although both types of tRNA have the same anticodon, UAC.

Usually, a single mRNA molecule is translated into several polypeptide chains by different ribosomes that proceed simultaneously along the mRNA at a distance of several nucleotides from each other. Each ribosome that attaches to the mRNA synthesizes a polypeptide. The synthesis of a polypeptide is concluded when a ribosome encounters a "terminator" codon in the mRNA. There are three terminator codons, UAA, UGA, and UAG, none of which codes for any amino acid.

One outstanding achievement of molecular genetics has been the "breaking" or deciphering of the genetic code. This was accomplished using different approaches by several groups of investigators, most notably by M. Nirenberg and by G. Khorana, and their collaborators. Table 2-1 shows the correspondence between the 64 possible triplets in mRNA and the 20 amino acids or the signals for termination of protein synthesis. The code is said to be "degenerate" because a given amino acid may be specified by more than one codon. Apparently, however, different codons for a given amino acid are sometimes recognized by a single species of tRNA. Pairing takes place because some "wobble" is possible in the third position of a codon once the correct base pairs have formed in the first two positions. However, as stated above, several tRNA species may exist for a given amino acid. There is no simple one-to-one relationship between the number of tRNA species and the number of different codons in the mRNA's of an organism.

The genetic code shown in Table 2-1 is universal. A given codon is always translated into the same amino acid in different organisms, with perhaps only minor exceptions. Lane, Marbaix, and Gurdon (1971) have provided the most impressive evidence that the genetic code is the same in different organisms. Purified mRNA coding for hemoglobin was extracted from rabbits and injected

TABLE 2-1. The genetic code, showing the amino acids (or termination signals) specified by each of the 64 nucleotide triplets (codons) in messenger RNA. The abbreviated nucleotide bases are **A**denine, **C**ytosine, **G**uanine, and **U**racil. The abbreviated amino acids are **Ala**nine, **Arg**inine, (**Asn**) Asparagine, **Asp**artic acid, **Cys**teine, **Gly**cine, **Glu**tamic acid, (**Gln**) Glutamine, **His**tidine, (**Ile**) Isoleucine, **Leu**cine, **Lys**ine, **Met**hionine, **Phe**nylalanine, **Pro**line, **Ser**ine, **Thr**eonine, **Tyr**osine, (**Trp**) Tryptophan, and **Val**ine.

First Letter	Second Letter: U	Second Letter: C	Second Letter: A	Second Letter: G	Third Letter
U	UUU, UUC } Phe UUA, UUG } Leu	UCU, UCC, UCA, UCG } Ser	UAU, UAC } Tyr UAA, UAG } Stop	UGU, UGC } Cys UGA Stop UGG Trp	U C A G
C	CUU, CUC, CUA, CUG } Leu	CCU, CCC, CCA, CCG } Pro	CAU, CAC } His CAA, CAG } Gln	CGU, CGC, CGA, CGG } Arg	U C A G
A	AUU, AUC, AUA } Ile AUG Met	ACU, ACC, ACA, ACG } Thr	AAU, AAC } Asn AAA, AAG } Lys	AGU, AGC } Ser AGA, AGG } Arg	U C A G
G	GUU, GUC, GUA, GUG } Val	GCU, GCC, GCA, GCG } Ala	GAU, GAC } Asp GAA, GAG } Glu	GGU, GGC, GGA, GGG } Gly	U C A G

into frog oocytes; rabbit hemoglobin was synthesized there. The information in the rabbit mRNA was properly recognized by the tRNA's of the frog oocytes, although the oocytes are normally never involved in hemoglobin synthesis. The universality of the genetic code suggests that the genetic code in its present form became fixed in some primordial form of life from which most (or all) living organisms have evolved.

GENOTYPE AND PHENOTYPE

A gene is transcribed unambiguously into a specific RNA sequence. For structural genes this is an mRNA that becomes translated into a single polypeptide chain. This "one gene—one enzyme" correspondence, first advanced as a hypothesis by Beadle and Tatum (1941), is now well established in its modified form as "one gene—one polypeptide." Only a fraction of all structural genes are engaged in transcription in any one cell at any one time. In multicellular organisms all cells carry identical sets of genes, but different subsets are active in the cells of different tissues (Bonner, 1965a; Davidson, 1968; Galau *et al.*, 1976). Cell and tissue differentiation occur as a consequence of the activity of different groups of genes. Which genes become activated depends on interactions between molecules in the cells, between neighboring cells, and between the cells and the external environment.

The detailed mechanisms that regulate the activity of genes are far from fully understood. For bacteria the generally accepted model of gene regulation is the "operon" proposed by Jacob and Monod (1961). An operon consists of an operator and several structural genes, all adjacent or in close proximity to each other. The operator, interacting with the cell environment through the products of regulatory genes, determines when the structural genes of the operon will be transcribed. Little is known about regulation of gene activity in higher organisms, but Britten and Davidson have proposed an interesting model (1969, 1971; Davidson and Britten, 1973). A fraction of the DNA of higher organisms consists of relatively few short sequences, each repeated from a few hundred to about a million times. Many of these sequences are about 300 nucleotides in length and are interspersed with structural genes throughout the genome (Davidson and Britten, 1973). According to Britten and Davidson, these highly repetitive short sequences of DNA are regulatory genes that control the transcription of structural genes. (This model of gene regulation is further discussed in Chapter 8.)

The development of an organism is determined by complex networks of interactions between gene products. The expression of a gene varies depending on what other genes are associated with it. Genes may interact with each other and with the environment to determine a given character. Interactions between genes at different loci are called epistatic. Dominance, codominance, and recessivity are interactions between genes at the same locus. Often a single gene may affect several traits, a phenomenon known as pleiotropism. There is no simple one-to-one relationship between genes and traits.

Johannsen (1909) introduced the important distinction between genotype and phenotype. The *phenotype* of an organism is its appearance—its morphology, physiology, and ways of life—what we can observe. The *genotype* is the sum total of hereditary materials of the organism. The phenotype changes continuously throughout the life of an organism, from the moment of conception to its death. The genotype, however, remains constant except for occasional mutations.

Because of the complex interactions between genes with each other and with the environment, the genotype of an organism does not unambiguously specify its phenotype. Rather, the genotype determines the range of phenotypes that may possibly develop; this range is called the *range of reaction,* or norm of reaction, of the genotype. Which phenotypes will actually be realized depends on the environments in which development takes place (see for example Figure 5-5, p. 149).

Lamarck, Darwin, and many others believed in the inheritance of "acquired" characteristics. According to Lamarck, the adaptation of organisms to their environments could be explained by use and disuse of their organs and other features. Consistent use of an organ would develop, strengthen, and modify it; disuse would gradually lead to atrophy and eventual obliteration. Such modifications due to use or disuse could become hereditary. Other authors have attributed adaptation to direct influences of the environment that would impress upon the organism heritable modifications to fit the characteristics of the environment.

Convincing evidence against the inheritance of acquired characteristics accumulated in the late nineteenth century and first half of the twentieth century through the work of Weismann and others. The advances of molecular genetics during the last three decades have unambiguously determined that the relationship between genotype and phenotype is unidirectional. The sequence of nucleotides in DNA is transcribed into RNA, which, for structural genes, is translated into proteins. Interactions between gene products and the environment determine the phenotype. There is no mechanism by which the process could be reversed. Phenotypic modifications resulting from use or disuse, or from the complex interactions between organisms and their environments, do not modify the hereditary information encoded in the DNA of the organism. The evolution of organisms depends on changes in the DNA, which occur through the processes of gene mutation and chromosomal change discussed in Chapter 3.

POPULATIONS AND GENE POOLS

Living matter exists in the form of discrete units, individuals. In unicellular organisms each cell is an individual. Multicellular individuals consist of many cells thoroughly integrated and interdependent, many of which die and may be replaced by other cells throughout the life of the individual. Populations are arrays of individuals. The term "population" has a variety of meanings even in scientific language. A demographer may speak of the "U.S. population," referring to all individuals living in the United States; or he may speak of the "human population," referring to all men, women, and children in the world. An ecologist may speak of the rodent population in a meadow, which might include mice, pocket gophers, and individuals of several other species.

A Mendelian population, or *reproductive community,* is a "community of individuals of a sexually reproducing species within which matings take place" (Dobzhansky, 1970, p. 310). The lowest level of a reproductive community is represented by *panmictic populations,* in which individuals mate at random. A local population is sometimes called a *deme;* this term is applied to individuals of the same species whether or not they reproduce sexually. Local populations of sexually reproducing organisms may be discontinuous, but population boundaries are often not well defined. Consider populations of the wild slender oat, *Avena fatua,* along the foothills of the Coast Ranges of California. Local clusters of individuals exist, and clusters of clusters can also be recognized. Often, however, individuals can be seen more or less spaced between the clusters. Sharp discontinuities in the distribution of populations exist, either because the intermediate territory is uninhabitable for the organisms or they are excluded by competitors, predators, or parasites. Even so, some continuity among the separate populations may exist because of migrant individuals.

The most inclusive reproductive community is the *species*. As a rule, the genetic discontinuities between species are absolute; sexually reproducing organisms of different species are kept from interbreeding by isolating mechanisms discussed in Chapter 6.

Evolutionary change occurs in populations, not in individuals. The individual is born, grows, and eventually dies. Although individual organisms change more or less throughout their lifetimes, their genetic constitution remains constant. The genetic constitution of a population, on the other hand, may change from generation to generation by the processes of genetic mutation, migration, drift, and natural selection, which are discussed throughout this book. From the evolutionary point of view the individual is ephemeral; only populations persist through time. The continuity derives from the mechanism of biological heredity.

The sum total of the genotypes of all individuals in a reproductive community, or Mendelian population, may be conceived as the *gene pool* of the population. For diploid organisms, the gene pool of a population of N individuals will consist of $2N$ genomes, that is $2N$ genes for each gene locus and $2N$ chromosomes for each of the chromosomes in the haploid complement. The exceptions are sex-linked genes and X chromosomes which exist in single dose in the heterogamic sex (the males in most animals, but the females in birds, butterflies, and others). The number of genes and chromosomes of each kind is $4N$ for tetraploids, $6N$ for hexaploids, and so on.

Genes at a given gene locus may exist in a variety of allelic forms. For example, two alleles exist at a locus controlling eye-color differences in man; one allele determines blue, the other brown-colored eyes. Description of a gene pool at a given locus requires specification of the kinds of alleles present and their frequencies. Allelic variants at different loci often are not associated at random, particularly when they are closely linked within the same chromosome. The associations between alleles at different loci are important for the understanding of the genetic structure of populations. Evolution consists of changes in the constitution of gene pools; that is, changes in allelic frequencies as well as changes in the associations between alleles at different gene loci. Evolution may also occur through changes in the amount and organization of the genetic material, described elsewhere in this book.

GENETIC VARIATION AND EVOLUTION

In *Origin of Species* Darwin supplied ample evidence that biological evolution has occurred. Most important, however, he provided a causal explanation of the evolutionary origin of living beings, namely, the theory of natural selection.

The starting point of Darwin's argument is the occurrence of hereditary variation. For Darwin this was an incontrovertible fact, although he did not know the processes by which hereditary variation arises. Darwin argued that some natural variations must be more advantageous than others for the survival and reproduction of their possessors. Organisms having advantageous variations are more likely to survive and to reproduce than organisms lacking them. The process leads to the spread of useful variations and to the elimination of harmful or less useful ones.

Today we have a more complete, and more profound, understanding of the processes of organic evolution. Natural selection remains the fundamental process directing evolutionary change. However, natural selection can only occur if there is hereditary variation. The more genetic variation there is in a population, the greater the opportunity for the operation of natural selection.

R. A. Fisher demonstrated mathematically in *The Genetical Theory of Natural Selection* (1930) that the amount of genetic variation with respect to *fitness,* i.e., relative reproductive efficiency, is directly correlated with the rate of evolutionary change by natural selection. This correlation is stated in the Fundamental Theorem of Natural Selection: *The rate of increase in fitness of a population at any time is equal to its genetic variance in fitness at that time.* The mathematical demonstration of the Fundamental Theorem may be found in Fisher's book and in textbooks of population genetics, such as Li (1955) and Crow and Kimura (1970). The strict formulation of the Fundamental Theorem applies to allelic variation in fitness at a single gene locus, and assumes that the environment remains constant. Nevertheless, changes in allelic frequency due to natural selection generally continue as long as there is hereditary variation in a population, and the rate of change is directly proportionate to the amount of variation.

Ayala (1965a, 1968a) has provided experimental evidence of the positive correlation between the amount of genetic variation and the rate of evolutionary change. *Drosophila serrata* is a species endemic in eastern Australia, New Guinea, and New Britain. Laboratory populations were set up with flies descended from various natural populations. Two types of experimental populations were established. "Single-strain" populations had founders descended from flies collected in a single locality; "mixed" populations were established by crossing flies collected in two different localities. Two sets of experiments were conducted, one at 25°C and the other at 19°C. Under the experimental conditions there was intense competition among the flies for food and space, which stimulates rapid evolutionary change. The adaptation of a population to the experimental environment was measured by the number of individuals in the population. The results of two experiments are shown in Table 2-2 and Figure 2-4. These populations were kept for 500 days or about 25 generations. Over the experimental period the average population size of the mixed population (Sydney × Popondetta) was significantly larger than that of the single-strain populations from either Sydney or Popondetta. All populations gradually increased their adaptation to the experimental environment. At 25°C the average number of flies in the Popondetta population

TABLE 2-2. Correlation between the amount of genetic variation and the rate of evolution in experimental populations of *Drosophila serrata* exposed to a new environment. The "mixed" populations initially have greater amounts of genetic variation than the "single-strain" populations, because the former are established by intercrossing two single-strain populations. [Data from Ayala, 1965b.]

	Mean number of flies in population	Mean number of flies increased per week
	Experiment at 19°C	
Single-strain (Popondetta)	1724 ± 58	8.4 ± 3.3
Mixed (Popondetta × Sydney)	2677 ± 102	20.4 ± 4.6
	Experiment at 25°C	
Single strain (Popondetta)	1862 ± 79	10.5 ± 4.6
Mixed (Popondetta × Sydney)	2750 ± 112	19.5 ± 5.8

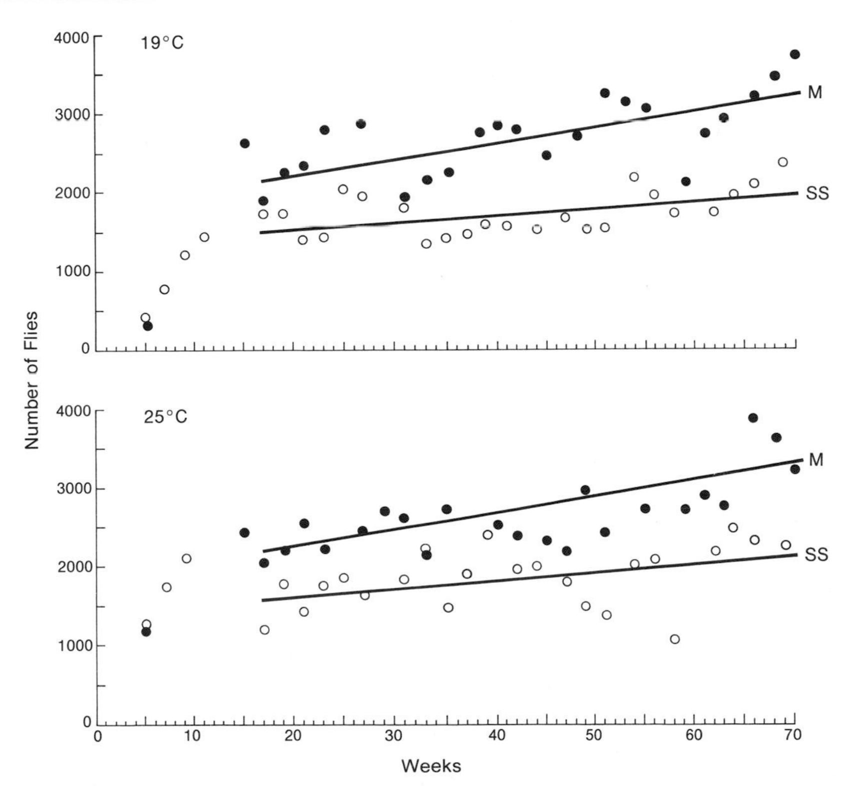

FIGURE 2-4
Change in population size due to natural selection in experimental populations of *Drosophila serrata* at two temperatures. (SS, single-strain population from Popondetta; M, mixed population obtained by crossing Popondetta × Sydney strains.) The experimental period corresponds to about 25 generations. Throughout the experiment the populations gradually increase in size, but the average rate of increase (represented by the slope of the solid regression lines) is about twice as large in the mixed populations (which are genetically more variable) than in the single-strain populations. [See Ayala, 1965b.]

increased at a rate of 10.5 ± 4.6 flies per week; the Sydney population increased at about the same rate. The Sydney × Popondetta population, however, increased in size at a rate of 19.5 ± 5.8 flies per week. This population had from the start about twice as much genetic variation as each of the two single-strain populations since it was started by mixing their gene pools and evolved at a much faster rate than either one.

In a series of related experiments, Ayala (1966, 1969a) increased the genetic variation of populations of *Drosophila serrata* and a closely related species, *D. birchii,* by exposing them for three generations to moderate doses of X-rays. Thereafter, both irradiated (experimental) and non-irradiated (control) populations were started. Over many generations the rate of adaptation to the experimental environment was significantly greater in the irradiated populations. In these studies, as in the experiments with mixed populations described above, the laboratory environment presented the experimental populations with new challenges. The process of natural selection was accelerated by stern competition for food and space. In all cases populations with greater initial stores of genetic variation evolved at faster rates.

TWO MODELS OF POPULATION STRUCTURE

Two models of the genetic structure of populations have been proposed, the so-called "classical" and "balance" hypotheses (Dobzhansky, 1955). The classical hypothesis proposes that the gene pool of a population consists at each gene locus of a wild-type allele with a frequency approaching one. Mutant alleles in very low frequencies may also exist at each locus. A typical individual would be homozygous for the wild-type allele at most gene loci; at a very small proportion of its loci the individual would be heterozygous for a wild and a mutant allele. Except in the progenies of consanguineous matings, individuals homozygous for a mutant allele would be extremely rare. The "normal," ideal genotype would be an individual homozygous for the wild-type allele at every locus.

According to the classical model, mutant alleles are continuously introduced in the population by mutation pressure, but are generally deleterious and thus are more or less gradually removed from the population by natural selection. Occasionally a beneficial mutant allele might arise, conferring higher fitness upon its carriers than the preexisting wild-type allele. This beneficial allele would gradually increase in frequency by natural selection to become the new wild-type allele, while the former wild-type allele would be eliminated. Evolution would thus consist of the replacement at an occasional locus of the preexisting wild-type allele by a new wild-type allele.

The historical origin of the classical model of the genetic structure of populations must be traced to the early decades of the twentieth century. It derives to a large extent from the laboratory experience with mutants, like those studied in *Drosophila* by Morgan and his school. These mutants were selected for study precisely because their drastic effects on the phenotype permitted the unambiguous classification of their carriers. Laboratory mutants almost invariably impair the fitness of their carriers, at least in homozygous condition. The most articulate exponents of the classical model are H. J. Muller and some of his students. As recently as 1966 Muller and Kaplan defended the classical model, arguing that a human individual is not likely to be heterozygous at more than 80 gene loci, and maybe on the average at only eight.

The proponents of the balance model of population structure include Dobzhansky (1970, for a review) and his students in the United States, and the school of ecological geneticists in Great Britain, notably E. B. Ford (1971). The balance hypothesis derives from experience with the direct study of natural populations. According to the balance model, there is generally no single wild-type or "normal" allele. Rather, the gene pool of a population is envisioned as consisting at most loci of an array of alleles in moderate frequencies. A typical individual is heterozygous at a large proportion of its gene loci. There is no "normal" or ideal genotype, only an adaptive norm consisting of an array of genotypes that yield a satisfactory fitness in most environments encountered by the population.

The proponents of the balance model argue that the ubiquitous allelic polymorphisms are maintained in populations by various forms of balancing natural selection (to be discussed in Chapter 4). The fitness conferred on its carriers by an allele depends on what other alleles exist in the genotype at that and other gene loci. It also depends, of course, on the environment. Gene pools are coadapted systems; the sets of alleles favored at one locus depend on the sets of alleles that exist at other loci. Evolution occurs by gradual change in the frequencies and kinds of alleles at many loci. As the configuration of the set of alleles changes at one locus, it also changes at many other loci.

The classical and balance models of gene-pool structure are contrasted in somewhat extreme form in Figure 2-5. The genotypes of three typical individuals are schematically represented according to each hypothesis. Gene loci are symbolized by capital letters. The wild-type allele of the classical model is represented by a plus sign over the letter representing the locus; allelic variants are represented by numbers. According to the classical model, a typical individual is homozygous for the wild-type allele at the great majority of loci, but at an occasional locus it is heterozygous for the wild and a mutant allele. In the balance model a typical individual is heterozygous at a substantial proportion, perhaps the majority, of loci. Different individuals are heterozygous at different sets of loci. At those loci at which two individuals are both either homozygous or heterozygous, they often are so for different alleles. Of course there are also unconditionally deleterious alleles, but these are kept at low frequencies by natural selection, and they very rarely appear in homozygous condition.

The balance model of the genetic structure of populations has now become definitely established, although some controversy remains concerning the processes maintaining the ubiquitous polymorphisms (Ayala, 1974a; Lewontin,

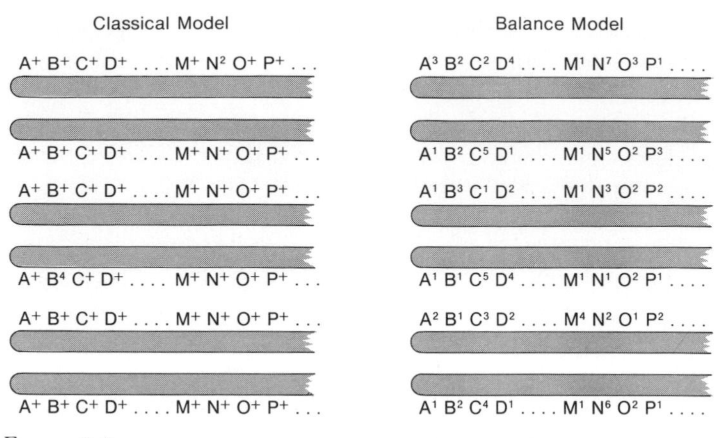

FIGURE 2-5
The classical and balance models of the genetic structure of populations. The hypothetical genotypes of three typical individuals are represented according to each model. The capital letters symbolize gene loci. The plus sign (superscript) represents the wild-type allele; each number represents a different allele. According to the classical model, individuals are homozygous for the wild-type allele at most loci, but an individual may be heterozygous for the wild allele and a mutant allele at an occasional locus (N in the first individual, B in the second). According to the balance model, individuals are heterozygous at a large proportion of the gene loci.

1974). Evidence has accumulated over the years showing that genetic polymorphisms are widespread. During the last decade the application of molecular genetic techniques has permitted estimation, at least to a first approximation, of the proportion of loci that are polymorphic in a given population, or that are heterozygous in a typical individual. We shall review these studies later in this chapter.

ALLELIC VARIATION AT INDIVIDUAL GENE LOCI

Morphological variation is conspicuously apparent in human populations in the facial features, body build, skin pigmentation, and other traits of people. It has also been recorded in virtually every class of animal or plant. Discontinuous variants (often called "morphs" in animals but "sports" in plants) are sometimes so strikingly different from the "normal" type as to have been mistakenly described as different species. Among birds, morphs were erroneously described as separate species in more than 100 instances (Mayr, 1963, 1970).

The genetic analysis of morphological variation has been carried out in only a few cases where the materials are favorable. These include color and pattern polymorphisms in snails, butterflies, grasshoppers, ladybird beetles, and birds;

dextral versus sinistral winding of the shell in snails; flower and seed color and pattern as well as growth habit in many plants. The genetics of color pattern variation in mice has been analyzed by Dunn and his collaborators. Many genetically determined blood groups are known in man, as well as in other mammals and birds; more are discovered every year (reviews in Manwell and Baker, 1970; Lewontin, 1974).

Dubinin and collaborators (1937) examined 129,582 individuals of *Drosophila melanogaster* collected in three successive years from several localities in southern Russia. Visible variants were found in 2,700 flies (2.08 percent of the total). About one-third of these flies were demonstrably homozygous for recessive mutants, or heterozygous for dominant ones. The aberrant phenotypes were for the most part identical with well-known laboratory mutants, including modification of bristles, eye and body color, and so on.

Direct examination of individuals collected in nature may not reveal much discontinuous morphological variation determined by single genes, since mutant genes are often recessive and therefore not detectable in heterozygotes. Hidden genetic variation can be detected by examining the inbred progenies of wild specimens (Chetverikov, 1926). In the F_1 progenies of inseminated *Drosophila* females one may discover the dominant mutants carried by these females and their mates; recessive sex-linked mutants carried by the females are manifested in their male progenies. Inbreeding the F_1 individuals permits detection of recessive autosomal mutants in the F_2 and F_3 generations. In monoecious (hermaphroditic) plants, selfing uncovers recessive mutants, since each gene carried in heterozygous condition becomes homozygous in one fourth of the progeny.

Spencer (1957) studied hidden genetic variation in *Drosophila mulleri* by obtaining laboratory progenies of females collected in nature. On the average, the females carried about one mutant gene with visible effects per wild fly. Studies by Alexander (1949, 1952) led to the estimation that there are between 0.56 (*Drosophila novamexicana*) and 2.38 (*D. hydei*) mutants with visible effects per wild fly. Boesiger (1962) studied a large number of wild *Drosophila melanogaster* flies collected in southern France. After two generations of inbreeding the number of detected mutants per wild fly ranged from none to 12. Most wild flies carried between two and five recessive mutants; the averages in different collections ranged from 2.29 to 4.88 mutants per fly. Pentzos, Boesiger, and Kanellis (1967) conducted similar studies with *Drosophila subobscura* flies collected in Greece, and estimated that a wild fly carries on the average from 3.5 to 4.6 recessive mutants with visible effects.

Several authors have studied recessive allelic variation with visible effects in a variety of organisms, particularly Diptera. In the house fly, *Musca domestica*, inbreeding revealed from zero to six mutants per fertilized female (Milani, 1967). In the mosquito *Aedes aegypti* the average number of mutants per female ranges from 0.72 to 2.96 in different samples (Craig and Hickey, 1967).

Crumpacker (1967) has summarized the literature concerning plant mutants causing chlorophyl and other deficiencies detectable in seedlings after selfing. Between 12 and 67 percent of the individuals tested proved to be heterozygous for one or more mutants in two species of rye grass *(Lolium perenne* and *L. multiflorum),* in timothy *(Phleum pratense),* orchard grass *(Dactylis glomerata), Festuca rubra,* clover *(Trifolium repens),* and cherry *(Prunus avium).* The most extensive data available are for Indian corn *(Zea mays),* with an overall frequency of 21.3 percent of heterozygotes for mutants causing chlorophyl-deficient seedlings. It must be born in mind that chlorophyl deficiencies are of several kinds, and may be caused in maize by mutations in at least 180 different gene loci. An individual with normal chlorophyl may be heterozygous for two or more non-allelic mutants.

The extensive literature on mutations causing various physical or mental disabilities in man has been concisely reviewed by Dobzhansky (1970). About 2.5 percent of all children are born with genetically determined malformations detectable by the naked eye (Stevenson, 1959, 1961). McKusick (1975) lists 1,218 dominant, 947 recessive, and 171 sex-linked traits in man determined by single-gene mutants, most of which result in mild or severe physical disabilities. Genetic predisposition is also involved in the etiology of many mental disorders. When physical and mental disabilities are all considered, the total incidence of congenital defects in human populations is about 4.5 percent, although official records generally report lower values (Kennedy, 1967). More than 40 different blood group polymorphisms, each determined by one or more gene loci, are known in man. The loci range from slightly polymorphic to very polymorphic, like the well known ABO and MN blood groups.

GENETIC VARIATION AND ARTIFICIAL SELECTION

The pervasiveness of genetic variation in natural populations is evident from the success of artificial selection for a great variety of traits in many different organisms. In artificial selection those individuals that exhibit the greatest expression of the desired characteristics are chosen to breed the next generation. Heritable changes over the generations in the phenotypic distribution of a population with respect to the selected trait indicate that the population has genetic variance for the selected trait.

Artificial selection has been successful for an innumerable variety of commercially desirable traits in many domesticated animal and plant species, including cattle, swine, sheep, poultry, corn, rice, and wheat (Falconer, 1964; Brewbaker, 1964; Lerner, 1958, 1968). The changes obtained by artificial selection are often staggering. Lerner (1958, 1968) has reported increases in the average egg production of a White Leghorn flock from 125.6 eggs per hen per year in 1933, to 249.6 eggs per hen per year in 1965. Artificial selection is sometimes practiced in two opposite directions in different lines derived from the same population. Woodworth and co-workers (1952) selected for high and low protein and oil content in corn for 50 generations. Protein content changed from 10.9 to 19.4 percent in the high line, and from 10.9 to 4.9 percent in the low line. Oil content changed from 4.7 to 15.4 percent and from 4.7 to 1.0 percent in the high and low lines, respectively. Throughout the 50 generations of selection there was little or no decline in the effectiveness of the selection process, and in some lines at least no plateau was ever reached. Genetic variance for protein and oil content was still present in the populations at the end of the experiment, in spite of the stupendous improvements already achieved.

Successful artificial selection has been carried out for at least 51 different traits in *Drosophila melanogaster.* The traits selected include body and wing size, number of abdominal and sternopleural bristles, resistance to DDT, rate of development, fecundity, and certain behavioral traits, such as phototaxis and geotaxis (for review see Lewontin, 1974). Selection in *Drosophila* has often been successful in two opposite directions; for example, increased and decreased number of bristles (Robertson, 1959), and positive and negative geotaxis or phototaxis (Dobzhansky and Spassky, 1967).

There are few reports of unsuccessful attempts at artificial selection. Dickerson (1955) failed to change viability and egg production in a closed flock of chickens; Maynard Smith and Sondhi (1960) failed to modify the symmetry of the ocelli in *Drosophila* flies. Of course, the possibility exists that failed attempts at artificial selection may be reported less often than successful ones. Nevertheless, the fact remains that genetic changes have been achieved by artificial selection for a great diversity of traits in a large number of different species. The general success of artificial selection indicates that genetic variation for all sorts of traits is widespread in natural populations. Darwin reviewed the success of breeders and animal fanciers to show the potentialities of selection acting on preexisting variability to produce permanent heritable changes in populations.

CONCEALED GENETIC VARIATION AFFECTING FITNESS

Natural selection acts on alternative genetic variants affecting *fitness* (reproductive efficiency), such as viability, fecundity, and rate of development. Many allelic variants modifying fitness characters may not be expressed in typical wild individuals, where they are likely to exist in heterozygous associations. Techniques to study recessive genetic variation concealed in "normal" individuals have been developed in *Drosophila.* The techniques essentially consist of obtaining numerous flies all homozygous for a single chromosome sampled from a natural population. Recessive allelic variants present in the wild chromosome will be expressed in the homozygous flies. The joint effect on fitness of all variants in the chromosome can be measured as the average performance of the set of homozygous flies relative to "normal" flies, i.e., flies heterozygous for random combinations of wild chromosomes.

The basic experimental procedure is outlined in Figure 2-6. Special "balancer" stocks are used. A balancer stock usually carries at least one dominant mutant

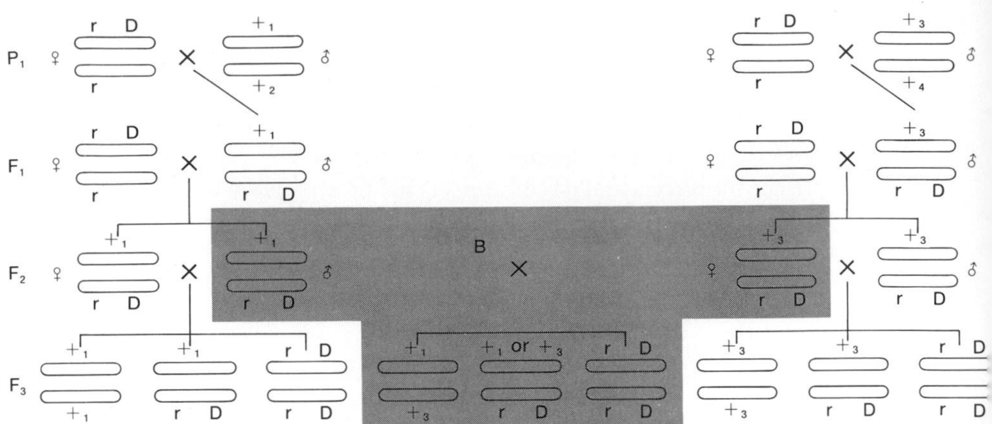

FIGURE 2-6.
Crosses used to obtain large numbers of *Drosophila* flies homozygous for a chromosome sampled from a natural population. **A.** In the P generation one wild male (carrying two wild chromosomes, either $+_1$ and $+_2$, or $+_3$ and $+_4$) is crossed to females of a balanced marker stock. The balancer stock contains one chromosome with multiple inversions (to inhibit recombination) and two mutant markers, one dominant and the other recessive; the second chromosome contains the same recessive mutant marker. In the F_1 generation a single male heterozygous for one wild chromosome and the balanced marker chromosome is crossed to females of the balancer stock. In the F_2 generation males and females heterozygous for the same wild chromosome and the balanced marker chromosome are intercrossed. Three kinds of progeny are expected in the F_3 generation: one-fourth should be homozygous for the wild chromosome, one-half should be heterozygous for the wild and the balanced marker chromosome, and one-fourth should be homozygous for the balanced marker chromosome (which also carries a recessive lethal gene, so that these flies die). Deviations between the observed and expected numbers of flies in the F_3 generation are used to measure the effects of the wild chromosome on the viability of homozygous flies. **B.** Control crosses are made by intercrossing F_2 flies heterozygous for the balancer chromosome and different wild chromosomes. The wild flies in the F_3 generation of these crosses are heterozygous for different wild chromosomes, and thus are genetically similar to wild flies.

in each chromosome, a recessive lethal gene, and one or more chromosomal inversions. The dominant mutant permits identification of heterozygotes carrying the balancer chromosome; the inversions inhibit genetic recombination in the heterozygotes. Flies homozygous for a balancer chromosome die because of the recessive lethal.

Males collected in a natural population are individually crossed to females of the balancer stock. A single F_1 male progeny from each cross is mated with one or several females of the balancer stock. The one wild chromosome carried by this F_1 male is multiplied without recombination in the F_2 flies heterozygous for the wild and balancer chromosomes. These F_2 heterozygous flies are intercrossed to produce in the F_3 three kinds of zygotes: homozygotes for the wild chromosome sampled from the natural population; heterozygotes for the wild and balancer chromosomes; and homozygotes for the balancer chromosome, which die before completing development. Any recessive genes in the wild chromosome will be expressed in the wild homozygotes. The homozygotes for the wild chromosome and the heterozygotes for the wild and balancer chromosomes are expected in the ratio 1:2, if there are no differences in viability between them. Departures from the expected ratio allow measuring the homozygous effects on viability of the wild chromosome.

The marker chromosome may affect the viability of its heterozygous carriers. This effect may be corrected for by means of the control crosses indicated in Figure 2-6. Crosses may be made in the F_2 generation between males from one line and females of another. The F_3 wild flies from these matings are heterozygous for two different wild chromosomes, and thus similar to wild flies from the natural population. Deviations from the expected 1:2 ratio in the F_3 progenies of these control crosses give an estimate of the viability of flies heterozygous for wild and balancer chromosomes relative to "normal" wild individuals.

Typical results of this kind of experiment are those obtained by Dobzhansky and Spassky (1963), who studied a large number of wild second chromosomes from a natural population of *Drosophila pseudoobscura.* Figure 2-7 shows the distribution of the homozygote viabilities of the chromosomes relative to that of flies heterozygous for random combinations of two wild chromosomes. The mean viability of these random heterozygotes is made equal to 100 percent. In homozygous conditions 28 percent of the wild chromosomes are effectively lethal or semilethal, since few or no flies emerge in the F_3 generation. A majority of the chromosomes (72 percent) are called "quasinormals." These have homozygous viabilities approaching that of random heterozygotes, but significantly lower on the average.

The variance in a distribution like that shown in Figure 2-7 has several causes. Some variance is due to sampling errors, since only 200 to 300 flies are counted in each F_3 generation; some may be due to uncontrollable environmental differences between cultures; and some to segregation in the rest of the genotype. Methods exist to separate these various effects from those due to genetic differences between the chromosomes tested (Wallace and Madden, 1953; Dobzhansky, 1970; Lewontin, 1974). After this is done, the wild chromosomes are classified as being lethal, semilethal, subvital, quasinormal, or supervital in homozygous condition, as shown in Figure 2-7 and Table 2-3.

Appropriate laboratory stocks to carry out the crosses shown in Figure 2-6 exist only in five species: *Drosophila melanogaster, D. pseudoobscura, D. persimilis, D. willistoni,* and *D. prosaltans.* Some of the numerous studies that have been carried out in these species are summarized in Table 2-4. The proportion of homozygous lethal chromosomes ranges from 10 percent for third chromosomes of *D. prosaltans* from Bahia, Brazil, to 60 percent for third chromosomes of *D. melanogaster* from Napa County, California. The average viability of quasinormal chromosomes ranges from 96.6 percent for third chromosomes of *D. prosaltans* from Bahia to 75.0 percent for second chromosomes of *D. pseudoobscura* from Stanislaus Forest, California.

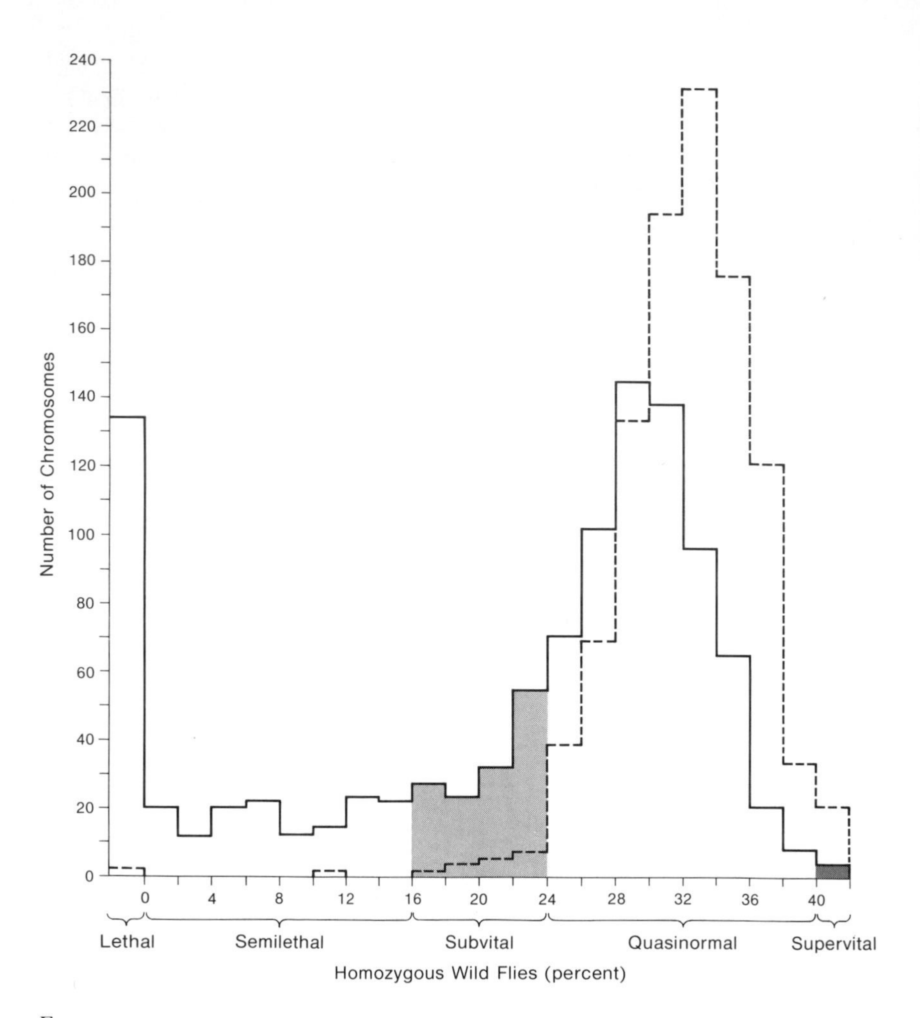

FIGURE 2-7.
Mean viability of *Drosophila pseudoobscura* homozygous (solid line) or heterozygous (broken line) for wild second chromosomes sampled from a single population. The mean viability of the heterozygous flies is 32.3 percent and the standard deviation 4.0 percent. A total of 1,063 wild second chromosomes were tested in homozygous condition: 134 chromosomes (13 percent of the total) had lethal effects, another 158 (15 percent) were semilethal, i.e., reduced the viability of homozygotes below half the deviations from the heterozygote mean (but above half the mean) are called *subvitals* (light grey); those increasing viability above two standard deviations from the heterozygote mean are called *supervitals* (dark grey). Chromosomes resulting in viabilities between two standard deviations around the mean are called *quasinormal*. [Data from Dobzhansky and Spassky, 1963.]

TABLE 2-3. Frequencies of wild chromosomes of *Drosophila pseudoobscura* (from Stanislaus National Forest, California) with different effects on the viability of homozygous flies. [After Dobzhansky and Spassky, 1953.]

Effect on viability	Viability relative to heterozygous flies	Chromosome		
		Second	Third	Fourth
Lethal and semilethal	Less than 50 percent	33.0%	25.0%	25.9%
Subvitals	More than 50 percent but below two standard deviations from the mean	62.6	58.7	51.8
Quasinormals	Between two standard deviations below and two standard deviations above the mean	4.3	16.3	22.3
Supervitals	Above two standard deviations from the mean	<0.1	<0.1	<0.1

TABLE 2-4. Frequency of chromosomes causing lethality or semilethality in homozygous condition in several species of *Drosophila*.

	Chromosome	Locality	Percent lethals and semilethals
D. melanogaster	II	Wisconsin	25.1%
	II	Long Island, New York	20.5
	II	Napa Valley, California	36.0
	III	Napa Valley, California	60.0
D. pseudoobscura	II	Bogota, Colombia	18.3
	II	Stanislaus Forest, California	33.0
	II	San Jacinto, California	21.3
	III	Mexico and Guatemala	30.0
	III	Stanislaus Forest, California	25.0
	III	Death Valley, California	15.0
	IV	Stanislaus Forest, California	25.9
	IV	San Jacinto, California	25.5
D. persimilis	II	Stanislaus Forest, California	25.5
	III	Stanislaus Forest, California	22.7
	IV	Stanislaus Forest, California	28.1
D. willistoni	II	Brazil (various locations)	41.3
	II	São Paulo, Brazil	24.0
	III	Brazil (various locations)	32.1
D. prosaltans	II	Brazil (various locations)	32.6
	III	Brazil (various locations)	9.5

The technique depicted in Figure 2-6 can be employed to detect concealed genetic variants affecting other fitness parameters besides viability. Fecundity, rate of development, and other fitness parameters can be examined in the homozygous flies of the F_3 generation, provided of course that the chromosome is not homozygous lethal. The technique can also be used for detecting recessive genes with visible effects, but it is unnecessarily cumbersome for that purpose.

Marinković (1967a) has studied the fecundity of flies homozygous for second chromosomes from a natural population of *Drosophila pseudoobscura*. The mean number of eggs laid per female between the sixth and twelfth day after emergence from the pupa is 260.4 for homozygous females, versus 322.2 for females heterozygous for two wild chromosomes. The fecundity of homozygous relative to heterozygous females is 260.4/322.2, or 80.7 percent. (For a similar study in *D. melanogaster* see Temin, 1966.) Marinković (1967b) also studied the effects of second chromosomes of *D. pseudoobscura* on rate of development. Homozygous flies require a longer time to complete development from egg to adult than do heterozygous flies.

The studies just reviewed of concealed genetic variation affecting fitness indicate that in homozygous condition most chromosomes reduce one or more fitness traits. However, there are a number of chromosomes that do not deviate from "normality" for one or another fitness trait. Are these chromosomes "normal," in the sense predicted by the classical model of population structure, that is, because at every locus they carry the "wild type" or "normal" allele? The answer is in the negative as shown by the following experiment.

Dobzhansky and co-workers (1959) selected the quasinormal chromosomes from a distribution like that shown in Figure 2-7. Genetic recombination may occur in females heterozygous for pairs of such chromosomes. The recombinant chromosomes were then made homozygous by the procedure outlined in Figure 2-6, and their viability effects measured. The results are shown in Table 2-5. The distribution of homozygous viabilities of these recombinant chromosomes extends over the same spectrum as the original chromosomes. The variance in viability among the recombinant chromosomes is 25 to 43 percent as large as that of the original wild chromosomes. Chromosomes with quasinormal

TABLE 2-5. Variation in viability obtained through recombination between chromosomes having "quasinormal" viability in homozygous condition.

Species	Lethal chromosomes	Variance of viability: Among natural chromosomes (a)	Among recombinant chromosomes (b)	Recovered variance (b/a)
Drosophila pseudoobscura	4.1%	140	60	42.9%
Drosophila prosaltans	5.7%	200	50	25.0%

homozygous viabilities are very different from each other in their genetic contents, although the interactions between their respective sets of alleles have similar effects on viability.

In the studies reviewed so far, the homozygous effects of chromosomes are measured with respect to one or another fitness parameter. It is preferable, however, to study their effects on total fitness. Sved and Ayala (1970) have devised a method to compare the fitness of chromosomal homozygotes relative to chromosomal heterozygotes. Chromosomes are obtained from a natural population, using the method shown in Figure 2-6. The flies recovered in the F_3 generation are used to establish experimental populations, where the course of natural selection can be studied over many generations. Since the balancer chromosome inhibits recombination, only two kinds of viable zygotes can exist at any time—those homozygous for the wild chromosome and those heterozygous for the wild and balancer chromosomes. All zygotes homozygous for the balancer chromosome die before completing development. If the homozygotes for the wild chromosome have lower fitness than the balancer heterozygotes, a stable equilibrium will eventually be established between the two types of flies. Their relative fitnesses can be directly calculated from the zygotic equilibrium frequencies. If the balancer heterozygotes have lower fitness than the chromosomal homozygotes, the balancer chromosome will gradually decrease in frequency. The relative fitness of the two kinds of flies can be estimated from the rate of elimination. Control experimental populations are set up with all the wild chromosomes and the balancer. In these control populations flies are either heterozygous for different wild chromosomes or heterozygous for one wild chromosome and the balancer. The former have genetic constitutions comparable to flies in a natural population. The control populations permit an estimation of the fitness of balancer heterozygotes relative to wild heterozygotes. This estimate of fitness can be used to estimate the fitness of chromosomal homozygotes relative to flies heterozygous for random combinations of wild chromosomes.

With the method of Sved and Ayala, overall fitness rather than a specific fitness trait is measured under population conditions. Under these conditions, the distribution of fitnesses of chromosomal homozygotes is very different from that shown in Figure 2-7. All chromosomes become either lethal or semilethal. Even when only chromosomes with quasinormal viabilities are selected, their overall fitness becomes extremely low. Figure 2-8 shows the fitness distribution of 23 second chromosomes of *Drosophila melanogaster* (Tracey and Ayala, 1974). Although the method is extremely laborious, studies have been conducted for *D. pseudoobscura* second chromosomes (Sved and Ayala, 1970); *D. melanogaster* second and third chromosomes (Sved, 1971; Tracey and Ayala, 1974); and *D. willistoni* second and third chromosomes (Mourão, Ayala, and Anderson, 1972). The results are summarized in Table 2-6.

The data reported in Table 2-6 are uniform. In all three species chromosomal homozygotes are lethal or nearly so when overall fintess is considered. The class of quasinormal chromosomes disappears. The number of gene loci

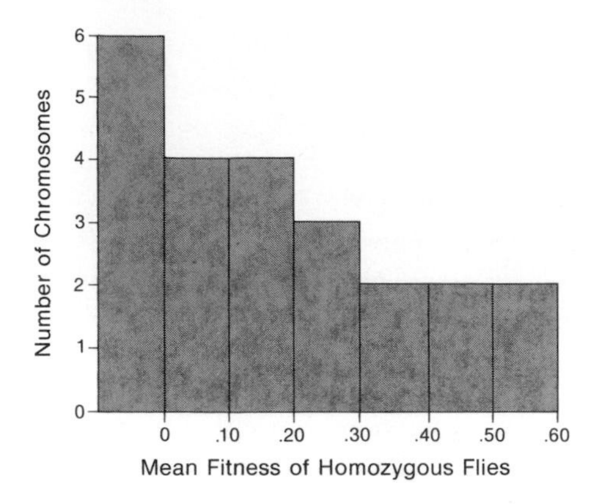

FIGURE 2-8
Fitness of *Drosophila melanogaster* homozygous for second chromosomes sampled from a natural population. The relative fitness of flies homozygous and heterozygous for wild chromosomes is measured in experimental population cages (see Fig. 4-6, p. 113). All the chromosomes had quasinormal viability using the tests shown in Figures 2-6 and 2-7. When all fitness components are taken into consideration, virtually all chromosomes have lethal or semilethal effects in homozygous flies.

TABLE 2-6. Fitness of *Drosophila* flies homozygous for second chromosomes with quasinormal viability obtained from natural populations.

		Fitness of homozygous flies		
	Chromosome	Mean	Standard error	Range
D. melanogaster	II	0.192	0.029	0.00–0.60
	III	0.318	0.037	0.06–0.73
D. pseudoobscura	II	0.366	0.030	0.10–0.58
D. willistoni	II	0.339	0.023	0.15–0.61

polymorphic for alleles modifying fitness cannot be determined from these studies. Yet one consideration indicates that many loci are involved. If only a few loci had allelic variants affecting fitness, one might expect the variance in fitness among chromosomal homozygotes to be large, and some chromosomes to be free or nearly so of any deleterious allele. However, all chromosomes substantially decrease fitness, and no chromosome confers on its homozygous carriers a fitness approaching that of heterozygotes for wild chromosomes. According to Tracey and Ayala (1974), more than one thousand polymorphic loci may be maintained in natural populations of *D. melanogaster* by balancing selection.

MOLECULAR TECHNIQUES TO QUANTIFY GENETIC VARIATION

The evidence reviewed in the previous sections indicates that large stores of genetic variation exist in natural populations. Yet the evidence does not permit an estimate of how many genes or what proportion of all genes are polymorphic, nor how polymorphic they are. The degree of genic variation in a population can be estimated only if one can detect allelic variants in single genes representing a random sample of the total genome (Lewontin and Hubby, 1966).

Studies of artificial selection and concealed genetic variation modifying fitness indicate that allelic variation is a pervasive phenomenon. Yet there is no way to determine precisely how many gene loci are variable, how polymorphic these are, or how many other invariant gene loci exist in the population. Allelic variants can be identified in single-gene loci modifying certain morphological traits or determining blood groups. For many such genes there is a one-to-one correspondence between genotype and phenotype. Yet a serious methodological handicap remains. In Mendelian genetics the presence of a gene is ascertained by studying segregation in matings between individuals. Only genes with allelic variation can be shown to exist. Therefore it is not possible to obtain a sample of the genome random with respect to variation, since only variable genes can be studied. Another difficulty arises because in Mendelian studies of single genes individuals must be assigned to discrete classes readily distinguishable. Yet most of the genetic variation relevant to evolution affects characteristics with continuous expression, such as fecundity, viability, longevity, rate of development, and size.

Recent developments in molecular genetics have made it possible to obtain a random sample of the genome, and to detect allelic variation in individual loci. The genetic information encoded in the nucleotide sequence of the DNA of structural genes is translated into a sequence of amino acids making up a polypeptide chain. Most enzymes and proteins are the products of individual genes, although some are made up of polypeptides coded by two or more genes. It is possible, to a first approximation, to equate variation in enzymes or proteins with variation in genes. A sample of proteins random with respect to their genetically determined variation may be obtained. The genes coding for such proteins represent a random sample of the genome. Variant as well as invariant gene loci can be surveyed. Variation in the amino acid sequence of a protein implies allelic variation in the gene coding for it. If we ignore the problem of redundant codons, an invariant protein implies lack of allelic variation in the gene coding for it.

To obtain the amino acid sequence of a protein is at present a delicate and time-consuming operation. To obtain the amino acid sequence of many proteins in each of many individuals is prohibitive because of the cost and time involved in purifying and sequencing the proteins. However, some relatively simple techniques that permit detection of amino acid substitutions in individual proteins have recently been developed. These are gel electrophoresis and enzyme assay. The procedure is schematically represented in Figure 2-9. Tissue samples from organisms are individually homogenized to release the proteins and enzymes. The homogenate supernatants are placed in a gel made of starch, agar, polyacrylamide, or some other jellylike substance. The gel with the tissue samples is then subjected for a given length of time to an electric current.

A. Tissue homogenates are placed in a gel made of starch, agar, polyacrylamide, cellulose, or some other substance providing a homogenous matrix.

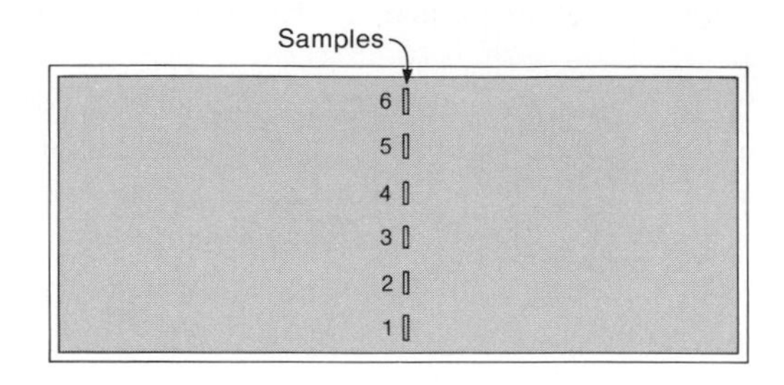

B. The gel with the samples in it is placed in an electric field for several hours.

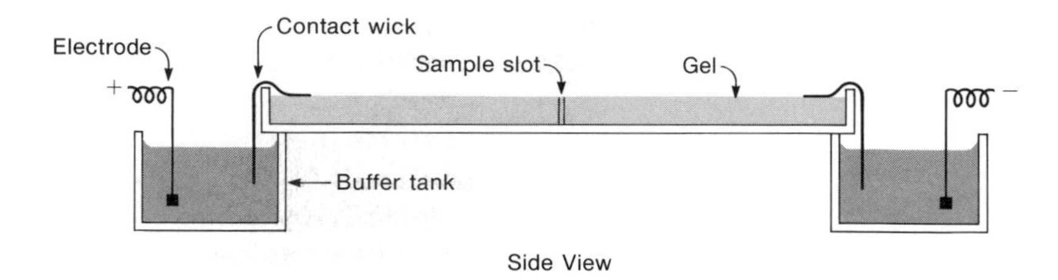

C. The gel is placed in a solution with an appropriate substrate for the enzyme to be assayed and a salt that produces a colored band when reacting with the product of the reaction catalyzed by the enzyme.

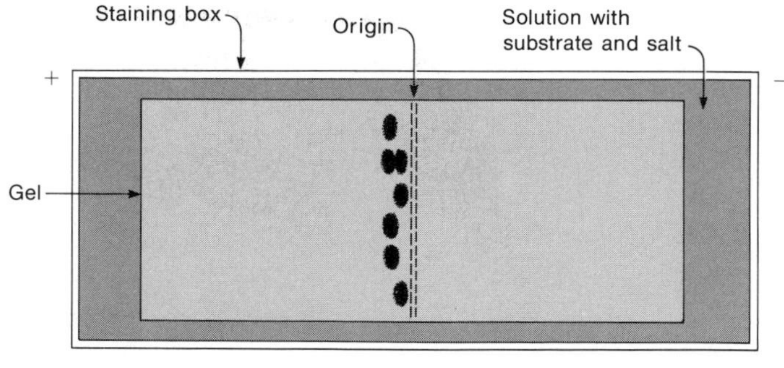

FIGURE 2-9.
Schematic representation of the techniques of gel electrophoresis, used to ascertain genetic variation in natural populations.

Each protein in the gel migrates in a direction and at a rate that depend on net electrical charge and molecular size. After removing the gel from the electric field, it is treated with appropriate chemical solutions to visualize the position of proteins or specific enzymes.

The appearance of two illustrative gels is shown in Figures 2-10 and 2-11. The gel in Figure 2-10 contains samples of *Drosophila pseudoobscura* female flies assayed for the enzyme phosphoglucomutase. The first two individuals

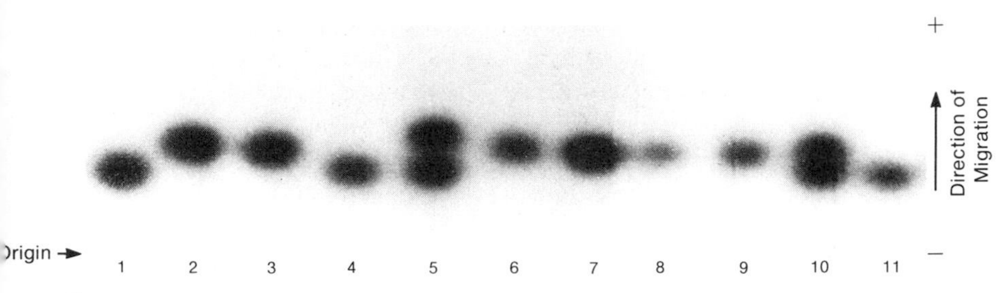

FIGURE 2-10.
Genetic variation at the *Pgm-1* locus in a sample of 11 *Drosophila pseudoobscura* females from a natural population. Crude homogenates of each fly were placed along a slot (the site marked "origin") in a gel. The gel was subject to electrophoresis for five hours and then stained for the enzyme phosphoglucomutase. Homozygous flies exhibit only one band; heterozygous flies exhibit two bands. The first fly (far left) is homozygous for the allele *100*, the second is homozygous for the allele *104*, the fifth is heterozygous for the alleles *100* and *104*, and so on.

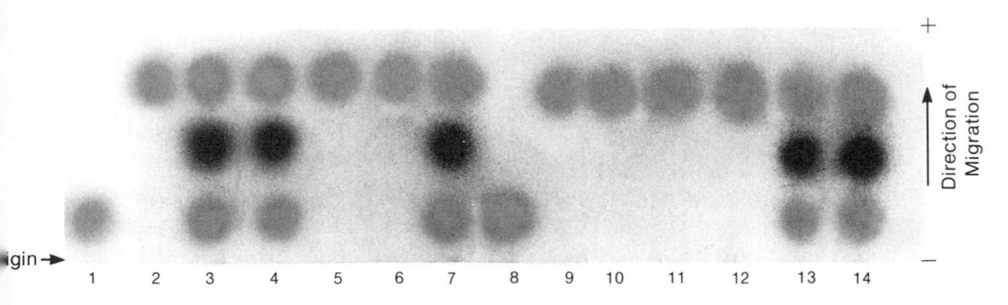

FIGURE 2-11.
Genetic variation at the *Acph* locus in a sample of 14 *Drosophila equinoxialis* flies. After electrophoresis the gel was stained for the enzyme acid phosphatase. Homozygous flies exhibit only one band; heterozygous flies exhibit three bands because the enzyme is a dimer. The first fly is homozygous for the allele *88*, the second is homozygous for the allele *100*, the third is heterozygous for alleles *88* and *100*, and so on.

have enzymes with different electrophoretic mobilities and thus different amino acid sequences. We shall represent the gene coding for phosphoglucomutase as *Pgm*, and the alleles coding for the enzymes of the first two individuals as Pgm^{100} and Pgm^{104}, where the superscripts reflect the difference in electrophoretic mobility of their enzyme products. Since the first two individuals each exhibit only one band, we infer that they are homozygotes with genotypes $Pgm^{100/100}$ and $Pgm^{104/104}$, respectively. The fifth individual in Figure 2-10 exhibits two phosphoglucomutase bands, which correspond with the bands of the first two individuals. We infer that the fifth individual is heterozygous at this locus, with the genotype $Pgm^{100/104}$. (The *Pgm* locus is sex-linked in *D. pseudoobscura.* The flies assayed in Figure 2-10 are females; males exhibit only one band for this enzyme because they have only one X chromosome.)

Figure 2-11 represents a gel with samples of *Drosophila equinoxialis* flies assayed for the enzyme acid phosphatase. The first two individuals again exhibit only one band each. If we symbolize the gene locus coding for acid phosphatase as *Acph,* their genotypes can be written as $Acph^{88/88}$ and $Acph^{100/100}$. The active form of acid phosphatase is, however, a dimer consisting of two units. The active enzymes in the first two individuals may be represented as (88 + 88) and (100 + 100) respectively. A heterozygote with the genotype $Acph^{88/100}$ will have three active forms of the enzyme, which may be represented as (88 + 88), (88 + 100), and (100 + 100). The third individual in Figure 2-11 has such a genotype. The enzyme (88 + 100) has an electrophoretic mobility intermediate between (88 + 88) and (100 + 100). If the two peptides associate at random, the heterodimer (88 + 100) will be twice as abundant as each homodimer. Tetramer and even higher-order enzymes are known; the genetic interpretation of electrophoretic phenotypes is based on the same basic principles just illustrated.

Electrophoretic techniques permit detection of allelic variants in individual genes. Variant as well as invariant gene loci can be identified. A random sample of genes with respect to variation is possible. Proteins and enzymes for which the appropriate assay techniques exist can be chosen for study without knowing a priori whether they are variable or how variable. A moderate number of proteins studied in a moderately large number of individuals is sufficient to estimate the amount of variation over the entire genome in a population. However, not all allelic variants are detectable by gel electrophoresis. Generally, only those amino acid substitutions that alter the net charge of the protein will change its mobility in an electrophoretic gel. Considerations of the genetic code and of electrical properties of amino acids suggest that only one third of all amino acid replacements are detectable by gel electrophoresis (Lewontin, 1974). Consequently, the amount of allelic variation is underestimated.

The basic information obtained from electrophoretic surveys of protein variation consists of the genotypic or the allelic frequencies at each locus in a given population. For simplicity, only the allelic frequencies are usually given. Generally there are fewer alleles than genotypes, never more. If there are n different alleles at a locus, the number of possible different genotypes is $n(n + 1)/2$. A variety of measures can be used to express in a single statistic the amount of genetic variation in a population. For a random mating population the most informative measure is the overall incidence of heterozygosity, which itself can be expressed in a variety of ways. The proportion of polymorphic loci in a population is another commonly used measure, but one which is less precise and informative.

In random mating populations the expected frequency of heterozygotes (H) at a locus can be directly calculated from the allelic frequencies. If there are n alleles with frequencies $f_1, f_2, f_3 \ldots f_n$, the expected frequency of homozygotes is simply, $f_1^2 + f_2^2 + f_3^2 \ldots + f_n^2$. The expected frequency of heterozygotes is $H = 1 - (f_1^2 + f_2^2 + f_3^2 + \ldots + f_n^2)$. A related measure is the effective number of alleles, n_e, which is the reciprocal of the frequency of homozygotes. The overall amount of variation in a population is estimated by the average frequency of heterozygotes per locus ($\overline{H}$). This is simply obtained by averaging H over all loci sampled. $\overline{H}$ may be expressed with its standard error, which reflects the amount of heterogeneity among the loci sampled.

The heterozygosity of a population can also be expressed as the average frequency of heterozygous loci per individual ($\overline{H}_i$). This is estimated by averaging (over all individuals) the proportion of heterozygous loci observed in each individual. The values of $\overline{H}$ and $\overline{H}_i$ are the same but their variance and standard errors are generally different. The variance of $\overline{H}_i$ reflects the heterogeneity among individuals for the set of loci studied. The variance of $\overline{H}$ measures heterogeneity among loci. In a random mating population $\overline{H}_i$ is normally distributed and, relatively speaking, small. There is no a priori reason, however, why $\overline{H}$ should be normally distributed, and generally it is not. The variance of $\overline{H}$ is generally larger than the variance of $\overline{H}_i$ (compare Figures 2–12 and 2–13). To estimate the amount of genetic variation in a population *over the entire genome*, $\overline{H}$ with its standard error is preferable since this statistic reflects the great amount of heterogeneity among loci, and may vary in value substantially when different sets of loci are sampled.

Genetic variation can also be measured by the proportion of polymorphic loci (P) in a population. This statistic is to a certain extent arbitrary and imprecise. It is arbitrary because it must first be decided when a locus will be considered polymorphic. Two criteria are commonly used: (1) the frequency of the most common allele in the population is no greater than 0.95, and (2) it is no greater than 0.99. The first criterion is more restrictive than the second, since every locus polymorphic by the first is also polymorphic according to the second, but not vice versa. P is also imprecise because for each locus it establishes only whether or not it is polymorphic, not how polymorphic it is. A locus with two alleles with frequencies 0.95 and 0.05, and a second locus with 10 alleles each with a frequency of 0.10, contribute equally to P, although the second locus has more genetic variation. If several populations of a species are studied, the average proportion of polymorphic loci per population ($\overline{P}_p$) can be calculated as the average of P over all populations. Alternatively, the proportion of populations in which a locus is polymorphic may be calculated first, and the average, $\overline{P}_l$, estimates the average proportion of populations in which a locus is polymorphic. Generally, $\overline{P}_p$ has a smaller variance than $\overline{P}_l$.

QUANTIFICATION OF GENETIC VARIATION IN POPULATIONS

Electrophoretic techniques for estimating genetic variation in populations were first applied in man by Harris (1966), in *Drosophila ananassae* by Johnson and co-workers (1966), and in *D. pseudoobscura* by Lewontin and Hubby (1966). Many populations of different organisms have been surveyed since the time of those pioneer studies.

An extensive survey of genetic variation has been conducted in natural populations of the Neotropical species *Drosophila willistoni*. The kind of information obtained for each locus is illustrated in Table 2-7, which shows genetic variation at three loci in each of five populations of *D. willistoni*. The three loci have been chosen as examples because *Lap-5* is extremely polymorphic, *Est-5* is moderately so, and *To* exhibits little or no genetic variation. The alleles are symbolized by numbers that refer to the electrophoretic mobilities of the enzymes relative to a standard. The number of genomes sampled establishes the reliability of the estimates of allelic frequencies. In diploid organisms the number of genes sampled is twice the number of individuals, except for sex-linked genes in the heterogametic sex (the males in *Drosophila*). The *To* is sex-linked; the other two loci in Table 2-7 are autosomal. The expected

TABLE 2-7. Variation at three gene loci coding for enzymes in five natural populations of *Drosophi* *willistoni*. *Lap-5* codes for a leucine aminopeptidase enzyme, *Est-5* codes for an esterase, and *T* codes for tetrazolium oxidase. *H* is the expected frequency of heterozygotes.

		Localities				
Gene locus	Alleles	Puerto Rico	Dominican Republic	Tame, Colombia	Santarem, Brazil	São Paulo, Brazil
Lap-5	96	0.00	0.005	0.01	0.02	0.01
	98	0.04	0.03	0.12	0.14	0.07
	100	0.60	0.74	0.28	0.39	0.25
	103	0.34	0.22	0.55	0.43	0.57
	105	0.02	0.005	0.04	0.02	0.09
	H	0.527	0.408	0.600	0.649	0.618
Genomes sampled:		320	620	192	492	1,806
Est-5	98	0.00	0.00	0.00	0.01	0.01
	100	0.13	0.18	0.02	0.06	0.04
	102	0.86	0.81	0.95	0.91	0.93
	104	0.01	0.006	0.04	0.01	0.02
	H	0.252	0.307	0.102	0.159	0.136
Genomes sampled:		636	320	190	82	1,916
To	86	0.01	0.00	0.00	0.00	0.000
	98	0.00	0.00	0.01	0.00	0.004
	100	0.99	1.00	0.97	0.98	0.995
	102	0.00	0.00	0.01	0.02	0.001
	H	0.012	0.000	0.025	0.032	0.010
Genomes sampled:		508	260	161	244	1,038

frequencies of heterozygotes are also given in Table 2-7. These are calculated on the assumption of random-mating equilibrium, according to the formula given above. In these populations of *D. willistoni* the observed and expected frequencies of heterozygotes are in good agreement.

A summary of genetic variation at the 36 enzyme loci studied in *D. willistoni* is given in Table 2-8. Some loci, such as *Adk-1, Est-7,* and *Lap-5,* are extremely polymorphic, with more than 50 percent heterozygotes. Other loci, such as *Idh, Mdh-2,* and *Me-1,* have little allelic variation. Most loci fall between these two extremes. A great range in the frequency of heterozygosity per locus is generally found in most organisms studied. There is considerable genetic variation in populations of *D. willistoni*. On the average, 18.0 ± 2.8 percent of the individuals are heterozygous per locus. The average proportion of polymorphic populations per locus is 53.0 ± 5.9 percent according to criterion 1, and 80.7 ± 3.9 percent according to criterion 2.

Figures 2-12 and 2-13 summarize a study of genetically determined protein variation in deep-sea invertebrates. Figure 2-12 is a histogram of the distribution of the frequency of heterozygotes per locus in six species. The mean value of H is 16.6 ± 5.4 percent. Populations of these organisms contain nearly as much genetic variation as *D. willistoni*. Figure 2-13 shows the distribution of the proportion of heterozygous loci per individual in the brittle star *Ophiomusium lymani*, illustrating the typical normal distribution of this statistic. The mean value of H_i is 16.4 ± 0.5 percent. As pointed out above, the variance of H_i is usually substantially less than that of H.

Genetic variation has been studied using the techniques of gel electrophoresis in a large number of organisms. New surveys appear every year in the literature. Table 2-9 summarizes the results obtained in 125 animal species and 8 plant species in which a fairly large number of loci have been assayed.

A great deal of genetic variation exists in most organisms studied. A most conspicuous exception is *Rumina decollata*, a self-fertilizing snail introduced into the United States before 1822, and in which no genetic variation has been detected in U.S. populations (Selander and Kaufman, 1973). No genetic variation has been observed in the elephant seal, *Mirounga angustirostris* (Bonnell and Selander, 1974), and little in the two species of starfish, *Asterias vulgaris* and *A. forbesi* (Schopf and Murphy, 1973). These snail, seal, and starfish populations have probably passed through recent periods when they consisted of few individuals, which may account for their reduced genetic variation. In general, vertebrates have less genetic variation than invertebrates. For the 57 invertebrate species in Table 2-9 the average heterozygosity ($\bar{H}$) has a mean value of 13.4 percent; for the 68 vertebrate species the average value of $\bar{H}$ is 6.0 percent. The average heterozygosity in man, 6.7 percent (Harris and Hopkinson, 1972), is very similar to the vertebrate average.

The amount of genetic variation in most organisms is staggering. Consider man, with an average heterozygosity of 6.7 percent, and 28 percent of the loci polymorphic according to criterion 2. If we assume that the genome of man consists of 100,000 structural gene loci (McKusick, 1975), an individual would be heterozygous at about 6,700 structural genes, and about 28,000

TABLE 2-8. Variation at 36 gene loci coding for enzymes in *Drosophila willistoni*. Data are based on the study of more than 5,000 genomes sampled from more than 100 different localities. [From Ayala *et al.*, 1974a, and other sources.]

Gene locus	Enzyme coded	Frequency of polymorphic populations*		Frequency of heterozygous individuals
		1	2	
Acph-1	Acid phosphatase	0.56	0.95	0.107
Acph-2	Acid phosphatase	0.83	0.85	0.148
Adh	Alcohol dehydrogenase	0.28	0.46	0.075
Adk-1	Adenylate kinase	1.00	1.00	0.572
Adk-2	Adenylate kinase	0.55	0.75	0.094
Ald-1	Aldolase	0.65	0.88	0.157
Ald-2	Aldolase	0.63	1.00	0.132
Ao-1	Aldehyde oxidase	1.00	1.00	0.334
Ao-2	Aldehyde oxidase	1.00	1.00	0.482
Aph	Alkaline phosphatase	0.85	1.00	0.243
Est-2	Esterase	0.67	0.95	0.112
Est-3	Esterase	0.75	1.00	0.107
Est-4	Esterase	0.35	0.48	0.152
Est-5	Esterase	0.24	0.75	0.089
Est-6	Esterase	1.00	1.00	0.285
Est-7	Esterase	1.00	1.00	0.601
Fum	Fumarase	0.50	0.69	0.092
Got	Glutamate-oxaloacetate transaminase	0.00	0.50	0.038
αGpd	Alpha-glycerophosphate dehydrogenase	0.02	0.25	0.019
G3pd	Glyceraldehyde-3-phosphate dehydrogenase	0.65	1.00	0.123
G6pd	Glucose-6-phosphate dehydrogenase	0.75	1.00	0.201
Hbdh	Hydroxybutyrate dehydrogenase	0.00	0.35	0.015
Hk-1	Hexokinase	0.36	1.00	0.054
Hk-2	Hexokinase	0.78	0.95	0.183
Hk-3	Hexokinase	0.02	0.70	0.042
Idh	Isocitrate dehydrogenase	0.06	0.55	0.030
Lap-5	Leucine aminopeptidase	1.00	1.00	0.537
Mdh-2	Malate dehydrogenase	0.24	0.70	0.073
Me-1	Malic enzyme	0.03	0.75	0.056
Me-2	Malic enzyme	0.85	1.00	0.321
Odh-1	Octanol dehydrogenase	0.47	0.95	0.132
Odh-2	Octanol dehydrogenase	0.16	0.36	0.106
Pgm-1	Phosphoglucomutase	0.62	1.00	0.143
To	Tetrazolium oxidase	0.15	0.58	0.072
Tpi-2	Triose phosphate isomerase	0.06	0.64	0.023
Xdh	Xanthine dehydrogenase	1.00	1.00	0.532
	Average frequency of polymorphic populations per locus:			
	Criterion 1	0.530 ± 0.059		
	Criterion 2	0.807 ± 0.039		
	Average frequency of heterozygous individuals per locus	0.180 ± 0.028		

*Two criteria are used to decide whether a locus is polymorphic in a given population: under criterion 1 a locus is considered polymorphic when the frequency of the most common allele is no greater than 0.950; under criterion 2, when it is no greater than 0.990.

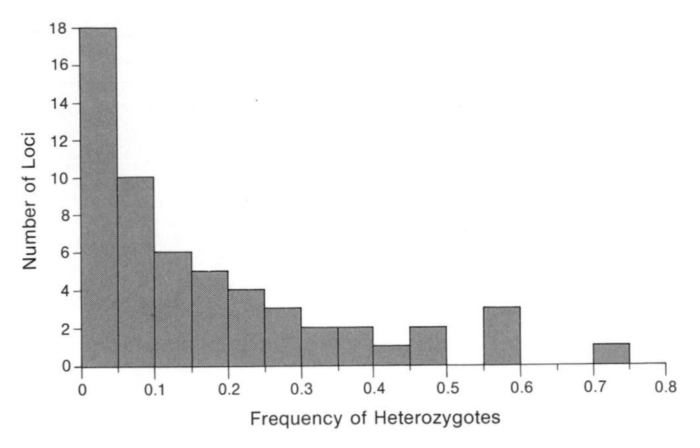

FIGURE 2-12.
Distribution of the frequency of heterozygous individuals per gene locus in deep-sea organisms. A total of 57 gene loci coding for enzymes were sampled in six different species (the brachiopod *Frieleia halli*, the brittle star *Ophiomusium lymani*, and four starfish: *Nearchaster aciculosus*, *Pteraster jordani*, *Diplopteraster multipes*, and *Myxoderma sacculatum ectenes*). The mean proportion of heterozygous individuals per gene locus is 16.6 ± 5.4 percent. The distribution is unimodal, with the largest class at the lower end of the distribution. [See Ayala and Valentine, 1974.]

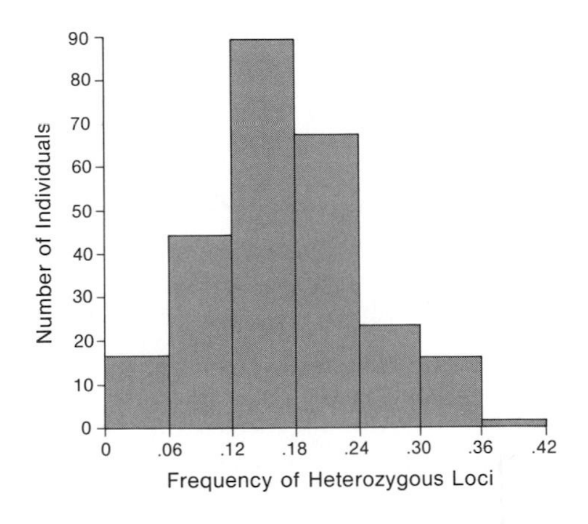

FIGURE 2-13.
Distribution of heterozygous gene loci in 257 individuals of the brittle star *Ophiomusium lymani*. Fifteen loci coding for enzymes were studied. The mean frequency of heterozygous loci per individual is 16.4 ± 0.5 percent. The distribution is approximately normal. [See Ayala and Valentine, 1974.]

gene loci would be segregating in a typical population. An individual heterozygous at 6,700 genes can potentially produce $2^{6,700}$ or $10^{2,017}$ different kinds of gametes. Even if we assume that the number of structural gene loci in man is only 30,000, the number of different kinds of gametes that can potentially be produced by a typical individual is 10^{605}.

We must now recall that electrophoretic studies underestimate the amount of genetic variation, since not all amino acid replacements result in proteins with different electrophoretic mobilities. Bernstein and co-workers (1973), using heat denaturation techniques, have found about 1.7 more alleles than

TABLE 2-9. Genic variation in natural populations of some major groups of animals and plants. The data are for continental species with large population sizes. [After Selander, 1976.]

	Number of species	Mean number of loci per species	Proportion of polymorphic loci per population*	Proportion of heterozygous loci per individual
Invertebrates:				
Drosophila	28	24	0.529	0.150
Haplodiploid wasps	6	15	0.243	0.062
Other insects	4	18	0.531	0.151
Marine	14	23	0.439	0.124
Land snails	5	18	0.437	0.150
Vertebrates:				
Fish	14	21	0.306	0.078
Amphibians	11	22	0.336	0.082
Reptiles	9	21	0.231	0.047
Birds	4	19	0.145	0.042
Mammals	30	28	0.206	0.051
Plants, outcrossing	8	8	0.464	0.170
Mean values:				
Invertebrates	57	21.8	0.469	0.134
Vertebrates	68	24.1	0.247	0.060
All animals	125	23.0	0.348	0.094

*The criterion of polymorphism is not the same for all species.

had been detected by electrophoresis at a gene locus in *Drosophila virilis*. The estimates of $\overline{H}$ given in Table 2-9 may have to be multiplied by a factor of perhaps two or three to estimate the true degree of allelic variation in the gene loci studied.

A note of caution must be advanced. For the animal species in Table 2-9, the number of gene loci that have been studied ranges from about 15 (in wasps) to 71 (in man). These numbers may represent less than one per thousand of all structural genes in these organisms. Moreover, most of the loci assayed code for enzymes and other soluble proteins. Genes coding for nonsoluble proteins and for regulatory genes are generally not included in the surveys. At present there is no way to ascertain whether the kinds of gene loci studied are a fair random sample of the total genome. Therefore, the estimates in Table 2-9 should be accepted with caution, although little doubt remains that natural populations of most organisms contain large stores of genetic variation.

3

The Origin of Hereditary Variation

CLASSIFICATION OF MUTATIONS

Heredity is a conservative process, although not perfectly so. The information encoded in the nucleotide sequence of the DNA is, as a rule, faithfully reproduced during replication, so that each replication results in two DNA molecules identical to each other and to the parental one. Occasionally, however, "mistakes" in the process of replication lead to different nucleotide sequences in parental and daughter DNA molecules. Changes in the hereditary materials are known as *mutations*. More than two million species of organisms live on earth, descendants of one or only a few primordial forms of life. Without the occurrence of hereditary changes, life could not have evolved or diversified—the same kinds of organisms that existed three billion years ago, and no others, would be living today, providing they had not become extinct in the meantime.

The term "mutation" is used to designate both the processes by which hereditary changes arise and the outcomes or end products of such processes ("mutants"). It is usually clear from the context whether the term refers to the process of change, to the changed hereditary materials, or to both. In the most inclusive sense, genetic mutations are changes in the hereditary materials not due to genetic recombination or to the independent assortment of chromosomes characteristic of the sexual process. They can be classified in one of two major categories: *gene* (or point) *mutations,* which affect only one or a few nucleotides within a gene; and *chromosomal mutations* (or aberrations), which affect the number of chromosomes, or the number or the arrangement of genes in a chromosome. Chromosomal mutations can be classified as follows:

1. Changes in the *number of genes* in chromosomes.

A. *Deficiency* or *deletion.* A segment of DNA containing one or several genes is lost from a chromosome.

B. *Duplication.* A segment of DNA containing one or more genes is present more than once in a set of chromosomes. Duplications often occur in tandem, i.e., the two duplicated segments lie adjacent to each other in the same chromosome.

2. Changes in the *location of genes* in the chromosomes.

A. *Inversion.* The location of a block of genes is inverted within a chromosome. When the rotated segment includes the centromere, the inversion is called *pericentric;* otherwise, the inversion is *paracentric.*

B. *Translocation.* The location of a block of genes is changed in the chromosomes. The most common forms of translocations are *reciprocal*, involving an exchange of blocks of genes between nonhomologous chromosomes. A chromosomal segment may also move to a new location in a different chromosome without reciprocal exchange, or within the same chromosome; these changes are sometimes called *transpositions* (Dobzhansky, 1970).

3. Changes in the *number of chromosomes*. These are of four kinds; the first two do not affect the total amount of hereditary material but the other two do.

A. *Fusion.* Two nonhomologous chromosomes fuse into one. This involves the loss of a centromere.

B. *Fission.* One chromosome splits into two. An additional centromere must be produced, otherwise the new chromosome would be lost when the cell divides.

C. *Aneuploidy.* One or more chromosomes of the normal set may be lacking or present in excess. The terms *nullosomic, monosomic, trisomic, tetrasomic,* etc., refer to the occurrence of a given chromosome zero times, once, three times, four times, etc., in a diploid organism.

D. *Haploidy* and *polyploidy.* The number of *sets* of chromosomes is other than two. Most organisms are diploid, i.e., they have two sets of chromosomes in their somatic cells, but only one set in their gametic cells. However, some organisms are normally haploid, i.e., they have only one set of chromosomes. Both haploid and diploid organisms exist among certain social insects, such as the honeybee, where haploid males develop from unfertilized eggs. Polyploid organisms have more than two sets of chromosomes; the organism is said to be *triploid* if it contains three sets of chromosomes, *tetraploid* if it contains four sets, and so on. Polyploidy occurs in many species in some groups of plants, but is much rarer in animals. The evolutionary role of polyploidy is discussed in Chapter 7.

HISTORICAL OVERVIEW

The paleontologist Wilhelm Waagen applied the term "mutation" to morphological discontinuities in a temporal series of fossil ammonites. Hugo de Vries in Holland and William Bateson in England described mutations as discontinuous hereditary variations causing major, easily recognizable changes. De Vries argued that the building blocks of evolutionary change are sudden mutational changes rather than the gradual "individual variability . . . [that]

cannot lead to a real overstepping of the species limits even with the most intense steady selection." De Vries based his mutation theory of evolution on the results of segregation in progenies of crosses of the evening primrose, *Oenothera*. Actually, the "mutants" described by de Vries were due to a variety of hereditary changes, particularly chromosomal aberrations. Richard Goldschmidt and others argued that there are two kinds of mutations: some produce variation among individuals, modifying their adaptation to the environments, while others produce new species, genera, families, etc. The mutations allegedly responsible for drastic changes were called "systemic mutations," "macromutations," and the like.

The modern concept of mutation is due to Thomas Hunt Morgan and his associates who worked with the fruit fly, *Drosophila melanogaster*. Mutations are defined as changes in single genes, with effects ranging from barely detectable to very drastic. Some mutations cause morphological variations, but others result in behavioral changes or modify the viability, fertility, or rate of development of their carriers. Mutations are said to be "spontaneous," i.e., due to unknown naturally occurring agencies. H. J. Muller showed that mutations arise with definite regularity, at rates that can be measured. He also discovered that the frequency of mutation is increased in the progenies of flies treated with X rays. Other high-energy radiations were later shown to be also mutagenic. Charlotte Auerbach was the first to demonstrate unambiguously that a chemical, namely mustard gas, had mutagenic properties. It was later shown that exposure to a variety of chemical agents, or to higher temperatures, increases the frequency of mutations.

Since the 1940s, mutations have been extensively studied in the mold *Neurospora*, in yeast and other unicellular organisms, in bacteria such as *Escherichia coli*, and in viruses. Morphological mutations modifying the appearance of colonies were discovered in these organisms, but major contributions to genetic knowledge and to the understanding of the mutation processes were made through the study of so-called *auxotroph* mutations. In contrast to prototrophic (wild-type) organisms, auxotrophic organisms require special nutritional supplements for growth and reproduction. Other types of mutations commonly studied in haploid organisms are those causing resistance or sensitivity to certain drugs, such as penicillin or streptomycin, or to infection by specific viruses. In the viruses themselves, mutations were discovered that modify the viruses' ability to infect their bacterial hosts or to multiply in them.

The discovery of the double-helical structure of DNA, and later of the genetic code, opened the way to an understanding of the process of mutation and its causes in physicochemical terms.

GENE MUTATIONS

A gene or point mutation occurs when the DNA sequence of a gene is altered and the new nucleotide sequence is passed to the offspring. The change may be due to the substitution of one or a few nucleotides for others, or to the addition or deletion of one or a few nucleotides. Nucleotide substitutions can be either transitions or transversions. *Transitions* are replacements of a purine by another purine (A by G, or vice versa), and of a pyrimidine by another pyrimidine (C by T, or vice versa). *Transversions* are replacements of a purine by a pyrimidine, and vice versa (C or T by either G or A, and vice versa).

Substitutions in the nucleotide sequence of a structural gene may result in changes in the amino acid sequence of the polypeptide encoded by the gene, although this is not always the case owing to the degeneracy of the genetic code. Consider the triplet AAT in DNA (corresponding to UUA in messenger RNA), which codes for the amino acid leucine. If the first A is replaced by G the triplet will still code for leucine, but if it is replaced by C it will code for valine instead (see Table 2-1, p. 28). A nucleotide substitution in the DNA that results in an amino acid substitution in the corresponding polypeptide may or may not severely affect the biological function of the protein. Nucleotide substitutions that change a triplet coding for an amino acid into a terminating triplet are likely to have severe effects. If the second A in the AAT triplet is replaced by T, the resulting triplet will code in messenger RNA for UAA, which is a terminator codon; the following triplets in the DNA sequence will not be translated into amino acids. Point mutations that result in the replacement of one amino acid for a different one are called *missense* mutations; when a triplet coding for an amino acid changes to a terminating codon, the mutations are called *nonsense* mutations.

Additions or deletions of nucleotide pairs in the DNA sequence of a structural gene often result in a very altered sequence of amino acids in the coded polypeptide. The addition or deletion of one or two nucleotide pairs shifts the "reading frame" of the nucleotide sequence from the point of the insertion or deletion to the end of the molecule. For example assume that a DNA segment is read as . . . CAT-CAT-CAT-CAT-CAT. . . . If a nucleotide base, say T, were inserted in the second position of this segment, it would then be read as . . . CTA-TCA-TCA-TCA-TCA. . . . The polypeptide segment corresponding to the original DNA sequence consists of five valine amino acids; the segment corresponding to the altered sequence consists of one aspartic acid and four serines. From the point of the insertion onwards the sequence of amino acids becomes altered. However, if a total of three nucleotide pairs is either added or deleted, the original reading frame is restored in the rest of the sequence. Additions or deletions of nucleotide pairs in numbers other than three or multiples of three are called *frameshift* mutations.

If one nucleotide is inserted at some point and another is deleted at some other point, the original reading frame and the corresponding amino acid sequence will be restored after the second mutational change. An instance of this situation has been worked out in the bacteriophage *T4* by Terzaghi and co-workers (1966) for the gene coding for lysozyme, an enzyme that digests bacterial walls. Some mutant strains of *T4* lack a functional lysozyme gene. Back mutations that restore the function of the lysozyme occasionally occur. The back-mutated strains are called "pseudo-wild" because the amino acid sequence in the lysozyme is different from that of wild strains. As shown in

Wild-type lysozyme	—	Thr	Lys	Ser	Pro	Ser	Leu	Asn	Ala	—
Wild-type mRNA		ACX	AAY	AGU	CCA	UCA	CUU	AAU	GCX	
			Delete					Insert		
Pseudo-wild mRNA		ACX	AAY	GUC	CAU	CAC	UUA	AUF	GCX	
Pseudo-wild lysozyme	—	Thr	Lys	Val	His	His	Leu	Met	Ala	—

FIGURE 3-1.
The effect of two compensating frame-shift mutations in the gene coding for lysozyme in the bacteriophage *T4*. Only a fraction of the messenger RNA is shown. A nucleotide deletion has occurred in the third codon, and an addition in the seventh codon. As a consequence, the amino acid sequence is changed from the third to the seventh amino acid, but the normal sequence is restored after the seventh amino acid shown. X = A, G, C, or U; Y = A or G. [After Terzaghi *et al.*, 1966.]

Figure 3-1, the sequence of amino acids in the wild and pseudo-wild strains can be explained if it is assumed that there is a nucleotide deletion in the mutant strain and that an insertion has occurred a few positions later in the pseudo-wild strain.

Considerable progress has been made in recent years toward understanding the causal processes of mutations. Transitions due to so-called *tautomeric shifts* (changes from one to another isomer) in the DNA nitrogen bases may occur. (Isomers are chemical compounds that contain the same numbers of atoms of the same elements but differ in structural arrangement and properties.) The accuracy of base-pairing during replication depends on the fact that the purines and pyrimidines of DNA usually exist in particular isomeric states. Occasionally, however, certain hydrogen atoms in the bases migrate to different unstable positions, modifying the pairing properties of the bases: C with the pairing properties of T, and vice versa; and A with the pairing properties of G, and vice versa.

Ultraviolet radiations may induce mutations by causing the formation of dimers between identical pyrimidines adjacent in the same DNA chain. The dimerized bases become displaced and are unable to form hydrogen bonds with the opposing purines in the complementary chain. During DNA replication, gaps occur in the complementary chain and these gaps may then be filled with the wrong bases. Transversions as well as transitions can result from this process.

Tautomeric shifts are transitory conditions of the DNA nitrogen bases. Stable changes may also occur in the bases, which then pair with the "wrong" bases during replication. Chemicals known to cause permanent alterations of the nitrogen bases are hydroxylamine, nitrous acid, and a variety of alkylating agents, such as mustard gas and the epoxides (dimethyl and diethyl sulfonate, methyl and ethyl methanesulfonate, and nitroguanidine). Nitrous acid deaminates cytosine into uracil, which pairs with adenine, and deaminates adenine yielding hypoxanthine which has the pairing properties of guanine. Hydroxylamine causes transitions from GC pairs to AT pairs. Alkylating agents alter the properties of the nitrogen bases, often leading to mispairing.

Base analogs are nitrogen bases resembling those normally occurring in DNA. Instead of thymine, the analog 5-bromouracil may be incorporated, but it undergoes tautomeric shifts and may pair with guanine upon replication. The base analog 2-aminopurine is a purine that may pair with either cytosine or thymine.

Some aromatic compounds, known as acridines, cause additions and deletions of one to more than 20 nitrogen bases in the DNA. Acridines become intercalated between the stacked base pairs of the double helix, but the process by which they result in insertions and deletions is not well known. One model suggests that the acridines stabilize DNA "buckles" formed by dimers of adjacent bases in the same chain. Insertions or deletions may occur during DNA replication whenever buckles exist in one of the chains.

EFFECTS OF MUTATIONS

The effects of point mutations on the ability of organisms to survive and reproduce may range from negligible to severe. In a missense mutation the amino acid substitution may or may not affect the essential biological function of the coded protein. In the latter case it may have little or no detectable effect on the organism, although the activity of the protein at certain ranges of temperature or pH, as well as some other biochemical properties, may be altered. Most allelic variants detected by electrophoretic techniques (see Chapter 2) are missense mutations differing from one another by one or a few amino acid substitutions with different electric charge. Twelve different alleles have been detected by electrophoresis at the *Est-5* locus in *Drosophila pseudoobscura* (Prakash, Lewontin, and Hubby, 1969). None is lethal or severely detrimental in homozygous condition.

The protein cytochrome *c* has been examined in scores of organisms ranging from bakers yeast to man. It consists of a single polypeptide chain containing a number of amino acids ranging from a minimum of 103 (in tuna) to a maximum of 112 (in wheat). The amino acids of the cytochromes of different organisms are sometimes different but the amino acids are the same in about 20 positions (nearly 20 percent of the total) in all organisms examined. Cytochromes *c* with different amino acid sequences are able to perform similar biological functions, although they may differ in their affinities for certain ions (Margoliash, 1972).

Missense mutations may have severe effects on the organisms when the amino acid substitution either occurs in the active site of an enzyme or modifies in some other way an essential biological function of a protein. Hemoglobin A is the most common hemoglobin in humans after birth. It is a tetramer consisting of two α and two β chains; the α and β chains are coded by two different genes. Normal β chains have an electrically charged amino acid, glutamic acid, in the sixth position. The mutant allele *S* of the β gene codes for a chain with valine, which is electrically neutral, in position six. The replacement of glutamic acid by valine can be explained by a single nucleotide substitution in the β gene (see Table 2-1). Normal hemoglobin A and hemoglobin S differ in net electric charge and therefore are easily distinguishable by electrophoresis. The substitution of valine for glutamic acid also alters the solubility of the polypeptide; at low oxygen tensions red blood cells with hemoglobin S have a characteristic sickle shape rather than the disc shape of cells with normal hemoglobin A (Figure 3-2). The deformed cells are defective in oxygen transport. Individuals homozygous for the *S* allele suffer from sickle cell anemia, a severe disease that usually results in death during childhood.

McKusick (1975) has compiled a catalog of traits in man determined by single genes. Mutations at most of the 2,336 gene loci included in his catalog produce various abnormalities, ranging from mild to severe. Deficient forms of specific enzymes have been demonstrated in 103 human hereditary disorders. Mutant genes that cause lethality or sterility, or decrease viability and fertility, are well known in organisms other than man. In natural populations of *Drosophila* about 20 percent or more of all chromosomes carry at least one mutation that is lethal in homozygous condition, as shown in Chapter 2 (see Figure 2-7). Mutations with less drastic effects occur with much greater frequencies than lethal mutations in natural populations. Lethal mutations or those

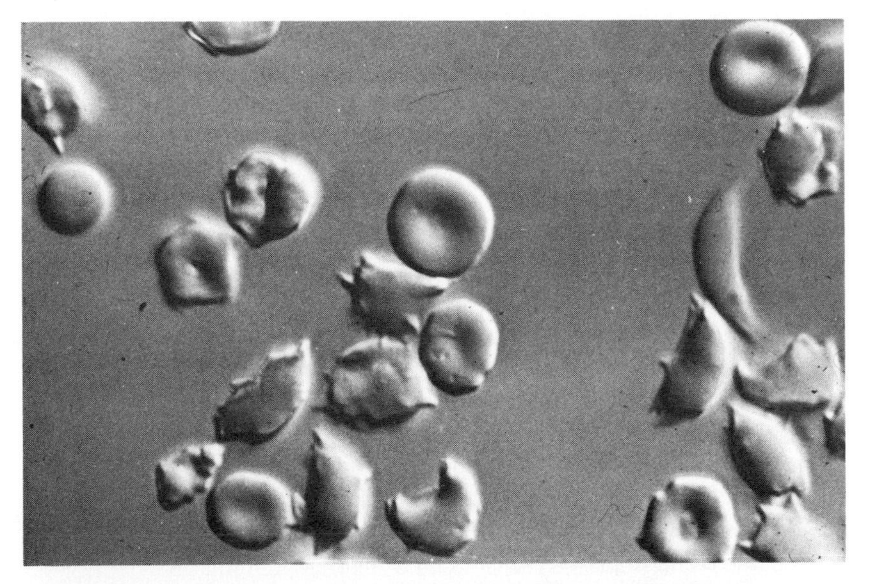

FIGURE 3-2.
Deoxygenated red blood cells from a person suffering from sickle-cell anemia. The abnormal cells exhibit a variety of bizarre shapes, including the characteristic sickle shape (upper right) for which the disease was named. Normal red blood cells exhibit typically a disc shape similar to that of the cell in the top center. Most of the irregularly shaped sickle cells regain normal shape in the presence of oxygen. [From A. Cerami and J. Manning, *Proc. Nat. Acad. Sci. U.S.A.*, 68: 1180–1183 (1971).]

affecting viability may express their effects at various stages of development, from early embryogenesis throughout the life of the individual. They may affect different tissues, organ systems, behavioral patterns, or metabolic processes.

The lethal or deleterious effects of mutations often depend on particular environmental conditions. In *Drosophila* a class of mutants studied by Suzuki and others are known as "temperature-sensitive lethals." At standard temperatures of 20° to 25°C, flies homozygous for these alleles survive and reproduce more or less normally; but at temperatures about 28°C or higher these flies become paralyzed or die, although wild-type flies can function normally. The auxotrophic mutants commonly studied in microorganisms also depend on the environment for their expression. If the nutrient that a mutant strain is unable to synthesize is present in the culture medium, the microorganisms survive and reproduce; without the nutrient they die or fail to reproduce. In man phenylketonuria (PKU) is a severe disease caused by homozygosis for a recessive gene. Phenylketonurics are unable to metabolize the amino acid phenylalanine, and suffer from mental retardation and other defects. Individuals with the PKU genotype may become effectively normal if they feed on a diet free of phenylalanine.

The effects of mutations on organisms may also depend on the genetic background in which the mutations occur. Epistatic (intergenic) interactions are common, as was pointed out in Chapter 2. An extreme example is intergenic suppression. The best-studied case is the gene *su3*$^+$ in *Escherichia coli,* which suppresses the expression of nonsense mutations. Strains carrying the *su3*$^+$ gene have a tRNA carrying tyrosine with an AUC anticodon. This tRNA recognizes the terminator codon UAG and there inserts tyrosine, thereby eliminating the chain-terminating effect of this triplet and allowing polypeptide synthesis to continue wherever such a nonsense mutation has occurred.

MUTATION AND ADAPTATION

Mutations are often said to be accidental, undirected, random, or chance events. It is worthwhile to clarify the different meanings attributed to these terms and in what senses they can be applied to the mutation process.

Mutations are accidents or chance events in the sense that they are rare exceptions to the regularity of the process of DNA replication, which normally involves precise copying of the hereditary information encoded in nucleotide sequences. Mutations are also accidents or chance events in the sense that they are unintended; naturally occurring mutations are not the results of any agent actively intending to produce them. Nor is there any known natural process tending towards producing nucleotide changes at specific sites in specific genes.

Also, mutations are accidental, random, or chance events because there is no way of knowing whether a given gene will mutate in a particular replication of a DNA molecule. We cannot predict which individuals will have a new mutant, and which ones will not at a given gene locus. This does not imply that no regularities exist in the mutation process. The regularities are those associated with stochastic processes to which probabilities can be assigned. There is a probability (although it may not have been ascertained) that a certain gene will mutate in a given number of replications, and thus there is a probability that a new individual will have a new mutation at a certain gene locus. For example, aniridia (absence of iris) in man results from a dominant gene that arises by mutation from the normal gene at a rate of one per 200,000 gametes. Therefore, the probability that a child of normal parents will be born with aniridia is one in 100,000 births. There is also a probability that a specific nucleotide pair may be replaced by another during replication. And it is possible, at least in principle, to estimate the probability that a gamete, or a zygote, will have a new mutation in one or more genes when all genes are collectively taken into consideration (see below).

Mutations are not random in the sense that any mutation would be equally likely to occur as any other mutation. Transitions and transversions may have different probabilities of occurrence; nucleotide substitutions may not be equally likely for different nucleotide pairs, for nucleotide pairs in different positions, or for different genes; nucleotide substitutions and frameshift mutations have different probabilities of occurrence. A gene cannot arise from just any other gene. The cytochrome *c* gene of man cannot arise by a mutation from the gene coding for hemoglobin β in man, or from the gene coding for cytochrome *c* in wheat.

Mutations are accidental, undirected, random, or chance events in still another sense very important for evolution; namely, in the sense that they are unoriented with respect to adaptation. Mutations occur independently of whether or not they are adaptive in the environments in which the organisms live. Microbiologists have known for a long time that confronted with adverse environmental conditions, bacterial cultures give rise to new genetically stable strains able to cope with the unusual environment. The regularity of this result inclined some bacteriologists to believe that the environment could induce specific mutations favorable in that environment. Ample evidence now exists showing that mutations arise with certain probabilities independently of whether or not they are favorable in the environment where they arise. The elegant experiments of Lederberg and Lederberg (1952) using "replica plating," as well as experiments by Luria and Delbrück (1943), Demerec and Fano (1945), and others, have settled this question definitely.

Newly arisen mutations are more likely to be deleterious than beneficial to their carriers. It is easy to see why this should be. The genes occurring in a population have been subject to natural selection. Allelic variants that occur in substantial frequencies in a population are therefore adaptive; they increase the fitness of their carriers relative to other variants that are kept at low frequencies or have been eliminated by natural selection. New mutations are likely to have also arisen during the past history of the population; if new mutants do not already exist in substantial frequencies it is because they have been eliminated or kept at low frequencies by natural selection owing to their harmful effects on the organisms. This may be illustrated with an analogy. Assume that we have a sentence written in English that has been chosen because it makes sense. If single letters or words are replaced with others at random, most substitutions will destroy the sense of the sentence. Mutations are random changes because they occur independently of whether they are beneficial or harmful, and therefore they are a disordering process.

Occasionally, however, a newly arisen mutation may increase the adaptation of the organism. The probability of such an event is greater when organisms colonize a new habitat, or when environmental changes present a population with new challenges. In these cases the adaptation of the population is less than optimal and there is greater opportunity for new mutations to be adaptive. Needless to say, there are mutations that destroy essential functions and thus are harmful in any environment. This brings about the important point that mutations are not beneficial or harmful in the abstract, but rather with respect to some environment. The same mutation may be harmful in some environments while adaptive in others. A mutation increasing the density of hair in a mammal may be adaptive in a population living in Alaska, but it is likely to be selected against in a population living in the tropics. Increased melanin pigmentation may be beneficial to men living in tropical Africa, where dark skin protects from the sun's ultraviolet radiation, but not in Scandinavia, where the intensity of sunlight is low, and light skin facilitates the synthesis of vitamin D. Mutations to drug resistance in microorganisms are clear-cut examples. In the unicellular alga *Chlamydomonas reinhardi,* mutations from streptomycin sensitivity (gene *str-s*) to streptomycin resistance (*str-r*) occur with a frequency of one for 10^6 cell divisions. In the absence of streptomycin the gene *str-r* is selected against, but if streptomycin is present in the culture medium the only cells that survive and reproduce are those carrying the *str-r* mutation. In general, mutations with small effects are more likely to be adaptive than mutations with large effects. Missense mutations involving single nucleotide substitutions are more likely to be adaptive than frameshift or nonsense mutations.

There is ample experimental evidence that some mutations may increase the fitness of multicellular organisms. Dobzhansky and Spassky (1947) started an experiment with seven strains of *Drosophila pseudoobscura,* each homozygous for a different chromosome. The strains were maintained in laboratory cultures for 50 generations, after which significant increases in viability had occurred in five of the seven strains. The initial viability of the strains was substantially below that of wild flies; at the end of the experiment the normal level had been recovered in at least three strains.

Ayala (1966, 1969a) has shown that new mutations may be adaptive in populations of wild flies when these are exposed to new environments. One experiment was conducted with *Drosophila birchii,* a species endemic to northeastern Australia, New Guinea, and New Britain. Pairs of experimental populations were established with the descendants of a large number of flies collected from a natural population. One population of each pair served as control; the males of the other population received 1000 r of X-rays during each of the first three generations so as to induce a large number of mutations. All populations were maintained under identical experimental conditions, with food and space severely limited to facilitate the operation of natural selection. During the initial generations the number of flies was less in the irradiated than in the control populations, owing to the elimination of deleterious mutations induced by radiation; but the irradiated populations evolved at a faster rate than the controls throughout the two and a half years (30 to 40 generations) of the experiment (Figure 3-3). Table 3-1 shows that the mean number of flies in the populations, as well as the rate at which the number of flies increased through

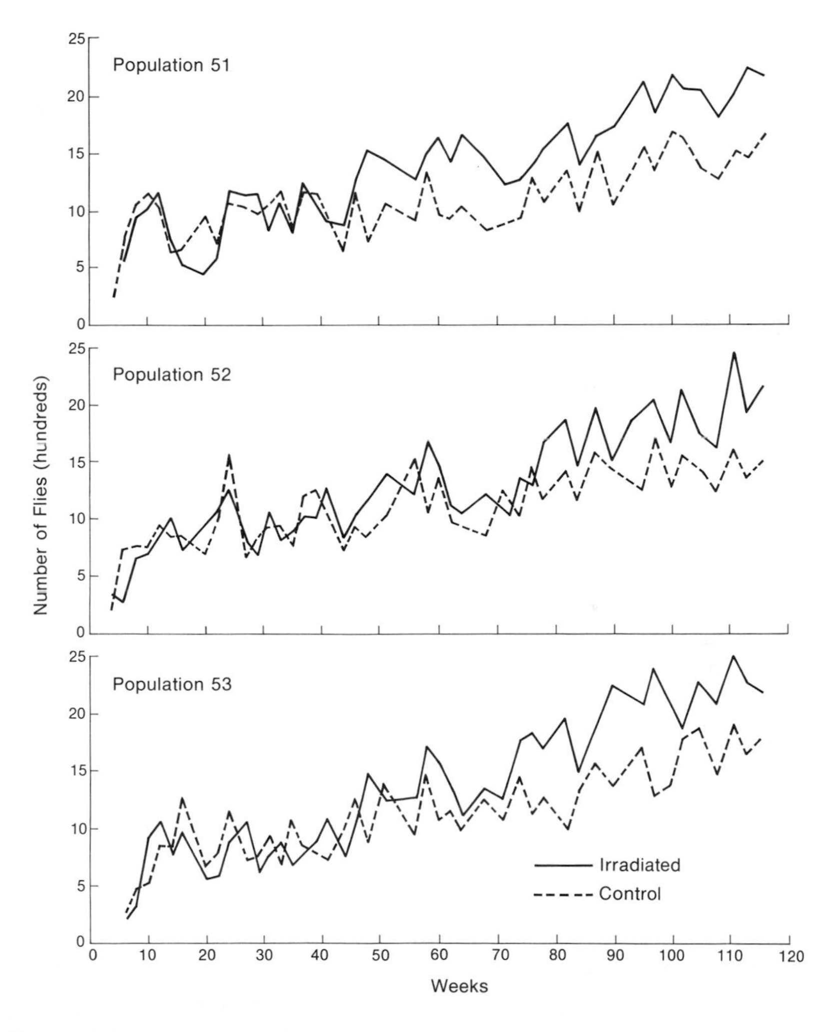

FIGURE 3-3.
Rate of evolution in irradiated and nonirradiated experimental populations of *Drosophila birchii*. Each of three large populations was divided into two: one was subject to irradiation for three generations, the other was the control. All populations became increasingly adapted to the experimental environment, as reflected in the gradual increase in size. The irradiated populations, however, increased at a faster rate than the controls, evincing that some mutations induced by radiation may be favorable to their carriers. [After Ayala, 1969a.]

TABLE 3-1. Rate of evolution of irradiated and nonirradiated populations of *Drosophila birchii* in the experiment in Figure 3-3. [From Ayala, 1969a.]

Population		Mean number of flies in population	Increase in number of flies per week
51	Control	1093 ± 47	6.9 ± 2.0
	Irradiated	1337 ± 73	11.5 ± 3.0
52	Control	1106 ± 53	8.3 ± 2.0
	Irradiated	1283 ± 86	15.2 ± 3.2
53	Control	1121 ± 70	6.4 ± 2.0
	Irradiated	1334 ± 103	12.9 ± 3.2

time, was greater in the irradiated than in the control populations. Although the average effect of radiation-induced mutations is deleterious, a minority proved to be favorable and permitted the populations to exploit more efficiently the experimental environment.

RATES OF MUTATION

Mutation rates of single genes are difficult to measure because of their low frequencies. Estimation is less difficult in microorganisms than in higher organisms, since millions or even billions of individuals can be easily and rapidly produced in microbial cultures. Table 3-2 gives a representative sample of mutation rates in a variety of organisms. Known mutation rates range from less than 10^{-9} to more than 10^{-4}; even within the same organism mutation rates for different genes may differ by three orders of magnitude. In viruses, bacteria, and unicellular organisms recorded mutation rates per gene per cell division range from less than 10^{-9} to 10^{-6}. In higher organisms mutation rates per gamete range from 10^{-6} to 10^{-4}. The highest mutation rate recorded in man is about 2×10^{-4} for neurofibromatosis, a syndrome characterized by spots of abnormal pigmentation in the skin and numerous tumors in the nervous system; at the other extreme, mutations for Huntington's chorea arise at a rate of 10^{-6}.

The genes for which mutation rates have been reported in multicellular organisms are for the most part those having clear-cut phenotypic effects, or in the case of man those responsible for infirmities of various kinds. There are few studies of mutation rates for allelic variants detectable by electrophoresis, which are so often found in natural populations of all kinds of organisms. Tobari and Kojima (1972) studied ten gene loci coding for enzymes in *Drosophila melanogaster* and observed an average mutation rate of 4.5×10^{-6} per locus per generation. In the same species, Mukai (1970) observed an average mutation rate of 4×10^{-6} per locus per generation in three loci. These mutation rates are about one order of magnitude less than those reported for morphological mutants (see Table 3-2).

Rates of mutation for genes modifying viability have been measured for entire chromosomes in various species of *Drosophila*. Some estimates of recessive lethal mutations per generation are: 0.0019 for the X chromosome of *D. melanogaster* (Dubinin, 1966); 0.0100 and 0.0178 for the second chromosomes of *D. pseudoobscura* and *D. persimilis*, respectively (Dobzhansky, Spassky, and Spassky, 1954); and 0.0098 and 0.0089 for the second and third chromosomes, respectively, of *D. willistoni* (Dobzhansky, Spassky, and Spassky, 1952; see Table 3-3 here). Mukai (1964) has estimated that second chromosomes of

TABLE 3-2. Mutation rates of specific genes in various organisms.

Organism and trait	Mutations per cell or gamete
Bacteriophage *T2* (virus)	
Host range	3×10^{-9}
Lysis inhibition	1×10^{-8}
Escherichia coli (bacterium)	
Streptomycin resistance	4×10^{-10}
Streptomycin dependence	1×10^{-9}
Sensitivity to phage *T1*	2×10^{-8}
Lactose fermentation	2×10^{-7}
Salmonella typhimurium (bacterium)	
Tryptophan independence	5×10^{-8}
Chlamydomonas reinhardi (alga)	
Streptomycin resistance	1×10^{-6}
Neurospora crassa (fungus)	
Adenine independence	4×10^{-8}
Inositol independence	8×10^{-8}
Zea mays (corn)	
Shrunken seeds	1×10^{-6}
Purple seeds	1×10^{-5}
Drosophila melanogaster (fruit fly)	
Electrophoretic variants	4×10^{-6}
White eye	4×10^{-5}
Yellow body	1×10^{-4}
Mus musculus (mouse)	
Brown coat	8×10^{-6}
Prebald coat	3×10^{-5}
Homo sapiens (man)	
Huntington's chorea	1×10^{-6}
Aniridia (absence of iris)	5×10^{-6}
Retinoblastoma (tumor of retina)	1×10^{-5}
Hemophilia A	3×10^{-5}
Achondroplasia (dwarfness)	$4\text{–}8 \times 10^{-5}$
Neurofibromatosis (tumor of nervous tissue)	2×10^{-4}

TABLE 3-3. Combined frequencies of lethal and semilethal mutations in second chromosomes of *Drosophila willistoni*. [After Dobzhansky, Spassky, and Spassky, 1952.]

Geographic origin	Sample size	Frequency per chromosome per generation	Confidence limits
Mogi, São Paulo	1806	0.0171	0.0120–0.0244
Pirassununga, São Paulo	2201	0.0076	0.0047–0.0123
Belem, Parà	2004	0.0057	0.0032–0.0102

D. melanogaster acquire mutations modifying viability at a rate no less than 0.1411 per generation; mutations with small effects on viability are approximately 20 times higher than those causing lethality.

Rates of mutation are under genetic control. Strains with different mutation rates have been demonstrated in *Drosophila melanogaster* (Dubinin, 1966) and *D. willistoni* (Dobzhansky, Spassky, and Spassky, 1952). The mutation rate of recessive lethals in second chromosomes of *D. willistoni* is ostensibly three times higher in strains from Mogi in south Brazil than in strains from Belem in the Amazon delta (Table 3-3). Genes that increase mutation rates have been identified in *Escherichia coli* and in *Drosophila*. The *mutT* gene of *E. coli* selectively increases the occurrence of AT → GC transversions (Treffers, Spinelli, and Belser, 1954). In *D. melanogaster* the second chromosome genes *mu-F* (Demerec, 1937) and *hi* (Ives, 1950) increase the frequency of lethal mutations in the X chromosome. The gene *mu* located in the third chromosome increases the mutation rates of lethals as well as of genes with morphological effects, apparently by affecting the mechanisms of chromosome repair (Green, 1970, 1973).

Since there is genetic variation affecting mutation rates, these may be subject to natural selection. May we infer that present levels of mutation rates are optimal? There is no definitive answer to this question. Mutations are the ultimate source of genetic variation; they provide the raw materials for evolutionary change. However, a great deal of genetic variation exists in most natural populations of organisms, and the large majority of newly arisen mutations are deleterious. It would seem, therefore, that natural selection would favor a gradual decrease of mutation rates. Why have mutation rates not been reduced to zero altogether? A possible answer is that present mutation rates are the minimal ones that can be achieved, i.e., that there is no way of reducing mutation rates below their present levels.

Another possible answer is that populations with very low mutation rates (approaching zero) may eventually become genetically depauperate by gradual loss of genetic variation, and become extinct when faced with environmental changes to which they are unable to adapt. If this were the case, would higher mutation rates be favored by natural selection? The experiments discussed above (Ayala, 1966, 1967) demonstrate that a temporary increase of mutation rates by radiation may lead to an increase in the rate of adaptation of populations facing new environments. On the other hand, other experiments have shown that when radiation is continuously administered, the adaptation of populations decreases because natural selection cannot offset the increased burden of the deleterious mutations continuously induced (Carson, 1964; Sankaranarayanan, 1964, 1966). Benado, Ayala, and Green (1976) have studied experimental populations of *Drosophila melanogaster* with and without the third chromosome mutator gene, *mu*. The adaptation of populations carrying the *mu* gene decreased throughout the 20 to 25 generations of the experiment. It seems likely that in natural populations the *mu* gene would be eliminated by natural selection.

RATES OF MUTATION AND RATES OF EVOLUTION

Mutations are either rare or ubiquitous events depending on how we choose to look at them. The mutation rates of individual genes are low, but each organism has many genes, and populations consist of many individuals. Mukai (1964) estimates that the rate of mutations modifying viability is at least 0.1411 per generation for each second chromosome of *Drosophila melanogaster*. A second chromosome has approximately 20 percent of the genetic material of the complete genome of *D. melanogaster*. Extrapolating Mukai's results we may estimate that the mean rate of viability mutations is no less than $0.1411/0.20 = 0.7055$ per individual per generation.

The haploid chromosomal complement of man contains about 3.2×10^{-12} grams of DNA, or 2.9×10^9 nucleotide pairs. Assuming an average of 2,000 nucleotide pairs per gene, there is enough DNA in man for one and a half million pairs of genes. *Drosophila melanogaster* contains five percent of the DNA in man, enough for about 75,000 pairs of genes. However the genetic information may be contained in only a fraction of the nuclear DNA. Assume that there are at least 100,000 pairs of genes in man (McKusick, 1975), and 10,000 in *Drosophila*, and that the average mutation rate per gene per generation is 10^{-5}. The average number of mutations arising per generation would then be estimated as at least 2×10^5 genes $\times$ 10^{-5} mutations per gene $= 2$ mutations for a human zygote, and as 2×10^4 genes $\times$ 10^{-5} mutations per gene $= 0.2$ mutations for a *Drosophila* zygote.

There are about 4×10^9 humans in the world. With two mutations per individual, the total number of mutations newly arisen in the present population of mankind is 8×10^9. The median number of individuals per insect species is estimated to be about 1.2×10^8. If we assume that 0.2 new mutations are acquired by each individual, there would be 2.4×10^7 new mutations per generation in a species of median size. Species of other groups of organisms, including vertebrates, may consist of fewer individuals than insect species, but even such species will acquire large numbers of mutations each generation.

When whole species are considered, many mutations occur every generation even at a single locus. If the average mutation rate per gene per generation is 10^{-5}, an average of about 80,000 new mutations arise every generation at each locus in mankind (4×10^9 individuals $\times$ 2 genes $\times$ 10^{-5} mutations per locus). In an insect species of median size, the average number of mutations per gene per generation may be about 2,400. It is not surprising that different populations of the same species and different species often become adapted to specific environmental challenges. Many species of insects have developed resistance to DDT in different parts of the world where spraying has been intense (Crow, 1957; Brown, 1967). Industrial melanism has evolved in many species of moths and butterflies in industrial countries. If the appropriate genetic variants to face an environmental challenge are not already present in the population, they are likely to arise soon by mutation. The potential of the mutation process to generate new variation is indeed enormous.

The large number of variants arising in each generation by mutation represents only a small fraction of the total amount of genetic variability present in natural populations. We have seen in Chapter 2 that roughly half the loci are polymorphic in populations of sexually reproducing organisms, and that an individual is usually heterozygous at 5 to 20 percent of its loci. An individual heterozygous at 10 percent of, say, 30,000 gene loci will have two allelic variants at each of 3,000 gene loci. The same individual will have acquired an average of about 0.6 new allelic variants owing to newly induced mutations. Thus, one may estimate that the amount of variation present in a population is about 5,000 times greater than that acquired each generation by mutation. Although these calculations are very rough, they show that newly arisen mutations represent only a small fraction of the genetic variation present in populations at any one time.

It follows that rates of evolution are not likely to be closely correlated with rates of mutation. Besides mutation, natural selection and migration help maintain high levels of genetic variation in natural populations. Even if mutation rates would increase by a factor of 10, newly induced mutations would represent only a very small fraction of the variation present at any one time in populations of outcrossing, sexually reproducing organisms. Organisms reproducing asexually, by apogamy or parthenogenesis, may be on occasion more dependent on newly arisen mutations. However, recent studies have shown that a great deal of genetic variation is present in bacteria, e.g., *Escherichia coli* (Milkman, 1973), and in organisms reproducing mostly vegetatively, e.g., *Neurospora intermedia* (Spieth, 1975), or parthenogenetically, e.g., the beetles *Otiorrhynchus scaber* and *Strophosomus capitatus* (Suomalainen and Saura, 1973).

EVOLUTION OF GENOME SIZE

Around 1950 it became well established that the DNA content per nucleus is constant in the various somatic cells of an organism, and that sperm and egg cells contain about one half as much DNA as somatic cells (Vendrely and Vendrely, 1948, 1949; Mirsky and Ris, 1949, 1951; Davidson *et al.*, 1950). Mirsky and Ris (1951) were the first to measure the amount of DNA in the cell nucleus of different organisms, including various invertebrate groups, fish, amphibians, reptiles, birds, and mammals. Among invertebrates they found lesser amounts of DNA per cell in sponges and coelenterates than in echinoderms, annelids, crustaceans, and molluscs; among molluscs, less DNA was found in limpets, snails, and chitons than in the more advanced squids. They suggested that the amount of DNA per cell may increase from primitive to more highly developed invertebrates. This correlation did not hold among vertebrates. The lungfish and some amphibians had 20 or more times as much DNA as men and rats, which had somewhat more DNA per cell than reptiles, and about twice as much as birds.

At present the DNA content has been measured by diverse methods in close to 2,000 different species, including about 1000 species of plants, nearly 300 species of fish, and more than 100 species of molluscs (for review see Hinegardner, 1976). The information is summarized for many organisms in Figure 3-4. There is a great range in the amount of DNA per cell throughout the whole diversity of life.

The lowest amounts of DNA are found in some viruses, about 10^4 nucleotide pairs per virus (RNA viruses have even less nucleic acid content than DNA viruses, the former range from 1.3×10^3 to 2×10^4 nucleotide pairs per virus). The largest amount is found in some polyploid primitive vascular plants (*Psilopsida*) and in two multinucleated protozoa (*Chaos chaos* and *Urostyla caudata*), somewhat more than 10^{12} nucleotide pairs per cell. Other organisms with very large amounts of DNA include the alga *Gonyaulax polyedra* (4×10^{11} nucleotide pairs per haploid chromosomal complement), the amoeba *Amoeba*

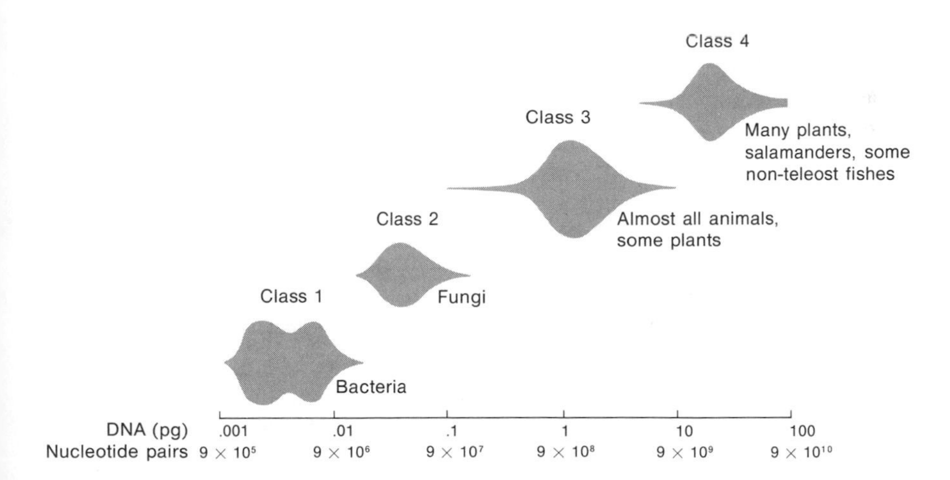

FIGURE 3-4.
Organisms classified according to their amounts of DNA. Organisms within each of the four broad classes have similar amounts of DNA; in each group the amount of DNA in most organisms varies over only one order of magnitude, i.e., differs by a factor from one to ten. For all organisms, from bacteria to plants and animals, the amounts of DNA vary over five orders of magnitude. [After Hinegardner, 1976.]

dubias (7×10^{11} pairs), the angiosperm *Sprekelia formosissima* (3.5×10^{11} pairs), the South African lungfish *Lepidosiren paradoxa* (1.1×10^{11} pairs), and the salamander *Amphiuma means* (Congo eel, 1.9×10^{11} pairs).

A wide range of four orders of magnitude of DNA content exists among algae, from 1.2×10^7 to 4.0×10^{11} nucleotide pairs; among gymnosperms the range extends from 8.4×10^9 (*Ephedra fragilis*) to 1.4×10^{11} (*Pinus resinosa*); among angiosperms, the lowest known value of DNA content is 2×10^9 nucleotide pairs in *Arabidopsis thaliana*, while several species have more than 10^{11} pairs. Among animals one of the lowest amounts of DNA occurs in *Drosophila melanogaster* (1.7×10^8 nucleotide pairs), while other insects, such as the grasshopper *Melanoplus differentialis* and the cricket *Grillus domesticus*, have more than 10^{10} nucleotide pairs per haploid chromosomal complement. A wide range exists also among amphibians, from 2.1×10^9 nucleotide pairs in the frog *Scaphiopus holbrookii holbrookii* to more than 10^{11} pairs in the Congo eel, the mudpuppy (*Necturus maculosus*), and others; urodela (salamanders) have higher DNA contents than anurans (frogs and toads). The range among teleost fishes extends from about 8×10^8 in *Tetraodon fluviatilis* to 8.8×10^9 in the catfish *Coryodoras aeneus*. The amount of DNA per cell varies relatively little from species to species among reptiles, birds, and mammals. Most reptiles have around 4.5×10^9 nucleotide pairs, most birds about $2–3 \times 10^9$, and most mammals about 5.7×10^9 per haploid complement.

Certain cell organelles are known to contain DNA with genetic information. The amount of DNA in mitochondria ranges from about 5×10^4 nucleotide pairs (sea urchin and cow's heart) to 1×10^6 (in turnip and mung bean), which correspond to intermediate size DNA viruses and small bacteria, respectively. The amount of DNA of individual chromosomes is very variable, but the smallest chromosomes of eukaryotes have about 6×10^7 nucleotide pairs, a number similar to that of the larger bacteria.

Considerable change in the amount of DNA per cell has occurred in the evolution of life. The amount of DNA increases from viruses to prokaryotes to eukaryotes. Among eukaryotes, the fungi generally have the least DNA, but few evolutionary trends can be identified. There is a general trend in the invertebrates toward greater DNA cell contents from simpler to more complex animals, although there are exceptions, such as *Drosophila melanogaster*, an advanced invertebrate with less DNA than much simpler organisms. Increases as well as decreases in DNA content have occurred in the evolution of the vertebrates; reptiles have more DNA than birds but less than mammals. The broad range found in the amphibians encompasses fishes, reptiles, and mammals, but many birds have less DNA than any amphibians. The evolutionary significance of these differences in DNA amount is unclear. Indeed, it appears that much DNA may be genetically "silent," i.e., may carry no hereditary information. The function of genetically silent DNA is unknown, although some authors have suggested that it may play some structural role.

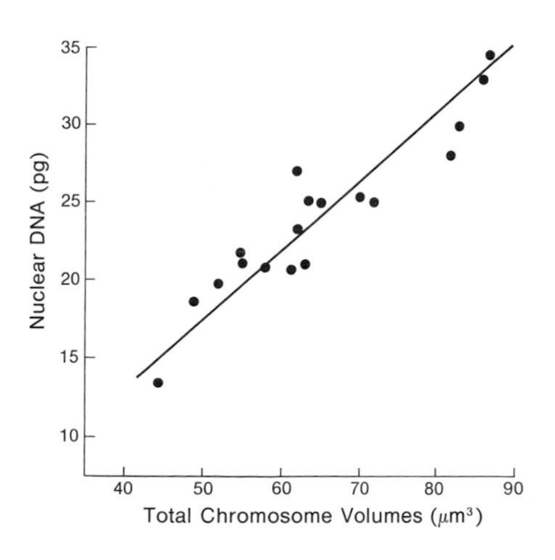

FIGURE 3-5.
The correlation between the amount of DNA and total chromosome volume in 18 species of the sweet pea, *Lathyrus*. DNA is measured in picograms (1 pg = 10^{-12} g); chromosome volume is measured in cubic micrometers (1 μm = 0.001 mm). All 18 species have the same number of chromosomes, 2n = 14. [After Rees, 1974.]

In certain groups of organisms large differences in DNA content are sometimes found even between species of a single family or genus. For example, the sweet pea genus, *Lathyrus*, includes 18 species all with a diploid number of 14 chromosomes. The amount of DNA per cell, however, is three times as large in *L. sylvestris* as in *L. angulatus*, while the other 16 species have intermediate amounts (Rees and Hazarika, 1969). Figure 3-5 shows that in these *Lathyrus* species the amount of DNA per cell is well correlated with the size of the chromosomes. As pointed out above, in several classes of vertebrates, i.e., mammals, birds, or reptiles, there is little variation between species in the amount of DNA per cell. The range among fishes extends over more than one order of magnitude, but members of the same family have virtually identical DNA amounts. Much less uniformity exists among amphibians. Among toads of the genus *Bufo* some species have twice as much DNA as others, and among the tree-frogs of the genus *Hyla* some species have more than four times as much DNA as other species.

CHROMOSOMAL DELETIONS AND DUPLICATIONS

Evolutionary changes in the amount of DNA may be due to a variety of processes, including polyploidy, polyteny, and duplications and deletions. Polyploidy is a common phenomenon in the evolution of most groups of plants (see Chapter 7), but it is a much rarer phenomenon in animals, since polyploidy cannot become easily established in species with separate sexes and regular outcrossing. Polyploid species occur among hermaphroditic animals, such as earthworms and planarians, and also among animals with parthenogenetic reproduction, including some beetles, moths, sow bugs, shrimps, fishes, and salamanders.

Polyteny, i.e., multiplication of the number of DNA strands within a chromosome, occurs in certain animal tissues, e.g., the salivary glands, the gut, and the Malpighian tubes of Diptera, but several sources of evidence indicate that polyteny is not a general phenomenon for the increase of DNA amounts in evolution (Rees, 1974; Bachmann, Goin, and Goin, 1974). Deletions and duplications of relatively small DNA segments are the most general processes by which evolutionary changes in the amount of DNA have taken place. Bachmann, Goin, and Goin (1974) have observed that when genome sizes of many teleost, anuran, or placental species are arranged in a frequency diagram, they conform to a logarithmic normal distribution around a single mode. This indicates that evolutionary changes in genome size are numerous and individually small, as would be the case with duplications and deletions, rather than large and discontinuous, as would be expected if they were due to polyploidy or polyteny.

The gradual character of change in DNA amounts can also be observed within a single genus, such as *Bufo*. The DNA content has been determined in 19 of the 250 known species. Figure 3-6 shows that the DNA content of these toad species ranges from about 7 to 15 $\times$ 10^9 nucleotide pairs, with a mode around 10 $\times$ 10^9; the distribution fits extremely well the expectations

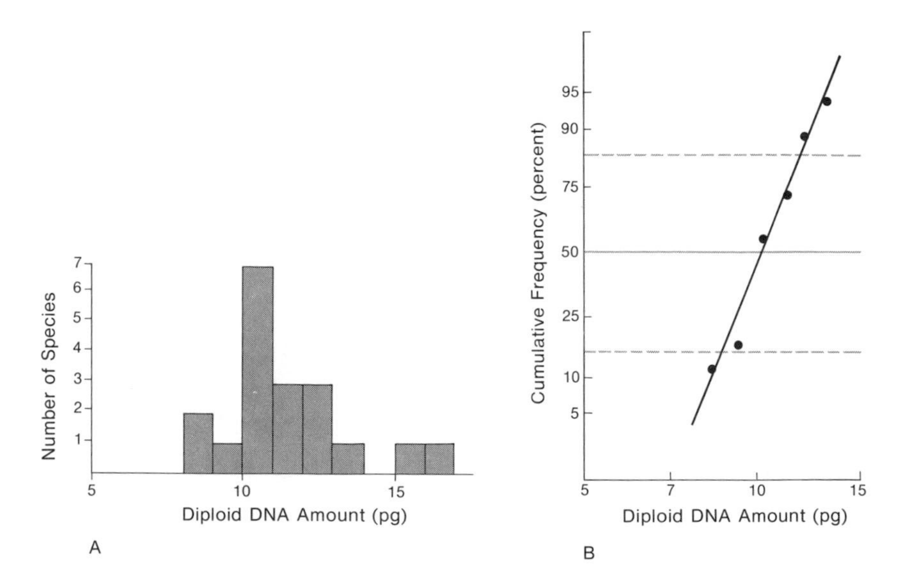

FIGURE 3-6.
The amount of DNA per nucleus in 19 species of the toad genus *Bufo*. **A.** The amount of DNA per nucleus ranges from about 7.2 $\times$ 10^9 to about 15.4 $\times$ 10^9 nucleotide pairs (or from 8 to 17 pg), with a mode about 9.5 $\times$ 10^9 nucleotide pairs. **B.** The cumulative frequencies of species plotted against the amounts of DNA on a logarithmic scale fall very nearly in a straight line, which represents the expectations of a logarithmic normal distribution. This good fit supports the hypothesis that the amount of DNA per nucleus evolves by small increments or decrements that are approximately proportional to the preexisting amount of DNA. [After Bachmann, Goin, and Goin, 1974.]

of a logarithmic normal distribution. All 19 species have 22 chromosomes per nucleus. The transition from about 7×10^9 nucleotide pairs to about twice that amount has occurred not by polyploidy, but by cumulative addition and subtraction of small amounts of DNA, resulting in a fairly smooth and continuous distribution. Similar results have been obtained for other genera in which a number of species have been studied (Bachmann, Goin, and Goin, 1974; see also Figure 3-5).

Chromosomal deletions or deficiencies are sometimes detected under the microscope in the large polytene chromosomes of Diptera by observing that one or several bands are missing in a chromosome. Whenever fairly large chromosomal segments are involved, deletions can be detected in normal, nonpolytene chromosomes with the light microscope. In recent years the development of special staining procedures, e.g., the acetic-saline-giemsa method, and the use of fluorescent dyes, e.g., quinacrine mustard, have greatly increased the number of deletions discovered in a variety of organisms. In man the *cri du chat* syndrome, which includes severe mental retardation and other abnormalities, has been associated with a deficiency in the short arm of chromosome 5. Leukemia patients are often carriers of the "Philadelphia chromosome," a chromosome 22 with a deleted segment.

Chromosomal deficiencies are often lethal in homozygous condition, as would be expected if genes responsible for some essential functions in the physiology or development of the organism were missing. The *Notch* phenotype in *Drosophila melanogaster* exhibits small incisions in the edges of the wings; it is due to any one of a series of heterozygous deletions in the X chromosome near the *white*-eye locus. In homozygous condition or in the males even the smaller *Notch* deletions are lethal. Large deletions are often lethal or at least seriously deleterious even in heterozygotes, presumably owing to developmental imbalance. Deletions of small chromosomal segments might conceivably not be lethal in homozygotes whenever the deleted genes had been previously duplicated and their function could still be carried out by the remaining genes. Also, deletions might not be lethal, and might perhaps be favored by natural selection, in the evolution of saprophytic or parasitic ways of life, when the function of certain groups of genes may not be needed; but there is little experimental evidence to support this suggestion. It is, however, well established that deletions, and not only duplications, have occurred in the evolution of life (see Hinegardner, 1976).

Duplications of genetic material, followed by divergence of the duplicated segments towards fulfilling different functions, have played a major role in evolution (Ohno, 1970). The ancestral form(s) of life of all DNA-carrying organisms probably contained a short DNA double helix consisting of only one or a few genes. The tens of thousands of different DNA genes found in, say, the human genotype are descendants of that ancestral short segment through multiple duplications and modifications. Visible evidence of gene duplication in phylogeny has been obtained in the study of the giant polytene chromosomes of various species of *Drosophila*. Repeats of groups of one or several bands are often found either contiguous to each other or in different parts of the chromosomes. The homology of the repeated segments is manifested not only by the visible similarity of the bands, but also by the tendency of the repeats to pair with each other in the polytene nuclei.

A classic example of a chromosomal duplication is the *Bar* phenotype in *Drosophila melanogaster*, which makes the eye abnormally narrow by reducing the number of facets in it. *Bar* flies have a small segment of the X chromosome duplicated in tandem; flies with three tandem repeats of the segment have a more exaggerated ultra-*Bar* phenotype. However, duplications of chromosomal material do not always have conspicuous effects on the phenotype of their carriers. *Clarkia unguiculata*, an annual flowering plant, may have whole chromosomes present three times, rather than the normal two, without exhibiting any detectable change in its phenotype or in its vigor.

Considerable information has accumulated in recent years on the evolutionary role of duplications of DNA material. For convenience, we may distinguish three general classes of gene duplications.

1. Duplications of single loci followed by divergent evolution toward different functions.
2. Genes that exist in several copies within each genome but which remain essentially identical to each other in DNA sequence and in function. The presence of several copies of a single gene allows the organism to obtain large amounts of the gene product in short time intervals.
3. Short sequences of DNA in eukaryotes that are repeated from a thousand to more than a million times in the genome, although not all copies may be exactly identical. These highly repetitive sequences of DNA may be involved in the regulation of gene activity.

We shall discuss these three classes of duplications in turn.

EVOLUTION OF DUPLICATED GENES

The genes coding for hemoglobins in vertebrates have evolved by a series of gene duplications followed by gradual divergence toward different but related functions. The genes coding for hemoglobins, as well as for myoglobins, can be traced to a single gene that became duplicated some 650 million years ago in the ancestral line of the vertebrates. Both myoglobins and hemoglobins are involved in oxygen transport—myoglobins in muscle, hemoglobins in blood. Myoglobins consist of a single polypeptide chain arranged in a complex three-dimensional structure with a heme group—a protoporphyrin ring with an iron atom—in the center. Myoglobins have a molecular weight of about 17,000. Hemoglobin molecules generally consist of four subunits: two polypeptide chains of one kind and two of another, each with a heme group in its center. With the exception of certain cyclostomes (fishes without jaws, like lampreys and hagfishes) all vertebrate hemoglobins are apparently made up that way. Hemoglobin molecules have a molecular weight of about 67,000. In normal human adults there are two types of hemoglobins, A and A_2. Hemoglobin A, the most common, consists of two α and two β polypeptide chains (symbolized as α_2 β_2); hemoglobin A_2, which makes up only about 2 percent adult hemoglobin, consists of two α and two δ chains (α_2 δ_2). A different hemoglobin is found in human embryos, the so-called fetal hemoglobin, which consists of two α and two γ chains ($\alpha_2\gamma_2$). The α, β, γ, and δ polypeptide chains are each coded by a different gene.

The phylogeny of the hemoglobin genes is shown in Figure 3-7. The gene ancestral to the modern hemoglobin genes became duplicated some 380 million years ago. One of the genes specialized in coding for α chains, the other diverged into a gene coding for a somewhat different chain. This duplication made possible the development of the tetramer structure consisting of two different chains found in the hemoglobins of the higher vertebrates. This structure enhances the possibility of heme–heme interactions and results in more efficient oxygenation and deoxygenation. Another duplication occurred around 150

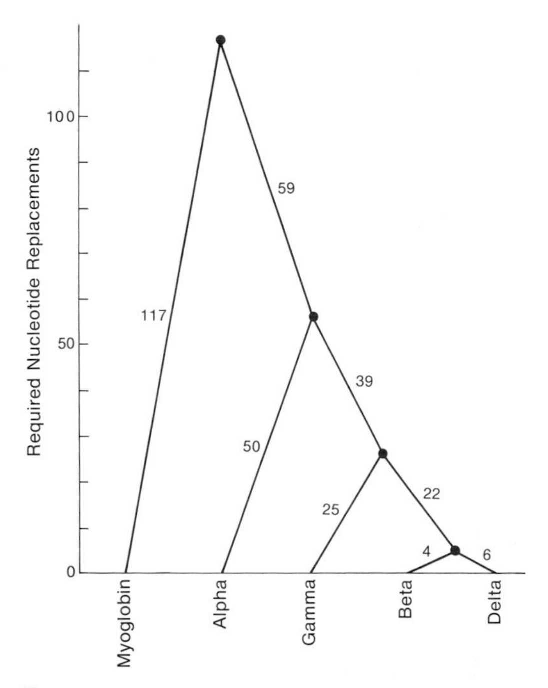

FIGURE 3-7.
Phylogeny of the globin genes. The black dots indicate probable duplication of ancestral genes giving rise to a new gene line. Numbers on segments are the nucleotide replacements required to account for the descent of the genes coding for the five globin molecules from a common ancestral gene (see Chapter 9). [See Fitch and Margoliash, 1970.]

million years ago in the ancestral evolutionary line leading to the higher primates. One of the genes continued coding for the β chain, the other evolved into the gene coding for the δ chain (Zuckerkandl and Pauling, 1965; Zuckerkandl, 1965). The δ chain made possible the appearance of a new kind of hemoglobin, A_2, in adult men and anthropoid apes. The β–δ duplication occurred after the separation of the hominoids from the other primates, since the lower primates do not have hemoglobin A_2.

The duplications of the α–β and the γ–β genes were accompanied or followed by chromosomal translocations, since at least in man the α, β, and γ genes are located in different chromosomes. Nevertheless, the β and δ genes remain closely linked, as evinced by the occurrence of the Lepore hemoglobins. At one end of the polypeptide chains of Lepore hemoglobins are peptides normally found in the β chain, and at the other end peptides typical of the δ chain. The genes coding for Lepore hemoglobins apparently arise by asymmetrical crossing-over between β and δ genes, the resulting gene having a DNA sequence consisting of part of the β and part of the δ gene.

During the evolution of hemoglobins, nucleotide substitutions, as well as nucleotide additions and deletions, have taken place. Human myoglobin consists of 153 amino acids, the α chains have 141 amino acids, and the β, γ, and δ chains consist of 146 amino acids each. Figure 3-7 indicates the minimum number of nucleotide replacements in the DNA required to account for the descent of the myoglobin and the various hemoglobin polypeptides of modern man from a single ancestral gene. Each globin gene has also changed in different lines of descent through evolutionary time. The α chains of man and chimpanzee have identical amino acid sequences, but the α chains of man and horse differ in 18 amino acids, while those of man and carp differ in 68 amino acids.

Other duplications of hemoglobin genes are known, and some must have occurred in recent evolutionary history. The highly inbred strain of mice known as SEC, used in laboratory studies of various sorts, contains two α genes probably arranged in tandem (the crossover frequency is less than 0.005 percent). The two genes code for different amino acids at position 68, one for serine and the other for threonine. This duplication must have arisen very recently since it has not been found in other laboratory strains of mice (Popp, 1969). Man has three and perhaps four copies of the γ gene, all clustered together.

Another instance of gene duplication in evolution involves two genes, *Pgi-1* and *Pgi-2*, coding for two forms of the dimeric enzyme phosphoglucose isomerase in bony fishes (Avise and Kitto, 1973). This duplication is present in all higher teleost fishes but is absent in the more primitive Holostei and Chondrostei, and thus it must have occurred after the lineages leading to the modern teleosts diverged from the holosteans. The duplication probably occurred during the Jurassic, some 150 million years ago, in the leptolepiforms, a now extinct group of fishes from which all modern teleosts have descended (Figure 3-8). The polypeptides coded by the *Pgi-1* and *Pgi-2* genes differ by at least one amino acid substitution, and the enzymes differ in their thermal stabilities and other properties. The *Pgi* enzymes all catalyze the same basic reaction

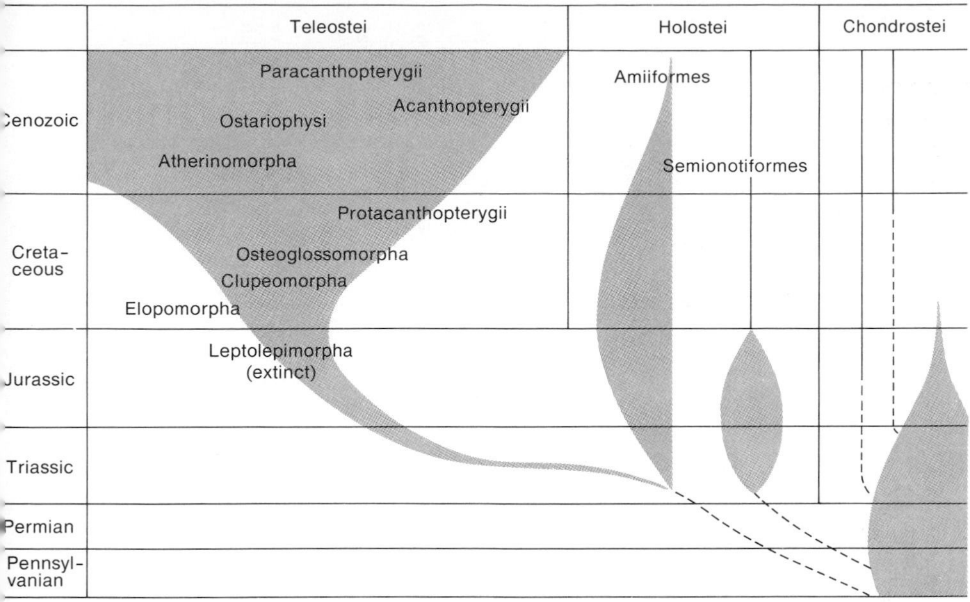

FIGURE 3-8.
Phylogenetic reconstruction of the evolutionary history of the higher bony fishes. (The widths of the grey areas indicate approximately the relative abundance of the taxa over time.) The duplicated genes *Pgi-1* and *Pgi-2* are present in all teleosts but absent in the holosteans and chondrosteans. The duplication of these genes must have occurred after the lineages leading to the teleosts and holosteans separated from each other; it probably occurred in the leptolepiforms during the Jurassic, some 150 million years ago.

but have become adaptively differentiated. The *Pgi-1* enzyme is predominantly active in most tissues, but not in skeletal muscle, where the *Pgi-2* enzyme is more active. The heterodimer enzyme, made up of one *Pgi-1* and one *Pgi-2* polypeptide, is strongly expressed in the heart, often along with moderate activity of the *Pgi-1* and *Pgi-2* enzymes.

Another example of gene duplication involves the enzymes trypsin and chymotrypsin, which have been well studied in cattle (see Ohno, 1970). These two enzymes are produced in the pancreas in their inactive forms as trypsinogen and chymotrypsinogen and are released in the intestine, where they are involved in the digestion of protein. The inactive forms of both enzymes are activated by trypsin, which removes a group of amino acids from their amino terminals (six from trypsinogen and 15 from chymotrypsinogen). Trypsin acts by hydrolyzing the carboxyl end of the basic amino acids lysine and arginine, while chymotrypsin digests protein at the carboxyl end of the aromatic amino acids phenylalanine and tyrosine.

Trypsin and chymotrypsin have great similarities in amino acid sequences (Figure 3-9), particularly at their active sites. The two enzymes have in common 14 out of 17 amino acids in one active site and 8 out of 10 in the other. Since

CH	1	*Cys*	*Gly*	*Val*	*Pro*	*Ala*	*Ilu*	*Gln*	*Pro*	*Val*	*Leu*	*Ser*	*Gly*	*Leu*	*Ser*	*Arg*	Ilu	Val	Asn	Gly	Glu
TRP	1												*Val*	*Asp*	*Asp*	*Asp*	*Asp*	*Lys*	Ilu	Val	Gly
CH	21	Glu	Ala	Val	Pro	Gly	Ser	Trp	Pro	Trp	Gln	Val	Ser	Leu	Gln	Asp	Lys	Thr	Gly	Phe	His
TRP	10	Gly	Tyr	Thr	Cys	Gly	Ala	Asn	Thr	Val	Pro	Tyr	Gln	Val	Ser	Leu	Asn	Ser	Gly	Tyr	His
CH	41	Phe	**Cys**	**Gly**	**Gly**	**Ser**	**Leu**	**Ilu**	**Asn**	**Glu**	**Asn**	**Trp**	**Val**	**Val**	**Thr**	**Ala**	**Ala**	**His**	**Cys**	Gly	Val
TRP	30	Phe	**Cys**	**Gly**	**Gly**	**Ser**	**Leu**	**Ilu**	**Asn**	**Ser**	**Gln**	**Trp**	**Val**	**Val**	**Ser**	**Ala**	**Ala**	**His**	**Cys**	Tyr	Lys
CH	61	Thr	Thr	Ser	Asp	Val	Val	Val	Ala	Gly	Glu	Phe	Asp	Gln	Gly	Ser	Ser	Ser	Glu	Lys	Ilu
TRP	50	Ser	Gly	Ilu	Gln	Val	Arg	Leu	Gly	Gln	Asp	Asn	Ilu	Asn	Val	Val	Glu	Gly	Asn	Gln	Gln
CH	81	Gln	Lys	Leu	Lys	Ilu	Ala	Lys	Val	Phe	Lys	Asn	Ser	Lys	Tyr	Asn	Ser	Leu	Thr	Ilu	Asn
TRP	70	Phe	Ilu	Ser	Ala		Ser	Lys	Ser	Ilu	Val	His	Pro	Ser	Tyr	Asn	Ser	Asn	Thr	Leu	Asn
CH	101	Asn	Asn	Ilu	Thr	Leu	Leu	Lys	Leu	Ser	Thr	Ala	Ala	Ser	Phe	Ser	Gln	Thr	Val	Ser	Ala
TRP	89	Asn	Asp	Ilu	Met	Leu	Ilu	Lys	Leu	Lys	Ser	Ala	Ala	Ser	Leu	Asn	Ser	Arg	Val	Ala	Ser
CH	121	Val	Cys	Leu	Pro	Ser	Ala	Ser	Asp	Asp	Phe	Ala	Ala	Gly	Thr	Thr	Cys	Val	Thr	Thr	Gly
TRP	109	Ilu	Ser	Leu	Pro	Thr	Ser	Cys	Ala	Ser	Ala	Gly	Thr	Gln	Cys	Leu	Ilu	Ser	Gly	Trp	Gly
CH	141	Trp	Gly	Leu	Thr	Arg	Tyr	Thr	Asn	Ala	Asn	Thr	Pro	Asp	Arg	Leu	Gln	Gln	Ala	Ser	Leu
TRP	129	Asn	Thr	Lys	Ser	Ser	Gly	Thr	Ser	Tyr	Pro	Asp	Val	Leu	Lys	Cys	Leu	Lys	Ala	Pro	Ilu
CH	161	Pro	Leu	Leu	Ser	Asn	Thr	Asn	Cys	Lys	Lys	Tyr	Trp	Gly	Thr	Lys	Ilu	Lys	Asp	Ala	Met
TRP	149	Leu	Ser	Asn	Ser	Ser	Cys	Lys	Ser	Ala	Tyr	Pro	Gly	Gln	Ilu	Thr	Ser	Asn	Met	Phe	Cys
CH	181	Ilu	Cys	Ala	Gly	Ala	Ser	Gly	Val	Ser	Ser	**Cys**	**Met**	**Gly**	**Asp**	**Ser**	**Gly**	**Gly**	**Pro**	**Leu**	**Val**
TRP	169	Ala	Gly	Tyr	Leu	Glu	Gly	Gly	Lys	Asn	Ser	**Cys**	**Gln**	**Gly**	**Asp**	**Ser**	**Gly**	**Gly**	**Pro**	**Val**	**Val**
CH	201	**Cys**	Lys	Lys	Asn	Gly	Ala	Trp	Thr	Leu	Val	Gly	Ilu	Val	Ser	Trp	Gly	Ser	Ser	Thr	Cys
TRP	189	**Cys**	Ser	Gly	Lys	Leu	Gln	Gly	Ilu	Val	Ser	Trp	Gly	Ser	Gly	Cys	Ala	Gln	Lys	Asn	Lys
CH	221	Ser	Thr	Ser	Thr	Pro	Gly	Val	Tyr	Ala	Arg	Val	Thr	Ala	Leu	Val	Asn	Trp	Val	Gln	Gln
TRP	209	Pro	Gly	Val	Tyr	Thr	Lys	Val	Cys	Asn	Tyr	Val	Ser	Trp	Ilu	Lys	Gln	Thr	Ilu	Ala	Ser
CH	241	Thr	Leu	Ala	Ala	Asn															
TRP	229	Asn																			

FIGURE 3-9.
The amino acid sequences of chymotrypsinogen A (CH) and trypsinogen (TRP) from cattle. When the amino acids in italic (at the beginning) are split off, the enzymes become activated. The segments in bold face are the active sites of the enzymes. At one active site the two enzymes have 14 out of 17 amino acids in common, at the other 8 out of 10, suggesting that the two enzymes are encoded by genes that arose one from the other by duplication. [After Ohno, 1970.]

trypsin is the activating molecule for both enzymes it seems likely that the gene coding for trypsinogen is the ancestral one.

Many other instances are known of gene duplications in evolutionary history. There is evidence, however, to suggest that most structural genes, i.e., genes translated into proteins, exist in each genome in single copies (Goldberg *et al.*, 1973; Davidson and Britten, 1973). This suggests that although gene duplications may not be rare events on the evolutionary time scale, they do not become continuously established. In general, enough time passes between duplications of a given gene to allow for evolutionary divergence of the duplicated genes through point mutations.

GENES IN MULTIPLE COPIES

The second class of duplicated DNA sequences consists of genes repeated from a few to several hundred times in each genome, all copies of which are presumably identical in nucleotide sequence and thus transcribed in identical RNA sequences. This class includes the genes coding for ribosomal RNA and transfer RNA, both of which are transcribed but not translated, and the genes coding for histone proteins and for antibodies.

Ribosomes are involved in protein synthesis. They consist of two sub-units, one small and one large, made up of three kinds of rRNA and some 50 kinds of ribosomal protein. During protein synthesis ribosomes become associated with messenger RNA and mediate the sequential codon–anticodon recognition between the mRNA and the tRNA's. The ribosomes also make possible the formation of peptide bonds between the amino acids brought in by the tRNA's. In eukaryotes the three kinds of rRNA are designated 5S, 18S, and 28S, the larger numbers indicating larger RNA molecules. The small ribosomal subunit contains an 18S rRNA molecule; the larger subunit contains a 5S and a 28S rRNA molecule.

In eukaryotes the gene coding for the 18S and 28S rRNA is located in the nucleolar organizer (NO), a chromosomal region lying in the nucleolus. Each gene is a sequence of DNA whose coded RNA product is split after transcription into the two types, 18S and 28S. The gene coding for the 18S and 28S rRNA is replicated a variable number of times in different organisms. In *Drosophila melanogaster* the NO contains about 130 sequences of this gene (Ritossa and Spiegelman, 1965; Ritossa, Atwood, and Spiegelman, 1966); in the African toad, *Xenopus laevis*, each NO carries about 400 repeated sequences of this rRNA gene (Brown and Gurdon, 1964). The repeated copies of this rRNA gene are arranged in tandem, although they are separated by short interspersed ("spacer") sequences of DNA (Davidson and Britten, 1973). The genes coding for the 5S rRNA are also replicated many times in eukaryotic organisms, although they are not located next to the genes coding for 18S and 28S rRNA. In *Drosophila melanogaster* the NO is in the X and Y chromosomes, while the genes coding for 5S rRNA are in one of the autosomes.

Natural selection may have favored the multiplication of the rRNA genes, since cells require large numbers of ribosomes for protein synthesis. Each molecule of mRNA is "read" by a group of several ribosomes, called polyribosomes, proceeding sequentially along the mRNA. The demand of a cell for ribosomes is great because many proteins are synthesized at the same time. Generally about 80 to 85 percent of all RNA in a cell is rRNA. Evidence of the great demand for rRNA is provided by the fact that the five to 10 genes coding for rRNA in *Escherichia coli* undergo nearly continuous transcription. In *Drosophila melanogaster*, bobbed mutants, which lack the NO in the Y chromosome, have reduced viability and fertility and develop more slowly than normal flies.

The five to 10 copies of rRNA genes in *E. coli* represent about 0.4 percent of its total DNA content; the 130 copies of rRNA genes in *D. melanogaster* represent about 0.3 percent; and the 400 copies of *Xenopus* represent about 0.2 percent. The relative proportion of the genome (0.2–0.4 percent) dedicated to the rRNA genes is approximately the same in these organisms, although it decreases somewhat in the more complex ones, which can therefore devote a higher proportion of the DNA to functions other than protein synthesis.

Transfer RNA molecules are involved in protein synthesis as the carriers of specific amino acids to the mRNA–polyribosome complexes, where the anticodon site of a tRNA molecule "recognizes" a corresponding codon in mRNA. The active forms of tRNA are relatively small molecules with about 70 to 80 nucleotides and a molecular weight of about 30,000; their precursors have about 40 additional nucleotides, which are eventually cleaved off to yield the functional tRNA molecules. There are 30 to 40 different tRNA molecules in *E. coli* and about 60 in higher organisms. Since only 20 different amino acids are used in protein synthesis, it follows that two or more tRNA's may have affinity for a single amino acid. For example, in *E. coli* there are five species of tRNA associated with the amino acid leucine. This species apparently has only a single copy of each of its 30 to 40 tRNA genes, but eukaryotes have multiple copies of each tRNA gene. In yeast there are about 400 tRNA genes, each of 61 different genes being repeated five to seven times. In *Drosophila melanogaster* each tRNA gene exists in about 13 copies, with a total of somewhat more than 700 tRNA genes.

Histones are basic proteins containing a relatively high proportion of the amino acids lysine and arginine; their molecular weight ranges from 11,000 to 21,000. There are five types of histones, distinguished by the relative amounts of lysine and arginine. Histones and DNA occur in about equal amounts by weight in the chromosomes of eukaryotes. Apparently most or all the DNA in the nucleus is associated with histone and forms a structure resembling a string of beads; the "beads" are made up of DNA coiled around a histone core and are connected by short DNA segments. Histones may be involved in gene regulation by altering the transcriptional properties of the DNA. The genes coding for the histones exist in the form of tandem clusters of several hundred copies of each gene (Kedes and Birnstiel, 1971; Davidson and Britten, 1973). Other genes coding for protein (structural genes) and known to exist in eukaryotes in multiple copies include the genes coding for antibodies (Hood *et al.*, 1975).

HIGHLY REPETITIVE DNA

A fraction of the DNA in eukaryotes, but not in *Escherichia coli* and other prokaryotes, consists of relatively few sequences, each replicated many times. Although the values obtained depend to a certain extent on experimental conditions, such as the length of the DNA fragments used in the reannealing procedures and temperature, there is little doubt that from 10 to 80 percent of all DNA in eukaryotes consists of multiple, virtually identical copies of each of a few DNA sequences. Data obtained from three species (an invertebrate, amphibian, and mammal), using the same experimental procedures, are given in Table 3-4. In man about 70 percent of the DNA consists of unique sequences; the rest consists of sequences of various lengths each repeated from 300 to 300,000 times (Britten and Davidson, 1971). In *Drosophila melanogaster* about

TABLE 3-4. Frequency of unique and repetitive DNA sequences in the genomes of three organisms. Data obtained from 450 nucleotide-long fragments using hydroxyapatite and 0.12M phosphate buffer at 60°C. [After Davidson and Britten, 1973.]

	Percent DNA	Number of copies per sequence	Complexity
Sea urchin (*Strongylocentrotus purpuratus*)	38%	1	3.0×10^8
	25	20–50	1.0×10^7
	27	250	1.0×10^6
	7	6,000	1.3×10^4
	3	*	*
African toad (*Xenopus laevis*)	54	1	1.6×10^9
	6	20	1.5×10^7
	31	1,600	6.0×10^5
	6	32,000	6.0×10^3
	3	*	*
Calf (*Bos taurus*)	55	1	1.5×10^9
	38	60,000	1.7×10^4
	2	1,000,000	60
	3	*	*

*This DNA fraction binds virtually instantaneously.

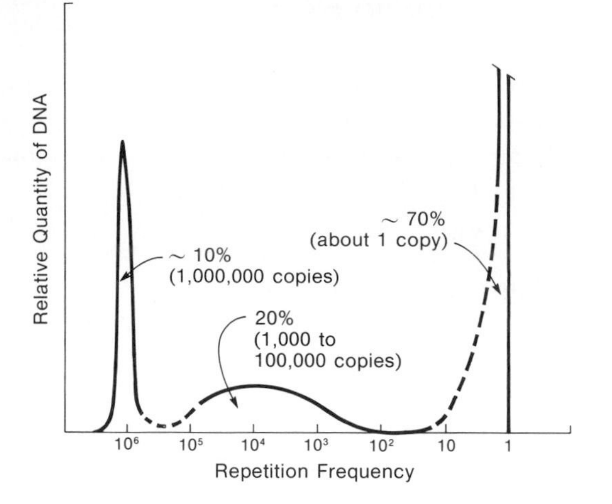

FIGURE 3-10.
Proportion of the total DNA with a given level of sequence repetition in the mouse. The scale on the abscissa is logarithmic. About 70 percent of the total DNA is made up of sequences that are each present only once in the haploid genome. About 20 percent of the DNA consists of sequences that are each repeated from 1,000 to 100,000 times. About 10 percent of the DNA consists of sequences that are each repeated about one million times; this fraction represents the satellite DNA. The graph is meant to show that there are various degrees of repetition of DNA sequences, but the actual configuration is largely conjectural. The spikes at the two ends (repetition frequencies 1 and 10^6) are probably good approximations to reality, but the shape of the curve in the intermediate points (repetition frequencies between 10 and 10^5) is not known. [After Britten and Kohne, 1968.]

90 percent of the DNA consists of unique sequences and 10 percent of a few sequences replicated many times. Results obtained with mouse DNA are shown in Figure 3-10.

Table 3-4 introduces the term *complexity*, which refers to the total number of nucleotide pairs in all DNA sequences when repeated sequences are counted only once. The complexity of a set of unique sequences is the total length of all such sequences. The complexity of a series of identical copies of a single sequence is the number of nucleotides in one of the copies, but there is some ambiguity, since DNA fragments very similar but not identical in nucleotide sequence may or may not appear as repetitive sequences, depending on the experimental conditions.

Little is known about the function of the highly repetitive sequences of DNA. Recent studies carried out primarily with the African clawed toad (*Xenopus laevis*) and the purple sea urchin (*Strongylocentrotus purpuratus*) indicate that much of repetitive DNA is made of a set of sequences about 300 nucleotides in length, each replicated many times. These sequences appear to be distributed throughout the genome, interspersed between single-copy sequences. Britten and Davidson (1969, 1971; Davidson and Britten, 1973) have suggested that these repetitive sequences may be involved in gene regulation. Their model is discussed in Chapter 8.

There is only one copy of most structural genes in each haploid complement, but recent evidence indicates that only a fraction of single-copy DNA codes for proteins (Galau, Britten, and Davidson, 1974; Levy and McCarthy, 1975). At the gastrula stage of sea urchin embryos, 28.5 percent of all single-copy DNA is transcribed, but only about 10 percent of those sequences (2.7 percent of the total single-copy DNA) is translated into protein. That is, about 90 percent of the transcribed sequences of single-copy DNA do not leave the nucleus, or at least do not become associated with ribosomes for translation. Little is known about the actual role of this large fraction of single-copy DNA (Britten and Davidson, 1971).

Extrapolation from the above results leads to some tentative speculations about the maximum number of structural genes in various organisms. We may assume that all single-copy DNA is transcribed at one or another stage of the life cycle, although there is no convincing evidence that such is the case. If about 10 percent of the transcribed DNA sequences become involved in translation, then about 10 percent of the total single-copy DNA would consist of structural genes. The human genome contains about 2.9×10^9 nucleotide pairs; 70 percent of that total consists of single-copy DNA. Ten percent of the single-copy DNA is about 2×10^8 nucleotide pairs, the fraction that we shall assume to make up the structural genes. The mean length of a structural gene is estimated to be about 1,800 nucleotide pairs. Therefore, the maximum number of structural genes in man would be about 110,000. Similar calculations estimate a maximum of 17,000 structural genes in the sea urchin and 83,000 in the cow.

Drosophila melanogaster has about one-twentieth as much DNA as man, but 90 percent of it, or 1.3×10^8 nucleotide pairs, consists of single sequences. If 10 percent of the single copy DNA codes for proteins, there would be about 7,250 structural genes in *D. melanogaster.* About 5,000 to 6,000 bands are observed in the polytene chromosomes of this species, and it has been suggested that each band may represent a single structural gene. Other estimates based on biochemical studies also suggest that the number of structural genes in *D. melanogaster* may be about 7,000 (Levy and McCarthy, 1975).

INVERSIONS AND TRANSLOCATIONS

Chromosomal inversions and translocations are rearrangements of the genome without addition or deletion of hereditary materials. Inversions are 180° rotations of chromosomal segments. If the gene sequence of a chromosome is represented as ABCDEF, the sequence ABEDCF represents the same chromosome with the segment CDE inverted. In inversion heterozygotes, synapsis of the homologous chromosomes requires the formation of a loop involving the inverted segments (Figures 3-11, 3-12, and 3-13). Heterozygous inversions can be recognized by the presence of such loops in preparations of cells at the pachytene stage of meiosis. Similar loops are also observed in the polytene chromosomes found in the salivary glands and other tissues of Diptera

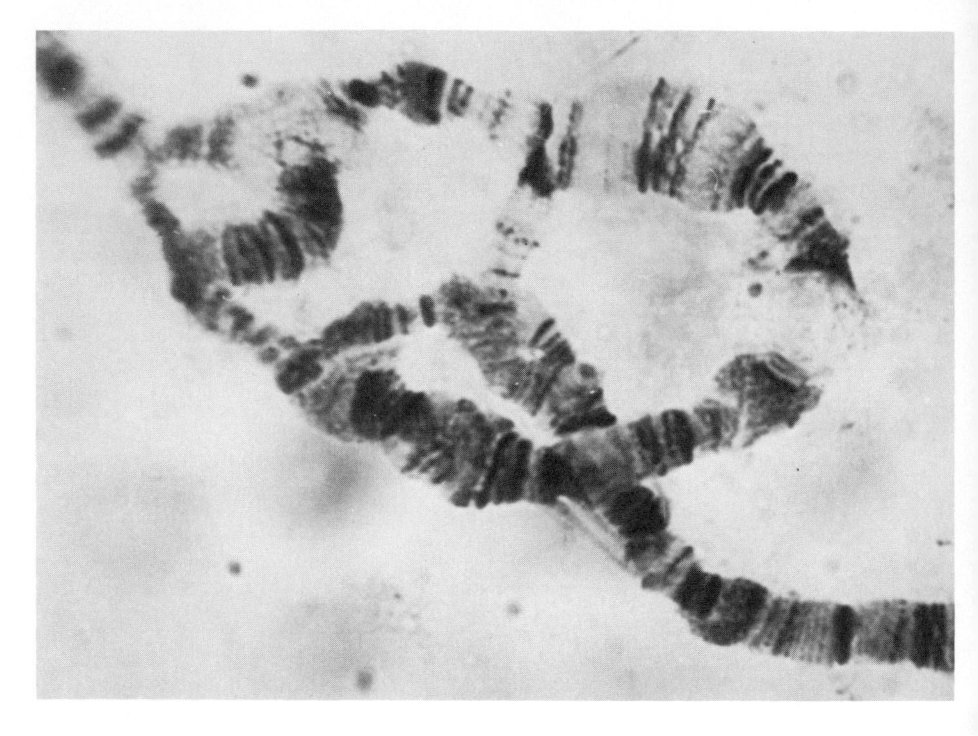

FIGURE 3-11.
A segment of the third chromosome of a *Drosophila pseudoobscura* fly heterozygous for different chromosome sequences. The polytene chromosomes of diptera result from the pairing of the two homologous chromosomes, each replicated many times. The loops are characteristic of inversion heterozygotes.

(Figure 3-11). The presence of inversions can also be detected because they effectively suppress recombination in heterozygotes.

Genetic recombination is suppressed in the progenies of inversion heterozygotes. The effects of crossing-over in an individual heterozygous for a paracentric inversion are shown schematically in Figure 3-12. Of the four chromosomes resulting from the meiotic divisions, one has two centromeres, one has none, and two are normal non-crossover chromosomes. The only gametes that can give rise to viable progeny are those containing the non-crossover chromosomes.

It would seem that heterozygotes for paracentric inversions should have reduced fertility, but such is not always the case. In *Drosophila* and related families of flies crossing-over does not occur during male meiosis; thus, males heterozygous for inversions do not exhibit reduced fertility. In females one of the normal chromosomes is always included in the egg nucleus, while the other three are eliminated in the polar bodies; thus, females heterozygous for paracentric inversions have undiminished fertility. In other Diptera, such as the midge *Chironomus*, chiasmata are formed and crossing-over takes place in male meiosis. Yet the fertility of male inversion heterozygotes is not appreciably

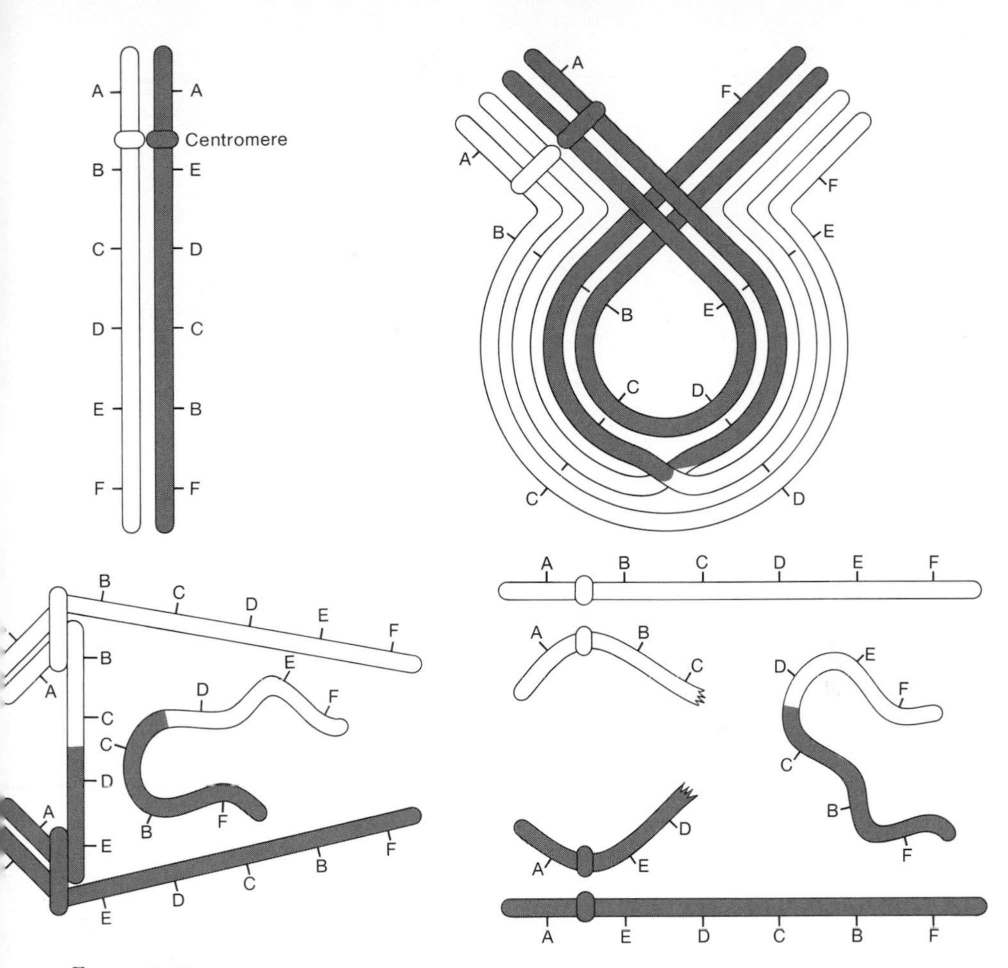

FIGURE 3-12.
Crossing-over in a heterozygote for a paracentric inversion. *Upper left:* Two chromosomes in the body cells of the individual. The chromosomal segment from B to E is inverted in one chromosome relative to the other; since the centromere is not included, the inversion is called *paracentric. Upper right:* Crossing-over between two nonsister chromatids. *Lower left:* Separation of the chromosomes at early anaphase of the first meiotic division. One chromosomal segment has two centromeres and will eventually break up as the centromeres move away from each other; another chromosomal segment has no centromere and will be lost. *Lower right:* The resulting chromosomes. Only two chromosomes have complete sets of genes; these are noncross-over chromosomes with the same gene sequences as the two original chromosomes.

diminished, apparently because spermatozoa containing abnormal chromatids fail to function. Heterozygotes for paracentric inversions also occur in many species of plants, where they have been detected because the dicentric chromatid appears at the first anaphase of meiosis as a "bridge," while the acentric fragment lags between the disjoining groups of chromosomes (see Figure 3-12).

The effects of crossing-over in a heterozygote for a pericentric inversion are shown in Figure 3-13. Of the four chromosomes resulting from the meiotic

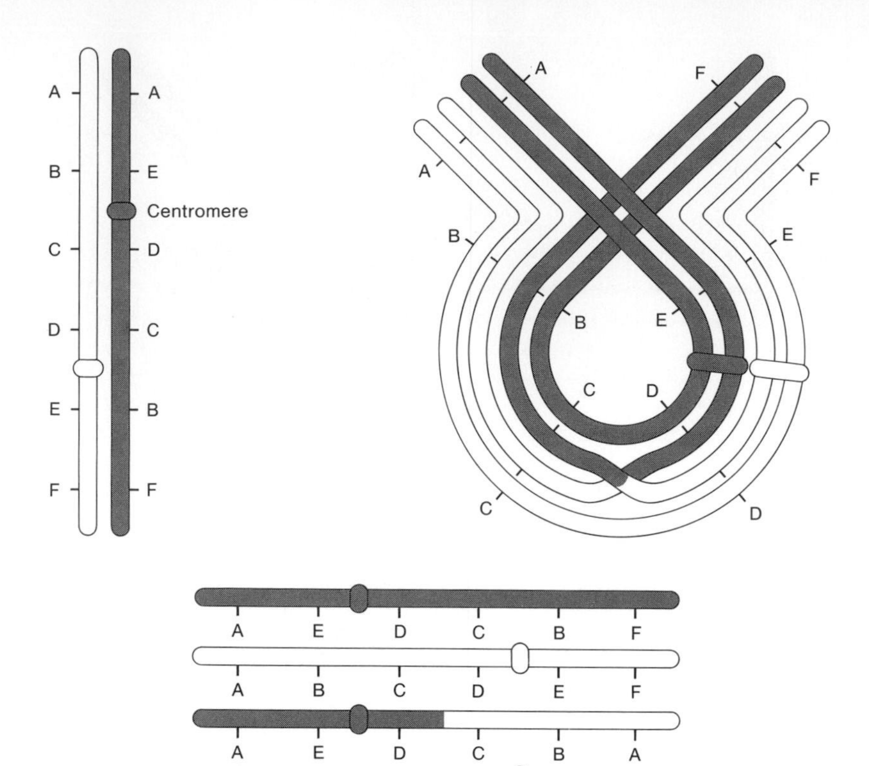

FIGURE 3-13.
Crossing-over in a heterozygote for a pericentric inversion. *Upper left:* The two chromosomes. The inverted segment from B to E includes the centromere, and therefore the inversion is called *pericentric. Upper right:* Crossing-over between two nonsister chromatids. *Below:* The four resulting chromosomes. Only the two top chromosomes have complete sets of genes; they are noncrossover chromosomes with the same gene sequences as the two original chromosomes.

divisions, two are the original noncrossover chromosomes, but the other two have either duplications or deficiencies. The gametes carrying the crossover chromosomes usually cannot produce viable progeny. Therefore, as with paracentric inversion heterozygotes, no recombination occurs in the progenies of heterozygotes for pericentric inversions. Moreover, females heterozygous for pericentric inversions are often partially sterile.

Pericentric inversions may produce chromosomes very different in appearance from the original ones. An originally metacentric (V-shaped) chromosome, after a pericentric inversion asymmetric around the centromere, may look like a submetacentric (J-shaped) or acrocentric (rod-shaped) chromosome; the reverse is also possible. Changes in the appearance of chromosomes may also be due to other chromosomal rearrangements, such as the translocation of a chromosomal segment to another position in the same chromosome (transposition). Whenever changes in chromosome shape occur, genetic and cytological studies may permit ascertaining whether or not a pericentric inversion has occurred in a particular case.

Reciprocal translocations involve the interchange of blocks of genes between nonhomologous chromosomes. If the gene sequences in two nonhomologous chromosomes are represented as ABCDEF and GHIJKL, the sequences ABCDKL and GHIJEF represent translocated chromosomes. In a translocation heterozygote, pairing during meiotic prophase results in a cross-shaped configuration represented schematically in Figure 3-14. Segregation at meiosis may occur in a variety of ways, as shown in the figure. Of the six types of

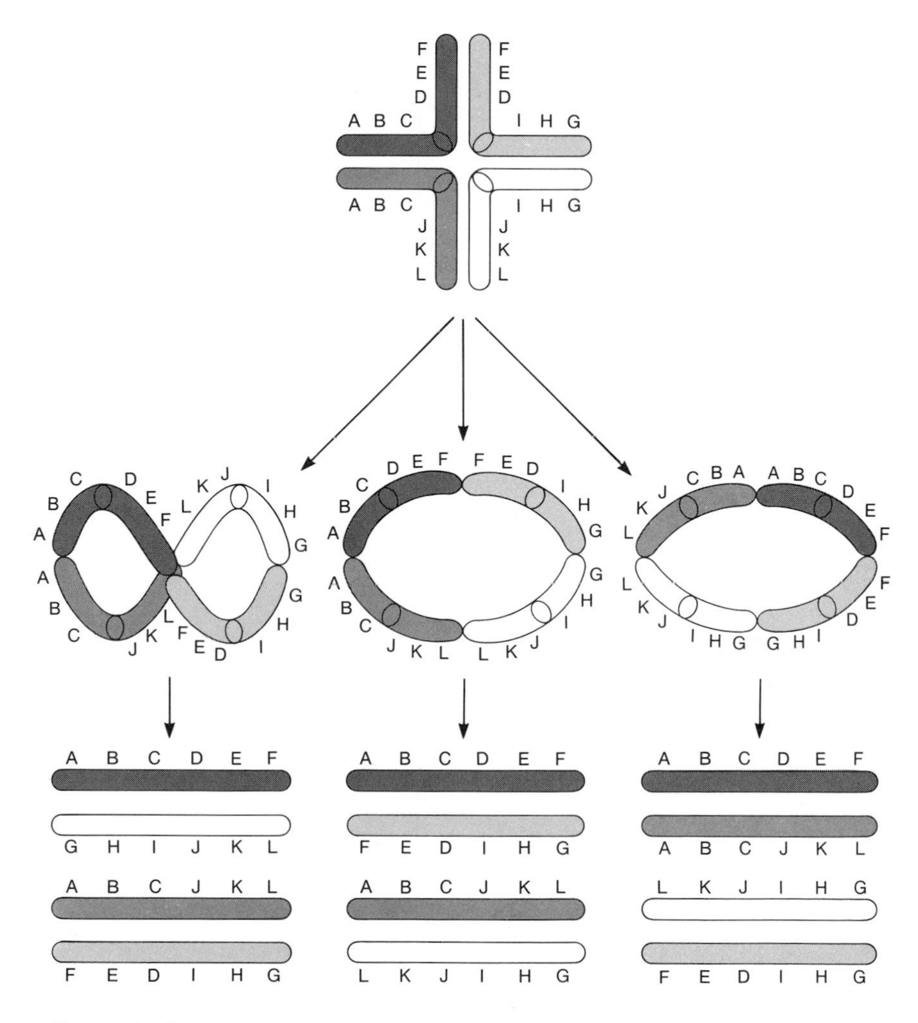

FIGURE 3-14.
Meiosis in a translocation heterozygote. At the top is the cross-shaped configuration formed when the chromosomes pair at the beginning of meiosis. The second row shows the three configurations that may occur at metaphase: a twisted ring and two open rings. The two lower rows show the six types of gametes formed. Only the two types on the left contain a complete set of genes; the other four contain some duplicated, and some missing, chromosomal segments. (For convenience, the chromosomes are shown unduplicated, without indicating the chromatids.)

gametes that can be formed, only the two on the left contain all of the chromosomal parts once and only once. All other gametes have some chromosome parts duplicated and some missing, and therefore cannot result in normal progeny. Since the only gametes producing normal progeny are those having either both nontranslocated chromosomes or both translocated chromosomes, it follows that the two chromosomes will behave as a single linkage group in translocation heterozygotes.

In plants, pollen grains and ovules with duplicated or deleted chromosomal segments are usually aborted. Animal gametes with duplicated or deleted chromosome parts may function, but the zygote formed by the union of one such gamete and a normal one usually dies or develops into an abnormal individual. Translocation heterozygotes are therefore semisterile. Translocation heterozygotes are very rare in animals but have been found in natural populations of many plants. The bizarre genetic systems evolved in *Oenothera lamarckiana* and other evening primroses are based on translocation heterozygosis (see Chapter 7).

CHROMOSOMAL FUSIONS AND FISSIONS

Chromosomal *fusion* occurs when two nonhomologous chromosomes fuse into one, thus reducing the number of chromosomes in the karyotype. Chromosomal *fission* takes place when a chromosome splits into two, thereby increasing the number of chromosomes in the karyotype. Fusion and fission of chromosomes are sometimes called "Robertsonian changes," after W. R. Robertson, who was the first to postulate fusion as a mechanism for reduction in chromosome number. Chromosomal fissions are called "dissociations" by some writers (White, 1973).

Considerable controversy exists concerning the precise mechanisms of chromosomal fusion and fission. Some workers postulate that chromosomal fusion and fission do not involve fusion or splitting of the centromeres, but rather are special kinds of translocations (White, 1973). However, studies with light and electron-microscopes indicate that the centromere of a biarmed chromosome may contain twice the material in the centromere of a telocentric chromosome (Lima-de-Faria, 1956; Comings and Okada, 1970). This suggests that chromosomal fission may, at least sometimes, occur by a splitting of the centromere, and chromosomal fusion may involve actual fusion of the centromeres.

The haploid chromosome numbers in most animals lie between six and 20, but the range extends from one (the nematode *Parascaris equorum* var. *univalens*) to about 220 (the butterfly *Lysandra atlantica*). In plants the gametic number of chromosomes may be as high as 631 (the fern *Ophioglossum reticulatum*, which is almost certainly a polyploid). Chromosomal fusions as well as fissions have occurred in the evolution of eukaryotes, but chromosomal fusions are thought to be more common than chromosomal fissions. Nevertheless, it is often difficult to ascertain in a particular group of organisms whether fusions or fissions are involved, since it may not be known which karyotypic configuration is the ancestral one. Evidence of chromosomal fusions exists for virtually all major groups of plants and animals (White, 1973). Increases of chromosome numbers by fission are well established in some cases, e.g., in *Anolis* lizards. The ancestral diploid chromosome number found in primitive species of this Neotropical genus consists of 36 chromosomes (12 metacentric macrochromosomes and 24 microchromosomes). *Anolis monticola*, a species considered to be advanced on the basis of morphology and other evidence, has 48 chromosomes (24 telocentric macrochromosomes and 24 microchromosomes). The diploid number of 48 observed in *A. monticola* has been derived from the primitive 2n = 36 karyotype by fission of the 12 metacentric macrochromosomes (Webster, Hall, and Williams, 1972).

The haploid chromosome complements of five species of the subgenus *Sophophora* of the genus *Drosophila* are shown in Figure 3-15. The ancestral condition for the genus appears to be five pairs of acrocentrics and one pair of dot-like chromosomes, as in the European species *D. subobscura* (Sturtevant and Novitski, 1941). Its close relative *D. pseudoobscura* has a new X chromosome, corresponding to the X and one of the autosomes of *D. subobscura*. The four pairs of acrocentric autosomes are fused into two pairs of metacentrics in

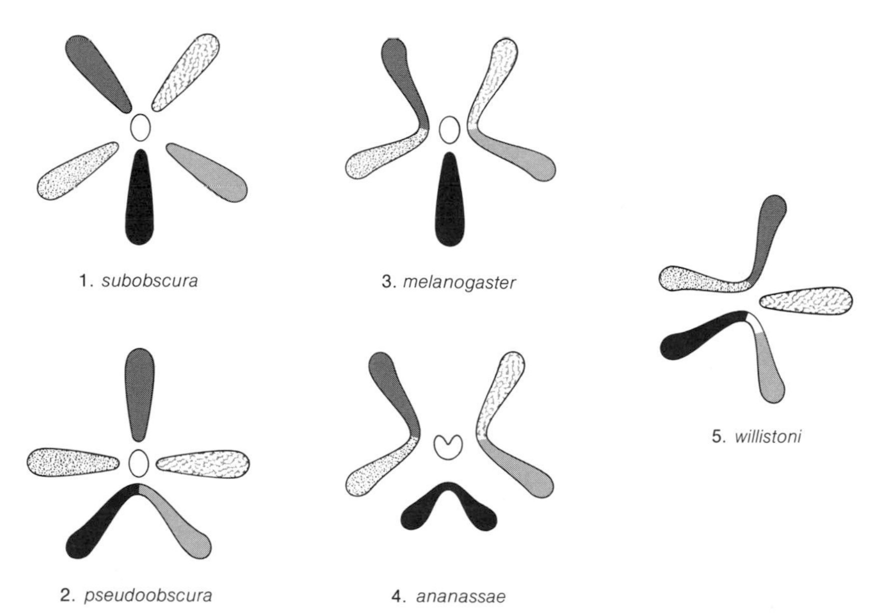

FIGURE 3-15.
The haploid chromosome complements of five *Drosophila* species. Homologous chromosome arms can be identified in the figure by the shading. The ancestral condition for the genus appears to be five pairs of acrocentric (rod-like) chromosomes and one pair of dot-like chromosomes, as in *D. subobscura*. The other conditions can be derived from the primitive one through various chromosomes fusions. The X chromosomes (black) of *D. ananassae* and *D. melanogaster* differ by a pericentric inversion that has changed the position of the centromere from the end to the middle of the chromosome in *D. ananassae*.

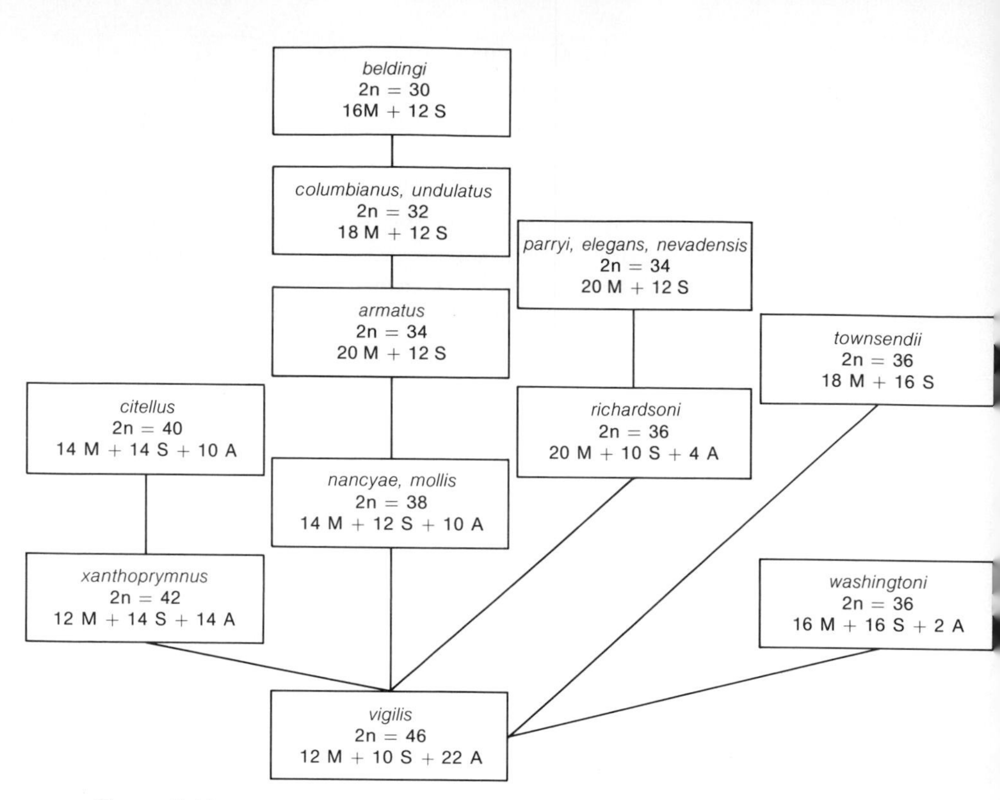

FIGURE 3-16.
Chromosomal evolution in ground squirrels of the genus *Citellus*. The total diploid number of chromosomes and the number of metacentric (M), submetacentric (S), and acrocentric (A) chromosomes of each species are shown in the boxes. The lines connect species that may have derived from each other by a combination of chromosomal fusions, fissions, and pericentric inversions. [After Liapunova and Vorontsov, 1970.]

D. melanogaster and *D. ananassae*, but in the latter species a pericentric inversion has transformed the acrocentric X chromosome into a small metacentric. In *D. willistoni* there are only three pairs of chromosomes, the ancestral dot-like chromosome being incorporated into the X chromosome.

Figure 3-16 shows the evolution of the karyotype in another group of animals, the ground squirrels of the genus *Citellus*. Modern cytogenetic techniques, like quinacrine fluorescent staining, have made it possible to compare the structures (banding patterns) of the chromosomes of man with those of his closest animal relative, the chimpanzee (*Pan troglodytes*). Man has 46 chromosomes and the chimpanzee 48. Thus, at least one chromosomal fusion, or one fission, has occurred in their evolution from a common ancestor. Other chromosomal rearrangements have taken place, including translocations—e.g., the long arm of chromosome 9 of the chimpanzee is homologous to the long arm of human chromosome 5—and probably inversions. Overall, however, the chromosomes of man and chimpanzee are fairly similar; Lin and co-workers (1973) estimate that about two thirds of the chimpanzee chromosomes have banding patterns very similar to human chromosomes.

4

Natural Selection

More than a million and a half living species have been described and named. No less remarkable than the sheer number of species is the immense diversity of their ways of life. Green plants build their bodies of water, carbon dioxide, some mineral salts, and sunlight energy. Animals require organic compounds derived ultimately from plants. Some organisms live in seawater, others in fresh water, still others on land, including arid deserts where water is scarce. Some require warmth, others resist arctic and antarctic colds. Some need light, others live in permanent darkness. Some are fairly omnivorous, others are narrowly specialized predators, or external or internal parasites living on a single host species. Some are consumers of dead bodies of other organisms or of the products of their decomposition.

The structure, function, and behavior of every species are adapted to its particular mode of life. In a way this statement is a truism, for an organism lacking adaptedness would not be alive. Yet the efficient and often astonishingly complex adaptations of living beings to their environments have fascinated biologists since Aristotle. Living beings appear to be contrivances designed for the purpose of survival and reproduction.

Purposefulness, or teleology, does not exist in nonliving nature. It is universal in the living world. It would make no sense to talk of the purpose of adaptation of stars, mountains, or the laws of physics. Adaptedness of living beings is too obvious to be overlooked. To be sure, artifacts made by man, such as furniture or machines, are made to serve some needs. Nevertheless, artifacts have purposes *imposed* on them by their creators; Ayala (1968b; also see Chapter 16 here) speaks of artifacts having an *external* or artificial teleology. Artifacts are made not only by man, but by some nonhuman animals as well. Thus, birds, termites, and ants build nests for housing and protecting their progeny.

Living beings have an *internal,* or natural, teleology. Organisms, from the smallest bacterium to man, arise from similar organisms by ordered growth and development. Their internal teleology has accumulated in the evolutionary history of their lineage. On the assumption that all existing life is derived from one primordial ancestor, the internal teleology of any organism is the outcome of approximately three and a half billion years of organic evolution.

The origin of organic adaptedness, or internal teleology, is a fundamental, if not the most fundamental problem of biology. There are essentially two alternative approaches to this problem. One is explicitly or implicitly vitalistic. Organic adaptedness, internal teleology, is considered an intrinsic, immanent, constitutive property of all life. However, like all vitalism, this is a pseudo-explanation; it simply takes for granted what is to be explained. The alternative approach is to regard internal teleology as a product of evolution by natural selection. Internal teleology is not a static property of life. Its advances and recessions can be observed, sometimes induced experimentally, and analyzed scientifically like other biological phenomena.

THE IDEA OF NATURAL SELECTION

Really great ideas in science are sometimes remarkably simple, so much so that they occur to more than one person independently. Natural selection is a case in point. The idea is basically that healthy and vigorous individuals have better chances of surviving and leaving progeny than ailing and frail ones. Empedocles (5th century B.C.) speculated that living things first arose as disjointed body parts—as heads, trunks, and limbs. These parts then combined at random, with only viable combinations surviving. Lucretius Carus (1st century B.C.) reiterated the myth of Empedocles, but added some less bizarre biological reflections. Between 1835 and 1837 Edward Blyth in England discussed some examples of what we would now call normalizing natural selection (see Eiseley, 1959). Although Darwin's failure to acknowledge Blyth as a precursor of his theory has puzzled some modern writers, we think the explanation is simple. Normalizing natural selection is a conservative rather than a creative factor in evolution. Darwin was interested in what changes biological species, Blyth in what keeps them stable.

In 1831 Patrick Matthew published a work in which natural selection (of course, not so named) was described as a factor of evolution. However, the title of the work was *Naval Timber and Arboriculture*, and the discussion of natural selection was overlooked by most readers. In 1844, Robert Chambers published *Vestiges of the Natural History of Creation* anonymously because he feared, justifiably, adverse reaction from the Victorian establishment. This was the most widely read pre-Darwinian statement of evolutionary views (the book went through several editions between 1844 and 1853). However, Chambers followed Lamarck in ascribing a greater role as an evolutionary factor to the inheritance of acquired traits, rather than to natural selection.

Darwin and Wallace considered natural selection, i.e., the differential survival and reproduction of organisms, the chief directing agent of evolutionary change. So do most evolutionists today. However, biology has advanced during the century and more since the first announcement of the Darwin–Wallace theory, and it is important to understand how the theory of natural selection has developed since then.

Darwin started his first notebook by compiling evidence on the transmutation of species in 1837. At that time he had no satisfactory explanation for transmutation. An explanation was suggested to him in September 1838, when he read "for amusement" the work of the sociologist T. R. Malthus, *An Essay on the Principle of Population.* In this essay, first published in 1798, Malthus argued that since human populations can increase exponentially, they must inevitably outgrow their means of subsistence. The increase is sooner or later checked by hunger, disease, or war. Darwin saw that the potentiality for exponential population growth is quite universal in the living world. Yet most of the time populations of most species remain approximately constant in numbers. It follows that only a part of the progeny survive, and the rest are eliminated by death. Is it a matter of chance alone what survives and what dies? No. In every species some individuals are stronger or in some ways better adjusted to the environments in which they live than other individuals. The offspring of the stronger are more likely to be represented among the following generations than the offspring of weaker individuals. The result of this differential survival is, in Darwin's words, that "natural selection is daily and hourly scrutinizing, throughout the world, the slightest variations, rejecting those that are bad, preserving and adding up all that are good; silently and insensibly working, whenever and wherever opportunity offers, at the improvement of each organic being in relation to its organic and inorganic conditions of life."

The same insight was suggested to Wallace in 1858 by the same essay of Malthus, under unlikely circumstances—when he was suffering from a malarial fever on the island of Halmahera in Indonesia. Although Darwin's insight antedated that of Wallace by almost 20 years, their discoveries were announced together in 1858. Darwin's great work, *Origin of Species*, was published in 1859; the sixth edition revised by Darwin appeared in 1872.

Some of the concepts of the Darwin–Wallace theory require comment. Artificial selection, practiced by breeders of agricultural plants and domesticated animals, has commonly been used as a model of the action of natural selection. However, in Lerner's words, "Natural selection has no purpose. . . . For any given generation, natural selection is a consequence of the differences between individuals with respect to their capacity to produce progeny. . . . Artificial selection, in contrast, is a purposeful process. It has a goal that can be visualized" (1958). Natural selection can and does take place in domesticated and laboratory organisms, and in mankind, under all sorts of natural and artificial conditions. Artificial selection is man-made, however. Natural selection has no selector, it is a self-generated outcome of interactions between organisms and their environments.

Darwin described natural selection as a consequence of the "struggle for life" or "existence." Although he carefully pointed out that the word "struggle" was used metaphorically, Darwin's phrase easily lent itself to abuse. So-called social Darwinists (of whom Darwin was not one) claimed that Darwin's theory justified war, aggression, and hostility between races and classes, and unrestrained economic competition. "Nature red in tooth and claw" was an adage that appealed to many in the heyday of capitalism and imperialism. Yet in nature the struggle for life does not necessarily take the form of actual combat between individuals of the same species.

Among higher animals combat within species is often ritualized. Victory or submission may be achieved without the opponents inflicting physical harm on each other (Lozenz, 1966). Plants "struggle" against aridity by developing devices to protect against excessive loss of water, not by sucking water from each other. It is no paradox to say that under many circumstances the most effective "struggle" for life is mutual help and cooperation. This idea was advanced by K. F. Kessler in 1880 and by Peter Kropotkin in 1902.

Herbert Spencer described natural selection as the "survival of the fittest." Not without hesitation, Darwin accepted this phrase in one of the later editions of *Origin of Species*. The use of the superlative is gratuitous, however. What survives is not the "fittest" individual, but simply tolerably fit individuals. Who are the fit, or fittest, in the process of natural selection? This question could not be answered scientifically until the middle of this century and the development of population genetics, which made possible a rigorous definition and measurement of Darwinian fitness.

Darwin freely admitted that some critically important parts of his theory could not, in his day, be fully clarified. He took it for granted that individuals composing a species vary in their chances to be winners in the struggle for life. What is the source of the variations on which natural selection operates?

Darwin did not always distinguish between what we now call hereditary or genotypic variation and environmentally induced phenotypic variation. Although in his day it seemed possible that both the environment and selection may change the hereditary materials, it was not until 1909 that the Danish geneticist W. Johannsen formulated the concepts of genotype and phenotype (see Chapter 2). Johannsen experimented with pure lines of beans, obtained by self-pollination from a single progenitor. Most individuals belonging to a pure line are genotypically identical, but show some phenotypic variation, e.g., in seed size and seed weight. Yet the averages of the progenies of large and small beans in the same pure line were identical. Thus the selection was without effect. By contrast, the progenies of large beans from a mixed population were on the average large, those of small beans were likewise small. The selection (artificial in this case) extracts from a mixture of genotypes the desired ones. The source of the genetic variability is the mutation process (see Chapter 3).

It was believed in Darwin's time that the heredities, or "bloods," of the parents mix and blend in the offspring. This "blood theory" of heredity created another difficulty, pointed out to Darwin in 1867 by Fleeming Jenkin, by training an engineer rather than a biologist. Suppose that some useful variation, for example a lighter coloration, appears in a species composed of dark individuals. The light individual will likely mate with a dark one (unless many light variants appear simultaneously). The progeny will be intermediate in coloration, and in turn will mate mostly with dark ones. In a few generations the light variant will disappear as completely as a drop of ink in a sea.

The "blood theory" of heredity makes sexual reproduction a destroyer of genetic variability. In fact the opposite is true, but this could not be understood until the "blood theory" was invalidated by Mendel's discoveries. The heredities of parents are not bloods that mix, but arrays of genes that segregate during the processes of sex-cell formation.

THE GENE POOL AND THE HARDY–WEINBERG EQUILIBRIUM

In 1908 Hardy and Weinberg independently drew a deduction from Mendel's law that became a foundation of evolutionary and population genetics. Consider two strains of a sexual and cross-fertilizing species that differ in a single gene, one strain being AA and the other aa. Let us place both strains in a previously uninhabited territory, for example an island, where they are equally fit to live. Suppose further that no immigration or emigration takes place, and that no mutations of the gene alleles A and a occur. Let the proportions of AA and aa individuals in the population be p and q, respectively, so that $p + q = 1$. The population is panmictic, that is, the carriers of the genes A and a mate at random, showing no preference for mating with partners who are either like or unlike themselves with respect to the genes A or a. With random mating, the two homozygous and one heterozygous genotypes will appear in the next generation with the following frequencies:

$$p^2\,AA + 2pq\,Aa + q^2\,aa = 1$$

What will the gene and genotype frequencies be in the following generations? One can consider all possible matings of the three kinds of individuals and the proportions of the different genotypes that they will yield in the progenies. A simpler and no less accurate way is to envision that every individual contributes equal numbers of gametes to the gene pool of the population (see Chapter 2) and that these gametes combine at random. Each homozygote yields one kind and the heterozygote two kinds of gametes in equal proportions. Therefore, the A and a gametes will have the following frequencies:

$$A = p^2 + pq = p(p + q) = p$$
$$a = q^2 + pq = q(q + p) = q$$

The gametic frequencies will be unchanged, and so will the zygotic frequencies of the homo- and heterozygous genotypes. The equilibrium proportions, $p^2 AA + 2pq Aa + q^2 aa$, will be maintained generation after generation. This invalidates the objection made by Jenkin to Darwin.

The Hardy–Weinberg equilibrium shows that the Mendelian mechanism is a conservative factor. The conservation of the gene frequencies makes the composition of the gene pool of a population or a species stable. Yet evolution is change. It demands that the frequencies of some genetic variants be altered, and processes that cause such alterations are known. Let us consider them one at a time, and then in interactions.

Gene frequency changes due to mutation alone (Chapter 3) are generally slow, at least in higher organisms in which the generation time is long. Suppose that the gene allele A mutates to a at a rate u per generation. If a population consists originally of AA individuals, the frequency of a in the gene pool will start increasing at first in proportion to the number of the generations elapsed. In man as well as in *Drosophila*, mutation rates (u) of the order of 10^{-5} (1:100,000) per gene per generation are common. If in one generation the gene frequency increases from zero to 0.00001, it will take 10 generations for it to grow to 0.0001. In man this means between 200 and 300 years. The rate of increase of the frequency of a becomes slower when this allele grows common in a population. Let p_0 stand for the frequency of A in the population at a certain time, p_n for the frequency n generations later, and u for the mutation rate. Then $p_n = p_0 (1 - u)^n$. Of course, after a long time and very many generations the allele A will become rare or disappear entirely. We shall see in the next chapter that this process may be speeded up in populations consisting of small numbers of individuals.

A mutation may be reversible. Suppose that the allele A of a gene mutates to a at a rate u, and a to A at a rate v. A stable equilibrium may eventually be reached when the frequency of A becomes equal to $v/(u + v)$, and at a to $u/(u + v)$. Such mutational equilibria (polymorphisms) are probably not uncommon in rapidly reproducing microorganisms. Among higher organisms the time required may be so long that factors other than mutation (selection, genetic drift, migration) are likely to intervene.

DARWINIAN FITNESS OR SELECTIVE VALUE

The carriers of a certain genotype, say AA, may be more viable—may survive, become adults, and reproduce more frequently—than those of an alternative genotype, aa. Or the carriers of one genotype may be more fertile and beget more progeny than the carriers of another. The carriers of a genotype may also develop faster and reach the reproductive age sooner than another genotype. Among animals the carriers of a genotype may be sexually more active and mate more frequently than the carriers of another genotype. Among plants a genetic variant may have flowers more attractive to insect pollinators than another variant. Any one or a combination of several advantages like the above may confer upon the carriers of a genotype a Darwinian fitness higher than that of another. The Darwinian fitness, also called the selective or adaptive value, usually symbolized as w, is a relative rather than an absolute measure. If the carriers of a certain genotype transmit their genes to the next generation at a rate we may denote as unity ($w_1 = 1$), the carriers of another genotype in the same population may pass their genes at a lower or a higher rate, $w_2 = 1 - s$ or $1 + s$. The value s is the selection coefficient.

Note that the components of the Darwinian fitness need not be correlated. Thus, the carriers of one genotype may be more viable but less fertile than those of another. The Darwinian fitness is a composite, a product of various advantages and disadvantages. A fitter genotype may nevertheless have some deficits. For example, two decades ago it was believed that in Western societies the more intelligent people, i.e., those with higher IQ, raised fewer children than the less intelligent ones (new data indicate this differential fertility was a temporary condition, which tends to disappear). Now, to the extent that human intelligence has a genetic component, one would have to say that people of low intelligence had a higher Darwinian fitness than those with high intelligence. The "fittest" turns out not to be a superman but merely the parent of the largest number of children reaching adulthood.

Suppose that a gene in a sexual diploid population is represented by two alleles, A and a, with frequencies p and q, respectively. A is dominant and a recessive. The Darwinian fitnesses of AA and Aa are both one, and that of the recessive homozygote aa is $1 - s$. We then have:

	AA	Aa	aa
Fitness, w	1	1	$1 - s$
Frequency before selection	p^2	$2pq$	q^2
Frequency after selection	p^2	$2pq$	$q^2(1 - s)$

The frequency of A will increase and that of a will decrease in the next generation by an increment $\Delta p = spq^2 / (1 - sq^2)$. Of course, the change will be greater with strong selection than with weak selection. Let us consider what happens with selection as weak as $s = 0.1$ (the homozygote aa has a one percent disadvantage), and as strong as $s = 1$ (the homozygote aa is lethal or does not reproduce). If the alleles A and a are equally frequent at the start, $p_0 = q_0 = 0.5$, the frequency p_1 after one generation of selection will be:

Selection coefficient, s	0.01	0.02	0.1	0.5	1.0
Frequency of A, p_1	0.50125	0.5025	0.5128	0.574	0.67

Even with weak selection, the frequency of A will gradually increase and eventually displace a. However, this may take a very long time, since selection against recessives becomes slower as the recessives dwindle in frequency. This is so even for a complete lethal ($s = 1$). Starting with a frequency $q = 0.5$, it takes eight generations to reduce q to 0.1, 50 generations to make $q = 0.02$, 100 generations to depress it to $q = 0.01$, and 1,000 generations to $q = 0.001$.

For an organism with a generation time as long as man (say, 25 years), this means that centuries or millennia may be needed to produce radical changes of gene frequencies.

Selection is more efficient if there is no dominance, or if the dominance is incomplete. The coefficient of dominance is symbolized by h ($h = 0.5$ when there is no dominance, $h = 1$ or $h = 0$ when one or the other allele is completely dominant). We then have:

	AA	Aa	aa
Fitness, w	1	$1 - hs$	$1 - s$
Frequency before selection	p^2	$2pq$	q^2
Frequency after selection	p^2	$2pq(1 - hs)$	$(1 - s)q^2$

After one generation the frequency of A will increase (and that of a will decrease) by an increment $\Delta p = hspq/(1 - hsq)$. To compare the efficiency of selection with and without dominance, the tabulation below shows the numbers of generations needed to effect a given frequency change in a gene discriminated against by selection $s = 0.02$, $hs = 0.01$.

From *To*	*Recessive*	*No Dominance*
0.25 → 0.10	710	110
0.10 → 0.01	9,240	240
0.01 → 0.001	90,231	231

The greater efficiency of selection when dominance is absent or incomplete is not surprising. When a gene allele is rare in a population, the heterozygotes that carry it greatly outnumber the individuals homozygous for it. Therefore, even a weak selection that acts on the heterozygotes will have an appreciable effect within a reasonably small number of generations.

NORMALIZING NATURAL SELECTION

Of considerable genetic and evolutionary significance is the interaction of mutation and selection. A mutation is basically an accidental change in some component of the genetic system, comparable to a misprint or error in copying some word or paragraph. It is not surprising that mutations are rarely useful to their carriers when the latter live in the "normal" or usual environments of their species. One may conjecture that most useful mutations have already occurred and have become incorporated in the species gene pool (Chapter 3). Useful mutations are more likely to be found when a population is transferred to a novel environment. Some mutations are favorable in heterozygous but unfavorable in homozygous condition. A majority of mutations range from neutral to deleterious to lethal. The accumulation in the gene pool of mutants lowering the fitness is impeded by normalizing natural selection.

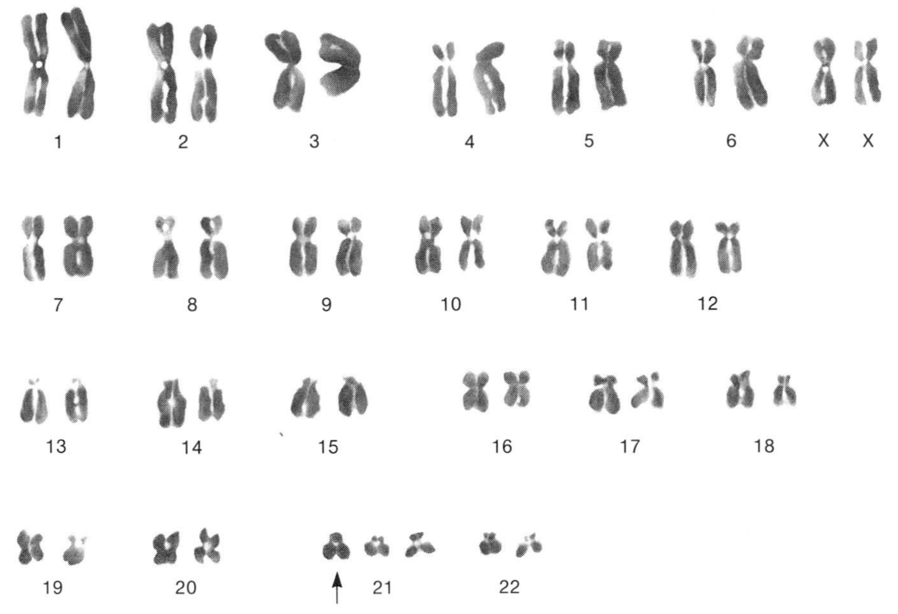

FIGURE 4-1.
The karyotype of a woman with Down's syndrome, in which three rather than the usual two chromosomes number 21 occur.

The situation is simplest for dominant mutations. A carrier of a dominant mutant shows its effects in the phenotype, and is therefore exposed to the action of natural selection. Let the mutation rate to a dominant condition be u per gamete per generation, and the Darwinian fitness of a carrier of the mutant gene $1 - s$. The equilibrium incidence of the dominant gene in the gene pool of the population will then be u/s. For a dominant lethal ($s = 1$) this will equal the mutation rate, $p = u$.

Mongoloidism, or Down's syndrome, in man is close to being a dominant lethal. Although most affected individuals survive, they seldom reproduce—their Darwinian fitness is very low. Mongoloidism is due to duplication of chromosome 21, so that most affected individuals have 47 chromosomes in their cells instead of the usual 46 (Figure 4-1). The incidence of Down's syndrome at birth is about 15 per 10,000 (it is greater among children of older mothers). The mutation rate (u) is 7.5×10^{-4}, and is opposed by selection not far from $s = 1$. The frequency of Down's syndrome phenotypes in a population is $2U/s$ (since one mutant gamete is enough to produce an affected diploid individual).

Chondrodystrophic (achondroplastic) dwarfism in man is one of the many dominant mutants that appreciably reduce Darwinian fitness, but not to zero level. The dwarfs have heads and trunks of normal size, but relatively short legs and arms (Figure 4-2). A study made in Denmark has recorded the birth of eight achondroplastic infants among some 94,070 infants in families in which

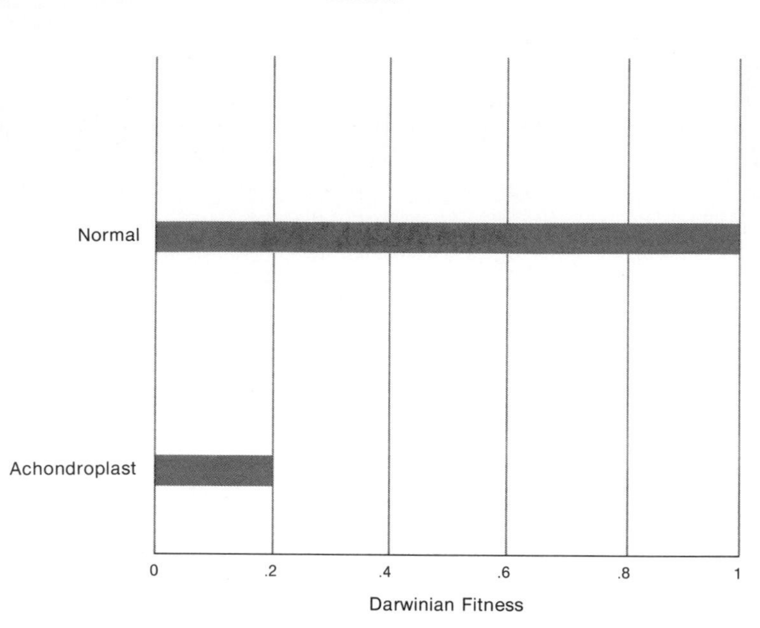

FIGURE 4-2.
Relative fitnesses of normal and achondroplastic humans. Achondroplastic dwarfs have heads and trunks of normal size, but relatively short legs and arms.

neither parent was an achondroplast. The mutation rate for achondroplasia is thus estimated approximately $8/(94{,}070 \times 2)$ or 4.2×10^{-5}. In families in which one parent is a dwarf, the number of children is on the average 0.25, while for normal families it is about 1.27. The Darwinian fitness of the achondroplasts is thus $0.25/1.27 = 0.2$, and the selection coefficient discriminating against achondroplasia $1 - 0.2 = 0.8$. The frequency of achondroplasts in the population should be $2u/s = (8.4 \times 10^{-5})/0.8$, or approximately 1 per 10,000, which is about what is observed.

The situation is more complex with deleterious recessive mutations. A recessive mutant that is deleterious or even lethal when homozygous is sheltered from natural selection when it is carried in a heterozygote with a "normal" dominant allele. Recessive mutants will therefore accumulate in the population, until matings of heterozygous carriers are frequent enough to start yielding the ill-adapted homozygotes. An equilibrium between mutation and normalizing selection will then be reached when the number of the mutant alleles eliminated equals that of new mutants arising per generation. This will happen when $sq^2 = u$, or $q = \sqrt{u/s}$ (where s is the selection coefficient against the recessive homozygote, q the frequency of the deleterious recessive in the gene pool, and u the mutation rate).

Consider phenylketonuria and albinism, two examples among hundreds of more or less deleterious recessive mutants found in human populations. Both of them are rare—between 1:10,000 and 1:20,000—except that in some small and isolated populations albinism is more frequent. Phenylketonuria is a disorder of the metabolism of the amino acid phenylalanine. If untreated, it results in brain damage, gross mental retardation, and a Darwinian fitness of zero. If discovered early, however, and the affected infant is placed on a diet nearly free of phenylalanine, serious damage can be avoided, and the fitness increased to near normal. This is a fine example of dependence of Darwinian fitness on the environment. Albinism causes failure to synthesize pigment in the hair, skin, and iris of the eyes. Albinos are somewhat handicapped compared to normally pigmented persons, but most of them survive and reproduce.

Why do phenylketonuria and albinism persist in most human populations? The simplest hypothesis is that they arise repeatedly by mutation, but that normalizing selection prevents their undue accumulation. If the frequency of a recessive homozygote is $q^2 = 0.0001$ (one per 10,000), the frequency of the recessive gene in the gene pool of the population will be $q = 0.01$, and that of heterozygous carriers (according to Hardy-Weinberg formula) $2pq = 2 \times 0.99 \times 0.01 = 0.0198$. That is, almost two percent of healthy persons in the population are carriers of the phenylketonuria gene, and two percent carriers of the albinism gene.

The fitness of untreated phenylketonurics is zero, hence the selection against them is $s = 1$. Assuming that the fitness of heterozygous carriers is one, the same as for noncarriers, we can estimate the mutation rate from the normal to the phenylketonuric allele, using the equation $u = sq^2$. This is $1 \times 0.01^2 = 0.0001$, or 10^{-4}. The fitness of homozygous albinos is not known; assuming it is about 90 percent normal ($s = 0.1$), we obtain an estimate of the mutation rate $u = 0.1 \times 0.01^2$, or 10^{-5}. Both estimates have a margin of uncertainty because we have assumed the fitness of the heterozygous carriers of phenylketonuria and albinism to be one. If it were lower, normalizing selection would be more important against the heterozygotes than against the homozygotes because the former are much more numerous than the latter. Our estimates of the mutation rates would then be too high. Similar problems arise also with respect to other deleterious recessive mutants, which are individually rare and yet are encountered in most human populations.

GENETIC BURDENS OR LOADS

Some examples of deleterious genetic traits in human populations have been used in this chapter to illustrate the action of normalizing natural selection. Other examples of such traits in animals and plants have been given in Chapter 2 as instances of pervasiveness of genetic variability. Indeed, genetic variability and normalizing selection are universal in the living world. Mutants arise in any species and many or most of them are deleterious. Some of these mutants may persist in the populations, i.e., one or several generations intervene between the origin and the elimination of a deleterious genetic variant. Hence, populations of any species, including the best adapted and most successful ones, carry burdens or loads of genetic defects. Unhappily, mankind is no exception.

One can distinguish between expressed (overt) and concealed genetic loads. The former comprise individuals with diseases, malformations, or constitutional weaknesses that are manifestly genetic. Concealed genetic burdens are due to recessive genes or gene complexes carried in heterozygous condition in individuals who are themselves healthy or "normal." When homozygous, however, these genes or gene complexes are also debilitating or lethal. Concealed loads are uncovered when heterozygous carriers of the same recessive defect mate; a part of their progeny will be homozygous and show the defect. Inbreeding increases the probability of uncovering concealed genetic malformations. In human populations some rare recessive diseases are found mainly in progenies of parents who are relatives. It is an open question whether incest taboos were first instituted because of experience with the genetic risks. There is no doubt, however, that avoidance of incest is genetically a sound practice.

Inbreeding experiments have been made with many species of normally outbreeding animals and plants. The uncovering of deleterious recessive mutants has been described in Chapter 2. Inbred lines of corn (*Zea mays*) have been studied most extensively. Almost without exception progenies obtained after several generations of self-pollination are strikingly less vigorous than outbred progenies. In genetically well-studied species of *Drosophila* techniques are available for uncovering and recording the frequencies of the constituents of concealed genetic loads. The essence of these techniques is a series of crosses designed to obtain individuals that carry in double dose (are homozygous for) certain chromosomes extracted from natural or experimental populations (see Figure 2-6, p. 40). The results obtained are qualitatively similar, although quantitatively variable, in all species. When chromosomes that are present in "normal," healthy flies are made homozygous, many of them become lethal (kill all the homozygotes), semilethal (kill more than 50 but less than 100 percent of the homozygotes), subvital (kill fewer than 50 percent), cause sterility of the females or males, or induce visible structural defects or abnormalities. Table 2-3 (p. 43) summarizes the results obtained for one species, *Drosophila pseudoobscura.*

Very few, if any individuals in natural populations are free of chromosomes that produce more or less serious defects when in double dose. Only a minority of the chromosomes give "normal" homozygotes, and a very small minority give supervital homozygotes. In point of fact, special experiments have shown that even these minorities are really handicapped. Normality (the adaptive norm) is defined as the average viability, fertility, development rate, etc., of heterozygous individuals. The heterozygotes conserve their normality when exposed to variations of temperature and other conditions. They possess developmental homeostasis, which makes them able to adjust to environmental changes. By contrast, the normal and supervital homozygotes deteriorate and become subvital, even semilethal, in changed environments. They are deficient in homeostatic ability to adjust to changed environments.

Components of concealed genetic loads become expressed not only as a result of inbreeding but also of accidental mating of not closely related individuals who happen to be heterozygous for the same recessive defect. The recessivity of genes that are lethal or grossly deleterious when homozygous may be incomplete; the heterozygous carriers may then be weakened compared to noncarriers. Extensive data on expressed genetic loads, though not as precise as one would like them to be, are available for human populations. According to Stevenson (1961) and Trimble and Doughty (1974), in British populations some 12 to 15 percent of pregnancies that continue more than four or five weeks end in spontaneous abortion, and two percent in stillbirths. An appreciable part, although doubtless not all, of the abortions and stillbirths are genetically caused. Schull and Neel (1965) found 3.55 and 3.42 percent childhood mortalities in Hiroshima and Nagasaki, respectively, in families in which the parents were not known to be related. The figures are 6.12 and 5.25 percent in families where the parents are first cousins. What proportion of morbidity in human populations stems from genetic causes is even more difficult to determine than for mortality. Stevenson estimates that in the population of Northern Ireland some 7.9 percent of all consultations with medical specialists and 6.4 percent of those with general practitioners are sought by genetically handicapped persons, and 26.5 percent of all hospital beds are occupied by such individuals.

What, if any, biological function do genetic burdens serve in living species? Genetic variability is a prerequisite for evolutionary change, and the mutation process is the ultimate source of this variability. You may deplore the fact that most mutations are deleterious, instead of only useful ones being produced exactly where and when they are needed. One must realize, however, that to do so a living species would have to anticipate its future needs and be able to act purposefully to satisfy them. Only mankind can do that, and, lamentably, only to a very limited extent. Deleterious mutations are the price the species pays for having some useful ones.

Moreover, deleterious and useful mutants are not fixed categories. What is deleterious in one environment may be useful in another. Insecticide resistance is imperative for survival when an insect population is being treated regularly with an insecticide, but mildly deleterious in the absence of any treatment. A gene may be useful in combinations with some, neutral with others, and harmful with still other genes. Biological species usually differ in numerous genes. We shall see in Chapter 6 that species formation involves the emergence of constellations of genes that work together harmoniously. In the short run, the store of genetic variability carried in the gene pool of a population is a genetic burden. Yet in the long run, genetic variability is a treasure house enabling the population to adapt to changing environments.

POLYMORPHISM AND BALANCING NATURAL SELECTION

We have seen that normalizing natural selection is preeminently a conservative force. It purges the gene pool of a population of deleterious genetic variants and thereby tends to keep the species constant. Its efficiency is far from perfect, however, since populations carry genetic defects, which lower the

Darwinian fitness of homozygous individuals. Although these defects are of many kinds, most are rare. Among hundreds or even thousands of genetic diseases and malformations known in man, most have an incidence of 1 : 10,000 or less. Taken all together, these deficits nevertheless add up to a genetic load which is by no means negligible.

Sometimes two or more genetic variants occur in a population with such frequencies that they cannot be regarded as anomalies. Such populations and species are called *polymorphic* (see Chapter 2). Species in which at least two phenotypically or genotypically discrete classes cannot be distinguished are *monomorphic*. Some polymorphisms are more apparent than others. For example, polymorphisms for eye and hair color in human populations are clearly visible, whereas blood group, enzyme, and taste-blindness polymorphisms require special tests to be revealed.

Polymorphism and continuous genetic variation are distinct in reality, although not in theory. Thus, variations in stature, weight, or head shape in human populations, though they doubtless have genetic foundations, are not usually described as polymorphisms. Tall and short, fat and slim, longheaded and roundheaded people do not form discrete classes. Not only do intermediates exist, but they are more frequent than the extremes. Nevertheless, if we could carry genetic analysis of these traits down to the gene level, we would find discrete variants in the gene pool underlying the continuous observed variation of the phenotypes. Eye color polymorphism, which at first sight seems discontinuous, is to a certain extent an instance of continuous variation, due to one gene with major and several with minor phenotypic effects. The more detailed the study of genetic variation in a species, the more likely it is to disclose polymorphisms.

What genetic mechanisms are responsible for the origin and maintenance of polymorphisms? This is a difficult and controversial problem in evolutionary genetics. Several mechanisms can be considered. A polymorphism can be balanced, so that two or more gene alleles or chromosomal variants are kept in a population at more or less stable equilibrium frequencies. Several forms of balancing natural selection can bring such a situation about. A transient polymorphism may be a stage about midway between the origin of a new useful variant by mutation and its eventual fixation in the population by directional natural selection. In addition, polymorphisms may have nothing to do with selection, i.e., the genetic variants concerned may be neutral, or equal in Darwinian fitness. They would then be due to mutation and random genetic drift (see Chapter 5). It is certain that all these mechanisms exist in reality; what is still controversial is their relative frequency in nature and their roles in the evolutionary development of the living world.

Consider the effects on Darwinian fitness of two gene alleles, A_1 and A_2. If the fitnesses of the homozygotes A_1A_1 and the heterozygotes A_1A_2 are equal, but either lower or higher than that of the homozygotes A_2A_2, the allele A_1 is said to be fully dominant and A_2 fully recessive. If the heterozygote A_1A_2 is intermediate in fitness between A_1A_1 and A_2A_2, the dominance is incomplete or absent. The heterozygote A_1A_2 may also be superior, or inferior, in fitness to both homozygotes. If it is superior, we have a case of overdominance, or hybrid vigor, or *heterosis.* Inferior heterozygotes betoken some degree of hybrid inviability or hybrid sterility (the term "negative heterosis" has also been proposed, but it is rather misleading and should be avoided).

Let the superior fitness of a heterotic heterozygote be one, and the fitnesses of the inferior homozygotes $1 - s_1$ and $1 - s_2$, where s_1 and s_2 are selection coefficients. In a sexual and outbreeding population there will be:

	A_1A_1	A_1A_2	A_2A_2	*Total*
Fitness, w	$1 - s_1$	1	$1 - s_2$	W
Frequency before selection	p^2	$2pq$	q^2	1
Frequency after selection	$p^2 - s_1p^2$	$2pq$	$q^2 - s_2q^2$	$1 - s_1p^2 - s_2q^2$

Traditionally, since Darwin's day, one has usually assumed that natural selection eliminates what is less fit and puts in its place what is more useful. Overdominance means however that neither A_1 nor A_2 is best by itself, because the fittest heterozygote, A_1A_2, has them both. The outcome of selection will therefore be a balanced equilibrium, conserving the polymorphism in the population. The frequencies of A_1 and of A_2 at equilibrium will be:

$$p = s_2/(s_1 + s_2) \qquad q = s_1/(s_1 + s_2)$$

If the advantage of the heterozygote is the same over both homozygotes ($s_1 = s_2$), the variants A_1 and A_2 will be equally frequent at equilibrium ($p = q = 0.5$). This works regardless of whether the advantage is small (for example $s_1 = s_2 = 0.01$) or great. If both homozygotes are lethal or sterile ($s_1 = s_2 = 1$) we have what is called a *balanced lethal system.* In many laboratory strains of *Drosophila* both chromosomes of a homologous pair have different mutant genes, which may or may not produce viable changes in the appearance of the flies when heterozygous but which are lethal in double dose. The mating A_1A_2 ♀ × A_1A_2 ♂ yields only A_1A_2 progeny; the strain is said to "breed true," but it can be shown that one half of the eggs deposited fail to develop to the adult stage. Similar in principle are some species of evening primroses (*Oenothera*) and other plants (see Chapter 2). They are "permanent heterozygotes" for one or several translocations between different chromosomes; all viable seeds are translocation heterozygotes like the plants that produce them.

One of the homozygotes may have a fitness greater than the other, while both are less fit than the heterozygotes ($A_1A_2 > A_1A_1 > A_2A_2$). At equilibrium, A_1 will be more frequent than A_2 in the population. The classical (and some people think, the only well-established) case of balanced polymorphism in man is that of sickle-cell anemia, a disease that is fairly common in some populations in Africa and in populations of African and Asiatic origin (Figures 4-3 and 4-4). The anemia is due to homozygosis for a gene s, which produces an abnormal hemoglobin S_1 instead of the normal hemoglobin A produced by the "normal"

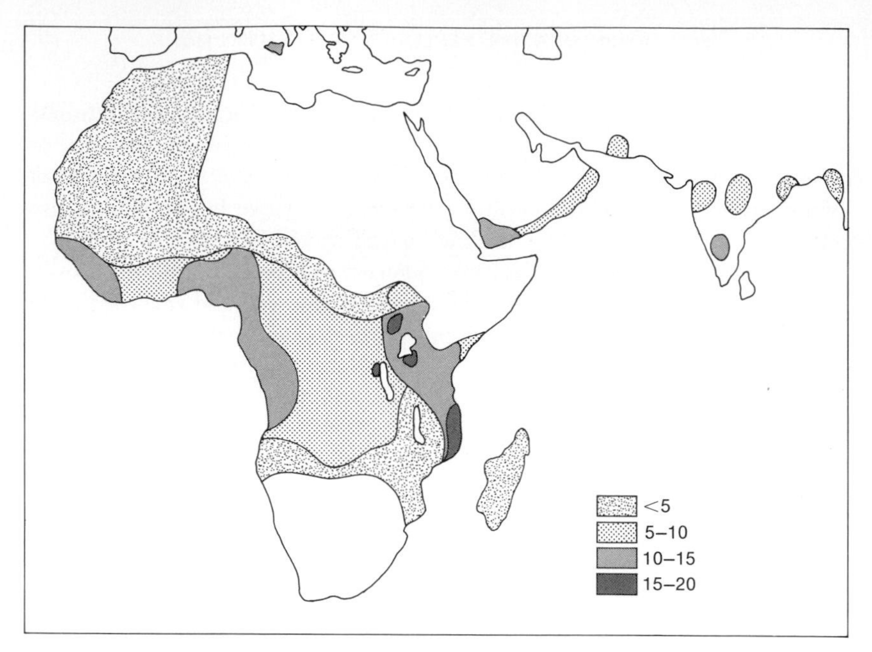

FIGURE 4-3.
Geographic distribution of the sickle-cell gene, *s*, which in homozygous condition is responsible for sickle-cell anemia. The frequency of *s* is high in certain parts of the world where falciparum malaria is endemic (see Figure 4-4). This association may be due to the high resistance to malaria of individuals heterozygous for the *s* gene. [After Cavalli-Sforza and Bodmer, 1971.]

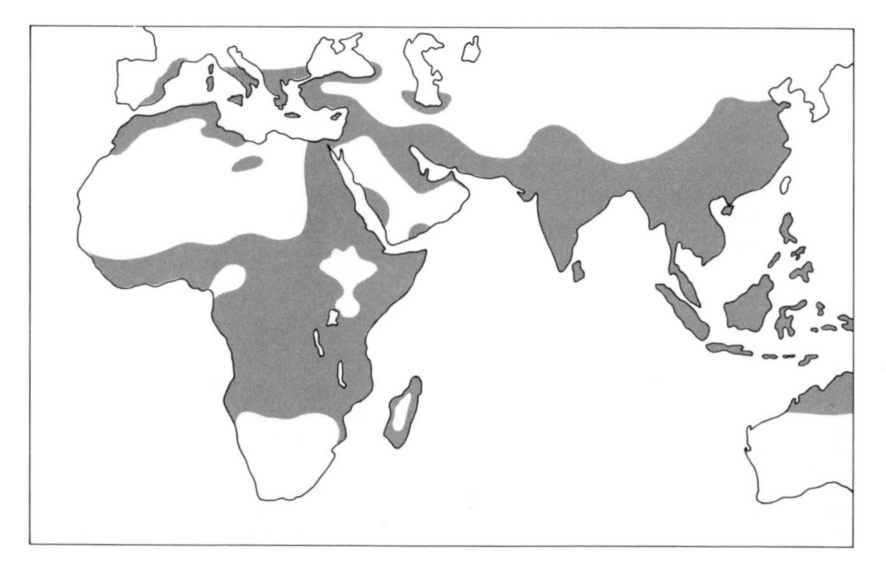

FIGURE 4-4.
Old World distribution of malignant malarias caused by the parasite *Plasmodium falciparum.* [After Cavalli-Sforza and Bodmer, 1971.]

allele *S*. The two hemoglobins differ in a single amino acid, glutamic acid in hemoglobin A being replaced by valine in position 6 in the beta chain of hemoglobin S_1. The homozygotes *s*/*s* usually die before reaching sexual maturity; their Darwinian fitness is only slightly greater than zero. Let us take it to be $w = 0.1$, so that the selection coefficient against *s*/*s* homozygotes is 0.9. The heterozygotes *S*/*s* not only will survive, but, as first shown by Allison, will be relatively more resistant to, or less affected by, a form of malaria widespread in the tropics (*Plasmodium falciparum*) than are the "normal" homozygotes *S*/*S*. The heterozygotes will thus be more fit than the homozygotes *S*/*S* in countries where the falciparum malaria is widespread; where this malaria is absent there is no advantage in the heterozygotes.

Let us suppose that in some country where malaria is moderately common the fitnesses of the homo- and heterozygotes are $S/S = 0.8$, $S/s = 1$, $s/s = 0.1$. At equilibrium, the heterotic balancing selection will establish the frequencies of the alleles $S = p = 0.9/(0.9 + 0.2) = 0.82$, and $s = q = 0.2/(0.9 + 0.2) = 0.18$. In every generation $q^2 = 0.18^2$, that is, 3.24 percent of the infants born will be the semilethal homozygotes *s*/*s*, most of whom will die before adulthood. However, in a malarial environment the average fitness of the population will be greater than that of a population consisting only of "normal" homozygotes, *S*/*S*. This is because $2pq = 2 \times 0.82 \times 0.18 = 0.295$; 29.5 percent of the individuals born will be heterozygotes that are relatively resistant to malaria and have a fitness of one, whereas the fitness of "normal" homozygotes in malarial environment is only 0.8. If the population migrates to a malaria-free country, or if malaria is eradicated, a population having the *S* gene in its gene pool will be slightly less fit than one free of *S*. Normalizing natural selection will begin to act to reduce the frequency of *S*, and eventually eliminate it entirely. On the other hand, if malaria becomes more frequent, balancing natural selection will tend to increase the frequency of *S*.

Although it has not been proven, some hereditary diseases and defects may be maintained in human populations by balancing selection if the fitness of the heterozygous carriers is greater than that of the noncarriers. The same holds for some of the constituents of the genetic loads in *Drosophila* and other organisms. Reliable evidence bearing on this problem is difficult to obtain because small selective advantages or disadvantages require impractical amounts of experimental or observational data.

Suppose that the Darwinian fitnesses of the three genotypes A_1A_1:A_1A_2:A_2A_2 are 0.99:1:0, i.e., the heterozygous carriers of A_2 have a one percent advantage over the noncarriers, while the homozygote A_2A_2 is lethal. How shall we go about discovering this one percent advantage? It may be due to a slightly lower childhood mortality, greater fecundity, earlier marriage, longer reproductive life, better social adjustment, or a combination of these and other factors. Moreover, the advantages may be pronounced in some physical or cultural environments and absent in other environments. Finally, the heterozygotes may have had a fitness advantage in the past, in environments before the advent of civilization and modern technology, but may no longer be advantageous. For all these reasons, pinpointing small selective advantages and disadvantages in man is a formidable task. It is laborious even with laboratory animals and plants. The question of how many recessive genetic defects are maintained in human populations by balancing selection, rather than by mutation pressure, is wide open.

Heterotic balancing selection can be demonstrated and measured in situations where selective advantages and disadvantages are large. Populations of some animal and plant species are polymorphic for variants of the chromosome structure, inversions, and translocations (see Chapters 2 and 3). As a rule these chromosomal variants produce no visible effects on the external appearance of their carriers, but they can be detected by microscopic examination of the chromosomal complements. This is easiest in many representatives of Diptera, which have giant banded chromosomes in the cells of larval salivary glands. The inversion heterozygotes (heterokaryotypes) have two chromosomes of a homologous pair with gene arrangements that differ in one or more chromosome segments (blocks of genes) placed in inverted orders—ABCD and ACBD. The pairing of such chromosomes gives rise to an easily visible loop (Figure 3-11, p. 88).

Drosophila pseudoobscura is one of many species the natural populations of which are polymorphic for chromosomal inversions. For reasons that are still unknown, most of the inversions occur in only one of the five chromosomes the species has—the third chromosome. Around 1940 the surprising observation was made that the relative frequencies of the inversions in some populations change from month to month. These changes are not only quite significant statistically but are cyclic with the seasons of the year, i.e., are repeated in successive years (Figure 4-5). The hypothesis was framed that the changes

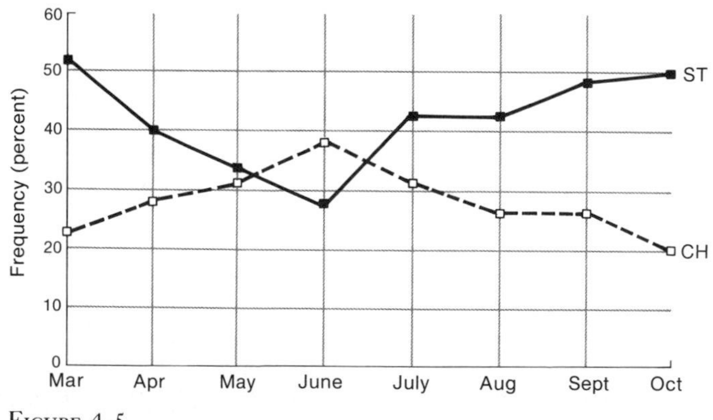

FIGURE 4-5.
Frequencies of two gene arrangements in a natural population of *Drosophila pseudoobscura*. The frequencies of the two chromosomal arrangements, CH and ST, change throughout the year. CH has its highest frequency at the beginning of summer, when the frequency of ST is lowest.

FIGURE 4-6.
A laboratory cage used in the study of experimental populations of *Drosophila*. The 15 food containers in the bottom are usually replaced at a rate of one every two days. The cage becomes an enclosed universe in which evolutionary changes in a population can be studied over many generations. Adult flies may be seen resting on all surfaces of the cage.

are due to the action of natural selection. To test the hypothesis, experimental populations are set up in the laboratory in so-called population cages (Figure 4-6). A population is started with known frequencies of two or more gene arrangements differing in inversions. The population is left to breed freely for as many generations as needed. At desired intervals, samples of larvae are taken, and the chromosomes in the salivary glands examined.

Typical results are shown in Figure 4-7. In the early generations (a generation takes about a month under the conditions of this experiment) the frequencies of the inversions change rapidly. Later the changes slow down, and eventually an equilibrium is reached when the frequency of the inversions remains constant. None of the competing inversions are eliminated and none reach 100 percent frequencies. This is prima facie evidence that a balancing natural selection is acting. The most probable explanation is that the fitness of the heterokaryotypes is higher than that of the homokaryotypes. Taking into consideration the slope of the selection curve and the equilibrium values, one can compute the adaptive values of the karyotypes. These turn out to be one for the heterokaryotypes and 0.89 and 0.41 for the homokaryotypes.

This is very strong selection, considering that the selection is not for some kind of genetic disease but for "normal" constituents of natural populations.

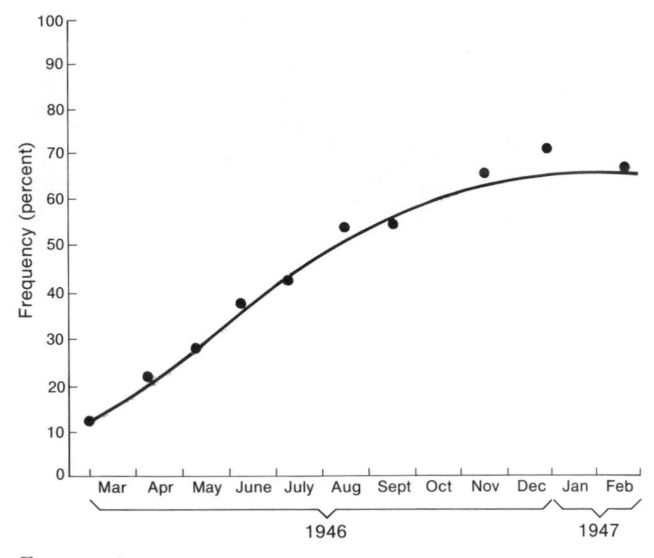

FIGURE 4-7.
Changes in chromosomal frequencies in an experimental population of *Drosophila pseudoobscura*. Two chromosomal arrangements, CH and ST, compete in the cage. The frequency of ST gradually increases from its initial frequency of 12 percent, to reach an equilibrium frequency around 70 percent. Correspondingly, CH decreases from about 88 percent to an equilibrium frequency around 30 percent.

One of the two homokaryotypes has a fitness less than one half that of the other! To be sure, the fitness is remarkably sensitive to environment (temperature, nutrition, etc.). The results in Figure 4-7 are from an experiment done at 25°C. Lowering the temperature by only 10°C renders the adaptive values of the karyotypes so nearly equal that no significant changes in the chromosome frequencies are observed for about 10 generations.

These experiments throw some light on cyclic seasonal changes in the frequencies of different chromosomes in nature. Such changes are expected if one of the homokaryotypes (ST/ST in Figure 4-5) is superior to the other (CH/CH) in summer (June–September), but inferior in spring (March–May). Just what physiological and ecological properties of the karyotypes are responsible for the changes is not well known. Some experiments suggest that ST is superior under conditions of crowding, and CH superior in uncrowded populations. Other characters, such as viability and fecundity, are probably also involved. Nevertheless, the evidence shows unambiguously that *Drosophila*, an organism that produces several generations per year in at least some of its natural habitats, undergoes genetic, evolutionary alterations under the influence of seasonal climatic or biotic changes in its environments. There is also evidence that natural selection induces genetic alterations in different years, presumably because of the year-to-year environmental fluctuations.

Another question is why an inversion of a block of genes should change the Darwinian fitness, either in homo- or heterozygous condition. After all, a chromosome with an inversion carries the same gene loci, though arranged in a different sequence, that it had before the inversion. The answer is suggested by the fact that gene recombination is reduced or suppressed in chromosomes in which the two members of a pair differ by an inversion. Suppose that a pair of chromosomes carry gene alleles $A_1B_1C_1D_1$ and $A_2C_2B_2D_2$, respectively. The two gene combinations interact favorably, giving heterosis in the heterokaryotype $A_1B_1C_1D_1/A_2C_2B_2D_2$. Other combinations of the same alleles, for example $A_1B_2C_1D_2$, have adaptively less favorable effects. An inversion that suppresses recombination guards against disintegration of a favorable gene pattern, and is therefore favored by the balancing natural selection.

Inversion polymorphism is really an especially clear case of a much more general phenomenon of linkage disequilibrium (Lewontin, 1974). Genes do not act in development independently of other genes. They interact in concert, and natural selection favors placing together in the same chromosome an assemblage of gene alleles that confers a high fitness on the possessors. Such assemblages of genes are frequently referred to as supergenes (Ford, 1971). The role of chromosomal inversions in evolution may then be conservation of adaptively valuable supergenes.

How widespread inversion polymorphisms are in organisms other than *Drosophila* and its relatives is uncertain. In the absence of giant chromosomes the detection of inversion is difficult—only those inversions that change the position of the centromere or those that produce certain unusual configurations at meiotic divisions are detected. Nevertheless, inversion heterozygotes are known in many animals, e.g., species of grasshoppers (White, 1973); in plants they are found chiefly in interspecific hybrids (Stebbins, 1971). In human populations inversions seem to be rare, but this may only mean that their discovery is uncommonly difficult. By contrast, polymorphism for translocations is found in populations of many plant species and relatively few animals. The biological function of translocation and inversion polymorphisms is probably similar: establishment of adaptively valuable linkage disequilibria and supergenes.

The adaptive superiority of heterozygotes, heterosis, is the most frequently discussed, but not necessarily the most widespread, genetic mechanism underlying polymorphisms in natural populations. Frequency-dependent selection may be at least equally or more important. In the foregoing discussion we tacitly assumed that the adaptive values and selection coefficients are constant for any given genotype, no matter what its frequency in the population relative to other genotypes. This simplifies the mathematical treatment of selection processes but it is often unrealistic. After all, the adaptive value, Darwinian fitness, is a function of the environment in which an organism lives, and other organisms of the same and different species living in the same neighborhood are important constituents of the environment.

Suppose that the fitnesses of two competing genotypes, say A_1A_1 and A_2A_2, are high when a given genotype is rare in a population, but diminish when that genotype becomes frequent. Natural selection will enhance the frequency of the initially rare genotype. As its frequency increases, its fitness dwindles, while the fitness of the alternative genotype grows as it becomes less frequent. It may be that at certain frequencies the carriers of A_1A_1 and A_2A_2 are equally fit. If so, a stable equilibrium will be established at these frequencies, and the population will become stably polymorphic (reviews in Wright, 1969; Ayala and Campbell, 1974). Very extensive experimental evidence on frequency-dependent selection in some species of *Drosophila* has been obtained by Kojima and his colleagues (see for example Kojima, 1971). The evidence is even more abundant in cultivated plants (for review see Allard and Adams, 1969). The yield of a planting of a proper mixture of genotypes may exceed that of equal areas planted to each of these genotypes singly.

There are at least two mechanisms that bring about frequency dependence. One is diversity of environments. Some environments may be more suitable, or preferred, by carriers of one genotype and others by competing genotypes. This may lead to diversifying natural selection (see below). Another mechanism is interaction of the carriers of different genotypes. The interaction may be mutually beneficial (facilitation, synergism), or may result in one or both genotypes being harmed (disoperation).

Dawood and Strickberger (1969), as well as Anxolabehere and Periquet (1972), have adduced evidence of such interactions between larvae of different mutants and species of *Drosophila*. It should be noted that heterotic and frequency-dependent balancing selections are not mutually exclusive—they can operate simultaneously. In fact, one of the first instances of frequency dependence was found in experimental population cages of *Drosophila pseudoobscura* polymorphic for chromosomal inversions (Levene *et al.*, 1954). We have seen above that polymorphisms for inversions are consequences of superior fitnesses of heterokaryotypes; nevertheless, analysis of the selection process in experimental populations has shown that the relative fitnesses of the karyotypes is dependent on the kinds and frequencies of the chromosomes present.

DIVERSIFYING NATURAL SELECTION

Biologists often speak about the environment in which a species or a population lives. Yet even carefully controlled laboratory environments are hardly ever completely uniform. Natural "environments" are mosaics of more or less similar or disparate habitats or subenvironments. Some of these habitats may be more suitable for some genotypes, and others for competing genotypes. How is a species to cope with the variety of habitats that it meets both in space and time? One strategy is natural selection favoring a genotype in which the carriers possess a well-developed homeostasis, such that they have well-adapted phenotypes in all, or at least in most, subenvironments. In a way, this strategy is universal. Not only every species but also every individual meets with a range of environments; if it were specialized to live in a single environment it would not live long. Another strategy is polymorphism. Natural selection diversifies the gene pool in such a way that two or several classes of genotypes having optimal adaptiveness in different subenvironments are produced. The two strategies are of course not mutually exclusive. Conditions under which each of them is expedient have been investigated mathematically by Levins (1968).

The converse of diversifying selection (also called "disruptive selection") is stabilizing selection. If the environment is rather uniform in space and time, it may be advantageous for a population to limit the range of its variability, so that only, or mostly, phenotypes tested by natural selection are produced and deviations eliminated. In a sense, stabilizing selection is related to normalizing selection, and so understood it is really universal. Every species has on the average a certain body size, weight, and proportions. It is doubtful that giants above seven feet tall or dwarfs below five feet would fare well in human populations. In fact, neonatal deaths are more frequent among abnormally heavy and abnormally light infants than among infants of average weight.

Diversifying selection may be closely related to frequency-dependent selection. Suppose that a population inhabiting a certain territory has two or several groups of genotypes and meets two or several subenvironments. A genotype that is rare but is well adapted to a certain habitat will be promoted and increased in frequency by selection when its habitat is not fully occupied. Conversely, it will no longer be promoted when its habitat is saturated. A frequent genotype may saturate its optimal environment, and its excess population may spill over into subenvironments that are stressful for them. Diversifying selection will then establish a balanced polymorphism. Frequencies of the different genotypes will be reached such that the exploitation of the available mosaics of subenvironments will be as complete and efficient as possible. Levene (1953), Maynard Smith (1970), and others have explored the conditions under which the polymorphism will be stable. Powell (1971) and McDonald and Ayala (1974) have shown experimentally that laboratory populations of *Drosophila* placed in environments deliberately made heterogeneous support more genetic polymorphisms than do populations in more homogeneous environments.

Diversifying selection is a process that enhances the adaptedness of populations that live in heterogeneous environments. By and large, genetic uniformity is a drawback, except in ideally constant environments, and diversity is an advantage. Does diversifying selection play any role in mankind? Although upbringing and training are the chief factors that qualify persons for different occupations and social roles, individual genetic predispositions do facilitate training for one or another vocation. Insofar as this is so, diversifying selection renders important services in the human species.

SEXUAL SELECTION

In many species of animals females and males are strikingly dimorphic. Males are often (though not always) larger, stronger, more brightly colored, and possessing various "adornments," some of which may seem to be handicaps rather than advantages to their possessors. Females and males always differ in behavior to some extent, and sometimes strikingly so. Could sexual dimorphisms originate in evolution by natural selection? Darwin found this improbable as a general explanation, and proposed, in 1871, sexual selection as a more plausible solution. In his words, "sexual selection depends not on a struggle for existence in relation to other organic beings or to external conditions, but on a struggle between the individuals of one sex, generally the males, for the possession of the other sex."

During the century since it was proposed by Darwin, the theory of sexual selection has met with even more opposition than the theory of natural selection. At present, sexual selection is usually regarded as a special case of natural selection (for a review see Mayr, 1972). After all, Darwinian fitness is reproductive fitness. The carriers of some genotypes transmit more of their genes to the following generations than do other genotypes. Higher fitness may be achieved because of some advantage in the "struggle for existence in relation to other organic beings or to external conditions"; it may also be achieved because of greater proficiency in securing mates. Components of fitness may or may not be correlated. Greater sexual prowess may go together with lower survival rate, or vice versa. Darwinian fitness is a resultant summation of advantages and disadvantages. Too conspicuous a coloration, overgrown antlers, or horny outgrowths, present in some recent and fossil species of mammals, may have been nuisances to their possessors, but if they enabled them to sire more offspring they were promoted by selection.

Sexual selection has been experimentally demonstrated. For example, yellow body color, one of the classic mutants in *Drosophila*, is selected against in experimental populations containing mixtures of yellow and normally pigmented flies. The lower fitness of yellow is due largely to the fact that yellow males are seldom accepted by normal females, while normal males are accepted about equally by both yellow and normal females. Sexual selection in nature is a complex issue. Do females really discriminate in favor of more "handsome" males? In some species, among both vertebrates and invertebrates, males engage in agonistic displays or actual fights, and the winner gets the female as the "prize." Here the females play passive roles. Persistence in courting females varies, and the more persevering males are accepted on the average more often than more sluggish ones.

Some modern authors have broadened the concept of sexual selection to mean any departure from random mating among inhabitants of the same neighborhood. So defined, sexual selection is certainly widespread. In the first place, different species, even closely related ones, usually exhibit ethological isolation, i.e., a preference for mating with conspecific individuals (this will be discussed in more detail in Chapter 6). To exercise such a preference, both females and males must be able to recognize the species and sex of individuals whom they meet. The recognition marks may be visual, auditory, olfactory, or kinesthetic. Selection is certainly involved in development of the recognition marks. The bright colorations and other adornments of males may then be understood either as appealing to "aesthetic perceptions" in the females or as species recognition marks. The fact that such adornments are found more often in males than in females is consistent with the more active sexual role observed of males in many species.

Moreover, deviations from randomness of mating are by no means uncommon among conspecific individuals in the same neighborhood. A most interesting situation, combining sexual and frequency-dependent selections, was discovered independently by Petit and by Ehrman in several species of *Drosophila* (see Petit and Ehrman, 1969). In Ehrman's experiments equal numbers of females and males of two kinds of *Drosophila pseudoobscura* were introduced into an observation chamber, but the proportion of flies of each kind varied from experiment to experiment. Whichever males were in a minority had a relative advantage in mating with females of either kind. In one experiment the two kinds of flies differed in geographic origin—the ancestors of one strain were collected in a locality of California, and those of the other in Texas. Table 4-1 shows the results.

TABLE 4-1. Observed and expected numbers of matings of two strains of *Drosophila pseudoobscura* in mating chambers with different proportions of the two kinds of flies. C, California; T, Texas. [After Petit and Ehrman, 1969.]

	Males mated		Females mated		
Proportions	C	T	C	T	Chi-square for males
12C:12T					
Observed	55	49	50	54	0.39
Expected	52.0	52.0	52.0	52.0	
20C:5T					
Observed	70	39	84	25	16.96
Expected	87.2	21.8	87.2	21.8	
5C:20T					
Observed	39	65	30	74	19.91
Expected	20.8	83.2	20.8	83.2	
23C:2T					
Observed	77	24	93	8	34.75
Expected	92.9	8.1	92.9	8.1	
2C:23T					
Observed	30	70	12	88	65.76
Expected	8.0	92.0	8.0	92.0	

It can be seen in Table 4-1 that when the numbers of males of California and Texas origins in the mating chamber were equal, the matings occurred in random proportions. This was not the case when either California or Texas males were a minority. The observed number of matings per minority male turned out to be greater than expected on the assumption of randomness, while the majority males mated less frequently than expected. The differences between the observed and expected number of males mating was statistically quite significant—a chi-square greater than 6.63 is expected to occur by chance in fewer than one per 100 trials. By contrast, the numbers of majority and minority females mating were nearly in proportion to their abundance. Mating advantages of minority males were also observed by Ehrman when the two kinds of the flies in the chamber differed in chromosomal inversions, in mutant genes, or in having developed at different temperatures. Some observations indicate that mating advantages of rarity occur also in *Tribolium* beetles and in certain butterflies. Nevertheless, this phenomenon can hardly be universal. Grossly deleterious mutants, such as hereditary diseases or malformations, surely do not confer mating advantages.

Mating advantages of minority genotypes may lead to stably balanced polymorphisms. Of course, a mating advantage tends to increase Darwinian fitness. However, the advantage decreases as the frequency of a genotype increases. At some frequencies the competing genotypes mate at random, and an equilibrium is reached. It is worthwhile to reiterate that different kinds of selection may operate simultaneously. It may be that the frequencies of chromosomal polymorphisms in natural populations of *Drosophila* (see above, and Chapter 5) are determined by interactions of heterotic and frequency-dependent balancing selections.

DIRECTIONAL SELECTION

Physical, e.g., climatic, and biotic environments of a species do not remain constant for long periods. Environmental changes do not necessarily induce genetic changes; the genetic variants composing a species may be sufficiently homeostatic so that adaptedness remains adequate in the altered environments. On the other hand, the Darwinian fitnesses of variant phenotypes and genotypes may be shifted, some in positive and others in negative directions. A *directional natural selection* then comes into operation. It transforms the gene pool of a species or population toward the highest level of adaptedness (given the genetic materials available) that can be reached in the new environment. Apart from environmental change, origin of new and favorable mutant genes or supergenes brings into action directional natural selection, even if the external environment continues without change.

Rates of changes in gene frequencies under directional selection depend of course on the relative Darwinian fitnesses of the favored genotypes. In principle, the rates are the same as under normalizing selection (see above). The similarity of the selection rates under normalizing and directional selections is not surprising. If selection favors a certain allele of a gene, it necessarily discriminates against the other alleles of the same gene.

Experimental models of directional selection have been made. A known mixture of different genotypes is introduced in a population cage, or sown or planted on an experimental field, and left to breed freely. In every generation, or at other convenient intervals, samples of the population are taken, and the frequencies of some genetic variants determined. From the observed rates of changes in these frequencies, the adaptive values of the variants can be computed. It should be stressed that what is observed in such experiments is natural selection, no matter how "artificial" the experimental environments may be.

Artificial selection, practiced by breeders of domestic animals or plants, can be used as a model of directional natural selection. It was so used by Darwin in *Origin of Species*. Experiments on artificial selection have also been made by geneticists on a variety of characteristics of many species. Perhaps the most notable result of these experiments is that at least some change in the desired direction can be obtained with almost anything one selects (see Chapter 2). In Lewontin's words, "There appears to be no character—morphogenetic, behavioral, physiological, or cytological—that cannot be selected in *Drosophila*" (1974). The significance of this must be appreciated. Natural populations, at least of sexual and outbreeding species, contain stores of genetic variability much greater than classical geneticists, and even Darwin, dared to suppose. Furthermore, this variability is pervasive, in the sense that it affects every aspect of the organism's constitution. This disposes of the disingenuous objection against natural selection as a directing agency of evolution that alleges that to progress in a given direction selection has to wait for the occurrence of novel mutations. To be sure, to quicken the progress of artificial selection practical breeders sometimes take steps to augment the genetic variance of the materials with which they work. Yet striking, even spectacular improvements in the productivity of agricultural plants and animals have been achieved by artificial selection without conscious steps to enhance the genetic variances. Call to mind the far-reaching differences between domesticated species and their wild ancestors!

Examples of the action of directional selection in nature can be seen most easily in genetic changes that occur in nondomesticated species because of transformation of their environments by man. The evolution of resistance of insect species to pesticides has in recent decades been a spectacular demonstration of the ability of living species to undergo genetic changes in response to challenges of the environment (for reviews see Brown, 1967; Georghiou, 1972). The story is always the same: when a new insecticide is introduced, a certain relatively small amount or concentration is sufficient to achieve a satisfactory control of the insect pest abundance; the necessary concentration gradually increases, until it becomes totally ineffective or economically impractical. In 1947 Saccá was the first to report that a population of the house fly, *Musca domestica*, had become resistant to DDT. Since then, resistance to one or more insecticides has been recorded in at least 225 species of insects and other arthropods. Usually the resistance is specific to one or a group of chemically related insecticides, but occasionally instances of multiple resistance are reported. An interesting parallel is found in a mammal species, the grey rat. A substance known under the trade name of Warfarin was destroying rats, chiefly through its action as a blood anticoagulant and interference with vitamin K. Warfarin-resistant rats appeared first in a locality in England, and have since been spreading and replacing the ordinary Warfarin-sensitive rats.

In a few species resistance to insecticides has been induced experimentally by selection in the laboratory. It has been ascertained that resistance can be due either to a single gene with a major effect or to a co-acting set of genes with individually minor effects. The ubiquity of genes that confer adaptedness to synthetic chemicals only recently invented by man may seem strange or even mysterious, however, the mystery can be dispelled by studies of the physiological bases of specific resistances. It is usually found that in the resistant strains the insecticide is detoxified by enzymes and metabolic processes that also exist in nonresistant strains, although with a lower level of activity. The emergence of pathogenic and other bacteria resistant to antibiotic and chemotherapeutic agents is due to directional selection quite analogous to that for insecticide resistance. Resistance in pathogenic bacteria has ushered in practical problems that are even more troublesome than those created by insect resistance to insecticides, inasmuch as in the former problems it is human health and even survival that are involved.

Directional natural selection has also been observed in the spread of so-called industrial melanisms in many species of moths, best studied in England (for reviews see Kettlewell, 1961; Ford, 1971). Since the middle of the last century the appearance and spread of darkly pigmented variants of moths has been recorded in industrial regions where the vegetation is blackened because of pollution from soot and other wastes. In some localities the dark varieties have almost completely replaced the formerly "normal" lightly pigmented forms, while in nonpolluted areas the light ones still occur. Altogether about 100 species have shown industrial melanisms. In a majority of the species examined genetically, the difference between the dark and light variants is due to a single gene, the black usually being dominant. Interestingly enough, further evolution of the melanics has also selected a system of gene modifiers with relatively minor phenotypic effects, which intensify the phenotypic effects of the main gene. Observations and experiments have shown clearly that predation of the moths by birds is the main selective factor promoting industrial melanism. The melanic varieties are protectively colored when they rest on blackened vegetation; they are conspicuous on nonpolluted vegetation, where light varieties are protected (Figure 4-8).

Plants have also produced some clear-cut instances of directional selection. Bradshaw, Gregory, Antonovics, and their colleagues have studied varieties of several species of plants able to grow on soils contaminated with heavy metals (copper, zinc, lead). Such contaminated soils are often found near old mines

FIGURE 4-8.
Above: The peppered moth, *Biston betularia,* and its melanic form, *carbonaria,* at rest on a soot covered oak trunk near the industrial city of Birmingham, England. The *carbonaria* form is much less conspicuous than the typical peppered form, which is very conspicuous. *Below:* The same forms resting on a lichened tree trunk in unpolluted countryside. The typical form is much less conspicuous than the melanic. [Courtesy of Dr. H. B. D. Kettlewell.]

where ores of these metals were exploited. The resistant varieties maintain themselves on soils with concentrations of the pollutants that are lethal or at least stressful for ordinary members of the same species. A conspicuous feature is that the resistant and nonresistant varieties may grow at distances only a few meters from each other, but rigorously confined to their respective contaminated and uncontaminated soils.

NATURAL SELECTION AND ADAPTEDNESS

Darwinian fitness is a function of reproductive efficiency. The carriers of some genotypes transmit their genes to the next generation more often than do carriers of other genotypes. By definition, natural selection promotes Darwinian fitness. However, is high Darwinian fitness always beneficial to an individual or population? Does it always go together with adaptedness, good health, improved survival ability, enhanced adaptability in varied environments? By and large it does; if it were otherwise, life on earth would have become extinct long ago. One must nevertheless be wary of projecting the vernacular meanings of "fitness" onto Darwinian fitness. Natural selection is not a benevolent genie, but a consequence of certain combinations of physical and biological variables. A hypothetical situation in which natural selection could have mischievous consequences is easily imaginable.

If intelligent people had fewer children then less intelligent ones, and provided that human intelligence is in part genetically conditioned, the average intelligence would exhibit a downward trend. In our system of values high intelligence is a positive quality, and low intelligence a negative quality. Moreover, deterioration of the intelligence may conceivably endanger the survival of the species. High Darwinian fitness may thus be a property of genotypes that are not only socially undesirable but injurious to the species as well.

Genetic research has uncovered several instances of hereditary variants promoted by natural selection despite their being, at least in the long run, harmful to the population. Dunn and his collaborators found most unusual mutants at the *t*-locus in the house mouse. A male mouse heterozygous for a *t*-mutant and for its "normal" allele generates more, sometimes many more, functioning spermatozoa with the *t*-mutant than with the normal allele. This deviation from the usual 1:1 segregation ratio confers upon the mutant alleles a selective advantage. If not counterbalanced by some disadvantage, the mutant *t*s will spread and crowd out the normal alleles. Actually, most mutant *t*s are lethal when homozygous. For this and perhaps other still undiscovered reasons, *t* alleles do not outnumber normal alleles, although they are found in many mouse populations.

Another example of a genetic variant that has subverted the normal processes of meiosis and sex cell formation, and thus gained a selective advantage, is the so-called chromosomal "sex ratio" found in natural populations of several species of *Drosophila*. A male with a "sex-ratio" X chromosome produces functioning spermatozoa most or all of which carry X chromosomes, and few or none with Y chromosomes. Crossed to any female, such males give progenies consisting of daughters and few or no sons. Ordinary males pass their X chromosomes to only one half of their offspring, because the other half are males that inherit Y chromosomes from their fathers and X chromosomes from their mothers. Unopposed by some other agency, the "sex ratio" X chromosomes would take over the population; in a species incapable of parthenogenesis, a population consisting only of females would become extinct. In fact, however, this does not happen in nature in populations in which "sex ratio" chromosomes are found. The reason for this is still conjectural.

Species now living are descendants of only a minority of the species that existed in the geological past, and increasingly smaller minorities as we look back to more distant times. Indeed, the usual fate of an evolutionary line is extinction (see Chapter 11). Yet the evolution of extinct lines is also controlled by natural selection. The solution to this apparent paradox is rarely found in the appearance of "subversive" mutants, like those of the *t*-locus in the mouse or the "sex ratio" in *Drosophila*. Extinctions may often be due to the fact that natural selection promotes genetic variants with high Darwinian fitnesses, here and now, regardless of what their fitnesses may be in future environments. Artificial selection is goal-directed, but natural selection is not. Having no information about the future, natural selection is opportunistic and blind. A genotype does not have high or low Darwinian fitness in the abstract; its fitness is a function of the environments in which it is placed or to which it can migrate.

A genotype only passably successful now may flourish in some future environments. Such a genotype is often said to be "preadapted" to what it will meet in the future. The concept of preadaptation is a fallacy if it is taken to mean that evolution somehow contrives adaptations for use in as yet nonexistent environments. Nevertheless, in building adaptedness for present environments, natural selection can only operate with what is inherited from the past. Bipedal gait was surely adaptive in mankind's remote ancestors, and it was selected for the benefit of those ancestors. Yet its upshot was the development of hands capable of manipulating tools, machines, and surgical instruments. This ability made possible the development of technology and culture, and is highly adaptive in human environments created much later by culture (see Chapter 14).

GROUP AND KIN SELECTION

In a population under selection the frequencies of some genes increase and those of others decrease as generation follows generation. The reproductive efficiency of individuals who carry the "successful" genes is higher than that of the carriers of "unsuccessful" ones. What survives or dies—remains sterile or engenders offspring—are however not genes themselves but living individuals, who are their carriers. The question can nevertheless be asked whether, in addition to natural selection on the individual level, there can be *group selection*, i.e., selection of groups of individuals, such as Mendelian populations. Related species compete for resources that both are in need of, and one species may outbreed and crowd out another. Thus, species introduced from foreign countries sometimes become pests or weeds, and eliminate or greatly reduce the abundance of native species. This is particularly striking on oceanic islands. Island endemic species often seem to have reduced competitive abilities, and they are given short shrift by introduced forms.

A more complex and interesting problem is whether group selection is also important intraspecifically. Wright, in his classic essay "Evolution in Mendelian Populations" (1931), devised a model of a species subdivided into semi-isolated colonies. The gene exchange between the colonies is presumed to be rare enough that the colonies may diverge genetically, owing to differential selection and random genetic drift (see Chapter 5). He compared the colonies to scouts or trial parties "exploring" the field of gene combinations. Evolutionary changes in some colonies may result in superior adaptedness, and in other colonies in inferior adaptedness and even extinction. The colonies that flounder and fail may then be replaced by migrants from the successful ones. The whole species will then rise to a higher level of adaptedness.

Variants of this model were later utilized to explain some evolutionary phenomena that are difficult to account for by the more usual selection of individual contributions to the gene pool of following generations. Consider, for example, altruistic behavior. For the purposes of the present discussion altruism can be defined as a behavior that benefits other individuals at the expense of the altruist. If altruistic behavior is genetically conditioned, the Darwinian fitness of the altruists will be diminished; that of individuals who refuse to act altruistically will, on the contrary, be augmented. It would seem that natural selection would discriminate against altruism and favor self-preservation and egotism. Yet a population, colony, or tribe that includes many altruists may prosper more than a population burdened with egotists. Altruism and related behaviors may be promoted by group selection. But they may also be promoted by natural selection when the altruistic behavior of an individual benefits its close relatives. This form of natural selection is known as *kin selection.*

It is arguable whether it makes sense to speak of altruism and selfishness except in species able to choose freely between different possible courses of action. The only such species is man (Chapter 14). Many kinds of behavior are equivalent to altruism, whereas they are really automatic and stereotyped. The behavior of honeybees, ants, termites, and other social insects has excited the curiosity and admiration of many observers for a long time (for an excellent review see Wilson, 1971). Individual "workers" toil selflessly and are ready to sacrifice themselves for the benefit of the colony to which they belong. The evolutionary development of this ostensible altruism is easily understood. The "workers" and "soldiers" do not reproduce. They are sterile females, or males with underdeveloped sex organs. The perpetuation of an insect society is the task of a sexual caste, the so-called queens and fertile males, which are a minority, sometimes single individuals, in the colony. Sexual individuals do little or no work after the colony is founded, and evince no particular readiness for self-sacrifice. The workers share genes with the sexual members produced in the colony. The result of this arrangement is that natural selection operates not in individuals but in colonies consisting of reproductive and sterile castes as units. The ostensible altruism of the workers enhances the probability of transmission to the next generation of the genes these workers share with the sexual caste of their colony.

Behaviors analogous to human altruism are by no means confined to social insects. Production of progeny exacts, as a minimum, an "investment" of nutrient materials on the part of the parents to produce sex cells. Where parental care is developed the progeny is, for a shorter or longer time, fed by the parents. More than that, parents defend their progeny from aggressors and predators, and in so doing expose themselves to dangers of injury or death. Among herd-living animals the large and powerful males may protect other members of the troop, particularly females with young, from predators, and do so at a risk to themselves. One should not expect natural selection to promote selfish behavior in every individual under all conditions. Hamilton (1964), Trivers (1972), and others advanced a theory of kin selection, which provides a simple and reasonable explanation of ostensibly altruistic behavior in animals. Altruism may promote the "inclusive Darwinian fitness" of the altruist, instead of his individual well-being. Indeed, each child shares 50 percent of his genes with siblings and with each parent. Therefore, a parent who is disabled or killed defending his progeny may "save" more of his genes than if he remained alive while his progeny perished. This is especially true of older parents, whose reproductive life is finished or nearly so. The closer the genetic relationship between the altruist and the beneficiaries of his conduct, the greater the biological, selectional warrant for altruistic behavior. Members of a herd, e.g., baboons or wild horses, are in general more or less close relatives. Altruistic and cooperative behaviors are, as a rule, more developed the closer the genetic relationships.

Group selection has also been invoked to explain some forms of behavior that do not fit the kin-selection model. In some species, physiological and behavioral phenomena result in decreasing reproductive rates of members of populations that have grown beyond optimal numbers. Wynn-Edwards (1962) sees in this the result of group selection that benefits populations rather than individuals. Indeed, excessive population growth may destroy the food supply and cause extinction of the population. Some species have evolved social hierarchies of dominant and subordinate individuals within populations. One alleged consequence of this is greater efficiency in the utilization of food. Another is that some weaker or "surplus" individuals are excluded from the population, and effectively condemned to die from starvation or predation. Group selection supposedly makes the surplus individuals stoically accept their fate because it benefits the species, even though at their expense. Most evolutionists find such explanations questionable. There is clearly much to be learned about the roles of different kinds of group and kin selection.

5

Populations, Races, Subspecies

A traveler in foreign lands need not be a biologist or anthropologist to see that the inhabitants of different countries are more or less distinct, both physically and culturally. A traveler who is also a biologist will notice that the rule of geographic diversity applies to animals and plants as well as to humans. In different parts of the distribution area that a species inhabits, its representatives usually show some more or less striking average differences. This local, or regional, or racial variation is more pronounced in some species than in others. Species that are geographically clearly differentiated are called polytypic; mankind is an example of a polytypic species. Polytypism is the variability between populations or groups. It should not be confused with polymorphism, the variability *within* populations (see Chapters 2 and 4). Yet we shall see below that polymorphism may serve as a store of genetic raw materials for origination of polytypism.

Although pioneer biologists could not fail to observe polytypism, it was interpreted differently by various writers. Linnaeus believed species to be created entities, and in his early writings he counseled botanists "not to bother with quite insignificant varieties." Yet in 1759 he went so far as to suggest "that many species belonging to the same genus were once a single species." Buffon was quite impressed by intraspecific variation, which he ascribed to environmental, particularly climatic, influences (1749, 1753). Thus, "The heat is the main cause of the black coloration. When heat is extreme, as in Senegal and Guinea, people are completely black . . . finally, when [heat] is quite moderate, as in Europe and Asia, people are white." It was left to Lamarck (1809) and Darwin (1859, 1871) to state that varieties are the materials and species the products of the evolutionary process. Although their interpretations of the causes of evolution were different—Lamarck ascribed genetic modifications to "habits required by the conditions," Darwin to natural selection—they agreed that varieties were precursors of species (for a critical discussion of this history see Boesiger, 1974).

After Darwin, in the late nineteenth and early twentieth centuries, many biologists attempted to sort out the genetic polymorphisms and polytypisms from phenotypic variations due to environmental influences. Much confusion resulted until Johannsen developed the concepts of genotype and phenotype (Chapter 2). The genotype does not determine unit characters, but a range of reactions to the environment during the process of development. It is therefore naive to ascribe one character to heredity and another to environment. The potentiality of any characteristic is of course always genetic, but the realization of this potentiality is contingent upon such environment in which development can occur.

Lamarck and his followers, the so-called Neo-Lamarckists, attempted to explain evolution as the inheritance of traits acquired by the organism during its lifetime. Such traits were due to the use or disuse of organs, or to environmental (or as we would now say, phenotypic) modifications. Although Lamarck took this idea for granted, the central idea of his theory was that evolution comes from an inner drive for perfection inherent in all life. This view was later called autogenesis, and autogenetic theories still persist as a minority view (for example see Grassé, 1973). Darwin also accepted the idea that acquired characters can be inherited, but only as a process subsidiary to natural selection. Weismann in 1883 challenged the idea that any acquired characters are inherited. He became the leading Neo-Darwinist, claiming natural selection to be the main cause of evolution (the theory of mutation was developed by de Vries later, around 1900). Nevertheless, the controversy over inheritance of acquired traits continued until the 1930s, and persists even now in some quarters.

The present book is essentially an exposition of the modern theory of evolution, called the "biological" or "synthetic" theory of evolution (because it is based on data from all biological sciences). The fundamentals of this modern theory were arrived at, largely independently, by Chetverikov, Fisher, Haldane, and Wright between 1926 and 1932; but the theory was developed by many authors, starting in the 1930s (see Chapter 1). In a nutshell, the theory maintains that mutation and sexual recombination furnish the raw materials; that natural selection fashions from these materials genotypes and gene pools; and that, in sexually reproducing forms, the arrays of adaptively coherent genotypes are protected from disintegration by reproductive isolating mechanisms.

Some eminent biologists have recently asserted that the problem of evolution is now resolved, except for minor details. This assertion is erroneous. To be sure, there is no reasonable doubt that the living world is a product and outcome of the three to four billion years of the earth's evolutionary history. However, the causes of evolution, and the patterning of the processes that bring it about, are far from completely understood. We cannot predict the future course of evolution except in a few well-studied situations, and even then only short-range predictions are possible. Nor can we, again with a few isolated exceptions, explain why past evolutionary events had to happen as they did. A predictive theory of evolution is a goal for the future. Hardly any competent biologist doubts that natural selection is an important directing and controlling agency in evolution. Yet one current issue hotly debated is whether a majority or only a small minority of evolutionary changes are induced by selection (for a discussion of this issue see Lewontin, 1974). The term "non-Darwinian evolution" has been coined for changes due not to natural selection but to random processes (some of which are discussed in this chapter and in Chapter 9). This term is ambiguous at best, because several old theories, beginning with Lamarck's, ascribed little or no role to natural selection, and thus were "non-Darwinian."

CLONES, PURE LINES, POPULATIONS

Group variability is a function not only of mutation and natural selection but also of the reproductive biology of the species concerned. Evolutionary patterns in organisms with different kinds of reproductive biologies will be treated in more detail in Chapter 7. Here we need only to be reminded of some elementary facts. Many organisms, especially prokaryotes but also some plants and some lower animals, reproduce asexually. Simple fission, budding, asexual sporulation, etc., give rise to clones. Unless mutation intervenes, a clone derived from a single progenitor consists of individuals with genotypes identical to that of the progenitor and to each other. Selection, natural or artificial, is without effect within a clone. But the population of a locality may consist of genetically distinct clones. A "locality" may be the gut of an animal harboring diverse bacterial clones, or a meadow with clones of an asexually reproducing plant, or a pond with clones of a blue-green alga. Selection acts to enhance the frequencies of the better-adapted clones and to depress the frequencies or eliminate the less well adapted ones.

Clones of some species of cultivated fruit trees (oranges, apples, pears) are made artificially by grafting scions of desirable varieties onto stocks that have other desirable properties. For example, scions of choice European grapevine varieties are regularly grafted onto stocks of American grapes because the latter are resistant to the attacks of a serious pest, the aphid *Phylloxera*. Barring mutation, many thousands of individuals are genotypically identical members of a clone. A practical breeder works in two ways. First, he chooses from among available clones those that do best in a particular climate and soil. Second, he looks for an opportunity to start new clones that may arise owing to mutation or to Mendelian segregation following occasional spontaneous or artificial hybridization.

Some plants and animals reproduce by forms of parthenogenesis or apogamy that also give rise to clones of genotypically identical individuals. Thus, some species of insects, fish, and lizards consist of females only, although closely related species are bisexual and incapable of parthenogenesis. Parthenogenetic clones may consist of genotypically identical individuals that are heterozygotes for many genes; recombination of these genes may be prevented by parthenogenesis. Self-pollination in plants and self-fertilization in some hermaphroditic animals may lead to formation of pure lines. Derived by generations of selfing from a single ancestor, a pure line, like a clone, may be composed of genetically identical individuals. Clones consisting of homozygous individuals may also result from certain varieties of parthenogenesis (White 1973).

Selection, both natural and artificial, proceeds by sorting out from among existing pure lines or clones those suitable for particular environments. It may also proceed by detecting new pure lines and clones that originate by mutation or, rarely, cross-fertilization. Indeed, many predominantly asexual or self-pollinating species are capable of producing progeny by sexual crossing. When such crossing takes place, Mendelian recombination in the hybrid progenies gives rise to swarms of new clones or pure lines. From these, selection picks out and propagates the ones suitable for the available environments. The work of Allard and his students on the wild oat, *Avena barbata*, has disclosed elegant examples of natural selection among pure lines (Hamrick and Allard, 1972). In some localities in California the populations of this almost exclusively self-pollinated species are mixtures of pure lines, the relative frequencies of which are correlated with environmental conditions, notably with moisture or dryness. Striking frequency differences may occur at distances of only tens or hundreds of meters.

Sexual reproduction with cross-fertilization is the prevalent mode of reproduction in the living world at large. It is overwhelmingly the most common mode in higher animals (with few exceptions) and higher plants (where the exceptions are more numerous). It is common in species with separate sexes as well as in hermaphroditic ones. Even though some hermaphrodites, e.g., pulmonate molluscs and many plants, are capable of self-fertilization, most progeny comes from cross-fertilization when more than a single individual is available.

The genetic and evolutionary consequences of sexual reproduction and outbreeding are of cardinal importance. An individual of a species consisting of clones or pure lines has a single parent, one grandparent, one greatgrandparent, etc. A species may be represented symbolically as a forest of parallel or branching lines of descent. In a sexual species an individual has two parents, four grandparents, eight great-grandparents, etc. Yet some of the ancestors were relatives. In mankind, owing to incest taboos, close relatives do not marry. However, if one could construct a complete pedigree of mankind, or of any other sexual species, one would find a network in which every individual is multiply related to every other individual. A sexual species is a reproductive community, all members of which are connected by ties of mating, parentage, and common descent. The reproductive community has a common gene pool, from which the genes of every individual are derived and to which they return unless the individual dies childless.

Possession of a common gene pool makes a sexual outbreeding species an inclusive Mendelian population. More precisely, it is an array of subordinate Mendelian populations interconnected by regular or occasional gene flow. The adjective "Mendelian" is necessary because "population" is applied also to assemblages of individuals who do not constitute reproductive communities (Chapter 2). Thus, one speaks of "populations" of diverse species of trees or birds in a forest, and also of "populations" of school students, soldiers, or prisoners. The Mendelian population is a form of supraindividual integration. An isolated individual of a sexual nonhermaphroditic species is biologically a dead end; individuals of a clone or a pure line are not so. A clone or a pure line is a "pure race" because it consists of genotypically identical individuals. There are no pure races in mankind or in other outbreeding sexual species, nor have they ever existed, because the occurrence of two or more individuals with identical genotypes in such species is highly improbable (monozygotic twins can be considered members of a clone.). Clones and pure lines are independent evolutionary lineages. On the other hand, a Mendelian population is an evolving supraindividual system. Of course, natural selection acts on individual phenotypes, and through them on individual genotypes. It is an individual who survives or dies, begets progeny or remains childless. Yet it is the collectivity, the gene pool of a Mendelian population, that increases or diminishes the probability of the next and the following generations being in harmony with their environments.

In many species sexuality, asexuality, and selfing are not obligatory but facultative methods of reproduction. In some, chiefly marine, invertebrates sexual and asexual or parthenogenetic generations regularly alternate. In aphids several parthenogenetic generations are followed by a sexual one, this latter usually coinciding with the advent of an unfavorable season. In plants, as already pointed out above, a part of the progeny may originate by selfing and other parts by crossing or by parthenogenesis. Species that possess two or more reproductive modalities may be Mendelian populations composed of numerous small clones or pure lines. Such species exploit the advantages of sexuality, which generates numerous genotypes, and of parthenogenesis or selfing, which permit well-adapted genotypes to multiply before they are returned to the melting pot of sexual reproduction and Mendelian recombination (see Chapter 7).

MICROGEOGRAPHIC RACES

Geographic races are Mendelian populations of a species (or arrays of clones or pure lines) that inhabit different territories and that differ in the incidence of some gene alleles or other genetic variants in their gene pools. In the study of the wild oat, *Avena barbata*, referred to earlier (Hamrick and Allard, 1972), samples were taken at seven locations on a hillside, 183 meters being the distance between the most remote locations. The location at the foot of the hill was the most mesic (humid) and at the top the most xeric (dry). The oats from the different locations differed quite appreciably in frequencies of alleles at six gene loci; five loci coded for enzyme variants distinguishable by electrophoresis (Chapter 2), and the sixth determined the color, black or grey. The differences in the gene frequencies were consistent with the hypothesis that they were induced by natural selection in response to the conditions of the habitat.

Certain microgeographic races of plants have developed a tolerance to salts of heavy metals. Tolerant races of the grasses *Agrostis*, *Festuca*, and *Anthoxanthum* grow well on soils with concentrations of copper, zinc, or lead that are lethal or nearly so to intolerant forms of the same species (Bradshaw, 1971). The tolerant races are found on the tailings of mineworks in which the ores of the respective metals have been quarried. The areas occupied by the tolerant races are small—seldom more than 500 meters across, and often only 100 meters or less. The boundaries between the tolerant and the surrounding normal, i.e., intolerant, races are remarkably sharp. "If transects are made across the boundary between contaminated and normal areas the tolerance of the populations sampled is found to change precisely at the boundary" (Bradshaw, 1971). Yet some of the species that have evolved tolerant races are wind-pollinated, so there must be some gene flow between the tolerant and the intolerant populations. McNeilly studied this problem in *Agrostis tenuis* in a locality where the prevailing wind direction is known. The adult population at a copper mine is uniformly copper tolerant, but some of its seeds give less tolerant plants. Conversely, some plants in the surrounding intolerant population produce seeds that yield plants exhibiting a tolerance. The natural selection that fixes the boundaries of the tolerant microraces must be very strong.

Chromosomal polymorphisms found in many species of *Drosophila* (see Chapters 3 and 4) give rise to some clear-cut examples of microgeographic races. Figure 5-1 shows the composition of populations of *Drosophila pseudoobscura* that live at different elevations of the Sierra Nevada range in the Yosemite region in California. The three commonest gene arrangements in the third chromosomes of this species are ST (stippled squares in Figure 5-1), AR (white squares), and CH (hatched squares). At the lowest elevation ST is most frequent, followed by AR and by CH. As one ascends the slope, AR grows and ST decreases in frequency. At elevations of 8000 feet and above, AR is much the commonest and ST the least common of the three kinds of chromosomes. The horizontal distance between the lowest and highest localities is about 60 miles. In contrast to the sharp boundaries of the metal-tolerant plant races discussed above, the transition between populations of *Drosophila* that live at low and at high elevations is gradual. What we observe are geographic, altitudinal gradients, or clines, in chromosome frequencies. It is most probable that these altitudinal gradients are induced by natural selection in response to the particular environments in which the flies live at different elevations. Exactly what environmental variables are responsible has not been ascertained. However, we do know that the same chromosomes that show the altitudinal

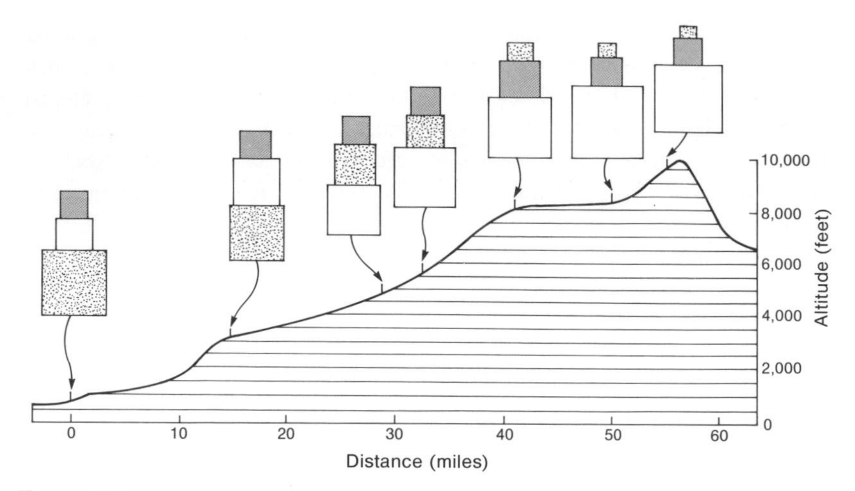

FIGURE 5-1.
The relative frequencies (indicated by the sizes of the squares) of three different kinds of chromosomes in altitudinal populations of *Drosophila pseudoobscura* in the Sierra Nevada near Yosemite Park, California. Stippled squares = ST; white = AR; hatched = CH. The fly populations that occur at different elevations are racially distinct.

gradients also show seasonal changes in relative frequencies, and some of these have been reproduced in laboratory experiments (see Figures 4-5 and 4-7, pp. 112 and 114).

GEOGRAPHIC RACES OR SUBSPECIES OF ANIMALS AND PLANTS

No sharp distinction can be drawn between microgeographic and the more commonly observed macrogeographic races. All gradations exist. Geographic races that differ in easily observable characteristics are often described and given Latin names by many zoological and rather fewer botanical taxonomists. A race given a Latin name is a subspecies. Giving names to races is a rather arbitrary procedure—one can bestow names on two, or several, or many races distinguishable in a polytypic species. As we shall see below, the number of races that are distinguished is a matter largely of convenience.

The microgeographic races of *Drosophila pseudoobscura* considered above occur within a total distance of only 60 miles. The species, however, lives in a vast territory, from British Columbia in Canada to Guatemala, and from the Pacific Ocean to the Rocky Mountains and Texas. Table 5-1 summarizes the data on the incidence of third chromosomes with different gene arrangements in populations in 12 localities. Of the 22 gene arrangements known in the species only the nine most common are included.

TABLE 5-1. Relative frequencies of third chromosomes with different gene arrangements in populations of *Drosophila pseudoobscura* in various localities. Each gene arrangement is designated by two capital letters. [After Powell, Levene, and Dobzhansky, 1973.]

	ST	AR	CH	PP	TL	SC	OL	EP	CU
Methow, Washington	70.4%	27.3%	0.3%	—	2.0%	—	—	—	—
Mather, California	35.4	35.5	11.3	5.7%	10.7	0.9%	0.5%	0.1%	—
San Jacinto, California	41.5	25.6	29.2	—	3.4	0.3	—	—	—
Fort Collins, Colorado	4.3	39.9	0.2	32.9	12.3	—	2.1	7.2	—
Mesa Verde, Colorado	0.8	97.6	—	0.5	—	—	—	0.2	—
Chiricahua, Arizona	0.7	87.6	7.8	3.1	0.6	—	—	—	—
Central Texas	0.1	19.3	—	70.7	7.7	—	2.4	—	—
Chihuahua, Mexico	—	4.6	68.5	20.4	1.0	3.1	0.7	—	—
Durango, Mexico	—	—	74.0	9.2	3.1	13.1	—	—	—
Hidalgo, Mexico	—	—	—	0.9	31.4	1.7	13.5	1.7	48.3%
Tehuacan, Mexico	—	—	—	—	20.2	1.1	—	3.2	74.5
Oaxaca, Mexico	—	—	10.3	—	7.9	—	0.9	1.6	71.4

Several items merit attention in Table 5-1, especially because their analogs will also be relevant when we consider human races. The population of each locality differs in composition from every other locality. Yet the differences are mostly quantitative, not qualitative. That is to say, the populations differ in the relative frequencies of the same genetic variables in which individuals within a population, specifically siblings and their parents, may also differ. Certain chromosomes are very common in some populations and completely absent in others. Thus, CU chromosomes are found only in southern Mexico, ST mainly on the Pacific Coast of the United States, CH reaches high frequencies in north-central Mexico, etc. No population has the full assortment of the chromosomes, and no chromosome is found everywhere. Does knowing the chromosomes permit determining the geographic origin of a fly (the flies look outwardly alike everywhere)? If we have a population sample of several hundred flies from any one of the 12 regions listed in Table 5-1, the incidence of chromosomal forms will usually tell where the sample came from. The two chromosomes of a single fly may only permit a rough designation. Thus, a fly having one ST and one AR chromosome could be found anywhere in western United States but not in Mexico; a fly with one CU and one TL chromosome probably comes from southern Mexico or Guatemala, not from the United States and not from northern Mexico, etc.

Into how many races can one subdivide the species *Drosophila pseudoobscura* by their chromosomes? No division is possible if it is required that no two individuals of different races have the same chromosomes. Two races—one in the United States and northern Mexico and the other in southern Mexico and Guatemala—would contain fewer than five percent, usually fewer than one percent, of ambiguous individuals. Relaxing the requirement of distinctiveness, one can classify five races: Pacific Coast, Great Basin states, Rocky Mountains and Texas, northern Mexico, and southern Mexico.

Figure 5-2 shows the frequencies of some of the same gene arrangements listed in Table 5-1 in a number of localities along the border between the

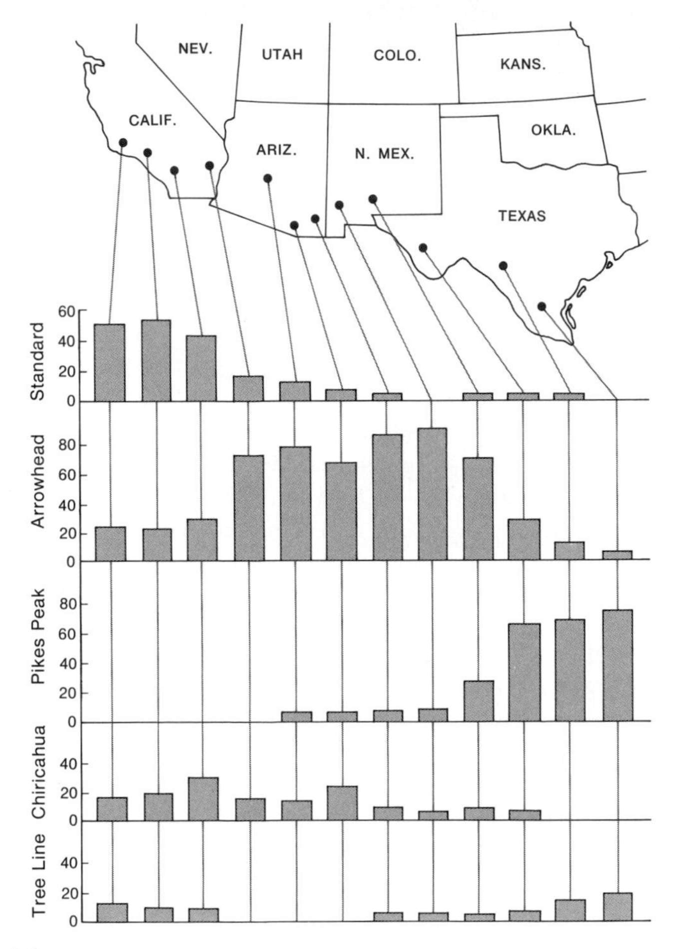

FIGURE 5-2.
Frequencies of five different gene arrangements in third chromosomes of *Drosophila pseudoobscura* in southwestern United States. The Standard chromosomal arrangement decreases in frequency from west to east. The Pikes Peak arrangement increases in frequency from west to east. The Arrowhead arrangement has low frequencies in the east and west but high in the center. The Chiricahua and Tree Line chromosomal arrangements have less conspicuous frequency changes.

United States and Mexico. ST decreases in frequency from west to east; PP does the opposite; AR is commonest in Arizona and New Mexico and dwindles eastward as well as westward. Furthermore, one notices that the geographic gradients of the chromosome frequencies are not uniform. The gradients become steep as one moves from coastal California to the arid intermountain zone, and again between the Rocky Mountains and Texas. The microgeographic variation shown in Figure 5-1 can be said to represent a magnified picture of the same steep gradients seen in Figure 5-2, but observed in the Sierra Nevada mountains. If the geographic gradients were uniform, any division of the species into races would be quite arbitrary. A division becomes feasible only when the places where the gradients steepen are chosen as race boundaries.

The distinctiveness of the races into which a species is subdivided varies greatly from species to species. We have deliberately considered first races differing in a single or few genes or chromosomal variants, and also having boundaries blurred by geographic gradients. Some (chiefly zoological) taxonomists have agreed that subspecies should be given taxonomic (Latin) names only if a majority, at least 75 percent, of individuals can be recognized as belonging to one or the other subspecies. According to Mayr (1970) there are approximately 8,600 species of birds divided into about 28,500 subspecies, or 3.3 subspecies per species on the average. Of course, some species are monotypic and others highly polytypic. Thus, the song sparrow *Passerella melodia* has 34 identified subspecies. Nevertheless, no matter how many races one may distinguish, no two individuals of any subspecies are either phenotypically or genotypically identical. As a rule, individuals and populations transitional between subspecies are found at or close to the geographic boundary at which subspecies meet. Such transitions, or geographic gradients in the boundary zones, may be narrowly localized or broad. This depends on whether or not the boundary coincides with some topographic or climatic barrier. Figure 5-3 shows subspecies of the golden whistler (*Pachycephala pectoralis*) living on islands or island groups of the Solomon Archipelago in the tropical Pacific Ocean. These birds probably could fly from island to island rather easily, but in fact they do so rarely if at all. Races of species living on continents rather than on islands often have no distinct boundaries. One observes rather zones of intergradation, inhabited by highly variable populations in which all intermediates between the races are found. The intergradation zones may be narrow or wide.

To recapitulate: the concept of race or subspecies refers neither to individuals nor to single genotypes, but to Mendelian populations or arrays of genotypes that inhabit parts of the distribution area of a polytypic species. Individuals of a race are both genotypically and phenotypically variable. Race differences may be small or large. They range all the way from populations differing in frequencies of one or few genes or chromosomal variants to populations in the process of evolving reproductive isolation, and thus standing on the brink of becoming separate species (Chapter 6).

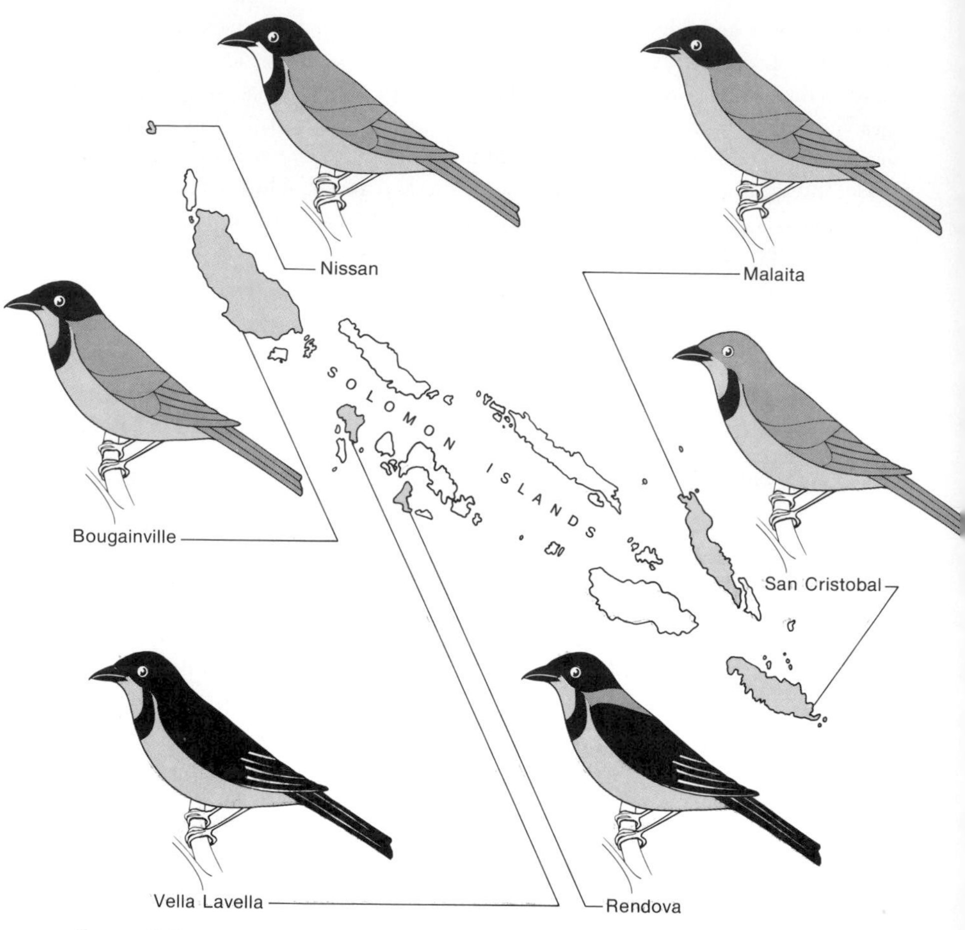

FIGURE 5-3.
Varieties of the golden whistler (*Pachycephala pectoralis*) of the Solomon Islands constitute races, which are kept distinct principally by geographic isolation. They differ in their black and white and colored markings. The dark gray areas represent green; light gray areas are yellow. [After Dobzhansky, 1950.]

HUMAN RACIAL VARIATION

Homo sapiens is a polytypic species. Its evolutionary pattern is in some very important respects unique, as will be discussed in Chapter 14. Here we are concerned with features of its racial variation, which are also somewhat unusual. As pointed out above, racially distinct human populations were originally allopatric, like races of most sexually reproducing species in nature. Geographic separation limited the opportunities for interbreeding and gene exchange, and thus encouraged genetic divergence. A new situation arose with the development of cultures. Some racially distinct populations became sympatric. The gene exchange between them increased compared to the allopatric past, although usually not enough for the races to merge into single variable populations. Thus, the traditional castes in India have descended from several successive waves of racially distinct invaders. They still show some remnants of genetic differences, after two or more millennia of sympatric coexistence. In man, as in some domesticated animal and plant species, one finds races that are partly allopatric and partly sympatric. In the species as a whole genetic divergence changed to convergence. Even the most distinct races of man are genetically less divergent than are the subspecies of many wild animal and plant species. The process of race formation often culminates in an ancestral species breaking up into two or more derived species (see Chapter 6). This is quite unlikely to occur in mankind. Racial variation in mankind is worth studying, in the first place because anything concerning man is of interest. Moreover, since man has been investigated more than any other species, understanding human racial variation helps to understand the processes of race formation in general, especially among sexually reproducing and outbreeding species.

Quite naturally, the study of human races had a long and checkered history. Linnaeus recognized four "varieties" of man (American, European, Asiatic, and African), while Blumenbach (in 1775) named five "races" (white or Caucasian, yellow or Mongolian, black or Ethiopian, red or American, and brown or Malayan). The skin color is evidently not the only character in which human populations differ. The shapes of the hair, nose, lips, and other traits are not only diverse but often not correlated with the skin color or with each other. Most classical anthropologists and biologists adhered to "typological" modes of thought (see Chapter 14), which require that individuals of a race (or species) should all conform to a racial (or specific) "type" or "essence," from which only accidental or pathological deviations are allowed. The situation was not remedied by naming more and more races. Different authorities disagreed about the number and also the delimitation of the "races." Some speculated that in a remote past there were unambiguously distinguishable "pure" races, but hybridization, or "mongrelization," mixed up everything. Attempts were made to distinguish not race populations but racial types. Several such "types" were alleged to coexist in the same population. Thus, the frequent racial types in the population of Ireland were supposed to be the Celtic, Nordic Mediterranean, Dinaric, and Nordic Alpine, while the pure Nordic, predominantly Nordic, East Baltic, and pure Mediterranean types were found with lesser frequencies. The assumption of racial types led to absurd conclusions, as for example when siblings were found to belong to racial types different from their parents as well as from each other. What was radically wrong was the basic assumption: races are not individuals, not types, but genetically different populations.

A new era opened in 1900 when Landsteiner discovered the four "classical" blood groups: O, A, B, and AB. In the following decades it was demonstrated that the incidence of individuals belonging to these groups is different in different human populations; that the blood groups are inherited according to the Mendelian rules; and that in addition to the "classical" there exist several other gene loci, each with two or more alleles, which produce discrete blood group systems. A competent laboratory technician takes a sample of the blood of a person and is able to determine which blood group phenotypes, and usually the genotypes over a range of loci, the person tested carries. By 1962 as many as 33 blood group loci were known in human populations, about one third of which are distinctly polymorphic (Lewontin, 1974). Studies of the proteins in the blood cells and the blood plasma disclosed about 45 further genes that are more or less polymorphic (Cavalli-Sforza and Bodmer, 1971; Harris and Hopkinson, 1972). A majority of these latter polymorphisms have been revealed by electrophoresis techniques (Chapter 2).

A pioneering attempt to classify the racial groups of mankind according to the incidence of known blood groups was made in 1950 by Boyd. He distinguished the European or Caucasoid, African or Negroid, Asiatic or Mongoloid, American Indian, Australoid, and Early European races. Only the last named was an innovation; it included the Spanish and French Basques, which have high frequencies of the *r* (rhesus) negative allele. More recent authors do not regard the Basques distinct enough to be treated as a race equivalent to the others. Traditional anthropologists opposed Boyd for a different reason: why classify men by their blood groups rather than by more easily visible traits? This was certainly a misconception. The advantage of the blood groups and enzyme polymorphisms is their genetic simplicity (most of them are coded by single genes), the feasibility of determining which gene alleles any individual carries, and therefore the possibility of describing populations in terms of the frequencies of certain genes. The gene-frequency method is a new paradigm, a new model for racial studies. It is intended to clarify and supplement rather than replace the description of all kinds of genetic differences in human populations.

Figure 5-4 shows the frequencies in human populations of the allele I^B. This allele is present in individuals with B and AB blood groups. The gene frequency gradients seen in Figure 5-4 are, on a world-wide scale, similar to the chromosome frequency gradients shown in Figure 5-2, on a geographically more restricted scale in *Drosophila* flies. The I^B allele is most common in Mongolia, central Asia, northern India, and in some aboriginal peoples of Siberia. Its frequency declines as one travels from Mongolia westward, eastward, and southeastward. It is altogether absent among American Indians and aboriginal Australians unmixed with European invaders. With these exceptions, it is present in most human populations, albeit in different frequencies. The differences between populations are quantitative rather than qualitative.

Frequency gradients are observed also with other alleles and other blood group genes. We cannot consider them here one by one; instead, Table 5-2 gives a summary for five major racial groups and for seven blood group genes. Not only the major races but also local populations within a race often differ in gene frequencies. For this reason, where many population samples within a racial group have been examined, Table 5-2 gives frequency ranges. Four alleles

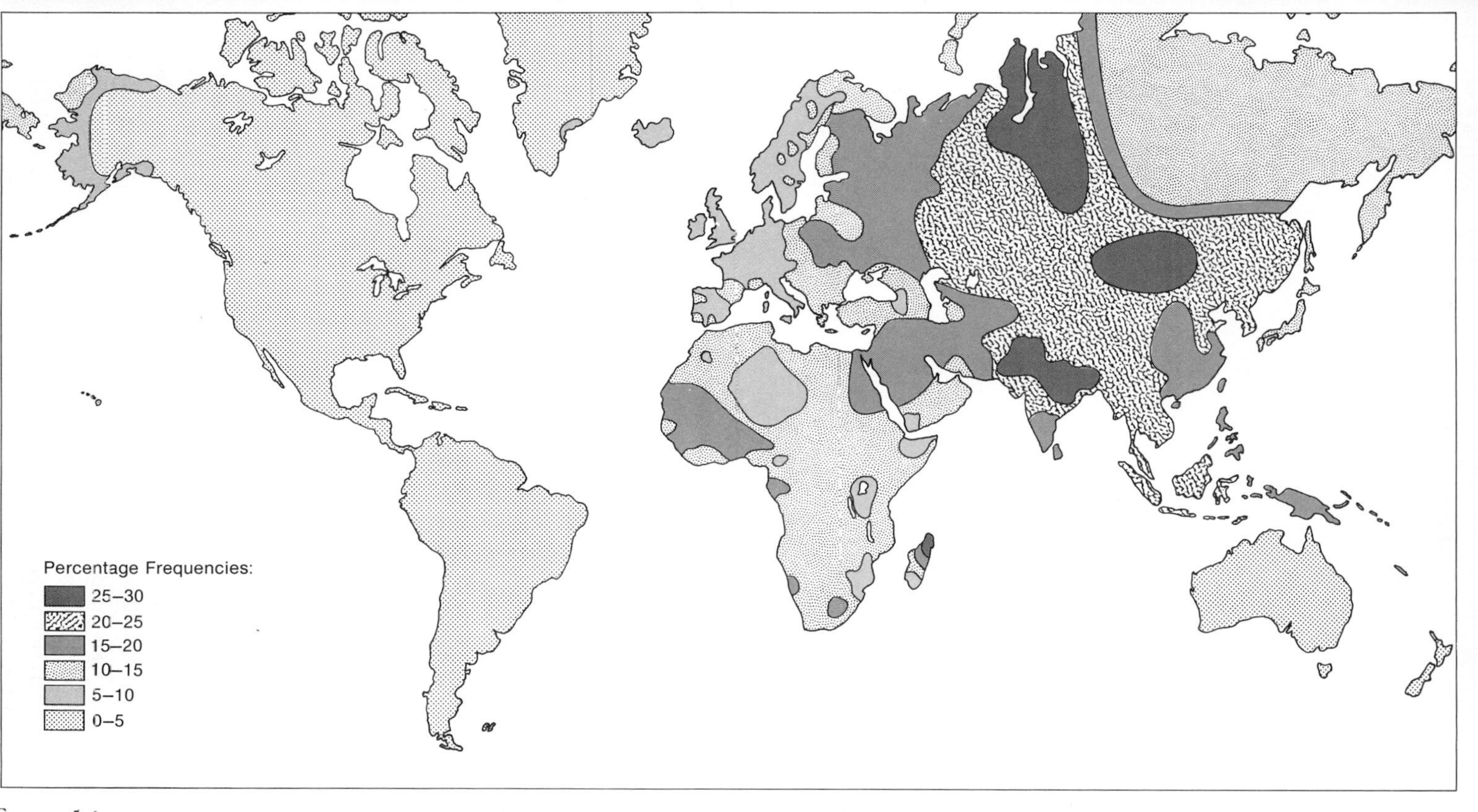

FIGURE 5-4.
Frequencies of the blood group gene I^B in aboriginal populations of the world. The I^B gene is most common in Mongolia, Central Asia, northern India, and in some aboriginal peoples of Siberia. Its frequency decreases west, east, and southeast from Mongolia. The gene is altogether absent among American Indians and aboriginal Australians unmixed with European invaders.

TABLE 5-2. Frequencies, in percent, of various blood group alleles in five racial groups of mankind. [After Stern, 1973.]

	Caucasoids	Negroids	Mongoloids	Amerindians	Australoids
I^{A1}	20–30%	10–20%	15–25%	0–55%	20–45%
I^{A2}	4–8	5	0	0	0
I^{B}	5–20	10–20	15–30	0	0
L^{MS}	20–30	7–20	4	15–30	0
L^{Ms}	30	30–50	56	50–70	26
L^{NS}	5–10	2–12	1	2–6	0
L^{Ns}	30–40	30–50	38	5–20	74
r	30–40	10–20	0–7	0	0
r'	0–2	0–6	0	0–17	13
r''	0–2	0–1	0–3	0–3	0
R^0	1–5	40–70	0–5	0–30	9
R^1	30–50	5–15	60–76	30–45	56
R^2	10–15	6–20	20–30	30–60	20
P	40–60	50–80	15–20	20–60	?
Lu^a	2–5	0–4	?	0–10	0
Fy^a	40	< 10	90	0–90	?
Di^a	< 1	0	1–12	0–25	0

are distinguished in the "classical" blood groups: I^{A1} or I^{A2} are present in persons with A bloods, I^B in those with B bloods, and both I^A and I^B in those with AB bloods (which are always heterozygotes). The fourth allele is the recessive i, for which persons with O bloods are homozygous (the frequencies of i are not shown in the table). American Indians have mostly O bloods (ii homozygotes). However, the tribe of Blackfeet in northcentral United States and adjacent Canada has the world's highest frequency of A bloods (55 percent of the allele I^{A1}). O bloods are also prevalent among Australian aboriginals and western Europeans. The distribution of I^B has already been discussed (Figure 5-4). Four alleles of the L gene yield the so-called M–N blood groups. The L^{MS} allele is fairly common among Caucasoids, Negroids, and Amerindians, but rare among Mongoloids and Australoids. The L^{Ns} allele is predominant among Australoids, fairly common among Caucasoids, Negroids, and Mongoloids, but rather rare among Amerindians.

The "rhesus" gene is complex but interesting. As many as 30 alleles of this gene are known; a majority of them are very rare, and only six are listed in Table 5-2, namely r, r', r'', R^0, R^1 and R^2. The r small, or rhesus-negative alleles, may cause a hemolytic disease of newborns or fetuses, when a mother homozygous for rr is pregnant with a fetus carrying R. The highest frequencies of r are found among the Basques (see above) and other Europeans, while this allele is rare or absent among Mongoloids, American Indians, and Australoids. R^0 is primarily an African gene, but notice that it occurs occasionally in all other races. Does this mean that the African admixture, owing to ancient or recent hybridization, is world-wide? Not necessarily; polymorphisms involving this allele may have always been species-wide. Likewise, R^1 is found most often in Asia, but also elsewhere throughout the world.

The lower part of Table 5-2 summarizes the data on frequencies of some alleles of the *P*, Lutheran (*Lu*), Duffy (*Fy*), and Diego (*Di*) genes. Most interesting is the Fy^a allele, which is prevalent among Mongoloids and some Amerindians, less so among Caucasoids, and least among the Africans. Di^a is also rather frequent among Mongoloids and Amerindians, and almost absent elsewhere. These facts are consistent with the generally accepted theory that American Indians are descendants of the Mongoloid immigrants who crossed the Bering Strait perhaps 20,000 years ago. The prevalence of the O blood type (*ii*), the absence of B (the I^B allele), and the rarity of A bloods in most unmixed American Indians do not agree well with the theory, but neither do these facts necessarily contradict the theory. One possible explanation is that the number of immigrant individuals who were the ancestors of the Amerindians was very small, and by chance did not include persons with B or A bloods. Another possibility is that there were some ancient Mongoloid peoples in northeastern Asia among whom A and B bloods were rare. Still another possibility is that B and A bloods were for some reason eliminated by natural selection among the founders of the Amerindian peoples.

Of course, many traits that differentiate human populations are genetically more complex than blood-group differences. The most externally conspicuous difference is that between black Africans and light-skinned Europeans. An indication of how incomplete is our present knowledge of human genetics is that the genetic basis of this difference is far from established. Various researchers have postulated anywhere from two to a dozen or more genes. Stern (1973) gives a model of five genes, each with two alleles, one of which produces the melanin pigment while the other does not. Let a dark-skinned African be $A_1A_1B_1B_1C_1C_1D_1D_1E_1E_1$ and a light (but not albino) European $A_2A_2B_2B_2C_2C_2D_2D_2E_2E_2$; they have 10 and 0 pigment genes, respectively. A first generation hybrid (mulatto) has five pigment genes. $A_1A_2B_1B_2C_1C_2D_1D_2E_1E_2$; recombination in the second and further generations may produce 3^5 or 243 genotypes, with numbers of pigment genes ranging from 0 to 10. Genotypes with intermediate numbers of pigment genes, say 4 to 6, and consequently with intermediate pigmentations, will be most common. Those with one or two, or with nine or 10 pigment genes, and consequently with very light or very dark pigmentation, will be rare. Note also that individuals with similar pigmentation phenotypes may have the same number of pigment genes but at different pigment loci, say $A_1A_1B_1B_1C_1C_2D_2D_2E_2E_2$ and $A_2A_2B_2B_2C_1C_2D_1D_1E_1E_1$. Similar phenotypes may be achieved on the basis of different genotypes.

Even if the above model was securely established, which it is not, one would still find it enormously difficult to analyze the racial variation in pigmentation in terms of geographic gradients of gene frequencies, as is being done with blood groups. Skin pigmentation is appreciably variable among "pure" blacks and "pure" whites; part of the variation is environmental (sun tanning) and part genetic. This proves no more than that no race is "pure." Yet it is an open question whether individual pigment variations are due to a scattering within populations of the same alleles that when combined produce the striking inter-populational differences. Perhaps alleles of the same gene loci but with intermediate potencies are involved. What of the racial groups intermediate between the darkest Africans and the lightest Europeans? Are the same or different pigment genes involved? Suppose one could in a series of generations select and marry the darkest individuals among the whites, or the lightest individuals among the African blacks; would one obtain a population of European descent with pigmentations as dark as in Africa, and vice versa?

Racial variation in skin pigmentation may at first sight appear qualitative, and that in blood groups predominantly quantitative. After all, indigenous populations of tropical Africa contain no white-skinned individuals (except albinos), nor do indigenous populations of Europe produce a sprinkling of blacks. Yet this is merely a consequence of several genes with additive effects being involved in pigmentation, whereas the various alleles of the blood group genes are usually identifiable one by one in each individual tested. One can "enumerate" blood types and enzyme alleles in a population, whereas one can only estimate the mean and variance of pigmentation (for a discussion of this difference see Lewontin, 1974). Classical anthropologists who engaged in race studies and classifications dealt mainly with characteristics of the second kind, i.e., polygenic ones. Skin color, stature, cephalic index, hair form, body build, etc., are all determined by several, but not precisely established, numbers of genes. In addition, most of these traits are subject to considerable environmental modification. One may hope that genetic analysis will eventually lead to a resolution of polygenic complexes into their gene components. Meanwhile, the gene frequency approach helps to clarify the nature of racial differences, rather than establish definite racial classifications.

RACE CLASSIFICATION

An issue frequently debated is whether races or subspecies are "real" or arbitrary subdivisions of species, invented by classifiers for their convenience. Understandably, the debate is most animated when the subject focuses on human races. Racists are eager to exploit race classifications for their own purposes; they would have individuals treated according to their racial origins rather than their particular qualities. Some opponents of racism propose that racism be countered with the argument that mankind has no races, only ethnic groups—as if ethnic prejudice were not as insidious as race prejudice. Is it not more sensible to have people understand the scientific concept of race?

One should distinguish two aspects of the problem of "reality" of race—race as a category of classification and racial differences as data of observation.

We have seen that populations of *Drosophila pseudoobscura* in parts of the United States and Mexico (Table 5-1) and even at high and low elevations of the Sierra Nevada range (Figure 5-1) differ in the relative frequencies of the gene arrangements in their third chromosomes. These are racial differences. We have also seen that human populations of different countries have diverse frequencies of certain blood groups (Figure 5-4, Table 5-2). These are likewise racial differences. There is nothing arbitrary or unreal about these differences. However, *classifying* races or subspecies is an altogether separate issue. One could affix racial names, in English or Latin, to populations of *Drosophila pseudoobscura* that inhabit the Pacific Coast of the United States, the Rocky Mountains, and northern and central Mexico. But this would hardly serve a useful purpose; saying that a given population is of such-and-such geographic origin is good enough. To most students of human racial variation racial labels seem convenient, but they are not usually put in Latin. However, Latin subspecific names are preferred by many investigators working with mammals, birds, butterflies, and other animals.

The number of names that one wishes to have is again a matter largely of convenience. In Table 5-2 we have listed five racial groups. An anthropologist working on populations of, say, Caucasoids may wish to subdivide them into several smaller entities—Nordics, Alpines, Mediterraneans, Armenoids, etc. One convenient criterion is whether or not the populations composing a race are reasonably uniform among themselves but distinct from other races. As pointed out above, if the geographic gradients of the gene and phenotype frequencies were all uniform, any subdivision would be arbitrary and probably inconvenient. In reality, the gradients are often steeper in some places than in others. For example, people look rather different north and south of the Sahara Desert in Africa, although in the valley of the Nile all intermediates occur. The same is true north and south of the Himalaya Mountains in Asia. This is not a matter of chance. Natural environments are dissimilar north and south of the Sahara, and north and south of the Himalayas. Furthermore, travel across the Sahara and the Himalayas was difficult, at least until the relatively recent invention of modern aircraft. Environmental disparities select different genes, and impediments to travel reduce the gene diffusion between populations. Regions in which the gene frequency gradients are steep or interrupted facilitate drawing race boundaries, even though such boundaries are rarely perfectly sharp.

A race or subspecies classifier often faces a dilemma. If only a few subdivisions are recognized there will be too much diversity within some of them. This is the case with the five racial groups of man (Table 5-2). There is certainly a great variety of genetically distinct populations among Caucasoids, Negroids, and Mongoloids, as well as among Amerindians and Australoids. On the other hand, with finer splitting the differences are less and less striking, and the boundaries more and more often blurred. Garn (1961) has proposed a subdivision of the human species into nine "geographical races" or 34 "local races" (Table 5-3).

TABLE 5-3. The nine "geographical races" and 34 "local races" proposed by Garn (1961).

Geographical races		
Amerindian	Melanesian-Papuan	Indian
Polynesian	Australian	European
Micronesian	Asiatic	African
Local races		
1. Northwest European	13. Lapp	25. Negrito
2. Northeast European	14. North American Indian	26. Melanesian-Papuan
3. Alpine	15. Central American Indian	27. Murrayian
4. Mediterranean	16. South American Indian	28. Carpentarian
5. Hindu	17. Fuegian	29. Micronesian
6. Turkic	18. East African	30. Polynesian
7. Tibetan	19. Sudanese	31. Neo-Hawaiian
8. North Chinese	20. Forest Negro	32. Ladino
9. Classic Mongoloid	21. Bantu	33. North American Colored
10. Eskimo	22. Bushman and Hottentot	34. South African Colored
11. Southeast Asiatic	23. African Pigmy	
12. Ainu	24. Dravidian	

RACE DIFFERENCES AND NATURAL SELECTION

The biological functions of polytypism and polymorphism are most easily understood as results of natural selection. Race differences are, then, adaptations to diverse environments in different parts of the distribution area of the species; polymorphisms are adaptations to environmental heterogeneity in a territory inhabited by a population. If the environment in which a species lives were uniform, race formation would be unnecessary. The hypothesis just stated does not require that any and all racial traits must have a role to play in the adaptedness of their possessors. In the first place, some traits may have been selected in environments that no longer prevail. Such traits may be regarded as functionless evolutionary relics. This is particularly plausible for racial traits of mankind; surely, our present environments are radically different from those in which our early ancestors lived. Second, some evolutionary changes may be neither favorable nor unfavorable to their possessors; they may be neutral, i.e., adaptively irrelevant. The adaptedness of racial traits must be proven by evidence; it cannot be taken for granted.

Some examples of racial traits that are adaptive have been given in the present and in preceding chapters. As could be expected, most of them are adaptations to environmental changes deliberately or inadvertently created by man. Classical examples are "industrial melanisms" in many species of moths. Populations with a high incidence of darkly pigmented individuals have evolved in regions where the vegetation is polluted by industrial wastes. Parallel situations in plants are the microgeographic races growing on the tailings of mineworks where the soils are polluted with salts of heavy metals. Insect populations resistant to insecticides have evolved in many species of pests, and resistance has been induced experimentally in laboratory populations. Rats resistant to Warfarin is an analogous case in a species of mammals. The wild oat, *Avena barbata*, is a weed introduced in California by man, where it has evolved microgeographic races differentially adapted to various more or less "natural" environments. *Drosophila pseudoobscura* is a species rarely found in man-modified habitats. Its races, which differ in the incidence of chromosomal variants, are doubtless adapted to different climatic and biotic factors. However, we cannot specify which factors are responsible for the differentiation.

Adaptive significance of geographically varying traits can often be inferred, though seldom demonstrated beyond doubt, if several or many species show parallel genetic modifications when they come to live in environments that are similar in some respects. Several "ecogeographical rules" were arrived at by zoological taxonomists in the nineteenth century, and codified in the present century, particularly by Rensch (for reviews see Mayr, 1963, 1970). Thus, races of many species from colder climates tend to be larger than races of the same species from warmer climates (Bergmann's rule). Among warm-blooded animals, protruding body parts (ears, tails, and legs) tend to be shorter relative to body size in races that live in cold climates than in those living in warm climates (Allen's rule). Both rules are connected presumably with heat conservation or heat dissipation. The volume, or weight, of the body increases as the cube, and the body surface as only the square of the linear dimensions. Other things being equal, large bodies but relatively short protruding parts are favorable in the cold, whereas smaller and more slender, linearly built bodies are advantageous in the heat.

Gloger's rule states that races of warm-blooded vertebrate species are more darkly pigmented in warm and humid areas than in cool and dry ones. Among some insects a variant of Gloger's rule applies: dark races are found in humid as well as cold areas, and light ones in warm as well as dry areas. The physiological basis of Gloger's rule is conjectural. It may have to do with temperature regulation and protective colorations. It should be noted, however, that the above "rules," although they hold in many species, are not universal. (Indeed, exceptions are not unexpected in any but the most naively simplistic theory.) Environments of different countries usually differ in many more variables than just temperature and humidity. Races of more or less widespread species have to become adapted to many biotic as well as climatic factors.

In 1922 the Swedish botanist Turesson proposed the term "ecotype" for local populations of plant species "arising as a result of the genotypical response" to particular kinds of habitats. He found that several species in Sweden have ecotypes corresponding to sand dunes, sea cliffs, forest shade, swamp, etc. These habitats recur mosaic-like in the same country, and so do the corresponding ecotypes. It is an open question whether the same ecotype arises independently in many places or spreads from one place to suitable locations elsewhere. Some ecological geneticists have used the term "ecotype" for what could otherwise be called races or subspecies. Thus, in the classical work of Clausen, Keck, and Hiesey (1940 and later) alpine, mid-elevation, and sea-level ecotypes or races of *Potentilla glandulosa* were found to differ in external appearance as well as climatic requirements. Individual plants from each type of environment were cut in three parts, and the cuttings were replanted in experimental gardens at sea level, mid-altitude, and alpine zones of the Sierra Nevada mountains of California. In many instances the transplants died or grew poorly in foreign habitats. However, in some instances the transplants, freed from competition with other vegetation, became even larger than in their native habitats. The evidence as a whole clearly shows that the altitudinal races are not only different in appearance but are adaptively fittest in their native habitats (Figure 5-5).

What physiological and ecological roles do racial traits in man play in the present environments, and what roles have they played in environments of the past? Far from having been studied in detail, reliable information about these roles is surprisingly scant. Consider a trait that catches the eye—skin color. As pointed out above, Buffon in the eighteenth century surmised that dark skin is adaptive in hot and light in cool climates. This notion is based on the well-known fact that dark-skinned peoples live (or lived) mainly in the tropics, and light-skinned in temperate and cold climates. There are, to be sure, exceptions to this rule. Indians in the American tropics are not particularly dark. The explanation that these people have not lived long enough in their tropical habitats to become more heavily pigmented is rather unconvincing, but no better explanation is available.

The difference between "black" and "white" skin is due to additive effects of several, but not precisely determined, number of genes. These genes determine the quantity of the pigment formed in the pigment cells, rather than the number of these cells. The amount of the pigment depends also on the exposure of the skin to sunlight, or more precisely to a part of the ultraviolet spectrum. The fact that sun-tanning protects from sunburns led to the hypothesis that dark pigmentation was selected as a safety device in lands with abundant sunshine. Skin cancers arise more frequently in light-skinned than in dark-skinned people, who are constantly exposed to sunlight. The selective role of this is probably minimal, not only because skin cancers are on the whole rare, but also because they appear mainly in postreproductive ages.

More plausible is the hypothesis that the genes for light pigmentations are selected in countries with scarce sunshine because the synthesis of vitamin D (the anti-rickets vitamin) is facilitated by lack of the pigment. By contrast, dark pigmentation prevents excessive amounts of this vitamin being synthesized (Williams, 1973). Still another suggestion is that dark skin colors serve as camouflage; dark-skinned naked hunters were less conspicuous than light-skinned ones to the game they stalked. In sum, it is probable that natural selection promoted dark pigmentation in some places and light pigmentation in others. Precisely which factors were effective in particular situations is however quite inadequately understood.

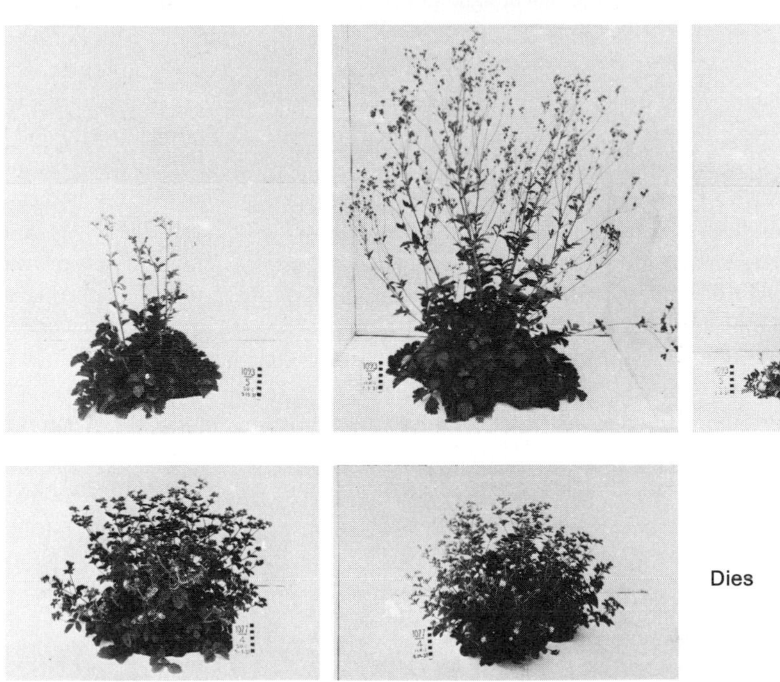

FIGURE 5-5.
Plants of three altitudinal races of *Potentilla glandulosa* in different environments. The differences among the plants in the columns are genotypic, due to selection and adaptation of their ancestors to different altitudes. The differences among the plants in the rows represent responses of the same genotype to different environments. [Original photographs courtesy of Dr. M. A. Nobs.]

Even less is known for certain about the adaptive significance of eye and hair colors and hair forms. The pigmentation of hair and eyes is generally correlated with skin color. The correlation is however not perfect. Persons with brown eyes and light skin, or blue eyes and skin of intermediate tints, are not rare. Eyes with a heavily pigmented iris are alleged to be less sensitive to sun glare than are blue eyes. This is possibly true but not well established. There is not even a plausible hypothesis about the selective factors responsible for differences in hair form. Yet in different races, and also among individuals of the same population, hair form varies greatly. The extremes as well as all intermediates exist between straight and coarse, straight and soft, wavy, curly, wooly, and "peppercorn" (this last name refers to spiral tufts of hair found particularly in Bushmen and Hottentots in Southern Africa).

Somewhat better understood are variations in stature and body shape. Tall people are found mainly in cold and temperate countries, while inhabitants of warm and hot countries tend to be short. This is in conformity with Bergmann's rule (see above). One of the striking exceptions are Watusi and related people, living on the upper Nile not far north of the Equator, who are among the tallest people (mean 176 cm) on earth. To be sure, African Pigmies (mean 142 cm) live not very far from the Watusi, though their mode of life is quite dissimilar. Allen's rule also applies to some human populations: linear, lanky bodies with long extremities are found among inhabitants of hot countries, heavy set chubby ones among inhabitants of cold countries. Figure 5-6 shows the correlation of body weight and mean annual environmental temperature. The correlation is significantly but not strongly negative. As could be anticipated, body weight, the question of nutrition aside, is positively correlated with stature and globose body shapes. Roberts (1953) gives a formula $W = 0.07\,S - 0.199\,T - 48.1$, where W is weight in kilograms, S stature in millimeters, and T the mean annual temperature in degrees Fahrenheit.

Racial climatic adaptedness was probably more important in the past than it is now, since people now protect themselves from extreme temperatures by wearing heavy, light, or no clothing, and live in protective dwellings. But it would be a mistake to think that genetic adaptedness no longer plays a role at all. Frostbite and sunstroke remain hazards. Though the body may be warmly covered, the face may still be exposed to cold and to wind. A slim and slender body build facilitates the dissipation of heat, which is advantageous in hot countries. Conservation of heat is aided by stocky and squat physique, which is adaptive in cold climates. Fat-padding and the facial features of the inhabitants of arctic countries, especially Mongolians, aboriginal Siberians, and Eskimos, have been interpreted as climatic "engineering" by natural selection. Similarly, the shape of the nose (nasal index = breadth : length of the nose) is correlated with frequently encountered humidity and temperature of the inspired air. A correlation of + 0.82 is found between mean absolute humidity and the nasal index.

A fascinating example of an interaction of genetic and cultural variables is that of milk drinking habits with respect to the presence or deficiency of the enzyme lactase. Strangely enough, the situation has only rather recently begun to be studied, and its genetic basis has not yet been fully clarified (for a review see McCracken, 1971). Lactose is the principal carbohydrate present in milk, and lactase splits the lactose molecule into glucose and galactose. These latter are absorbed in the small intestines, whereas lactose is not. Lactase is present in the great majority of infants; its absence, apparently due to a rare recessive,

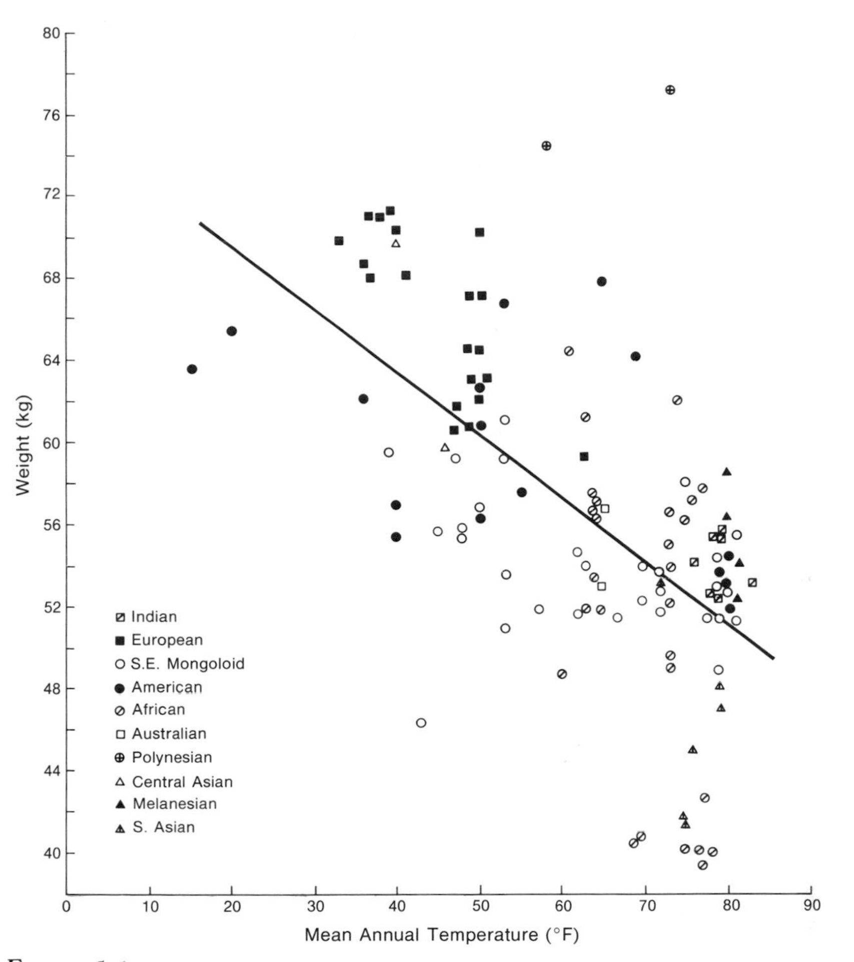

FIGURE 5-6.
The negative correlation between body weight and mean annual temperature of habitat in human populations. Although great variation exists, people living in colder climates tend to be heavier (and taller) in conformity with Bergmann's rule. [After Roberts, 1953.]

is usually lethal. But while in some persons lactase is present in adults as well as in children of all ages, in other persons it disappears, probably gradually, in late childhood. A probable, though not securely established genetic explanation is that the retention of lactase is due to a dominant gene and its disappearance to homozygosis for a recessive.

Dairy animals—goats, cattle, sheep, horses—were domesticated in western Asia in the sixth millennium B.C. or even earlier. That utilization of milk was an early development has been supported by the discovery at Ur in Iraq and in Egypt of ancient drawings of cows being milked. Domestication of animals for food and transportation spread through most of the Old World, and eventually to the New World and Australia. The role of milk and milk products in the diets of adult people is however far from uniform. It is very important in some but almost nil in other cultures. The adaptive value of the gene or genes that enable lactase to be maintained in adults is likewise varied. Natural selection enhanced the frequencies of these genes among habitual milk consumers, but little or not at all where milk is seldom utilized. Table 5-4 and Figure 5-7 give samples of the data.

TABLE 5-4. Lactose intolerance in different peoples. [After Gottesman and Heston, 1972.]

	Individuals examined	Percent intolerant
Europeans in Australia	160	4%
American Caucasians	245	12
Finns	134	18
Australian Aborigines	44	85
African Bantu	59	89
Chinese	71	93
American Blacks	20	95
Thais	134	98
American Indians	24	100

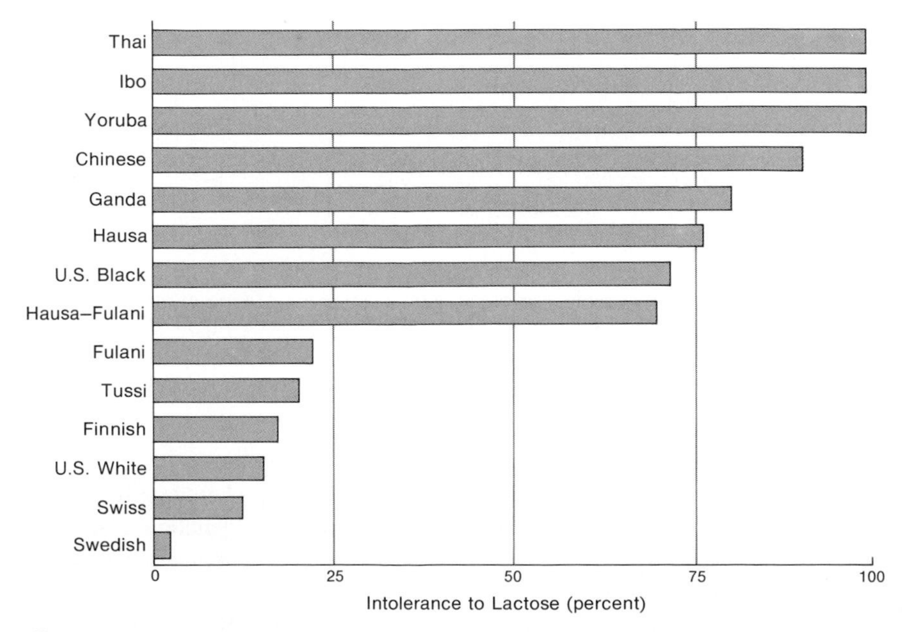

FIGURE 5-7.
Intolerance to lactose in various human populations. The proportion of individuals tolerant of this milk sugar is highest in peoples that have used milk and milk products in their diets for millennia. Populations that traditionally do not use milk as food are nearly 100 percent intolerant to lactose. [From Kretchmer, 1972.]

While the presence of lactase is indispensable for life in infants, one may conjecture that its retention past weaning age became adaptively valuable only after the domestication of animals and the use of milk. Table 5-4 shows that the lactose tolerance is common in cultures that habitually use milk in their diets, and that intolerance is common in cultures using milk and milk products rarely or not at all. Gottesman and Heston (1972) calculated that in a population with only a 0.01 selective advantage of lactose-tolerant phenotypes, the difference in the incidence of tolerant and intolerant phenotypes observed in populations of milk users could be produced in about 400 human generations. If milk consumption is about 6,000 years old, a somewhat larger selection coefficient is required.

ADAPTIVELY NEUTRAL TRAITS

Many examples of different forms of natural selection have been given in this and the foregoing chapters. There is no doubt that numerous polymorphisms within, and racial differences between, populations have originated and are being maintained by natural selection. However, selection may not be responsible for all polymorphisms and polytypisms. Many polymorphic and racial traits seem to be without adaptive significance; that is, they are neither useful nor harmful, but ostensibly neutral. At the same time, a trait may only appear to be neutral because it has not been studied in sufficient detail. Small differences in fitness can escape detection unless large-scale, often laborious and expensive, experiments are undertaken to reveal them. Nevertheless, some authors contend that neutral traits do in fact exist, and with high frequencies. Some believe that a majority of the differences between individuals, populations, species, and higher categories are neutral traits. This view is a radical deviation from the theory that evolutionary changes are governed by natural selection. What is involved here is nothing less than one of the major unresolved problems of evolutionary biology.

Perhaps the simplest example of presumably neutral genetic changes are those stemming from the "degeneracy" of the genetic code (see Chapter 2). Some amino acids are coded by two to six nucleotide triplets. Mutations are most frequently substitutions of single nucleotides in DNA. Suppose that a mutation transforms one triplet into another coding for the same amino acid. The protein that the mutated gene produces will be the same as that coded by the original, unmutated gene. It will have the same sequence of the same amino acids. The mutation, although it has altered a segment of the DNA chain of a gene, would seem to be of no consequence to the organism. Yet even here the neutrality of the genetic change is not absolutely certain. Transfer RNA's needed to translate the nucleotide sequences into amino acid sequences are specific for each nucleotide triplet. It is conceivable, though not proven, that the RNA's for synonymous triplets are not equally available and abundant.

Seemingly more persuasive arguments for neutrality of many genetic variants come from consideration of the mechanisms that maintain genetic burdens and polymorphisms in natural populations. We have seen in Chapter 4 that one source of genetic burdens is the occurrence of deleterious mutations. The input of deleterious mutations into the gene pool is opposed by normalizing natural selection, which works to eliminate such mutations. The opposing forces lead eventually to an equilibrium, when the numbers of deleterious mutations entering the gene pool per generation are equalled by the numbers of eliminations. Some geneticists have called the elimination of a deleterious variant "genetic death." Users of this term should bear in mind that a "genetic death" does not necessarily mean that an individual dies. The term applies to lethal genes that cause abortion or deaths after birth, as well as to genes producing viable but sterile individuals, and to those that reduce the number of offspring. Note that lethal mutations cause as many eliminations or genetic deaths as mutations that are only mildly subvital. In either case the number of eliminations must at equilibrium equal the number of mutations. This is because drastic deleterious mutations persist in the population on the average for fewer generations than mildly deleterious ones. The result is that, in a population at equilibrium, the number of "genetic deaths" is approximately the sum of the mutation rates of all genes yielding deleterious genetic variants. (Some eliminations may of course involve individuals with two or more deleterious mutants.)

The key difficulties are obviously to determine the sum of mutation rates and the percentage of mutations that are deleterious. Recent studies have shown that mutations are more frequent than classical geneticists dared to assume. Mukai (1969) found that about 15 percent of the second chromosomes of *Drosophila melanogaster* acquire new mutants in every generation (see Chapter 3). Mukai's technique detected only mutations that caused decreases of the viability of the flies, at least when homozygous. Neutral mutations, or mutations changing components of fitness other than viability (for example, fecundity or development rate), were not detected. Most mutations were individually small in their effects, but the aggregate effect was easily detectable. The second chromosome contains about two fifths of the total chromosomal material. Assuming that the genes in the other chromosomes are as mutable as those in the second, easily 30 percent of the sex cells produced by a "normal" *Drosophila* transport a new deleterious mutation. Roughly the same figure was arrived at by Muller in studies of man, on the basis of much less precise inferential data and considerations. If each mutation has to result in a "genetic death," where and how is this stunning mass of defective heredity eliminated?

Some mutations may be useful in heterozygous conditions, or in some environments a population encounters. Many mutations are of course unconditionally harmful, when homozygous as well as heterozygous, and seemingly in all possible environments. Yet the work of Wallace (1958 and later) showed that the viability effects of some mutations are modulated by the genetic constitutions of their carriers. The net effect of mutations induced in *Drosophila* by small doses of gamma ray treatments is a loss of viability in homozygotes. However, *under certain conditions,* the same mutations increase viability when heterozygous. If one irradiates flies made homozygous for their chromosomes (by the technique illustrated in Figure 2-6), the average viability of the progeny heterozygous for the induced mutations increases. No such increase is observed in the progeny of irradiated flies from an ordinary genetically variable population. The evidence suggests that, in contrast to complete homozygosis, a certain amount of heterozygosis confers hybrid vigor. But above a certain level, additional heterozygosis no longer accentuates the vigor. The physiological mechanisms underlying this phenomenon are at present only conjectural.

Even if only some new mutations are accepted in a population, on account of their favorable heterozygous effects, the problem of genetic burdens is not solved. In fact, it is aggravated. The phenomenon of heterotic balancing selection has been discussed in Chapter 4. With a heterozygote being superior in Darwinian fitness to both homozygotes, natural selection establishes an equilibrium condition in which heterozygotes and homozygotes are maintained with predictable frequencies. This happens when homozygotes are either only slightly or greatly inferior, or even lethal. Hence, the maintenance of heterotic balanced polymorphisms imposes a persistent genetic load on the population. This is inescapable with sexual reproduction and outbreeding, because a population cannot consist entirely of heterozygotes. Relatively ill-adapted homozygotes will be produced in every generation. With random mating in a large population, the total loss of fitness (L) is easily estimated. If the selection coefficients that operate against the two homozygotes are s and t respectively, $L = st/(s + t)$.

Suppose that the homozygotes for the two alleles of a polymorphic gene are only slightly inferior in fitness to the heterozygote, for example $s = t = 0.02$ (a two percent disadvantage of the homozygotes). The fitness of a polymorphic population at equilibrium will then be 0.99, taking the fitness of the heterozygotes to be 1.00. The loss of one percent of fitness can probably be tolerated by populations of most organisms. However, how many such polymorphisms can a population afford to maintain? A seemingly reasonable simplifying assumption is that the gains or losses of fitness produced by polymorphisms of different genes or chromosomal variants are independent. If one polymorphism reduces the fitness by one percent, from 1.00 to 0.99, then n polymorphisms will reduce it to 0.99^n. With $n = 100$ polymorphic genes, the fitness of the population will be 0.99^{100}, or 0.366; about two-thirds of the population are sacrificed to genetic deaths. If the disadvantages of the homozygotes are greater, the loss of fitness is also greater. It becomes doubtful that any species, even one generating very large numbers of progeny, can withstand such losses.

No wonder that partisans of the classical theory of genetic population structure, of whom H. J. Muller was the leading figure, believed heterotic balanced polymorphisms to be rare and of little significance in evolution (for example, see Muller, 1950; and Chapter 2 here). Most individuals of a species or race were supposed to be homozygous at most of their gene loci. At a minority of the loci there were heterozygotes for deleterious alleles from mutation, which normalizing natural selection worked to eliminate or to keep at the lowest possible frequencies. Balanced polymorphisms could not be denied completely, but were believed to be rare and, in the long run, temporary expedients to be corrected by natural selection. A new allele arising by mutation should eventually be found that would be as fit in homozygous condition as are the heterozygotes under balanced polymorphism.

The growing information that showed polymorphisms to be widespread led to the formulation of the balance theory of population structure (Chapter 2). Individuals of sexual outbreeding species are heterozygous for a considerable proportion of their genes, and populations are polymorphic for a majority of the gene loci. Some of this variability is maintained by mutation pressure generating deleterious mutants opposed by natural selection. This much is acknowledged by both the classical and balance theories. But a greater part of the genetic variability is sustained by various forms of balancing selection, or else represents a residue of the genetic variability selected in the past and now neutral or nearly so. The heterotic balance is, of course, only one of several known forms of balancing natural selection. Diversifying, frequency-dependent, sexual, and other forms of selection (Chapter 4) may be even more important in populations that live in heterogeneous environments. The term "annidation" was proposed by Ludwig in connection with diversifying selection operating in environments that have a variety of ecological niches (see also Levins, 1968).

The continuing controversies between followers of the classical and balance theories have been most fruitful in that they have stimulated a great deal of research in evolution. In 1966 three groups of investigators opened a new phase of this research by their electrophoretic studies of allozyme polymorphisms in *Drosophila* and man (Chapter 2). These studies have been extended by many authors to many groups of organisms. A hitherto unsuspected amount of genetic variability has been revealed. Depending on the criteria applied, between 30 and 80 percent of the gene loci are polymorphic in populations. Minimum estimates of the proportions of genes heterozygous per individual are on the average 13.4 percent in species of invertebrates, and 6.0 percent in the vertebrate species studied (Table 2-9). The classical theory is clearly invalid. The problem of the maintenance of the seeming deluge of genetic variability in populations is however far from solved by this invalidation. It becomes an ever stronger challenge to evolutionists.

Several authors resolved to cut the Gordian knot. King and Jukes (1969), Crow and Kimura (1970), Kimura and Ohta (1971), Nei (1975), and others proposed that a majority of the mutational changes at the molecular level, and a majority of polymorphisms found in natural populations, are neither useful nor harmful to their possessors, but simply neutral. Natural selection neither promotes nor discriminates against neutral variants. Their fate in populations is determined by chance. For this reason, populations can without constraint accumulate the immense amounts of genetic variability, the discovery of which proved fatal to the classical theory. Theories that recognize natural selection as the only guiding process of evolution are sometimes labelled "panselectionist." Theories assuming neutrality of most genetic variants can then be called "panneutralist."

RANDOM GENETIC DRIFT

Panneutralism may or may not be right—this is an unsettled issue—but it is a theory that is self-consistent and reasonable. It has long been realized that gene frequencies can be changed both by deterministic (directed) and stochastic (random) processes. Wright (1955) listed recurrent mutation, recurrent migration, and selection as deterministic processes. The most interesting stochastic process is random genetic drift, due to accidents of sampling, especially in small populations. Fluctuations in mutation, migration, and selection rates, as well as "unique events," such as a novel favorable mutation, the development of a new population from a single or very few immigrants (founder effect), and unique selective incidents or unique hybridizations, are also stochastic phenomena. If the environment is constant, and one knows the parameters of the deterministic processes, such as mutation opposed by selection, one can, at least in principle, predict the gene frequencies that will be reached. With stochastic processes, mathematical models and predictions can also be advanced. In recent years a great number of studies on stochastic phenomena have been carried out by theoretical population geneticists and mathematicians. Experimental tests of the theoretical models lag far behind.

Hardy and Weinberg have shown that in the absence of mutation, migration, and selection, the gene frequencies in a population remain constant (Chapter 4). Complete constancy can however be expected only in ideal, infinitely large populations. In reality no population is infinitely large and some consist of small numbers of individuals. A new generation arises from a sample of sex cells taken from the gene (or gamete) pool of the preceding generation. In a large population the sample will contain very nearly the same gene frequencies that are present in the gene pool. In a small population the sampling errors may cause the gene frequency in the next generation to be higher, or lower, than in the preceding one. The smaller the population the more appreciable the sampling errors will be. Moreover, in a succession of generations the sampling errors are likely to accumulate. Suppose that we have many small populations, all of which have initially the same gene frequency, or that in one population the frequencies of many genes are initially uniform. In the next generation the distribution of the gene frequencies will have a certain statistical variance; the variance will grow in proportion to the number of generations elapsed, even though the mean gene frequency for all the populations (or for all the loci) considered together may be similar or identical to the original mean. This random genetic drift, or random walk, will eventually result in the formerly variable genes reaching fixation; one or the other of the alternative gene alleles or chromosomal variants initially present will reach the frequency of unity (fixation) or zero (loss).

It stands to reason that random drift can diversify an array of genetically initially identical but isolated populations (for example, on an archipelago of islands) without intervention of natural selection. The key variable is the effective population size, N. The effective size cannot be greater than the number of gametes that gave rise to the adult individuals composing a popu-

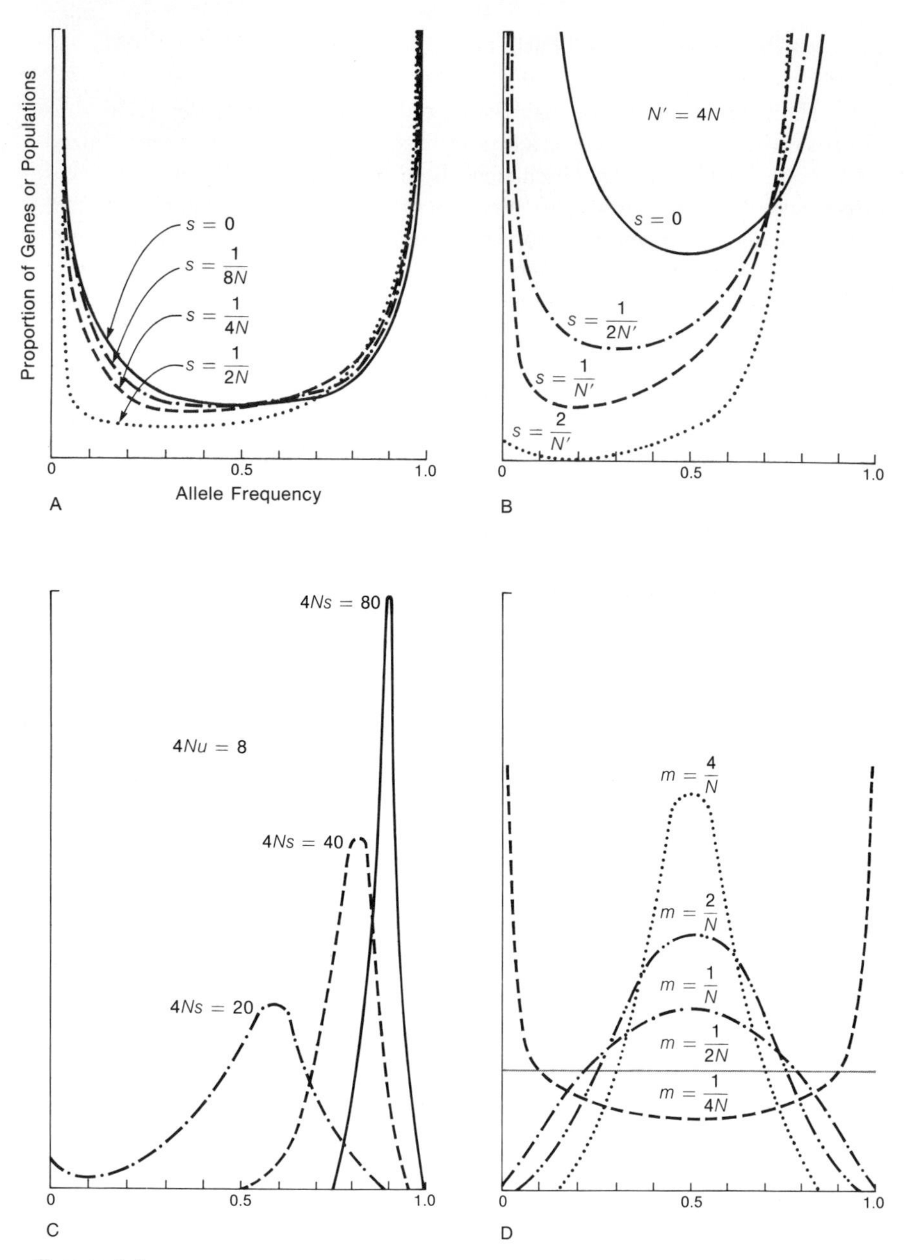

FIGURE 5-8.
Distribution of gene frequencies in populations of different sizes under different conditions of selection, mutation, and migration. The abscissa represents the frequency, q, of one of two alleles. The ordinate refers to different subpopulations or gene loci in a single population. **A**. Equilibrium distribution in small populations when the selection coefficients, s, are small and range in value from 0 to $1/2N$. **B**. Equilibrium distributions when the selection coefficients are the same in absolute value as in A but the populations are four times as large. Selection has greater effect in B than in A. **C**. Equilibrium distributions under simultaneous mutation and selection pressures. The mutation rate is

lation, but it can be much smaller. Of course, only individuals who produce surviving progeny count. If a population expands and contracts in different generations, N is closer to the minimum than to the maximum size. Variations in the number of surviving progeny of different parents reduce N below that of a population in which the survivorship is uniform. The drift is more rapid with small than with large N's, but given enough generations it will lead to a diversification of gene frequencies, and to eventual fixation of gene or chromosome variants. Kimura and Ohta give an interesting theoretical model. Suppose that in a population of N individuals no two gene alleles at a certain locus are identical, but all are equivalent with respect to the fitness of their carriers. Over a series of generations some of the initially present alleles will be lost, but others will be represented by several or many copies of the originally unique allele. After numerous generations one allele will be fixed and all others lost. For neutral alleles, the chances of fixation of any one of the $2N$ original alleles are equal. How many generations will be needed for fixation? The number is, on the average, $4N$. If N is one million, this means four million generations! Of course, related species or isolated races of the same species will become perceptibly different in gene frequencies much sooner. Whether or not this long-term drift explains the difference between species and higher groups in ostensibly neutral traits (particularly biochemical traits) will be considered in Chapter 9. Here we are interested in the possibility of adaptively neutral racial differences.

It should be stressed that random genetic drift does not always act in isolation from deterministic evolutionary forces. A genetic variant that is neutral with respect to fitness in one environment may be slightly or appreciably favorable or deleterious in other environments. This consideration is obviously important when gene frequency changes taking place during long successions of generations are surveyed. A model assuming subdivision of a species into completely isolated colonies is rarely realistic; the colonies exchange migrants regularly or occasionally. Wright's classical diagram of interactions of genetic drift with deterministic factors is reproduced in Figure 5-8. Without selection ($s = 0$) genetic drift results in a U-shaped distribution of the gene frequencies in different colonies, or at different, variable loci in a population. That is to say, most colonies reach fixation at allele frequencies $q = 0$ or $q = 1.0$; intermediate frequencies grow progressively rare. Graphs A and B in Figure 5-8 show that a U-shaped distribution is reached more rapidly in a small population (N) than in a larger population ($4N$). Natural selection that is very small (s ranging from

---

twice the reciprocal of the effective population size ($4Nu = 8$, or $u = 2/N$). The selection coefficients are the same in absolute value as in A, but the populations are 10 times larger. **D.** Equilibrium distributions with migration between subpopulations when the average frequency of the gene is $q = 1/2$. Small migration rates, such as $m = 4/N$ (or four migrant individuals per generation), overcome the effects of random drift so that no population becomes fixed. Virtually no population becomes fixed even when migration is as low as one migrant individual every two generations ($m = 1/N$), but many fixations take place if there is only one migrant individual every four generations ($m = 1/4N$). [After Wright, 1931.]

$1/8N$ to $2/N$) makes the gene frequency distributions skewed in the direction favored by the selection. At $4Ns = 80$ (or $s = 20/N$, still a very small value if N is in the thousands or larger) the loss of alleles favored by selection will be negligible (graph C in Figure 5-8). Gene diffusion from colony to colony easily overcomes the effects of the genetic drift (graph D in Figure 5-8). With gene flow between colonies of a species as rare as $m = 1/2N$ (one migrant every two generations) random fixation of genetic variants is rare, and with higher migration rates it becomes negligible.

ADAPTIVELY NEUTRAL RACE DIFFERENCES

Nothing is easier than to give examples of genetic differences between populations of a species for which no positive or negative utility is apparent. This may mean nothing more than that studies to discover such utility have never been done or were not commensurate with the difficulties encountered. Some variable traits in human populations have however been studied by many investigators, and the information accumulated is not negligible. The "classical" blood groups, O, A, B, and AB, are involved both in intrapopulational polymorphism and in interpopulational differences (see Figure 5-4 and Table 5-2). Individuals enjoying good health, or sickly individuals, may belong to any of these blood groups; a person in a polymorphic population cannot guess his blood group until his blood is tested. Several investigators claimed "incompatibility" in certain matings in which the parents differ in blood groups, particularly with O mothers and A or B fathers. Hemolytic disease of the newborns is observed in a small proportion of the progeny from such matings (as it is observed in a greater proportion of matings of rhesus-negative mothers with rhesus-positive fathers, see above). Curiously enough, O–A and O–B incompatibilities *decrease* the likelihood of the hemolytic disease in matings incompatible for the rhesus alleles. Some investigators also found "pregnancy wastage" in OAB incompatible matings, but other investigators failed to confirm this (for a review see Cavalli-Sforza and Bodmer, 1971).

Much better established are differences in the relative incidence of certain diseases among carriers of different blood groups. The following figures for the ratio of incidence of some diseases in carriers of certain blood groups are taken from the above mentioned review by Cavalli-Sforza and Bodmer:

Duodenal ulcer	O : (A + B + AB)	1.4
Stomach ulcer	O : (A + B + AB)	1.8
Stomach cancer	A : O	1.25
Cancer of pancreas	A : (O + B)	1.27
Pernicious anemia	A : O	1.5

Whether these associations between blood groups and disease, alone or together with the blood group incompatibilities and pregnancy wastage, can explain either the maintenance of the blood group polymorphisms or the racial differences in the incidence of the blood groups is most doubtful. The diseases listed are rather rare or occur mostly in people of postreproductive age. There is no evidence whatever that the Amerindian tribes having only O and those having many group A individuals differ in the incidence of these diseases. Nor is such a correlation indicated for the peoples of the Old World. Some researchers claimed that the geographic variation of blood group frequencies is related to the prevalence, in historical or prehistorical times, of infectious diseases—plague, smallpox, and syphilis. Such relationships could be powerful selective factors, but the claim has so far not been substantiated.

The burgeoning field of allozyme studies has disclosed an enormous, hitherto unsuspected abundance of polymorphisms and species differences (see Chapters 2, 4, 6, and 9). It was this unexpected abundance of polymorphisms that forced the former adherents of the classical model of population structure to switch to panneutralism. What part of the allozyme differences are controlled by natural selection? Most mutations are presumed to be due to substitution of single amino acids in protein molecules. Such substitutions in certain parts of the molecule change the functional properties of the protein, and may even be lethal. However, other substitutions seem not to change the enzyme function, at least in vitro. Far more work is needed to discover the selective forces, if any, that control the allozyme variants in populations. Among the scanty data available, one can mention the findings of Powell (1971) and McDonald and Ayala (1974), who showed that more allozyme polymorphisms can be maintained in laboratory populations of *Drosophila willistoni* and *Drosophila pseudoobscura* living in diversified than in those living in uniform environments. Koehn and Rasmussen (1967) found wide variations in the frequencies of certain esterase alleles in the fish *Catastomus clarkii* correlated with the temperature of the habitat. The correlation found in natural populations was confirmed in laboratory experiments. In another fish, *Anoplorchus*, Johnson (1971) observed habitat-correlated variations in lactate dehydrogenase alleles among populations of Puget Sound in the northwestern United States.

In some widely distributed species the frequencies of gene alleles responsible for allozyme, serum, and blood group differences show wide geographic variations. Mankind is an outstanding example of such geographic variation. So are the mice *Mus musculus* and *Peromyscus polionotus* (Selander *et al.*, 1969, 1971). Selectionists entertain the hypothesis that geographic variation is due to differential selection in different environments. Neutralists ascribe it to divergence owing to genetic drift. Yet in most species of *Drosophila* allozyme frequencies vary little throughout the species distribution area, whereas the frequencies of chromosomal variants in the same species are greatly diversified. This is particularly striking if one compares the continental population of *Drosophila willistoni* in Central and South America with populations of the same species on the oceanic islands of the Lesser Antilles (Ayala *et al.*, 1971). The allozyme polymorphisms are similar on the continent and the islands, while the chromosomal inversion polymorphisms are far more limited on the islands. These islands are of volcanic origin, and were never a part of any continent;

their inhabitants are descendants of accidental introductions of very few founders by wind or other agencies. The same allozyme allele can arise repeatedly by mutation, and reach the same frequencies on the continent and the islands if it is favored to the same extent by natural selection. By contrast, the same chromosomal inversion is unlikely to originate more than once. Even if such a monophyletically arisen inversion is favored by natural selection, it can occur on the islands only if it was present among the immigrant founders. Other evidence that allozyme variants may be subject to selection is given by Ayala and co-workers (1974a) and Marinković and Ayala (1975a, b).

Cavalli-Sforza (1973) formulates the problem of genetic differences between human populations as follows: "What variation would be expected if drift alone (but not selection) operates?" An important issue in this connection is the choice of genetically determined characters to study. Cavalli-Sforza and Bodmer (1971) state that "when the aims of an analysis of racial differentiation are carefully defined, the best characters to work with are genetic polymorphisms." Classical anthropologists worked mostly with easily visible external traits—skin, hair and eye color, body size and build, head form, and facial features such as nose, lips, eyelids, etc. These traits are polygenic, the separate gene effects being difficult or impossible to track in isolation, and their phenotypic manifestation is often subject to environmental modifications. The usefulness of polymorphisms for clarification of the concept of race has been pointed out above, and yet it is quite possible that different categories of traits are influenced to different extents by selection and drift. Cavalli-Sforza and his collaborators have shown that genetic heterogeneities observed between populations of villages in the Parma Valley in Italy are in agreement with predictions based on genetic drift with gene diffusion, and not selection. They have also shown that the classification and the phylogenetic relationships of some 15 human populations from different parts of the world, arrived at through statistical analysis of the blood group gene frequencies, is in reasonable agreement with those accepted by classical anthropologists.

Lewontin (1972) has compared the magnitudes of the genetic diversity found among individuals within human populations, between populations of the same race, and between races. He used the Shannon information measure $H = -\sum_{i=1}^{n} p_i \log_2 p_i$ to quantify the polymorphisms and the interpopulational differences observed with nine blood group systems and seven enzyme and blood serum variants. Seven races (Caucasians, Black Africans, Mongoloids, South Asian Aborigines, Amerindians, Oceanians, and Australian Aborigines), each represented by as many populations as have been studied for the genetic variants chosen, are included in the data. The diversity between individuals within populations accounts for 85 percent of the total diversity in the human species; diversity between populations within a race accounts for 8.3 percent and diversity between races for only 6.3 percent.

Lewontin concludes that "based on randomly chosen genetic differences, human races and populations are remarkably similar to each other, with the largest part by far of human variation being accounted for by the differences between individuals." And further: "Human racial classification is of no social value and is positively destructive of social and human relations. Since such racial classification is now seen to be of virtually no genetic or taxonomic significance either, no justification can be offered for its continuance." This opinion is questionable. Races, species, genera, and other categories are needed above all for the pragmatic purpose of communication. One must be able to indicate what kind of organisms one is talking or writing about. Lewontin himself has classified the populations he has studied into seven racial groups. A random collection of genetic differences is desirable for some types of investigation, and differences chosen on the basis of some criteria are desirable for others. After all, it is possible that some traits are adaptively neutral and others are governed by heterotic, diversifying, or sexual selection. It would certainly be interesting to discover which differences—individual, interpopulational within a race, and interracial—belong to these classes. The data gathered by anthropologists studying racial variation in mankind should not be underrated.

HALDANE'S DILEMMA

Consider a population in which a gene A_1 confers on its carriers a Darwinian fitness greater than in the carriers of A_2. Natural selection acts to enhance the frequency of A_1 and to reduce that of A_2. This may happen because the progeny of A_1 survive more frequently than of A_2, or because the former have a greater fecundity, sexual activity, longevity, or any combination of these and other advantages. Whatever the cause, one may say that carriers of A_2 are eliminated by "genetic deaths." Substitution of more favorable for less favorable alleles by natural selection occurs at a "cost," and imposes upon the population a "substitutional" genetic load. The concept of substitutional load has a paradox at its core. Imagine a population in which every member has a high Darwinian fitness; a new and still more favorable mutation arises; now every member except the carrier of the mutant has a new genetic load that must be eliminated for the population to reach a still higher level of fitness.

In 1957 Haldane analyzed the consequences of this situation. During the passage of a favorable mutant from its origin to fixation many individuals have to suffer genetic death; the number of such individuals is generally much greater than the number of individuals alive in any one generation. Crow and Kimura (1970) give the following example of gene substitution: "if the typical allele has an initial frequency of 10^{-4}, a population of one million individuals will have to have nine million genetic deaths each generation if it is to substitute an average of one allele per generation. Or more probably, if there is to be a gene substitution every 100 generations, the average fitness will be lowered by 0.09." Now, in evolution many genes must be changed to transform one species into another. Granted that most living species produce numbers of progeny far in excess of those needed to have the population survive, it is difficult to understand how evolution can happen at such an enormous cost in genetic deaths. Haldane saw clearly that he was confronted by a dilemma. In his words, "I am quite aware that my conclusions will probably need a drastic revision. But I am convinced that quantitative arguments of the kind here put forward should play a part in all future discussions of evolution."

Several authors (Brues, 1964; Sved, 1968; Crow and Kimura, 1970; Lewontin, 1974) have examined various ways to escape the dilemma. The simplest and surest way is to postulate that only a few genes are substituted in racial, specific, and transspecific evolution by natural selection. The bulk of genetic changes in evolution involve neutral variants, substituted by random genetic drift. Substitution of neutral genes entails no genetic deaths and no substitutional load. This is perhaps the strongest argument in favor of the panneutralist position. But there are other alternatives.

The assumption made in many mathematical models of evolutionary processes is that the substitution at each gene locus induces its quota of genetic deaths independently of other loci. The effects of different loci are merely additive. As pointed out above, if one locus reduces the fitness of a population from 1 to 0.99, then n independently acting loci with the same selective disadvantage will reduce the fitness to 0.99^n. This assumption of the independence of the gene effects is however not a necessary one. Genes frequently interact. The model of truncation selection assumes that several or many deleterious genes must be present in the same individual to induce a "genetic death." Each one by itself can be tolerated. One "genetic death" will thus eliminate several deleterious genes at once. The truncation selection model is, in principle, applicable to the issue of the maintenance in the population of many balanced heterotic polymorphisms, as well as to that raised by Haldane's dilemma. Sved, Reed, and Bodmer (1967) have shown that "a large number of selectively balanced polymorphisms may exist [in the same population] even if the fitnesses of optimum genotypes are not unduly high in comparison to the population mean." Sved (1968) found that, granting certain assumptions, there is no obvious upper limit for the rate of gene substitution by natural selection. To what degree these assumptions and models are realistic is not possible to decide on the basis of the presently available information. The relative importance of natural selection and genetic drift, and more generally of the deterministic and stochastic factors in evolution, remains the most important unsolved problem in our understanding of the mechanisms that bring about biological evolution.

6

Species and Their Origins

Classification is a necessity. A telephone directory would be worthless if it listed entries in a random sequence. The names of the clients must be arranged alphabetically or according to some other system. Biological classification makes the immense variety of living organisms manageable. Plinius, in ancient Rome, formulated a threefold classification of animals: those living on land, in water, and in the air. This proved inconvenient. Dolphins, fishes, and many invertebrates all live in water, but their body structures are more similar to those of some land animals than to each other. Linnaeus classified animals and plants according to their body structures, a system he regarded as "natural." He devised a hierarchy of systematic categories—species, genus, order, class, and kingdom. Among these, he rightly considered species fundamental. However, he also considered them permanent and immutable entities. In 1737 he wrote, "Species tot sunt, quot diversas formas ab initio produxit Infinitum Ens" (There are as many species as the Infinite Being has produced at the beginning). As mentioned in Chapter 5, in his later days Linnaeus admitted the possibility that some species change. Lamarck and Darwin showed that species are transformed into new species in time, and that an ancestral species may split into two or more derived and contemporaneous species.

Much evidence led Lamarck and Darwin to the conclusion that species are constantly evolving rather than created from whole cloth. Perhaps the weightiest evidence was the difficulty encountered by zoological and botanical systematists delimiting species in many animal and plant groups. Indeed, some species are quite distinct in appearance, and forms transitional between them, e.g., mankind, the chimpanzee, gorilla, and orang, are altogether absent. But other species are outwardly very similar or identical (sibling species). Still others have the dividing lines between them blurred by intermediates or hybrids. Finally, some species are hardly more distinct than races or subspecies of a single species.

Hence, Darwin concluded that "species are only strongly marked and permanent varieties, and that each species first existed as a variety." This conclusion was not a solution of the species problem, but it was a seminal idea that served as guiding light in further study.

Darwin's use of the word "variety" lacked precision. Geographic races of wild species and breeds of domesticated ones, intrapopulational polymorphs, and individual abnormalities (presumably of mutational origin) could all be referred to as "varieties." M. Wagner (1889), Kleinschmidt (1900), K. Jordan (1905), Semenov-Tian-Shansky (1910), Rensch (1929), Mayr (1942) and other systematists set forth the view that species-splitting occurs through gradual divergence of geographically separate populations (see Mayr, 1942, 1963, for critical reviews). Races and subspecies may be "incipient species." This does not mean, of course, that every race will some day turn into a separate species, only that some of them may do so if the genetic divergence proceeds far enough. This became known as the theory of geographic, or allopatric, species formation (speciation). On the other hand, Bateson (1894), de Vries (1901), Goldschmidt (1940), and others argued that species arise not by gradual divergence but rather by abrupt changes. To de Vries, a mutant belonged to a species distinct from its ancestors. Some researchers, although they did not insist that the change must be abrupt, argued that a geographic separation of diverging races is not a necessary condition for speciation. A population meeting different environments in the same vicinity may split into distinct species. This is the theory of sympatric speciation. The theories of speciation have served to stimulate a great deal of important research. There was, and still is, much polemics between the proponents of the different theories. Yet it is becoming more and more apparent that speciation can occur in different ways.

WHY SHOULD THERE BE SPECIES?

Living beings vary in size from microbes as small as 100–250 millimicrons in diameter to giants like blue whales and sequoia trees. They live on unlike foods, in unlike environments—in water, in humid jungles, and in dry deserts—and at temperatures as low as −23°C and as high as 80–85°C. Yet it is probable that the living world is monophyletic. If life arose only once, then all existing organisms are products of divergent evolution. The evolution was gradual; sudden radical transformations did not occur. Therefore, if all organisms that ever lived were preserved as fossils, one could find all transitions between all now existing forms of life. But only a minuscule fraction have been fossilized (see Chapter 10). Anyway, the living world is radically discontinuous. Not only does life present itself to our observation in the form of discontinuous quanta called individuals, but individuals usually occur in discontinuous arrays. Some examples of such arrays are mentioned above—mankind, the chimpanzee, gorilla, and orang. It is legitimate to ask, What is the meaning, the biological function, of the discontinuous arrays we call species?

It has been mentioned repeatedly in the foregoing chapters that the mechanisms of Mendelian segregation and recombination generate a profusion of gene constellations. One parent heterozygous for n genes has a potentiality of producing 2^n genetically different gametes; two parents heterozygous for the same n genes have the potentiality of producing 3^n, and for n different genes 4^n genetically diverse zygotes. Wright (1932) pointed out that these potentialities are far in excess of what can be realized. Even with overconservative estimates of the number of genes and gene alleles, the number of potentially possible genotypes in the living world is many orders of magnitude greater than the number of atoms and subatomic particles estimated by cosmologists in the universe (see Chapter 2). What exists is an almost infinitesimally small sample of what could exist.

Arrays of the gene combinations that exist have been winnowed by natural selection. Inviable or poorly viable combinations were not perpetuated, even if they ever appeared. This is not to say that all possible organisms that could live are living or have lived. Natural selection has no pre-established program to bring into being all that could live. It is useless to speculate what organisms might have emerged that did not emerge. The situation of the living world can conveniently be envisaged in the form proposed by Wright, a symbolic "map" of adaptive "peaks" and "valleys" (Figure 6-1). Such "maps" should

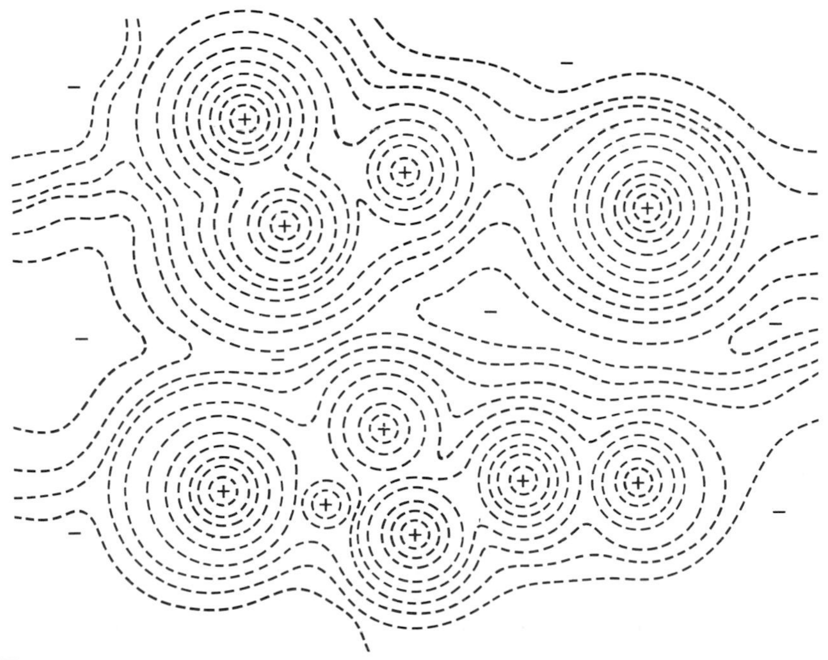

FIGURE 6-1.
Adaptive peaks and valleys in the field of gene combinations. The contour lines encompass genotypes with comparable adaptive values. Populations occupy adaptive peaks (+); the valleys (–) represent genotypic combinations with low adaptive values.

have as many dimensions as there are genes with two, or several, or many alleles. With two dimensions, the two axes symbolize the allelic variants of two gene loci. The Darwinian fitness (adaptive value) of the combinations of these alleles is symbolized by contours, like in a topographic map.

One can imagine a world where life exists in only one ecological niche, uses only one kind of food, and subsists in only one environment. A single species would then be enough. Perhaps a situation of this sort was realized when the first life arose from inanimate matter. But since then life has conquered and spread into many environments. Life has responded to the diversity of environments by evolving genotypes adapted to inhabit them. It has mastered many adaptive peaks. In the field of potentially possible gene combinations some combinations are adaptively coherent, while others are unfit to survive and reproduce. A species is an array of related gene combinations that occupy one of the adaptive peaks represented in Figure 6-1. The summit of the peak is held by the genetic "elite," the genotypes that are fittest in the environments in which the species lives. The carriers of the less-fit genotypes can be said to occupy the slopes of the adaptive peak. The least fit, e.g., lethals and other constituents of the genetic burden (see Chapters 4 and 5), perish in the "adaptive valleys." The question of why there should be species can thus be answered: because there are many adaptive peaks.

DISCONTINUITY OF ORGANIC VARIATION

Figure 6-1 shows several adaptive peaks, some of which are higher than others. High peaks symbolize species of high adaptedness, and lower ones are held by forms less well adapted but still able to survive. There can also be unoccupied peaks, i.e., opportunities for living that are not exploited. Some unoccupied peaks may in time be settled if genotypes able to occupy them eventually evolve. Others may remain unoccupied if suitable gene combinations fail to emerge. How do we know that unoccupied peaks exist? There is good evidence that animals and plants introduced to new countries sometimes become highly successful pests or weeds, much more so than in their native countries. For example, the gypsy moth (*Porthetria dispar*) does relatively little harm to the vegetation in its native Europe, but in New England, where it was introduced, it has multiplied so much that it has denuded forests. There evidently was an unoccupied adaptive peak that the gypsy moth occupied quite successfully. One may speculate that man could well have lived in the early Tertiary or even Mesozoic periods; however, it was only in the late Pliocene and Pleistocene that the human species finally appeared. The human peak, this Everest of the adaptive landscape, was unoccupied for tens or even hundreds of millions of years. Evolution can be described as a series of occupations of previously vacant adaptive peaks.

The adaptive peaks in Figure 6-1 are shown grouped in clusters, separated by some shallow and some deep adaptive valleys. This symbolizes the two fundamental characteristics of organic diversity—its discontinuity and the hierarchial ordering. Instead of all possible gene combinations generated at random, one observes species between which, with exceptions to be discussed below, intermediates are absent. For example, no individual has ever been seen about which one could have the slightest hesitation as to whether it belongs to the species humankind or the species chimpanzee. However, the genetic and structural gaps between species of one genus are on the average smaller than between genera, smaller between genera than between families, and so forth. Classifiers of animals and plants utilize discontinuities to achieve a natural and convenient system. The system is "natural" in so far as the discontinuities are observable realities.

Any number of examples could be given to illustrate the discontinuous and hierarchical character of organic variation. We shall use the cat-like animals and other carnivores because they are familiar to most readers. All breeds or races of domestic cats belong to the species *Felis cattus*. They mate freely and produce fertile hybrid progeny. The European wildcat, *Felis sylvestris*, is a related but by now nearly extinct species. Hybrids of wildcats and domestic cats have been obtained in zoos, although their fertility has not been tested. The bobcat (*Felis lynx*) is sometimes treated as a genus (*Lynx*) separate from *Felis*; crosses of bobcats with domestic and wildcats have been reported in zoos, but their viability and fertility are uncertain. The lion, tiger, leopard, and jaguar were formerly considered species of the genus *Felis*, but are now placed in a separate genus *Panthera* (*P. leo, P. tigris, P. pardus, P. onca*). No hybrids between these species occur in nature, but they do mate and produce hybrids in zoos. There are reports of hybridization of lions and tigers; the hybrids ("tiglons") are viable and fertile, at least as females. The crosses lion × leopard, lion × jaguar, leopard × jaguar, and leopard × tiger have all been obtained, but the fertility of the hybrids is insufficiently studied (Gray, 1972). Crosses of species of *Panthera* with domestic cats are impossible on account of differences in size, but it would be interesting to try to achieve hybridization by artificial insemination. *Felis*, *Panthera*, and some other genera make up the cat family (Felidae). Together with the dog family (Canidae), bears (Ursidae), weasels (Mustellidae), and mongooses (Viverridae), these families constitute the order Carnivora.

Let it be reemphasized that discontinuity and a hierarchical ordering are universal in the living world. They are consequences of the fact that only a minority of the potentially possible gene combinations can give rise to viable organisms. Discontinuities and group hierarchies are observed among microorganisms, even viruses, as well as among lower and higher animals and plants. The very universality of these phenomena, while it is a godsend to systematists, raises some knotty problems. Are species of, say, bacteria self-same or even analogous biological phenomena to species of grasses, trees, insects, birds, sexually reproducing and asexual, recent and fossil organisms? Is it merely a matter of convenience whether we place certain species in the same or in different genera? Is it arbitrary to treat genetically distinct populations as races of the same species or as separate species? We have seen above, in the example of the carnivore mammals, that some species put in one genus can be placed in different genera.

REPRODUCTIVE ISOLATION

Reproduction is the primal and indispensable function of all life. However, it is achieved by a variety of methods. Species and speciation assume different forms depending upon the reproductive biology of the organisms concerned. In the remainder of this chapter our attention will be concentrated on species of sexually reproducing organisms. These constitute the majority of the living world, although asexual reproduction and parthenogenesis occur in many groups, especially among lower organisms. There are also many varieties of sexual reproduction. The common feature is the union (syngamy) of the nuclei of two cells (gametes) to form the nucleus of a single cell, the zygote. The two uniting gametes are usually very different in appearance and size, one of them being designated female and the other male. Female and male gametes may be produced in different individuals (bisexuality, dioecy) or in the same individual (hermaphroditism, monoecy). In unicellular organisms the cells that act as gametes may be indistinguishable in appearance. In some bacteria, instead of two cells fusing together, one cell donates a chromosome, or a part of it, to the partner cell. In some organisms arrangements exist that reduce the probability or preclude self-fertilization (in hermaphrodites) or mating of close relatives. In some other forms, however, self-fertilization is a frequent, or the only, method of reproduction.

Notwithstanding the variations, the essence of sexual reproduction that is relevant to evolutionary problems can be simply stated as follows. The nucleus of a zygote contains the sum of the chromosomes and genes present in the gametes whose union gave rise to the zygote. The formation of the zygote is followed, either immediately or at some later stage of the life cycle, by meiosis. Recombination of the parental chromosomes and their parts during meiosis leads to a multitude of gene combinations among the gametes that give rise to the next generation. This may be adaptively favorable or unfavorable to the species, depending on the circumstances. Some newly produced genotypes may have a fitness superior to the parental genotypes, especially if environments are changing. By contrast, other gene constellations may be inferior and even lethal. The situation is resolved by a compromise. On the one hand, a species holds fast to the genotypes the adaptedness of which has been tested by natural selection and found satisfactory in existing environments. On the other hand, improvisation of new genotypes "explores" the field of possible gene combinations, and may lead to the "discovery" of unoccupied adaptive peaks (Figure 6-1). "Exploration," however, entails a "cost." Many, or most, or even all novel gene combinations produced may be inferior in adaptedness. The losses incurred by too much gene recombination may be excessively heavy and endanger the continuation of the species or population.

Species are a compromise between too much and too little adaptive conservatism and wasteful innovation. In sexually reproducing organisms species can be defined as Mendelian populations, or arrays of Mendelian populations, between which the gene exchange is limited or prevented by reproductive isolating mechanisms. An isolating mechanism is any genetically conditioned impediment to gene exchange between populations. It is convenient to classify reproductive isolating mechanisms as premating (prezygotic) and postmating (zygotic). The former impede or prevent hybridization of members of different populations, and hence the production of hybrid zygotes. The latter reduce the viability or fertility of hybrids that have arisen. Premating and postmating mechanisms serve ultimately the same function—they stave off the gene exchange between species populations. The prezygotic isolating barriers are:

1. Ecological or habitat isolation, which occurs when species occupy different habitats (biotopes) in the same territory.
2. Seasonal or temporal isolation, which is found between populations in which the members reach sexual maturity, or flowering, at different times, usually at different seasons of the year.
3. Ethological or sexual isolation, which is a result of weakening, or absence, of sexual attraction between females and males of different species.
4. Mechanical isolation, which comes about because of different structures of flowers, impeding pollen transfer, or different sizes or shapes of the genitalia, making copulation and sperm transfer difficult or impossible. Among flowering plants, different species may be specialized to attract different insects as pollinators.
5. Gametic isolation, which results from female and male gametes failing to attract each other. Or else, spermatozoa or pollen of one species may be inviable in the sexual ducts, or stigmas and styles of flowers, of another species.

Postmating or zygotic isolating mechanisms are as follows:

1. Hybrid inviability, which eliminates many or all hybrid individuals before they reach sexual maturity. The elimination may take place at any stage of the life cycle, promptly after fertilization or shortly before the advent of the reproductive age.
2. Hybrid sterility, which disrupts the processes of gamete formation in the hybrids, so that the latter fail to produce functioning sex cells.
3. Hybrid breakdown, which reduces the viability or fertility in the progenies of hybrids, i.e., in F_2 or in backcross generations.

EXAMPLES OF PREZYGOTIC ISOLATING MECHANISMS

There are no two species between which gene exchange is prevented by all the prezygotic and zygotic isolating mechanisms listed above. On the other hand, careful study of a pair of species usually discloses not a single but two or several isolating mechanisms reinforcing each other's action. Moreover, and this is both interesting and meaningful, the assemblages of isolating mechanisms vary greatly among species of different animal and plant groups, and even among pairs of species of the same genus. Although reproductive isolating mechanisms are quite heterogeneous, their evolutionary functions are basically similar, namely, limitation or prevention of the gene exchange between species. At least in the short run it is immaterial whether species are kept separate by ecological, ethological, or hybrid sterility barriers. It should also be kept in mind that reproductive isolation is not an all-or-none phenomenon, but admits of degrees. Thus, ethological isolation may be complete, or it may amount to only slight preferences of females for males of their own rather than of foreign species. A hybrid may be wholly sterile, or it may have its fertility reduced below that of the parental species.

Many species of plants occur only or mainly in localities with a certain kind of soil, or only in xeric (dry) or mesic (humid) habitats. Habitat specialization may act as an isolating mechanism. An excellent example is species of bushes *Ceanothus* in California studied by Nobs (1963). *Ceanothus jepsonii* is found exclusively on basic serpentine soils, while *C. cuneatus* and *C. ramulosus* have broader ranges of tolerance to soil type. Although the latter species often grow near *C. jepsonii,* natural hybrids are extremely rare. However, the species have been crossed artificially and produced fertile hybrids.

Habitat separation plays a part in the reproductive isolation of sibling species (see below) of mosquitoes of the *Anopheles maculipennis* group. Though *Anopheles* was implicated in the transmission of malaria at the turn of the century, it was shown some 30 years later that only certain species of this genus are vectors of human malaria. The situation was particularly puzzling because *Anopheles* was present in many places where malaria did not occur. The pioneering works of de Buck, Hackett, Missirolli, Swellengrebel, and others, confirmed and extended by more recent investigators, showed that there are at least six species in Europe and western Asia that are morphologically identical as adults, but which were confused under the name *A. maculipennis.* They are however distinct in the characteristics of their egg "floats," in the gene arrangements in their chromosomes, in their courtship rituals, production of fertile or sterile hybrids when crossed, ability or inability to transmit malaria, and what is particularly interesting for us here, in their habitat preferences. Some (*A. labranchiae, A. atroparvus*) live in brackish water, others in running fresh water (*A. maculipennis*), or stagnant (*A. melanoon, A. messeae*) fresh water. Although the habitats of the different species may be not far distant from each other, the different preferences of these species contribute to their reproductive isolation. Habitat isolation is found also among sibling species of tropical mosquitoes of the *Anopheles gambiae* complex (Kitzmiller *et al.*, 1967; Davidson *et al.*, 1967; Coluzzi, 1970).

A good example of seasonal isolation of plant species are the pines *Pinus radiata* and *P. muricata,* which grow together on the Monterey Peninsula in California (Stebbins, 1950; Mirov, 1967). The former species sheds its pollen in early February, the latter in April. Trees intermediate between the two species in appearance, presumed hybrids, do occur but "form only a fraction of one percent of the stand and appear less vigorous and productive than typical *P. radiata* and *P. muricata*." Among many related animal species that differ in breeding season, the toads *Bufo americanus* and *B. fowleri* have been studied most carefully (Blair, 1941; Cory and Manion, 1955). Though *B. americanus* breeds earlier in spring than *B. fowleri*, the breeding seasons overlap slightly and fertile species hybrids are produced. In all probability, what kept the species apart was a combination of seasonal and habitat isolations, mutually reinforcing each other. *Bufo americanus* prefers to live in forested areas and *B. fowleri* in grasslands. In territories unmodified by human activities the species are well separated, but the destruction of forests and utilization of land for agriculture has created man-made habitats acceptable for both toad species. As a result, in some places these species have intercrossed and formed intermediate hybrid populations.

Sexual (ethological or behavioral) isolation is the most powerful agency keeping apart related sympatric species in many, though not all, groups of the animal kingdom. In species with separate sexes, females and males must search for each other, come together, perform the often spectacularly complex courtship and mating "rituals," and finally copulate or release their sex cells in sufficient proximity to make the formation of zygotes possible. Even in many hermaphroditic animals, in which egg cell and spermatozoa are formed in the same individual, self-insemination is an exception rather than the rule. Thus, in pulmonate snails pairs of individuals mutually inseminate each other, although an isolated individual may give progeny by self-fertilization. The chain of search—courtship—mating reactions may diverge in different species at one or more points, preventing the formation of hybrid progeny.

Ethological isolation of species of *Drosophila* can be easily observed in the laboratory. One technique is to introduce virgin females of two species and males of one of them in a container with nutrient medium, leave them for some hours or days, dissect the females, and record the presence or absence of sperm in their sperm receptacles by examination under the microscope. Except when the species are closely related, all or a part of the conspecific females are inseminated, while the females of the other species remain virgins. With related species one usually finds that the males inseminate a greater proportion of the females of their own than of the foreign species. For example, in an experiment by Levene and Dobzhansky females of *Drosophila pseudoobscura* and *D. persimilis* were kept for four days with males of *D. persimilis*. Of the 178 *D. persimilis* females, 79.2 percent were inseminated, while only 22.5 percent were inseminated among 173 *D. pseudoobscura* females.

Another technique used to measure ethological isolation is to place females and males of two species of *Drosophila* in an observation chamber and record the matings that take place. In *Drosophila* and other animals in which no permanent bond is established between the mates, and in which the parents, or at least the males, do not feed or guard the progeny, one usually finds that the males are promiscuous. They approach and start courting females of their own as well as of foreign species. The females are more discriminating; they accept conspecific males after more or less prolonged courtship, and reject males of other species. This difference in the behavior of the two sexes is understandable, since the parental "investment" is greater for females than for males. The eggs produced are far fewer than spermatozoa. The more females a male inseminates the greater his Darwinian fitness is likely to be; he can afford an occasional "mistake." A female must choose a male whose progeny will be vigorous and fertile. Dobzhansky, Ehrman, and Kastritsis observed the following numbers of matings in chambers with equal numbers of females and males of two species, *D. bifasciata* and *D. imaii:*

bifasciata ♀ × *bifasciata* ♂	229
bifasciata ♀ × *imaii* ♂	13
imaii ♀ × *bifasciata* ♂	9
imaii ♀ × *imaii* ♂	375

Exactly what stimuli cause *Drosophila* females to reject the courtship of males of foreign species is not well known. Olfactory (chemical), visual, auditory, and tactile signals may all be involved in different species combinations. In some moths and butterflies chemical stimuli are most important, and specific substances (pheromones) have been chemically identified (Figure 6-2). Littlejohn (1965) and others have shown that among frogs and toads auditory signals ("calls") are species-specific. Their effectiveness as isolating mechanisms are experimentally demonstrable. Females are attracted to the calls of conspecific males recorded and reproduced by loudspeakers, and may not react to calls of other species.

As long ago as 1844 the entomologist Dufour expounded the lock-and-key theory. The genitalia of the females and males of the same species fit each other so exactly, that deviations in shape or size make copulation impossible. The basis of this theory is that closely related species, particularly insects, are often more securely distinguishable by their genitalia than by external appearance. Dufour believed that the lock-and-key relationship of the genitalia makes species separate and unchangeable. This is undoubtedly an exaggeration. Not only large and small individuals of the same species but sometimes different species with clearly distinguishable genitalia copulate successfully. Nevertheless, it has often been observed, particularly in laboratory experiments, that interspecific copulation results in injury or even death of the participants. An example of this is copulation of *Drosophila pseudoobscura* females with *D. melanogaster* males. Females and males of dog breeds very different in size may be attracted to each other but unable to copulate.

Analogous situations are found in flowering plants. Some flowers are "promiscuous" in the sense that they can be pollinated by many species of insects or other animals. Others are specialized to be pollinated by a few related or even a single species. The specialization may be expressed in the flower structure, which admits only some species of pollinators or which makes the

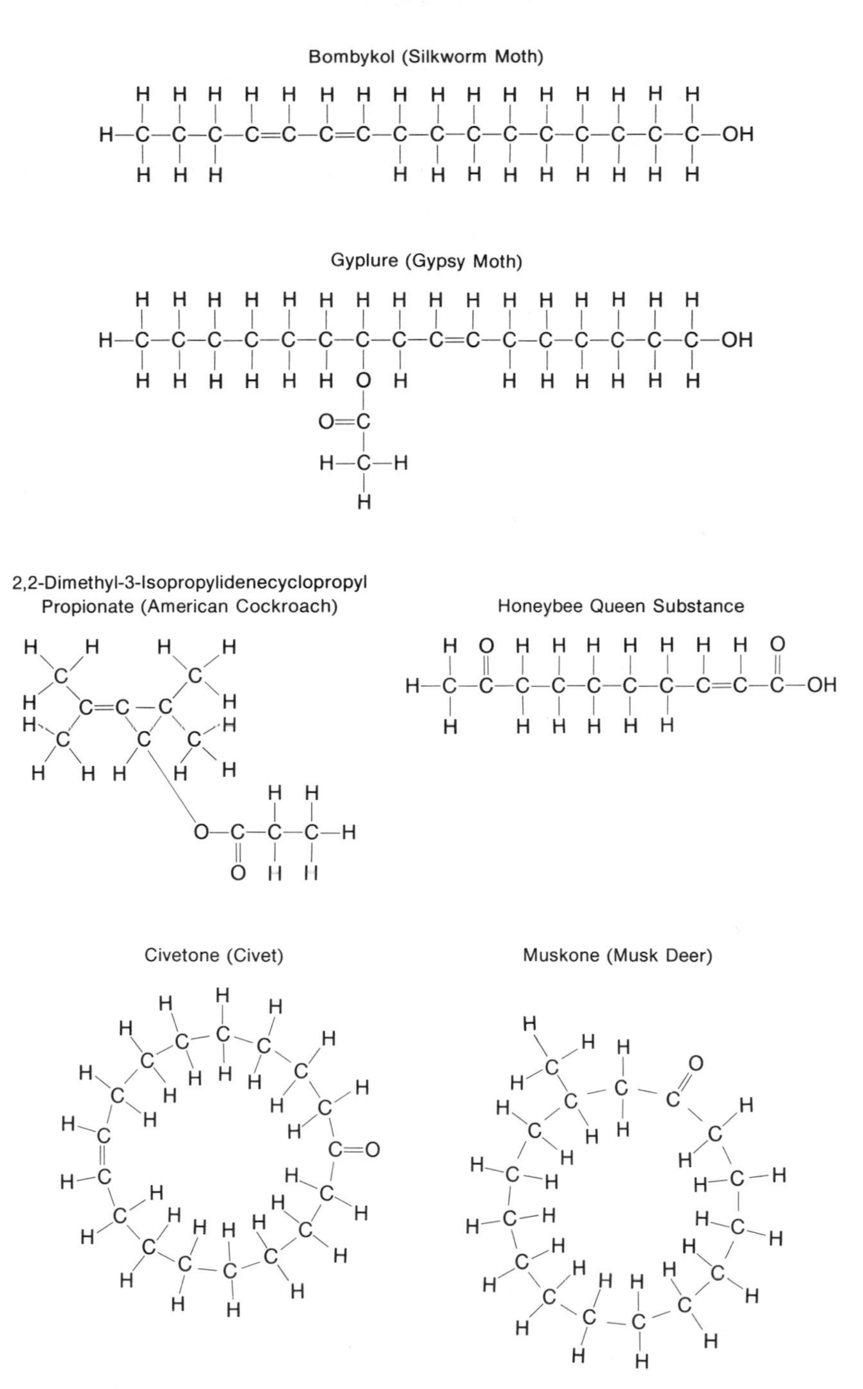

FIGURE 6-2.
Sex pheromones from four insect species and two mammals. The high molecular weight of most sex pheromones accounts for their narrow specificity and high potency. [After Wilson, 1963.]

flowers attractive to some but not to other visitors (Grant and Grant, 1965; Grant, 1971). Two interesting examples are the sage species, *Salvia apiana* and *S. mellifera* (Grant and Grant, 1964). The distribution areas of these species overlap widely in California, but the gene exchange between them is restricted by a combination of habitat, seasonal and mechanical isolation, as well as by a partial hybrid sterility. The flowers of *S. mellifera* are so constructed that at least 12 species of small and medium-sized bees (including the honey bee) and one species of fly effectively pollinate them. The flowers of *S. apiana* require visits of larger carpenter or bumble bees to successfully pollinate them. A most extraordinary but apparently quite efficient isolating mechanism is found in several species of orchids, whose flowers mimic in shape the females of certain species of wasps and bees (Dodson, 1967). Male wasps or bees are attracted to these flowers and engage in "pseudocopulation" with them, thus receiving and transferring the pollen sacs.

Gametic isolation may be important in some marine animals, which discharge their egg cells and spermatozoa into ambient water and let the fertilization take place outside the parental bodies. The attraction between the eggs and the sperms of different species may be reduced or absent. Moreover, the release of the sexual products in water may be stimulated by chemical substances produced by conspecific individuals, which induce sexually mature individuals of a given species to discharge their sex cells almost simultaneously. In animals with internal fertilization the spermatozoa of foreign species may be incapacitated in the sexual ducts of alien females. This occurs, for example, in some species of *Drosophila* that have the so-called insemination reaction, i.e., swelling of the vagina by secretion of a fluid. It is interesting that while the insemination reaction is very strong in some *Drosophila* species, other species show not a trace of it. At least superficially similar phenomena in the plant kingdom are failure of the pollen grains of one species to germinate on the stigma of another, or slowness of growth of the pollen tube in styles of different species. Thus a flower pollinated with a mixture of the pollen grains of the same and a foreign species may fail to produce hybrid seed (Smith, 1970).

We have deliberately not included geographic isolation among the prezygotic isolating mechanisms. There is of course no doubt that races or subspecies of sexually reproducing forms can be kept apart by the simple fact of being allopatric, i.e., living in different territories. The same is true of allopatric species, some of which easily hybridize when brought together in a laboratory or experimental field. However, geographic isolation is not a reproductive-isolating mechanism stemming from the genetic constitution of populations. Geographically isolated populations (on islands, for example) may be genetically identical. Or if they are genetically different, the differences may be consequences rather than causes of the isolation. A rigorous test of efficient reproductive isolation is the ability of populations to coexist sympatrically (in the same territory) with little or no gene flow between them. Species, not races, of sexually reproducing and outbreeding forms can coexist sympatrically without gene exchange and eventual fusion. Only in the human species does one encounter sympatric races. They may, at least temporarily, coexist with little or no gene exchange, owing to cultural rather than to genetic isolating agents (Chapter 14).

HYBRID INVIABILITY, STERILITY, AND BREAKDOWN

If a gene exchange between species leads to production of offspring of diminished fitness, prezygotic isolating mechanisms are the most economical means to avert fitness losses. However, isolating mechanisms are often not foolproof; hybrid zygotes may at least occasionally be produced. Thus, *Drosophila pseudoobscura* and *D. persimilis* occur together in many localities, despite their fairly distinct environmental preferences. Ethological isolation, which is much stronger in nature than in experimental cultures, appears to be the principal agent keeping these species apart. Among some 27,000 individuals examined from localities where both species are present, only a single hybrid was encountered.

A hybrid zygote may be aborted soon after its formation, or at any stage of the life cycle. Hybrid individuals may be frail—able to survive in protected laboratory cultures but not outdoors. Of course, some species hybrids not only survive but exhibit luxuriance (somatic hybrid vigor). A classic example of luxuriance among species hybrids is the mule, progeny of horse × ass crosses. At the opposite extreme are goat × sheep crosses; fertilization takes place but the hybrid embryos die in early developmental stages. Moore found various degrees of inviability among hybrids between closely related species of the leopard frog (*Rana pipiens*) complex. Sonneborn and his students observed conjugation between sibling species (which these investigators call "varieties") of infusoria of the *Paramecium aurelia* group. If an exchange of nuclei between the conjugating pairs takes place, the hybrids may die (hybrid inviability); if they survive, their progenies may be in various degrees inviable (hybrid breakdown). Numerous examples of reduced viability among species hybrids in plants have been recorded by botanists (Clausen, 1951; Grant, 1971; Stebbins, 1950).

Partial or complete hybrid sterility occurs in numerous species hybrids, and sometimes in hybrids of races or strains within species. It may or may not be a form of hybrid inviability, since it is often found in hybrids that are quite vigorous somatically. The mule is the classical example of a vigorous but completely sterile hybrid (fertile female mules have occasionally been claimed, but the evidence is unconvincing). Sterility is due to disturbances in the processes of germ-cell formation. These disturbances are of many kinds in different hybrids.

The sex organs in some sterile hybrids appear to be quite normal, and microscopic examination reveals that the processes of gamete formation proceed normally through the pre-meiotic stages. At meiosis a part or all of the chromosomes of the two parental sets may fail to pair and form bivalents. For example, in the classic work of Karpechenko (1927) the common radish (*Raphanus sativus*) was crossed to cabbage (*Brassica oleracea*). Both species have 18 chromosomes in their diploid cells, and so does the F_1 hybrid. Nine pairs are regularly formed at meiosis, whereas in the parental species the hybrid shows no pairing (18 univalents) or only an occasional bivalent. The distribution of the chromosomes at meiosis is almost random, the division products have varying numbers of chromosomes (mostly between six and 12), no viable pollen or ovules are formed, and the hybrid plants are nearly sterile. However, some cells come to include the entire complement of 18 chromosomes, and the union of such unreduced gametes gives rise to a tetraploid hybrid with 36 chromosomes. This hybrid has a regular meiosis with 18 bivalents (nine pairs formed by radish and the other nine by cabbage chromosomes), normally viable pollen and ovules, and is consequently fertile. It represents an artificially created allopolyploid species or even genus, *Raphanobrassica* (Figure 6-3).

The sterility of the diploid radish × cabbage hybrids clearly stems from an inability of the radish chromosomes to "recognize" their homologs among cabbage chromosomes, and vice versa. This is an example of chromosomal (segregational) sterility, which is alleviated or removed by the chromosome doubling that provides a homolog to every chromosome in hybrid cells undergoing meiosis. In many sterile interspecific hybrids, such as mules, disturbances of the sex-cell formation become apparent before meiosis. In still other hybrids the meiotic pairing and the meiotic divisions occur normally, but subsequently irregularities set in, and no functional gametes are generated.

The male hybrids of *Drosophila pseudoobscura* and *D. persimilis* are excellent examples. Depending on the strains crossed, the full number of chromosomal bivalents (five) are formed in the primary spermatocytes, or some of the chromosomes remain unpaired. Regardless of complete or incomplete pairing, the cell-division machinery grossly malfunctions, and nothing like normal

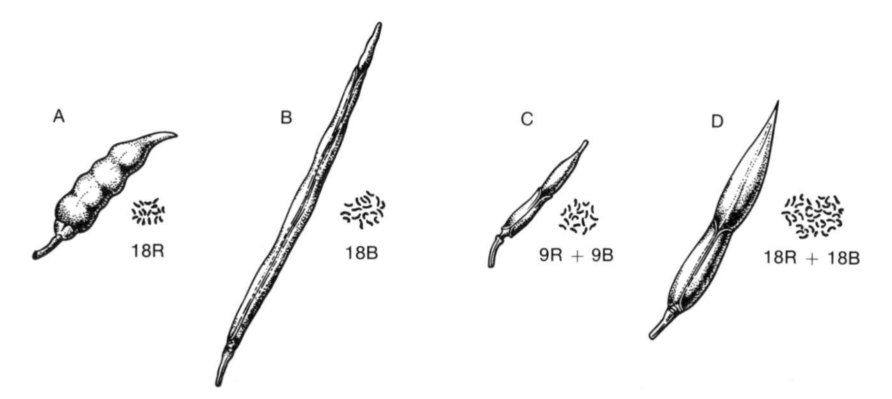

FIGURE 6-3.
Radish (A) and cabbage (B) are crossed to produce a diploid hybrid (C). An artificially created allotetraploid species, *Raphanobrassica* (D), is obtained by duplication of every chromosome of the diploid hybrid.

spermatozoa are formed. However, female hybrids between the same *Drosophila* species are fertile and can be backcrossed to males of both parental species. Furthermore, groups of tetraploid cells (20 chromosomes) are occasionally formed in the testes of the hybrid males. While chromosome doubling restores normal chromosome pairing and fertility in the tetraploid *Raphanobrassica*, it does not do so in *D. pseudoobscura* × *D. persimilis* hybrids. The cell division apparatus malfunctions in the tetraploid spermatocytes as much as it does in the diploid ones. This is because the sterility is genic (developmental) rather than chromosomal. Genic sterility results from failures of the gene complements of distinct species to function harmoniously in the hybrids. Although no disharmonies may arise in the somatic cells, the processes of germ-cell formation may be disrupted. In such a case the hybrid is somatically vigorous but sterile.

Genic sterility is known in hybrids between species with very similar or even identical gene arrangements. Chromosomal sterility indicates that so many alterations of the gene arrangement have taken place in the evolutionary history that the chromosomes of the crossed species no longer find pairing partners in hybrid cells. With genic sterility the germ-cell formation may be disrupted before, during, or after meiosis; chromosomal sterility manifests itself at and after meiosis. Of course, genic and chromosomal sterilities are not mutually exclusive, and some interspecific hybrids owe their sterility to combinations of both. By and large, genic sterility is more frequent in animals, and chromosomal sterility in plants (see Chapter 7).

Some interspecific hybrids are partially or completely fertile in one or in both sexes. If so, it still does not necessarily follow that the gene exchange between the populations of such species can take place without impediment. Hybrid breakdown may occur in the second or the backcross generations. Its cause is the formation of genotypes of low fitness, owing to recombination of the gene complements of the parental species. The F_1 hybrid females from *Drosophila pseudoobscura* × *D. persimilis* crosses are vigorous and deposit as many eggs as nonhybrid females. Yet when these females are backcrossed to males of either parental species, individuals of both sexes in the offspring are weak, and many of them are sterile. The degree of the hybrid breakdown varies depending upon the strains of the parental species used to produce hybrids.

Many instances of hybrid breakdown have been observed in plants. The hybrids between cotton species *Gossypium barbadense, G. hirsutum,* and *G. tomentosum* have been thoroughly studied by Stephens (1950). With some exceptions, these cotton species give vigorous and fertile F_1 hybrids, but F_2 hybrids die in seed or seedling stages, or produce "unthrifty" plants.

EVOLUTIONARY ORIGIN OF REPRODUCTIVE ISOLATION

It is appropriate to reiterate at this point that species in nature are kept genetically apart by diverse isolating mechanisms. Many species hybrids are sterile, but it would be far from the truth to suppose that the production of sterile hybrids is a criterion of species distinction. Many species, even those belonging to different genera, yield fertile hybrids when crossed. Likewise, ethological isolation exists between many, although not all species. When brought together artificially, some species show not a trace of aversion to interbreeding. The evolutionary process is out-and-out opportunistic; whatever kind of reproductive isolation is workable between a given pair of species is likely to become established. One should also avoid the naive mistake of supposing that species manifesting no reproductive isolation in laboratory or field experiments lack isolation also in their natural habitats. Many species hybrids that do not exist, or are exceedingly rare in the field, have been obtained in experiments.

Different isolating mechanisms are, physiologically considered, quite heterogeneous. Indeed, there seems to be little in common between, say, ecological, ethological, and hybrid sterility barriers to interbreeding. Yet all of them can serve the same function: partial or complete blockage of the gene exchange between populations. Why is this important? Because species are arrays of genotypes occupying different adaptive peaks (Figure 6-1), and a breakup and recombination of their genotypes is likely to yield a swarm of genetic endowments of low fitness.

The above argument should not be overstated. Some genotypes that might result from mixtures of distinct species can be equal or even superior in adaptedness to the parental species themselves. This is why introgressive hybridization, recombinational speciation, and allopolyploidy all play some roles in evolution (see Chapter 7). Yet we are also familiar with hybrid breakdown (see above). A superior genotype emerging from mixtures of the genes of distinct species may often be like the proverbial needle in a haystack.

Two, not mutually exclusive theories have been advanced to explain the origin and development of reproductive isolation between genetically diverging populations. One regards the isolation as an accidental by-product of genetic divergence. As populations grow genetically more and more dissimilar, it is less and less likely that their genotypes will interact harmoniously in a hybrid. The other theory, adumbrated by A. R. Wallace in 1889 and developed by some modern authors, regards reproductive isolation as a product of natural selection. Suppose that hybrids are less fit than the parental species that produce them. Individuals of a given species that mate with others of the same species will have a greater number, or more vigorous, or more fertile progeny than individuals involved in interspecific matings. Genetic characteristics that enhance the likelihood of intraspecific matings will thus be promoted, and those permitting interspecific unions will be discriminated against by natural selection. The selection can be quite opportunistic. It does not matter, at least in the short run, whether interspecific matings are eliminated by seasonal, mechanical, or other isolating mechanisms, or by cooperation of two or more mechanisms.

It is most likely that postmating isolating mechanisms arise mainly as by-products of genetic divergence, and premating ones are induced or enhanced by natural selection. The number of gene differences that must accumulate to induce hybrid inviability, sterility, or breakdown is still conjectural (however, see Ayala, 1975a, and Table 6-1 here for estimates of the genetic distances between races and species). Examples have long been known of strains of the same species giving viable hybrids and other strains giving inviable hybrids when crossed to another species. Thus, some strains of the plant *Crepis tectorum* carry a gene that produces no visible effects in that species, but which is lethal in hybrids with another species, *C. capillaris*. Of course, this does not mean that these species differ in a single gene; their hybrids are sterile, even if they do not receive the gene lethal to hybrids from *C. tectorum*. The hybrid lethality is a result of interaction of the hybrid genotype (which may be heterozygous for hundreds or thousands of genes) with a particular gene variant contributed by *C. tectorum*. Two is probably the minimal number of genes that can result in some reproductive isolation between populations. Single genes that induce lethality or sterility are frequent components of genetic burdens (Chapter 4). However, such genes are likely to be eliminated by natural selection rather than serve as foundations of postmating reproductive isolation between species. Reproductive isolation arises rather when one population incorporates a gene or a gene complex A, and another population a gene complex B, such that A interacting with B causes loss of fitness due to inviability, sterility, or breakdown of the hybrids.

Once postmating isolating mechanisms have arisen, it becomes advantageous to the species involved to minimize the loss of fitness resulting from interspecific matings. Experimental demonstration of selection enhancing or weakening ethological (sexual) isolation has been achieved in laboratories using various strains and species of *Drosophila*. The results of Koopman (1950) and Kessler (1969), working with *D. pseudoobscura* and *D. persimilis*, are particularly elegant. As stated above, these species display an appreciable ethological isolation in laboratory environments, and a much stronger one in nature. Females and males of both species were placed together; individuals that failed to engage in interspecific matings (high selection line) and those that mated with individuals of other species (low selection line) were used as parents of following generations. After 18 generations of such selection, the following percentages of intraspecific and interspecific matings were observed:

Lines	*pseudoobscura* ♀ *pseudoobscura* ♂	*pseudoobscura* ♀ *persimilis* ♂	*persimilis* ♀ *pseudoobscura* ♂	*persimilis* ♀ *persimilis* ♂
Unselected	49%	6%	4 %	41%
High lines	39	2	0	59
Low lines	27	21	1.4	51

Other things being equal, the greater the loss of fitness among hybrids between two species, the greater the pressure of natural selection to restrict or suppress the chances of such hybrids being produced. A feedback enhances the premating isolating mechanisms as the postmating ones are intensified. This leads to the phenomenon of so-called character displacement. Where the geographic distribution areas of two species overlap, the differences between these species are likely to be greater than where only one species lives by itself. Littlejohn (1965) found that the mating calls of the tree frogs *Hyla ewingi* and *H. verreauxi* are similar where these species are allopatric, but quite distinct where they are sympatric. Vaurie (1950) found an equally striking situation with two species of nuthatches (*Sitta*). Where these two bird species are allopatric they are rather similar in bill size; where both species occur sympatrically one of them has a bill appreciably larger than the other.

It is obviously in the zone of the geographic overlap, where two species are exposed to the challenge of hybridization and production of progeny of low fitness, that the building-up of premating reproductive isolation by natural selection will be most intense. Sooner or later this isolation spreads over the whole geographic area of the two species. Once reproductive isolation is complete, the species will have embarked on separate and independent evolutionary courses. One of them may flourish and the other become extinct, or both may live side by side. However, no matter how advantageous a new mutation or gene combination may be in one species, say man, it cannot benefit another species, such as the chimpanzee, or vice versa. Before reproductive isolation is complete, incipient species may converge and fuse instead of diverging; the two toad species *Bufo americanus* and *B. fowleri* are examples of this situation. These species sometimes hybridize when they become sympatric, and fuse to form a "hybrid swarm."

SIBLING SPECIES AND SEMISPECIES

Linnaeus and other pioneer zoologists and botanists distinguished species mostly by easily visible external traits. Anatomical traits were eventually added in some groups, e.g., insect genitalia, and physiological and biochemical traits in other groups (microorganisms, lower fungi). But it soon became apparent that visible differences between species are far more striking in some organisms than in others. Species of pheasants, especially the males, look so different that many of them were originally placed in separate genera. The same is true of birds of paradise and some groups of flamboyantly colored butterflies. Yet some of these apparently sharply distinct species are genetically so close that they not only cross easily in experiments, but give fertile and seemingly vigorous F_1 and F_2 hybrids. By contrast, species of some genera are isolated reproductively but differ only in some apparently minor and not easily visible details, or are identical in outward appearance. These are called *sibling species.*

Conservative taxonomists have opposed recognition of sibling species. They are understandably averse to dealing with species they cannot identify by inspection of single individuals. What then is the evidence that sibling species are species and not races or variants of the same species? The answer is simple. Sibling species are reproductively isolated and have all the earmarks of distinct species, although they lack outwardly visible recognition marks. Sibling species other than those already mentioned are the mosquito species of the

FIGURE 6-4.
Geographic distribution of six sibling species of *Drosophila* living in the American tropics. *D. willistoni, D. paulistorum, D. equinoxialis,* and *D. tropicalis* are widely distributed. *D. insularis* and *D. pavlovskiana* are narrow endemics, the former on some islands of the Lesser Antilles and the latter in Guiana. All six species are morphologically virtually indistinguishable but reproductively completely isolated.

Anopheles maculipennis and *A. gambiae* complexes. They differ in geographic distribution, preferred habitats, the type of blood they feed on (human blood versus that of other animals), the ability or inability to produce sterile or fertile hybrids, gene arrangements in their chromosomes, and finally as carriers and noncarriers of malaria. The sibling species can usually be diagnosed by their egg floats, though not in pinned specimens of adult insects in entomological collections.

Biologists working with *Drosophila* flies have uncovered many examples of sibling species. *Drosophila pseudoobscura* and *D. persimilis* have been mentioned above. Six sibling species related to *D. willistoni* live in the tropics of the Western Hemisphere. Four of them are widespread, as shown in Figure 6-4; two others (*D. insularis* and *D. pavlovskiana*) live in very restricted areas—the former on some islands of the Lesser Antilles and the latter in Guiana. Figure

6-4 shows that in many places, two, three, and even four sibling species are sympatric. The ability of the species to coexist sympatrically suggests that reproductive isolation is strong enough to interdict all or most gene exchange. Nevertheless, the species are quite similar in appearance; males show very slight but diagnostic differences in the genitalia, while females are indistinguishable. In laboratory experiments the species exhibit strong but incomplete ethological isolation, although the rare interspecific inseminations occasionally result in viable hybrid progenies (sterile males, some fertile females). No hybrids have been found in nature.

Why is it that species differences in some groups are strikingly visible, while in other groups sibling species are so difficult to diagnose? One should not assume that sibling species are necessarily genetically very close and represent races or subspecies not having attained the stage of full species. More likely, visual recognition of potential mates is important in some animals, while olfactory or auditory stimuli are important in the courtship and mating in other animals. After all, females and males of *Drosophila pseudoobscura* and *D. persimilis* recognize each other in nature almost without fail, in light and in darkness, and do not need visible traits for this purpose. Another possibility is that in some groups the formation of species specialized for diverse ways of life proceeds by modification of physiological, ecological, and behavioral traits, leaving the outwardly visible characteristics unchanged.

Mayr (1970) defines *semispecies* as "showing some of the characteristics of species and some of subspecies," and Grant (1971) as "populations [that] are neither good races nor good species but are connected by a reduced amount of interbreeding and gene flow." In other words, semispecies are borderline situations between races and species. Such borderline situations are bound to exist if the process of species formation is gradual rather than instantaneous. *Drosophila paulistorum*, one of the sibling species related to *D. willistoni*, is a complex of six semispecies. No outwardly visible characteristics by which one could distinguish the semispecies have been found. In laboratory experiments they manifest strong preferences for mating within their own semispecies. The ethological isolation, however, is less strong than between the sibling species. Inter-semispecific insemination results in viable hybrids of both sexes; the hybrid females are fertile, the hybrid males are completely sterile.

Figure 6-5 shows that each semispecies has a geographic distribution area different from that of every other semispecies. Nevertheless, in many places two or even three semispecies are sympatric. No hybrids between them have been found in nature. The sympatric coexistence demonstrates that at least in some places some semispecies have attained a degree of reproductive isolation characteristic of full species. Yet the fertility of hybrid females suggests that the possibility of gene exchange between the semispecies is not eliminated altogether. What is clear beyond doubt is that the semispecies of *D. paulistorum* have diverged less than the sibling species of the *D. willistoni* group, one of which is *D. paulistorum*.

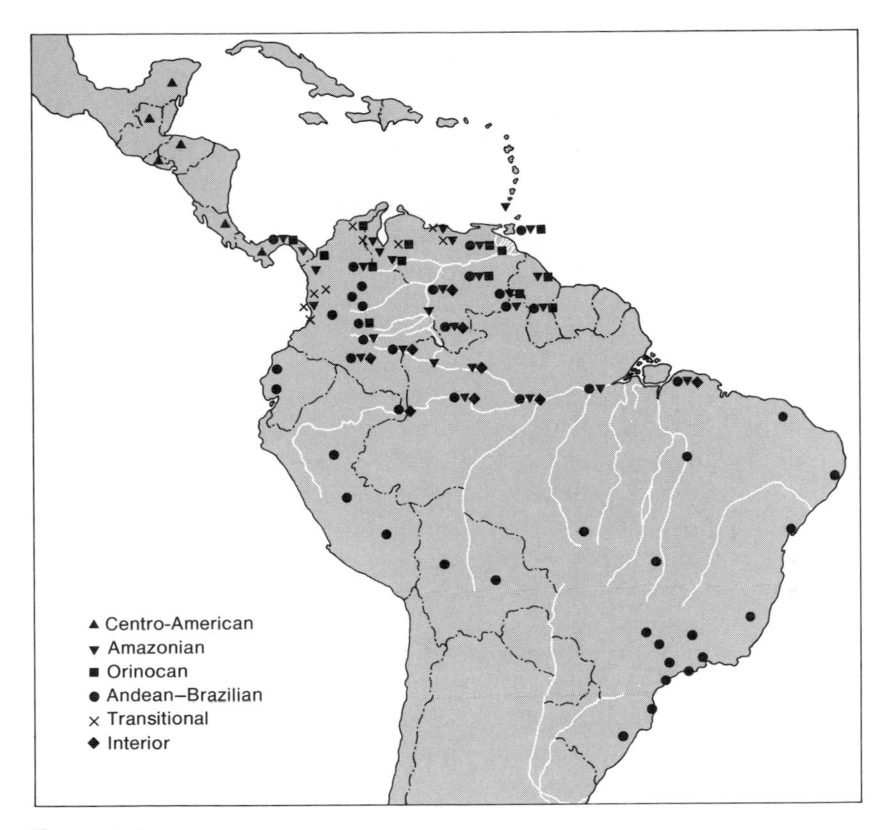

FIGURE 6-5.
Geographic distribution of the six semispecies of *Drosophila paulistorum.* The semispecies represent advanced stages of the process of speciation. Male hybrids between different semispecies are sterile. Ethological (or sexual) isolation between semispecies is quite pronounced, and is complete in habitats where two or three semispecies coexist.

An example of a group of semispecies that at first sight seem a case utterly different from those of *D. paulistorum* are the birds of paradise *(Astrapia)* in New Guinea (Mayr, 1963). Instead of being indistinguishable in appearance, these magnificently colored and adorned birds are so diversified that they were described as belonging not only to different species but to different genera. And yet careful studies have shown that the genetic differences between these phenotypically so strikingly multifarious birds are maintained chiefly by geographical separation. Where their distribution areas come in contact the semispecies readily hybridize. We may have to conclude that identical appearing semispecies of *D. paulistorum* are genetically more, not less, divergent than the birds of paradise!

SPECIATION ON OCEANIC ISLANDS

In 1835 Darwin visited the Galápagos Islands, some 600 miles off the west coast of South America. There he encountered a number of endemic species not found elsewhere in the world. The Galápagos are oceanic islands that in the geologic past never had a land connection with any continent or other group of islands. Inhabitants of these islands are descendants of ancestors introduced by passive long-distance transport, such as wind, sea, and currents, from more or less remote places. Why have they evolved into new endemic species on the islands? Ever since Darwin it has been believed that studies of speciation on islands may throw light on the general processes of species origin. Of course, there is the possibility that island speciation may be a special case, different from speciation in widespread continental species. Be this as it may, studies of speciation on islands are a fascinating chapter of evolutionary biology.

The faunas and floras of oceanic islands are characteristically unbalanced. Many kinds of animals and plants commonly found on continents, or on islands that were parts of continents in geologically recent past, are absent on oceanic islands. By contrast, some groups are over-represented by surprisingly large numbers of very diverse species. Thus, as many as 14 species of so-called Darwin's finches live on the Galápagos, while only six species of all other passerine birds and one species of cuckoo are found in these islands. Why so many finches and relatively few species of other land birds? A probable answer is that the ancestor of Darwin's finches arrived on the Galápagos earlier than other birds did, and encountered an abundance of unoccupied ecological niches (or adaptive peaks). The finches would thus have undergone extensive adaptive radiation, evolving a variety of species able to exploit opportunities for living which in balanced continental faunas are exploited by other kinds of birds. Figure 6-6 illustrates the great diversity of feeding habits of these species of finches, and also the variety of their beak shapes. The size and conformation of the beak are adaptively adjusted to the kind of food a bird is dependent on. Beaks are usually more similar in related continental species than in species of Darwin's finches, which is understandable since in the Galápagos the finches occupy a variety of ecological niches that elsewhere are settled by other bird families (Lack, 1947; Bowman, 1961).

The archipelago of Hawaii is a group of volcanic islands far from any continent. There are no Darwin's finches in Hawaii. A quite different group of birds, Drepanididae, or honeycreepers, underwent adaptive radiation there. Several species, each with beaks suitable for procuring different kinds of food, have evolved. These species have correspondingly different ways of life (see Chapter 8, Figure 8-6). Even more spectacular is the adaptive radiation of drosophilid flies in the Hawaiian islands. Of the world fauna of *Drosophila*, about 1,250 described species, about one-third of the total are endemic to Hawaii (see Table 9-1, p. 273). Yet the total area of the islands of this archipelago is about equal to that of the state of New Jersey. Species of *Drosophila* in Hawaii are not only numerous but extraordinarily diverse in body structures, food habits, and

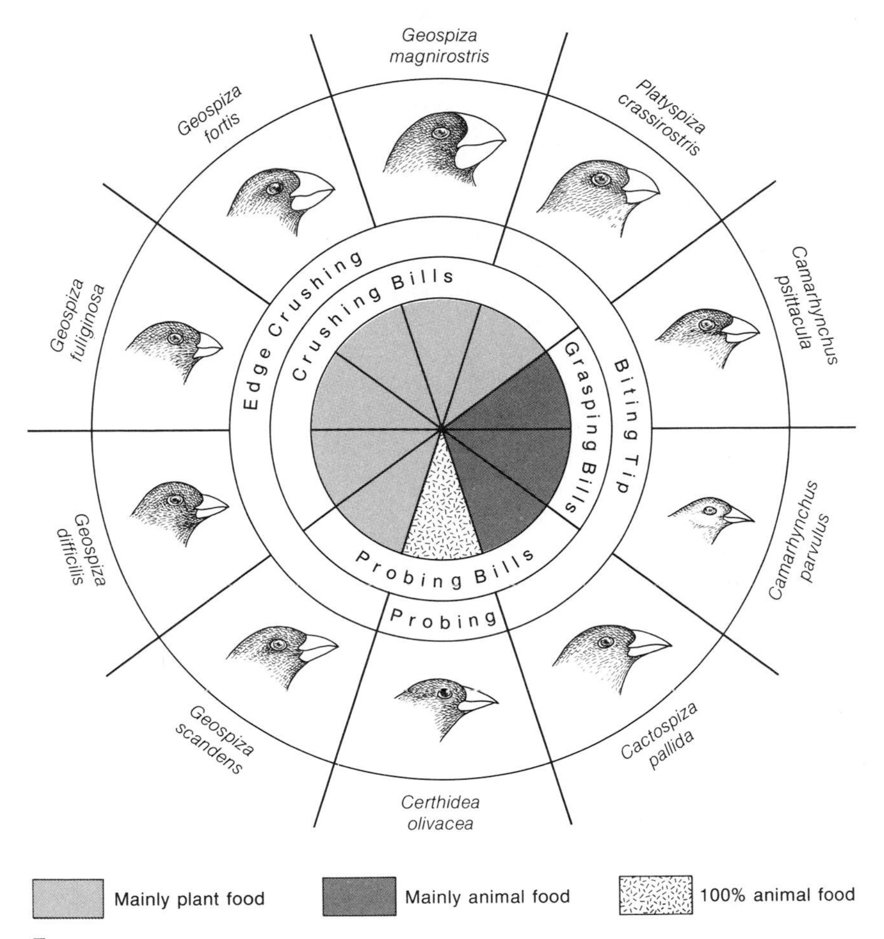

FIGURE 6-6.
Darwin's finches. Ten species of Geospizinae from Indefatigable Island in the Galápagos. Different species feed on different foods and have evolved beaks adapted to their feeding habits.

sexual behavior. Very elaborate mating rituals are found among Hawaiian *Drosophila.* This is as expected on theoretical grounds; with so many related species a fly must insure that it mates with a right partner (Carson *et al.,* 1970).

The geological age of the islands of the Hawaiian archipelago occupied by *Drosophila* is between five and less than one million years. All Hawaiian drosophilids are descendants of one or two ancestral species that reached the islands by some accidental long-distance transport. The hundreds of species now living there must have evolved within an interval of time that is short on the geological time scale. By and large, the oldest islands have the most ancient species, whereas the youngest islands have more recently arisen species. What factors have led to this rapid and profuse speciation? Mayr, Carson, and some other students of speciation on oceanic islands infer that a species may arise from very small numbers of "founders." Suppose that a single fertilized female is transported by wind or some other agency from one island to a new island where it was not previously found. This "founder" will evidently bring only a small fraction of the gene pool of the population from which it came, and natural selection will begin its work on the new island with that limited supply of genetic variability. However, the progeny of the founder will be inbred for several generations, and may come to contain gene combinations that are rarely, if ever, produced on the ancestral island. The population on the newly colonized island may undergo rapid genetic change and become a new species. Of course, such rapid speciation may not be restricted to oceanic islands. It may conceivably occur in colonies on the periphery of geographic distribution areas of continental species. This hypothesis is a long way from being proven conclusively, but it is a fruitful notion because it suggests many experiments and observations to verify it.

INTERSPECIFIC HYBRIDS AND CHROMOSOME COMPARISONS

Another unsolved problem that is being actively studied at present is the extent of genetic differences between species. This problem has a long history. As pointed out earlier, de Vries at the turn of the century believed that a single mutation gave rise to a new species. This is true only of a special kind of mutation, allopolyploidy, which can give rise to a new species in a single generation (see Chapter 7). Mutations observed by Morgan were not new species but variant forms of the ancestral one. Morgan rightly concluded that species differ in several or many gene changes, derived ultimately from mutations. However, not even rough estimates of the numbers of these changes were obtainable in Morgan's pioneering days.

Some light on the problem came from studies of recombination in F_2 and later generations of fertile interspecific hybrids. Most of these studies were made on plants, because fertile hybrids between animal species are relatively rare. Spectacular results were obtained by Lotsy and by Baur with hybrids of species of snapdragons, *Antirrhinum maius* and *A. glutinosum*. There appeared not only a great mass of recombinations of traits of parental species, but allegedly (the results have been questioned by other investigators) traits present in neither parent. Such new traits could arise from epistatic interactions of the parental genes. Baur's minimal estimate of the number of gene differences between the snapdragon species was 100, and this estimate was evidently limited to genes with externally visible effects. Figure 6-7 shows the flowers of the violets *Viola arvensis* and *V. rothomagensis*, and of their F_1 and F_2 hybrids studied by Clausen (1951). The impressive variety of recombinants obtained in progenies of species hybrids lured some biologists into extravagant speculations.

FIGURE 6-7.
Flowers of *Viola arvensis* (top left), *V. rothomagensis* (top right), their F_1 hybrid (top center), and their F_2 progeny (four lower rows). Great phenotypic variety is observed in the F_1 individuals as a consequence of recombination between traits of the two parental species. [Courtesy of Dr. M. C. Nobs.]

Thus, Lotsy believed that evolution can be accounted for by recombination of different but unchanging genes—that mutation was unnecessary.

As early as the nineteenth century, pioneer cytologists found that species differ frequently, though by no means always, in chromosome numbers and shapes. An enormous number of observations on chromosomal differences has accumulated since. Here are a few examples. Man has 46 chromosomes, while related anthropoid apes (chimpanzee, gorilla, orang) have 48. Species of *Drosophila* have from 3 to 7 chromosome pairs. Numerous species of grasshoppers of the family Acrididae all have 12 pairs of chromosomes.

The discovery of the mutational origin of chromosomal aberrations (deficiencies, duplications, translocations, and inversions; see Chapter 3) attracted the attention of investigators to the role played by such mutations in the origin of chromosomal polymorphisms within, and of chromosomal differences between species. Rediscovery of the giant chromosomes in the salivary glands of fly larvae in 1933 (they were seen but not understood in 1881) furnished a relatively easy and precise technique for comparison of the gene arrangements in the chromosomes of related species in organisms that have such giant chromosomes. The sibling species *Drosophila pseudoobscura* and *D. persimilis*, mentioned above, differ usually in at least four inverted sections on different chromosomes. A less closely related species, *D. miranda*, differs from these two in many large and small inversions in every chromosome, as well as in a translocation which has joined the Y-chromosome to one of the autosomes. A minimum estimate of the number of chromosome breaks needed to transform the chromosomes of *D. pseudoobscura* into those of *D. miranda,* or vice versa, is of the order of 100.

Figure 6-8 shows the comparative structure of the chromosomes in 9 species of the *virilis* species group of *Drosophila*. Taking the gene arrangements in the chromosomes of *D. virilis* as an arbitrary standard, the letters symbolize the different inversions of blocks of genes that have taken place in evolution. For example, the chromosomes of *D. flavomontana* differ from *D. virilis* in 25 inversions of gene blocks. With still less closely related species, the gene arrangements in the chromosomes are changed so radically that even the giant chromosomes of the salivary gland cells do not permit full decipherment of all the changes in the gene orders that have occurred in the evolutionary descent. And yet one should not conclude that rearrangements of the gene orders in the chromosomes are necessary for speciation. Carson and his collaborators have made detailed comparisons of the giant chromosomes in the salivary gland

---

FIGURE 6-8 (opposite)
The haploid chromosome complements of nine species of the *Drosophila virilis* group. The chromosome arms are labeled X, 2, 3, 4, 5, and the dot-like chromosome 6. The letters represent inversions of chromosomal segments that have taken place in the evolution of these species. *D. virilis* is probably the ancestral species from which the others have derived, as indicated by the arrows.

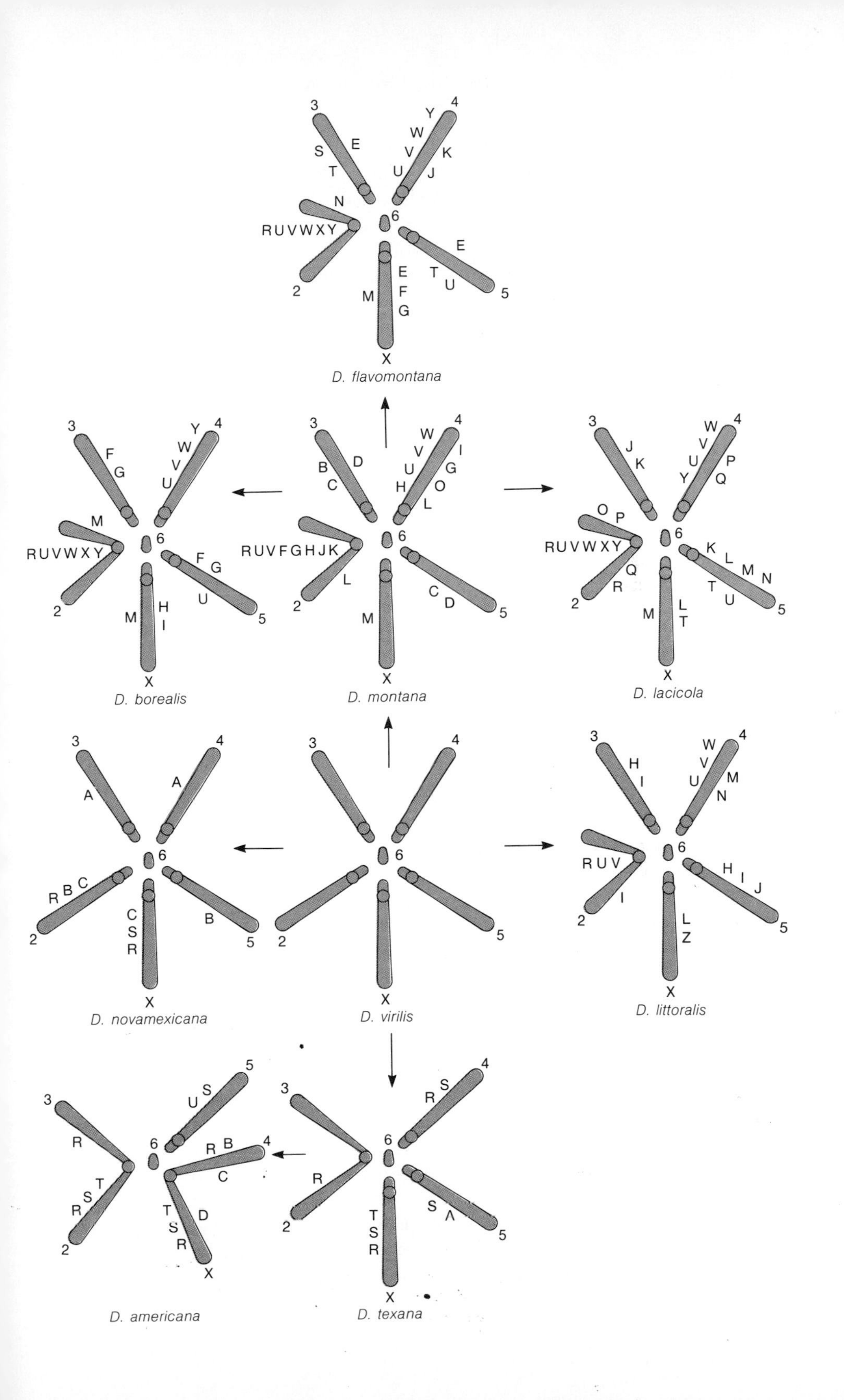

D. flavomontana
RUVWXY
D. borealis
RUVWXY
D. montana
RUVFGHJK
D. lacicola
RUVWXY
D. novamexicana
D. virilis
D. littoralis
RUV
D. americana
D. texana

cells of 96 species of *Drosophila* inhabiting the Hawaiian archipelago. Several examples of pairs or groups of related species were found to be homosequential, i.e., having identical gene arrangements in all their chromosomes.

SPECIES DIFFERENCES IN ALLOZYMES

Mendelian genetics deals with gene differences. We know that there are genes determining blood groups or eye color in man because we observe segregations of these characteristics in families in which there appear individuals with different blood groups or different eye colors. From the observed segregation of characters we infer the existence of segregating genes. If all humans had the same blood antigens and the same eye color, the existence of these genes would escape detection. This imposes a constraint on Mendelian genetics; one may, with enough labor, count the genes in which species differ but not the genes in which they are alike. The constraint has been partly relaxed since 1966, when the techniques for studying enzymes and other soluble proteins in populations by means of gel electrophoresis were developed (see Chapter 2). One can record the proteins represented by variants (allozymes) with different electrophoretic mobilities in individuals, races, species, etc. One can also register the proteins with similar mobilities. Estimates are thus obtained of the numbers of genes that are different and similar in individuals or populations studied.

Hundreds of studies of allozyme variation have been carried out in diverse groups of plants and animals, including man, under the leadership of Allard, Ayala, Harris, Lewontin, Selander, and others. The aspect of this work that interests us here is the estimation of the genetic differences and similarities between species (for a review see Ayala, 1975a). Of course, the magnitudes of these differences and similarities are not expected to be, and are not, constant for all species. This is because, first, the formation of species is a gradual process, and some species have only recently emerged from races, while others had time to accumulate gene differences. Second, speciation occurs in different ways in different groups of organisms, and some of these ways entail more genetic divergence than others (see Chapter 7).

The most detailed studies of allozyme variation between closely related species have been made in the *Drosophila willistoni* group of sibling species. Five levels of evolutionary divergence are represented among these flies:

A. Local populations of a species, which intercross easily and yield fertile hybrids.

B. Geographically separated subspecies, showing only a trace or no ethological isolation, but producing sterile male hybrids in at least one of the reciprocal crosses.

C. Semispecies, geographically allopatric or sympatric, showing strong but not complete ethological isolation, and producing in laboratory experiments fertile female but completely sterile male hybrids.

D. Sibling species, similar or identical in externally visible traits, yet giving rise to no viable hybrids in nature, and exhibiting strong ethological isolation.
E. Reproductively fully isolated and morphologically distinguishable species.

Average values of genetic identity, I, and of genetic distance, D, between the groups at various levels of genetic divergence can then be calculated. As explained in more detail in Chapter 9, the value D can be interpreted as the average number of electrophoretically detectable codon substitutions per gene that have accumulated in the populations studied since they diverged from a common ancestor. The value I is the proportion of genes the products of which are not distinguishable by their electrophoretic behavior. The sum of D and I is greater than unity, for the simple reason that two or several codon substitutions may have occurred in some genes. The values of genetic identity and genetic distance for populations of the *Drosophila willistoni* group at various levels of genetic divergence are shown in Table 6-1.

The data in Table 6-1 require some comment. Recall that the information obtained by electrophoretic techniques underestimates the genetic differences. Gene alleles coding for different allozymes with similar electrophoretic mobilities are classified as identical. Even so, only about 80 percent of the genes seem to be identical on the average in different subspecies or different semispecies. The proportion of ostensibly identical genes decreases to 56 percent in sibling species and to only 35 percent in morphologically distinguishable species belonging to the same subgenus and species group in *Drosophila*. Clearly, then, the proportions of genes represented by similar alleles rapidly decrease (and the proportions of those represented by different alleles increase) during the process of species formation and divergence. The part of the genetic divergence due to natural selection and the contribution of the random genetic drift are, of course, quite different problems.

One point deserves special emphasis. Table 6-1 shows that the average genetic distances are the same between subspecies and between semispecies. Yet the semispecies in the *Drosophila willistoni* group have reached a degree of reproductive isolation permitting some of them to coexist sympatrically. Subspecies show mere rudiments of reproductive isolation. It would however be erroneous to conclude that reproductive isolation arises independently of genetic

TABLE 6-1. Genetic identity and genetic distance between various levels of evolutionary divergence in the *Drosophila willistoni* group. [After Ayala *et al.*, 1974c.]

Level of divergence	Identity	Distance
A. Local populations	0.970 ± 0.006	0.031 ± 0.007
B. Subspecies	0.795 ± 0.013	0.230 ± 0.016
C. Semispecies	0.798 ± 0.026	0.226 ± 0.033
D. Sibling species	0.563 ± 0.023	0.581 ± 0.039
E. Full species	0.352 ± 0.023	1.056 ± 0.068

divergence. What the data show is rather that given enough genetic divergence, superimposition of reproductive isolation may occur through a limited number of gene differences responsible particularly for isolating mechanisms.

The genetic distances found between subspecies or between species of the *Drosophila willistoni* group (Table 6-1) need not be the same as between subspecies or species in other groups of animals or plants. Let it be emphasized again that the processes of speciation occur in different ways in different organisms. The amount of genetic divergence that precedes and follows the advent of reproductive isolation is not everywhere constant. The most impressive demonstration of this comes from comparisons of men with anthropoid apes, particularly the chimpanzee (King and Wilson, 1975). The genetic distance between man and chimpanzee turns out to be 0.62, and genetic identity 0.54. Consulting Table 6-1 we find that these values are matched by sibling species of *Drosophila*! While sibling species are indistinguishable in external appearance, the bodily structures of man and chimpanzee are great enough to place them in different zoological families; the differences in mental abilities are more radical still (see Chapter 14). The limited divergence in the structural genes coding for allozymes and other proteins is, in the case of man and chimpanzee, far exceeded by divergence in the regulatory genes, which are not readily detected by electrophoretic techniques.

7

Patterns of Speciation

In Chapter 6 biological species were defined as arrays of Mendelian populations between which gene exchange is limited or prevented by reproductive isolating mechanisms. Once this definition becomes accepted, the exploration and analysis of reproductive isolating mechanisms becomes a major task for the evolutionist.

The origin of isolating mechanisms is associated, often intimately, with morphological and ecological diversification. Nevertheless, many of these mechanisms are *sui generis*. They are related to, but different from the primary trends of differentiation. As pointed out in the previous chapter, in animals behavior patterns as well as some other isolating mechanisms evolve through direct selection against the disharmonious mixing of divergent gene pools. In most plants and microorganisms behavior divergence cannot develop, so that other differences are the primary basis of reproductive isolation. In the most specialized flowering plants relationships with animal pollen vectors may become as specific as the behavioral patterns responsible for ethological isolation in animals. More often, however, reproductive isolation is based upon chromosomal repatterning and disharmonious gene action in development. Both of these phenomena will be analyzed further in the present chapter.

In plants and microorganisms neither divergences responsible for genic disharmony in hybrids nor chromosomal divergences that give rise to hybrid sterility are inevitable by-products of morphological and ecological divergence. This fact became well established by the earliest explorations of diversity within plant species, particularly the research of Turesson and that of Clausen, Keck, and Hiesey (Stebbins, 1950). By the 1930s and 1940s plant evolutionists had become aware of the fact that many plant species, particularly trees and shrubs, can diverge greatly from each other, without becoming separated by permanent barriers of reproductive isolation. They can remain morphologically distinct even when sympatric over large areas, but in localized regions, often under the influence of habitat disturbance, they can form hybrid swarms of such size and complexity that the parental gene complexes become nearly or quite unrecognizable. Grant (1971) has given an excellent account of such groups of species, which in the words of Mayr (1963) "pass the test of sympatry" in some parts of their range but "fail" it in others. Grant has designated this stage of evolution as that of the "semispecies," a use of the term that is different from the original definition of Mayr (1931, 1942). Among the best-known examples are species of oaks (genus *Quercus*), which have been carefully analyzed by Tucker (1952) and others.

At the other end of the spectrum are plant genera that, like *Drosophila* and many other animal genera, contain numerous groups of sibling species that are indistinguishable or barely distinguishable morphologically, but are nevertheless separated by strong barriers of reproductive isolation. A well-known example is the composite genus *Holocarpha* (Clausen, 1951). Most groups of plant species lie between these extremes (for a review see Grant, 1971).

ECOGEOGRAPHIC FACTORS AFFECTING THE ORIGIN OF SPECIES

The origin of a new species requires a succession of events that are usually distributed over a considerable time span, and so is difficult or impossible to reproduce experimentally. Nevertheless, each of the separate processes necessary for speciation has been duplicated under laboratory or garden conditions, so that this crucial step of evolution is open to analysis by experimental and quantitative methods. The entire sequence, including the establishment in nature of a newly arisen, reproductively isolated population as a spontaneous, self-perpetuating entity, has not yet been duplicated.

Our understanding of this sequence in nature must be based upon extrapolation from experimental data. The situations most favorable for such analysis by extrapolation are those in which geological and other evidence indicates that a particular species population has evolved recently.

Analyses of variation within species, including the differentiation of races (Chapter 5), supports paleontological evidence indicating that compared to variation within species, the origin of a new species is a relatively rare event. It takes place only when both ecological conditions and the genetic structure of a population are particularly favorable for it. It is an important task for the evolutionist to define the conditions that are most favorable for speciation.

The classical opinion, still widely held, is that in outcrossing populations speciation requires, first, spatial isolation of a subpopulation from the remainder of the ancestral species, and second, divergent selection pressures that establish new barriers of reproductive isolation, based upon a large number of genetic differences. This *geographical theory of speciation* is outlined in Figure 7-1. Its

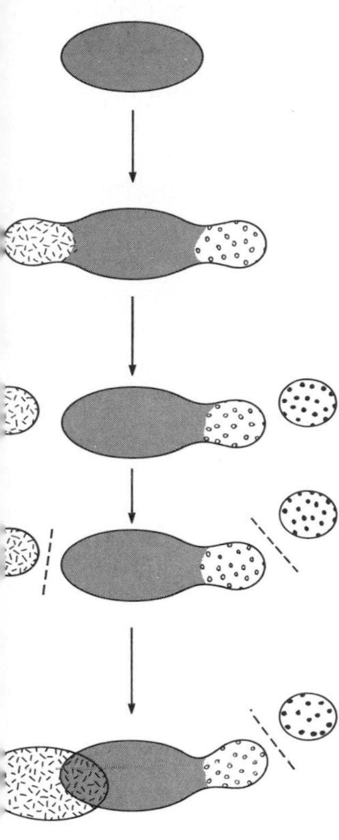

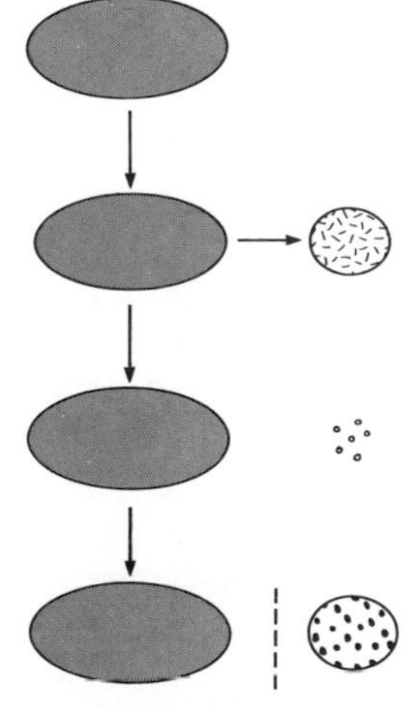

FIGURE 7-1.
Origin of a new species according to the conventional and quantum models of geographic speciation.

history, going back to naturalists of the late nineteenth century, particularly M. Wagner, has been outlined by Mayr (1963). The principal arguments in favor of it are as follows.

1. Numerous conditions intermediate between subspecies and full species are known among both animals and plants. Some of these have been described in Chapter 6, others will be discussed later in this chapter. They are often designated "semispecies."

2. Nearly all full-fledged species, even closely related ones like *Drosophila pseudoobscura* and *D. persimilis*, are separated from each other by more than one, often several, kinds of reproductive isolating mechanisms, each of which is determined by differences at a large number of gene loci. Hence the divergence that produced the differentiation of these species must have involved many successive steps. This gradual differentiation process could not continue in the presence of gene flow from the parental population.

3. Most closely related species differ from each other by morphological and physiological characters that are only quantitatively, not qualitatively, different from the characters that separate fully interfertile subspecies and partially interfertile semispecies. No evidence exists for the postulate that special genetic differences must arise before species can evolve, except for those that contribute to reproductive isolating mechanisms.

Ever since the speciation process became a separate topic in the 1930s, alternative methods of speciation have been proposed. The principal ones are *quantum speciation*, proposed by Simpson (1944) and discussed by Grant (1963, 1971) and Carson (1973, 1975), and called "saltational speciation" by Ayala (1975a); *parapatric speciation*, discussed most recently by Murray (1972) and Bush (1975); and *sympatric speciation*.

Quantum speciation is defined by Grant (1971, p. 114) as "the budding off of a new and very different daughter species from a semi-isolated peripheral population of the ancestral species in a cross fertilizing organism." The most characteristic example in plants is *Clarkia lingulata*, a narrowly endemic species of the central Sierra Nevada, clearly derived from the more widespread *C. biloba*, and differing from it in both floral morphology and karyotype (Figure 7-2). Other examples are known in various plant genera, often associated with the shift from predominant outcrossing to facultative or nearly obligate self-fertilization. The association of quantum speciation with translocations and other chromosomal changes, which occurs generally in the genus *Clarkia* (Lewis, 1966), is not a necessary accompaniment of the process. It has not taken place in some speciation events among Hawaiian species of *Drosophila*, described below.

The differences between quantum speciation and conventional geographical speciation are as follows.

1. Quantum speciation is rapid, requiring only a few generations.

2. The ancestors of new species do not include a large proportion of the populations belonging to the preexisting one, and may consist of only one or

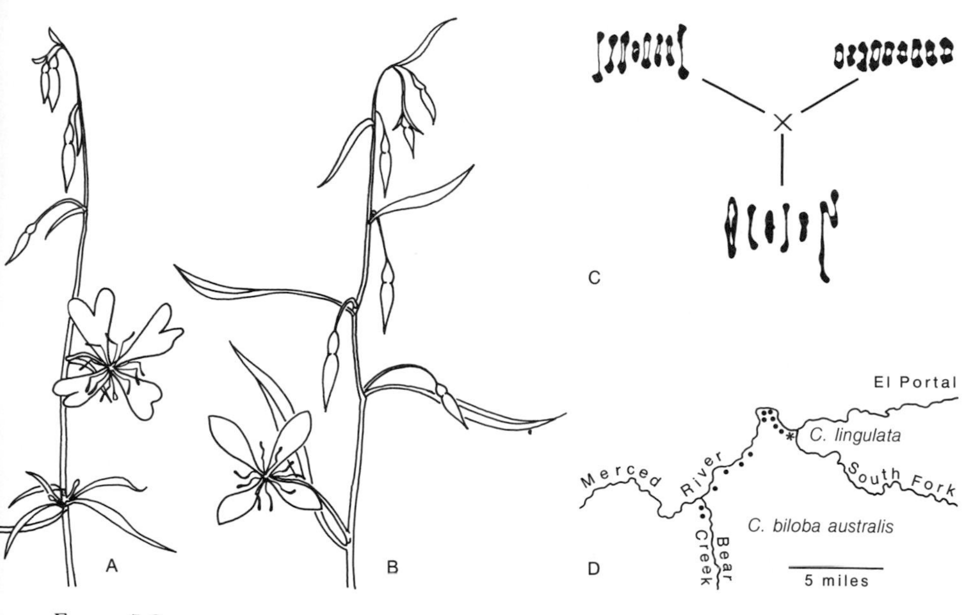

FIGURE 7-2.
The morphological, cytological, and geographical relationships between two species of *Clarkia* in the Sierra Nevada of California. **A.** Flowering branch of *C. biloba,* showing distinctive bilobed petal. **B.** Flowering branch of *C. lingulata,* showing similarity to *C. biloba* except for petal shape. **C.** Paired chromosomes, at meiotic metaphase, of *C. biloba* ($n = 8$) and *C. lingulata* ($n = 9$), which when crossed produce a hybrid (*C. biloba* $\times$ *lingulata*) showing strong homology between chromosomes of the parental species. The 17 chromosomes in the hybrid (8 from *biloba,* 9 from *lingulata*) are shown at meiotic metaphase forming four pairs, a ring of four, and a chain of five, indicating that the parental species differ from each other with respect to at least two reciprocal translocations of chromosome segments. **D.** Map of a small portion of the Merced River Canyon, in the Sierra Nevada of California, west of Yosemite Valley. The black dots (at left) indicate the southernmost populations of *C. biloba;* the asterisk (at right) indicates the two known populations of *C. lingulata.*

a few individuals. Conventional speciation is a process of splitting; quantum speciation is a budding process.

3. Conventional speciation may be promoted by drastic reductions in population size, but this is not necessary, as it is for quantum speciation.

4. Conventional speciation in its entirety is either guided by or is a by-product of natural selection. Quantum speciation usually and perhaps always includes one or more stochastic or chance events.

Two other models have been proposed that in some respects resemble quantum speciation as defined by Grant. The first is speciation via systemic or species-specific mutations, as proposed by Goldschmidt for animals, and by Lamprecht, Willis, and others for plants. The theoretical objections to these models are well known (Dobzhansky, 1941; Stebbins, 1950). Factual evidence against them comes from analyses of the genetic basis of reproductive isolating mechanisms in the progeny of partially fertile interspecific hybrids, such as *Drosophila pseudoobscura* × *persimilis* and *Phaseolus multiflorus* × *vulgaris* (Wall and York, 1957). In these and several other examples reproductive isolation is based upon differences in several different gene loci. The "species-specific mutation," postulated earlier for *Phaseolus*, could not be recognized. No genetic evidence exists for the monofactorial basis of a reproductive isolation barrier between two valid species.

Another type of model has been proposed by Carson (1973, 1975), on the basis of his analyses of speciation in Hawaiian *Drosophila*. The 500 or so species have almost certainly evolved during the last five to six million years, and a large proportion of them, endemic to the largest and most recent island, Hawaii, during the last 700,000 years. Each species is relatively homogeneous. Subspecies, which according to the classical theory should be very numerous in a group undergoing rapid speciation, are rare or absent. Morphological and cytological comparisons of closely related species on the same and on neighboring islands permit reconstruction of probable routes of migration that were associated with events of speciation. Although inversion polymorphism occurs within some of the species, it is less frequent than in many continental species, and poorly or not at all correlated with species differences. Comparisons of salivary chromosomes indicate that some species are homosequential and have similar banding patterns.

Carson interprets these facts as follows. Each species contains in its gene pool two kinds of variability, open and closed. Elements of the open system recombine freely via meiosis and fertilization in these outcrossed species, are regularly subject to natural selection, and are present in different frequencies in populations adapted to different environments. Their variation is similar to that which in most species, as in woody plant genera of Hawaii like *Metrosideros* and *Acacia*, gives rise to different races and subspecies. The closed variability system consists of elements that "under the usual conditions which prevail when natural selection is applying steady pressure . . . cannot be segregated from one another in such a way that a viable fertile organism of high fitness is produced." This condition results from the presence of blocks of gene loci, or "supergenes" (see Chapter 4). Genes making up supergenes are genetically linked and also interact to produce fitness in such a way that substitution of a new allele at any one of these loci, through either mutation or crossing over, drastically reduces fitness.

The closed system of variability can be altered only by means of a "genetic revolution" associated with a rapid increase in population size, followed by a crash, and subsequently a slower increase based upon survivors of the crash. During the flush—crash part of the cycle natural selection is greatly relaxed, and chance alterations of gene frequency can play an important role. The few survivors of the crash possess closed systems of variability that are different from those found in the original population.

Carson's genetic and ecological observations on Hawaiian populations of *Drosophila* are well explained by this hypothesis. Speciation often accompanies migration from one island to another or colonization of an area that has been devastated by a volcanic eruption. Under these conditions a few immigrants may enter an area unoccupied by competitive *Drosophila* populations, so that their descendant population can expand rapidly with relaxation of selection. As the population reaches a saturation level, competition within it may become intense and the food supply limiting, so that any slight modification of the environment can induce the crash. Among the survivors of this crash, a few may be able to recolonize the habitat. Most probably, however, they will be descendants of flies that during the first expansive phase underwent recombination of the genes that formed the previously closed system of variability. New mutations of genes belonging to this system may also have occurred. The recolonization process, in its later stages, will be subject to a renewal of natural selection, which will form new coadaptive relationships between gene loci, and thus erect a new closed system of variability incapable of combining with the old one. New barriers of reproductive isolation are thus erected, and a new species arises.

Geographical and quantum speciation are not mutually exclusive concepts. The quantum model is well suited to organisms that have low vagility and high fecundity, like small insects and annual plant species. It is much harder to visualize in large mammals, most birds, and forest trees, which have high vagility of either organisms or their propagules and cannot colonize a new habitat quickly and repeatedly. From the operational point of view, Carson's hypothesis has the advantage that it is stated in such definite terms, and involves such a small time span that it can be tested experimentally. Analyzing this and other speciation hypotheses by quantitative experimental methods is one of the most important tasks for evolutionists in the immediate future. The large body of data assembled by Ayala (1975) on genetic differentiation that accompanies speciation indicates that the acquisition of reproductive isolation may not require changes in a large proportion of the genome. This suggests that the "closed system of variability" postulated by Carson may be only a small segment of the total gene pool. If so, its rapid reorganization is not hard to visualize.

According to Bush (1975), parapatric speciation may occur "whenever species evolve as continuous populations in a continuous cline." It is virtually synonymous with stasipatric speciation as described by White (1973). It is visualized as a rapid process that, like quantum speciation, involves only relatively few individuals of the parental population. It differs from quantum speciation, however, in that no spatial isolation is required, and the erection of reproductive isolating mechanisms is believed to be guided entirely by natural selection. Like quantum speciation, it would involve a certain amount of inbreeding.

The strongest evidence for parapatric speciation found so far is the snail genus *Partula* on the island of Moorea (Murray, 1972). Since genetic data on these species are not available, no criteria exist for deciding which of two interpretations, parapatric speciation or quantum speciation similar to that postulated by Carson for Hawaiian *Drosophila*, is the most reasonable one. Two other examples postulated by Bush also lack genetic data: the morabine grasshoppers of Australia and the mole rats of the genus *Spalax* in the eastern Mediterranean region. Consequently, the question of whether or not parapatric speciation exists as a phenomenon different from quantum speciation must be left open.

Sympatric speciation, defined by Mayr as the "origin of isolating mechanisms within the dispersal area of the offspring of a single cline," has been proposed by various authors for a number of years, and appears to be supported by mathematical models as well as some laboratory experiments on disruptive selection. Other experiments have cast doubt on its probability or even possibility. Bush states that it "appears to be limited to special kinds of animals, namely phytophagous and zoophagous parasites and parasitoids." The sequence of events must be triggered off by transfer of a female to a new host.

Here also a distinction from quantum speciation will be hard to establish. If a female, by mutation or genetic recombination, can acquire adaptation to a new host and transmit it to her offspring, rapid colonization of this new habitat could take place. This would be inevitably followed by a crash as the host became saturated with parasites and moribund. Recovery would require migration to a new host, followed by natural selection for an entirely new adaptive complex of genes that would establish a self-perpetuating balance between host and parasite. In theory, therefore, there appears to be little to choose between quantum speciation and sympatric speciation within genera of parasites. Whether or not material differences exist between these two kinds of speciation cannot be determined until more experimental genetic and ecological data are at hand.

DIFFERENCES BETWEEN SPECIATION IN ANIMALS AND IN PLANTS

Morphological and cytogenetic patterns of variation are *prima facie* evidence of the more reticulate pattern of evolution in plants as compared to animals. At the level of the genus or even of groups of genera, plants' evolutionary "trees" may resemble a network of interwoven branches. In many groups of microorganisms the absence or rarity of sexual reproduction may cause the "tree" to consist of a mass of slender parallel or nearly parallel evolutionary lines, represented by related asexual clones, which anastomose (mix) from time to time when sexual reproduction or parasexual transfer of genetic material takes place.

Five conspicuous differences between animals and plants radically affect their respective patterns of speciation:

1. As Grant (1971) has pointed out, a plant is a much less complex organism than an animal. One way of quantifying this difference is to compare the number of different kinds of differentiated cells found in the adult body. Estimates made by Stebbins indicate that this number varies from 47 to 52 kinds of cells in

flowering plants, about 66 different kinds in an earthworm, 100 to 150 in an insect, and 200 to 250 in a mammal such as our own species. Moreover, the integration and delicate balance between organs required for motility and sense perception in animals are far greater than anything existing in plants. Finally, plants have an open system of growth, based upon embryonic tissues or meristems that occur at the ends of their shoots and are potentially immortal, while animal integration demands a finite body size and compactness, and youth, maturity, old age, and death are a necessary consequence. This difference affects speciation because when developmental patterns are relatively simple, the possibility that elements from two different patterns can be combined to make a functional intermediate is much greater than when they are highly complex.

2. Because of their highly developed sense organs, animals can easily develop barriers of reproductive isolation based upon divergent behavior patterns, i.e., ethological isolation. As shown in the previous chapter, such barriers are built up by natural selection with relative ease, and nearly always arise to reinforce isolation based upon other kinds of barriers. In many groups of plants, on the other hand, particularly those having promiscuous cross pollination by the wind, the possibility for building up prezygotic barriers of reproductive isolation is much less (Baker, 1959).

3. Because of their great capacity for vegetative growth, sterile hybrid plants in many groups may be virtually immortal. For instance, the Canadian pondweed (*Elodea canadensis*) was introduced into Europe as female plants about 1840; but no males of this dioecious species arrived. Nevertheless, between 1840 and 1880 it spread by vegetative means throughout the inland waters of Europe (Gustafsson 1946–1947, p. 46). Careful data on a land plant were obtained by Harberd (1961) with respect to the grass red fescue (*Festuca rubra*). By cultivating under controlled conditions samples taken from large quadrats and testing their self-incompatibility relationships, he could determine the limits of individual, asexually reproducing clones. One particular genotype, which had spread over an area more than 240 meters in diameter, was estimated to be hundreds and perhaps thousands of years old. More casual observations by numerous authors indicate that this situation is not exceptional, but is normal for rhizomatous plant species.

This means that sexual sterility is by no means as great a hindrance to occupation of habitat in plants as it is in animals. In perennial species, particularly those equipped with rhizomes or bulbs, a sterile hybrid can occupy indefinitely whatever habitat may be open to it. Moreover, even if it has a very low level of fertility, such that less than one per cent of the flowers can produce seed by sexual means, it can eventually produce many offspring that share its adaptive vegetative characteristics. An example of this situation is two species of perennial grasses in California, *Elymus condensatus* and *E. triticoides* (Stebbins, 1959). They are sympatric throughout much of northern California, but normally occupy different habitats: *E. condensatus* is typical of brushy hillsides, while *E. triticoides* grows in wet bottomlands, often along streams. Artificial hybrids between them are highly vigorous, but so sterile that among hundreds of florets examined no viable seeds have been found. Nevertheless, natural clones that resemble these hybrids in morphological characters and sterility are not uncommon and differ so greatly from one another that they must surely represent different genotypes. Some of these sterile intermediates occur several miles from the nearest individuals of either *E. condensatus* or *E. triticoides*, suggesting that they have been derived from occasional seeds produced by other hybrid clones.

4. For reasons that will be discussed later in this chapter, sterile hybrids of plants may acquire fertility by chromosome doubling or polyploidy more easily than animals. For this reason alone, reticulate evolution is likely to be much more common in plants than in animals.

5. Many plant species are both hermaphroditic and capable of self-pollination, so that uniparental reproduction by autogamy is for them a normal method of reproduction.

POLLINATION BIOLOGY AND SPECIATION IN PLANTS

The effect of efficient vegetative reproduction in blurring the boundaries between species should be clear from the facts presented in the previous section. On the other hand, two factors of pollination biology serve to increase the sharpness of species boundaries, and in some instances to multiply them so that the rate of speciation is sharply increased. These are predominant self-fertilization and specialized adaptation to insect pollinators.

Although relatively few species of plants are exclusively self-fertilizing, and species having predominant self-fertilization often contain an unexpectedly high amount of genetic variability (Allard and Kahler, 1972), nevertheless the genetic system of self-pollinators is different from that of cross-pollinating relatives. This tends to strengthen and multiply barriers of reproductive isolation, so that many groups characterized by predominant self-fertilization contain clusters of sibling species. This fact was first recognized by Baker (1959), who cited examples in *Primula* and *Armeria*. Grant (1971) has developed this topic further.

The shift from cross- to self-fertilization is not by any means the only way in which clusters of sibling species can arise in plants. They are well known in self-incompatible groups, such as *Holocarpha* and *Layia* (Clausen, 1951). The possible origin of such species clusters by means of quantum speciation has already been discussed.

Another way in which self-fertilization can promote the proliferation of microspecies is in association with permanent heterozygosity for translocations. North American species of *Oenothera*, subgenus *Euoenothera*, form a classic example of this condition, both because one of them, *O. lamarckiana*, introduced into Europe, served as the basis of de Vries's mutation theory, and because extensive work on the genus was carried out later by several geneticists (for reviews see Cleland, 1964; and Grant, 1971).

Some species of this subgenus, such as *Oenothera hookeri* of California and adjacent states, are large-flowered and mostly cross-pollinated. Their 14 chromosomes form seven bivalents at meiosis. Central and eastern North America, however, are populated by a swarm of "species" that are small-flowered, self-pollinating and are permanent translocation heterozygotes. At meiosis most of them exhibit rings of 14 chromosomes and no bivalents, although *O. lamarckiana* and a few others have a ring of 12 chromosomes and a single bivalent. In spite of their translocation heterozygosity they produce constant progeny because of the regular disjunction of the chromosomes at meiosis, with adjacent chromosomes in the ring going to opposite poles, and alternate chromosomes to the same pole (Figure 7-3). The chromosomes in the ring form two "complexes"; the members of each complex are transmitted together as effectively as though they were a single chromosome and the genes contained in them a single linkage group. The production of homozygous seeds is prevented by the peculiar mechanisms of two different nonallelic lethals, one being included in each complex. In some "species" these lethals act in early zygotes or embryos, but more often one of them eliminates the pollen and the other the embryo sac that carries it.

These species of *Oenothera* have succeeded in combining the advantages of heterosis, the assured seed set from selfing, and a genetic constancy that enables a well-adapted genotype to quickly occupy all habitats similar to the one in which it originated. They have, in fact, two grand supergenes, which combine only with each other to produce an adaptive heterozygote. Occasionally, recombinations between supergenes or new translocations can occur; these were some of the "mutants" that de Vries observed in his classical work.

Plants differ greatly among each other with respect to the specificity of their mechanisms for cross-fertilization. The least specific is wind pollination, since the pollen is produced in great masses, scattered at random, and except for isolation by distance is just as likely to reach flowers belonging to different species as the one from which it came. This condition has two effects. Divergence due to barriers of reproductive isolation is easily checked by gene flow, so that prolonged spatial isolation is required to build up effective barriers, and in some instances they apparently never arise. Many wind-pollinated species, particularly grasses, are self-incompatible, so that outcrossing is enforced. In others, such as most tree species inhabiting temperate climates, flowers are unisexual and separated on different branches or different trees.

Cross-fertilizing, wind-pollinated species usually have, or have had in the past, wide geographic ranges. Their populations, which inhabit different regions, are well differentiated from each other both morphologically and ecologically, so that subspecies, clines, and ecotypes are detected with relative ease, and reproductive isolation between them is often poorly developed. The condition of "semi-species," already mentioned, is common in them. Well-analyzed examples are *Pinus sylvestris, Plantago maritima* (Stebbins, 1950), and several species of grasses (McMillan, 1964).

An intermediate condition exists in species that are pollinated by a variety of unspecialized insects. Such species may have nearly or even as extensive geographic ranges as wind-pollinated species, and many of them contain clearly marked subspecies, clines, or ecotypes. This category includes the majority

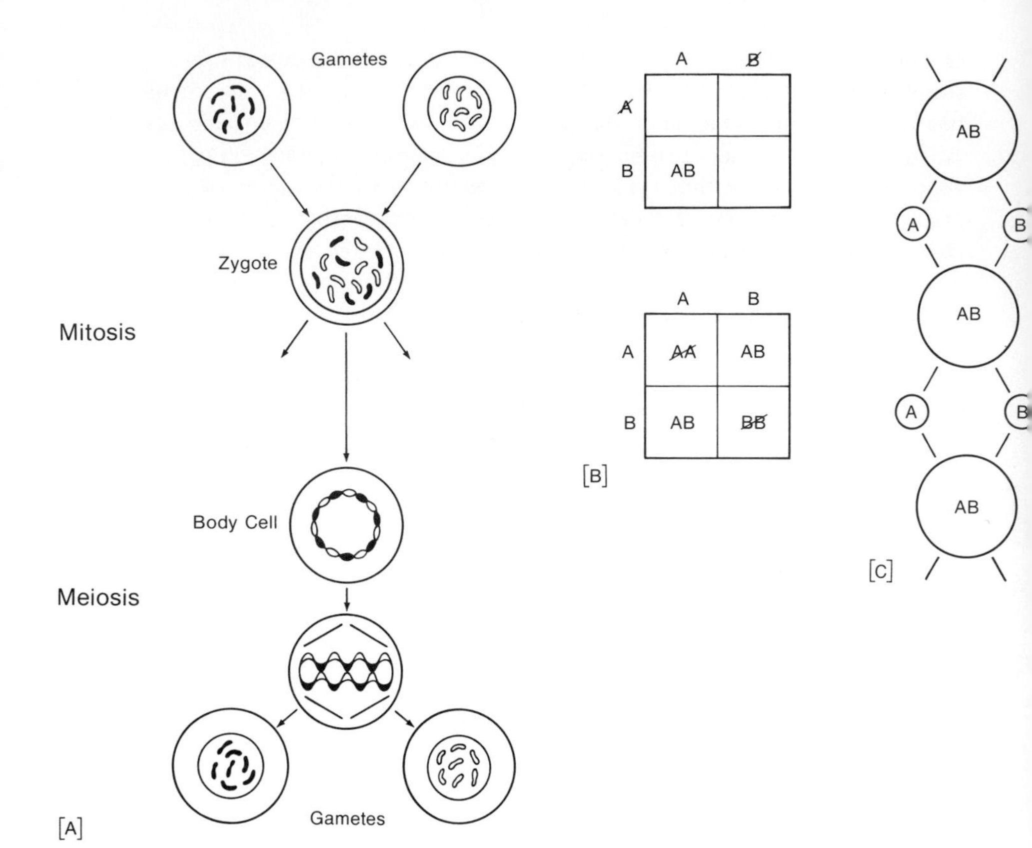

FIGURE 7-3.
Diagram showing how heterozygosity for translocations involving the entire chromosomal set plus a system of balanced lethals alters chromosomal segregation in evening primroses (genus *Oenothera*, subgenus *Euoenothera*). **A.** Gametes (eggs and pollen) containing chromosomes with different segmental arrangements form a zygote from which a new plant is derived. At meiosis, the chromosomes unite to form a ring of 14, which is oriented on the meiotic spindle in such a way that alternate chromosomes pass to the same pole, giving rise to only two kinds of gametes: one with a full set of chromosomes derived from the pollen parent, and the other with full set from the ovulate parent. **B.** In most species of *Oenothera*, balanced lethals act in such a way that one of the chromosomal complexes (B) is unable to direct the synthesis of pollen grains, causing them to abort, while the other (A) causes abortion of megaspores. Pollen is therefore 50 percent sterile, and all viable grains contain the A set of chromosomes. Since normally only one out of four megaspores can produce an embryo sac and egg cells, the abortion of megaspores containing the A set of chromosomes does not affect seed fertility: all gametophytes and egg cells are formed from one of the two megaspores that contain the B set. Thus all zygotes result from fertilization of an egg containing a B set and a pollen grain containing an A set, and have the constitution AB, generation after generation, as shown in **C.** [See Cleland, 1964.]

of species of several well-known families: pink (Caryophyllaceae), mustard (Cruciferae), buttercup (Ranunculaceae), rose (Rosaceae), parsley (Umbelliferae), aster (Compositae), and lily (Liliaceae). The best-known studies of ecotype differentiation have been conducted on species such as *Hieracium umbellatum, Senecio virgaurea, Potentilla glandulosa, Solidago sempervirens,* and the *Achillea borealis–Achillea lanulosa* complex (Stebbins, 1950), all of which are adapted to unspecialized insect pollinators.

The most extensive speciation based upon differential population biology occurs in those groups in which each species has flowers highly specialized for pollination by a particular vector, or a special method of transferring the pollen by a more generalized vector. The best known examples of this kind are in the orchid family. This family has evolved flowers having an architecture that is particularly susceptible to large numbers of variations on a single adaptive theme. They are bilaterally symmetrical, or zygomorphic because one of the six perianth members or tepals is modified into an elaborate structure known as the lip; the stamens and stigma are united to form a single structure, the column, which in each species produces a very specific relationship between the pollen-releasing anthers and the pollen-receiving stigma; and their sticky pollen mass cannot be transported by any means other than a particular vector (Figure 7–4). Unless the flower is visited by an insect or other visitor particularly adapted to one or a few related species, no seed is formed. On the other hand, all orchids are long lived perennials and produce enormous numbers of tiny seeds. A single successfully pollinated capsule of an orchid (*Cychnoches chlorochilon*) may contain as many as 3,770,000 seeds (Correll, 1950). Orchid populations can maintain their numbers even if a large proportion of their flowers produce no seed at all. Their remarkable vegetative persistence in suitable habitats, plus their amazing fecundity, permits them to rely upon rare visits by unusual pollinators and to pass through inadaptive bottlenecks from one pollination system to another. Although orchids are often regarded as rare plants, they are very common in the tropics, and in many tropical floras the orchid family exceeds all others in the number of species present.

The richness and diversity of orchid speciation depends upon both the variety of pollinators to which they have become adapted and the various ways in which pollination is effected. It is revealed by a wealth of description and elegant color photographs in the monograph by van der Pijl and Dodson (1966). The more primitive terrestrial genera are pollinated chiefly by solitary bees, but ichneumonid wasps and some species of Diptera are recorded for many species. The larger-flowered tropical species, from which the commercial kinds are derived, are pollinated chiefly by large carpenter bees (*Xylocopa* spp.). Among the numerous vectors of tropical species, many of which are not yet recognized, are hummingbirds (Trochilidae).

Among the more interesting vectors that orchids have added to their repertory are males belonging to various species of Hymenoptera. An example now familiar is the phenomenon of pseudocopulation, according to which the flowers attract sexually active males by mimicking females of a particular insect species

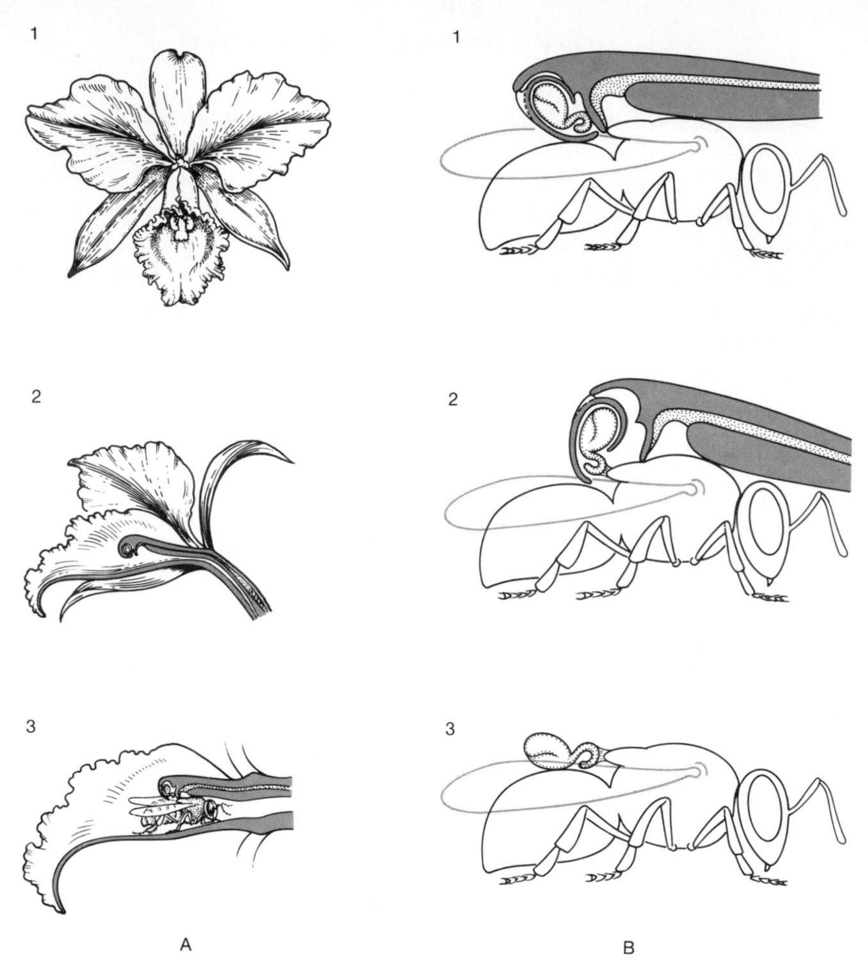

FIGURE 7-4.
Pollination of an orchid, a highly specialized and vector-specific flower. **A.** *A-1.* Face view of a well-known tropical species (*Cattleya*), showing the three narrower outer tepals and the three broader inner tepals, the lowest of which is the lip. The latter structure varies greatly in shape and ornamentation from one species of orchids to another. *A-2.* Longitudinal section of the flower. Partly enclosed in the lip is the stamen that contains pollinia, the stigma that receives pollen, and the column that supports them. *A-3.* A similar section, enlarged, showing the position of an insect visitor in the act of delivering pollen and receiving a new pollinium. As it moves back and forth, the insect (in this case a small bee) first deposits pollen on the stigmatic flap, then receives a pollinium that adheres to its thorax. **B.** Enlarged sectional drawings showing how the pollinium becomes stuck to the bee's thorax. [After van der Pijl and Dodson, 1966.]

with respect to the overall form and color pattern of the lip; and more particularly by subtle odors, produced by specific sesquiterpene lactones and hydrocarbons, as well as by coatings of hairs on the lip, which produce a tactile stimulus for their copulation reaction. Experiments have shown a great specificity of interaction between a species of orchid and its male hymenopteran pollinator (Kullenberg and Bergstrom, 1973). Species capable of forming fertile hybrids nevertheless maintain their distinctness, even when growing together sympatrically in large numbers, presumably because of pollinator specificity (Stebbins and Ferlan, 1956).

As was clearly demonstrated by Dodson and his co-workers (Williams and Dodson, 1972), male bees act as specific pollinators in another way. They gather fragrant compounds, transport them to a courtship area, and use them as sex attractants; the orchid flower serves as a sort of "beauty parlor" for the sexually ambitious male. Each attractant substance is specific to a particular flower and its pollinator. Moreover, Vogel (1972) has obtained evidence indicating that this kind of adaptation has arisen not only in tropical species adapted to male Euglossine bees, but also in the familiar temperate European genus *Orchis*, of which one species (*O. papilionacea*) produces substances attractive to male bees of the genus *Eucera*.

Numerous hybridization experiments (reviewed by Lenz and Wimber, 1959) have shown that almost the only barriers of reproductive isolation separating related species in many groups of orchids are of this kind. They have been carried out most extensively in the subtribe Laelieae, which includes *Cattleya* and others of the more showy kinds highly prized as corsages. Ten of the "genera" ordinarily recognized by taxonomists have been intercrossed in 20 "intergeneric" combinations. In other instances crossing has been possible between species classified into different subtribes. One might question the evolutionary validity of many of these "genera" and "subtribes." Whatever may be the taxonomic status of the species involved, the fact remains that floral isolations form the chief barriers in this family for establishing and maintaining distinct species. This condition, however, is not universal among orchids, as is shown by the presence of irregular meiosis in hybrids between species belonging to a single section of the genus *Vanda* (Kamemoto and Shindo, 1964).

Phenomena similar to those described for such highly specialized groups as the Orchidaceae and the genus *Ficus* exist to a lesser degree in many plant families. In the family Polemoniaceae, or phlox family, adaptive radiation for different pollen vectors has been clearly demonstrated (Grant and Grant, 1965; Stebbins, 1974a), and exists even among the subspecies of a single species, *Gilia splendens*. Levin (1970) has shown by means of a careful quantitative analysis that two species that are sympatric in Texas, *Phlox drummondii* and *P. cuspidata*, are kept distinct largely by the specificity of lepidopteran pollinators, which is aided by recognition signals inherent in the color pattern of the flower. In the genus *Penstemon* (Scrophulariaceae) a similar diversity of adaptations to pollinators has apparently been increased as a result of hybridization followed by spatial and ecological isolation of segregates, and selection for novel pollen vectors (Straw, 1956).

In another genus of Scrophulariaceae, the species *Pedicularis attollens* and *P. groenlandica* are kept distinct in spite of the fact that they are sympatric and are visited by the same species of *Bombus*. The architecture of their flowers is so different that the bee behaves very differently when visiting the different species, and so cannot transport pollen between different species. The most elaborate devices for differential pollination by the same vector are, however, in the genus

Asclepias, the milkweeds. Their sticky pollen is borne in sacs attached in pairs to clips, which can be lifted from a flower by the legs of a visiting insect and deposited in pockets next to the stigma of another flower. The fit between the sac of the pollen clip and the pocket of the stigma is so precise that although a pollen vector can carry on its legs the clips belonging to several different species, only those derived from a plant of the same species will fit the stigma pocket.

The effectiveness of these pollination mechanisms in producing the morphological diversity by which botanists recognize and separate species was shown by Grant (1949) in a survey of the diagnostic characters used by botanists for separating species in a great number of different genera. In genera of plants adapted to pollination by specialized animal vectors, a high proportion of the morphological characters upon which species distinctions are based are concerned with the perianth, stamens, and stigmas, which constitute the pollination mechanism. On the other hand, diagnostic characters in groups pollinated by less specialized vectors, or by wind or water, are based upon characters associated with vegetative adaptations or with seed development and dispersal (Figure 7-5).

POSTZYGOTIC ISOLATING MECHANISMS IN ANIMALS AND PLANTS

In Chapter 6 three kinds of postzygotic reproductive isolating barriers were described: hybrid inviability or weakness, hybrid sterility (either genic or chromosomal), and hybrid breakdown. The evolutionary role of these barriers depends largely upon their developmental basis. They are all expressions of the general phenomenon of disharmony of gene interaction: the genes derived from different species cannot interact with each other to produce a harmonious

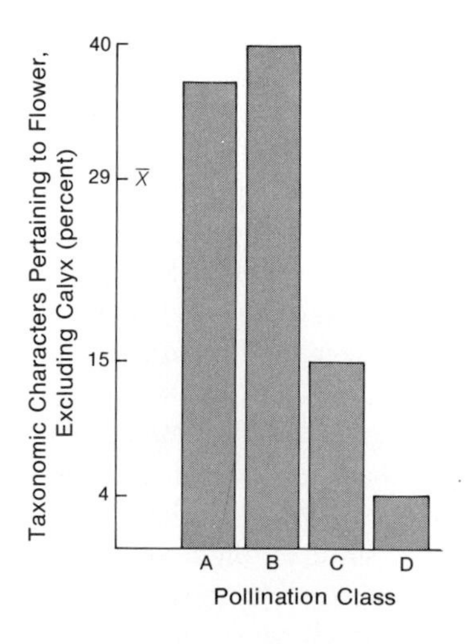

FIGURE 7-5.
The relative taxonomic importance of floral characters in flowering plants having different modes of pollination. In plants pollinated by birds (A), 37 percent of the characters used by taxonomists to separate species are differences in floral structure. In plants pollinated by bees and/or long-tongued flies, the proportion is 40 percent (B). In plants having flowers of a generalized type, which are pollinated promiscuously by unspecialized insects, the proportion is only 15 percent (C). In wind- and water-pollinated species the proportion drops to four percent (D). [After Grant, 1949.]

course of development. This disharmony is expressed in various ways. The primary distinction is between disharmony expressed in somatic tissues of developing F_1 hybrids and that which is manifest after meiosis in these hybrids. It may be delayed until gametophyte development (in plants) or even until the development of F_2 genotypes. Hybrid inviability or weakness is similar to "genic" hybrid sterility, the disharmony being manifest in somatic tissue of the F_1, while "chromosomal" hybrid sterility resembles hybrid breakdown in that the disharmony affects the formation of gametes and various later stages. For this reason, Stebbins (1966) has suggested that the terms *developmental sterility* (rather than "genic") and *segregational sterility* (rather than "chromosomal") are more appropriate. These terms will be used here.

Research directed toward a fuller understanding of these phenomena increases evolutionary knowledge in two ways. First, it promotes understanding of the causes of reproductive isolation, and second, analyses of the abnormal conditions brought about by genic disharmony during the development of an F_1 hybrid contributes to an understanding of regulation during normal development and differentiation. These conditions are best fulfilled by studies of early stages of development in F_1 individuals from wide interspecific or intergeneric crosses in animals. A review of earlier research (Stebbins, 1958) led to the conclusion that genic disharmony is largely stage-specific and due to the inability of the hybrid genome to carry out certain specific processes of differentiation. The three most critical stages are early cleavage, gastrulation, and the differentiation of gonads, including the meiocytes themselves. Disharmony at the latter stage is responsible for developmental hybrid sterility, and is manifest in the reduced testis size and abnormal spermatocytes of many hybrids, including those between closely related species or even subspecies, as discussed in the previous chapter.

The strongest barriers to hybrid development are disharmony in carrying out critical processes associated with gastrulation. Studies at the molecular level have shown that the development of eggs of echinoderms and amphibians up to the blastula stage is regulated chiefly or entirely by gene products transcribed from the maternal genes and present as messenger RNA in the cytoplasm of the unfertilized egg; massive activation of genes in the embryonic nuclei begins only at gastrulation. This activation has been studied in intergeneric hybrids of echinoderms by Denis and Brachet (1969a, b; 1970) in the lethal hybrid *Paracentrotus lividus* × *Arbacia lixula*, and by Whiteley and Whiteley (1972) in *Dendraster excentricus* × *Strongylocentrotus purpuratus*. In the former hybrid the amount of paternal DNA may be somewhat reduced, apparently because some paternal chromosomes are often lost during the mitosis of early cleavage divisions. On the other hand, the amount of RNA resembling that of paternal (*Arbacia*) embryos is considerably greater than the amount of this kind of RNA present in embryos containing complete genomes of the paternal species. This RNA is entirely of a kind produced in the parental embryos at many stages of development, including pre-gastrulation; RNA's specific to gastrulation stages are not present at all in hybrid embryos. One might infer from these experiments that regulation of both gene inhibition and activation is defective in these hybrids. On the other hand, up to gastrulation the *Dendraster* × *Strongylocentrotus* hybrids have apparently normal amounts of DNA and RNA, resembling that of both parents. Nevertheless, embryos at the final stage before blockage are deficient in several of the enzymes normally produced by the paternal genome (Whiteley and Whiteley, 1975). Either transcription is deficient for gene loci or, more probably, processes associated with migration of RNA's from nucleus to cytoplasm or with protein synthesis itself are blocked in these hybrids. In hybrids containing cytoplasm of loach (*Misgurnus*) and a haploid nucleus derived from sperm of goldfish (*Carassias*), RNA synthesis is likewise normal, but translation and protein synthesis are deficient (Neyfakh, 1971).

Additional biochemical evidence for the disturbance of gene regulation as a cause of hybrid inviability or weakness consists of the absence in adult hybrids of certain enzymes that are present in one or both parents (Ohno, 1969; Ornduff *et al.*, 1973), or the presence of novel compounds that appear through incomplete development of certain biosynthetic pathways (Levy and Levin, 1971; Ornduff *et al.*, 1973). In the sunfish family (Centrarchidae) this unbalance is evident in hybrids between species belonging to different genera or tribes, e.g., *Micropterus* × *Leopomis*, but not in hybrids between congeneric species (Whitt *et al.*, 1973).

Stage-specific hybrid inviability is much less marked in plants than in animals, but is nevertheless evident in early stages of embryo formation, in which the relationship between the embryo and the surrounding endosperm plays an important role. This kind of breakdown can often be avoided by extracting very young embryos and culturing them in vitro (Stebbins, 1950, 1958). It is expressed very differently in reciprocal hybrids, both between species having different levels of polyploidy (Stebbins, 1958) or between species having the same chromosome number but divergent developmental pathways, as in the genus *Primula* (Valentine and Woodell, 1963).

Up to now, information on genic disharmony has been obtained almost entirely from sexual hybrids. Nevertheless, some authors (Ephrussi, 1972; Levin, 1975) have pointed out that hybrid cell or tissue cultures may in the future contribute greatly to this problem. One adult amphidiploid hybrid produced by protoplast fusion is that between *Nicotiana glauca* and *N. langsdorffi* (Carlson *et al.*, 1972). It is indistinguishable from amphidiploids of the same combination produced by the usual sexual method. If hybrid organisms can be produced by this means from such wide crosses that sexual hybridization is impossible, even followed by embryo culture, this method will have exceptional value. Unsuccessful attempts to produce such hybrids by parasexual means will be valuable tools for analyzing hybrid weakness or inviability. If hybrid inviability is due to the inability of certain genes derived from different species to interact harmoniously during some particular developmental stage, tissue or organ cultures might be induced to circumvent this stage. If genes of one species are ineffective in directing development in the genic background of another species at many stages of development, the genomes of two species could be combined only in cells that metabolize and proliferate, but do not differentiate. Plant material is most favorable for testing such hypotheses, since adult individuals are in general more easily produced from tissue or organ cultures of plants than from cultured cells of animals.

Developmental hybrid sterility can be regarded as a special form of hybrid weakness in which the genic disharmony affects in particular the developing gonads or meiocytes. In animals it takes chiefly the form of reduced testes, which contain few or no primary spermatocytes, as in *Drosophila pseudoobscura* × *D. persimilis* (Dobzhansky, 1951, 1970). Both decreased viability and developmental sterility of hybrids tend to follow a generalization first recognized by Haldane and sometimes designated "Haldane's Law" (Stebbins, 1958): weakness or sterility is more strongly expressed in the heterogametic than in the homogametic sex. This result is entirely to be expected on the basis of genetic dominance and recessiveness. In the heterogametic sex, alleles of all genes located on the X chromosome are present in the hemizygous condition. Hence, deleterious recessives, which produce disharmony as a result of epistatic interaction with genes located on the autosomes, will always have this damaging effect. In the homogametic sex they are more likely to be "covered" by dominant alleles derived from the X chromosome of the opposite parent.

In some plant hybrids, such as those between certain species of *Elymus* and *Agropyron* (Stebbins, 1958), developmental sterility takes the form of much reduced stamens that lack pollen. In other instances, as in an interspecific hybrid of *Paeonia*, the stamens are replaced by rudimentary sterile carpels. In still other hybrids of both plants and animals, chromosome pairing is nearly or quite normal, indicating homology of the parental genomes, but disturbances of the spindles and other cytoplasmic structures of the meiocytes cause them to degenerate without forming gametes.

Both hybrid weakness or inviability and developmental hybrid sterility are much more common in animals than in plants. The condition found in many hybrids between subspecies of *Drosophila*, in which males have well-developed developmental sterility but females are entirely fertile (Ayala *et al.*, 1974b), is not known in any plant groups. This difference can easily be understood on the basis of the fact, stated earlier in this chapter, that developmental patterns are usually more complex in animals than in plants.

Segregational hybrid sterility is much more common in plants than in animals. It may be manifest in the form of reduced chromosome pairing, formation of multiple associations of chromosomes, or other abnormalities of chromosome behavior at meiosis, but this is not necessarily the case. After normal meiotic pairing, disharmonious combinations of genes may be segregated to the gametes or spores. The disharmony is produced by recessive genes having unfavorable epistatic interactions; such genes are absent from diploid somatic tissue since they are masked by dominants that do not have unfavorable effects. This kind of sterility is frequent in plants because the products of meiosis are not gametes but spores, which germinate and grow, giving rise to haploid gametophytic tissue. In higher plants this tissue is reduced on the male side to a two- or three-nucleate structure, the pollen grain, and on the female side to the eight-nucleate embryo sac. Nevertheless, the growth of pollen grains in particular requires the synthesis of many proteins and other macromolecules that are coded by genes in the haploid nucleus.

The operational definition of segregational as opposed to developmental sterility was made by Dobzhansky (1937) using the terms "chromosomal sterility" and "genic sterility." Hybrids having segregational sterility can be rendered fertile by doubling the chromosome number to produce amphidiploids or allopolyploids: those having developmental sterility remain sterile even when polyploid. This is because dominant genes having disharmonious epistatic interactions during gonad development will exhibit these effects equally effectively, or even more so, when present as duplicates. On the other hand, doubling the chromosome number of a hybrid affects chromosomal pairing in such a way that completely homologous chromosomes, derived from a single chromosome that entered the diploid hybrid, can pair preferentially with each other, thus eliminating chromosomal irregularities at meiosis (Stebbins, 1950, 1971). The result of preferential pairing is that most gametes of the doubled hybrid receive the same complement of chromosomes as that existing in the somatic cells of the undoubled hybrid (Figure 7-6). Since such chromosomal complements have already been screened for normal gene action during the development of the hybrid prior to meiosis, they can be expected to produce normal gametophytes.

Two separate hypotheses can be offered for the origin of developmental and segregational hybrid sterility. In both instances these barriers arise as by-products of natural selection for divergent adaptations, but the action of selection is different. Developmental hybrid sterility, as well as hybrid inviability and weakness, is a by-product of divergent selection for different developmental sequences, associated with adaptation to different climatic regimes or other environmental factors. A classic example is the group of frogs once regarded as

---

FIGURE 7-6 (opposite)
Preferential pairing in hybrid polyploids. **A.** The chromosome complements at somatic metaphase and at first metaphase of meiosis for two hypothetical species, each having the gametic number $n = 3$. Nonhomologous segments of chromosomes are indicated by grey in one species and black in the other; nonhomologous segments differ from each other with respect to a number of small rearrangements. Segments that are homologous in the two species are shown white. **B.** Somatic complement and bivalents at meiotic metaphase in an undoubled F_1 hybrid between the two species. All chromosomes are paired by union of homologous segments, although the pairing is less intimate, and fewer chiasmata are formed than in nonhybrid plants. **C.** Somatic complement and meiotic metaphase in a plant derived by doubling the chromosome number of the F_1 hybrid. At meiosis, only bivalents are formed, since the completely homologous chromosomes derived from a given parent pair with each other, preventing the partial homology of chromosomes derived from different chromosomes from expressing itself.

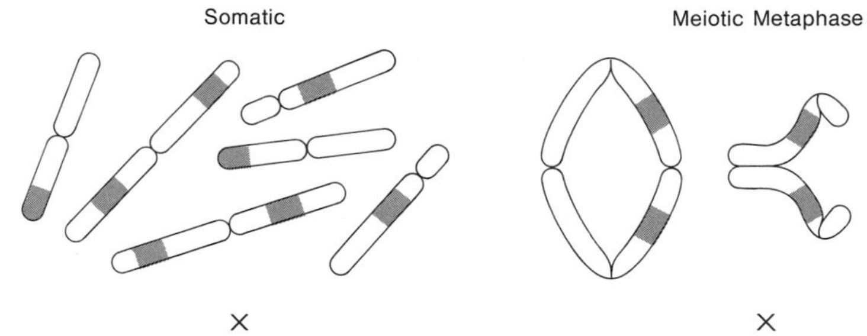
Somatic
Meiotic Metaphase
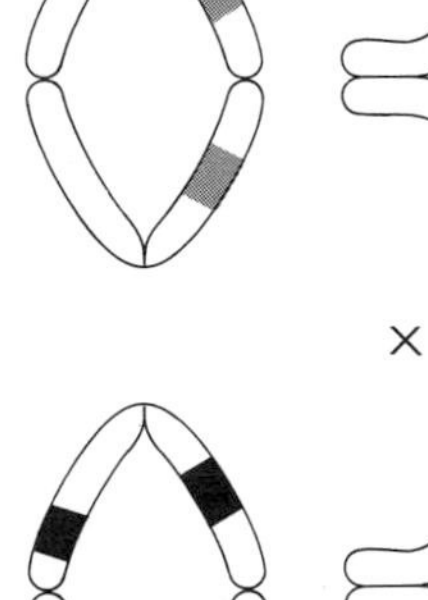
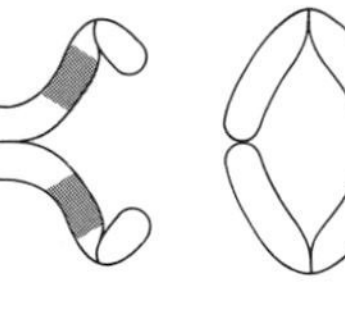
×
×
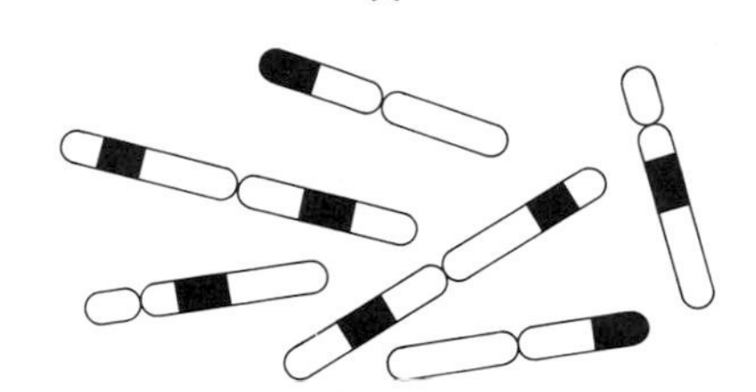
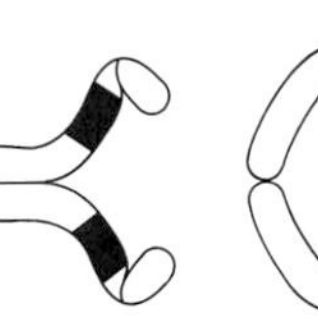

A
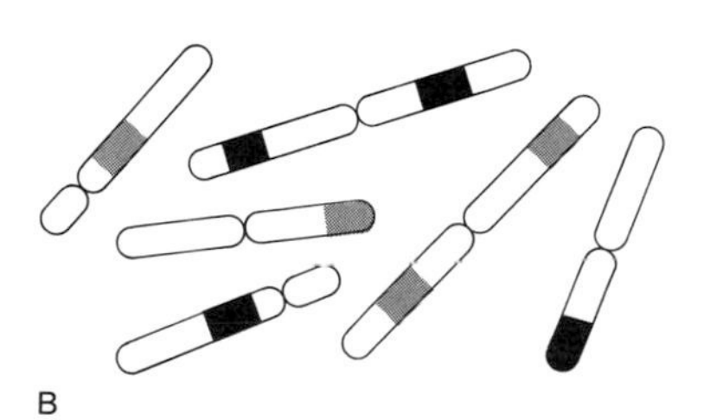
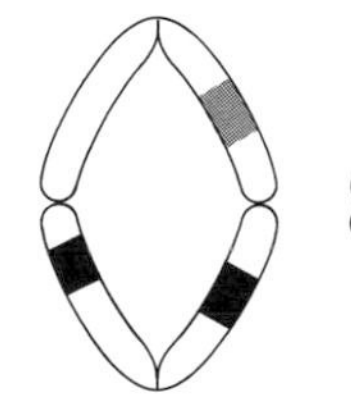
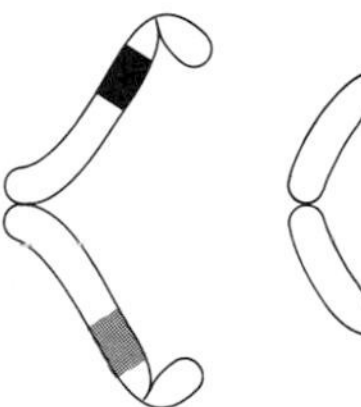
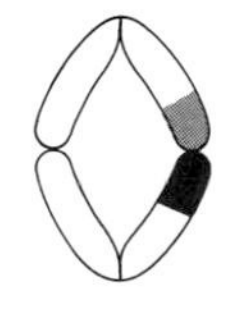
B
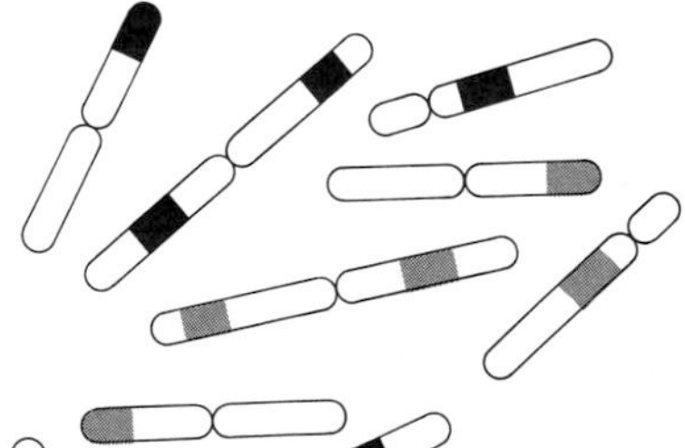
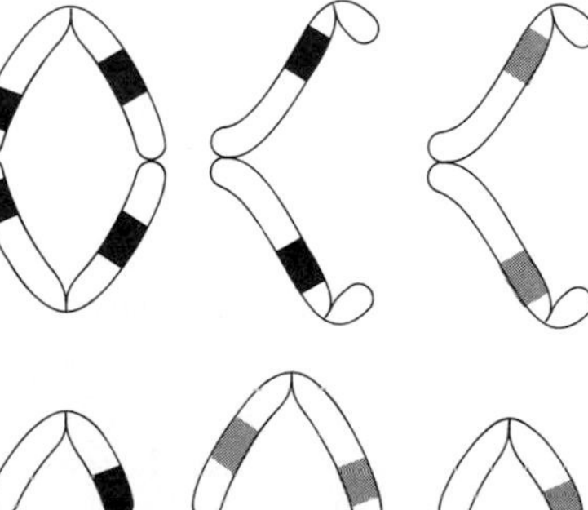
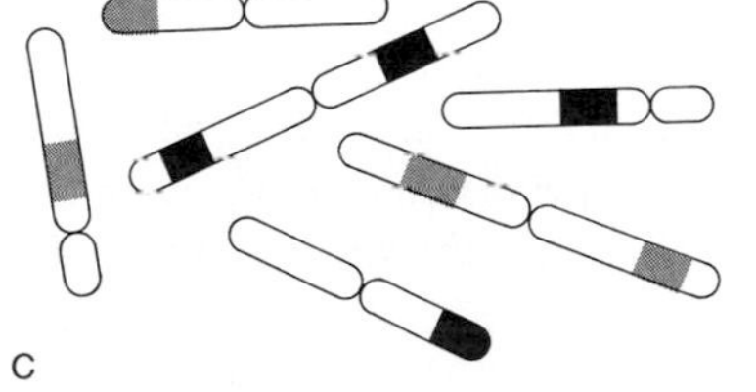
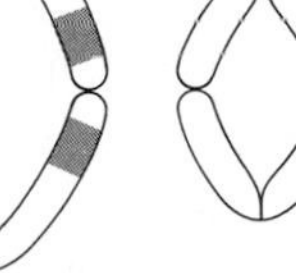

C

races or subspecies of *Rana pipiens* but now recognized as a cluster of sibling species (Fowler, 1964; Dobzhansky, 1970). If the species from Vermont is crossed with that from Florida, the F_1 embryos from Vermont eggs and Florida sperm have abnormally large heads and abortive, rudimentary tails, while Florida eggs fertilized by Vermont sperm give embryos with rudimentary heads and hypertrophied tails. These results can be explained on the basis of disharmony between regulator substances already present in the unfertilized egg, and those coded by genes of the hybrid during gastrulation and later developmental stages. Similar disharmonies between species having different rates of gonad maturation could give rise to developmental sterility, except that the difference would not be reciprocal, since in such examples the critical stages occur so late that they would not be affected by gene products in the unfertilized egg.

Segregational hybrid sterility, on the other hand, is based primarily upon differentiation with respect to the structural organization of the chromosomes, i.e., chromosomal repatterning. It originates in selective pressures favoring the establishment of a succession of new chromosomal arrangements. These can be inversions, translocations, or transpositions of small segments of chromatin from one region to another of the same chromosome. Any of these changes can interfere with either the normal pairing of chromosomes or the regular segregation of opposite alleles to the gametes, or both. An important principle in this connection is that the amount of segregational sterility in a hybrid depends chiefly upon the *number* of different rearrangements for which the hybrid is heterozygous, rather than the amount of material involved in a particular chromosomal rearrangement. Transposition or translocation of a segment containing no more than two or three vital gene loci can produce as much or more sterility as translocations involving whole arms. On the other hand, both the morphology of the karyotype and the nature of chromosome pairing at meiosis are more likely to be affected by one or a few conspicuous changes, so that these are the most likely to be noticed and to receive attention from cytogeneticists. The hypothesis known as cryptic structural hybridity (Stebbins, 1950) states that in many instances the chromosomal repatterning involves segments so small that individually they have little visible or no visible effect on meiosis, so that in spite of a high degree of pollen sterility, chromosome pairing is nearly or quite normal. Evidence for it is of three kinds: (1) differences between partly homologous chromosomes with respect to fine details of structure revealed by special techniques; (2) minor irregularities of pairing at pachytene in hybrids between closely related species; and (3) the preferential pairing of completely homologous chromosomes, which causes polyploids derived from many hybrids to form few or no multivalents, even though at the diploid level a complete set of bivalents is formed (for a review see Stebbins, 1971). In *Gossypium* the genetic effects of preferential pairing have been determined quantitatively, by recording abnormal segregation ratios for certain marker genes in artificial hybrids.

Two explanations have been offered for chromosomal repatterning. One of them is the "catastrophic speciation" hypothesis of Lewis (1966), which postulates that as a result of environmental stress a sudden repatterning of the chromosomes can occur, involving simultaneous rearrangements of many segments. The resulting individual gives rise to a new species sympatrically by uniparental reproduction. Lewis's examples are drawn from his extensive investigations of the genus *Clarkia* (Onagraceae), in which translocations are particularly common, even within individual populations. The species believed to have originated by this method are largely self-fertilizing. The theory would be hard to apply to a cross-fertilizing species, in which uniparental reproduction is impossible. However, one of the examples Lewis cites, *Clarkia franciscana*, has been reinvestigated on the basis of enzyme differences (Gottlieb, 1973), casting considerable doubt upon the "catastrophic" interpretation. Another series of examples, the *Clarkia unguiculata* complex, can just as easily be explained by postulating quantum speciation, accompanied by the establishment of successive chromosomal changes. Finally, Lewis and Bloom (1972) have found that still another localized species, *C. nitens*, which was described because it was thought to be reproductively isolated by a system of translocations, is now becoming merged with a related species, *C. speciosa*, apparently because the system of translocations has not been effective enough to keep the entities apart in the absence of other barriers. Lewis's hypothesis serves at best to explain certain special cases, and cannot account for the general phenomenon of chromosomal repatterning.

A second possibility, suggested by Darlington (1940) and Stebbins (1950), is that repatterning is based largely on divergent selection for "supergenes," or adaptive complexes of genetically linked loci. New linked combinations, and hence new arrangements of chromosome segments, are favored when populations enter new habitats. Some chromosomal divergence can be expected within a single species, but this may or may not be enough to bring about partial sterility. Effective barriers are the product of a long succession of environmental shifts accompanied by chromosomal rearrangements. This hypothesis implies two predictions. First, in groups undergoing active speciation imperfect barriers of segregational hybrid sterility will exist in some hybrids between populations of the same species. This condition has been observed in a number of plant groups, such as the Compositae, tribe Madiinae (Clausen, 1951; and see Figure 7-7), and the genus *Clarkia* (Lewis, 1966). Second, given proper techniques for detecting them, differences in pattern should be recognizable between chromosomes of different populations of a species, and these should have a regular ecological distribution. This condition is well demonstrated in the genus *Trillium*, both *T. kamschaticum* of eastern Asia and *T. ovatum* of Pacific North America (Stebbins, 1971). This hypothesis is the most plausible one to account for most examples of speciation involving principally segregational hybrid sterility. Some of these are likely to be rapid, and have the characteristics of quantum speciation. In that case, rearrangement of chromosomal segments that leads to segregational sterility can greatly help the reorganization of the closed system of variability, as postulated by Carson (1975).

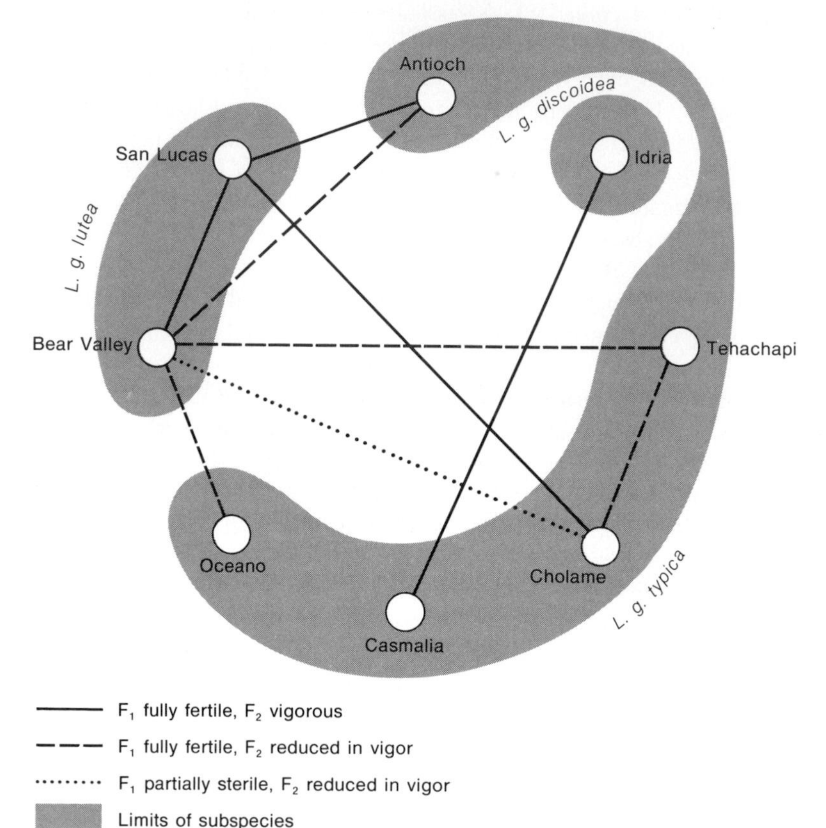

FIGURE 7-7.
Different levels of sterility in artificial hybrids between various populations of *Layia glandulosa,* an annual species of tarweed (family Compositae) native to California. The development of reproductive isolating barriers is only partly correlated with the morphological differences between subspecies. Moreover, two populations, from Antioch and Bear Valley, appear to be isolated from each other only if the direct hybrid between them is considered, but gene exchange between them is possible via an intermediate population, San Lucas. (All populations have a haploid complement of eight chromosomes.) The frequency of situations like this in genera analyzed by numerous hybridizations lends support to the classical hypothesis of speciation, according to which the formation of partially isolated populations form intermediate steps in the speciation process. [After Clausen, 1951.]

HYBRIDIZATION AS AN EVOLUTIONARY CATALYST

Hybridization between populations having different ecological norms or adaptations can have three kinds of results. One is the enrichment of the gene pool of a species, particularly in association with the colonization of a new habitat. This is most effective when followed by introgression, or backcrossing with subsequent fixation of backcross types. Numerous examples have been described in plants. One of the most striking is between two species of shrubs inhabiting the arid regions of western North America, *Purshia tridentata* and *Cowania stansburiana* (Stutz and Thomas, 1964). In this instance the parental species are placed by taxonomists in different genera, and the probable stabilized introgressant, which has a wide geographic distribution, is classified as a separate species, *Purshia glandulosa* (Figure 7-8).

Second is the production of such extensive hybrid swarms that boundaries between formerly distinct subspecies or even species are blurred out of existence. This has happened in the case of *Clarkia nitens* and *C. speciosa* (Lewis and Bloom, 1972) and is approached in populations of *Quercus dumosa–Quercus turbinella* in southern California (Tucker, 1952). This effect is most likely to occur as a result of habitat disturbance by man. In many instances the disturbance produces "hybridization of the habitat." A classic example is the clones of *Iris* inhabiting the Mississippi Delta. Before human disturbance two distinct sympatric species existed: *I. fulva* in well-drained forests along the edge of rivers and streams, and *I. hexagona* var. *giganticoerulea* in open, poorly drained swamps. Cutting of forests and draining of swamps have produced every possible combination of the two sets of conditions represented by the formerly distinct habitats. As a result, an enormous diversity of hybrids and hybrid derivatives have become established in these new habitats, exhibiting a great variety of color patterns and forms of flower (Anderson, 1949).

Based upon situations like these, Grant (1971) makes an important distinction between "hybridized" and "open" habitats. A hybridized habitat is one that combines characteristics of those occupied by the parental species, but may be subject to extensive competition from other species, and also limited in extent, so that the populations of hybrid origin may be restricted in both space and time. An open habitat is one in which interspecific competition is reduced; it may of course also have the characteristics of a hybridized habitat. The greatest opportunity for the success of hybrid derivatives is of course in habitats that are both "hybridized" or intermediate and open. Such habitats arise frequently as a result of human disturbance, particularly open fields and roadsides, but natural phenomena, such as landslides, river flooding, volcanic eruptions, and glaciation, can also bring them about. The progeny of hybrids are most likely to achieve success in habitats that either are intermediate or share certain characteristics of the habitats occupied by the parental species, and are open, with reduced competition. The success of *Purshia glandulosa*, mentioned above, is probably due to the fact that it has become adapted to a habitat of this kind. A similar example in animals is *Pipilo erythrophthalmus* × *ocai*, the towhees of Mexico

FIGURE 7-8.
Leaves, petals, and calyces (plus hypanthia or floral cups) of several plants belonging to two closely related genera of the rose family, *Purshia* and *Cowania,* which are xeric shrubs occurring in western North America. At left, *Purshia tridentata* (Pursh) DC; at right, *Cowania mexicana* var. *Stansburiana* (Torr.) Jeps. Right center: a natural hybrid between *P. tridentata* and *C. m.* var. *Stansburiana,* collected with its parents near Ephraim, Utah. This hybrid is highly sterile, but produces a few seeds when pollinated with pollen from *P. tridentata.* Left center: *Purshia tridentata* var. *glandulosa* Jones. In many characteristics this plant resembles the natural hybrid shown to the right of it, and in all its characteristics it either resembles *P. tridentata* or is intermediate between that species and *C. m.* var. *Stansburiana.* It is therefore regarded as the product of introgression between these two species. [After Stebbins, 1959.]

(Sibley, 1954), which have produced hybrid swarms well adapted to the intermediate and relatively open habitats provided by degraded oak forests in west central Mexico.

The third possible result of natural hybridization is hybrid speciation, as defined by Grant (1971). If hybridized and open habitats are available, a small proportion of the descendants of a partly fertile species hybrid may become stabilized and give rise to a population within which free gene exchange is possible, but which is reproductively isolated from the parental population by one or more barriers. Grant defines hybrid speciation as the origin of a new species directly from a natural hybrid. He includes stabilization of permanent heterozygotes, as in *Oenothera*; amphiploidy; hybrid speciation with external barriers; and recombinational speciation.

Hybrid speciation with external barriers occurs chiefly between species that are strongly isolated from each other with respect to prezygotic mechanisms, such as ecological isolation and differences in floral structure as well as in pollinators. Grant cites two well-known examples: the origin of *Delphinium gypsophilum* from *D. hesperium* × *recurvatum* and that of *Penstemon spectabilis* from *P. centranthifolius* × *grinnellii*. Particularly in the case of *Penstemon* the availability of an intermediate, open habitat plus a new pollinator were almost certainly essential for the successful establishment of the new species. There is good reason to believe that in plant groups having specialized pollination mechanisms, such as the Orchidaceae and the genus *Ficus*, hybrid speciation of this kind has had an important role in increasing the number of species. A preliminary study of two species of the genus *Ophrys* in the Mediterranean region shows that a recognized species, *O. murbeckii*, has originated in this fashion (Stebbins and Ferlan, 1956).

Recombinational speciation is the result of recombination, in the progeny of a semisterile hybrid, and of the factors responsible for postzygotic isolating mechanisms, usually segregational hybrid sterility. The diagram in Figure 7-9 shows how this can happen. Experimentally produced examples are certain lines derived from the artificial hybrid *Nicotiana langsdorffii* × *sanderae* and from *Gilia malior* × *G. modocensis* (Grant, 1971).

Grant has revealed a phenomenon that can be a strong deterrent to recombinational speciation: the linkage of genes affecting viability or fertility to those responsible for morphological differences. If such linkages are numerous, the only gametes likely to function and produce fertile derivatives are those bearing an almost complete set of alleles similar to those present in one of the parental species; therefore the only fertile derivatives of a hybrid will be those similar to one or the other of these species. Because of this phenomenon, recombinational speciation is most likely to succeed if the hybrids are only partly sterile and if the number of chromosome pairs, i.e., linkage groups, is high. Because of these restrictions, the occurrence of recombinational speciation is greatly favored by the availability of suitable open habitats, which enable the genetically stabilized derivative to migrate into an area separate from its parents and thus subject it to new selection pressures.

THE OCCURRENCE OF SEMISPECIES IN PLANTS

Examples of semispecies among animals have been described in Chapter 6. Because of the slowness with which barriers of reproductive isolation accumulate in many plant groups, semispecies are particularly common in higher plants. Some of these are similar to the birds of paradise mentioned in Chapter 6. They consist of completely allopatric populations, often separated on different continents, which are distinguished from each other by numerous, easily recognized,

	Species 1		Species 2
Parents	$A_1A_1B_1B_1C_1C_1D_1D_1$	$\times$	$A_2A_2B_2B_2C_2C_2D_2D_2$

$\downarrow$

F$_1$ Hybrid: $A_1A_2B_1B_2C_1C_2D_1D_2$

$\downarrow$

Gametes Produced By F_1 Hybrid				
	$A_1B_1C_1D_1$	$A_1B_2C_1D_1$	$A_1B_2C_1D_2$	$A_1B_2C_2D_2$
	$A_1B_1C_2D_2$	$A_2B_1C_1D_1$	$A_2B_1C_2D_1$	$A_2B_1C_2D_2$
	$A_2B_2C_1D_1$	$A_1B_1C_2D_1$	$A_2B_1C_1D_2$	$A_2B_2C_1D_2$
	$A_2B_2C_2D_2$	$A_1B_1C_1D_2$	$A_1B_2C_2D_1$	$A_2B_2C_2D_1$

(first column) Viable (25 percent); (remaining columns) Inviable (75 percent)

F_2

The four types of viable gametes produce 10 different F_2 combinations, four of which are fully fertile:

$A_1A_1B_1B_1C_1C_1D_1D_1$ $A_1A_1B_1B_1C_2C_2D_2D_2$

$A_2A_2B_2B_2C_2C_2D_2D_2$ $A_2A_2B_2B_2C_1C_1D_1D_1$

These two combinations are the two parental genotypes (first column). These two will produce partly sterile hybrids when crossed to either parent (second column).

Generalized Formulas

Assume parents differ with respect to n pairs of complementary gene loci:

Proportion of fertile gametes produced by F_1 hybrid: $\frac{1}{2^n}$

Proportion of balanced fully fertile F_2 combinations: $\frac{2}{2^n + 1}$

Proportion of fully fertile F_2 combinations that will produce partly sterile hybrids when crossed to either parent: $\frac{2^n - 2}{2^n}$

Proportion of fully fertile F_2 combinations that will have the genotype of one or the other parent: $\frac{2}{2^n}$

FIGURE 7-9.
Recombinational speciation in descendants of hybrids characterized by segregational sterility. Two parental species may differ from each other with respect to several sets of complementary genes, of which the products interact in such a way that products coded by Species 1 are not complementary to products coded by Species 2. For instance, A and B may be genes that code for polypeptide chains of a multimer enzyme of such a nature that the chain produced by allele A_1 produces a functional enzyme when combined with that produced by allele B_1, but not with that produced by B_2. Similarly, the products of A_2 and B_2, but not A_2 and B_1, may form a functional combination. In the figure it is assumed that there are two such complementary gene pairs segregating simultaneously. The genes A and B form one pair, the genes C and D form a different pair. The pedigree shows that the F_1 hybrid between Species 1 and Species 2 produces gametes, of which 25 percent are functional. There are 16 F_2 zygotic combinations

and constant morphological differences, but which form fully fertile F_1 hybrids when brought together and crossed artificially. Well-known examples are four species of the genus *Platanus* occurring, respectively, in the Mediterranean region (*P. orientalis*), eastern North America (*P. occidentalis*), Arizona (*P. wrightii*), and California (*P. racemosa*) (Stebbins, 1950; Grant, 1971), Similar examples exist in several other genera, particularly of woody plants, such as *Catalpa* and *Castanea*. Other examples consist of pairs of species or subspecies, sympatric over a large part of their geographic distribution, and which in some regions exist side by side and retain their identity for indefinite periods of time, while in other regions they form extensive hybrid swarms. This condition can arise only when the species are well-differentiated ecologically, and the hybrid swarms are to be expected in newly colonized or disturbed areas. The list of examples of this kind is very large: Grant (1971) cites them in the genera *Gilia, Ipomopsis, Aquilegia, Diplacus* (= *Mimulus,* in part), *Juniperus, Geum, Iris, Betula, Salix, Melandrium, Nothofagus,* and *Quercus.* None of these examples involve polyploidy. The greater frequency of such examples in plants than in animals is probably due partly to the lack of ethological isolation between plant species, plus the fact that sympatric species of plants may often be confined to contrasting habitats. The frequent occurrence of semispecies is one reason why many botanists are skeptical of the value of the biological species concept.

Semispecies of plants are not always, and perhaps not usually, species *in statu nascendi.* A clear illustration is provided by *Quercus douglasii-dumosa-turbinella* in California. In the coast ranges of northern California the deciduous tree *Quercus douglasii* exists sympatrically with the evergreen and shrubby *Q. dumosa* and hybrids between them are rare. Furthermore, populations of the two species in this area show few signs of introgression (Tucker, 1952). By contrast, populations in the interior coast ranges of central and southern California are almost continuous hybrid swarms involving *Q. douglasii, Q. dumosa,* and *Q. turbinella.* The latter is a small evergreen tree that in its most typical form occupies arid mountain areas in Arizona. The regional difference can have two explanations: (1) habitats in northern California are older and more stable than in the southern area, and (2) the presence of *Q. turbinella* provides a bridge for transferring genes from *Q. douglasii* to *Q. dumosa.* The importance of this example lies in the fact that fossil leaves precisely similar to those of modern *Q. douglasii, Q. dumosa,* and *Q. turbinella* are well known

---

among the four kinds of viable gametes produced by the F_1 hybrid. Of these 16 combinations, 10 are different from each other. Four of the 10 will be fully fertile, i.e., all their gametes will be viable; the other six will produce some inviable gametes. Two of the four F_2 fully fertile zygotic combinations will form partly sterile hybrids when backcrossed to either parent. The bottom of the figure gives generalized formulas for the case in which parents differ with respect to n pairs of complementary gene loci.

from deposits of Pliocene and even Miocene age (Axelrod, 1939). The counterparts of the modern species have existed and remained distinct for at least 10 million and perhaps as many as 15 million years, without evolving perfect barriers of reproductive isolation. The same is true for other woody genera, such as *Platanus* and *Castanea.*

THE EVOLUTIONARY SIGNIFICANCE OF POLYPLOIDY

The most widespread and distinctive cytogenetic process of speciation in higher plants is polyploidy, the multiplication of entire chromosomal complements. Polyploid series such as the somatic numbers 14, 28, and 42 in wheats, 18, 36, 54, 72 and 90 in *Chrysanthemum*, and 26, 52 in cotton are now very well known and have been extensively studied. Such series exist in at least 30 per cent of genera of flowering plants, and are even more common among ferns. Moreover, evidence accumulated in recent years strongly suggests that most genera of flowering plants having basic gametic numbers of $x = 12$ or higher have been derived by ancient events of polyploidy (Stebbins, 1971). If this is true, then all of the modern species belonging to such prominent families as Magnoliaceae, Winteraceae, Fagaceae, Juglandaceae, Salicaceae, Ericaceae, and Oleaceae have been derived at least originally by means of polyploidy.

Why has polyploidy been of such overriding importance? In the first place, it contributes characteristics that by themselves have an intrinsic value, particularly larger flowers of firmer texture and larger seeds. However, it does not necessarily increase the overall vigor of the plant; in fact, many autopolyploids derived from a single diploid population are less vigorous and fertile than their diploid progenitors (Stebbins, 1950). Second, polyploidy results in the fixation of heterozygous gene combinations, particularly those derived by hybridization between races of the same species.

At equilibrium, a tetraploid population contains for each locus not the usual three genotypes in proportions that follow the Hardy–Weinberg law, but five genotypes, AAAA, AAAa, AAaa, Aaaa, and aaaa, at frequencies of p^4, $4p^3q$, $6p^2q^2$, $4pq^3$ and q^4. The proportion of heterozygotes to homozygotes at the diploid level is 1:1, and at the tetraploid level 7:1. Whenever heterozygotes have an adaptive advantage, polyploidy will be favored because it makes completely heterozygous populations easier to maintain by selective elimination of the relatively few homozygotes. On the other hand, given the same selective pressure, the response of the tetraploid population is only half as rapid as a diploid population containing the same allele frequencies. Polyploidy is therefore basically a conservative process.

These theoretical predictions are fulfilled by the experience of cytogeneticists during the 38 years since colchicine treatment made possible the large-scale production of polyploids. Despite numerous efforts, autopolyploids derived from a single diploid strain have in no instance succeeded widely and continuously as new varieties of cultivated plants, with the possible exception of triploid sugar beets and a few other clonal varieties that are not reproduced by seed. A similar lack of success has accompanied efforts to produce autopolyploids from diploids of wild species, and to establish them under natural conditions. During the 1940s autopolyploids from species of the grass genera *Stipa, Elymus, Phalaris,* and *Ehrharta* were planted in natural sites, together with their diploid progenitors. Of these, the only one that was comparable to the diploid in vigor and seed fertility was that of *Ehrharta erecta*. When planted simultaneously with the diploid in about 20 different sites, it performed more poorly and eventually disappeared in all but one of these, a steep hillside under oak trees near the Botanical Garden of the University of California, Berkeley. In this site both diploid and tetraploid races have persisted by natural reseeding for more than 30 years, and the ecological relationship between them has become clear. The tetraploid has persisted in two restricted areas, both characterized by deep shade of oak trees, and on a steep hillside with unusually good drainage. The diploid, on the other hand, has spread into a variety of sites, some of them more than a hundred meters from the original planting. The tetraploid continues to dominate the habitat in which it was first planted, but has been much less capable than the diploid of entering new habitats.

This unexpectedly disappointing experience of evolutionists and plant breeders with a process and technique, widely proclaimed in the 1940s as a major avenue for increasing dramatically the productiveness of crop plants, creates an apparent paradox. Natural polyploids have been most successful in more recent geological epochs as invaders of new and disturbed habitats, such as those vacated by the Pleistocene glaciers (Stebbins, 1971). This is exactly opposite to the behavior of the autotetraploid *Ehrharta erecta* in the experiment described above. The solution of this paradox lies in the high probability that all polyploids that have been successful invaders, even those that have cytogenetic characteristics of autopolyploids, are at least partly of hybrid origin. As Anderson (1949) demonstrated many years ago, hybridization between either subspecies or species may produce diploid genotypes having a high capacity for invading new habitats. In many instances, however, they are unstable either because of sterility or excessive segregation due to their highly heterozygous nature. Both of these defects can be corrected by polyploidy followed by artificial or natural selection for adaptive polyploid genotypes. Polyploid evolution, which has been a dominant process in many groups of higher plants, has involved a series of successful equilibria between the disruptive effects of wide hybridization (both between differently adapted populations of the same species and between different species) and the stabilizing or conservative effects of chromosome doubling.

This theory is amply supported by the experience of plant breeders with artificial polyploids. Fertile, economically valuable tetraploid garden varieties of *Antirrhinum majus* can be produced only by crossing autotetraploids derived from different varieties (Stebbins, 1950). The importance of intervarietal hybridization for economically valuable autotetraploids was shown by Demarly (1963) for lucerne (*Medicago sativa*) and by Wexelsen (1965) for red clover (*Trifolium pratense*).

Polyploidy has an even greater stabilizing effect on interspecific hybrids. Chromosomes brought in from different parental species are likely to differ with respect to their structural patterns, while the doubling process provides the complement of the polyploid with two sets of similar chromosomes, derived from the same species. At meiosis, *preferential pairing* takes place between these similar chromosomes (Figure 7-7; Stebbins, 1950), and segregation with respect to the interspecific differences is greatly reduced or prevented altogether. The frequency of preferential pairing increases according to the amount of chromosomal difference between the parental species.

Finally, polyploidy facilitates gene exchange between distantly related species. For instance, a diploid species containing a chromosome set or genome designated by the letter A, hence of the constitution AA, may hybridize and produce tetraploids separately with two other diploids, BB and CC, which are distantly related to each other. Nevertheless, the two tetraploids thus produced, AABB and AACC, can often hybridize with each other and exchange genes by introgression because of the buffering effect of the genome they have in common. Another way in which polyploids can increase their gene pool is by one-way introgression involving partly fertile triploids. They can, therefore, build up the complex entities known as *compilospecies* (de Wet and Harlan, 1966), containing genes derived from several different diploid ancestors.

The study of polyploidy has certain advantages. First, polyploid evolution is, to a large extent, irreversible. To be sure, individual tetraploids derived from a single diploid population can revert to their ancestor if little or no subsequent genetic or cytological modification has taken place. However, if any appreciable amount of evolution has taken place subsequent to the initial doubling, the "polyhaploids" derived from a tetraploid are weak, sterile, or both (Stebbins, 1970; Grant, 1971). Second, particularly with respect to polyploids of recent origin, the evolutionary steps that gave rise to a new species can be repeated experimentally, as was done many years ago by Müntzing with *Galeopsis tetrahit* and by McFadden and Sears with the bread wheat, *Triticum aestivum* (Stebbins, 1950). Finally, the possibility of forming new polyploids in the laboratory or garden and testing them under natural conditions offers one of the few opportunities available for conducting synthetic, rather than analytical, experiments on evolution.

In most treatments of the subject, polyploids are separated into two categories: autopolyploids, which are regarded as of nonhybrid origin, and allopolyploids, which are said to be derived from hybrids. There are two difficulties with this classification (Stebbins, 1971). In the first place, no single characteristic can be used to distinguish between these supposed categories; external morphology, using characters generally recognized by taxonomists, can be particularly deceptive. Second, two sharply defined categories do not exist. Instead, a thorough knowledge of all kinds of polyploids shows that the classic typology, in which polyploids derived from a single diploid population are contrasted with plants derived from such diverse ancestors as cabbage (*Brassica oleracea*) and radish (*Raphanus sativus*), tends to hide a whole spectrum of intermediate situations, into which most polyploids can be classified.

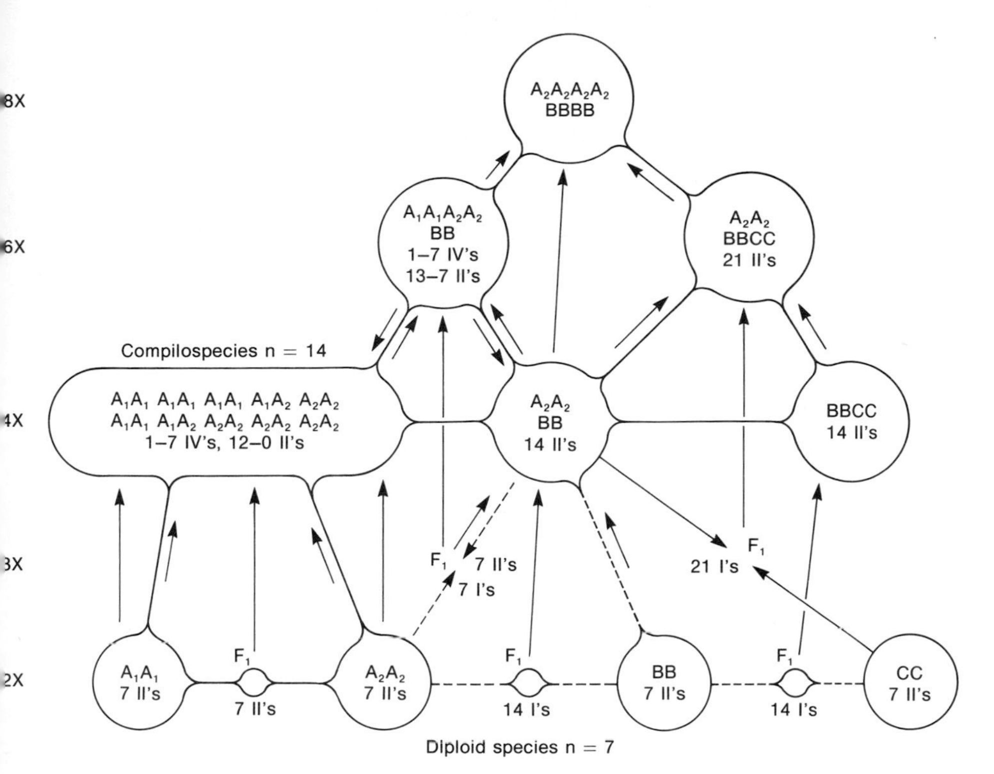

FIGURE 7-10.
A hypothetical polyploid complex with species having the basic gametic number n = 7. *Bottom row:* Of four diploid species, A_1A_1 and A_2A_2 are so closely related to each other that their chromosomes are partly homologous, and capable of pairing in F_1 hybrids. The other two, BB and CC, are more distantly related and contain different, nonhomologous or weakly homologous genomes. *Third row:* A series of tetraploid species is derived from the four diploids in the bottom row. A compilospecies (left) is derived from autotetraploids of A_1A_1 and A_2A_2 and from various hybrids between them; it has segregated into populations that have various combinations of the parental chromosomes and show various degrees of intermediacy between the parental species. Two allotetraploid species are derived, one from the genomes of A_2A_2 and BB (center), the other from BB and CC (right). *Second row:* Two hexaploid species are derived from the genomes of one tetraploid and the genome of one original diploid species. *Top row:* An octoploid species is derived by autopolyploidy from the tetraploid A_2A_2BB. Arrows indicate possibilities of gene exchange between species with different ploidy levels. Exchange can occur by formation of highly but not completely sterile triploid or pentaploid hybrids, followed by backcrossing and introgression of genes into the parental species having the higher chromosome number. [After Stebbins, 1971a.]

A more realistic way of studying polyploids from the evolutionary point of view is to recognize *polyploid complexes* (Figure 7-10), which consist of a series of different diploids and their polyploid derivatives, which have varying degrees of hybridity in their origin (Figure 7-9). In some such complexes, such as *Dactylis* (Stebbins, 1971), the diploids all behave cytogenetically like different races or subspecies of the same species, and the polyploids are all autopolyploids. In other complexes, such as *Gossypium* and certain sections of *Clarkia*, the diploid ancestors are highly distinct species, and various intermediate situations exist. Valuable information on the direction of plant migrations can be obtained by comparing the distributions of diploid and polyploid members of the same complex. Evidence obtained from statistical comparisons of the frequency of polyploidy in different geographical regions, a rather popular method in the 1920s and 1930s, is much less reliable. Contrary to earlier opinions, polyploids are not usually more successful than their diploid ancestors in regions having climates characterized by severe cold or drought; in many instances they occur in regions that are more mesic and favorable than some or all of their diploid ancestors. Polyploids do tend to be more frequent in regions that have been opened to plant colonization in relatively recent geological epochs, such as the glaciated regions of the Northern Hemisphere. With respect to human disturbance, the picture is less clear. When polyploids in many complexes, such as the grass genera *Aegilops* and *Avena,* are compared with their ancestral diploids, the greater success of the polyploids as weeds is quite evident. Nevertheless, in statistical comparisons of weedy and nonweedy floras, the frequency of polyploids is not significantly different. This is because in many groups in which polyploidy has not occurred at all, diploid species have been able to become weedy by other means.

At the level of transspecific evolution, polyploidy has been a conservative rather than progressive factor. This is evident from the fact that many of the most archaic genera of vascular plants—such as *Psilotum, Tmesipteris, Lycopodium, Equisetum,* and *Ophioglossum* among spore bearers, and *Tetracentron, Trochodendron,* and all species of Magnoliaceae and Winteraceae among angiosperms—have basic chromosomes so high that they appear to be of polyploid derivation. Groups of polyploids that have undergone considerable transspecific evolution at the polyploid level appear to have previously reverted to an essentially diploid genetic condition, by means of mutations and chromosomal changes, though retaining a high chromosome number. Based upon the cytogenetic behavior of both species and hybrids, this appears to be true of the apple tribe (Pomoideae) of the Rosaceae. Whatever progressive tendencies are possessed by polyploid complexes are probably derived from hybridization, chromosomal rearrangements at the polyploid level, or mutation.

In animals polyploidy has played a minor role for several reasons (Mayr, 1963; White, 1973). It disturbs the sex chromosome mechanism, so that successful polyploid animals must be either parthenogenetic, such as the brine shrimp *Artemia*, the moth *Solenobia*, and the weevils (Curculionidae), or hermaphroditic, such as the planarian genus *Dugesia.* If the sex chromosomes are not strongly differentiated from each other, and particularly if the Y chromosome contains dominant genes for maleness (or for femaleness in groups having the female as the heterogametic sex), this difficulty can be circumvented. The probability that species of Salmonidae are polyploid, which is considerably heightened by the discovery of multiple cistrons for lactate dehydrogenase (Massaro and Markert, 1968) may be explained on this basis. The same may be true of polyploid amphibians reported by Becak (1969). In bees (Apoidea), in which a polyploid frequency of 65 percent has been reported (Kerr and Silveira, 1972), the peculiar condition of haploid parthenogenetic origin of males may facilitate the regularization of the sex-determining process.

White (1973) has made the reasonable suggestion that one chief reason for the rarity of polyploidy in animals is that they are cross-fertilized, so that a single newly arisen polyploid animal is unable to reproduce by itself. This is true also of self-incompatible annual plants, and accounts for the rarity of polyploidy in plants of this kind, compared to self-compatible annuals (Grant, 1971). On the other hand, self-incompatibility, with its consequent need for cross-fertilization, is not by itself a serious barrier to polyploid evolution in perennial plants, particularly those that have efficient means of vegetative reproduction. A single polyploid plant of this kind can afford to wait decades or even centuries for the arrival of another polyploid having a similar genotype, particularly if it has a high degree of vegetative vigor and is well adapted to an available habitat.

Another source of weakness in polyploid animals is the upset of their delicately balanced developmental processes. For instance, increased cell size reduces the number of neurons that can exist in the brain and other nervous centers, thereby greatly diminishing the ability of artificially induced autopolyploids of amphibia to react to stimuli (Fankhauser, 1945). Finally, a very common category of polyploids in plants, those derived from diploids having very different genomes (commonly known as allopolyploids), is hardly to be expected in most animal groups, since hybrids formed from distantly related species are usually characterized by hybrid weakness or developmental sterility, which is not altered by chromosome doubling.

THE RELATIONSHIP BETWEEN POLYPLOIDY AND APOMIXIS

In many groups of plants and some animals polyploidy is associated with the abandonment of sexual reproduction, and the appearance of some form of zygote or seed production without fertilization, or apomixis. The various ways in which this can happen have been summarized recently (Stebbins, 1971). Most plant apomicts are clearly of hybrid origin, and apomixis has stabilized their hybrid condition to a large extent. For this reason, genera having large proportions of apomictic polyploids, such as *Rubus, Potentilla, Taraxacum, Poa,* and the *Bothriochloa–Dichanthelium* complex of Gramineae–Andropogoneae, are particularly difficult to classify. Species in the biological sense are confined to the sexual, diploid ancestors, and cannot be recognized by the taxonomist without extensive cytogenetic studies. When obligate apomixis has set in, the group has reached a "dead end," and produces new variants only by the slow process of mutation or by the origin of new apomicts from sexual species. In such groups the entire scope of morphological and ecological variation exists among the diploid sexual ancestors; the apomictic polyploids contain only innumerable variants on already established themes.

On the other hand, many apomictic complexes, such as *Rubus, Poa,* and the Andropogoneae, contain a high proportion of genotypes that are facultatively apomictic, being able to produce seeds either asexually or as a result of fertilization. Such groups have not lost their capability of generating new variation. They have, rather, a balance between temporary constancy and long-term variability not unlike that existing in facultative selfers. Based upon geographic distribution patterns, one can conclude that in each of these genera facultatively apomictic polyploids have existed ever since the Tertiary Period, and they show no sign of becoming less common. A safe prediction is that they will exist indefinitely, though the probability is low that they will produce any new adaptive themes.

Asexual reproduction by means of eggs produced without meiosis or fertilization is widespread in various animal phyla, and is called by some zoologists (White, 1973) "thelytoky," which is equivalent to the term "apomixis" as used by botanists. In addition to the well-known examples in earthworms, crustaceans (*Artemia*), weevils, moths (*Solenobia*), and other invertebrate groups, well-documented examples now exist in fishes (Schultz, 1969; Uzzell, 1970) and lizards (Darevsky, 1966). In animals, however, the forms are always either completely sexual or obligate apomicts. Facultative apomicts, which figure prominently in the differentiation of species in several genera of plants, are apparently absent. This situation is largely responsible for the fact that apomictic complexes in animals all have systematic and geographic distributions that indicate recent origin. They appear in about one out of every 1,000 species groups, flourish for a short time before reaching an evolutionary "dead end," and then become extinct.

THE SPECIES PROBLEM IN PRIMITIVELY ASEXUAL ORGANISMS

The concept of biological species is based upon gene exchange by means of sexual reproduction between populations belonging to the same species, and the absence or rarity of such exchange between populations belonging to different species. Apomictic groups descended from sexual ancestors are exceptions to the rule that each population belongs to a particular biological species. These exceptions, however, are not hard to understand, and a system that accommodates them can be constructed. In plants the framework of the *agamic complex* (Babcock and Stebbins, 1938), consisting of a series of distinct sexual species upon which are superposed a mass of asexually reproducing biotypes, largely or entirely of hybrid origin, serves well to clarify the evolutionary relationships of most genera in which apomixis occurs (Figure 7-11).

On the other hand, evolutionary relationships among microorganisms in which sexual reproduction is unknown cannot be understood in this way. The biggest problem is that of the prokaryotes (Chapter 12). In bacteria and blue-green algae, genetic recombination, when it exists, does not take place by the

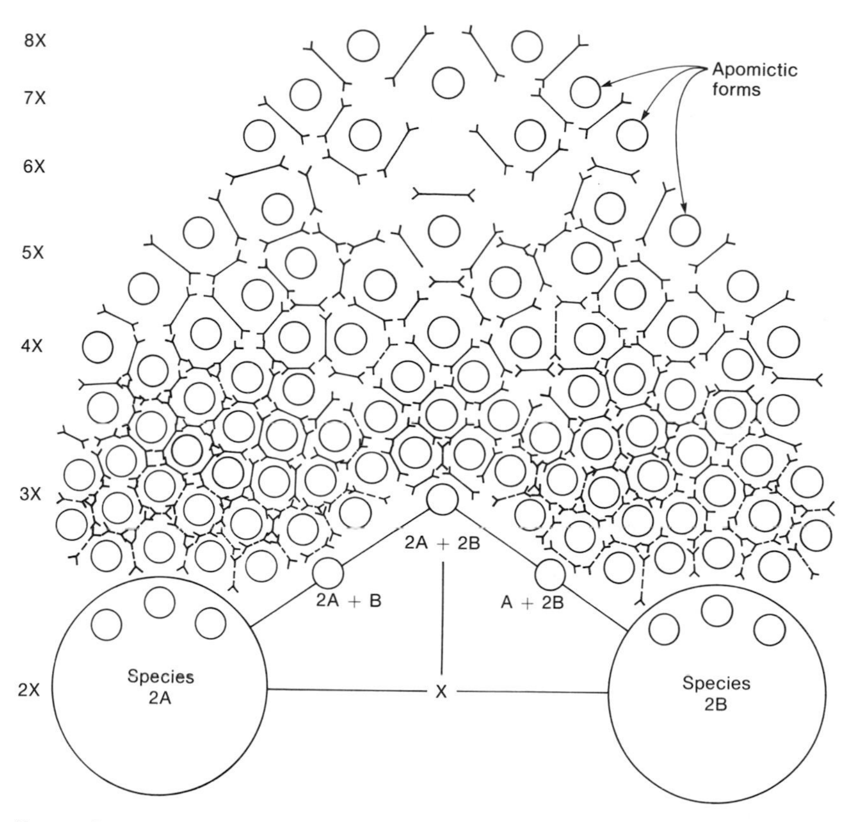

FIGURE 7-11.
The cytogenetic structure of an agamic complex, consisting of two sexually reproducing ancestral species plus a series of polyploid derivatives that have lost sexuality and reproduce by apomixis. Because of the loss of sexuality, these biotypes or clones cannot exchange genes with each other; hence, each one is represented by a circle that is surrounded by barriers. Since balanced chromosomal complements and regular behavior of chromosomes at meiosis are not required for seed production, many chromosomal combinations are tolerated that would not survive sexual reproduction. Occasional, facultative sexuality of many biotypes makes barriers incomplete; such barriers are indicated by broken lines. Due to the formation of gametes having the unreduced chromosome number, the products of hybridization between facultative apomicts often have higher chromosome numbers than either parent, so that higher levels of polyploidy can be built up. Although agamic complexes present a bewildering array of recombinations of the characteristics found in the parental species, they cannot generate any distinctively new characteristics. [After Babcock and Stebbins, 1938.]

formation of gametes by meiosis, followed by the union of two entire haploid complexes of genes to produce a zygote. Genes are recombined by any of various ways in which small pieces of DNA, containing only a few genes, are detached from the parental genophore or "chromosome" and transferred separately to another individual.

The commonest of these in nature is transduction, in which the transfer is effected by a temperate bacteriophage. Of the other two known methods, transformation is an artificial process, while bacterial conjugation, in which part of the chromosome is transferred, is known only in *Escherichia coli*.

The important difference between transduction and sexual recombination is that the former process never includes a stage in which two entire genomes derived from different parents must cooperate to produce a functional individual. Selection can operate upon each separate plasmid or episome, and incorporate into the recombinant individual only those genes that can cooperate with the existing genome to produce a harmonious and adaptive combination (Ferone *et al.*, 1970). The success of the plasmid in effecting recombination is largely independent of the nature of the genome from which it is derived. Consequently, while the possibility of genetic recombination between populations of prokaryotes is inversely correlated with the genetic distance between them, this relationship is not of the precise kind responsible for the barriers of reproductive isolation that separate species in sexual eukaryotes. Genes can be transferred across wide taxonomic and genetic gaps in the form of plasmids (Hedges, 1972). Description of prokaryotes must be based upon criteria entirely different from those used by the evolutionist to recognize species of sexual eukaryotes. The problems arising from this situation have been carefully discussed by Ainsworth and Sneath (1962) as well as by Mandel (1969). Mandel proposes the following definition: "a bacterial species is a type culture and those cultures resembling it." Though not very satisfactory, it is as good as any known to us.

A similar condition exists in many primitive eukaryotes, such as most flagellates, and in some multicellular organisms, such as the fungi imperfecti, which reproduce entirely by asexual methods. Some of these are probably derived from unknown and perhaps extinct sexual ancestors; others may be primitively asexual. In them, as in prokaryotes, description of species must be based upon whatever criteria are most useful for investigators.

8

Transspecific Evolution

The process of natural selection, acting upon the sources of genetic variability that reside in the gene pools of species, is clearly adequate to produce, preserve, and accumulate the sorts of changes that lead from one species to another. There is a voluminous body of theory and evidence to explain the origin of species through microevolution.

The differences between distantly allied species are profound, however. Contrast a species of arthropod, say a butterfly, with a species of mammal, such as man. Although the biochemical materials of which they are composed are extremely similar (see Chapter 9), the ways that these materials are organized and employed—the ways that they react in developmental processes and are fitted together to form cells, tissues, organ systems, and body plans—are very different. Indeed, the developmental pathways of butterflies and men are distinctive from the first cleavage of their fertilized eggs. The information in the mammalian genome must be very different from that in the arthropod genome. The differences between such species are in fact so impressive that some investigators have suggested that they have arisen through mechanisms distinct from the microevolutionary processes of adaptation. This has led to some of the major controversies in evolutionary theory. Before examining this problem, it will be useful to examine the ways in which organisms are classified into groups larger than species.

THE TAXONOMIC HIERARCHY

As discussed in Chapter 6, organisms are classified hierarchically: individuals are grouped into species, species into genera, genera into families, and so on up to the level of kingdom (Table 8-1). Each species, genus, family, or other

TABLE 8–1. The principal categories of the taxonomic hierarchy and examples of each category.

Categories	Taxa	
Kingdom	Animalia	Animalia
Superphylum	Metameria	Deuterostomia
Phylum	Arthropoda	Chordata
Class	Insecta	Mammalia
Order	Lepidoptera	Primates
Family	Danaidae	Hominidae
Genus	*Danaus*	*Homo*
Species	*Danaus plexippus*	*Homo sapiens*
Individual organism:	A monarch butterfly	A human

formal group of organisms is a *taxon*, and each level of the hierarchy is a *taxonomic category*. To avoid confusion, each taxon is given a name that is unique to its kingdom. The content of taxa, e.g., the species that comprise a genus, is not regulated by any generally accepted rules; different workers use different systems of criteria for grouping taxa of lower categories into those of higher categories.

Evolutionists have been particularly interested in two features of classification. First, they have preferred a classification that groups taxa according to their phylogenetic relations, since these indicate the evolutionary pathways that have been followed during life history. The establishment of phylogenetic relationships is very difficult; some of the techniques are described in the next chapter. Evidence is often inconclusive and opinions differ even among the taxonomists best acquainted with the groups in question. Therefore it is not uncommon to find more than one phylogenetic scheme for a given group of organisms, each being an hypothesis or model. It is hoped that evidence will eventually disprove those phylogenies that are incorrect.

If a classification is to reflect evolution, all the members of a taxon should be closely related and descended from a common ancestor. The application of this principle is commonly modified. Often, related organisms with similar morphologies and adaptations are retained within the same taxon, even if one of them happens to be the last common ancestor of lineages that are classed in a different taxon (Figure 8-1A). The last common ancestor is therefore not necessarily included in the new taxon, though it may be if it has evolved the main characters of the new taxon before giving rise to the descendant lineages that form the taxon (Figure 8-1B).

Taxa whose members are descended from a common ancestor are called *monophyletic*, and those whose members have come from diverse ancestral lineages are called *polyphyletic*. Strictly speaking, all members of a monophyletic taxon should be descended from the same parents, or at least from the same population or species. Any of these restrictions would exclude taxon II in Figure 8-1A, which would then be polyphyletic. However, the term "monophyletic"

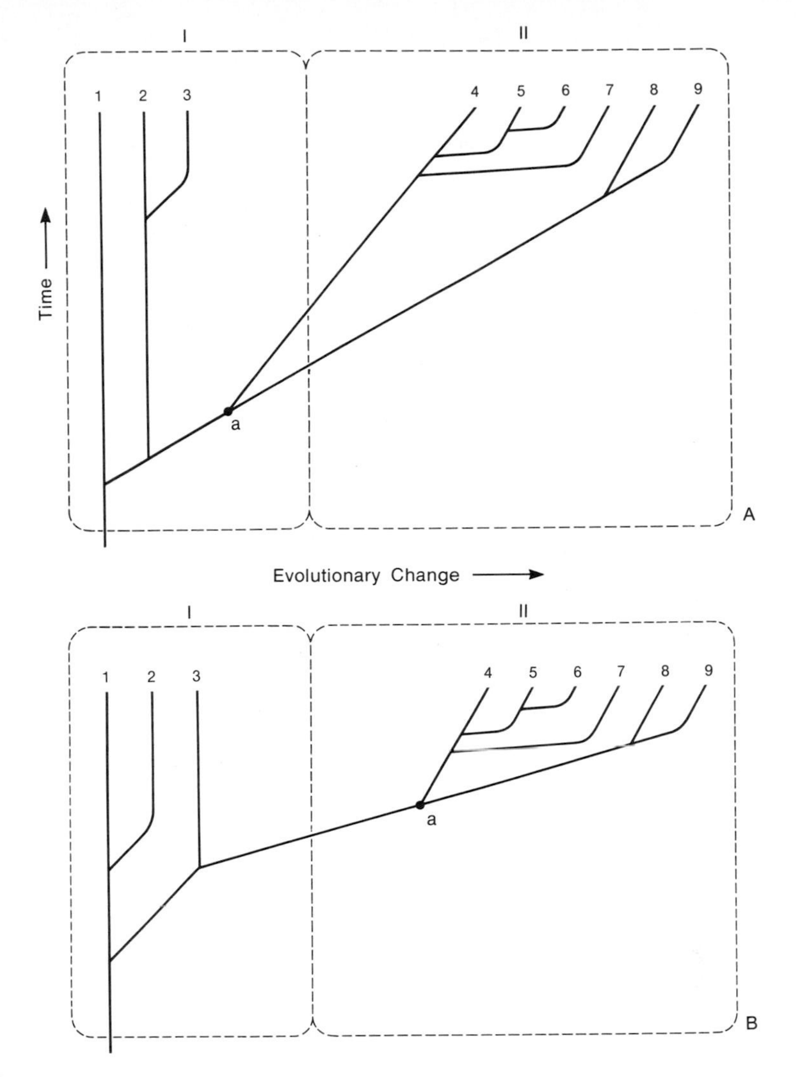

FIGURE 8-1.
Two cases of monophyly. **A.** Taxon I is composed of taxa 1 to 3, and taxon II of taxa 4 to 9. The last common ancestor of taxa 4 to 9 is taxon *a,* which closely resembles taxa 2 and 3, with which it shares a close common ancestor. Taxon II may be considered to be monophyletic, even though the last common ancestor of taxa 4 to 9 belongs to taxon I. **B.** The last common ancestor of taxa 4 to 9 is *a,* which is classed in taxon II.

is traditionally given a wider application. Simpson (1961) has framed a definition that expresses the situation in many traditional classifications: "Monophyly is the derivation of a taxon through one or more lineages, from one immediately ancestral taxon of the same or lower rank." In using the term "monophyly," it is helpful to mention the taxonomic level at which it is inferred. A class would be monophyletic at the family level if all lineages ancestral to the class originated in the same family.

Two sorts of evolutionary change are involved in the patterns of Figure 8-1; first, change in the characters of a lineage through time, represented in the figure by the movement of lineages towards the right, and second, splitting, or the creation of new branch lineages. Rensch (1954) has proposed the terms *anagenesis* to indicate evolutionary advance, and *cladogenesis* to indicate lineage splitting or branching. "Anagenesis" has come to be applied to nearly any kind of evolutionary change, whether leading to a marked advance or not, except for the branching pattern, and we shall use it here in this extended sense. A third case, in which lineages neither split nor otherwise change but merely persist in time, has been called *stasigenesis* (Huxley, 1957).

Anagenetic episodes commonly create organisms with novel characters and abilities, beyond those of their ancestors. Organisms with such new functional abilities are said to have achieved a new *grade*. The development of homoiothermy ("warm-bloodedness"), for example, created a new grade. Grades may be monophyletic or, as in homoiothermy, polyphyletic, since both birds and mammals have achieved this condition independently. Lineage branches that result from lineage splitting are called *clades*, and must be monophyletic, at least in the broad sense of the term (Figure 8-2).

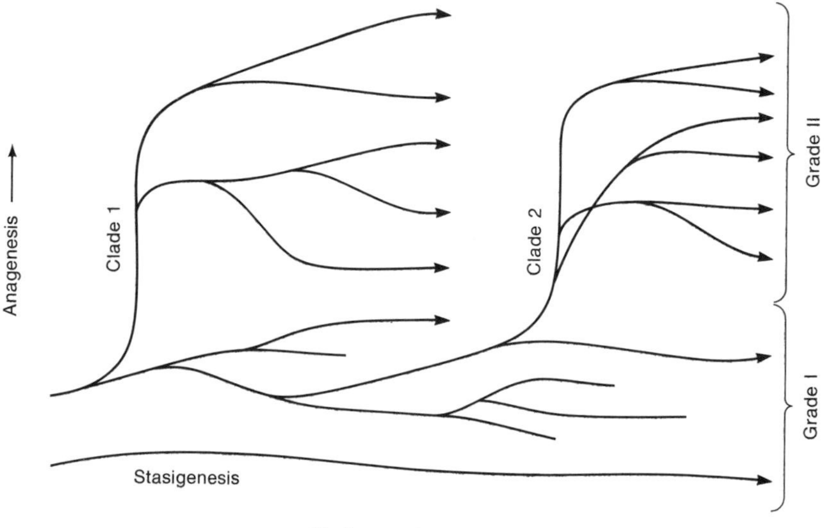

FIGURE 8-2.
Relationship between anagenesis (evolutionary advance or change), cladogenesis (branching of lineage), and stasigenesis (evolutionary persistence); between clades (branches) and grades (levels of functional or morphological complexity); and between monophyly and polyphyly (descent from single or multiple lineages, respectively). Time runs to the right and evolutionary change runs up or down. The branching pattern represents the evolution and splitting of lineages, some of which undergo periods of rapid anagenesis. Grade II has been achieved twice by independent lineages, each of which has then split repeatedly to form a clade. The clades are each monophyletic, but grade II is polyphyletic.

A second criterion of classification preferred by evolutionists is that the relative ranking of taxa reflect their biological significance. Taxa that have the more distinctive ground plans or adaptations (and therefore have undergone the most anagenesis), that play the more important roles in the biosphere, or that have undergone the more important radiations, should be given higher rankings than taxa that are of less significance in these respects. Not all taxonomists have constructed classifications with such considerations foremost, but a major tradition of classification has certainly followed these lines.

Ranking of taxa varies among traditional workers, since the assessment of biological significance is in part a subjective judgment. Some taxonomists are consistent "splitters," dividing the family tree into numerous nominal taxa, while others are "lumpers," employing few taxa for many species. For example, the monograptids are an important family of the extinct hemichordate group called graptolites. This family is split into about 19 genera that are widely recognized today by continental European workers. In Britain and America, however, the species of monograptids are usually lumped into six or seven genera (Bullman, 1970). Simpson (1961) gives a mammalian example among cats. Extreme splitters have erected 28 genera of living felids, while extreme lumpers make do with only one. Such differences in taxonomy are not fundamental and do not in themselves alter the phylogenetic model. However, it is clear that even though two taxa may be in the same category, they may have a different biological significance; families of one order may be different constructs than families in another order. In fact, this is the general rule. Although taxonomic categories roughly classify lineages resulting from separate evolutionary divergences according to their relative importance within their lineages, the class limits are imprecise and do not precisely correlate among different higher taxa.

Partly because of such difficulties as these, classifications have tended to be fluid, changing frequently as new data have come to light or as old data are reassessed by workers with novel approaches. The consequent instability of classifications is a vexing problem, for which a number of solutions have been proposed. One such suggestion is to classify organisms, not on the basis of their phylogenetic relationships, but exclusively according to their morphological resemblances. This *phenetic* approach to classification has a certain operational simplicity. Morphological characters may be measured or encoded numerically. The numerical scores of one taxon may then be compared with those of another for a great many characters, simply and efficiently by computer, and a numerical index of similarity may be calculated between taxa. Clusters of taxa that are relatively similar morphologically can then be grouped into higher taxa. In theory, at least, the entire biota can be arranged according to such a numerical taxonomy. (For an extended discussion of these techniques see Sneath and Sokal, 1973.)

A major difficulty with phenetic classification is that closely related species that have diverged morphologically are placed in different groups, while species that resemble each other fairly closely but happen to be relatively unrelated are

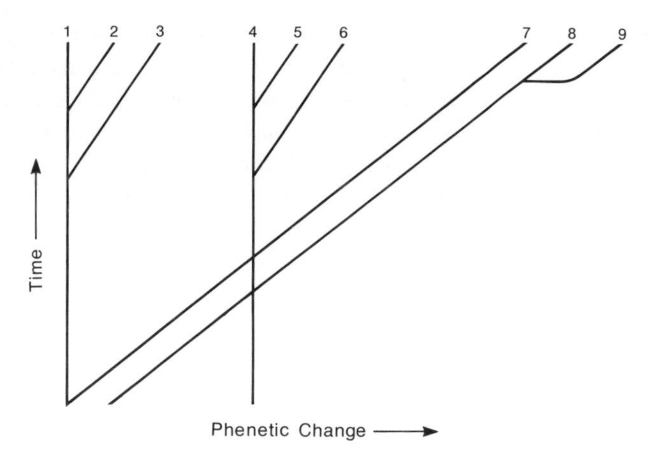

FIGURE 8-3.
Polyphyletic taxa arising from a phenetic classification. Nine taxa of the same rank (1–9) are grouped into the higher taxa 1-2-3, 4-5-6, and 7-8-9 by morphological (phenetic) criteria, although taxon 7-8-9 is a polyphyletic grouping.

placed in the same grouping (Figure 8-3). On the other hand, once the resemblances among taxa are quantified, their morphological relations are defined in a manner that is repeatable (barring new kinds of measurement, new samples, or new methods of comparison). Although phenetic classification leaves problems of ranking and inclusiveness of taxa, nevertheless there is a gain in objectivity and in taxonomic stability, but at the cost of reducing the evolutionary significance of the classification. For some taxa, phenetic relations in fact correspond to phylogenetic relations, but this is usually not the case. Since as evolutionists we are interested in the degrees of phylogenetic relationships among organisms, a phenetic classification does not serve our purpose well. Nevertheless, the techniques of numerical taxonomy certainly are of great use in assessing the facts of morphological relations, which together with other data may form the basis of phylogenetic inferences. The inferences may then be used to construct a classification that reflects evolutionary events.

Another approach to classification is called *phylogenetic systematics* by its chief founder, Hennig (1950, 1966). Since traditional taxonomists also espouse a classificatory system that is phylogenetic, they have objected to using the general term *phylogenetic* for a special taxonomic system, which they prefer to call *cladistic*. The cladists argue that the branching of lineages is the one critical event in evolution that provides an objective datum for classification. If a lineage evolves only by anagenesis, it forms a continuous succession of generally interfertile ancestral–descendant populations; there may be no point that is more appropriate than others at which to draw a taxonomic boundary, even though the anagenetic changes may eventually become profound. This problem is illustrated in Figure 8-1, wherein the boundaries between taxa I and II in both parts A and B are drawn arbitrarily. They could be moved in either direction at the discretion of the taxonomist. Not so in the cladistic approach, wherein cladogenesis automatically creates sister taxa of equal rank, taxonomically separated from the ancestral taxon. Furthermore, each taxon must be monophyletic in the restricted sense that the last common ancestor is included in the taxon. Thus in Figure 8-1A, assuming that taxa *a* and 1-9 are species, taxon II would have to include the last common ancestor of species 4 to 9, which is *a*. Indeed, species *a* must cease to exist when it branches. A cladistic taxonomy of the pattern in Figure 8-1A is shown in Figure 8-4A; a cladistic taxonomy of the pattern in Figure 8-3 is shown in Figure 8-4B.

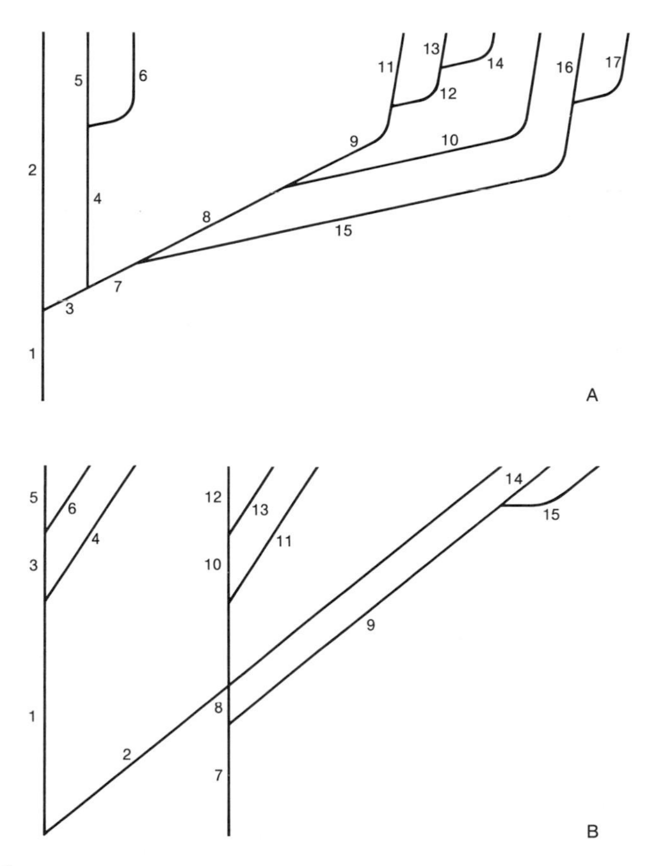

FIGURE 8-4.
Species assignments of two lineage patterns according to cladistics. Figure *A* is the same (monophyletic) pattern shown in Figure 8-1A, where the taxa are classified phenetically; figure *B* is the same (polyphyletic) pattern shown in Figure 8-3, where again the taxa are classified phenetically. Each branching episode produces sister taxa of the same rank, which are taxonomically separated from their ancestor. Thus, species 1 and 2 in figure *A* and 1, 3, and 5 in figure *B*, although morphologically and ecologically identical, are assigned to separate taxa because of branching episodes.

As an example, take the phylogenetic pattern depicted in Figure 8-5A. This is the pattern, much simplified, that is exhibited by birds (lineage A), reptiles including dinosaurs (lineages B to E), and mammals (lineage F). A cladistic classification requires that the clade ABCD be a different taxon from the clade EF, although these two clades must be in the same taxonomic category; that within ABCD, D be distinct from ABC; and that within ABC, C be distinct from AB. A cladistic classification of the lineages in Figure 8-5A is shown in Figure 8-5B, and the traditional classification is depicted in Figure 8-5C. In

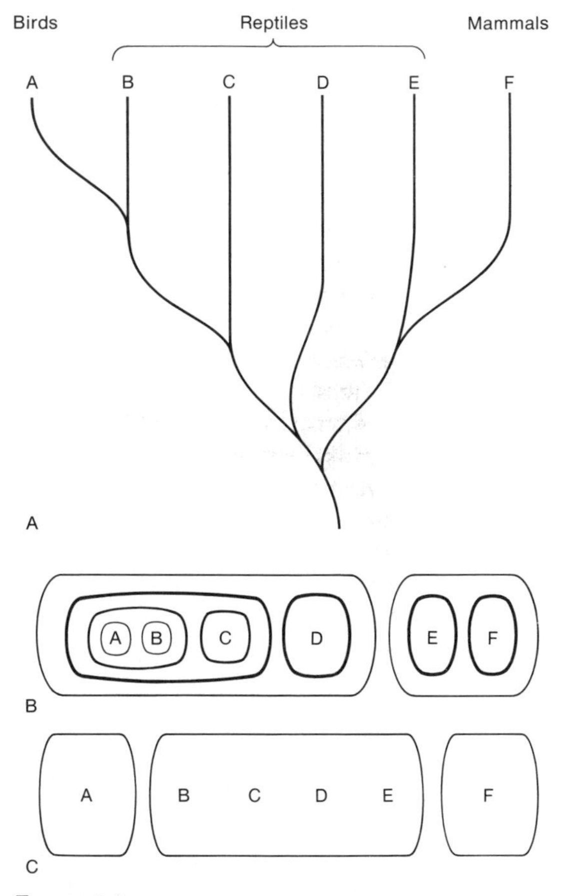

FIGURE 8-5.
Alternate classifications of taxa based on phylogenetic relationships. **A.** Diagrammatic phylogenetic tree of some vertebrates. **B.** Cladistic classification, in which the branching pattern is the sole criterion for grouping. **C.** Conventional evolutionary classification, which separates adaptively distinct organisms (birds, reptiles, mammals) into separate monophyletic groups. This approach is somewhat judgmental. [After Mayr, 1972.]

the latter, each group is monophyletic (at least in the diagram, and probably in history as well), and the fundamental distinctions between birds, reptiles, and mammals are given appropriate weight.

A cladistic taxonomy imparts a certain uniformity and simplicity to classifications, removing subjectivity from many taxonomic decisions, and it does display the phylogeny in its purest form. On the other hand, cladistic taxonomy often obscures the evolutionary and ecological significance of taxa. In Figure 8-1A, species *a* must be placed in taxon II cladistically because it is the stem ancestor of species 4–9, although it has not yet evolved the features that become characteristic of taxon II (and which undoubtedly underlie its subsequent successful radiation). Biologically and morphologically as well, it belongs with taxon I. The "objectivity" of cladistics, which is won by a strict formalism, can result in the grouping of adaptively incongruous forms. On the other hand, to dismember a lineage that reflects morphological and ecological stability (species 1, 3, and 5 in Figure 8-4B) merely because some marginal populations have branched off (to form species 4 and 6) is to split an adaptively and genetically congruous form on the basis of the same formalism. These are significant drawbacks (Mayr, 1974a).

Another aspect of cladistics involves the problem of ranking taxa, which is especially difficult since anagenetic change is virtually ignored in the classification system. One proposed solution is to rank taxa according to the time of origin of their stem ancestor (Hennig, 1966). Under this scheme, groups arising during the Precambrian are phyla; those arising during the Cambrian through Devonian are classes: those during the Carboniferous and Permian, orders; during the Triassic through Lower Cretaceous, families; during the Upper Cretaceous through the Oligocene, tribes; and during the Miocene, genera. The biological significance of the groups is not considered, but again a strict formalism is observed. Long-lived groups that have not changed much for long periods (those that are chiefly stasigenetic) are ranked higher than later groups that have undergone extensive anagenesis and radiation to become of major biological importance. Thus, *Lingula*, the long-lived brachiopod taxon traditionally ranked as a genus, must be placed in a higher category than the entire class Mammalia, simply because it arose earlier (Mayr, 1974a). It is true that most phyla arose early, while classes, orders, and lower taxa tend to have had the peaks of their appearances and diversities in successively later times. These points will be touched on again in Chapter 13.

The writings of Hennig and other cladists have caused taxonomists to examine more closely the role of cladogenesis in evolution, and cladists have been among the leaders in formulating criteria for separating ancestral and descendant lineages and recognizing sister taxa. This can only lead to increased understanding of patterns of descent. The rise of numerical taxonomy and cladistics exemplifies the continuing search for repeatability, objectivity, and stability in classification. As it becomes increasingly possible to measure the amount and kind of evolutionary change between taxa, it also becomes possible to approach these goals with a classification that reflects evolutionary history.

NONADAPTIVE MODELS

The phylogenetic patterns of branching and change have frequently been depicted in the form of tree-like diagrams, "family trees" of life. The first was published by Haeckel in 1866 (Figure 8-6). A tree showing all species would be impossibly complicated, even though the form of such a diagram is clearly conceivable. Species connected according to their evolutionary pathways would cluster to form genera, which in turn would cluster into families, families into orders, and so on. Divisions between the clusters become more profound (and of course biologically more significant) as we move up the levels of categories. When an attempt is made to trace the major branches back to their presumed common sources, as for example by following the history of two phyla back through the fossil record, we find that the branches are distinctive, and were already well separated when they first appeared as fossils (Chapter 13).

Each phylum possesses a distinctive anatomical architecture or ground plan, an integrated assemblage of characters that function well in concert—a property that Darwin called coadaptation. Some biologists have argued that gradual evolution of numerous harmoniously coadapted suites of distinctive characters is not plausible; no one character could be adaptive in its own right, and therefore all characters must have appeared more or less simultaneously. The morphological distance between a new type of organism and its ancestor would be crossed at a single jump, an hypothesis known as *saltation*. Accordingly, intermediates would never have existed, and the new form would not have been selected for its adaptiveness. This would also explain the sudden appearance of many new lineages as fossils (Schindewolf, 1950).

The saltation hypothesis of evolution dates back to Darwin's time, and was employed by Mivart (1871) and others who opposed the notion of natural selection. When the Mendelian aspects of heredity became generally known around the turn of the century, the appearance in progeny of unusual variations was attributed to mutation—some correctly and some incorrectly (Chapter 3). The phenomenon of mutation seemed to provide a mechanism to explain saltation. Such an idea has persisted among a few scientists. For example, Goldschmidt (1940, 1948, 1952) suggested that monstrosities occasionally arose via mutations with very large phenotypic effects. These "monsters" might prove on some occasions to be well adapted, and thus a new type of organism would appear suddenly. Of course, the probability that fitness would be enhanced by any truly large mutational effect is extremely small. The number of "hopeful monsters" that would have to arise in order that viable new types could be found is astronomical. There is no evidence of any such hordes of monstrosities, living or fossil, nor is any mechanism known that would produce them in appropriate quantities.

Evolution along trends that are fairly constant in direction over long periods of time, owing to inherent or immanent features of the organisms involved, is called *orthogenesis*. That evolution proceeded in this way was postulated in the

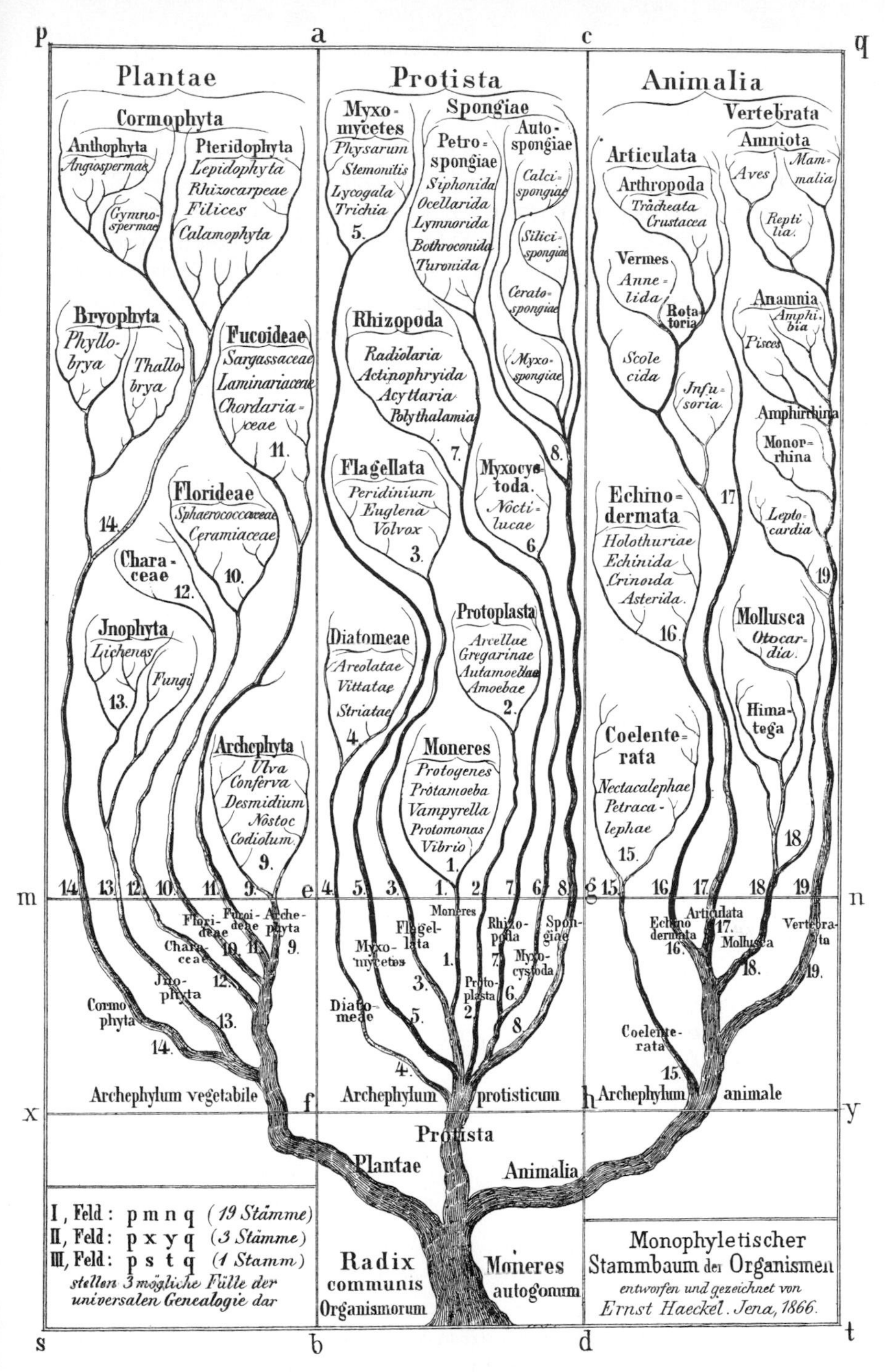

FIGURE 8-6.
The first phylogenetic tree of life, published by Ernst Haeckel in 1866. The modern arrangement is quite different from the structure shown here.

last quarter of the nineteenth century to explain seemingly gradual and continuous increases in body size in time series of fossils. Cope (1896) especially established that such trends persisted in many lineages. Some authorities even contended that orthogenetic trends might lead to disadvantageous conditions, yet would nevertheless persist because of the inherent tendencies that produced them. A common example given was the extinct Irish "elk," *Megaloceros*, actually a large cervine deer, which had massive antlers (up to 85 pounds mounted on a 5-pound skull) that were obviously cumbersome if used in fighting. It has been suggested that the orthogenetically predestined development of the antlers led to a lowering of adaptation and eventually to extinction (Lull, 1922).

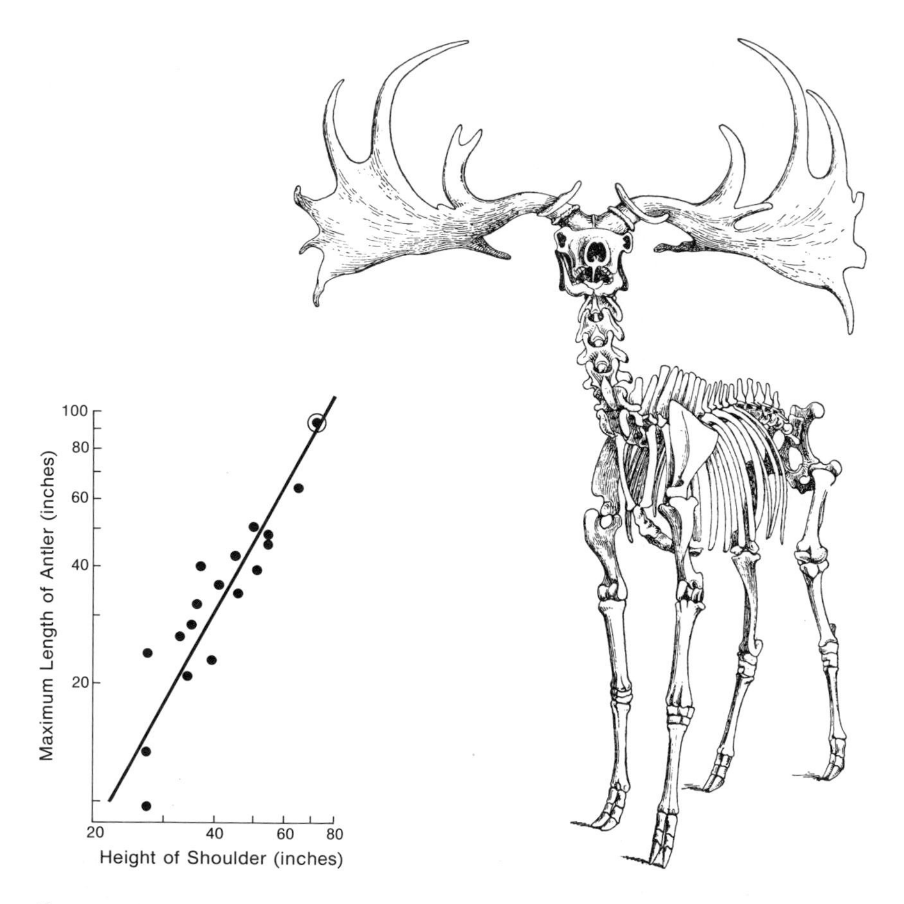

FIGURE 8-7.
Relation between body size and antler size in cervine deer. On the whole, the antlers of cervine deer become proportionately larger as body size increases. The largest known ratio of antler to body size belonged to the Pleistocene Irish "elk" *Megaloceros* (pictured). Although the antlers are quite large, they are not unexpectedly large, and in fact are precisely the size (circled dot in the graph) that could be predicted from the observed trend (the straight line). [After Gould, 1973.]

As fossil evidence has accumulated, many of the presumed orthogenetic trends have been shown to have been artifacts of a poor record, created by extrapolation between only a few points; many such features are now known to have evolved in different directions at different rates and at different times. Nevertheless, there is no doubt that some lineages have displayed long-term trends, and increases in body size are common examples. There is, however, every reason to believe that the observed trends are adaptive. Larger body size has many advantages: strength in capturing prey, fleetness in evading predators, and success in mating competition are examples. So long as increasing size does not entail offsetting disadvantages, such as an inordinate food requirement, it might be favored by selection.

Gould (1973) has shown that in cervine deer antler size increases loglinearly with respect to body size (a positive allometric relation), and that as *Megaloceros* has the largest recorded body size, it also has the largest antlers by far (Figure 8-7). Cervine deer frequently use their antlers in display to promote success in mating, rather than in combat, and it is reasonable that *Megaloceros* did also. Indeed, their large body sizes may have evolved partly because of selection for increased antler size through reproductive success for well-endowed stags (Gould, 1973).

ADAPTIVE MODELS

Difficulties in accounting for the origins of higher taxa of organisms are the most serious for various evolutionary models. Nonadaptive explanations, such as saltation, have been falsified. In natural selection, however, we have a well-established process that is known to cause evolution in adaptively favorable directions within species, as evidenced by innumerable field and laboratory studies (Chapters 4 to 7). To explain the origin of truly novel types through natural selection, then, we need a model that demonstrates the adaptive utility of novel structures and functions from their first appearance through their full development and elaboration into characteristic features of higher taxa.

An important contribution towards understanding of this problem was Sewall Wright's concept of adaptive peaks (Chapter 6), which explains the origin of discontinuities among contemporary species. Simpson (1944, 1953) has extended these ideas and developed a widely accepted conceptual framework to explain the origin of higher taxa through natural selection. He pointed out that the environment is naturally divided into regions within which certain characteristic modes of adaptation are required of the inhabitants—the land and the sea, the forest and prairie, and the rocky shore and sandy beach are some obvious examples. These regions are separated by discontinuities that act as adaptive thresholds. Nevertheless, the thresholds are crossed numerous times in the history of life. Even the threshold between the terrestrial and marine realms, a particularly profound environmental discontinuity, has been crossed several times in each direction by evolving lineages. Each of these realms contains many internal discontinuities of topography, climate, and other features that subdivide it into smaller regions, each of which presents characteristic adaptive problems. And these regions are themselves further subdivided. Indeed, the planetary environment is a mosaic of conditions on a series of scales. Each taxon has a particular adaptive mode and functional range, which permits it to inhabit a certain number of the mosaic patches. Simpson considers a taxon, together with the environmental regimes that it inhabits, to constitute an *adaptive zone.* Boundaries between adaptive zones tend to be found at environmental discontinuities between mosaic patches.

Lineages that are able to evolve across a zonal boundary are usually those that live near the boundary and are adapted to the peculiar conditions there. Such a lineage may be atypical of the taxon to which it belongs, and frequently appears to be peculiarly specialized (Mayr, 1963). It commonly possesses adaptations that, while evolved for life in its previous adaptive zone, prove to be especially useful in the neighboring zone as well. These adaptations are *preadaptations*, and their existence helps the lineage to inhabit the zonal border and invade the new zone. Even though evolution may lead the lineage across the zonal boundary, conditions at the border and in the new zone will be sufficiently different from those to which the lineage had been subjected that adaptation in these novel environments will usually be lessened. The size of the population might therefore be reduced. In these small populations genetic drift may play a role by permitting novel genotypes to become more frequent than selection would ordinarily permit.

Although extinction is probably common among lineages that explore novel environments to which they are not well adapted, selective pressures for adaptation to the new conditions would nevertheless be high, and competition would be minimal since the new zone would presumably be uninhabited by similar forms. Some pioneering populations manage to acquire the functions necessary to regain a high degree of adaptation. Alleles that formerly were of less adaptive value and that may not even have ordinarily been present in the ancestral gene pool of the lineage would become common; the harmonious coadaptation of these alleles within the genome might cause further changes.

The process of evolutionary adjustment to a newly occupied adaptive zone can be called *postadaptation* (Simpson, 1944). If the newly occupied environment is extensive, the pioneer lineage may well radiate into a number of descendant lineages through the processes of speciation outlined previously, usually with expansion and perhaps partitioning of the functional range of the pioneer taxon as it diversifies.

To summarize, anagenesis within a lineage leads to preadaptations that are fortuitous insofar as broaching of a zonal boundary is concerned, but which do in fact permit the invasion of a new zone. The invasion is commonly accompanied by accelerated anagenesis that permits the lineage to evolve along a pathway leading across the boundary into a new zone. This is followed by postadaptational evolution at more moderate rates, commonly accompanied or followed by extensive cladogenesis. A novel adaptive type is thus generated, distinct from its ancestors and represented by a number of allied lineages. Adaptation is maintained by the pioneering lineage throughout the zonal invasion, although it may (or may not) be lowered at times. At any rate, the evolution occurs by gradual steps along an *adaptive pathway* leading from part of one adaptive zone across an adaptive threshold to part of another zone. The extent of radiations in the new zone and their pattern—whether or not they are composed of distinctive clusters of radiating lineages—depends upon the nature of the discontinuities within the newly invaded environment, and upon the adaptive breadth attained by the new lineages (Chapters 10 and 13). The taxonomic category to which the new adaptive type is assigned depends upon the distinctiveness of form and function that it attains, and in some measure upon the extent of its subsequent diversification.

This model has proven to be a powerful one. There are some technical difficulties with the concept of the adaptive zone. Since the model relies heavily upon the interaction of organisms with their environments, ecological processes are emphasized. For ecological purposes it is most useful to separate environmental conditions from the functions of the organisms. A convenient conceptual model that accomplished this is a geometric one in which the environment is represented as a multidimensional space (Wright, 1932; Hutchinson, 1957). In this space, each distinct environmental parameter is represented by a separate dimension. Thus, the complete environmental model has as many dimensions as there are environmental parameters, and any point in the model represents some combination of parameters that is unique. The parameters include biotic environmental components that can function as food, predators, competitors, and so on, as well as physical features. The actual range of environmental parameters that can occur in any region of earth may be represented by a hypervolume within the model. If we change the region we are representing, the model hypervolume will change to represent the environmental regime in the new region; if we monitor a region through time, the model hypervolume will vary with temporal changes in the environmental regime there. The term *biospace* has been employed for this geometric representation of a potentially habitable environment (Doty, 1957; Valentine, 1969). A simple biospace model is depicted in Figure 8-8.

No organism is a complete generalist; each can exist only within a more or less narrow range of environmental states. This range forms a small hypervolume within biospace. The hypervolume occupied by an entire species is composed of the combined hypervolumes of all the individuals belonging to that species and will be considerably larger than the hypervolume of any individual. The hypervolume of a genus includes the hypervolumes of all its constituent species, and so on up the taxonomic hierarchy. A single general term for the hypervolume occupied by a group of organisms is *ecospace* (Figure 8-8; Valentine, 1969). Ecospace is defined by function. The ecospaces of some taxa, especially of highly specialized forms with narrow environmental tolerances, may be very small, while ecospaces of flexible, opportunistic forms with broad tolerances may be large. The ecospaces of populations or species represent

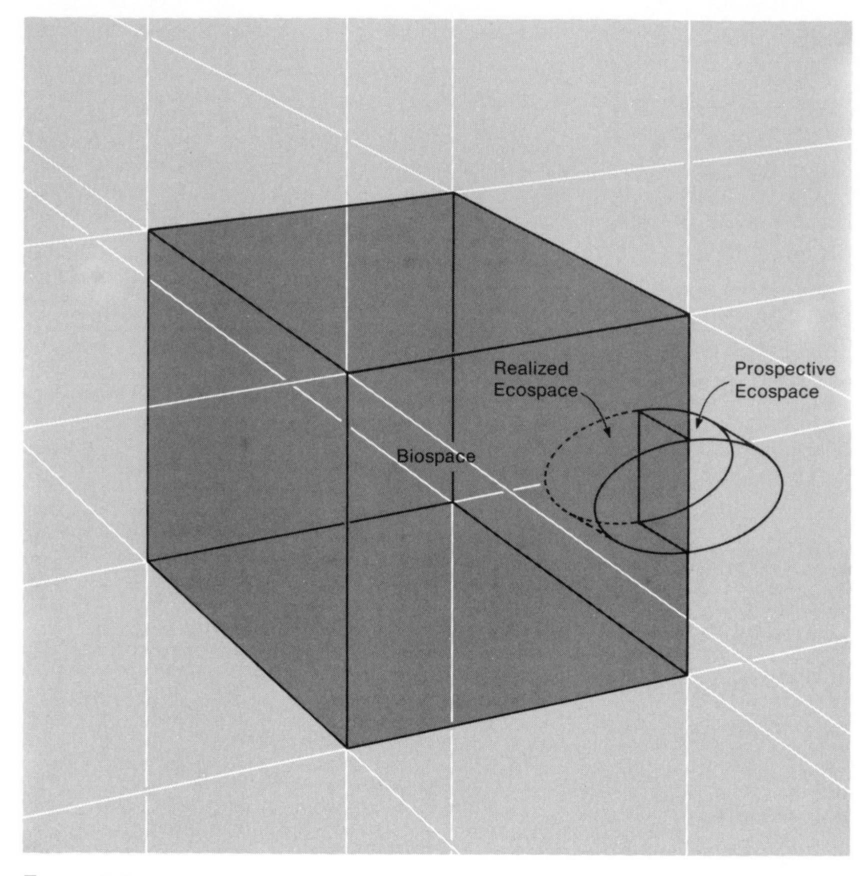

FIGURE 8-8.
An environmental hyperspace lattice, wherein any point represents a unique combination of environmental factors. The realized environment, within which organisms can live, is *biospace*. The region of biospace tolerated by some biological unit (an individual, species, community, etc.) is the *ecospace* of that unit. [After Valentine, 1969.]

adaptive peaks, and are usually called "niches" in ecology. We are avoiding that term because of its confused usage.

Examples of evolutionary events that fit this model well can be inferred from both the living biota and the fossil record. Some cases, such as Darwin's finches, have already been reviewed in Chapter 6. A similar case leading to supraspecific taxa is that of the Hawaiian honeycreepers, a family of birds (Drepanididae) that may have evolved from a single immigrant form (Amadon, 1950). There are two subfamilies, the Drepanidinae and the Psittirostrinae. The latter is far more rich in species and has been the subject of an evolutionary model by Bock (1970). The most primitive psittirostrine genus is *Loxops*, which probably evolved from an early drepanidine (Figure 8-9). The stem of this genus, which was evidently similar to the polytypic species *Loxops virens*, gave rise to four other congeneric species and also to a distinctive fifth species that has itself radiated. This second radiation produced three additional species to form a group that is considered to be a different genus, *Hemignathus*, and a fifth species that is distinctive and so is placed in a monotypic genus, *Pseudonestor*. In turn, a stem species evolved from *Pseudonestor* and radiated into six living species, producing the genus *Psittirostra*.

The origin of the features that characterize each genus is of special interest; these features seem to have arisen through the same processes that produced the separate congeneric species. The character of the bill and associated feeding apparatus is particularly diagnostic among the psittirostrines, as well as among Darwin's finches, because both radiations involved partitioning or expansion of food items and feeding habits. Bock (1970) has argued that the differences have arisen first through geographic isolation of separate bird populations on different islands until some degree of reproductive isolation was obligatory, and secondly through character displacement (Chapter 6) that occurred when such populations migrated so as to become sympatric.

Double invasions of new islands by distinct populations of the same species are envisioned in some cases. Differences in feeding habits that happen to have evolved within the previously allopatric populations became emphasized as the sympatric populations diverged in habit and morphology owing to inter-population competition. This produced two distinctive species, each with a characteristic bill morphology. One of the species produced in this way, *Hemignathus lucidus*, had a lengthened upper jaw, continuing a trend established for the subspecies *Loxops virens stejnegeri*, from which it may have evolved. A number of long-billed species radiated from *H. lucidus* or a similar form, producing a species group that has been classed as a genus. *H. lucidus* is not greatly different from *L. v. stejnegeri* and would probably be included in the genus *Loxops*, had not the subsequent radiation occurred. However, the generic separation of *Hemignathus* from *Loxops* has both a functional and morphological basis, since species of *Loxops* chiefly probe for insects in bark, open leaf buds, and legume pods, with only minor utilization of some forms of nectar. Species of *Hemignathus* also probe for insects, presumably in somewhat different manners, but rely on nectar to a greater degree.

Other genera of psittirostrines are also differentiated by feeding habits, morphology, and food items; they are heavy-billed (Figure 8-9) and phytophagous as well as insectivorous. The heavy bills evidently developed to open insect tunnels in wood, and were subsequently employed for feeding on plants themselves—a nice example of preadaptation. The evolutionary changes leading from the early forms of *Loxops* with small thin bills that probed for insects to forms of *Psittirostra* with heavy conical bills used in plant eating can be ascribed to many small steps that can now be reconstructed from the patterns of form and function in living species, although it is possible that actual ancestral steps were represented by extinct forms. Exploration of an adaptive zone (probing for insects in the *Loxops* way) led eventually to preadaptations that permitted the invasion of a new adaptive zone (phytophagy). Evidently the diversification of this subfamily has led to the partitioning of the ecospace of some primitive

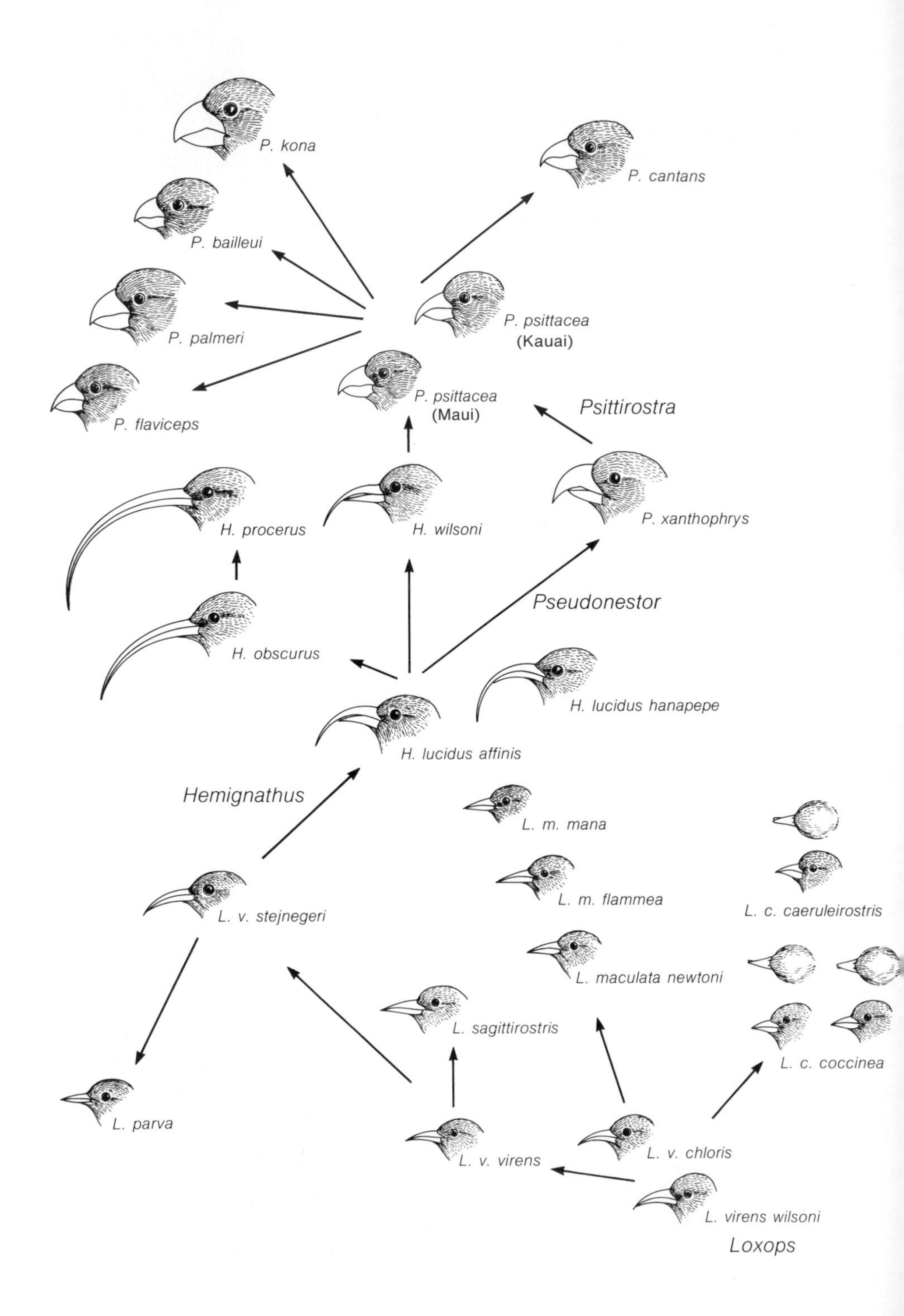

P. kona
P. cantans
P. bailleui
P. palmeri
P. psittacea
(Kauai)
P. flaviceps
P. psittacea
(Maui)
Psittirostra
H. procerus
H. wilsoni
P. xanthophrys
Pseudonestor
H. obscurus
H. lucidus hanapepe
H. lucidus affinis
Hemignathus
L. m. mana
L. m. flammea
L. c. caeruleirostris
L. v. stejnegeri
L. maculata newtoni
L. sagittirostris
L. c. coccinea
L. parva
L. v. chloris
L. v. virens
L. virens wilsoni
Loxops

forms among the descendant species, as well as to the deployment of some descendant lineages into new regions of biospace. At any rate, this example indicates that the rise of distinctive genera (and indeed of subfamilies) is plausible through the gradual evolutionary processes envisioned in Simpson's model. Since the Hawaiian Islands date from less than 10 million years, the psittirostrines have diversified within that time, perhaps considerably less time.

This example of generic evolution is particularly well studied. Furthermore, the island habitats clearly provide an ideal setting both for the occurrence of geographic isolation and for occasional subsequent interisland migration, so that both allopatry and subsequent sympatry may recur numerous times in different species. Even within continents or within the marine realm, however, there are numerous "habitat islands," such as mountain peaks, stream valleys, shallow embayments, or rocky continental shelf margins, which are generally isolated but between which migration can occasionally occur. Thus, evolutionary processes similar to those along the Hawaiian Island chain can occur in all major environmental realms.

Marine snails of the family Acmaeidae (limpets) are represented by as many as 16 sympatric species along the coast of southern California. Each species inhabits a distinctive habitat. Many live on rocks, but one lives mostly in the high intertidal zone, another mostly at mid-tide, and another in the low intertidal zone, while some live at protected and some at exposed sites. One species lives on the shells of trochoid gastropods, while three others live on kelp, surf grass, and eel grass, respectively. These limpets browse over their substrates for diatoms and blue-green algae or other minute plants, but they chiefly feed in different places and thus have somewhat separate trophic supplies. Even though all these species perform similar trophic functions, the total ecospace of the genus is greatly broadened by the adaptations for separate habitats.

There are at least three separate alliances of species, sometimes considered separate genera. *Acmaea* contains chiefly subtidal species that feed on coralline algae. The other two groups feed on diatoms and microscopic algae; *Colisella* is usually intertidal on rocks or shells, while *Notoacmaea* is most common on kelps and marine "grasses" (marine seed plants) but occurs on intertidal rocks

---

FIGURE 8-9 (opposite)
Phylogenetic relations inferred for living species of Hawaiian honeycreepers of the subfamily Psittirostrinae. A species of the genus *Loxops,* perhaps *L. virens,* founded the psittirostrines, and radiated into at least five species, *L. maculata, L. coccinea, L. sagittirostris, L. parva,* and *Hemignathus lucidus.* The latter is not greatly different from its ancestral subspecies, *L. virens stejnegeri,* but has itself radiated to produce three species with which it forms a rather distinctive adaptive group. Accordingly, this group is recognized as a genus, *Hemignathus.* Another species derived from *H. lucidus* is distinctive and is placed in a separate, monotypic genus; this is *Pseudonestor xanthophrys.* Six living species have radiated from *P. xanthophrys* and form a rather compact, distinctive group of their own, which has been given generic standing as *Psittirostra.* [After Bock, 1970.]

as well (McLean, 1969). It appears that anagenetic trends have created three distinctive stocks, each of which has undergone branching and perhaps anagenesis to extend its ecospace. The situation is quite reminiscent of the Hawaiian honeycreepers. Presumably these allied limpets have arisen allopatrically and are secondarily sympatric, and it is plausible that the narrow partitioning of trophic supplies between many of the species is due to functional displacement based on competition.

From these and other examples we can abstract the main patterns of evolution that are involved in the formation of genera. Figure 8-10 depicts the evolution of distinctive species groups that taxonomists may recognize as genera. The vertical scale, which represents ecological functions in the environment or functional aspects of biospace, is vastly oversimplified since multiple dimensions are here represented by only one. Species groups that are morphologically somewhat distinctive may originate when an anagenetic episode leads a lineage along an adaptive pathway into a biospace that is distinct from the ancestral biospace in some dimensions (usually food or habitat resources). Distinctive morphological features that evolved by this pioneering lineage to cross the adaptive threshold and persist in the new biospace serve to distinguish it and its descendant species (Figure 8-10A).

Other patterns can be discerned. Some distinctive clusters appear to result from the continuing radiation of a stem lineage and its descendants (Figure 8-10B). In such cases there is no particular adaptive threshold between species groups, which are morphologically distinctive because each has inherited the adaptations of its stem lineage, even though the stem lineages themselves were sister lineages. Partitioning of biospace, rather than radiation to occupy additional biospace, may have a similar result (Figure 8-10C). Descendant species may be placed into morphologically distinctive groups even though major ecological discontinuities and associated adaptive thresholds may not exist within the occupied biospace. Morphological gaps between species clusters may also result from extinction (Figure 8-10D). If a large cluster of allied species is decimated by extinction, surviving subclusters may be classed as distinct genera in the absence of a good fossil record. Taxonomic groupings can thus be created by patterns of extinction. This may often be the case among plants, which are poorly represented as fossils. As can be seen from the example of the Hawaiian honeycreepers, combinations of several of these patterns contribute to the formation of many species groups that are nominal genera.

Most families appear to have originated by a continuation of the same processes that result in the formation of genera: A distinctive phenotypic response to an ecological opportunity proves particularly successful, and an array of species results; these are grouped into several genera. If sufficiently distinctive, these may then be formed into a family. The marine snail genera *Acmaea, Colisella*, and allied limpets are grouped into the family Acmaeidae, which occupies a distinctive region of biospace. As the morphological gaps between families are usually rather profound, they are not likely to be due only to the historical factors represented in Figure 8-10A and B, but usually involve

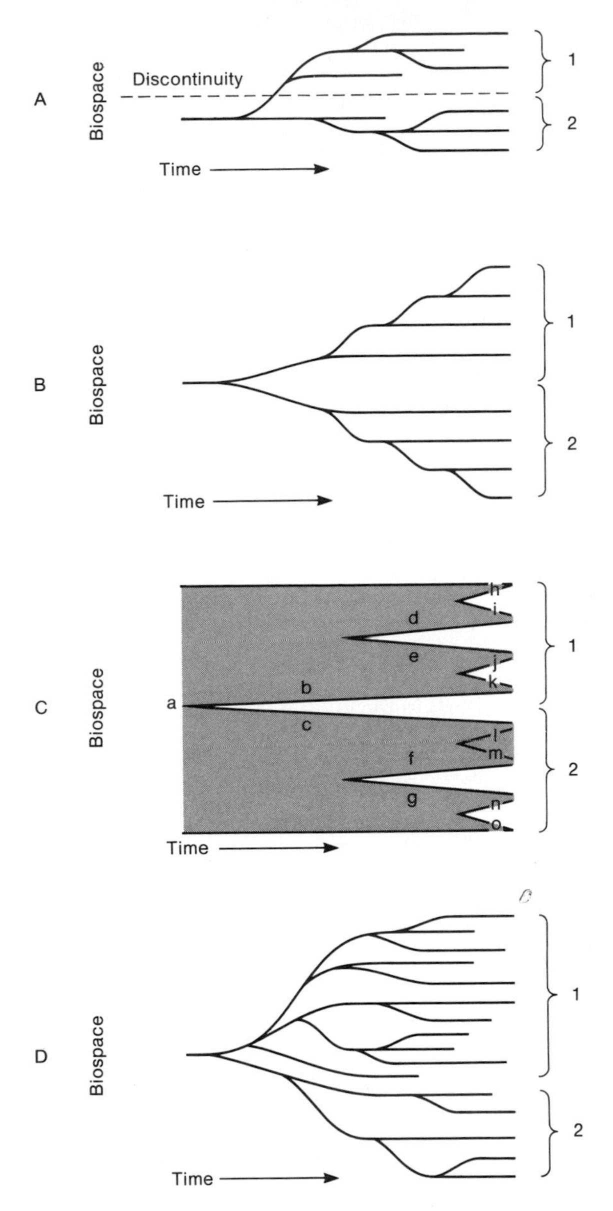

FIGURE 8-10.
Historical events producing monophyletic species clusters classified as genera. **A.** Biospace contains an ecological discontinuity, which forms an adaptive threshold; the species descended from the lineage that crossed the threshold share distinctive morphological traits. **B.** Species clusters result from divergences into distinctive subregions of biospace, with the functional range of the lineages remaining about equal; there is no major ecological discontinuity. **C.** Trend toward specialization associated with progressive partitioning of biospace among several species. Although there is no major ecological discontinuity, lineages *h* to *k* inherit characteristic traits of lineage *b,* while lineages *l* to *o* inherit characteristic traits of lineage *c,* and on this basis two genera are recognized. **D.** Repeated divergences coupled with extinctions of intermediate types produce two distinctive species clusters, recognized as genera.

more or less lengthy anagenetic episodes that lead to large morphological distance between their descendants and those of their parent stock or other generic clusters. Thus they tend to follow the pattern depicted in Figure 8-10C. This appears to be even more surely the case with still higher taxonomic levels.

For example, the series of changes that may have led from terrestrial dinosaurs to aerial birds has been inferred in terms of adaptive zones by Bock (1965). Some quadrupedal dinosaurs developed bipedal locomotion for fleetness, and this trend was accompanied by a lightening of the skeletal structures. Some of these light, active lineages must have taken to life in trees, and some arboreal forms became adept at leaping, with obvious advantages in escaping predators or in avoiding the ground when transferring from tree to tree in search of food. To prolong the time in the air, gliding is an obvious step, and once wing-like structures appeared, their further elaboration as true instruments of flight was possible. Following the development of active flight, modern birds have appeared, many lineages of which have very high proficiency in flight.

Birds could hardly have arisen in a single giant step from their ancestral dinosaurian progenitors; rather, they had to develop in a series of smaller steps that gradually altered the lineages but which added up to an important anagenetic episode. Evolution proceeded along adaptive pathways; each new adaptive mode along a pathway was clearly sufficiently well adapted that the lineage survived (although it is possible that competition with subsequent, descendant lineages may have contributed to the extinction of some earlier ones). Each mode grew naturally from a former mode, usually owing to a combination of preadaptations plus some adaptational novelties, the foundations for which were present in the ancestral populations. In turn, each mode provided access to a further mode, the next step along the pathway of adaptation.

GENETIC PATTERNS

As we have seen, the origins of higher taxa can be explained as adaptive responses to special ecological opportunities. Plausible adaptive models have been postulated for the origin of nearly all higher taxa. A number of these are reviewed in Chapter 13. The fossil record is adequate to verify only a few of these models, however. For most other taxa more than one plausible evolutionary pathway has been postulated. Additional fossil or molecular evidence will be required to test nearly all our current phylogenetic models.

Despite the plausibility of many of these models, with respect to adaptive pathways, there are still important general problems concerning the origin of higher taxa. One concerns the time required for evolution of vastly different biological architectures. Great morphological changes appear to have taken place during only relatively short periods of geological time; whole phyla seem to evolve within several tens of millions of years or less. Another problem concerns the differences between such relatively simple animal phyla as the Coelenterata and such relatively complex phyla as the Chordata, which must involve more genetic change than simply the progressive substitution of new structural alleles for old. Presumably, the more complex organisms require more genetic information than the less complex. Coelenterates have about seven cell types in their bodies, while chordates possess about 200 or so (Chapter 7; and see Bonner, 1965b). To the extent that this is a measure of complexity, the chordates are more complex by more than one order of magnitude. Yet in any given organism all cell types contain an identical genotype (except for gametes and some unimportant exceptions). Each cell type must have a distinctive developmental history, involving the operation of numbers of structural genes in harmoniously ordered sequences that differ from those of other cell types.

Developmental differences between cells probably arise from the kinds of genes that are activated and more especially from differences in the sequences and combinations of gene activations that occur. There must be a complex regulatory apparatus to control cell differentiation. Thus, we must develop hypotheses to explain rapid evolution of higher taxa and the evolution of genetic regulation. It is quite possible that these two problems are in fact linked (Valentine and Campbell, 1975).

Some aspects of genetic regulation in the single-celled prokaryotes have now been worked out, while knowledge of regulatory systems in the more advanced eukaryotes is still in the hypothetical stage. Several models of gene regulation of eukaryotes have been suggested (for example, Britten and Davidson, 1969, 1971; Kauffman, 1971; and Holliday and Pugh, 1975). They provide much insight into the ways in which evolution of the regulatory genome may occur, and the ways in which it may have influenced the course of evolutionary history.

Although there are several regulatory mechanisms present in prokaryotic cells, the best known were first described by Jacob and Monod (1961; also see Dobzhansky, 1970) and elaborated by other workers. Some *structural* genes (those that produce messenger RNA) are controlled by *operator* genes that lie next to them along a DNA strand. The operator genes are themselves inhibited by a class of proteins called *repressors*. Each repressor is specific for one operator. When it combines with this operator, the repressor prevents transcription of messenger RNA from the adjoining structural gene.

Repressors exist in two forms, active and inactive. In some systems the repressor is normally in the active form, except in the presence of certain substances to which it binds in preference to the operator. These systems are called inducible systems and the binding substance the *inducer*. In other systems the repressor is normally inactive and can only bind to the operator when it is complexed with a substance called a *co-repressor*. These systems are called *repressible systems*. Sometimes the inducer is the substrate of the enzyme or protein coded by the structural gene involved; the co-repressor is often the very enzyme or protein itself. In such cases the substrates and end products provide positive and negative feedback to insure that adequate but appropriate amounts of the structural gene product are maintained in the cell.

The number of distinctive enzyme activities that occur in higher organisms is probably not very much greater than those in lower organisms, but the way in which they are organized must be different, and the total number of gene activations must be vastly greater during the life cycle of a complex organism. Metazoan genomes clearly require more elaborate regulatory systems. For one thing, their genes are organized into chromosomes, so that regulatory genes must be able to control structural genes that may be on separate chromosomes. Furthermore, metazoans are multicellular and thus require some mechanisms for cellular integration. Additionally, they possess several different cell types, with different gene batteries employed in the different types. Thus, many cellular developmental pathways must be encoded within the same genome.

Britten and Davidson (1969, 1971; and Davidson and Britten, 1973) have formulated a general regulatory model for higher organisms that is consistent with the requirements of the genetic system and with our present knowledge of cell biology (Figure 8-11). As they emphasize, their model may well require modification as new knowledge comes to light, but some such regulatory system is justifiable. *Sensor genes* (SEN) receive a stimulus that results in the switching on of a battery of *structural genes* (S), which have products that function along a biosynthetic or metabolic pathway. The substance responsible for inducing activity in the sensor gene might be a hormone, perhaps acting through an intermediate agent. When the proper inducing agent is present, the sensor gene activates adjacent *integrator genes* (I), each of which produces a specific *activator RNA* molecule. These activator RNA molecules suffuse throughout the genome and complex with *receptor genes* (R).

The complexing is possible because the sequence of bases in the activator RNA molecule is complementary to that of receptor gene; therefore, the integrator gene from which the RNA was transcribed and the receptor gene must be similar or identical in sequence. The receptor gene lies in proximity to a structural gene, and when complexed with activator RNA, causes the structural gene to transcribe. This results in the elaboration of polypeptides at appropriate sites in the cell. In short, the need for a suite of gene products is detected by a sensor that causes an integrator to transcribe; the product of this transcription, activator RNA, is detected by a number of receptors, which cause structural genes to transcribe.

If many structural genes (which may be scattered widely through the genome) are associated with receptors that have identical (or similar) base sequences, then a single integrator can cause them all to transcribe. Such coordination would be salutary if the products of such a battery of genes were all required for integrated activity in a biochemical cycle or pathway. If the product of a given structural gene is required in more than one activity, that gene would be associated with more than one receptor, and would belong to more than one battery. The polygene batteries are thus not mutually exclusive in this model. Furthermore, if the products that result from genes controlled by a given integrator could stimulate one or more sensors, then additional gene batteries would be activated, and whole suites of integrated gene batteries brought into play. Thus, sets of gene batteries may be organized in a hierarchical pattern. The model depicted in Figure 8-11 implies that many repetitive DNA sequences must be present and interspersed with nonrepetitive sequences in the genome (see Chapter 3).

A possible variation on this model is that each structural gene has a unique receptor, and that batteries of structural genes are organized by multiple integrators, all of which (in any given battery) are activated by a given sensor.

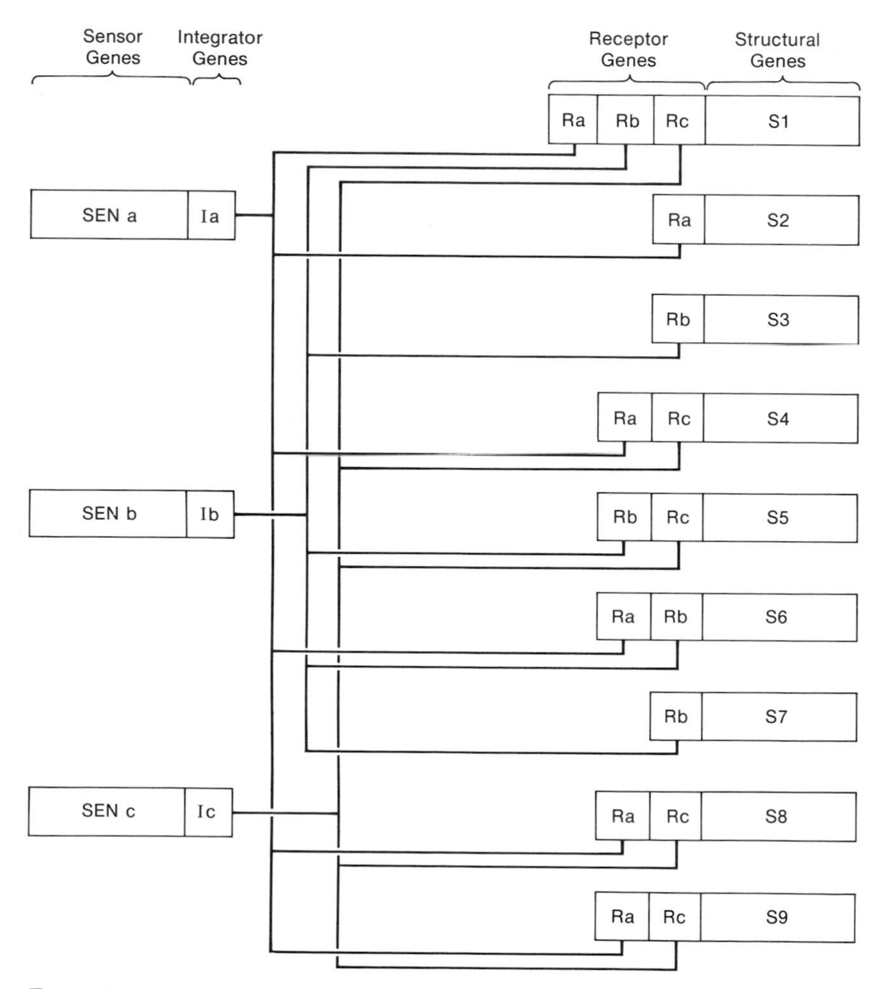

FIGURE 8-11
The Britten–Davidson model of gene regulation. Sensor genes (SEN) respond to the presence of some particular inducing substances (perhaps hormones) by activating adjoining integrators (I), which transcribe to produce an RNA, which diffuses through the genome and activates receptors (R). These in turn cause structural genes (S) to transcribe, eventually producing polypeptide gene products. Whole batteries of structural genes may be activated by a single sensor, and suites of sensors may be controlled by master sensors. [After Britten and Davidson, 1969, 1971.]

Another possibility is that integrators have protein products that are involved in activating the receptors. In any event, details of the organization of eukaryotic regulatory systems are not yet demonstrated, so we shall employ the simpler model described earlier.

One of the major evolutionary consequences of the existence of a regulatory system in the genome is that changes may occur not only among the structural genes but also among the regulatory genes. And just as a mutation may eliminate, reduce, alter, or even enhance the activity of a structural gene product, so may a regulatory mutation change the regulatory activity of the gene involved. Assuming that the sequence specificities need not be perfect at the various steps depicted in Figure 8-11, then a point mutation at, say, an integrator site might still permit activation of most receptors, although some might no longer accept the activating substances from the mutant integrator (Britten and Davidson, 1969).

For a given gene battery, effects of a mutation at a sensor site could range from deactivation of the battery to switching it to an entirely new activating substance so that the battery would function in new biochemical circumstances. Mutation of the integrator might also deactivate the battery, or it might change the composition of the battery, eliminating some and adding other structural genes. Mutation of the receptor could cause a particular structural gene to become part of one or more new batteries and eliminate it from some or all of its former gene batteries. Thus, gene activity could be extensively repatterned through mutations, whole batteries being activated under new circumstances, or reconstituted, or individual genes moved from one set of batteries to another. Even without any structural gene mutations, major changes in biochemical pathways can arise in a variety of ways, and if mutations occurred to more than one type of regulatory gene, the resulting changes could create a wide variety of novel patterns involving large numbers of structural genes. We assume that the novel patterns arise in small steps, since mutations with large phenotypic effects are very unlikely to increase fitness.

In addition to these mutations within the genes, chromosomal alterations could also create major changes in the regulatory patterns. The Britten–Davidson model implies that position is a major determinant of the function of different segments of DNA. For example, if a novel gene is emplaced next to a sensor, it will replace or supplement the former integrator and produce a novel activator substance. The effect of this new integrator would depend upon the extent to which appropriate receptors were present in the genome. Expansion of the regulatory apparatus may also occur, and would depend in part upon sources of new DNA, which, as we have seen in Chapter 3, are available to the genome.

The origin of novel organisms that achieve new levels of morphological complexity must surely involve an elaboration of the regulatory genome to accommodate the additional information required to encode the new developmental complications. Since new grades of organization arise during the appearance of new biological architectures or ground plans, some repatterning of the old regulatory system would ordinarily also occur to bring genes, gene batteries, or sets of gene batteries into new functional associations, producing novel arrangements of, for example, tissues and organs. At the same time, changes in the structural gene complement, perhaps including the addition of new genes, would be expected. The development of phyla of new grades, such as the coelomates from flatworms or the chordates from invertebrates, are examples of such evolutionary changes (see Chapter 13).

The radiations that so commonly follow the origin of a new adaptive type produce more or less distinctive variations on the new morphological theme, but at essentially the same grade of complexity. Such radiations may be based primarily upon repatterning of genes and gene combinations within a regulatory hierarchy. However, some expansion of the regulatory genome and evolution of structural gene complements must also occur. Examples of such radiations include the Mesozoic mammals, which diverged into a number of orders, and the radiation of the primitive echinoderms into numerous classes. In Figure 8-12 the architecture of several classes of the phylum Mollusca is depicted. Note that all the classes have a common morphological theme or ground plan, but that the development and arrangement of body parts differ from class to class. The extent of proliferation of cell types, the directional controls on cell growth and morphogenetic movements, and the timing of the appearance of organs

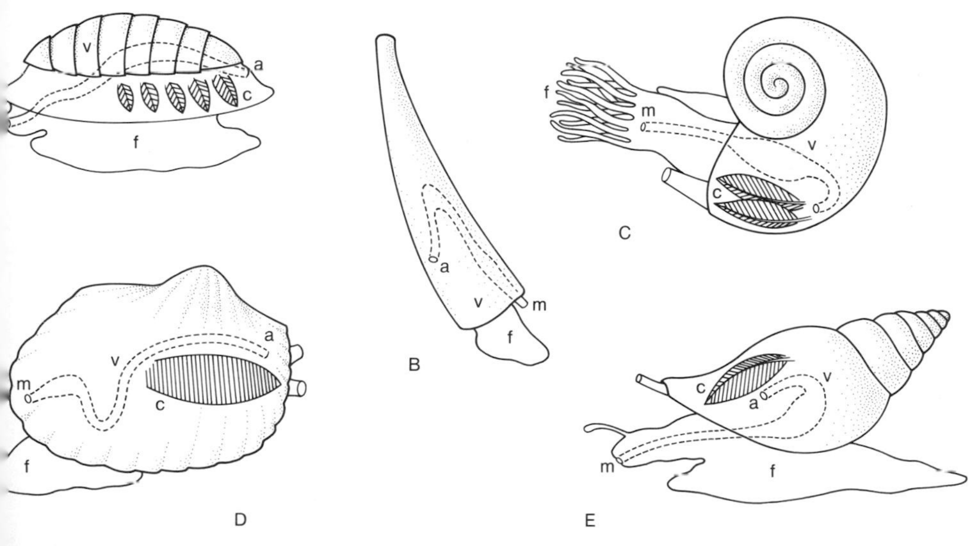

FIGURE 8-12.
Variations of molluscan architecture, which consists basically of a dorsal visceral mass *(v)* covered by a mantle and shell, a ventral foot *(f)*, and a head. Other features are the mouth *(m)*, anus *(a)*, and ctenidia or gill *(g)*. **A.** Amphineura (or Polyplacophora). **B.** Scaphopoda. **C.** Cephalopoda. **D.** Bivalvia (or Pelecypoda) **E.** Gastropoda. Monoplacophora not shown (see Figure 13-4).

during ontogeny may have different modes in the different classes. These are features that may evolve through changes in the pattern of gene regulation (see Chapter 13 for many additional examples).

Regulatory differences may account for many morphological differences between allied taxa even in lower categories. An example on the family level has been reported by King and Wilson (1975). They investigated genetic differences between chimpanzees and humans, and found that insofar as their structural gene complements are concerned they are no more different than sibling species of mammals or *Drosophila* (Chapter 6), whether measured by amino acid sequencing, electrophoretic, or immunological techniques (Chapter 9). Yet the morphological differences between humans and chimpanzees are impressive. Evidently, the chief genetic differences underlying them reside in the regulatory system. Regulatory changes may well underlie some speciation events as well (Wilson, Maxson, and Sarich, 1974).

There thus appear to be three major modes of evolutionary change as reflected by changes in the genome. One involves change in the frequency of structural genes and in the alleles of regulatory genes. A second involves repatterning of the genes. The third involves the growth of the genome, primarily through elaboration of the regulatory apparatus. All three modes must ordinarily operate to produce those series of anatomical changes that lead to the development of a wholly new animal type, and all three modes could operate to produce new species. Nevertheless, genome growth probably predominates during the evolution of new grades of complexity, gene repatterning during the radiation of major new variations within grades, and structural gene frequency changes must be especially important during phyletic evolution within species lineages.

It is possible that evolution of the regulatory genome is associated with a number of other evolutionary phenomena. For example, the shifting of characters from one part of an ontogeny to another may represent regulatory repatterning. The shifts that seem most common involve the retention of juvenile characters into later life, or *paedomorphosis*, and the shifting of reproduction into early ontogenetic stages, or *neoteny*. Such shifts, especially neoteny, have been postulated to explain the origin of taxa from subspecies to phyla.

The planktonic (floating) larvae of some primitive deuterostome invertebrates, especially of the tunicates (phylum Urochordata), resemble the theoretical *archetype* of the chordates, the simple common ancestor that can be postulated to have given rise to this phylum. Adult tunicates are sessile benthic suspension feeders. The embryologist Garstang conceived an explanation for the resemblance. He suggested that some primitive tunicate-like invertebrate larvae were subjected to large selective pressures for early sexual maturity—to shorten their lives before reproduction and thereby to increase fitness. It was also advantageous for them to remain long in the water column to achieve maximum dispersal. Eventually, reproduction was established before metamorphosis in some lineages, so that they became entirely planktonic, and the benthic portion of their life cycles dropped out. The adult planktonic creatures evolved into fish. This suggestion is not universally accepted, but demonstrates the possible advantages of neoteny. Shifting the activity of batteries of genes associated with reproductive functions into earlier parts of a life cycle might involve gradually switching the production of the substances that activate the appropriate sensors into gene batteries that operate in progressively earlier ontogenetic stages.

9

Phylogenies and Macromolecules

The diversity of life is staggering. About 400,000 species of plants and 1,500,000 species of animals have been described and named, but the census is far from complete. The diversity of the living world is apparent not only in the large number of species, but also in their heterogeneity. Organisms are extremely diverse in size, way of life, and habitat, as well as in structure and form.

Yet, despite their prodigious diversity, organisms share much in common. Oxygen, hydrogen, and carbon are common chemical elements in all organisms; together they account for about 98.5 percent by weight of any living being. Four kinds of macromolecules—proteins, carbohydrates, lipids, and nucleic acids—are the basic molecular constituents of all living processes. The genetic information of all organisms, from bacteria to man, is encoded in the double-helical structure of DNA. The processes of transcription and translation and the genetic code are essentially uniform throughout all life.

Certain similarities are shared by some, not all organisms. These similarities can be used to classify, i.e., to characterize, some groups of organisms and distinguish them from others. Centuries ago biologists noted that living things can be arranged in a hierarchic fashion, that some organisms (and groups of organisms) resemble each other more than they resemble other organisms (or groups). The basic process responsible for the hierarchy of similarities among living things is of course evolution—some organisms resemble each other more than others because they are more closely related by lines of descent.

No modern classification of life entirely ignores evolution. The three prevailing systems of classification, known as phenetic, cladistic, and evolutionary, were discussed in Chapter 8. In this chapter we shall discuss the methods used to infer relationships of evolutionary descent (phylogeny) on the basis of resemblance and dissimilarity.

HOMOLOGY AND ANALOGY

Phylogenetic relationships are ascertained on the basis of several complementary sources of evidence. First, we have remnants of organisms living in the past, the "fossil record" discussed in Chapter 10. In some cases the fossil record provides definitive evidence of the phylogenetic relationships among groups of organisms. However, the fossil record is far from complete, and is often seriously deficient. Second, comparative studies of living forms also provide information about phylogeny. Comparative anatomy is the branch of science that in the past has contributed the most information, although additional knowledge came from comparative embryology, cytology, ethology, biogeography, and other biological disciplines. In recent years the comparative study of informational macromolecules (proteins and nucleic acids) has become a powerful tool for the study of phylogeny.

Morphological similarities among organisms were probably always recognized by men. Among the Greeks, Aristotle (384–322 B.C.), and later his followers and those of Plato, particularly Porphyry (234–305 A.D.), classified organisms (as well as inanimate objects) on the basis of similarities. The Aristotelian system of classification was further developed by some medieval Scholastics, notably Albertus Magnus (1193–1280) and Thomas Aquinas (1225–1274). The modern foundations of taxonomy were laid in the eighteenth century by Linnaeus (1707–1778), and by the lesser-known botanist Adanson (1727–1806). Cuvier (1769–1832), another great taxonomist, suggested that similarities of form are related to similarities of function. Lamarck (1744–1829) dedicated much of his work to the systematic classification of organisms, and proposed that their similarities are due to ancestral relationships.

The modern theory of evolution, originating from Darwin (1859), provides a causal explanation of the similarities among living beings. Organisms evolve by a process of descent with modification. Changes, and therefore differences, gradually accumulate over the generations. The more recent the last common ancestor of a group of organisms, the less their differentiation; similarities of form and function reflect phylogenetic propinquity. For this reason, phylogenetic affinities can be inferred on the basis of relative similarity.

A distinction is necessary between resemblances due to propinquity of descent and resemblances only due to similarity of function. *Homology* is correspondence of features in different organisms due to inheritance from a common ancestor. The forelimbs of humans, dogs, whales and chickens are homologous; the skeletons of these limbs are all constructed of bones arranged according to the same pattern because they were inherited from an ancestor with similarly arranged forelimbs (Figure 9-1). *Analogy* is correspondence of features due to similarity of function but not related to common descent. The wings of birds and flies are analogous. Their wings are not modified versions of a structure present in a common ancestor, but rather have developed independently as adaptations to a common function, flying (Figure 9-2). The similarities between the wings of bats and birds are partially homologous and partially

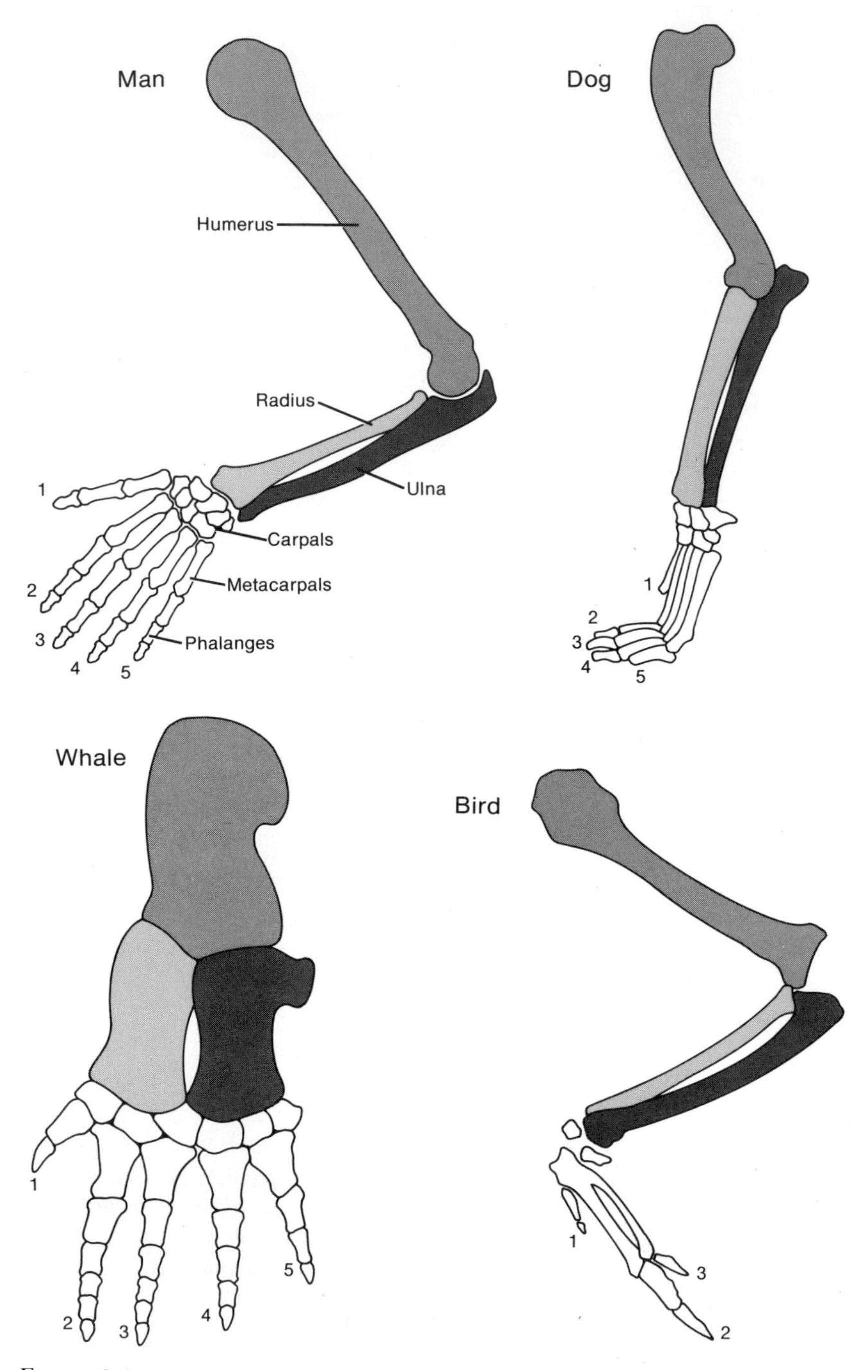

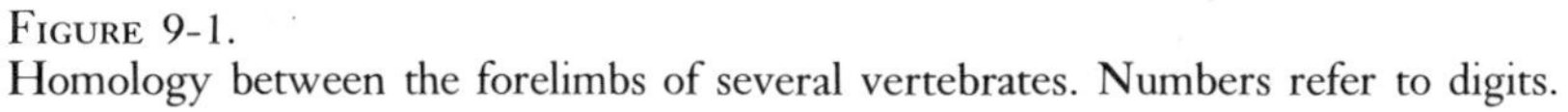

FIGURE 9-1.
Homology between the forelimbs of several vertebrates. Numbers refer to digits.

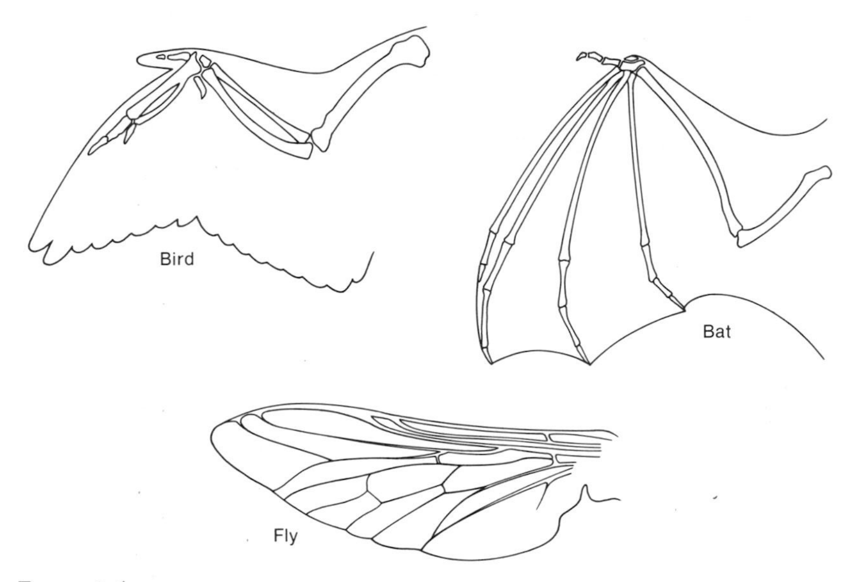

FIGURE 9-2
Analogy between the wings of birds, bats, and flies. The wings of flies and birds (or bats) are analogous; they have evolved independently in these organisms as adaptations to flying. The similarities between the wings of bats and birds are partially homologous and partially analogous. The skeletal structures are homologous owing to common descent from the forelimb of a reptilian ancestor, but the modifications for flying are analogous, having evolved independently.

analogous. The skeletal structure is homologous because of common descent from the forelimb of a reptilian ancestor, but the modifications for flying are different and independently evolved, and in this respect are analogous.

Features that have independently evolved into more rather than less similar ones are said to be *convergent*. Convergence often occurs due to similarity of function, such as the evolution of wings in the ancestors of birds, bats, and flies. The shark (fish) and the dolphin (mammal) exhibit similarities in external morphology due to convergence, since they have evolved independently as adaptations to aquatic life.

Taxonomists speak also of *parallel* evolution. Parallelism often involves similarities due to both homology and analogy. According to Simpson (1961), "parallelism is the development of similar characters separately in two or more lineages of common ancestry and on the basis of, or channeled by, characteristics of that ancestry." Parallelism and convergence are not always clearly distinguishable. Strictly speaking, convergent evolution occurs when descendants resemble each other more than their ancestors did with respect to some feature. Parallel evolution implies that two or more lineages have changed in similar ways, so that the evolved descendants are as similar to each other as their ancestors were. The evolution of marsupials in Australia paralleled the evolution of placental mammals in other parts of the world. There are Australian marsupials resembling true wolves, cats, mice, squirrels, moles, sloths, and anteaters (Figure 9-3). The evolution of these placental mammals and of the corresponding marsupials proceeded independently but in parallel lines due to their adaptation to similar ways of life. Some resemblances between a true anteater (*Myrmecophaga*) and a marsupial anteater (*Myrmecobius*) are due to homology—both are mammals. Others are due to analogy—both feed on ants.

Parallel and convergent evolution are also common in plants. New World cacti and African euphorbias are similar in overall appearance although they belong to separate families. Cacti and cactus-like euphorbias are succulent, spiny, water-storing, flowering plants adapted to the arid conditions of the desert. Their cactus-like morphologies have evolved independently as responses to similar environmental challenges. Cacti and cactus-like euphorbias are instances of convergent evolution.

Parallel and convergent evolution are particularly common in the reproductive structures of flowering plants. Because fossil remains of these structures are rare, it is often difficult to ascertain whether convergent or parallel evolution rather than homology is involved. A characteristically difficult situation occurs in the group of families having bell-shaped or tubular flowers with joined petals. Heaths (*Erica*), primroses (*Primula*), phloxes, mints (Labiateae), figworts (Schrophulariaceae), gentians, bluebells (*Campanula*), honeysuckles (*Lonicera*), sunflowers (Compositae), and others were all at one time grouped together in the superorder Sympetalae. Although some of these flowers may be strictly homologous, botanists are now certain that others, perhaps the majority, represent parallelisms or convergences.

EVALUATION OF HOMOLOGIES

So far we have discussed homology between different organisms. A different kind of homology may be recognized, namely between repetitive or serial structures of the same organism. This has been called *serial homology*. Serial homology exists, for example, between the arms and legs of man, or among the seven cervical vertebrae of mammals, or among the branches or leaves of a tree. The jointed appendages of arthropods are elaborate examples of serial homology. Nineteen pairs of appendages exist in the crayfish, all built according to the same basic pattern but serving diverse functions—sensing, chewing, food-handling, walking, mating, egg-carrying, and swimming (Figure 9-4).

Relationships in some sense akin to those between serial homologs exist at the molecular level between genes and proteins resulting from ancestral gene duplications. The α, β, γ, and δ hemoglobin chains and the genes coding for them are homologous; similarities in their amino acid sequences occur because they are modified descendants of a single ancestral sequence. Homologous

FIGURE 9-3.
Parallel evolution of placental and Australian marsupial mammals. [After Simpson and Beck, 1965.]

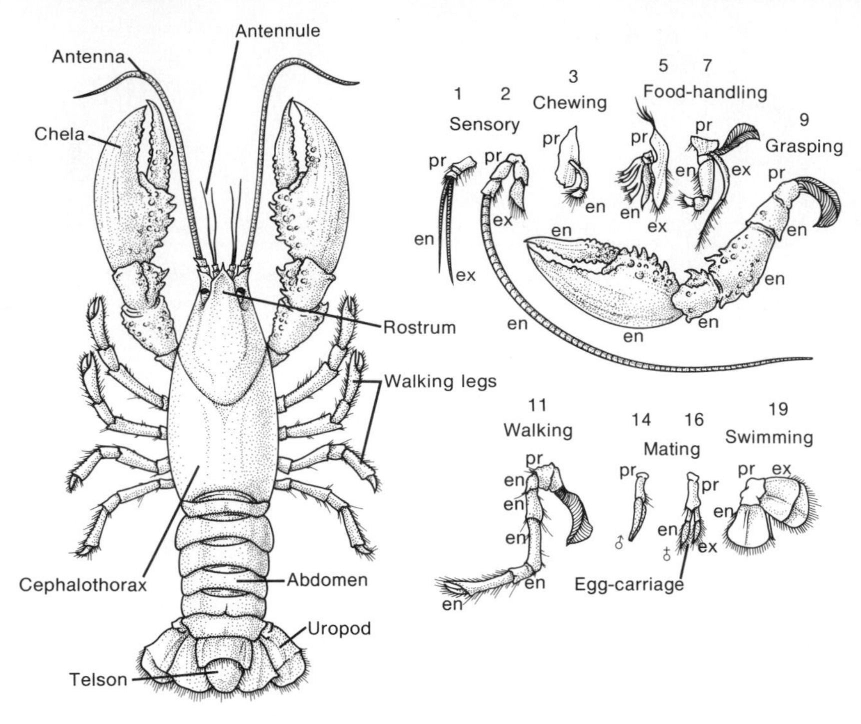

FIGURE 9-4.
Serial homology in the crayfish. The appendages are all modified versions of a basic pattern. This pattern has a basal structure, called the *protopodite (pr),* which may consist of more than one segment. There is also a two-fingered structure. The "finger" nearest the midline of the body is called the *endopodite (en),* and the lateral one is called the *exopodite (ex);* each consists of one or several segments. The appendages have derived through modifications, repetitions, and occasional loss of the elements of the basic pattern.

comparisons between hemoglobins can be of two kinds. We may compare the same hemoglobin, e.g., the α chain, of different species, such as man, chimpanzee, dog, and snake. Or we can make comparisons between the α, β, γ, and δ hemoglobins of a single species, such as man. Homologous genes (and proteins) can be either orthologous or paralogous (Fitch and Margoliash, 1970).

Orthologous genes are descendants of an ancestral gene present in the last common ancestor of the species carrying the orthologous genes. Therefore the evolution of orthologous genes reflects the evolution of the species in which they are found. Comparisons among orthologous genes serve to establish species phylogenies. For example, comparison of the α hemoglobins of different mammals leads to inferences about mammalian phylogeny. *Paralogous* genes are descendants of a duplicated ancestral gene. The evolution of paralogous genes reflects divergence that has accumulated since the genes duplicated. Homologies between paralogous genes of the same species, e.g., the α, β, γ, and δ hemoglobins in man, serve to establish gene phylogenies, i.e., the evolutionary history of duplicated genes within a given lineage. For example, the duplication leading to the α and β hemoglobins of man occurred further in the past than the duplication giving rise to the β and δ hemoglobins (see Chapter 3).

In establishing gene phylogenies comparisons should be made between paralogous genes (or proteins) of the *same* species. For example, myoglobin and the α and β hemoglobin chains of man should be compared with each other, and so on. Data obtained from different species should corroborate each other, since the phylogeny of duplicated genes is the same in all species carrying the same set. Differences between paralogous genes in *different* species (say the α hemoglobin in man and the β hemoglobin of dogs) reflect the phylogeny of the duplicated genes as well as the evolution of the species.

Whether or not the similarities between features of different organisms are due to homology cannot always be decided unambiguously, but may be a matter of probability. To determine phylogenetic relationships we must be able to distinguish similarities that are homologous from those that are analogous or simply accidental. Moreover, degrees of homology must be quantified in some way in order to determine the propinquity of common descent among species. Difficulties also arise in determining the degree of homology among various organisms. Consider the five forelimbs shown in Figure 9-1. Are the homologies greater between man and the horse than between man and the whale; or between man and the whale than between man and the chicken? The fossil record may sometimes provide the appropriate information, but as we have noted the fossil record is far from complete. The evidence from comparative studies of living forms and paleontological evidence must be examined together.

The comparative studies of proteins and nucleic acids presented later in this chapter provide quantitative estimates of the degree of homology between organisms. Methods used to distinguish homology from accidental similarity are discussed in this section. In general, two complementary criteria help to distinguish homology from other similarities (Simpson, 1961). (1) With respect to any given feature (a limb, organ, behavioral pattern, protein, etc.) *minuteness of resemblance* suggests homology. (2) The greater the *number of similar features* between any two organisms, the more likely it is that homology, and therefore common ancestry, is responsible for the similarities of any one feature.

The detailed correspondence between the parts of the forelimbs of man, dogs, horses, chickens, and whales, indicate that they are homologous. In all these forelimbs there are scapulas, humeri, radii, ulnas, carpals, metacarpals, and phalanges. The number of digits is five in man, dogs, and some whales, three in chickens, and only one in horses. But the correspondence in the structure and position of the forelimb bones is too precise to be accounted for by anything but common ancestry. Comparisons of other features, for example of the rest of the skeleton, show that many other traits are quite similar in the five animals shown in Figure 9-1. By comparing multiple features, the relative degrees of similarity among all five animals can be established. For example, it becomes apparent that man, dogs, horses, and whales are more similar to each other than any of them is to the chicken.

Inger (1958) has shown that simple streamlining and oral suckers have evolved independently in separate groups of tadpoles living in fast streams. Nevertheless, characteristic details of the oral discs permit identification of the phylogenetic groups to which the tadpoles belong. The homologies are more numerous and detailed among tadpoles of the same phylogenetic group than among tadpoles belonging to different groups but living in torrents. Examination of the detailed structure of convergent features usually makes it possible to detect analogy because resemblance rarely extends to the fine details of complex traits.

As a rule, similarities due to analogy, and in general to parallel or convergent evolution, lack the detailed correspondence of parts observed in cases of homology. Moreover, the number of similar features is limited. The Tasmanian wolf (*Thylacinus*) and the true wolf (*Canis*) shown in Figure 9-3 resemble each other, but their similarities are related to a particular habit of predatory adaptation. For characteristics not related to that adaptation, *Canis* exhibits greater and more detailed similarities with the cat (*Felis*) or the squirrel (*Glaucomys*) than with the Tasmanian wolf. Bats, birds, and flies (Figure 9-2) have evolved wings as adaptations to flying, but the correspondence in the morphology of the wings is far from precise. Moreover, bats, birds, and flies are quite different in many other features.

Assume that a group of organisms, say mammals, have been studied with the purpose of deciding their most likely phylogeny. A great variety of homologous features have been compared. How do we evaluate all this information, which might be conflicting in some cases? Traditionally, comparative anatomists and taxonomists have found from experience that within a group of organisms some features are more reliable indicators of phylogeny than others. That is, not all homologous features are weighted equally. An element of subjectivity is thus introduced; experts do not always agree on what traits should be given greater weight. To circumvent this problem, some taxonomists have suggested that as many characters as possible be studied and all be given equal weight. The data are then processed (usually with the help of an electronic computer) and "objective" taxonomic classifications are obtained. These procedures and the theory behind them are known as "phenetics" or "numerical taxonomy" (for a comprehensive treatment see Sneath and Sokal, 1973). The methods of numerical taxonomy yield "dendrograms"—organisms are connected in a branching arrangement in which organisms or groups of them are connected at nodes. Whether dendrograms should be interpreted as phylogenies depends on whether it can be assumed that the traits under consideration have evolved at approximately constant rates.

Numerical taxonomy, cladistics, and evolutionary taxonomy differ not only in their methods of classification, but also in the theories behind the methods. Numerical taxonomy has been severely criticized by some leading taxonomists and evolutionists, e.g., Simpson (1961) and Mayr (1969). Many taxonomists and comparative anatomists prefer to give different weight to different traits on the basis of accumulated knowledge and their experience rather than rely on the impersonal manipulations of data made by a computer. However that may be for anatomical, behavioral, and certain other traits, there is little doubt that the methods of numerical taxonomy are appropriate to evaluate the data obtained from the study of a single protein or DNA sequence.

Determination of phylogenetic relationships on the basis of anatomical, embryological, behavioral, and ecological knowledge requires intensive study of the biology of the organisms. There are specialized treatises in which such studies are presented, and we shall not review them here. In the following sections we shall, however, summarize certain methods that can be applied without much knowledge of the biology of the organisms under study. One method is based on the linear arrangement of genes in polytene chromosomes; its application is limited to Diptera. The other methods are based on the direct study of DNA and proteins. For the most part these latter methods have been developed only recently. They are powerful methods to ascertain phylogenetic relationships. Needless to say, decisions concerning phylogeny should be made on the basis of all available information, including biochemical, chromosomal, anatomical, embryological, ethological, ecological, biogeographical, and paleontological knowledge.

CHROMOSOME PHYLOGENIES

Changes in the number and shape of chromosomes have occurred in evolution through chromosomal fusions, fissions, translocations, inversions, duplications, and deletions (Chapter 3). Number, size, and shape of chromosomes have been used to establish phylogenetic relationships among related species (for a review see White, 1973). The giant polytene chromosomes of Diptera exhibit sequences of individually recognizable bands that permit detection of inversions and other chromosomal rearrangements. Ancestral relationships can sometimes be inferred from chromosomal rearrangements, particularly those involving *overlapping* inversions.

Overlapping inversions are chromosomal inversions that include part of previously inverted chromosome segments. Suppose that the sequences of bands in three different homologous chromosomes are ABCDEFGH, AEDCBFGH, and AEDFBCGH. The second sequence can arise from the first, or the first from the second, by a single inversion involving the segment BCDE. Similarly, the third sequence can arise from the second, or viceversa, by a single inversion of the segment CBF. However, the third sequence cannot arise from the first, nor the first from the third, by a single inversion. The second sequence represents a necessary intermediate step between the first and the third. If no other information is available, there is no way of knowing which one of the three sequences may have been the ancestral one, but only three phylogenetic relationships are possible: $1 \rightarrow 2 \rightarrow 3$, or $3 \rightarrow 2 \rightarrow 1$, or $1 \leftarrow 2 \rightarrow 3$. Any phylogenies involving the direct transitions $1 \rightarrow 3$ or $3 \rightarrow 1$ can be excluded.

Phylogenies have been established by means of overlapping inversions in organisms with polytene chromosomes, such as *Drosophila*, black flies (Simuliidae), midges (Chironomidae), and mosquitoes (Culicidae). Natural populations of *Drosophila pseudoobscura* and *D. persimilis* are polymorphic for a variety of arrangements in their third chromosomes (for a review see Dobzhansky, 1970). The phylogenetic relations inferred from the sequences of bands are shown in Figure 9-5. Any two chromosome arrangements connected by an arrow can be derived one from the other by a single inversion. The Standard arrangement is found in both species. The Hypothetical arrangement has never been found but has been postulated as a necessary "missing link" between

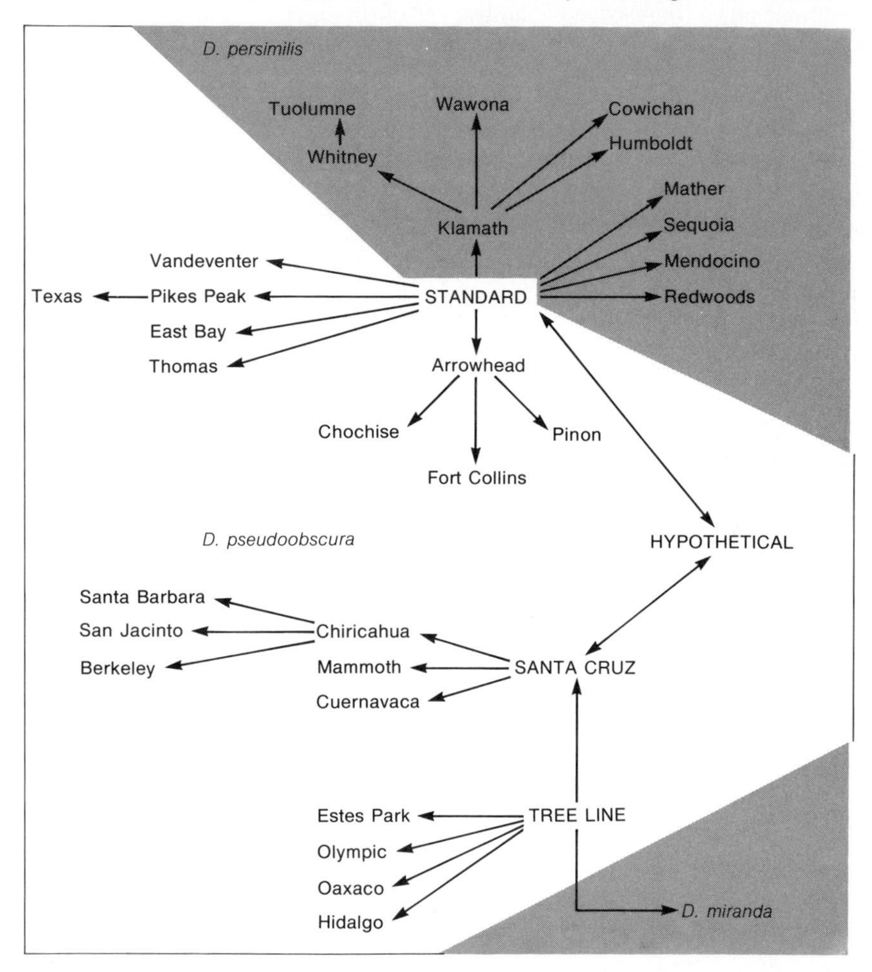

FIGURE 9-5.
Phylogeny of the gene arrangements in third chromosomes of *Drosophila pseudoobscura*, *D. persimilis*, and *D. miranda*. The phylogenetic relationships are inferred from the sequence of bands in polytene chromosomes of the salivary glands. The Standard chromosomal arrangement is found in both *D. pseudoobscura* and *D. persimilis;* all other sequences exist in only one species. The Hypothetical arrangement has never been found, but is postulated as a "missing link," perhaps extinct, between Standard and Santa Cruz. The pattern suggests that the ancestral arrangement is likely to be one of the four: Standard, Hypothetical, Santa Cruz, or Tree Line.

Standard and Santa Cruz. Conceivably the ancestral gene arrangement could be any one in Figure 9-5, or some other one as yet undiscovered. Nevertheless, the pattern shown in Figure 9-5 makes it likely that the four arrangements in the central portion (Standard, Hypothetical, Tree Line, and Santa Cruz) are ancestral, and the remainder are derived from them as shown by the arrows.

The phylogeny of the picture-winged group of Hawaiian *Drosophila* is a remarkable example of what can be accomplished studying polytene chromosomes. The archipelago of Hawaii contains a most interesting biota. Few animals or plants have colonized the archipelago, owing to its wide geographic separation from continents or islands. Some groups of organisms, such as mosquitoes, cockroaches, and mammals, never reached Hawaii until introduced by humans. Animals and plants that reached the Hawaiian islands found a propitious environment for evolutionary diversification; much of the native flora and fauna is endemic to the archipelago (Table 9-1; see Chapter 6). Colonizations were often followed by "adaptive radiations"—diversification into many species adapted to the various conditions existing in separate islands.

The total area of the archipelago is some 16,000 square kilometers, about half the size of Belgium or less than one-twentieth the size of California. Yet Hawaii contains around 500 endemic species of drosophilids, nearly one third of the number known for the entire world. Greater morphological and ecological diversity exists in Hawaiian *Drosophila* than in the rest of the world. A phylogeny of 101 species based on chromosomal arrangements is shown in Figure 9-6. The chromosomes do not provide information about the direction of ancestor–descendant relationships. Any one of the species in the diagram could be the ancestor from which all others derived through gradual accumulation of overlapping inversions. However, *D. primaeva*, shown at the bottom left of Figure 9-6, is probably the most similar to the ancestral species from which the species have derived through the steps indicated in the diagram. The chromosomal arrangements of *D. primaeva* show several homologies with continental species of the *D. repleta* group, from which Hawaiian *Drosophila* are thought to be derived (Carson *et al.*, 1970).

TABLE 9-1. Endemic flora and fauna in the Hawaiian archipelago. [After Carlquist, 1970, and other sources.]

	Genera		Species, subspecies, varieties	
	Total number	Percent endemic	Total number	Percent endemic
Ferns	37	8%	168	65%
Flowering plants	216	13	1729	94
Land molluscs	37	51	1064	99+
Insects	377	53	3750	99+
Drosophilids*	4	75	510	100
Birds	41	39	71	99

*Species introduced by man are not included.

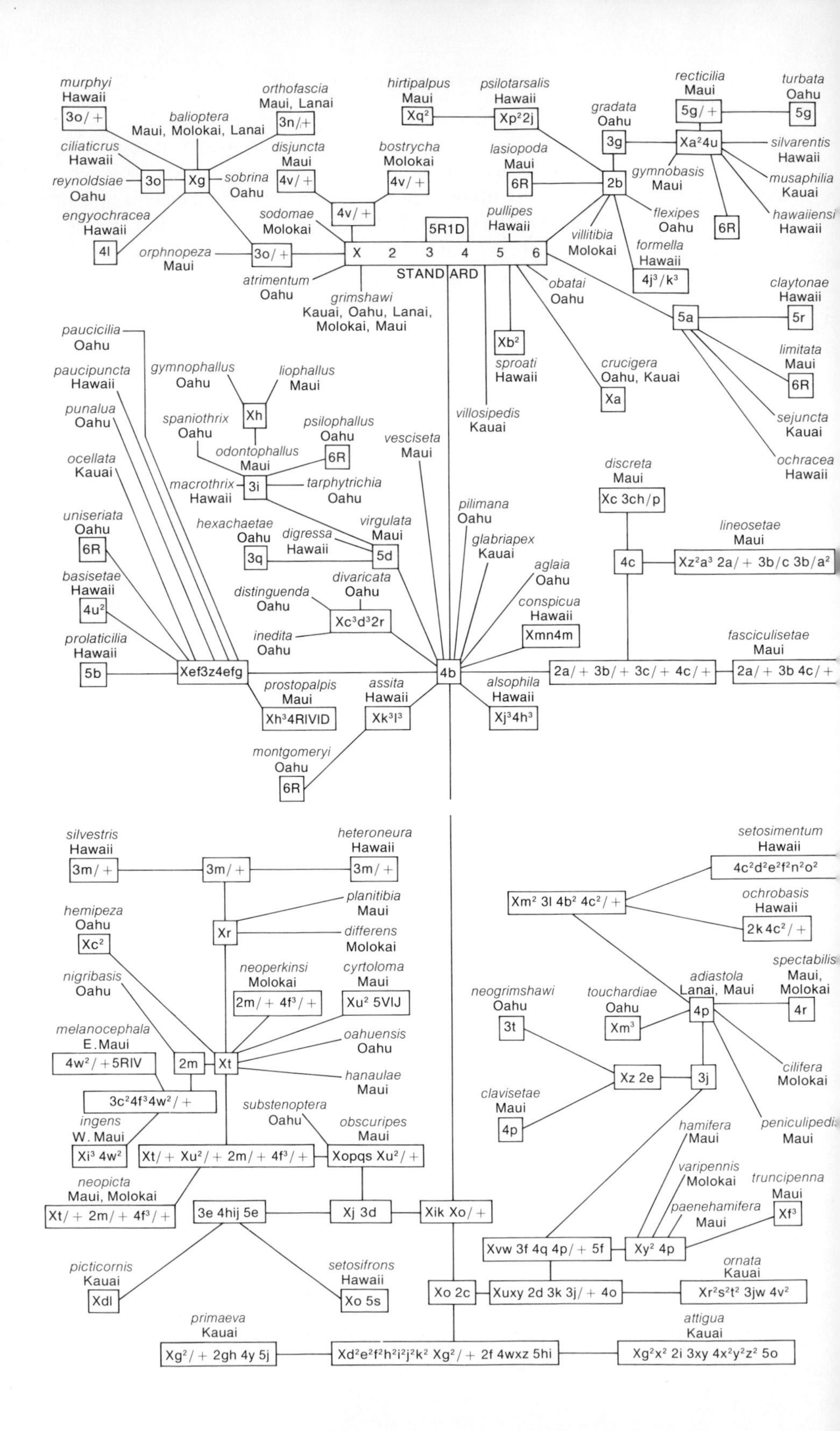

murphyi
Hawaii
3o/+
balioptera
Maui, Molokai, Lanai
orthofascia
Maui, Lanai
3n/+
hirtipalpus
Maui
Xq^2
psilotarsalis
Hawaii
Xp^{2}2j
recticilia
Maui
5g/+
turbata
Oahu
5g
gradata
Oahu
3g
Xa^{2}4u
silvarentis
Hawaii
ciliaticrus
Hawaii
3o
Xg
sobrina
Oahu
reynoldsiae
Oahu
disjuncta
Maui
4v/+
bostrycha
Molokai
4v/+
lasiopoda
Maui
6R
2b
gymnobasis
Maui
musaphilia
Kauai
engyochracea
Hawaii
4l
4v/+
sodomae
Molokai
5R1D
pullipes
Hawaii
flexipes
Oahu
6R
hawaiiensis
Hawaii
orphnopeza
Maui
3o/+
X 2 3 4 5 6
STANDARD
villitibia
Molokai
formella
Hawaii
4j^3/k^3
atrimentum
Oahu
grimshawi
Kauai, Oahu, Lanai, Molokai, Maui
obatai
Oahu
claytonae
Hawaii
5a
5r
paucicilia
Oahu
Xb^2
sproati
Hawaii
limitata
Maui
6R
paucipuncta
Hawaii
gymnophallus
Oahu
liophallus
Maui
crucigera
Oahu, Kauai
Xa
punalua
Oahu
Xh
spaniothrix
Oahu
psilophallus
Oahu
villosipedis
Kauai
sejuncta
Kauai
odontophallus
Maui
6R
vesciseta
Maui
ocellata
Kauai
ochracea
Hawaii
macrothrix
Hawaii
3i
tarphytrichia
Oahu
discreta
Maui
Xc 3ch/p
pilimana
Oahu
uniseriata
Oahu
6R
hexachaetae
Oahu
digressa
Hawaii
virgulata
Maui
glabriapex
Kauai
lineosetae
Maui
3q
5d
4c
Xz^{2}a^3 2a/+ 3b/c 3b/a^2
basisetae
Hawaii
4u^2
divaricata
Oahu
aglaia
Oahu
distinguenda
Oahu
Xc^{3}d^{3}2r
conspicua
Hawaii
inedita
Oahu
Xmn4m
prolaticilia
Hawaii
fasciculisetae
Maui
5b
Xef3z4efg
4b
2a/+ 3b/+ 3c/+ 4c/+
2a/+ 3b 4c/+
prostopalpis
Maui
assita
Hawaii
alsophila
Hawaii
Xh^{3}4RIVID
Xk^{3}l^3
Xj^{3}4h^3
montgomeryi
Oahu
6R
silvestris
Hawaii
3m/+
3m/+
heteroneura
Hawaii
3m/+
setosimentum
Hawaii
4c^{2}d^{2}e^{2}f^{2}n^{2}o^2
planitibia
Maui
Xm^2 3l 4b^2 4c^2/+
ochrobasis
Hawaii
hemipeza
Oahu
Xr
differens
Molokai
2k 4c^2/+
Xc^2
neoperkinsi
Molokai
cyrtoloma
Maui
spectabilis
Maui, Molokai
nigribasis
Oahu
2m/+ 4f^3/+
Xu^2 5VIJ
neogrimshawi
Oahu
touchardiae
Oahu
adiastola
Lanai, Maui
4p
4r
melanocephala
E. Maui
oahuensis
Oahu
3t
Xm^3
4w^2/+5RIV
2m
Xt
hanaulae
Maui
Xz 2e
3j
cilifera
Molokai
3c^{2}4f^{3}4w^2/+
substenoptera
Oahu
clavisetae
Maui
ingens
W. Maui
obscuripes
Maui
4p
peniculipedis
Maui
hamifera
Maui
Xi^3 4w^2
Xt/+ Xu^2/+ 2m/+ 4f^3/+
Xopqs Xu^2/+
varipennis
Molokai
truncipenna
Maui
neopicta
Maui, Molokai
paenehamifera
Maui
Xf^3
Xt/+ 2m/+ 4f^3/+
3e 4hij 5e
Xj 3d
Xik Xo/+
Xvw 3f 4q 4p/+ 5f
Xy^2 4p
picticornis
Kauai
setosifrons
Hawaii
ornata
Kauai
Xdl
Xo 5s
Xo 2c
Xuxy 2d 3k 3j/+ 4o
Xr^{2}s^{2}t^2 3jw 4v^2
primaeva
Kauai
attigua
Kauai
Xg^2/+ 2gh 4y 5j
Xd^{2}e^{2}f^{2}h^{2}i^{2}j^{2}k^2 Xg^2/+ 2f 4wxz 5hi
Xg^{2}x^2 2i 3xy 4x^{2}y^{2}z^2 5o

FIGURE 9-6 (opposite)
Phylogeny of 101 *Drosophila* species from Hawaii. The phylogenetic relationships are inferred from the sequence of bands in polytene chromosomes. The sequence taken as Standard is represented in the box at the center near the top; X, 2, 3, 4, and 5 represent the five large chromosome arms, 6 represents the small dot-like chromosome. The Standard sequences are found in *D. sodomae, D. atrimentum, D. grimsawi, D. villosipedis, D. obatai,* and *D. pullipes.* The chromosomal arrangements of all other species are derived from the Standard one through inversions. Inverted sequences in a given chromosome arm are represented by letters following the symbol for the chromosome, e.g., Xa represents inversion a in the X chromosome, 4b represents inversion b in the fourth chromosome arm. A letter followed by a slash and another letter (or the sign +) indicates that the species is polymorphic for two inverted sequences (or for the Standard and one inverted sequence), e.g., 2a/+ 3b/c represents a species that is polymorphic for the a inversion and the Standard sequence in the second chromosome, and for the b and c inversions in the third. The chromosome formulas are read additively, from the Standard through all intermediate steps. For example, the formula for *D. hawaiiensis* (top right hand corner) is Xa^2 2b 3g 4u, indicating that the chromosomes of this species differ from the Standard sequences by inversions a^2 in the X chromosome, b in the second, g in the third, and u in the fourth. The ancestral sequences could be those of any species in the figure or some others as yet undiscovered. It seems likely, however, that *D. primaeva* (bottom left hand corner) is the most similar to the ancestral species. [After Carson Kaneshiro, 1976.]

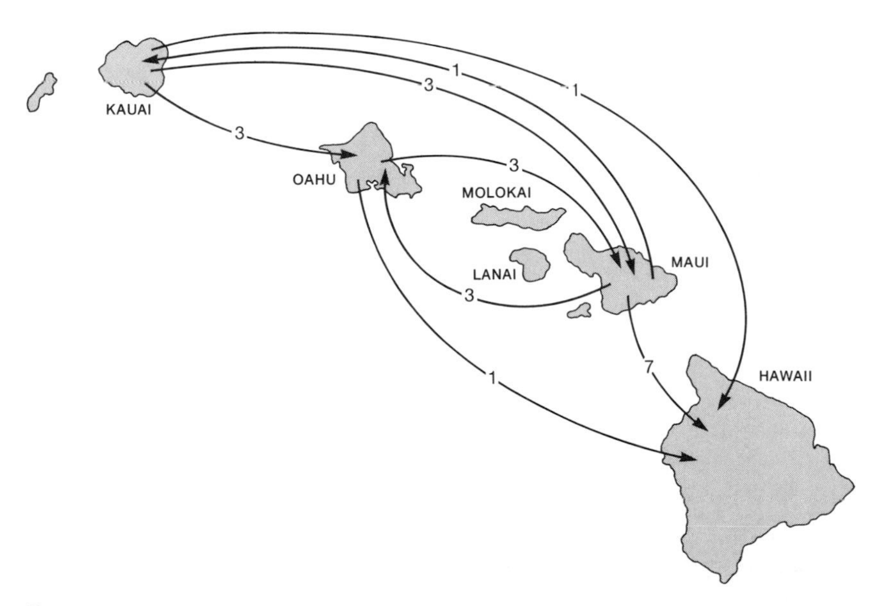

FIGURE 9-7.
Minimum number of colonizations postulated to account for the evolution of the picture-winged *Drosophila* species of Hawaii. The arrows indicate the direction of migration; the numbers in each arrow indicate the minimum number of separate colonizations postulated in each direction. [After Carson *et al.,* 1970.]

Comparisons of the chromosomal arrangements found in different islands of the Hawaiian archipelago have helped elucidate the pattern of colonizing migrations between islands. A minimum of 22 inter-island migrations are postulated to account for the evolution of the picture-winged species (Figure 9-7). For the most part, colonizations have occurred from geologically older to younger islands. Hawaii, the most recently formed island, has received colonizers from all other large islands, but it has not been the origin of any colonizing migrant. Kauai, the oldest of the larger islands, has been the source of colonizations to all other large islands, but has received only one successful colonization, from Maui (Carson *et al.*, 1970).

METHODS OF "HYBRIDIZING" DNA

Genetic information is encoded in the nucleotide sequence of DNA molecules. The degree of genetic differentiation between two species may be measured by the proportion of nucleotide pairs that are different in their DNA. This can be accomplished by making hybrid molecules between single strands of DNA from different organisms. Doty and his colleagues discovered that the two strands of DNA could be dissociated and reassociated in vitro, and succeeded in obtaining hybrid molecules made from strands coming from two different species of bacteria or viruses (Doty *et al.*, 1960; Schildkraut *et al.*, 1961, 1962). It was soon realized that these techniques could be used to measure DNA differentiation between species (Hoyer, McCarthy, and Bolton, 1964).

Two different measures of DNA differentiation can be obtained by hybridization: (1) the extent of the hybridization reaction, i.e., the fraction of the DNA of two species that forms hybrid molecules; (2) the proportion of nucleotide pairs that are complementary in the hybrid molecules. Single strands of DNA may reassociate into double-stranded molecules even though not all nucleotide pairs are complementary. The proportion of noncomplementary nucleotides is estimated by observing the temperature at which the two strands of the hybrid DNA molecule separate. A brief description follows of the methodology used to determine the degree of homology between DNA of different species.

Double-stranded DNA can be denatured or "melted" into single strands by heating it to about 100°C. The heat breaks the hydrogen bonds between the two complementary strands. If the solution is rapidly cooled, the strands remain separated. One common technique to measure the proportion of DNA sequences that are homologous in two species starts by trapping the single-stranded polynucleotide segments of one species, A, in a homogeneous matrix such as agar or nitrocellulose membrane filters. The agar or filter is sheared into small pieces containing single-stranded DNA fragments.

The filter- or agar-bound DNA of species A is then incubated at temperatures around 60°C in a solution containing single-stranded DNA segments of the same species, A, and a second species, B. The unbound DNA of species A and B in solution has been previously denatured and sheared into small segments about 500 nucleotides in length. The free single-stranded DNA of species A is labeled with a suitable radioactive isotope, such as tritium, ^3H, or phosphorous, ^{32}P. The solution contains a small constant amount of the radioactive DNA of species A, but the amount of DNA of species B is varied in separate experiments. The solution is incubated for several hours to permit association between the free and the bound DNA. The remaining free DNA is then washed out. The amounts of free DNA of species A and B that have formed duplexes with the bound DNA can then be determined, since the DNA of species A is radioactive.

The results of a typical experiment are shown in Figure 9-8. Denatured DNA from *Drosophila melanogaster* was immobilized in a nitrocellulose membrane filter, later cut into circular pieces 7 mm in diameter. The filters holding DNA were incubated in a solution containing 1 μg ($= 10^{-6}$ grams) of ^3H-labeled DNA from *D. melanogaster*, and variable amounts of DNA from some other species. In the control experiments the two DNAs in solution are from *D. melanogaster.* As the amount of unlabeled DNA increases, the amount of radioactive DNA that pairs with the filter-bound DNA decreases and gradually approaches zero.

When the solution contains 1 μg of labeled DNA from *D. melanogaster* and variable amounts of DNA from *D. simulans,* the results are different. As the amount of *D. simulans* DNA is increased up to about 100 μg, the amount of labeled DNA from *D. melanogaster* pairing with the filter-bound DNA

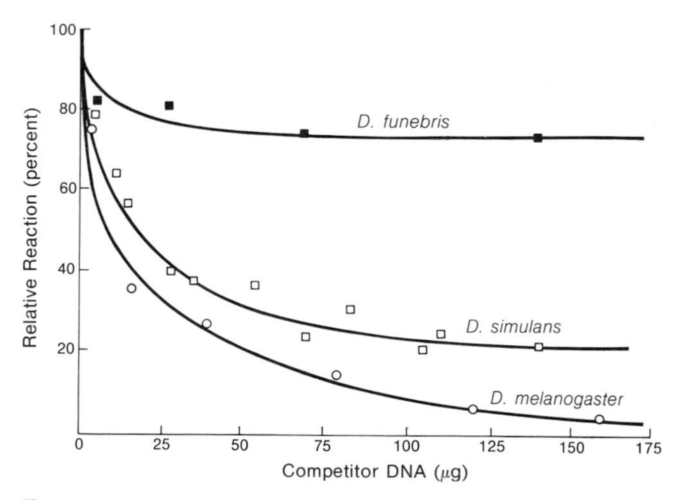

FIGURE 9-8.
Homology between the DNA sequences of *Drosophila melanogaster* and those of two other species. The experiment is carried out by hybridizing 1 μg of filter-bound DNA from *D. melanogaster* with 1 μg of radioactively labeled DNA from *D. melanogaster* and varying amounts of DNA from other species. All DNA's are first separated into single strands by heat denaturation. [After Laird and McCarthy, 1968]

gradually decreases to about 20 percent of the total paired DNA. But when the amount of *D. simulans* DNA in solution is increased beyond 100 μg, the proportion of labeled *D. melanogaster* pairing with the bound DNA is not reduced below 20 percent. This implies that about 20 percent of the filter-bound DNA of *D. melanogaster* has no complementary sequences in *D. simulans*. When the species tested is *D. funebris*, the proportion of labeled *D. melanogaster* DNA forming duplexes never decreases below 80 percent. Therefore, about 80 percent of *D. melanogaster* DNA has no homologous sequences in *D. funebris*. Armadillo and *D. melanogaster* DNA apparently have no homologous sequences (Laird and McCarthy, 1968).

The degrees of homology between DNA from rhesus monkeys and other primate species are shown in Figure 9-9; 130 micrograms of agar-bound rhesus DNA were incubated with 0.5 μg of ^{32}P-labeled rhesus DNA and variable amounts of DNA from various species in turn (Hoyer and Roberts,

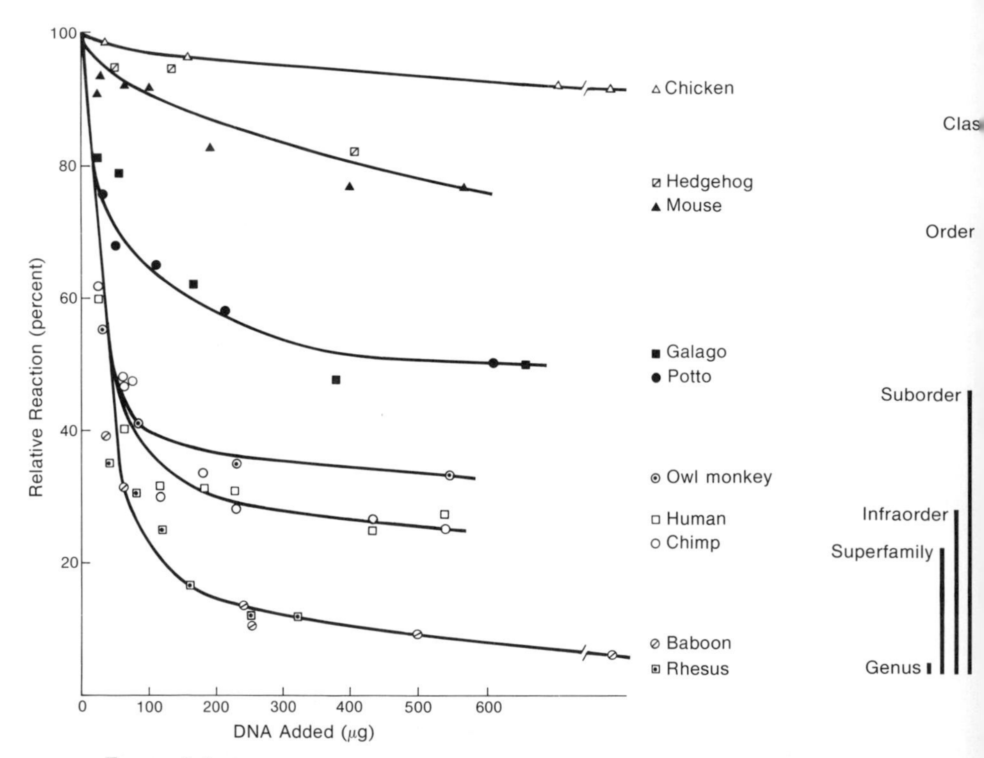

FIGURE 9-9.
Homology between DNA sequences of rhesus monkeys and those of various other species. The techniques used are basically similar to those in Figure 9-8. The amount of filter-bound DNA from rhesus monkey was 130 μg. The DNA in solution included half a microgram of labeled rhesus DNA and varying amounts of DNA from other species. The proportions of nonhomologous DNA are indicated by the bars at right, where the taxonomic separation between rhesus monkeys and the other species is indicated. [After Hoyer and Roberts, 1967.]

TABLE 9-2. Competition between 200 μg (or more) of DNA of various species and 0.5 μg of ^{14}C-labeled human DNA for binding with 0.5 g of agar-bound human DNA. [From Hoyer and Roberts, 1967.]

	Degree of taxonomic differentiation	Percent inhibition of human–human DNA binding
Man	–	100%
Chimpanzee	Family	100
Gibbon	Family	94
Rhesus monkey	Superfamily	88
Capuchin monkey	Superfamily	83
Tarsier	Suborder	65
Slow loris	Suborder	58
Galago	Suborder	58
Lemur	Suborder	47
Tree shrew	Suborder	28
Mouse	Order	21
Hedgehog	Order	19
Chicken	Class	10

1967). All the rhesus DNA molecules have homologous sequences in baboons; the baboon and rhesus DNA fragments pair equally efficiently with the agar-bound DNA from rhesus. The baboon and rhesus monkey cannot be differentiated by this technique. About 30 percent of rhesus DNA sequences have no homologs in either chimpanzee or human DNA; about 50 percent have no homologs in galago DNA, and so on.

Table 9-2 shows the results of an experiment using 0.5 g of agar-bound human DNA incubated with 0.5 μg of ^{14}C-labeled human DNA and variable amounts of DNA from other species (Hoyer and Roberts, 1967). Human and chimpanzee DNA are not distinguishable by this procedure. Man and chickens have 10 percent homologous sequences. The proportion of DNA from other species that has homologous sequences with human DNA ranges between 10 and 100 percent.

In the experiments described in the previous paragraphs the proportion of sequences forming duplexes with bound DNA depends on the temperature of incubation, the length of the reaction, and other experimental conditions (McCarthy and Church, 1970). Moreover, the sequences forming duplexes need not be complementary for every nucleotide. Sequences form whenever a major fraction of the nucleotides are complementary; some noncomplementary nucleotides may be interspersed between homologous segments. The proportion of noncomplementary nucleotides in interspecific DNA duplexes can be determined by the rate at which the DNA strands separate at increasing temperatures.

The techniques used to determine the proportion of noncomplementary nucleotides in hybrid DNA duplexes are conceptually simple. DNA duplexes are formed using bound DNA as described above, or simply placing two types of single-stranded DNA in free solution for several hours to allow hybrid double strands to form. The hybrid DNA is then heated by increasing the temperature at a rate of 1°C every few minutes. The duplex DNA gradually dissociates into single strands that are collected at regular intervals. The proportion of duplex DNA dissociated at each temperature is then plotted (Figure 9-10). The proportion of noncomplementary nucleotides is determined by comparing the dissociation curves of hybrid DNA and control DNA duplexes, the latter formed by re-association of DNA strands from a single organism. The critical parameter called thermal stability (*TS*) is the temperature at which 50 percent of the duplex DNA has dissociated. The difference (ΔTS) between the *TS* of hybrid DNA and the control is known to be approximately directly proportional

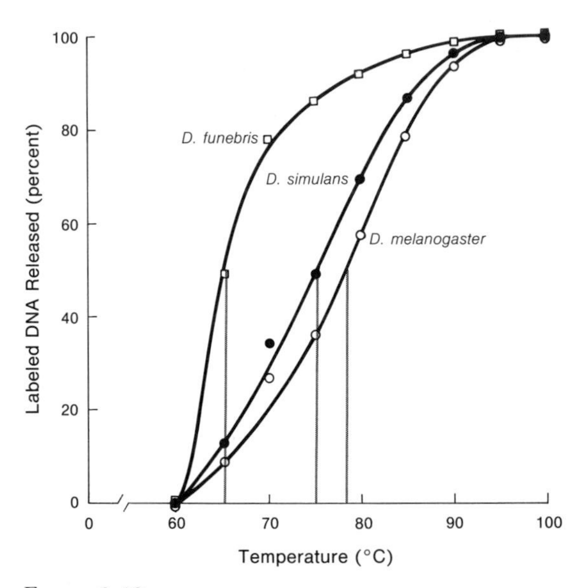

FIGURE 9-10.
Thermal stability profiles of DNA duplexes having one strand from *Drosophila melanogaster* and the other from the species indicated. The critical parameter called thermal stability *(TS)* is the temperature at which 50 percent of the duplex DNA has dissociated. The difference (ΔTS) between the *TS* of hybrid DNA and that of nonhybrid DNA *(D. melanogaster* with *D. melanogaster)* corresponds approximately to the percentage of mismatched nucleotide pairs in the hybrid DNA duplex. The *TS* for the nonhybrid duplex DNA is 78° C, for the *D. melanogaster–D. simulans* duplex DNA 75° C, and for the *D. melanogaster–D. fundebris* duplex DNA 65° C. Thus, the proportion of nucleotide pairs different from *D. melanogaster* is three percent for *D. simulans* and 13 percent for *D. funebris.* [After Laird and McCarthy, 1968.]

to the proportion of unpaired nucleotides in the hybrid DNA. For some time it was thought that each degree centigrade in ΔTS corresponded to 1.5 percent mismatched nucleotide pairs (Laird, McConaughy, and McCarthy, 1969). More recent experiments indicate that the correspondence may be $1°C\ \Delta TS =$ 0.65 percent mismatched nucleotides (McCarthy and Farquhar, 1974). These estimates are, however, subject to substantial experimental errors. For the time being, one may assume that the correspondence is approximately $1°C\ \Delta TS = 1$ percent mismatched nucleotides.

DNA PHYLOGENIES

The thermal stability profiles shown in Figure 9-10 were obtained using duplexes between ^{3}H-labeled DNA from *Drosophila melanogaster* and DNA of other species bound to nitrocellulose membrane filters as described above (see Figure 9-8). The *TS* for homologous duplexes of *D. melanogaster* DNA is 78°C. The *TS* for *D. melanogaster–D. simulans* hybrid DNA is 75°C and $\Delta TS = 3°C$. For *D. melanogaster–D. funebris* hybrid DNA, $TS = 65°C$ and $\Delta TS = 13°C$ (Laird and McCarthy, 1968). Therefore, the hybrid DNA's of *D. melanogaster* and *D. simulans* are noncomplementary in about 3 percent of their nucleotides, while those of *D. melanogaster* and *D. funebris* have about 13 percent mismatched nucleotides.

A phylogenetic tree of several genera related to the house sparrow (*Passer domesticus*) is shown in Figure 9-11. Thermal stabilities were determined using first the slate-colored junco (*Junco hyemalis*) and then the hermit thrush (*Catharus guttatus*) as references for DNA binding. The data from both sets of comparisons were combined to construct the phylogeny. The numbers in the tree branches are the differences in thermal stability (ΔTS), and thus represent approximately the percent of nucleotide replacements that have occurred in the evolution of each branch (Shields and Straus, 1975).

Thermal-stability profiles can be used to determine rates of nucleotide change per unit time. This requires that the time of divergence of the phyletic lines be known from paleontological or other data. Table 9-3 gives estimates of the nucleotide differences between various primates including man. The corresponding phylogenetic tree and the percent of nucleotide changes occurring in each branch are depicted in Figure 9-12. It is apparent in the last two columns of Table 9-3 that the rate of nucleotide change per unit time is not the same in the various lineages, even if we allow for some experimental error. For example, the nucleotide changes per year between man and galagos (5.2) or between the green monkey and galagos (5.0) are about three times as large as the nucleotide changes between man and chimpanzees (1.6) or between man and the gibbon (1.8). The rate of nucleotide change is more nearly the same in all lineages if it is calculated in terms of the generations elapsed rather than the number of years. The generation time is longer in large animals, such as man and the chimpanzee, than in small animals, such as the galago or the green monkey.

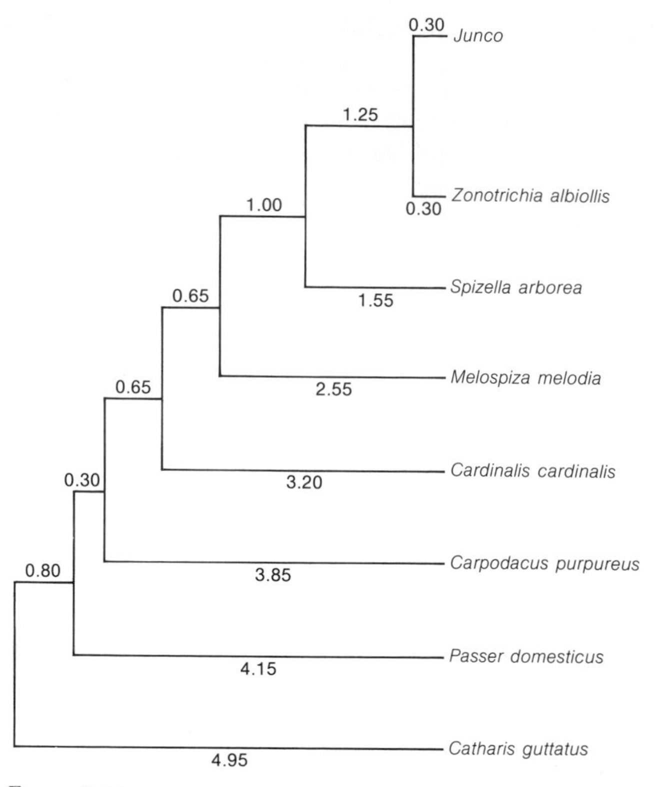

FIGURE 9-11.
Phylogeny of several species of birds related to the house sparrow *(Passer domesticus),* based on thermal-stability profiles of DNA hybrid duplexes. The numbers on the branches are estimated changes in the percentage of nucleotide-pair substitutions that have occurred in evolution. [After Shields and Strauss, 1975.]

Up to the present time DNA-hybridization studies have provided limited information on phylogenetic relationships. The techniques are laborious and expensive, due in part to the need to label the DNA of one species with a radioactive isotope. Difficulties also arise because the results obtained are very sensitive to the experimental conditions used. Inconsistent results are not rare and are sometimes observed in a single set of experiments. Hoyer and colleagues (1972) measured ΔTS between man and orangutan. They first used labeled human DNA and unlabeled orangutan DNA, and then labeled orangutan DNA and unlabeled human DNA. ΔTS was 2.9°C in the first experiment, but only 1.9°C in the second experiment, although identical procedures were used in both cases.

Most recently, techniques have been developed to obtain the nucleotide sequence of the messenger RNA coded by specific genes (see Fitch, 1976). Although extremely laborious at present, these techniques will provide precise measures of the differentiation of specific genes in separate species. They may become an important phylogenetic tool of the future.

TABLE 9-3. Rates of nucleotide change in the evolution of primate DNA. [After Kohne, Chiscon, and Hoyer, 1972.]

DNAs compared	Millions of years since divergence	Percent nucleotide changes	Percent changes per million years*	Nucleotide changes per year
Man and:				
Man	0	0%	–	–
Chimpanzee	15	2.4	0.08%	1.6
Gibbon	30	5.3	0.09	1.8
Green monkey	45	9.5	0.11	2.1
Capuchin	65	15.8	0.12	2.4
Galago	80	42.0	0.26	5.2
Green monkey and:				
Green monkey	0	0	–	–
Rhesus	15	3.5	0.12	2.4
Man	45	9.6	0.11	2.2
Chimpanzee	45	9.6	0.11	2.2
Gibbon	45	9.6	0.11	2.2
Capuchin	65	16.5	0.12	2.4
Galago	80	42.0	0.25	5.0

*The percentage of nucleotide changes per million years is obtained by dividing the percentage of nucleotide changes (column 2) by twice the number of years since divergence (colume 1). The number of years of independent evolution between any two extant species is twice the number of years since their divergence from a common ancestor. The authors of the table assumed that 1°C ΔTS = 1.5 percent mismatched nucleotide pairs.

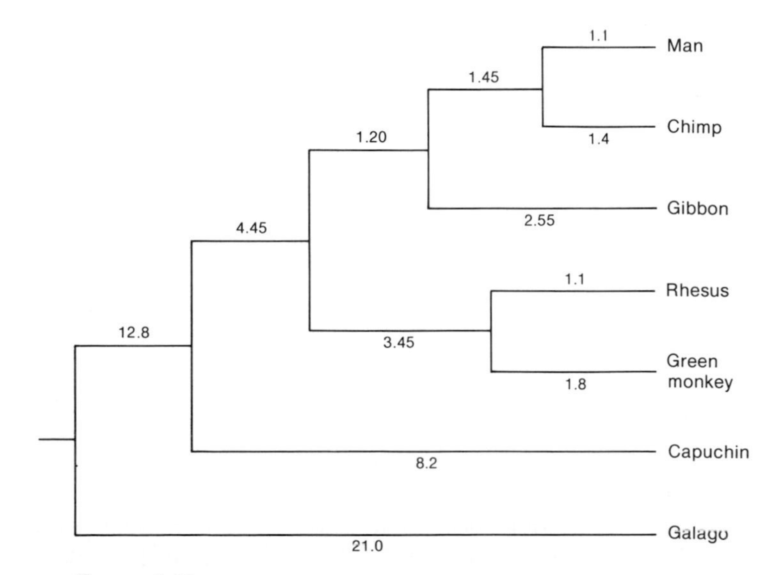

FIGURE 9-12.
Phylogeny of primate species, based on thermal-stability profiles of DNA hybrid duplexes. The numbers on the branches are estimated changes in the percentage of nucleotide-pair substitutions that have occurred in evolution. [After Kohne, Chiscon, and Hoyer, 1972.]

ELECTROPHORETIC MEASURES OF GENETIC DIFFERENTIATION

The information contained in the nucleotide sequence of structural genes is passed on to the amino acid sequence of proteins through the processes of transcription and translation. The genetic code is redundant in the direction of nucleic acid to protein—two or more nucleotide triplets may code for the same amino acid. The genetic code is ambiguous in the direction of protein to nucleic acid because individual amino acids may be coded for by any one of two or more nucleotide triplets. If the nucleotide sequence of a structural gene is known, the amino acid sequence (called the *primary structure*) of the coded protein is unambiguously determined. Knowledge of the primary structure of a protein does not allow unambiguous determination of the nucleotide sequence coding for it. Proteins different in their primary structures are encoded by different genes, although proteins with identical primary structures may or may not be coded by identical genes.

As explained in Chapter 2, proteins differing in their net electric charge can be separated by the techniques of gel electrophoresis. Not all amino acid changes in primary structure can be detected by gel electrophoresis, only those resulting in charge differences, and occasionally those changing the conformation (*secondary* and *tertiary structures*) of the proteins. For this reason, and also owing to the redundancy of the genetic code, not all allelic differences can be detected by electrophoresis of the corresponding proteins, but proteins differentiated by gel electrophoresis are always products of different genes. Thus, electrophoretic techniques underestimate the amount of genetic variation. With this important reservation, as was pointed out in Chapter 2, protein electrophoresis provides the means of estimating the amount of variation in natural populations. The rationale is that a number of genes can be selected for study without knowing a priori whether or not the genes are variable, or how variable they are. Therefore, a moderate sample of gene loci may be assumed to represent an unbiased or random sample of the genome of the organisms studied. Estimates of genetic variation obtained from the study of a few loci can be extrapolated to the whole genome.

The same rationale justifies the use of electrophoresis to estimate the degree of genetic differentiation between different populations. A moderate number of proteins is studied in two populations. The proteins are chosen without previous knowledge of whether or not they are different in the two populations. The average degree of genetic differentiation observed in such a random sample of proteins estimates the amount of genetic differentiation between the two populations over the whole genome. Gel electrophoresis is therefore a fairly simple method of measuring genetic differentiation between species. When several species are studied a dendrogram can be constructed that reflects the relative degrees of genetic differentiation among the species. Interpreting the dendrogram as a phylogeny requires the assumption that the amount of genetic differentiation averaged over a number of gene loci is approximately proportional to the time since their last common ancestor.

In recent years gel electrophoresis has been used to measure genetic differentiation among closely related species in various groups of organisms, from annual plants, through insects and other invertebrates, to fishes, reptiles, amphibians and mammals. These studies have provided much relevant, and in some cases critical, information about phylogenetic relationships.

The data obtained by electrophoresis are gene and genotypic frequencies for electrophoretically detectable alleles. The gene and genotypic frequencies observed in two different species are transformed into measures of genetic distance using any one of a variety of statistical methods. Two statistics widely used are "genetic identity," I, and "genetic distance," D, calculated as follows (Nei, 1972). Let A and B be two different populations, and l a given locus. The normalized probability that two alleles, one from each population, are identical is

$$I_l = \frac{\sum a_i b_i}{\sqrt{\sum a_i^2 \sum b_i^2}},$$

where a_i and b_i are the frequencies of the ith allele in populations A and B, respectively. I_l equals one when two populations have the same alleles in identical frequencies, and zero when two populations have no alleles in common. When several gene loci are considered, the *mean genetic identity* is defined

$$I = \frac{J_{AB}}{\sqrt{J_A J_B}},$$

where J_{AB}, J_A and J_B are the arithmetic means, over all gene loci sampled, of $\sum a_i b_i$, $\sum a_i^2$, and $\sum b_i^2$, respectively. The value of I may range from zero (complete genetic differentiation) to one (complete genetic identity).

The *mean genetic distance, D,* between two populations is defined as

$$D = -\log_e I.$$

D can range in value from zero (no genetic differences) to infinity. If it is assumed that nucleotide substitutions within a gene occur independently from each other, and that the number of nucleotide substitutions per locus follows a Poisson distribution, D may be interpreted as a measure of the average number of electrophoretically detectable nucleotide substitutions that have accumulated in the evolution of two populations since they separated from a common ancestral one. The statistics I and D were used in Chapter 6 to measure the amount of genetic differentiation during the speciation process.

ELECTROPHORETIC PHYLOGENIES

Assume that a set of loci have been studied by electrophoresis in each of a number of related species. The D values for all comparisons between pairs of species in the group can be used to construct a dendrogram that reflects genetic affinities among the species. Various mathematical methods may be employed to construct dendrograms based on measures of genetic distance. Different methods do not always yield identical dendrograms, although the results are generally fairly similar. Some of the numerical methods used to cluster species proceed, in essence, as follows. The two most similar species are first connected together as a cluster; the species most similar to the previous two is then added, and so on. To add a species to an existing cluster, one of two different criteria may be used: (1) the species to be added may be the one that is most similar to the *last* previously added species, or (2) one may choose the species that is the most similar, on the average, to *all* the species already clustered (Sneath and Sokal, 1973). There are methods for constructing dendrograms that do not require one to assume that evolutionary rates are homogeneous throughout a group of species (Farris, 1972). The latter methods are therefore preferable whenever dendrograms are intended to represent phylogenies.

The North American minnows are a group of fishes of the family Cyprinidae that consists of about 250 species. Genetic distances among nine species (each one classified in a different monotypic genus) of minnows from California are given in Table 9-4 (Avise and Ayala, 1975). The distances are based on electrophoretic studies of 24 gene loci. Two species, *Hesperoleucus* and *Lavinia*, are genetically very similar, $D = 0.055$. The two most different species are *Pogonichthys* and *Notemigonus*, $D = 1.118$. Figure 9-13 is a phylogenetic tree of the nine species based on their genetic distances. The dendrogram indicates probable phylogenetic relationships, as well as the amount of genetic change that occurred in each branch. The lineage going to *Notemigonus* experienced 0.610 electrophoretically detectable nucleotide substitutions per locus since it separated from the last ancestor common with the other eight species. The lineage leading to these latter species had 0.099 substitutions per locus before it split into two lineages, one giving rise to *Pogonichthys* and *Richardsonius*, the other to the remaining six species, and so on. The tree shown in Figure 9-13 is based on the matrix of genetic distances given in Table 9-4. If we had electrophoretic data for additional species, the resulting tree would be not only more detailed, but also the relationships among the nine species in Figure 9-13 might be somewhat changed.

TABLE 9-4. Genetic distances (above diagonal) and similarities (below diagonal) between nine specie of California minnows. [After Avise and Ayala, 1975.]

	1	2	3	4	5	6	7	8	9
1. *Hesperoleucus*	—	0.055	0.095	0.194	0.518	0.705	0.432	0.251	0.90
2. *Lavinia*	0.946	—	0.147	0.216	0.616	0.746	0.519	0.354	0.91
3. *Mylopharodon*	0.909	0.863	—	0.131	0.546	0.600	0.453	0.174	0.79
4. *Ptychocheilus*	0.824	0.806	0.877	—	0.541	0.600	0.526	0.333	0.98
5. *Orthodon*	0.596	0.540	0.579	0.582	—	1.079	0.776	0.518	1.09
6. *Pogonichthys*	0.494	0.474	0.549	0.549	0.340	—	0.519	0.679	1.11
7. *Richardsonius*	0.649	0.595	0.636	0.591	0.460	0.595	—	0.443	0.97
8. *Gila*	0.778	0.701	0.840	0.717	0.596	0.507	0.642	—	0.88
9. *Notemigonus*	0.406	0.399	0.454	0.372	0.335	0.327	0.377	0.413	—

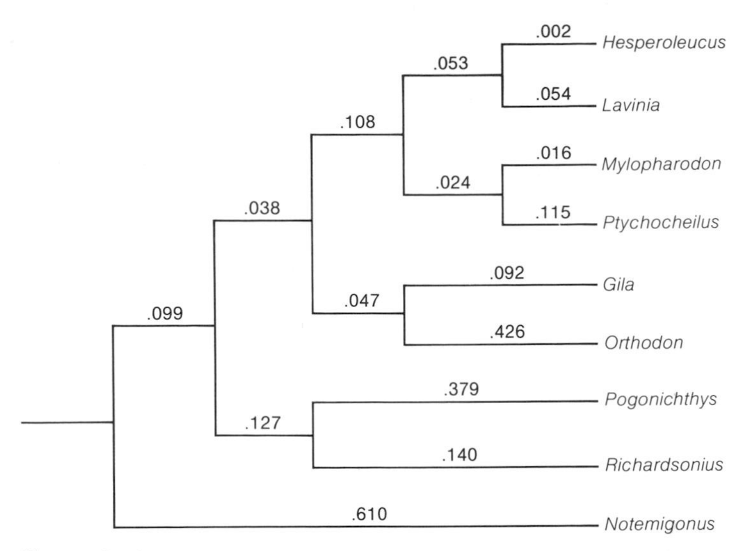

FIGURE 9-13.
Phylogeny of nine species of minnow fishes (each classified in a different monotypic genus), based on electrophoretic differences at 24 gene loci coding for enzymes and other proteins. The numbers on the branches are estimated nucleotide substitutions (detectable by electrophoresis) per locus that have taken place in evolution. [After Avise and Ayala, 1975]

Electrophoretic studies of genetic differences are particularly useful for establishing phylogenies of species that exhibit little or no morphological differentiation. The *willistoni* group of *Drosophila* consists of several closely related species endemic to the New World tropics (see Chapter 6, and Figures 6-4 and 6-5). Six species are siblings, morphologically nearly indistinguishable, although the species of individual males can be identified by slight but diagnostically reliable differences in their genitalia. Two sibling species, *D. willistoni* and *D. equinoxialis,* each consist of two allopatric subspecies with incipient reproductive isolation in the form of partial hybrid sterility when crossed in the laboratory. One species, *D. paulistorum,* consists of six semispecies, or incipient species, with more or less developed ethological isolation as well as sterility of male hybrids. Neither the subspecies nor the semispecies can be distinguished morphologically. Genetic differentiation at 36 gene loci has been studied electrophoretically in six sibling species and in *D. nebulosa*, a species morphologically distinguishable from the siblings but closely related to them (Ayala *et al.*, 1974c). Figure 9-14 shows the phylogenetic relationships among the taxa, as well as the amount of genetic differentiation in each lineage, estimated on the basis of the matrix of genetic distances. In this figure the vertical distances between neighboring taxa are approximately proportional to their genetic distances. The data in the matrix and in the figure also permit estimation of the amount of genetic differentiation at various stages of the speciation process (see Chapter 6).

To determine the phylogeny of a group, the results of electrophoretic studies should be incorporated with any other relevant evidence. Different studies provide complementary evidence, and are likely to agree on the whole. The

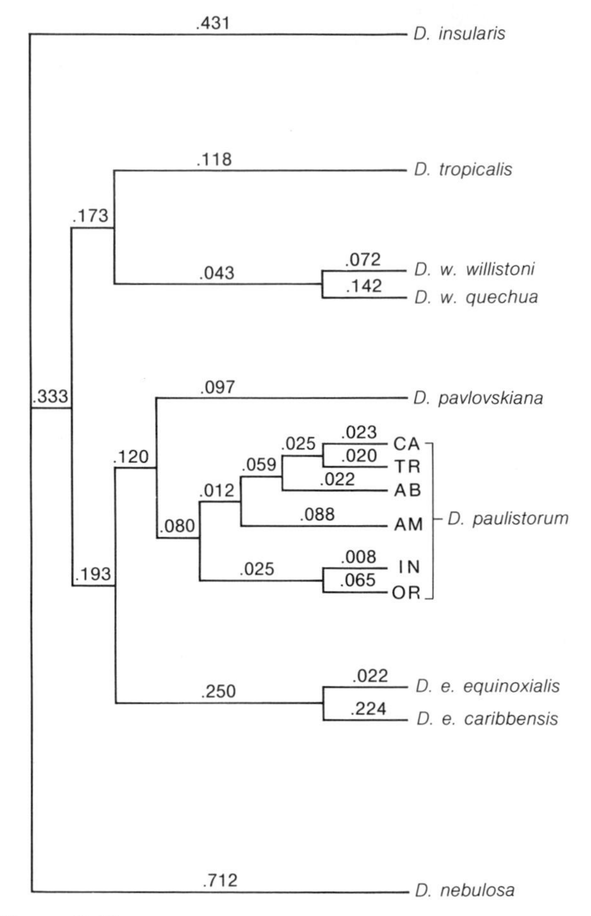

FIGURE 9-14.
Phylogeny of species related to *Drosophila willistoni,* based on electrophoretic differences at 36 gene loci coding for enzymes. The numbers on the branches are estimated nucleotide substitutions (detectable by electrophoresis) per locus that have taken place in evolution. The vertical distances between neighboring taxa are roughly proportional to their genetic differentiation. There are seven species in the phylogenetic tree. Two species, *D. willistoni* and *D. equinoxialis,* consist of two subspecies each. *D. paulistorum* represents a complex of six semispecies or incipient species, called Centroamerican (CA), Transitional (TR), Andean–Brazilian (AB), Amazonian (AM), Interior (IN), Orinocan (OR). [See Ayala *et al.,* 1974c.]

taxa of the *D. willistoni* group have been intensively studied with respect to their reproductive affinities, chromosomal gene arrangements, sexual behavior, geographic distribution, ecology, and statistical differences in morphological parameters. The phylogenetic relationships based on these studies are shown in Figure 9-15 (Spassky *et al.*, 1971). The relationships shown in this figure agree

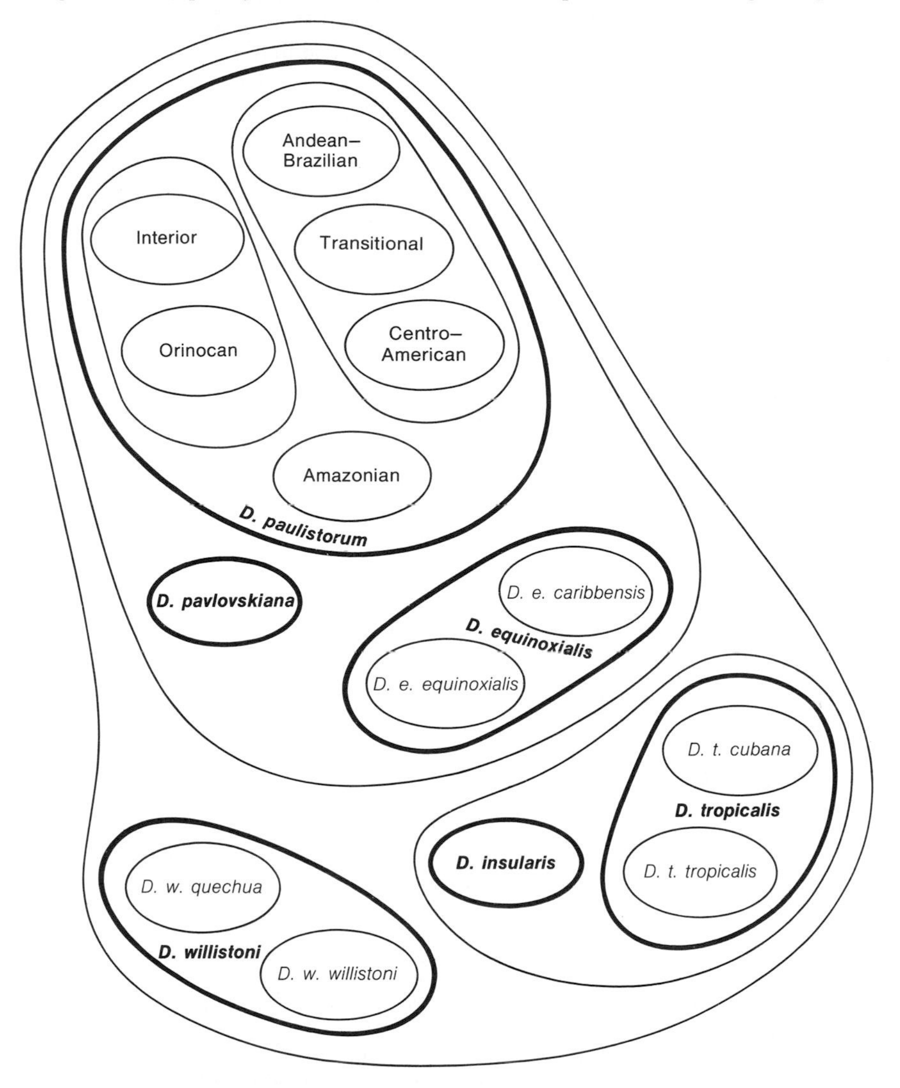

FIGURE 9-15.
Phylogenetic relationships of the sibling species of the *Drosophila willistoni* group, based on the study of reproductive affinities, chromosomal differences, sexual behavior, geographic distribution, ecology, and morphometric differences. The diagram represents a cross-sectional view of phylogenetic branches. The phylogenetic relationships are very similar to those shown in Figure 9-14, although they are based on different evidence. Note that this diagram includes two subspecies of *D. tropicalis* (only one is included in Figure 9-14), but does not include *D. nebulosa*. [See Spassky *et al.*, 1971.]

very well with those in Figure 9-14 based on electrophoretic data. One advantage of the genetic data obtained by electrophoresis is that the degree of differentiation between taxa or groups of taxa can be quantified.

The techniques of gel electrophoresis have been extensively used in recent years to study phylogenetic relationships among related species in a large variety of animal and plant groups (for a review see Ayala, 1975a). This methodology is useful only when the taxa are genetically not very different, i.e., when species of the same genus or closely related genera are compared. In electrophoretic gels two forms of a protein are judged to be the same if they have identical migrations. Whenever two proteins migrate at different rates we know that they are structurally different but we do not know whether they differ in one, two, or several amino acid replacements. The average number of amino acid replacements per protein is estimated from the proportion of proteins that are identical in two species, by assuming that the number of replacements follows a Poisson distribution. That is, estimates of genetic distance between taxa are based on the zero class (the proportion of unchanged loci) of the Poisson distribution. If two taxa are electrophoretically different in every protein, the zero class is empty and no estimates can be made of the average number of amino acid substitutions per protein (number of allelic replacements per locus). If the proportion of proteins with identical electrophoretic mobility (the zero class in the Poisson distribution) is very small, estimates of genetic distance are unreliable because they are potentially subject to large errors. Moreover, when genetically very different taxa are compared it is likely that proteins with ostensibly identical electrophoretic mobilities may in fact differ in amino acid composition, since not all amino acid replacements are detectable by electrophoretic techniques. Other biochemical methods described below make possible determination of phylogenetic relationships among taxa not closely related.

IMMUNOLOGICAL TECHNIQUES

Estimates of the degree of similarity between proteins can be obtained by immunological techniques, such as immunoelectrophoresis, immunodiffusion, quantitative precipitation, complement fixation, and turbidimetry (Reichlin, Hay, and Levine, 1964; Sarich and Wilson, 1966). In outline, the immunological comparison of proteins is performed as follows. A protein, say albumin, is purified from an animal, say man. The purified protein is injected into a mammal, such as a rabbit. The rabbit develops an immunological reaction and produces antibodies against the foreign protein (antigen). The antibodies produced by the immunized rabbit will thereafter react not only against the specific antigen used (human albumin in the example), but also against other related proteins (such as albumins from other primates). The greater the similarity between the protein used to immunize the rabbit and the protein tested, the greater the extent of the immunological reaction. The degrees of dissimilarity between the albumin of man (or whatever protein used in the original immunization) and albumins from different species are expressed as "immunological distances."

An efficient and sensitive immunological method that requires only small amounts of purified protein is *microcomplement fixation* (Champion *et al.*, 1974). "Complement" is a series of sequentially acting chemical substances found in vertebrate serum. When complement is added to antigens and antibodies under appropriate experimental conditions, it becomes fixed within the three-dimensional lattice of the antibody–antigen complexes. The amount of antigen–antibody reaction is measured by determining the amount of complement fixed in the reaction. Suitably prepared ("sensitized") red blood cells are added, and any complement not fixed by the antibody–antigen complexes is available to lyse the cells. The amount of lysed cells is determined spectrophotometrically. The number of lysed red blood cells is proportional to the amount of free (unfixed) complement. The greater the amount of lysed cells the lesser the extent of the antigen–antibody reaction. Immunological distances measured by microcomplement fixation are approximately proportional to the number of amino acid differences of related proteins (Champion *et al.*, 1974).

Microcomplement fixation of various proteins has been used to determine immunological distances in a variety of organisms, including mammals, birds, and amphibians. Table 9-5 shows the immunological distances between man, apes, and Old World monkeys calculated from data by Sarich and Wilson (1967). Antibodies were prepared independently against albumin obtained from man (*Homo sapiens*), chimpanzees (*Pan troglodytes*), and gibbons (*Hylobates lar*). These were reacted against albumins obtained from seven species of apes and six species of Old World monkeys (Cercopithecoidea: *Macaca mulatta, Papio papio, Cercocebus galeritus, Cercopithecus aethiops, Colobus polykomos,* and *Presbytis entellus*). The albumins of the two species of chimpanzee appear identical. The tests with antiserum prepared against man show that albumins from the African apes (chimpanzee and gorilla) are more similar to human albumin than the albumins from the Asiatic apes (orangutan, siamang, and gibbon); albumins from the Old World monkeys are most different. The antiserum to chimpanzee produced similar results; chimpanzee albumin is more similar to the albumins of man and gorillas than to those of the Asian apes, while the Old World monkeys are most different. The antiserum to gibbon albumin

TABLE 9-5. Immunological distances between albumins of various Old World primates. [Calculated from data in Sarich and Wilson, 1967.]

Species tested	Antiserum to		
	Homo	*Pan*	*Hylobates*
Homo sapiens (man)	0	3.7	11.1
Pan troglodytes (chimpanzee)	5.7	0	14.6
Pan paniscus (pygmy chimpanzee)	5.7	0	14.6
Gorilla gorilla (gorilla)	3.7	6.8	11.7
Pongo pygmaeus (orangutan)	8.6	9.3	11.1
Symphalangus syndactylus (siamang)	11.4	9.7	2.9
Hylobates lar (gibbon)	10.7	9.7	0
Old World monkeys (average of six species)	38.6	34.6	36.0

indicates that the albumins of gibbon and siamang are very similar; it also indicates that orangutan albumin is not much more different from albumins of the other Asian apes than the albumins of the African apes. A phylogenetic tree based on the albumin distances is depicted in Figure 9-16.

Lysozyme is an enzyme present in most animal species as well as in many plants and microorganisms. The immunological distances between lysozyme of man or the baboon and a variety of primates are shown in Table 9-6. Compared with human lysozyme, the lysozyme of chimpanzees appears to be identical (Immunological Distance = 0), that of orangutans very similar (I. D. = 1), but that of gorillas quite different (I. D. = 32). These results are not fully consistent with those obtained with albumin, since the albumin of gorillas is more similar to that of man than to that of orangutans. Another anomaly in Table 9-6 is that the Old World monkeys appear more closely related to the New World monkeys than to man and the apes (see the last column in the table). Yet, there is ample evidence indicating that the phyletic line leading to the New World monkeys separated from the lineage leading to the Old World monkeys before the latter separated from the lineage leading to man and the apes. One more inconsistency in Table 9-6 involves the immunological distance between man and the baboon; when anti-human serum is used the I. D. is 126, but when anti-baboon serum is used it is 66.

These inconsistencies do not invalidate the use of microcomplement fixation or other immunological methods for evolutionary studies. Rather, the inference to be drawn is simply that phylogenies should be based on all available evidence, not just on one single trait. Similar inconsistencies occur in other kinds of studies. For example, if gross external morphology alone is considered, dolphins and seals might appear more closely related to some fishes than to terrestrial mammals. In order to establish phylogenetic relationships, data obtained from the immunological study of a given protein should be combined with immunological studies of other proteins, with other biochemical evidence, and with morphological, behavioral, and any other relevant information.

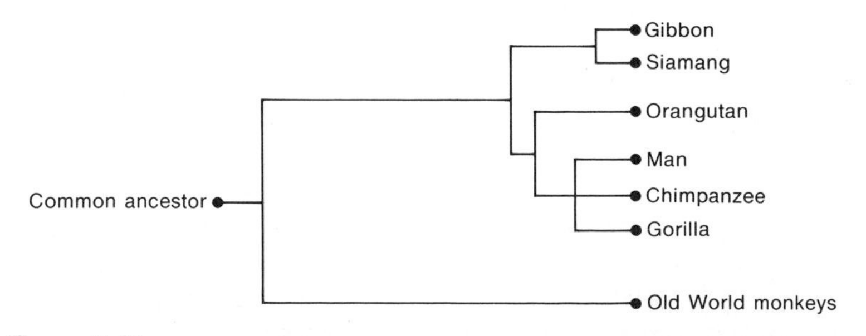

FIGURE 9-16.
Phylogeny of man, apes, and Old World monkeys, based on immunological differences between their albumin proteins. Man, the chimpanzee, and the gorilla appear more closely related to each other than either one of them is to the orangutan. [After Sarich and Wilson, 1967.]

TABLE 9-6. Immunological distances between the lysozymes of man or baboon and those of various primates. The lysozyme was obtained from milk. [After Wilson and Prager, 1974.]

Species tested	Anti-lysozyme to	
	Man	Baboon
Homo sapiens (man)	0	66
Pan troglodytes (chimpanzee)	0	67
Pongo pygmaeus (orangutan)	1	76
Gorilla gorilla (gorilla)	32	38
Old World monkeys:		
Cercopithecus aethiops (green monkey)	93	6
Cercopithecus talapoin (talapoin)	114	3
Macaca mulatta (rhesus)	122	1
Macaca speciosa (stump-tailed macaque)	124	2
Macaca fascicularis (crab-eating macaque)	130	2
Macaca radiata (bonnet macaque)	131	2
Papio cynocephalus (baboon)	127	0
New World monkeys:		
Saimiri sciureus (squirrel monkey)	127	36
Saguinus oedipus (tamarin)	134	37
Callithrix jacchus (marmoset)	137	36

AMINO ACID SEQUENCES OF PROTEINS

Proteins are polymer molecules made up of amino acids. The same 20 kinds of amino acids are found in all organisms. Proteins may consist of one, two, or more polypeptides. Each polypeptide is coded by one gene. The sequence of amino acids in a polypeptide is called its primary structure. Polypeptide chains have a helical structure, called the secondary structure of the protein. Polypeptide helices fold into three-dimensional configurations, called the tertiary structure. Under normal physiological conditions the various amino acids of a polypeptide interact through hydrogen bonds (H—H), disulfide linkages (S—S), and other kinds of bonds to produce a stable configuration. That is, the secondary and tertiary structures of a polypeptide are determined by its primary structure. Proteins consisting of more than one polypeptide have a *quaternary structure*, which refers to the three-dimensional topology of the associated polypeptides. Hemoglobin A, the most common form of hemoglobin found in human adults, consists of four polypeptides, two alpha and two beta chains. Many other proteins consist of two or more polypeptides.

The primary structure of a polypeptide is determined by the nucleotide sequence in the DNA coding for it. The number of differences between two homologous polypeptides or proteins reflects the number of differences in the corresponding genes. Genetic differentiation between species, and therefore probable phylogenetic relationships, can be inferred from the degree of differentiation in the primary structure of proteins.

The common procedure for establishing the amino acid sequence of a polypeptide requires, first, breaking the protein into small fragments of peptides, and then determining the amino acid sequence in each peptide. The polypeptide is broken into fragments with enzymes that hydrolyze the bonds between contiguous amino acids at specific sites. The resulting peptides are separated from each other using such procedures as column chromatography and two-dimensional paper chromatography. The amino acid sequence in each peptide is ascertained usually with the help of an apparatus known as a "protein sequencer" or "sequinator."

The procedure just described is carried out at least twice for each polypeptide to be sequenced, using two enzymes that hydrolyze the protein at different points. Trypsin and chymotrypsin are the most commonly used enzymes. Trypsin hydrolyzes the bonds between the carboxyl group of either lysine or arginine and the amino group of the contiguous amino acid. Chymotrypsin hydrolyzes polypeptides at the carboxyl ends of either tryptophan, phenylalanine, or asparagine. Two different sets of peptides are obtained with trypsin and chymotrypsin. The complete sequence of the protein is determined from the overlaps between the amino acid sequences of the two sets of peptides.

The first protein to be sequenced was insulin, which consists of 51 amino acids (Sanger and Thompson, 1953). During the early 1950s it was shown that the amino acid sequences of the insulins of cattle, pigs, sheep, horses, and sperm whales are identical except for replacements in three consecutive amino acids. The procedures to establish the primary sequence of proteins are extremely laborious. Yet more than 500 sequences or partial sequences are presently known through the efforts of scores of investigators working in many laboratories (Dayhoff, 1972). Many additional protein sequences are determined every year.

Evolutionary changes in proteins may involve not only amino acid replacements but also additions and/or deletions of amino acids. Homologous proteins may therefore differ not only in the sequence of amino acids but also in the length of the polypeptides. The alpha and beta chains of human hemoglobin are homologous—they are coded by genes that have arisen by ancestral duplication (Chapter 3). Yet the alpha chain consists of 141 amino acids, the beta chain of 146 amino acids. In order to determine the degree of similarity between homologous proteins, the possible occurrence of additions and deletions must be taken into account. The similarities between the alpha and beta hemoglobin chains are maximized if they are aligned over 148 positions (Figure 9-17). Gaps are assumed to exist at positions 2, 48, and 56 to 60 in the alpha chain, and at positions 19 and 20 in the beta chain. Each gap may be due to either a deletion in one chain or an addition to the other.

The occurrence of additions and deletions of amino acids in homologous proteins raises a problem. A typical protein consists of about 100 or more amino acids. Each of the 20 common amino acids is likely to occur several times in

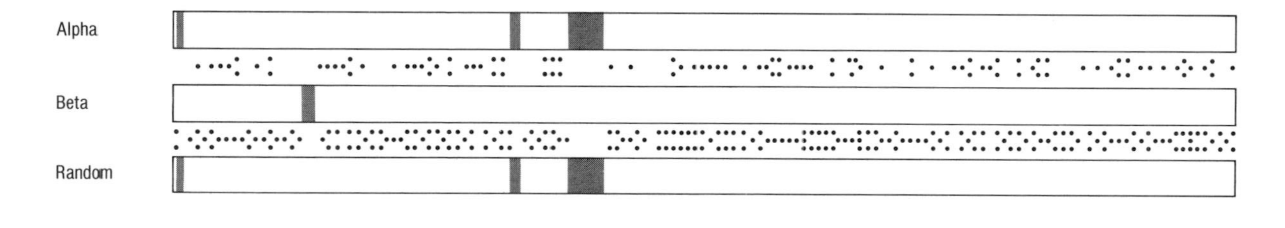

Beta	Val	His	Leu	Thr	Pro	Glu	Glu	Lys	Ser	Ala	Val	Thr	Ala	Leu	Try	Gly	Lys	Val	Asn	–	–	Val	Asp	Glu	Val
Alpha	Val	–	Leu	Ser	Pro	Ala	Asp	Lys	Thr	Asn	Val	Lys	Ala	Ala	Try	Gly	Lys	Val	Gly	Ala	His	Ala	Gly	Glu	Tyr

Beta	Gly	Gly	Glu	Ala	Leu	Gly	Arg	Leu	Leu	Val	Val	Tyr	Pro	Try	Thr	Gln	Arg	Phe	Phe	Glu	Ser	Phe	Gly	Asp	Leu
Alpha	Gly	Ala	Glu	Ala	Leu	Glu	Arg	Met	Phe	Leu	Ser	Phe	Pro	Thr	Thr	Lys	Thr	Tyr	Phe	Pro	His	Phe	–	Asp	Leu

Beta	Ser	Thr	Pro	Asp	Ala	Val	Met	Gly	Asn	Pro	Lys	Val	Lys	Ala	His	Gly	Lys	Lys	Val	Leu	Gly	Ala	Phe	Ser	Asp
Alpha	Ser	His	Gly	Ser	–	–	–	–	–	Ala	Gln	Val	Lys	Gly	His	Gly	Lys	Lys	Val	Ala	Asp	Ala	Leu	Thr	Asn

Beta	Gly	Leu	Ala	His	Leu	Asp	Asn	Leu	Lys	Gly	Thr	Phe	Ala	Thr	Leu	Ser	Glu	Leu	His	Cys	Asp	Lys	Leu	His	Val
Alpha	Ala	Val	Ala	His	Val	Asp	Asp	Met	Pro	Asn	Ala	Leu	Ser	Ala	Leu	Ser	Asp	Leu	His	Ala	His	Lys	Leu	Arg	Val

Beta	Asp	Pro	Glu	Asn	Phe	Arg	Leu	Leu	Gly	Asn	Val	Leu	Val	Cys	Val	Leu	Ala	His	His	Phe	Gly	Lys	Glu	Phe	Thr
Alpha	Asp	Pro	Val	Asn	Phe	Lys	Leu	Leu	Ser	His	Cys	Leu	Leu	Val	Thr	Leu	Ala	Ala	His	Leu	Pro	Ala	Glu	Phe	Thr

Beta	Pro	Pro	Val	Gln	Ala	Ala	Tyr	Gln	Lys	Val	Val	Ala	Gly	Val	Ala	Asn	Ala	Leu	Ala	His	Lys	Tyr	His
Alpha	Pro	Ala	Val	His	Ala	Ser	Leu	Asp	Lys	Phe	Leu	Ala	Ser	Val	Ser	Thr	Val	Leu	Thr	Ser	Lys	Tyr	Arg

FIGURE 9-17.
Below: The amino acid sequences of alpha and beta hemoglobin chains in man. The similarities between the two chains are greatest if they are aligned over 148 positions as shown. Deletions are assumed to have occurred at positions 2, 48, and 56 to 60 in the alpha chain, and at positions 20 and 21 in the beta chain (alternately, additions may have occurred at the same positions in the alternative chain). *Above:* The dots between the two chains represent the minimum number of nucleotide substitutions required to account for their differences. The dots below the beta chain represent the minimum number of nucleotide substitutions between the beta chain and a random sequence of amino acids similar in length to, and with the same gaps as, the alpha chain. The greater density of dots under the beta chain reveals how many more substitutions occur when unrelated sequences are compared.

A Alanine
C Cysteine
D Aspartic acid
E Glutamic acid
F Phenylalanine
G Glycine
H Histidine
I Isoleucine
K Lysine
L Leucine
M Methionine
N Asparagine
P Proline
Q Glutamine
R Arginine
S Serine
T Threonine
V Valine
W Tryptophan
Y Tyrosine

```
                          10        20        30        40        50
                 12345678901234567890123456789012345678901234567890123456789
Human            --------GDVEKGKKIFIMKCSQCHTVEKGGKHKTGPNLHGLFGRKTGQAPGYSYTAA
Rhesus monkey    --------GDVEKGKKIFIMKCSQCHTVEKGGKHKTGPNLHGLFGRKTGQAPGYSYTAA
Horse            --------GDVEKGKKIFVQKCAQCHTVEKGGKHKTGPNLHGLFGRKTGQAPGFTYTDA
Pig, bovine, sheep --------GDVEKGKKIFVQKCAQCHTVEKGGKHKTGPNLHGLFGRKTGQAPGFSYTDA
Dog              --------GDVEKGKKIFVQKCAQCHTVEKGGKHKTGPNLHGLFGRKTGQAPGFSYTDA
Gray whale       --------GDVEKGKKIFVQKCAQCHTVEKGGKHKTGPNLHGLFGRKTGQAVGFSYTDA
Rabbit           --------GDVEKGKKIFVQKCAQCHTVEKGGKHKTGPNLHGLFGRKTGQAVGFSYTDA
Kangaroo         --------GDVEKGKKIFVQKCAQCHTVEKGGKHKTGPNLNGIFGRKTGQAPGFTYTDA
Chicken, Turkey  --------GDIEKGKKIFVQKCSQCHTVEKGGKHKTGPNLHGLFGRKTGQAEGFSYTDA
Penguin          --------GDIEKGKKIFVQKCSQCHTVEKGGKHKTGPNLHGIFGRKTGQAEGFSYTDA
Pekin duck       --------GDVEKGKKIFVQKCSQCHTVEKGGKHKTGPNLHGLFGRKTGQAEGFSYTDA
Snapping turtle  --------GDVEKGKKIFVQKCAQCHTVEKGGKHKTGPNLNGLIGRKTGQAEGFSYTEA
Bullfrog         --------GDVEKGKKIFVQKCAQCHTCEKGGKHKVGPNLYGLIGRKTGQAAGFSYTDA
Tuna             --------GDVAKGKKTFVQKCAQCHTVENGGKHKVGPNLWGLFGRKTGQAEGYSYTDA
Screwworm fly    ----GVPAGDVEKGKKIFVQRCAQCHTVEAGGKHKVGPNLHGLFGRKTGQAAGFAYTNA
Silkworm moth    ----GVPAGNAENGKKIFVQRCAQCHTVEAGGKHKVGPNLHGFYGRKTGQAPGFSYSNA
Wheat            ASFSEAPPGNPDAGAKIFKTKCAQCHTVDAGAGHKQGPNLHGLFGRQSGTTAGYSYSAA
Fungus (Neurospora)  ----GFSAGDSKKGANLFKTRCAECHGEGGNLTQKIGPALHGLFGRKTGSVDGYAYTDA
Fungus (baker's yeast) ---TEFKAGSAKKGATLFKTRCELCHTVEKGGPHKVGPNLHGIFGRHSGQAQGYSYTDA
Fungus (Candida) --PAPFEQGSAKKGATLFKTRCAECHTIEAGGPHKVGPNLHGIFSRHSGQAQGYSYTDA
```

```
                         60                  70                  80                 -90                  100                 110
                         0 1 2 3 4 5 6 7 8 9 0 1 2 3 4 5 6 7 8 9 0 1 2 3 4 5 6 7 8 9 0 1 2 3 4 5 6 7 8 9 0 1 2 3 4 5 6 7 8 9 0 1 2
                  Human  N K N K G I I W G E D T L M E Y L E N P K K Y I P G T K M I F V G I K K K E E R A D L I A Y L K K A T N E
          Rhesus monkey  N K N K G I T W G E D T L M E Y L E N P K K Y I P G T K M I F V G I K K K E E R A D L I A Y L K K A T N E
                  Horse  N K N K G I T W K E E T L M E Y L E N P K K Y I P G T K M I F A G I K K K T E R E D L I A Y L K K A T N E
    Pig, bovine, sheep   N K N K G I T W G E E T L M E Y L E N P K K Y I P G T K M I F A G I K K K G E R E D L I A Y L K K A T N E
                    Dog  N K N K G I T W G E E T L M E Y L E N P K K Y I P G T K M I F A G I K K T G E R A D L I A Y L K K A T K E
             Gray whale  N K N K G I T W G E E T L M E Y L E N P K K Y I P G T K M I F A G I K K K G E R A D L I A Y L K K A T N E
                 Rabbit  N K N K G I T W G E D T L M E Y L E N P K K Y I P G T K M I F A G I K K K D E R A D L I A Y L K K A T N E
               Kangaroo  N K N K G I I W G E D T L M E Y L E N P K K Y I P G T K M I F A G I K K K G E R A D L I A Y L K K A T N E
        Chicken, Turkey  N K N K G I T W G E D T L M E Y L E N P K K Y I P G T K M I F A G I K K K S E R V D L I A Y L K D A T S K
                Penguin  N K N K G I T W G E D T L M E Y L E N P K K Y I P G T K M I F A G I K K K S E R A D L I A Y L K D A T S K
             Pekin duck  N K N K G I T W G E D T L M E Y L E N P K K Y I P G T K M I F A G I K K K S E R A D L I A Y L K D A T A K
        Snapping turtle  N K N K G I T W G E E T L M E Y L E N P K K Y I P G T K M I F A G I K K K A E R A D L I A Y L K D A T S K
               Bullfrog  N K N K G I T W G E D T L M E Y L E N P K K Y I P G T K M I F A G I K K K G E R Q D L I A Y L K S A C S K
                   Tuna  N K S K G I V W N N D T L M E Y L E N P K K Y I P G T K M I F A G I K K K G E R Q D L V A Y L K S A T S —
          Screwworm fly  N K A K G I T W Q D D T L F E Y L E N P K K Y I P G T K M I F A G L K K P N E R G D L I A Y L K S A T K —
          Silkworm moth  N K A K G I T W G D D T L F E Y L E N P K K Y I P G T K M V F A G L K K A N E R A D L I A Y L K E S T K —
                  Wheat  N K N K A V E W E E N T L Y D Y L L N P K K Y I P G T K M V F P G L K K P Q D R A D L I A Y L K K A T S S
    Fungus (Neurospora)  N K Q K G I T W D E N T L F E Y L E N P K K Y I P G T K M A F G G L K K D K D R N D I I T F M K E A T A —
 Fungus (baker's yeast)  N I K K N V L W D E N N M S E Y L T N P K K Y I P G T K M A F G G L K K E K D R N D L I T Y L K K A C E —
      Fungus (Candida)   N K R A G V E W A E P T M S D Y L E N P K K Y I P G T K M A F G G L K K A K D R N D L V T Y M L E A S K —
```

FIGURE 9-18.
Amino acid sequences of cytochrome *c* proteins from 20 different species. The proteins have been aligned to maximize their homology; there are differences in length at the two ends. The amino acid positions are numbered according to the wheat sequence, which as 112 amino acids. At 20 positions (gray) the same amino acid is found in all the sequences. Any two species, particularly closely related ones, share in common more than 20 amino acids, e.g. the cytochromes of man and rhesus monkey have 103 out of 104 amino acids in common.

any given protein. If we had the freedom to place gaps anywhere, any two proteins would be likely to have identical amino acids at a number of sites whether or not the proteins were homologous. Several methods have been devised to decide whether similarities in amino acid composition between two proteins are due to accidental coincidence or to homology (for a review see Fitch, 1973). One method makes all possible comparisons between fragments (spans) of two proteins, and uses statistical procedures to determine whether or not the amount of similarity is greater than that expected by chance alone (Fitch, 1966). The known sequence of each protein is divided into all possible spans of 30 consecutive amino acids. (The spans should be neither too long nor too short, and 30 amino acids have been found to be a convenient number). One such span would include all amino acids from position 1 to 30, a second from 2 to 31, a third from 3 to 32, and so on. A protein with x amino acids has $x - 29$ such spans of length 30. For example, $141 - 29 = 112$ and $146 - 29 = 117$ spans of 30 consecutive amino acids are possible for the alpha and beta hemoglobin chains, respectively. Comparisons are made, with the aid of an electronic computer, between every span of one protein and every span of the other protein. The number of sites occupied by identical amino acids in both proteins is determined for all comparisons. If the overall number of similarities is greater than expected by chance alone, the proteins are considered homologous.

The degree of similarity between presumed homologous proteins is often so large that homology can be inferred without carrying out any statistical tests. Cytochrome *c* is a protein involved in cell respiration, and in higher animals and plants is found in the mitochondria. The amino acid sequences of the cytochromes *c* of 20 diverse organisms are shown in Figure 9-18. The number of amino acids ranges from 103 in the bullfrog and the tuna to 112 in wheat; vertebrates generally have 104 amino acids. Homology is evident when all cytochromes *c* are aligned as in Figure 9-18. The cytochromes *c* of man and the rhesus monkey are different only in position 66, where man has isoleucine but the rhesus monkey has threonine. Even evolutionary distant organisms share in common many amino acids, e.g., man and baker's yeast have identical amino acids in 64 positions. All 20 sequences in Figure 9-18 have the same amino acids in 20 of the 112 sites.

The degree of differentiation, or distance, between any two homologous proteins of known amino acid sequence can be measured in two ways. First, one may simply count the number of amino acid differences between two given sequences. For example, the cytochromes *c* of man and the rhesus monkey differ by one amino acid (at site 66), those of man and the horse by 11 amino acids (at sites 19, 20, 23, 54, 55, 58, 66, 68, 91, 97, and 100), and so on.

A second procedure, which yields more information than the first, makes use of the genetic code (Table 2-1). The replacement of one amino acid by another may require, as a minimum, one, two, or three nucleotide substitutions in the corresponding DNA triplet. For example, in position 66 the cytochrome *c* of man has isoleucine, that of the rhesus monkey has threonine; the minimum number of nucleotide differences between the codons for these two amino acids is one. On the other hand, in position 20 man has methionine and the horse has glutamine; the corresponding codons differ by no less than two nucleotides. The minimum number of nucleotide differences is thus determined for each amino acid replacement between two proteins, and all such differences are added. This yields the minimum number of nucleotide changes that must have occurred in the separate evolutions of the two genes coding for the proteins compared. Table 9-7 gives the minimum numbers of nucleotide differences among the genes coding for the cytochromes *c* of 20 organisms. (Seventeen organisms are the same in Figure 9-18 and Table 9-7; the figure also includes gray whales, bullfrogs, and wheat, while the table includes donkeys, pigeons, and the yeast *Saccharomyces.*) The minimum number of nucleotide differences between the genes coding for two homologous proteins is sometimes called their *mutation distance.* The mutation distance between two genes is often greater than the number of amino acid differences between the corresponding proteins. Moreover, the actual number of nucleotide differences between two genes may be greater than their mutation distance, since the latter measures only the *minimum* number of nucleotide differences.

Whether the distance between two proteins is measured by the number of amino acid differences or by the number of nucleotide substitutions, a problem remains. The number of changes that have occurred in evolution may be greater than the number of present differences, since several intermediate conditions may have existed. For example, at a given position in a codon the changes might have been A→G→C→G. At present there would be only one difference, whereas three replacements would have occurred over time. Amino acid differences and mutation distances often underestimate the number of changes that have actually happened in evolution. The greater the time since two homologous genes or proteins had the last common ancestor, the greater on the average the error of estimation. However, methods exist to correct for this type of error.

PROTEIN PHYLOGENIES

Matrices of genetic distances such as those shown in Table 9-7 provide the data for constructing dendrograms using the clustering procedures mentioned earlier in this chapter, or other methods specifically designed to deal with primary structures of proteins (Dayhoff, 1969; Fitch, 1973). Figure 9-19 is a dendrogram based on the cytochrome *c* matrix shown in Table 9-7. Overall, the relationships shown in Figure 9-19 correspond fairly well with the phylogeny of the organisms as determined from the fossil record and other sources. There are disagreements, however. Chickens appear more closely related to penguins than to ducks and pigeons; the turtle, a reptile, appears more closely related to birds than to the rattlesnake; men and monkeys diverge from the other mammals before the marsupial kangaroo separates from some placental mammals. In spite of these erroneous relationships, it is remarkable that the study of a single protein yields a fairly accurate representation of the phylogeny of 20 organisms as diverse as those in the figure.

TABLE 9-7. Minimum numbers of nucleotide differences in the genes coding for cytochromes *c* in 20 organisms. The nucleotide differences are inferred from the amino acid sequences of the cytochromes *c*. [After Fitch and Margoliash, 1967.]

		1	2	3	4	5	6	7	8	9	10	11	12	13	14	15	16	17	18	19
1	Man																			
2	Monkey	1																		
3	Dog	13	12																	
4	Horse	17	16	10																
5	Donkey	16	15	8	1															
6	Pig	13	12	4	5	4														
7	Rabbit	12	11	6	11	10	6													
8	Kangaroo	12	13	7	11	12	7	7												
9	Duck	17	16	12	16	15	13	10	14											
10	Pigeon	16	15	12	16	15	13	8	14	3										
11	Chicken	18	17	14	16	15	13	11	15	3	4									
12	Penguin	18	17	14	17	16	14	11	13	3	4	2								
13	Turtle	19	18	13	16	15	13	11	14	7	8	8	8							
14	Rattlesnake	20	21	30	32	31	30	25	30	24	24	28	28	30						
15	Tuna	31	32	29	27	26	25	26	27	26	27	26	27	27	38					
16	Screwworm fly	33	32	24	24	25	26	23	26	25	26	26	28	30	40	34				
17	Moth	36	35	28	33	32	31	29	31	29	30	31	30	33	41	41	16			
18	*Neurospora*	63	62	64	64	64	64	62	66	61	59	61	62	65	61	72	58	59		
19	*Saccharomyces*	56	57	61	60	59	59	59	58	62	62	62	61	64	61	66	63	60	57	
20	*Candida*	66	65	66	68	67	67	67	68	66	66	66	65	67	69	69	65	61	61	4

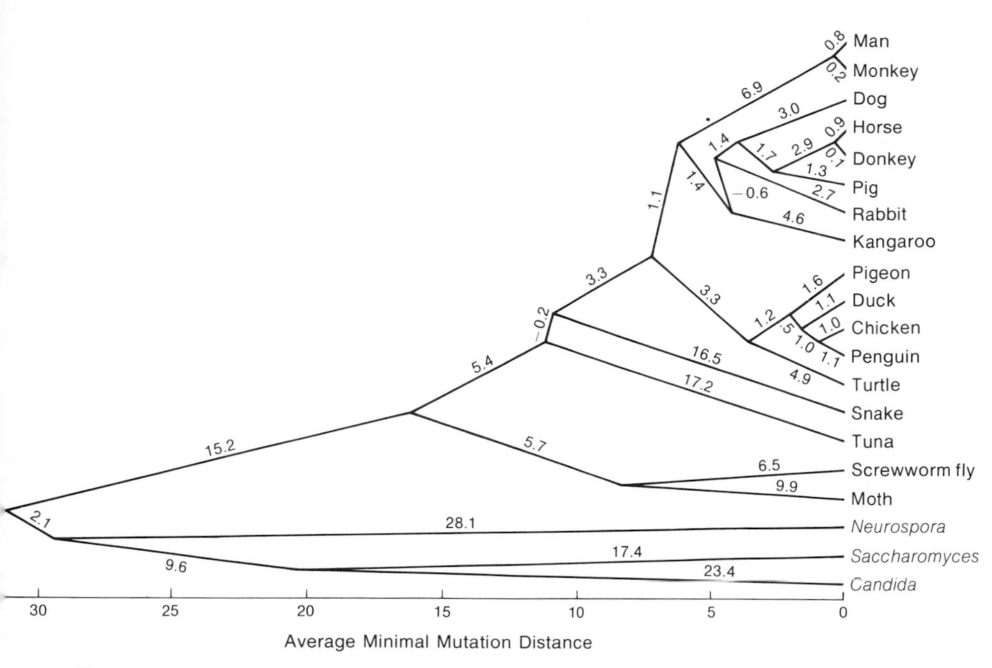

FIGURE 9-19.
Phylogeny of 20 organisms based on the cytochrome *c* matrix in Table 9-7. The phylogeny agrees on the whole fairly well with phylogenetic relationships inferred from the fossil record and other sources. This good agreement is remarkable since it is based on the study of a single protein and encompasses organisms ranging from yeast, through insects, fish, reptiles, amphibians, birds, and mammals, to man. The numbers on the branches are the estimated number of nucleotide substitutions that have taken place in evolution. [After Fitch and Margoliash, 1967.]

Unfortunately, protein-sequencing is a very laborious procedure, and the number of known sequences is as yet limited. As more and more proteins are sequenced in more and more organisms, our knowledge of phylogeny will improve considerably. Once again, however, we must bear in mind that the information obtained from the study of protein sequences should be used in conjunction with all other kinds of information relevant to phylogeny.

Cytochromes *c* are slowly evolving proteins. Organisms as different as man, the silkworm moth, and baker's yeast have in common a large proportion of amino acids in their cytochromes *c*. The evolutionary conservatism of these cytochromes makes it possible to establish the degree of ancestral propinquity among organisms only remotely related. This same conservatism, however, makes cytochrome *c* useless for determining phylogenetic relationships among closely related organisms, since these may have cytochrome *c* sequences that are completely or nearly identical. The primary structure of cytochrome *c* is identical in man and chimpanzee, which diverged 10 to 15 million years ago;

it differs by only one amino acid replacement between man and the rhesus monkey, whose most recent common ancestor lived 40 to 50 million years ago.

Fortunately, different proteins evolve at different rates. Phylogenetic relationships among closely related organisms may be inferred by studying the primary structures of rapidly evolving proteins, such as carbonic anhydrases and fibrinopeptides in mammals. Carbonic anhydrases are proteins physiologically important in the reversible hydration of CO_2 and in certain secretory processes (Maren, 1967). Two kinds of carbonic anhydrases, CA I and CA II, exist in all mammals including marsupials, while only one is found in fishes, birds, and invertebrates. The two carbonic anhydrases are the result of a gene duplication that must have occurred after the divergence of mammals and birds (some 300 million years ago), but before the divergence between marsupials and placental mammals (about 120 million years ago).

Mammal carbonic anhydrases consist of about 260 amino acids. The amino acids at 115 of the 260 positions in CA I have been identified in various primates including man. For these 115 amino acids the minimum numbers of nucleotide substitutions between man and each of five other species are as follows: chimpanzees, one; orangutans, four; vervet monkeys, six; rhesus monkeys, six; baboons, seven (Tashian *et al.*, 1972). A phylogenetic tree based on these estimates as well as some more recent data is shown in Figure 9-20. The overall configuration of the dendrogram conforms well with the known phylogeny of these animals.

Fibrinopeptides are polypeptide segments cleaved from fibrinogen when blood clots. The fibrinopeptides are another example of fast-evolving molecules. They have been used to study the phylogeny of closely related organisms,

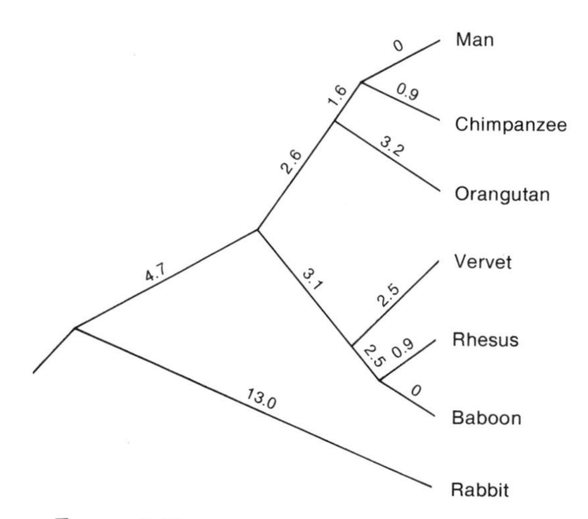

FIGURE 9-20.
Phylogeny of various primates, based on the sequence of 115 amino acids in carbonic anhydrase I. The numbers on the branches are the estimated numbers of nucleotide substitutions that have occurred in evolution. [After Tashian *et al.*, 1976.]

such as the primates (Wooding and Doolittle, 1972) and the artiodactyls or cloven-hoofed mammals (Doolittle, 1970).

Phylogenetic relationships among species are determined by comparing the primary structures of orthologous, not paralogous, proteins (see above). Orthologous genes and proteins start their separate evolutions at the time when the corresponding species have their last common ancestor. Paralogous genes evolve independently of each other from the time when the gene duplication takes place. Erroneous inferences concerning phylogenies of species can be made if genes or proteins thought to be orthologous are instead paralogous. Assume, for instance, that comparisons are made between the carbonic anhydrase CA II from man, CA II from rhesus monkeys, and CA I from chimpanzees. Among the 115 amino acids compared there are 85 differences between man CA II and chimpanzee CA I but only two between man CA II and rhesus CA II. If the three proteins were erroneously thought to be orthologous, man would appear much more closely related to the rhesus monkey than to the chimpanzee. The differences between man CA II and chimpanzee CA I have accumulated not since the time of divergence between the ancestors of man and chimpanzee, but since the duplication of the CA I and CA II genes about 200 million years ago. Grossly erroneous inferences would also be made in a phylogeny of mammals based on the amino acid sequence of the alpha hemoglobin chain of some species, the beta chain of some other species, and the delta chain of still others.

Some inconsistencies occasionally found in protein phylogenies may be due to comparisons between paralogous rather than orthologous proteins. For example, immunological studies of lysozymes show ducks and chickens closely related to each other, but very far removed from geese. This unreasonable relationship could be accounted for if the lysozymes studied in ducks and chickens were orthologous, but the lysozyme studied in geese were paralogous to the others. Recent evidence indicates that such may have indeed been the case (Arnheim and Steller, 1970; Fitch, 1973).

Comparisons between paralogous genes or proteins serve to determine phylogenies of *genes* rather than species. The phylogeny of the globin genes produced by the various duplications shown in Figure 3-7 (p. 79) was determined by comparisons between paralogous genes. Similar phylogenies can be established for other paralogous genes coding for proteins of known amino acid sequences, such as the carbonic anhydrases. It is in fact possible to construct a phylogeny that includes both paralogous and orthologous genes. Figure 9-21 depicts a phylogeny of globin genes that gives the number of nucleotide replacements between orthologous genes of different species as well as between duplicated genes.

THE NEUTRALITY THEORY OF PROTEIN EVOLUTION

The biochemical techniques described in this chapter and the methods of comparative anatomy lead to inferences of phylogenetic relationships based on resemblance. Similarities due to homology have to be distinguished from

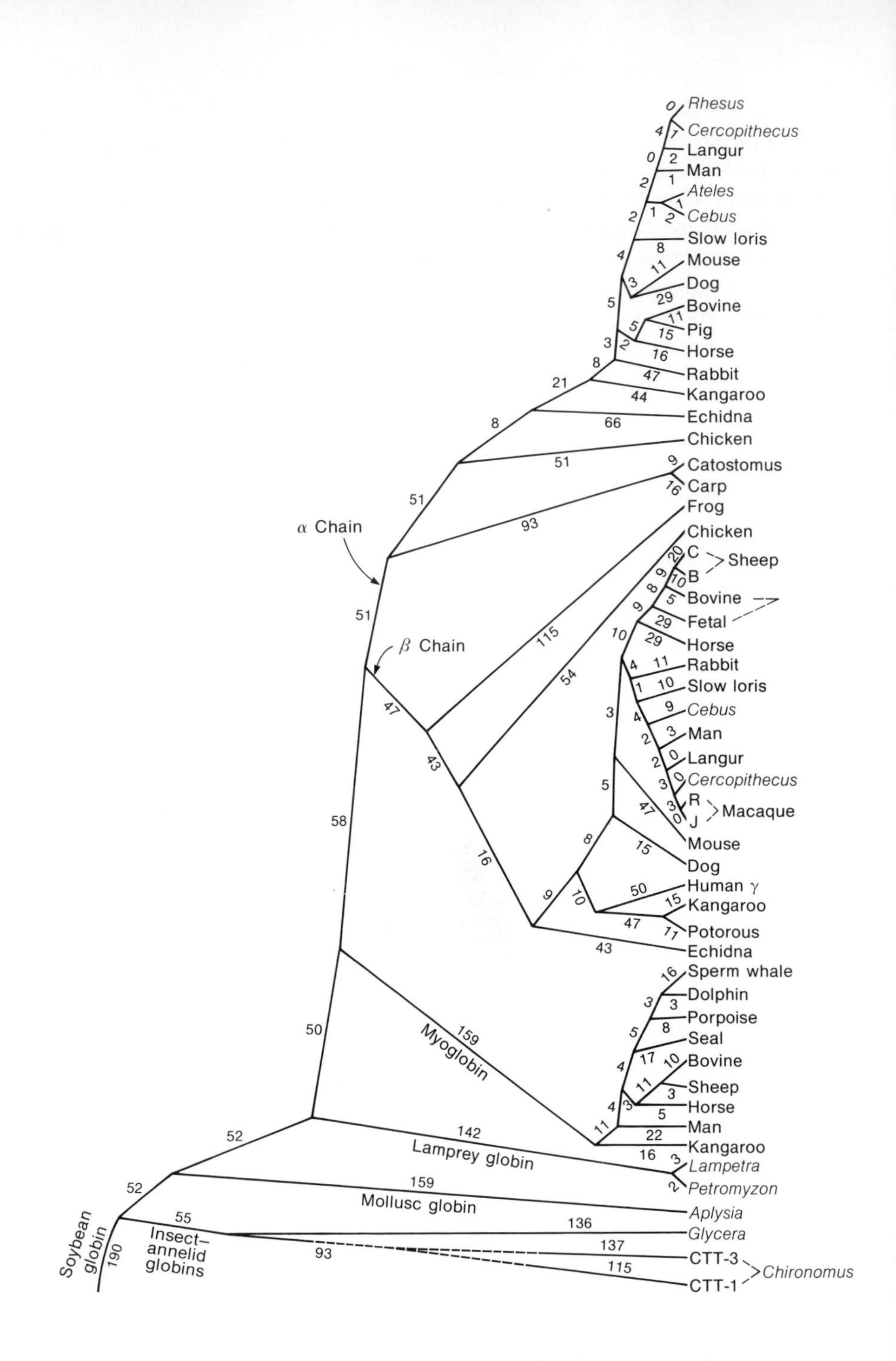

FIGURE 9-21.
Phylogeny of the globin genes and various organisms based on the amino acid sequences in 55 globins. The numbers on the branches are the estimated numbers of nucleotide substitutions that have occurred in evolution [After Goodman, 1976.]

those due to analogy, and convergent and parallel evolution must be taken into account. For similarities due to common ancestry, the assumption is made that degrees of similarity reflect degrees of phylogenetic propinquity. On the whole, this is a reasonable assumption, since evolution is a gradual process of change. However, differences in rates of evolutionary change among lineages may be a source of error. Assume that a species, S_1, diverged from the common ancestor of two other species, S_2 and S_3, before the latter diverged from each other. Assume also that a certain trait, say a protein, has evolved at a much faster rate in the lineage leading to S_3 than in the other two lineages. It might be the case that the primary structure of the protein would be more similar between S_1 and S_2 than between S_2 and S_3. The phylogeny inferred from the study of the protein might then be erroneous. Again, it is for that reason that all relevant sources of evidence must be taken into account in the construction of phylogenies.

The hypothesis has been recently advanced, however, that rates of amino acid replacements in proteins and nucleotide substitutions in DNA may be approximately constant because the vast majority of such changes are selectively neutral (see Chapter 5). New alleles appear in a population by mutation. If alternative alleles do not affect the fitness of their carriers, changes in allelic frequencies from generation to generation will be affected only by the random process of sampling—by genetic drift. Rates of allelic substitution would be "stochastically" constant, i.e., would occur with a constant probability for a given protein. That probability can be shown to be simply the mutation rate for neutral alleles (see below).

The neutrality theory of protein evolution has been championed by King and Jukes (1969), Kimura and Ohta (1971), and others. These authors admit that the evolution of morphological, behavioral, and ecological traits is largely governed by natural selection. They propose, however, that the evolution of most proteins, and of the genes coding for them, is for the most part due to chance. The neutrality theory of protein evolution assumes that, for any gene, a large proportion of all possible mutants are harmful to their carriers; these mutants are eliminated or kept at very low frequencies by natural selection. A large fraction of mutations, however, are assumed to be adaptively equivalent. Since these mutants do not affect the fitness of their carriers, they are not subject to natural selection. According to the neutrality theory, evolution at the molecular level consists for the most part of the gradual replacement of one amino acid sequence for another functionally equivalent to the first. The theory assumes that although favorable mutations occur, they are so rare that they have little effect on the overall evolutionary rate of amino acid substitutions.

Neutral alleles are not defined as adaptively identical in the mathematical sense. Operationally, neutral alleles are those whose differential contributions to fitness are so small that their frequencies change more due to the accidents of sampling through generations than to natural selection. The conditions for this can be simply stated. Assume that we have two alleles, A_1 and A_2, and that the adaptive value of one relative to the other is $1{:}1 + s$ (where s may be a positive or negative number). The two alleles are effectively neutral if, and only if

$$|N_e s| \ll 1, \tag{9-1}$$

where N_e is the effective size of the population or, approximately, the number of breeding individuals.

It is clear from the inequality that whether or not two alleles are adaptively equivalent depends not only on their differential effect on fitness, s, but also on the effective size of the population. Assume, for example, that $s = 0.001$. In a population with 100 breeding individuals, $|N_e s| = 100 \times 0.001 = 0.1$. Selection would have little effect; the allelic frequencies would change from generation to generation mostly by random drift. Assume now that the same two alleles are present in a population of effective size 10,000. Then $|N_e s| = 10{,}000 \times 0.001 = 10$. Changes in allelic frequencies would be determined for the most part by selection. The allele with the highest adaptive value would gradually replace the other. In a population with $N_e \geq 10{,}000$ individuals, two allelles will be effectively neutral only if $s < 0.0001$.

We want now to find out the rate of allelic substitution, k, per unit time in the course of evolution. Time units can be years, generations, or multiples thereof. In a panmictic (random mating) population with N diploid individuals

$$k = 2Nmx, \tag{9-2}$$

where m is the mutation rate of a gene per gamete per unit time (time is measured in the same units as for k), and x is the probability of ultimate fixation of an individual mutant. The derivation of Equation 9-2 is straightforward—there are $2Nm$ mutants per unit time, each with a probability x of becoming fixed.

Let us now consider the probability of fixation, x_n, of a newly arisen neutral mutant. Since there are $2N$ genes in the population, and all are assumed to have equal probability of becoming fixed, the probability of fixation of a newly arisen neutral mutant is simply

$$x_n = \frac{1}{2N}. \tag{9-3}$$

Replacing the value of x_n in Equation 9-2, we obtain the rate of allelic substitution, k_n, for neutral alleles as

$$k_n = 2Nm\left(\frac{1}{2N}\right) = m. \tag{9-4}$$

That is, the rate of substitution of neutral alleles is precisely the rate at which neutral alleles arise by mutation, independently of the size of the population and any other parameters. This is not only a remarkably simple result, but also one with momentous implications if it indeed applies to protein evolution. For a given protein, the rate of evolutionary change would occur with a constant probability if we assume that the mutation rate remains constant. Moreover, the evolutionary rate of allelic substitutions would provide an estimate of the rate of neutral mutations in the corresponding gene. For a neutral allele that eventually becomes fixed, the average number of generations, t_n, that elapse from its appearance by mutation until it becomes fixed can be shown to be $t_n = 4N_e$. Since the value of t_n is a function only of N_e, it is the same for all gene loci within a given population, independent of their mutation rates and other variables. In a population of effective size 10,000, any newly arisen mutant that eventually becomes fixed takes, on the average, about 40,000 generations until fixation.

We now consider the evolutionary rate of allelic substitution for alleles subject to natural selection. Assume that a newly arisen mutant, A_2, is advantageous relative to the preexisting allele, A_1, so that the fitness for A_1A_1 is 1, for A_1A_2 $1 + s$, and for A_2A_2 $1 + 2s$. If s is a positive number much smaller than one, the probability of fixation, x_s, of the new allele is approximately (Kimura and Ohta, 1972)

$$x_s = 2N_e s/N. \quad \text{9-5}$$

The probability of fixation of a new mutant as a function of the product $N_e s$ is shown in Figure 9-22. (When the value of $N_e s$ is close to zero, the probability of fixation becomes very nearly $1/2N$, as shown earlier). Substituting Equation

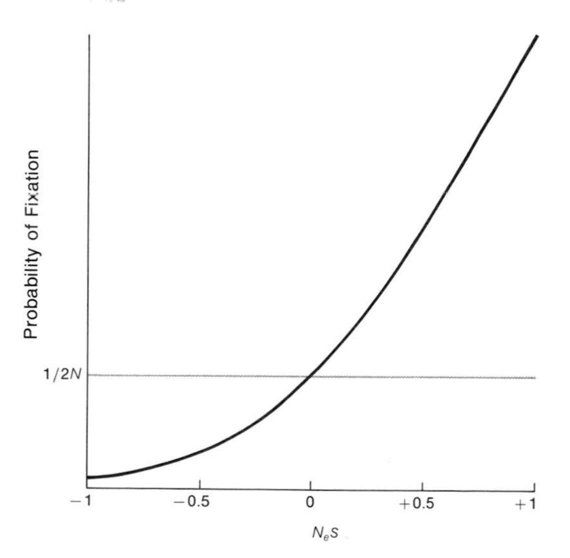

FIGURE 9-22.
Probability of fixation of a mutant gene as a function of the product of the effective population size (N_e) times the selection coefficient (s) of the gene. When $s = 0$, the probability of fixation is $1/2N$ and depends exclusively on genetic drift. When the product $N_e s$ is greater than 1 or smaller than -1, the probability of fixation (or elimination) of the mutant allele depends largely on selection, with drift having a relatively small effect. [After Kimura and Ohta, 1972.]

9-5 into Equation 9-2 we obtain the rate of allelic substitution, k_s, for a selective mutant without dominant effects:

$$k_s = Nm(2N_e s/N) = 4N_e sm. \qquad 9\text{-}6$$

The rate of substitution (k_x) of selective alleles depends on the effective size of the population, the selection coefficient, and the mutation rate. The greater the selective advantage of an allele, the faster the rate at which it will replace the disadvantageous allele. As pointed out above, selection has a significant effect whenever $|N_e s| > 1$. In such a case, $|4N_e sm|$ will be greater than m. In general, then, the rate of evolution is greater for selectively advantageous alleles than for neutral alleles.

The conclusions just reached need to be qualified, however. The value of s is unlikely to remain constant over long periods of evolutionary time. If s changes from positive to negative at different times, as might be the case, the rate of fixation could be considerably smaller than that given by Equation 9-6. This equation applies only to a simple case of selection with no dominance (i.e., the heterozygote has intermediate fitness between the two homozygotes). Other modes of selection exist. With heterozygote superiority or frequency-dependent selection, more or less stable polymorphisms may persist for long periods of time. Indeed, Equation 9-6 has little operational applicability.

THE MOLECULAR CLOCK OF EVOLUTION

If the neutrality theory of protein evolution were correct for a large number of proteins, the implications for the study of phylogeny would be momentous. Equation 9-4 shows that the expected rate of allelic substitution for neutral alleles is simply the rate at which neutral alleles arise by mutation. For a given gene, mutation rates are likely to remain fairly constant over long periods of evolutionary time. Allelic substitutions would therefore be expected to occur at a constant rate. Protein and gene evolution would serve as evolutionary clocks. First, the degree of protein differentiation among species would be a measure of their phylogenetic relatedness. Second, the actual "chronological" time of the various phylogenetic events could be estimated. Assume that we have a phylogeny such as that shown in Figure 9-19 based on cytochrome *c*. If the rate of evolution of cytochrome *c* were constant through time, the number of nucleotide substitutions that have occurred in each one of the "legs" of the phylogenetic tree would be directly proportional to the time elapsed. If the actual geological time of any one of the events in the phylogenetic tree were known from some outside source (such as the paleontological record), the times of all other events could be determined by a simple proportion. That is, once it is "calibrated" by reference to one single event, the molecular clock can be used to measure the time of occurrence of all other events in a phylogeny.

The molecular clock implied by Equation 9-4 is of course not a "metronomic clock," like timepieces in ordinary life that measure exact time with theoretically perfect regularity. The neutrality theory predicts, instead, that molecular evolution is a "stochastic clock," like radioactive decay. The *expected* rate of change is constant, although some variation occurs. Over fairly long periods of time a "stochastic clock" is nevertheless quite accurate. Moreover, it should be emphasized that each gene or protein would be a separate clock, providing an independent estimate of phylogenetic events and their time of occurrence. Each gene or protein would "tick" at a different rate (the mutation rate to neutral alleles, m, of the gene), but all of them would be timing the same evolutionary events. The joint results of several genes or proteins would provide a precise evolutionary clock.

Is the neutrality theory of molecular evolution correct? Are the rates of nucleotide or amino acid replacement stochastically constant for any given protein? Early studies of protein evolution seemed to suggest that they were stochastically constant. More detailed and sophisticated studies have recently shown that molecular evolution is not stochastically constant. This, however, does not imply that protein evolution cannot be used as an evolutionary clock. Because protein evolution is not stochastically constant, we cannot use any single protein as an evolutionary clock. But as we shall see below, the study of *many* proteins provides a fairly accurate evolutionary clock. The *average* amount of molecular change over many proteins in many organisms appears to be sufficiently constant to be used as an approximate clock of evolutionary events.

The phylogeny represented in Figure 9-23 provides an example of evidence originally thought to support the constancy of molecular evolutionary rates (Kimura and Ohta, 1972). The figure indicates the number of amino acid differences between chains of α hemoglobins. The most recent ancestor common to the carp, a fish, and the four mammals lived some 350 million years ago. Yet the number of amino acid differences between the carp and man (68) is about the same as between carp and mice (68), carp and rabbits (72), and carp and horses (67). Each one of these comparisons reflects the evolution of α hemoglobin for 700 million years—350 million years between the last common ancestor and carp, and 350 million years between the last common ancestor and each mammal. At first sight, the degree of protein differentiation appears remarkably similar in all four comparisons; but this similarity is deceiving. The most recent ancestor common to the four mammals lived some 70 million years ago. For 630 of the 700 million years of the evolution of the α chain (350 million years in the carp lineage and 280 million years in the lineage leading to the mammals) any amino acid replacement would affect all comparisons between any mammal and the carp. Only during 70 million years (or 10 percent of the total time involved) would heterogeneity of evolutionary rates affect the comparisons made above.

Information about the homogeneity or heterogeneity of evolutionary rates can be obtained by comparing the amino acid sequences of α hemoglobin of the four mammal species (Figure 9-23). The number of amino acid differences range from 17, between man and mice, to 28, between rabbits and mice. The

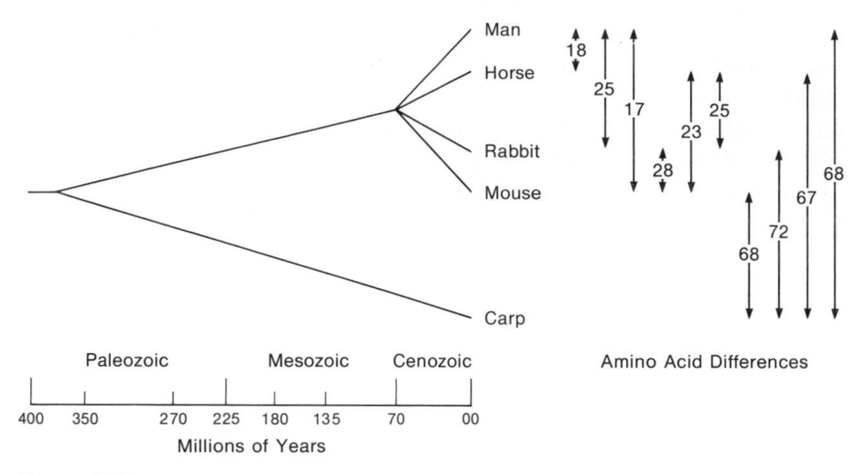

FIGURE 9-23.
Phylogeny of carp and four mammal species, indicating the number of amino acid differences between them in the alpha chain of hemoglobin. [After Kimura and Ohta, 1972.]

number of amino acid differences is 65 percent greater between mice and rabbits than between mice and man. Since one of the "legs" (the lineage going to mouse) is the same in these two comparisons, it follows that 11 more amino acid replacements have occurred in the lineage leading to rabbits than in the lineage going to man. The different rates of evolution in these two lineages might conceivably be accounted for if it were assumed that rates of amino acid replacement are constant *per generation* rather than per year. More generations have occurred in the rabbit lineage than in the human lineage. But even when rates of replacement are assumed to be constant per generation, the agreement is far from good. The number of differences between man and rabbits is 25, while between man and mice it is only 17. The lineage leading to man is the same in these comparisons. Therefore eight more replacements must have occurred in the rabbit lineage than in the mouse lineage, although approximately the same number of generations must have passed in both lineages, or if anything, fewer in the rabbit lineage.

Are the differences in evolutionary rates just discussed compatible with a *stochastically* constant rate? The neutrality theory predicts that the *probability* of replacement should be constant through time, but stochastic variation is to be expected. The question raised is far from definitely solved. Yet, as more and more homologous sequences are compared and as more sophisticated methods of analysis are used, evidence is gradually accumulating against the idea of constancy of evolutionary rates at the molecular level (Jukes and Holmquist, 1972; Tashian *et al.*, 1972; Langley and Fitch, 1974; Goodman *et al.*, 1975; Fitch, 1976; see also Ayala, 1976). As an example, we may consider the phylogeny of the globin molecules shown in Figure 9-21. Goodman, Moore, and Matsuda (1975) have calculated that for the first 380 million years of evolution of vertebrate globin (from the divergence between invertebrates and vertebrates to the divergence between birds and mammals), the hemoglobin genes evolved at an average rate of about 46 nucleotide replacements per 100 nucleotides per 100 million years (percent NTR/100 MY). However during the next 300 million years (from the mammal–bird split to the present) the average rate of hemoglobin evolution appears to be 15 percent NTR/100 MY, or only one third of the previous rate. Statistical tests indicate that the hypothesis of a constant rate of hemoglobin evolution has to be rejected.

Langley and Fitch (1974) have devised an ingenious method to test the constancy of evolutionary rates. They have applied the method to the amino acid sequences of four proteins in 18 species of vertebrates. The four proteins are hemoglobin α and β, cytochrome *c*, and fibrinopeptide A. The 18 species included one fish (carp), one frog *(Rana)*, one bird (chicken), and 15 mammals. The method is based on the minimum number of nucleotide substitutions required to account for the observed differences in amino acid sequence. The data for all four proteins are examined together by adding up the number of nucleotide replacements in each of the four. This method maximizes the amount of information that can be obtained from the data.

Two kinds of comparisons are made (Table 9-8). First, the total number of substitutions per unit time is examined for different times. The hypothesis tested is whether the *overall* rate of change is uniform over time. Because the departure from expectation is statistically significant, it cannot be attributed to stochastic variation alone. The conclusion follows that the proteins have not evolved at a constant rate. It might be possible to argue at this point that the rate of molecular evolution is constant *per generation* rather than per year. The heterogeneity in evolutionary rates might be due to variations in generation length through evolutionary time. However, this has to be rejected once a second type of comparison is taken into consideration. The hypothesis tested now is whether the *relative* rates of evolution are constant through time. If the heterogeneity in overall rates of evolution were due to changes in generation length through time, the relative rates of change among the proteins should remain constant, since all the proteins would change their rates proportionately. As shown in

TABLE 9-8. Statistical test of the constancy of evolutionary rates of four proteins (hemoglobins alpha and beta, cytochrome *c*, and fibrinopeptide A) in 18 vertebrate species. [After Fitch, 1975.]

	Chi-square	Degrees of freedom	Probability
Overall rates (comparisons among branches over all four proteins)	48.7	24	2×10^{-3}
Relative rates (comparisons among proteins within branches)	102.7	62	10^{-3}
Totals:	151.4	86	10^{-4}

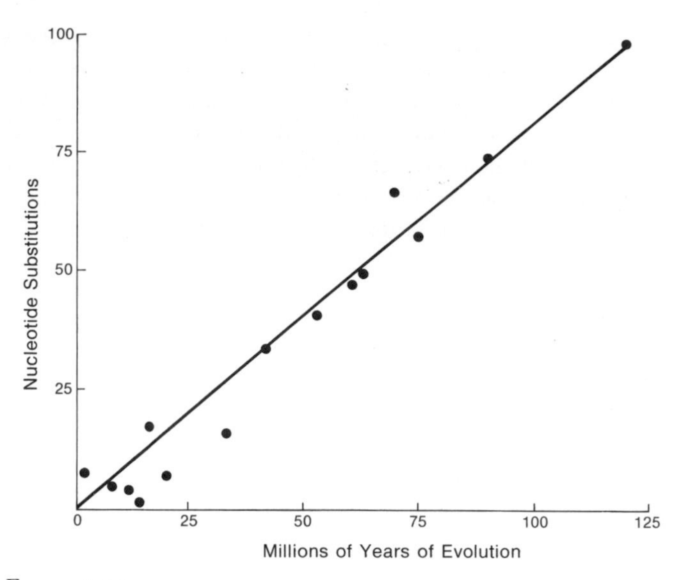

FIGURE 9-24.
Nucleotide substitutions versus paleontological time. The total nucleotide substitutions for seven proteins (cytochrome *c*, fibrinopeptides A and B, hemoglobins alpha and beta, myoglobin, and insulin C-peptide) have been calculated for comparisons between pairs of species whose ancestors diverged at the time indicated in the abscissa. The solid line has been drawn from the origin to the outermost point, and corresponds to a total rate of 0.41 nucleotide substitutions per million years (or 98.2 nucleotide substitutions per 2 × 120 million years of evolution) for the genes coding for all seven proteins. The fit between the observed number of nucleotide substitutions and the expected number (as determined by the solid line) is fairly good in general. However, in the primates (points below the diagonal at lower left) protein evolution seems to have occurred at a slower rate than in most other organisms. [After Fitch, 1976.]

Table 9-8, the relative rates of change are also not constant. More recently the analysis has been extended to seven proteins from 17 mammals (Fitch, 1976). The overall as well as the relative rates of protein evolution deviate significantly from constancy.

Proteins do not evolve at a stochastically constant rate. Nevertheless the *average* rates of evolution over long periods and over many proteins may be used as an approximate evolutionary clock (Figure 9-24). The minimum number of nucleotide substitutions that have occurred in all four proteins named above (α and β hemoglobins, cytochrome *c*, and fibrinopeptide A) is plotted against paleontological time. The overall correlation is fairly good for all organisms except the primates, which have evolved at a substantially lower rate than the average of the other organisms (Langley and Fitch, 1974).

In conclusion, informational macromolecules such as proteins and nucleic acids contain significant information about phylogeny. Rates of molecular evolution are not constant, yet protein changes may be used as an approximate evolutionary clock whenever they are averaged over many proteins and organisms. The application of protein-sequencing to the study of phylogeny is a fairly young scientific field. There is little reason to doubt that in the future the study of proteins and nucleic acids will make increasingly significant contributions to our knowledge of the evolutionary record.

10

The Geological Record

THE GEOLOGICAL TIME SCALE

The geological record is a mine of information on the environments and the life forms of the ancient past. Extending back well over 3.5 billion years—the earliest dated rocks on earth are about 3.8 billion years old—it forms our most direct evidence on the sequence of environmental changes that have swept across the face of the earth down through the ages, and also of the course of biological evolution.

Since natural selection involves interactions between organisms and their environments, the geological record permits us to compare environmental changes and concurrent evolutionary events over vast sweeps of time, which cannot be simulated in the laboratory. We are thus enabled to learn the historical patterns of evolutionary change and the rates at which these patterns have been produced. Additionally, we can trace the lines of descent of many groups of organisms from actual fossil remains, at least in outline.

In order to trace the sequence of environments or the sequence of descent in evolution, we must at least determine the relative order in which those events occurred, not only at any one place but also at many different places on earth. To determine that warm climates were widespread at some past time, for example, we must be able not only to discover criteria for warm temperatures in the geological record, but also to determine that warm temperatures were found in many parts of the earth contemporaneously. Much depends upon our ability to correlate rocks from place to place, that is, to determine that they were formed at about the same time.

When we attempt to infer the rates at which environments changed or organisms evolved, we require still more information. We need a measure of the absolute duration of time passed between separate events, such as between the first appearance and the full development of a new character or a novel type of organism, or between the appearance and disappearance of animal or plant groups, certain climatic regimes, sea-level changes, or some similar events.

Early in the nineteenth century geologists realized that major sequences of sedimentary rocks contained assemblages of fossil organisms that were distinct from assemblages in lower or higher strata. The lower sediments were naturally the first to be deposited, while younger sediments came to rest on top of older. Therefore, the stratigraphically lowermost rocks that contained fossils contained the oldest assemblages. Once the general sequence of fossils was learned, fossils could be used as indicators of the relative ages of the rocks in which they were found, according to their place in the known succession. This was usually true even for regions that were previously unexplored geologically.

Early geologists had no means of determining the absolute ages of the fossils at any point in their succession, so they divided the succession into an ordinal scale comprised of a hierarchical system of classes. Class divisions were chosen where the fossil assemblages changed their characters abruptly, especially when these changes corresponded with major changes in the character of the rock succession. This ordinal scale of geological time is shown on the left in Figure 10-1. Berry (1968) has written a clear account of the history and properties of this time scale. The correlation of events from one region to another thus depends upon the sequence of fossils, and in turn upon the course of evolution.

Today we are able to determine with reasonable accuracy the absolute ages of a wide variety of ancient rocks independently of fossil evidence. This is possible because many minerals, of which rocks are composed, contain radioisotopes, varieties of elements that contain a constant number of nuclear protons and neutrons, and that decay spontaneously into other isotopes. The decay rate for each radioisotope is constant; it is commonly expressed as a *half-life*, the length of time necessary for one-half of any given amount of the isotope to decay. If we know the half-life of a radioisotope and can determine how much of it was originally present in a rock, and how much of it is left, we can calculate how old the rock is. The half-lives of some radioisotopes commonly used to date rocks are listed in Table 10-1, along with some of their daughter decay products. In practice, the original amount of a radioisotope is usually calculated by measuring the amount of the remaining parent isotope plus the amount of its decay products that have accumulated in the rock.

All rocks are not suitable for these techniques; the best are igneous rocks that have cooled from the molten state, such as volcanic or granitic rocks, and have not been altered by subsequent events. Many rocks are altered so that isotopes are removed or added, destroying the accuracy of age measurements. Fossils do not occur in molten rocks, but in sediments that have accumulated on the land surface or on the sea floor—sands, muds, clays, and such. Thus,

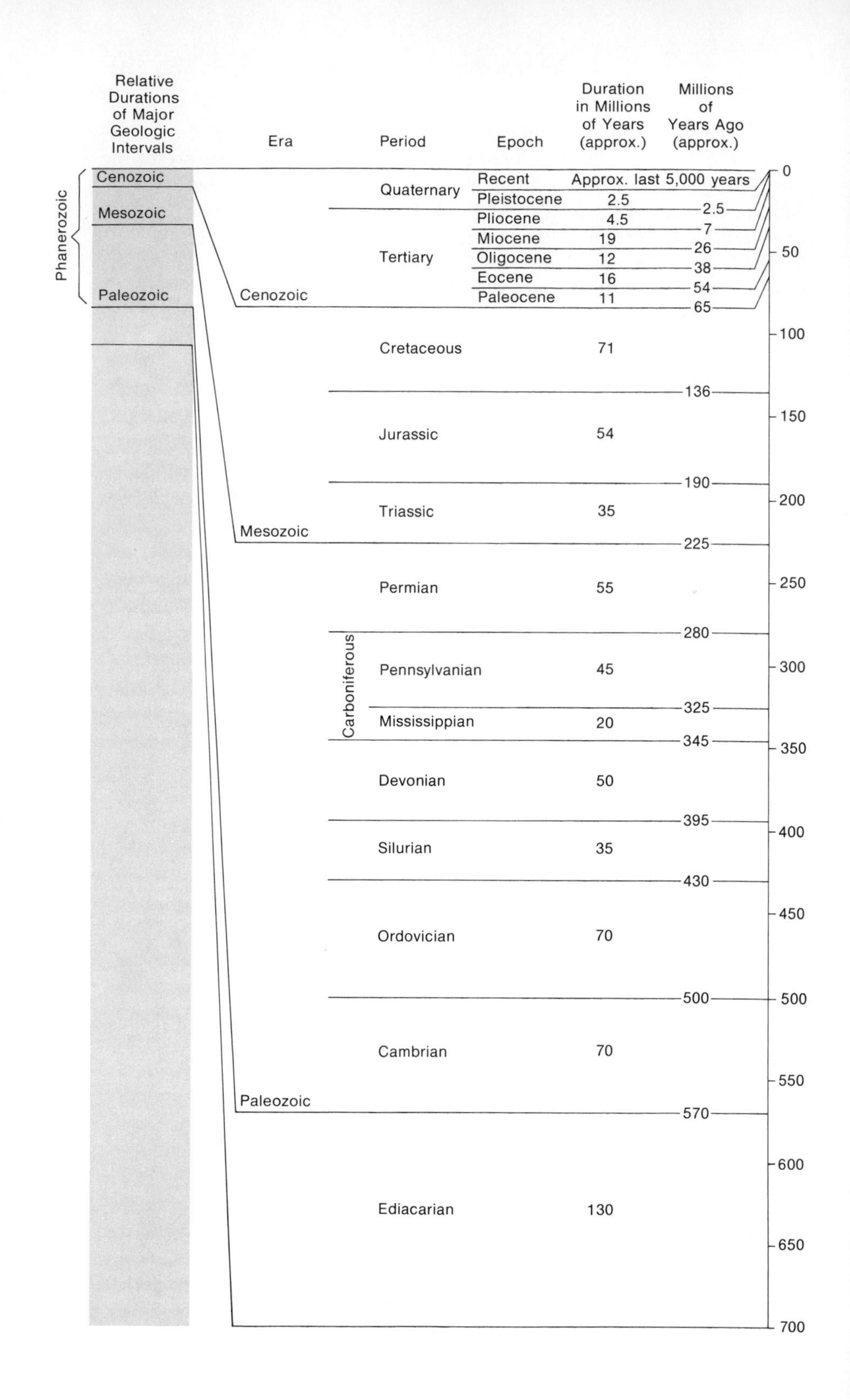

Relative Durations of Major Geologic Intervals
Era
Period
Epoch
Duration in Millions of Years (approx.)
Millions of Years Ago (approx.)
Phanerozoic
Cenozoic
Mesozoic
Paleozoic
Quaternary
Recent
Approx. last 5,000 years
Pleistocene
2.5
Pliocene
4.5
Tertiary
Miocene
19
Oligocene
12
Eocene
16
Paleocene
11
Cenozoic
2.5
7
26
38
54
65
Cretaceous
71
136
Jurassic
54
190
Triassic
35
Mesozoic
225
Permian
55
280
Carboniferous
Pennsylvanian
45
325
Mississippian
20
345
Devonian
50
395
Silurian
35
430
Ordovician
70
500
Cambrian
70
Paleozoic
570
Ediacarian
130
0
50
100
150
200
250
300
350
400
450
500
550
600
650
700

TABLE 10-1. Radioisotopes commonly employed in determining the ages of ancient rocks.

Parent	Half-life in millions of years	Daughter
Rubidium-87	47,000	Strontium-87
Uranium-238	4,510	Lead-206
Potassium-40	1,300	Argon-40
Uranium-235	713	Lead-207

fossils are difficult to date directly, and usually their ages must be inferred from dates on igneous rocks that are associated in some known way with the fossil occurrences. A few radioisotopes, such as uranium-238 and uranium-237, are actually deposited in skeletons. When they are well preserved and concentrated they may be used with their decay products to date the actual fossil that contains them, and the associated fossils as well. Unfortunately, many skeletons do not incorporate uranium in sufficient amounts.

One dating method sometimes applied directly to fossils, one that is quite accurate for relatively recent materials containing abundant carbon, is the carbon-14 (^{14}C) method. Carbon-14 is created by cosmic ray bombardment on the earth's upper atmosphere from nitrogen-14, a stable isotope that does not decay. The natural rate of production of ^{14}C appears to be fairly constant; once produced, the isotope acts like stable carbon isotopes and enters the CO_2 cycle, becoming assimilated into living organisms according to its proportionate representation. However, unlike stable carbon isotopes, ^{14}C decays; its half-life is only 5,730 years. As time goes on, the amount of ^{14}C in any organic substance gradually decreases, by one-half every half-life, until at an age of about 40,000 years or somewhat less, it is no longer detectable in amounts that permit dating. At ages younger than this, we can date many organic materials very accurately by merely measuring the amount of ^{14}C that is left. (For an introductory account of radiometric-dating techniques, see Eicher, 1968.)

FIGURE 10-1 (opposite).
The geological time scale, consisting of eras, divided into periods, divided into epochs. Rocks representing each time division contain characteristic suites of fossils, although the suites tend to grade into each other at the boundaries. The Paleozoic, Mesozoic, and Cenozoic eras are collectively called the Phanerozoic. In the figure the Phanerozoic is scaled to absolute time, as determined by radiometric dating. The column at the left indicates the scale of the Phanerozoic eras compared with all of geological time since formation of the earth some 4,600 million years ago. [After Eicher, 1968.]

There is always some error in the analysis of radioisotopes, and our knowledge of their half-lives is not precise, though it is generally accurate within two percent. For very great ages, such as billions of years, even a small percentage error represents much time; one percent of a billion years is 10 million years, which can be a long time in terms of evolutionary events. Further errors arise from imperfections in the preservation of parent and daughter isotopes in a decay series, and these can be difficult to detect. Therefore, the dates of our absolute time scale represent approximations only, subject to revision as new evidence comes to light. In general, the more recent parts of the geological record are dated more accurately than older parts. The earliest fossils seem to be primitive bacteria and blue-green algae, and date from over three billion years ago (Schopf, 1970, 1974). The earliest well-authenticated records of animal body fossils are not well dated as yet, but may be over 700 million years old. The approximate dates of the beginnings and endings of subdivisions of the geological time scale are indicated in Figure 10-1.

THE RECORD OF ANCIENT ENVIRONMENTS

The geological record is far from complete. In order to evaluate geological data it is necessary to have some appreciation of the quality of the record. Much of the record of past environments is inferred from sedimentary rocks, chiefly eroded from older rocks or accumulated from the remains of the skeletons of organisms. The sedimentary record is biased both in space and time. In any given region the accumulation of sediments is episodic, and there is nothing approaching a continuous sedimentary record of even the last 700 million years. Probably the worst record is of upland terrestrial environments; these are sites of erosion, so that any sediments that chance to come to rest there are usually destroyed relatively rapidly. However, terrestrial sediments often do accumulate in lowland basins and on plains, both in continental interiors and along coasts.

Exposed seacoasts tend to be sites of erosion, while sediments accumulate in bays, estuaries, and deltas, and at least episodically over wide stretches of the continental shelves. Sedimentary patterns in these environments shift with changing sediments and current patterns, related in turn to such factors as rainfall and runoff patterns, to changing topographic relief, and, most of all, to changes in sea level. Numerous times in the geological past swinging sea levels have raised and subsequently lowered the amount of continental platforms overlain by shallow seas. Sheets of shallow-water marine sediments have accumulated at times on the continents. Many of these have been subsequently eroded and lost during regressions of the sea; but parts remain, and since some of them are exposed on the continents, they are relatively accessible, and contain the best fossil records that we have.

In the deeper waters of the continental slopes and at abyssal depths, sedimentation is slow but much more constant than in shallow water. However, the deep-sea floor is being continuously destroyed. Old sea floor descends into the earth's interior down deep-sea trenches, as new sea floor appears at the surface along deep-sea ridges (see below). Most of the present deep-sea floor is younger than 100 million years (mid-Cretaceous), although there is one patch that is as old as about 180 million years (early Jurassic) (Figure 10-2). Thus, even in the deep sea, where sedimentation is most continuous, the record is relatively short. It is also of very difficult access.

The record of ancient environments is fragmentary. It must be inferred from many lines of evidence provided by numerous scientific disciplines. Local conditions at the time of formation of sediments can usually be inferred from the characters of the rocks themselves. Usually, terrestrial sediments can be distinguished from marine, and shallow-water sediments can be distinguished from deep-water by their textures, structures, and compositions. Under ideal

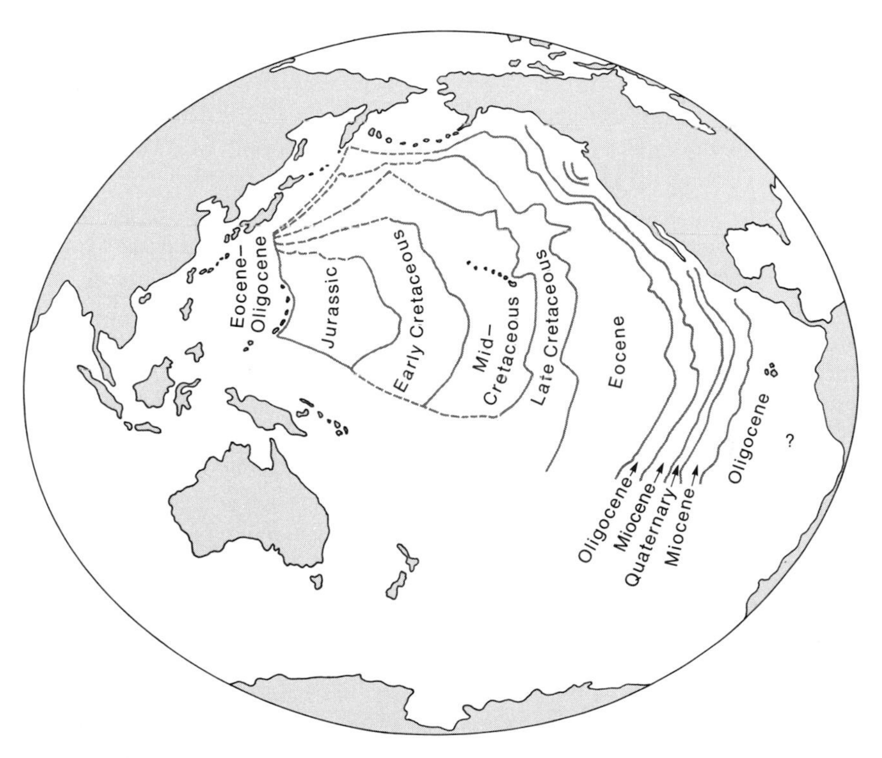

FIGURE 10-2.
The geological age pattern of basaltic rocks of the North Pacific sea floor, indicating the age of the oceanic crust there. A spreading ridge (the East Pacific Rise) lies beneath the strip of crust labeled "Quaternary." The crust becomes progressively older laterally away from this ridge. Presumably the oldest (Jurassic) crust, in the northwestern Pacific, originated at the ridge but spread westward as younger crust (Cretaceous, then Eocene, and so on) was generated at the ridge. North America has moved westward owing to spreading of the mid-Atlantic ridge (see Figure 10-4) and has partially overridden the east Pacific ridge system. [After A. G. Fischer *et al.*, 1970.]

conditions, sediments that have accumulated in different intertidal and subtidal zones can be distinguished. Weathering processes that operate in different climatic zones produce different mineral products, and these are incorporated into terrestrial and marine sediments alike, providing evidence of ancient climates. Careful mapping of sedimentary patterns in rocks now exposed on the continents often leads to reconstructions of the general environmental framework associated with the time of sedimentation. A vast body of work, employing hosts of criteria, has led to many marvelously detailed reconstructions of local environmental conditions. (For an introduction to the techniques of environmental reconstruction, see Laporte, 1968.)

By piecing together the results of such studies, a very generalized understanding of climatic history has been developed. Until recently, however, there had been no general explanations for the particular sequences of climates or environmental successions, no general theory of environmental change. Such a theory is now beginning to emerge. Since the direction of organic evolution must have been determined in large part by the kinds of environmental changes that have occurred, natural selection being chiefly an ecogenetic process, some understanding of the principal causes of these changes is relevant; in fact, an understanding of environmental history is necessary to achieve a full explanation of the history of life.

Our ability in principle to form a theory of environmental change developed in the 1960s with the rise of the theory of global tectonics. This latter theory, major elements of which have been extensively verified from geophysical and geological evidence, holds that the outer part of the earth, including the crust and upper mantle, is divided into a system of large *lithospheric plates*, of which there are six major and a few smaller ones at present (Figure 10-3). These plates grow along their margins, marked by broad ridges, such as the mid-Atlantic ridge. New material is added to the plates by the welling-up of magma (molten rock) from the earth's interior at the center of the ridge. Points on opposite sides of a ridge are on different lithospheric plates, and they spread away from each other at a rate commensurate with the rate of injection of magma into the ridge center. Common rates of spreading are from 1 to 6 cm per year on each side of a ridge; rates are usually the same on opposite ridge sides. Proceeding across a plate in the direction of spreading, we usually come to a locality where the plate is being consumed (*subducted*) and the material returned to the earth's interior, down a deep-sea trench (Figure 10-4). The amount of spreading and the amount of consumption probably balance out globally, so that the volume of the earth crust remains about the same. The magma that wells up beneath the ridges forms, upon cooling, a rock type called basalt, which is the major constituent of the oceanic crust. It is heavier on the average than the rocks that make up the continents, and thus tends to sink lower into the more plastic shells of the earth's interior. For this reason, and because oceanic crust is thinner than continental crust, ridges are essentially restricted to the ocean basins. In fact, the ocean basins are generated by the lateral spreading of the basalt layer (and of the lithospheric plate in general, which contains denser rocks beneath the crust proper).

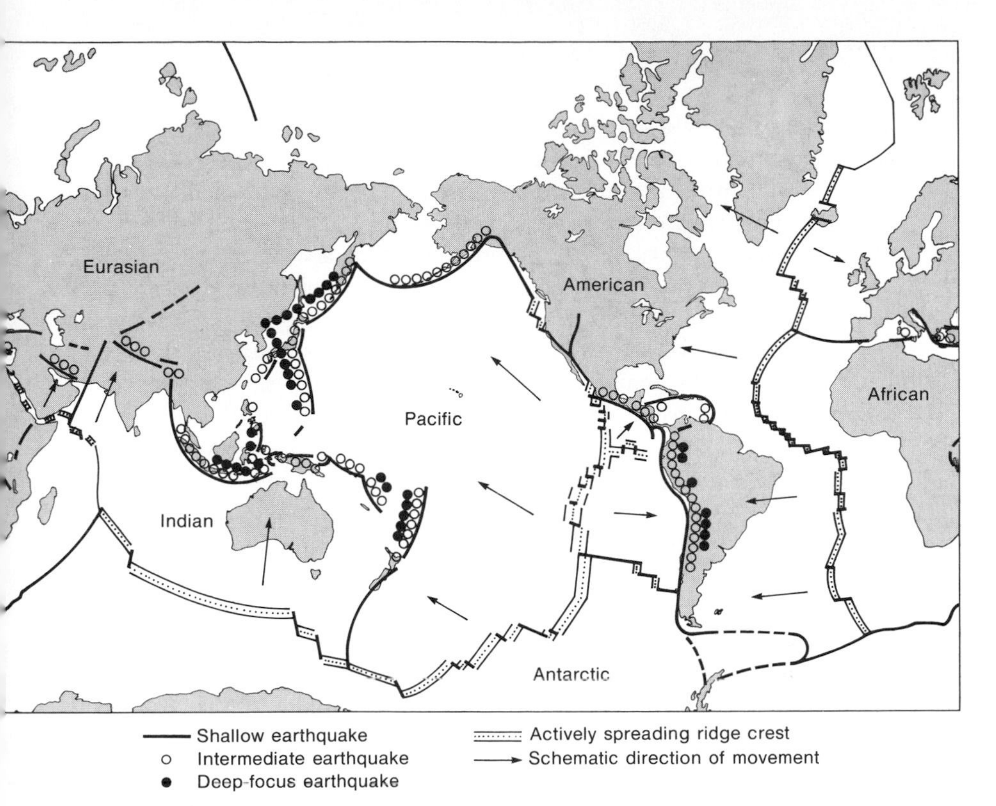

FIGURE 10-3.
The major lithospheric plates of the earth at present: Eurasian, Indian, Pacific, American, African, and Antarctic. They are formed at the ridges (double lines with dots) at rates from about 1 to 6 cm per year. The plates are consumed down subduction zones (solid lines bordered by circles, which indicate earthquakes). Note that three plates bear two continents each, two plates bear one continent each, and one major plate (Pacific) has no continents. [After Vine, 1969.]

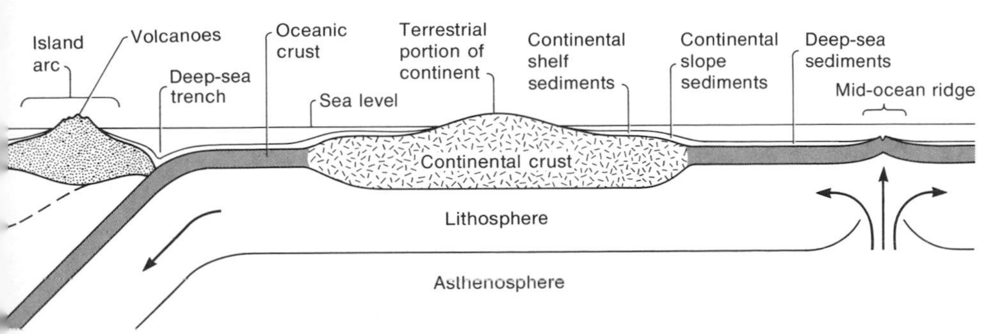

FIGURE 10-4.
Diagrammatic cross-section of the lithosphere and upper portion of the asthenosphere, showing a spreading ridge and a subduction zone. The continent is shown during a time of widespread shallow seas, the deposits of which are the best sources of fossils. [Partly after Dewey and Bird, 1970.]

Continents are relatively light parts of the earth's crust that, in a manner of speaking, float on denser subcrustal material; since continental crust is both less dense and thicker than the oceanic crust, continents stand higher than ocean basins. Continents are so light that they are not subducted down trenches, although they move with the plates in which they are imbedded, spreading away from active ridge margins. As one result, continents frequently have trenches near one of their margins, and at times continents actually collide over a trench site. The colliding margins are usually deformed into mountain belts, and the continents become sutured together. When this happens, there must be a change in the geometry of spreading, since the differential ridge motions that drove the plates containing the collided continents must stop. Thus, continents may be assembled into larger masses and plates may coalesce.

Continents also may fragment into smaller, daughter continents if a ridge springs up across them to create new spreading plate margins and thus subdivide a previous plate. Other features of plate-tectonic processes that are of biological importance include the volcanic arc systems that form along trenches on the plate that is not being consumed (Figure 10-4). The volcanoes represent more volatile material that is subducted and then rises to the surface after being heated in the earth's interior. Such activity may form mountain ranges on continents (as the Andes) or island arc systems in the ocean (as the Aleutian Islands).

Plate-tectonic processes like those of today appear to have operated throughout the Phanerozoic and probably during the late Precambrian. It is uncertain that similar processes occurred during the first two billion years of earth history, i.e., before about 2.6 billion years ago. Unfortunately, the evolution of the modern global tectonic style has not yet been traced. At any rate, the effects of Phanerozoic plate-tectonic processes upon the earth's environment have been fundamental and pervasive. Take, for example, the climate. As continents move into new regions of the earth, especially as they cross latitudes, their climates change; whole climatic zones sweep gradually across moving continental masses and their epicontinental seas. Moreover, the continents perturb global conditions within climatic zones and thus alter the climates themselves. Variations in topographic relief also arise principally from plate-tectonic processes, either from marginal continental deformations or from volcanism related to plate boundary conditions. The climatic changes are modified by variations in local topography and other factors into a vast complex of detailed sequences of conditions.

Important components of environmental change stem from variations in continental size. As continents collide and are sutured together into larger masses, greater climatic variety generally results, for continental interiors become increasingly remote from the sea. This leads to extreme seasonality and "continentality" of interior climates. At least at one time in the past all the continents became assembled into a supercontinent called Pangaea (Chapter 13). Conversely, continental fragmentation tends to reduce continentality. The precise patterns of climatic differentiation resulting from continental growth and fragmentation depend also upon the latitude of the continents, the orientation of mountain ranges, patterns of ocean currents, and even the proximity of other continents.

Significant changes in sea level are known to have occurred repeatedly during the geological past, and these have had important effects upon terrestrial and shallow marine environments. When continents are widely flooded by shallow seas, the moderating effects upon the climatic regimes must be important; continentality decreases and more maritime conditions prevail. Habitat space for shallow marine and terrestrial organisms varies with continental emergence and submergence. Variations in sea level probably arise from changes in ocean-basin volume (primarily a function of spreading ridge activity plus trench development and oceanic volcanism), as well as the trapping of water in glaciers. The interpretation of causes of sea-level change is beset by difficulties and has not yielded satisfactory historical models; nevertheless, the facts of continental emergence and submergence are well-established from empirical evidence.

Because of these and other geologic events, the environmental regime is more or less continuously changing; the regime may be mild or severe, heterogeneous or monotonous, stable or fluctuating, and it may change slowly or rapidly. The processes of weathering, erosion, and sedimentation constantly shift the local environmental mosaic, while on a broader scale the mosaic assemblages of whole lands and seas are altered in quality, size, and number. The planetary biota respond to these changes through evolution and extinction. The sequence of major environmental regimes can, in principle, be interpreted from paleogeographic reconstructions, which are becoming increasingly detailed. Major environmental changes should correspond with unusual events of extinction or diversification revealed by the fossil record. For example, major changes in the patterns of environmental variability may have played major roles in the origin of the major animal groups. These topics will be explored in Chapter 13.

THE RECORD OF ANCIENT LIFE

Like the geologic record, the fossil record is spotty, and is biased according to the preservability of different environmental regimes. For example, our best record of terrestrial organisms is of lowland forms entrapped in sediments washed onto plains, lakes, or swamps in coastal basins. In the sea, shallow-water forms from depths less than about 200 meters are by far the best represented. Because most marine species have lived in shallow water (about 90 percent of living marine species do) and because shallow-water marine sediments are in general better preserved than others, the fossil record of shallow marine communities is far better than for any other major environment. All animal phyla originated in the sea; the marine record includes a greater variety of higher taxa and extends farther back in time (to about 700 million years ago) than the terrestrial record.

Nevertheless, even the shallow marine realm is represented in the fossil record by only a fraction of its extinct species. Most fossils are remains of mineralized skeleton; the large number of organisms without durable skeletons, around two thirds of all animals, is largely lost. Many animals, including those without mineralized skeletons, burrow, track, or otherwise disturb the sediments on or in which they live, or they bore into lithified rocks. These marks are commonly preserved even when the sediments lithify, and are called *trace fossils*. They tell us something of the activities of animals whose bodies we may never see, and sometimes aid in interpretations of the paleoecology of rocks that contain no fossil skeletons. Nevertheless, we have little or no information on most unskeletonized animals. Even most animals with skeletons are not preserved; the skeletons weather or erode away, are dissolved by groundwater, or are otherwise destroyed as the sediments in which they are entombed become lithified.

We are not certain just how many species have lived or just what proportion of them are preserved. Typical estimates of the fraction of marine species that have actually been described as fossils, calculated by Valentine (1970) and Durham (1967), range from about 1 in 60 to 1 in 150. The percentage of fossil terrestrial vertebrate species that are known must be much smaller, and for insects it is vanishingly small.

Fortunately, we do not merely have every, say, hundredth species preserved, with regular gaps in between. Instead, the fossil record is very irregular. In some regions hundreds of millions of years of the record may be either missing, represented by unfossiliferous rocks, or hopelessly concealed beneath other rocks, while the record of some few million years may be quite good. At some localities virtually the entire skeletonized fraction of the original living communities appears to be preserved, and sometimes fairly continuous fossiliferous sedimentary successions may represent several million years.

At the places and for the times that the fossil record is most complete, we can study the pattern of change within lineages and attempt to extrapolate a history of life from these episodic data. The record is sufficiently good that we can discern a number of distinctive patterns of anagenesis, cladogenesis, and extinction, as occurring both among individual taxa as well as whole communities and provinces and at times among the entire planetary biota. On the other hand, it is not surprising that many critical gaps occur in the history of individual lineages. For example, only about 100,000 marine fossil species have been described. If there has been a total of 2,000,000 marine species with skeletons, then only about 5,000 pairs of ancestor–descendant species can be expected in a random sample of 100,000 species. If 5,000,000 marine species have had skeletons, then only about 2,000 pairs of ancestor–descendant species would be expected in the same sample (for methods of calculation see Mood, 1940). If these estimates embrace the actual figure, only between 0.25 and 0.04 percent of possible ancestor–descendant marine species pairs have been preserved, even among well-skeletonized forms. The percentage among the entire marine biota must be much smaller.

Most taxa at the higher levels, such as phyla, have contained a relatively large number of species and individuals throughout geologic time. Thus, even though the fossil record is spotty, there is a good chance that at least some members of a phylum will be represented in any extensive series of fossil samples. A relatively well-skeletonized, diverse phylum will be represented at many localities, and for most well-skeletonized phyla we probably have records that approach their true ranges in geologic time and geographic distribution in shallow-water environment. Of course, there are small soft-bodied phyla, e.g., the Pogonophora, that live in environments that do not have much fossil record, such as the deep sea.

Since, on the average, classes have fewer individuals than phyla have, they have less of a chance of being fossilized. This trend increases with lower and lower categories; most well-skeletonized orders are probably represented by fossils, but many families and a great many genera may be missing. As we have seen, an average species may stand only one chance in perhaps a hundred or more of being represented by fossils.

However, no phylum is known historically very completely; there are inevitably large temporal and geographic gaps in the record. In fact, no phylum can be traced from a preceding one in the fossil record, although this may be due to the very early origin of phyla. On the average, the known record of some classes is more complete than that of any phylum; some have lived in times, places, and environments that happen to be relatively well recorded. The antecedents of some classes (amphibians and mammals among the vertebrates, asteroids and ophiuroids among invertebrates) are fairly well documented. In proceeding to lower and lower categories, there are increasing numbers of taxa for which we have fairly good records of origins, distributions, regional and temporal variations, and times of extinctions.

The fossil record of species presents special problems. Partly this is because reproductive isolation cannot be determined from fossil material. In practice, fossil species are categorized by morphological distinctiveness; these units are called *morphospecies*. Some morphospecies doubtless include more than one biological species, e.g., sibling species, while in other cases groups of fossil populations regarded as distinct morphospecies may be polymorphic species. These problems are well understood and are simply accepted by paleontologists; they form an annoyance, but not a critical difficulty. Tracing phylogeny through time, then, involves tracing a sequence of morphologies.

Tracing species lineages has proven exceedingly difficult because of a second problem, namely, that nearly all species originate allopatrically from their main ancestral populations. If a new species subsequently expands its range and comes to occupy the same region as its ancestral species, it will usually replace or displace the latter, although it may co-exist with it in some circumstances. In such cases there will be a gap in the record of the lineage at localities that include both species, except for the (usually restricted) region where the descendant species first originated. Furthermore, many species have a standardized range of phenotypes that commonly do not change much, if at all, during their recorded histories. Thus, the fossil record of many species consists of the sudden appearance of a new morphospecies without a documented ancestor, the persistence of this new form often essentially unchanged, and then its eventual disappearance, sudden or piecemeal. Species phylogenies must usually be reconstructed by inference, with heavy reliance on stratigraphic and distributional as well as morphological data (see Hallam, 1962). Excellent examples of such reconstructions have been presented by MacGillavray (1968) and Eldredge (1971), and the problem has been reviewed by Eldredge and Gould (1972).

Differences in the fossil records of taxa of different levels mean that different methods of data collection and interpretation are required to deal with them. To study the evolution of phyla we must deal with very incomplete records, although there are many data points, and we interpolate to fill the gaps. On the other hand, at lower taxonomic levels we have some relatively complete examples, and the main task is to extrapolate from these few exemplars. Among lower taxa we can seldom trace lineages very far.

When the morphological changes undergone by evolving higher taxa are plotted through time, several distinctive patterns similar to those displayed in speciation are found (Figure 10-5). *Divergence*, the evolution of increasing morphological difference, is a common occurrence when lineages split and evolve along adaptive pathways that diverge in biospace; it may also occur when separate lineages follow different morphological solutions to the same problems. When a number of lineages split from some primitive form and deploy morphologically, usually to exploit a number of distinctive biospace regimes to which the primitive lineages are somewhat preadaptive, a multiple divergent pattern results, called *radiation* or *adaptive radiation*. The diversification of mammalian orders is a good example.

Parallel evolution results when two or more distinctive lineages evolve along similar lines, so that morphological change in one is paralleled by a similar change in the other, usually in response to similar environmental opportunities or requirements (see Chapter 9). Perhaps the most notable example of parallelism is found among the arthropods. A large body of evidence supports the

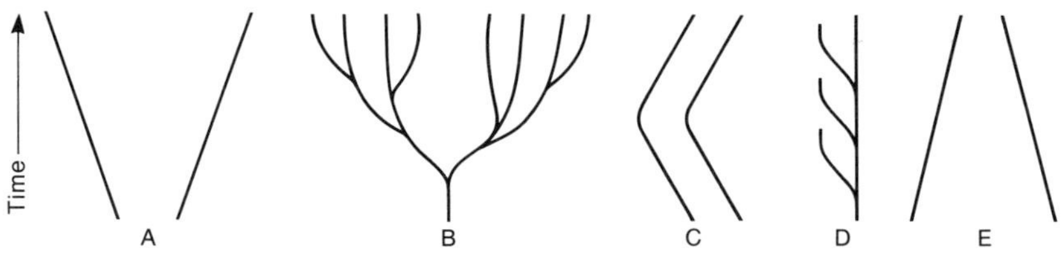

FIGURE 10-5.
Some common evolutionary patterns as indicated by morphological change among different lineages. **A.** *Divergence* of two lineages. **B.** *Radiation,* in which several lineages diverge from a common ancestor. **C.** *Parallel evolution,* in which two lineages change in similar ways in reaction to some environmental event. **D.** *Iterative evolution,* in which morphologically similar forms appear more than once from a root stock. **E.** *Convergence,* in which two lineages evolve toward morphological similarity, presumably under selection to adapt to similar environmental conditions.

view that the phylum Arthropoda is polyphyletic. Between two and four separate lineages of annelid-like worms have each undergone arthropodization to produce the assemblage of characters that represents the arthropod ground-plan (Teigs and Manton, 1958; Anderson, 1973; Cisne, 1974). The Trilobita, Chelicerata, and Crustacea form at least one monophyletic lineage, and the Uniramia (which includes centipedes and insects) forms another. In some morphological features the similarity of independently evolved characters extends right down to the level revealed by electron microscopy; this is true, for example, of the arthropod-style eyes in separate lineages. Such remarkable similarities are explained as evolving in lineages with rather similar genetic makeups and in response to identical functional requirements.

Iterative evolution involves the successive appearance of similar sequences of morphologies from a basic stock. Some of the classic examples of iterative evolution have been shown to be incorrect. An example is the coiled oyster, *Gryphaea*, which was once believed to have originated on several different occasions during the Jurassic, each time from flattish oysters, each time to become extinct. But Hallam and Gould (1975) have shown that Jurassic *Gryphaea* has had a continuing history. Nevertheless, some cases of iterative evolution seem genuine. Irregularly coiled (heterostrophic) ammonites have arisen three or more times from coiled ancestors (Wiedmann, 1969), and globigeriniform morphologies have been produced independently among planktonic foraminifera (see Loeblich and Tappan, 1964).

Very spectacular evolutionary patterns are produced by *convergent* evolution, when separate lineages become morphologically more similar (Chapter 9). Dolphins (mammals), some sharks (fishes), and some ichthyosaurs (reptiles) resemble each other closely in form, despite belonging to separate classes. When species of distinct lineages come to resemble each other closely in overall morphology, they are termed *homeomorphs*. Numerous examples of homeomorphy have been described among invertebrates, for example among fossil brachiopods.

THE FOSSIL RECORD OF EVOLUTIONARY RATES

The ability to obtain close approximations of the absolute ages of fossils has made possible calculation of the rates of change of particular taxa and whole faunas and floras. Simpson (1944, 1953) has made outstanding contributions to the study of evolutionary rates, and has drawn important conclusions about modes of evolution. Measurements of rates of evolution would be most accurate if they could be expressed as rates of change in genomes. It is possible to obtain estimates of average rates by measuring genetic differences between species that diverged at known times. However, this work is still in its infancy, although there are estimates of rate of change in some molecules (see Chapter 9). In the fossil record we must rely chiefly upon changes in morphology, which may be evaluated in various ways (Table 10-2). Measurements of the rate of change

TABLE 10-2. Some approaches to the estimation of evolutionary rates using the fossil record.

Kind of rate	What the rate measures
Morphological change	
Single character rate	Rate of morphological change of one character, which may not correspond with rates of other characters or of lineage as a whole.
Character complex rate	As the size of the complex increases to include most characters, this rate of increase tends to approximate the average rate of morphological change within a lineage.
Taxonomic change	
Time-phyletic rate	Rate of origin of morphologies sufficiently distinctive to merit recognition as new species within an evolving lineage.
Group rate: appearances	Rate of origin of new taxa.
Group rate: duration frequencies	Distribution or average of length of life of taxa.
Group rate: turnover rates	A sort of average of both appearance and disappearance rates of taxa.

of a morphological character, of a complex of morphological characters, or of the total morphological aspect of a lineage, in some known time, will provide estimates of evolutionary rates. Figures 10-6 and 10-7 present examples of measurements of this type.

Figure 10-6 depicts the range of variation in rib strength (the ratio of rib height to rib width) within a lineage of fossil Silurian brachiopods of the genus *Eocoelia* in Wales (Ziegler, 1966). The probable elapsed time is 10 million years, inferred from radiometric dating of rocks elsewhere. The relative duration of the species of *Eocoelia* is based upon the estimated durations of sequences of sediments containing certain fossil graptolites and brachiopods that are associated with *Eocoelia* in time (Cocks, 1971). Despite the great age of this lineage (415 to 405 million years old or so), the relative times involved are probably

FIGURE 10-6 (opposite).
Evolution of a single morphological character, rib strength, in the Silurian brachiopod *Eocoelia*. The time scale in the graph ranges from about 405 to 415 million years ago; rib strength is indicated on the horizontal scale. It remained rather constant until near 413 million years ago, when it changed rapidly for a while, and then continued to change more gradually until 405 million years ago. Rate of change in this character can easily be expressed as change in the ratio of rib height to rib width per million years. [After Ziegler, 1966.]

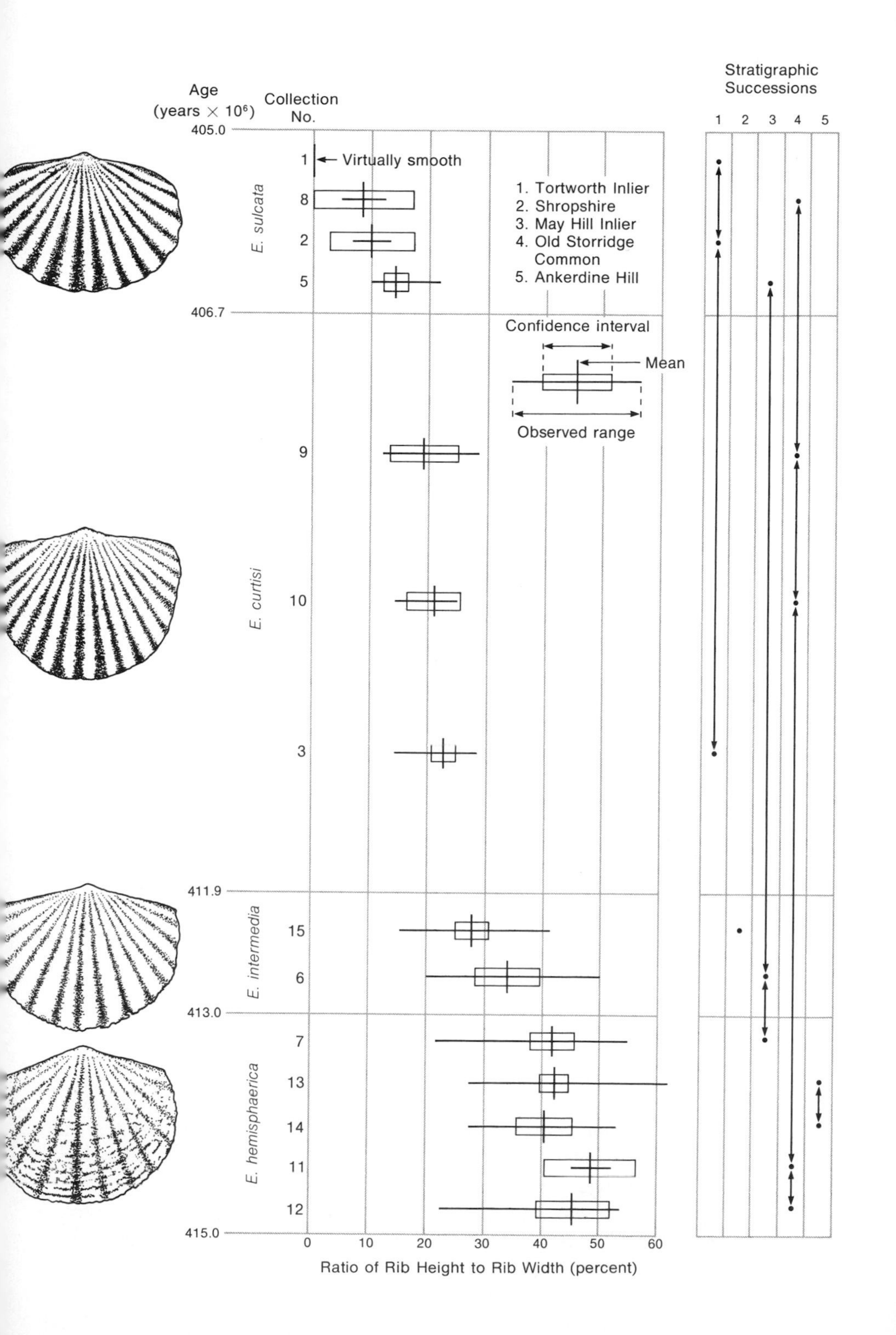

Age (years × 10⁶)
Collection No.
Stratigraphic Successions
1 2 3 4 5
405.0
406.7
411.9
413.0
415.0
E. sulcata
E. curtisi
E. intermedia
E. hemisphaerica
1
8
2
5
9
10
3
15
6
7
13
14
11
12
Virtually smooth
1. Tortworth Inlier
2. Shropshire
3. May Hill Inlier
4. Old Storridge Common
5. Ankerdine Hill
Confidence interval
Mean
Observed range
0 10 20 30 40 50 60
Ratio of Rib Height to Rib Width (percent)

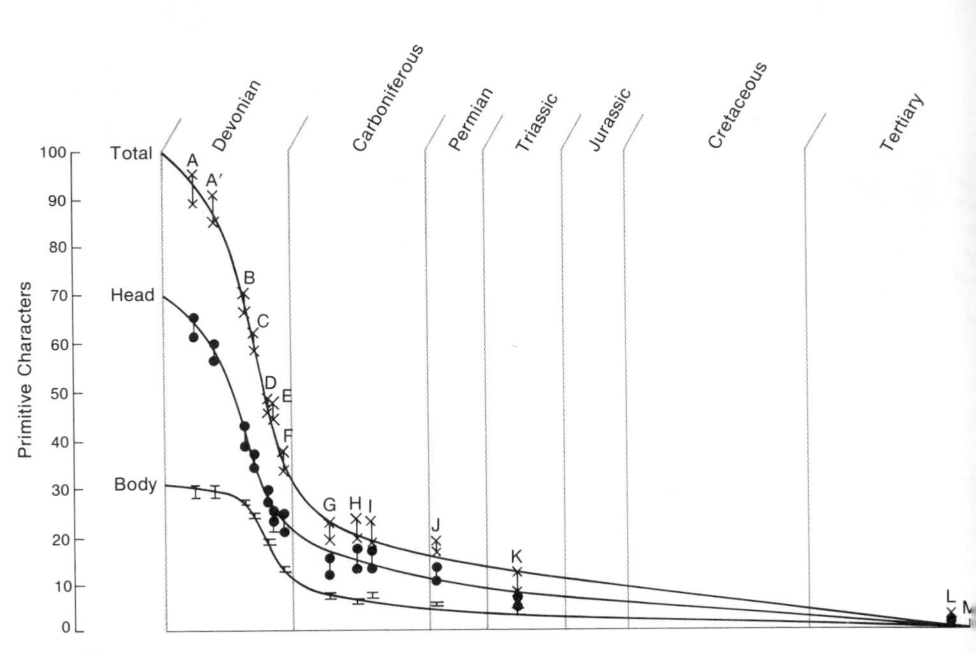

FIGURE 10-7.
Morphological evolution of character complexes in lungfish (Dipnoi). The curves indicate the rate of disappearance of primitive character complexes—for the head, the body, and for both combined. Evolution was most rapid during the Devonian and early Carboniferous. The letters refer to lungfish genera: A,A[1], *Dipnorhynchus;* B, *Dipterus;* C, *Pentlandia;* D. *Scaumenacia;* E, *Fleurantia;* F. *Phaneropleuron;* G, *Uronemus;* H, *Ctenodus;* I, *Sagenodus;* J, *Conchopoma;* K, *Ceratodus;* L, *Epiceratodus;* M, *Protopterus* and *Lepidosiren.* [After Westoll, 1949.]

rather accurately estimated. It can be seen that evolution of this character was most rapid between 413 and 412 million years ago. Other morphological characters changed during this interval as well. This morphological succession has been divided into four nominal morphospecies (Figure 10-6), although it actually represents the gradually changing sequence of ancestral and descendant populations in a single lineage.

Figure 10-7 is based upon studies of morphological change in 21 characters of lungfishes (Westoll, 1949). Each character is scored according to how closely it matches the primitive condition, which is employed as a standard. The score is totaled to give, for any given lungfish, a measure of the degree of change from the standard. When total scores of a series of fossil lungfishes are plotted against time, rates of change of whole complexes of characters result. Figure 10-7 illustrates the degree to which a time series of lungfishes departs from the primitive standard, with characters subdivided into those of the body and those of the head. For the body, rate of change is low at first (the line is nearly horizontal), accelerates greatly in the late Devonian, but is reduced again in the Carboniferous, remaining low until today. Characters of the head change rapidly at first, nearly completing the phase of rapid evolution by the late Devonian, after which the rate of change is low. The significance of this difference between head and trunk is that the diet of these animals shifted first, so that their dentitions altered from a biting or slashing to a grinding type, resulting in changes in many head characters. After these changes were well under way, a more sluggish locomotory habit evolved. Presumably the new food source permitted swimming to change from a swift type to a slower, ripple-movement type.

One conclusion of studies of this sort is that evolutionary rates vary greatly from lineage to lineage and, within a given lineage, from time to time. Rates may also vary among different characters or character-complexes within the same lineage at a given time. The latter situation implies that selection is acting differently upon different functions. This is called *mosaic evolution*, a common phenomenon. It is therefore difficult to reconstruct intermediate forms. For example, if we have older and younger forms within a given lineage, which display morphological differences, there is no assurance that a true intermediate ever existed, that is, that there ever was a form that stood halfway between them in all characters. Instead, one or more of the characters may have evolved first, and achieved their final state before other characters began to change. To reconstruct the state of the lineage at any intermediate time requires a hypothesis explaining which characters evolved when.

A spectacular example of a misjudgment of the effects of mosaic evolution is the "Piltdown man," a forged fossil that misled anthropologists for decades before being exposed as a hoax. The perpetrator joined the jaw of an orangutan to an old human cranium, and planted these remains in sediments contemporaneous with man's immediate ancestors. The forgery passed as representative of an intermediate, small-jawed, large-cranial stage in the evolution of man from ape-like ancestors. The intermediates were in fact just the opposite; they had human-size jaws but small crania (Chapter 14).

Estimates of evolutionary rates from character changes do not give absolute measures that are strictly comparable between different taxa, and especially not between taxa in different categories. The changes of the characters in lungfish imply vastly greater evolution than the changes of ribs in brachiopod lineages. The adaptive significance of changes in skeletal morphology must be considered in order to appreciate the underlying evolutionary changes. Even measures of the rates of change of total morphological distance are not comparable, for similar scores in different lineages may be due to quite different amounts of genetic change or have quite different levels of adaptive significance.

Other methods of measuring evolutionary change involve rates of taxonomic change (Table 10-2). These can be measured as a rate of change of taxa within a clade or of the number of taxa belonging to a clade at any time. The duration of taxa may be taken as an estimate of evolutionary rates; the shorter the duration, the faster the rate. Such estimates may be made at any taxonomic level. Because the fossil record is so poor at the species level, it is common to use genera or families for taxonomic rate studies in order to reduce the bias arising from incomplete preservation.

Species both originate and become extinct in two distinct ways. They can originate by splitting from a previous lineage or by developing from a previous lineage so that the later species is merely the temporal extension of the earlier. Species can disappear by extinction or by evolving into a descendant species. The cases of appearance or disappearance by evolution within a single lineage cause "pseudodiversification" and "pseudoextinction." From data on families it has been estimated that pseudoextinctions usually account for about 5 percent of all extinctions, and never more than 20 percent (Van Valen, 1973); pseudodiversifications have the same range. The *Eocoelia* lineage depicted in Figure 10-6 displays several pseudoextinctions and pseudodiversifications; for the time interval represented, there are three episodes of extinction and three of speciation, and a total diversity of four species. Actually there is only a single lineage at any time, and the average rate of appearance of species is 0.4 per million years. This sort of rate is called a *time-phyletic* rate (Table 10-2).

Time-phyletic rates cannot strictly be calculated for taxa in categories higher than species. Genera, for example, do not evolve from other genera, but from particular species of a genus. For evolutionary taxonomists there is no special reason why an older genus should become extinct merely because one of its species has given rise to a descendant that is classed in a new genus. It is possible for the new genus to develop an ecological range similar to its ancestral genus and eventually to replace it, but that is a different matter. Precise calculations of phyletic rates can be made for genera only when each ancestor–descendant series contains no more than one species. The same principle applies to higher taxa. Because the fossil record is so lacking in phylogenetic species sequences, the phyletic approach to evolutionary rates cannot be widely applied.

Taxonomic rates may be expressed as *group rates* rather than phyletic rates. An example is the rate of appearance of genera or of families within a given higher taxonomic group, such as a class or a phylum. Figures 10-8 and 10-9 show the rates of appearance and disappearance of families of Phanerozoic marine Bivalvia (Mollusca) and Cenozoic terrestrial placental mammals in the fossil record (Stanley, 1973). Assuming that appearances approximate origins and disappearances approximate extinctions, the bivalves have evolved much more slowly and have experienced much less turnover than the mammals. Furthermore, within each group the taxonomic rates of evolution have varied through time. In bivalves the rate of origin of families was particularly high in the early Mesozoic and early Cenozoic, while in mammals it was highest in the Paleocene and Oligocene. A probable reason for some of the apparent fall-off in rate of appearances of families in the late Cenozoic is that new genera that appeared during this time are still assigned to the old families from which they have sprung, but in time (millions of years) some of them will have diversified to the point of representing different families (Simpson, 1944).

Figures 10-8 and 10-9 also depict the approximate diversities, i.e., approximations of the total numbers present at any time, of marine bivalves and ter-

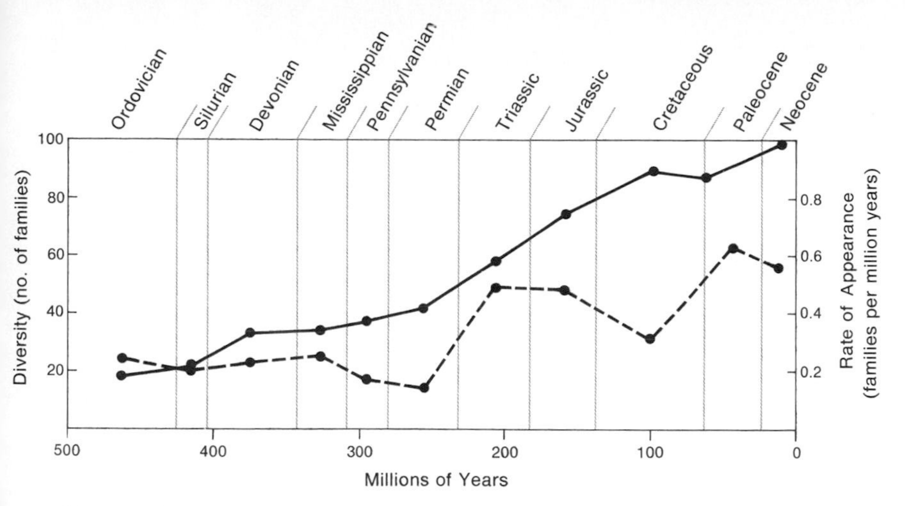

FIGURE 10-8.
Rates of appearance (broken line) of families of Bivalvia (Mollusca) and their standing diversities (solid line) during Phanerozoic time. Bivalves have never appeared at a rate much greater than about 0.6 family per million years. [After Stanley, 1973.]

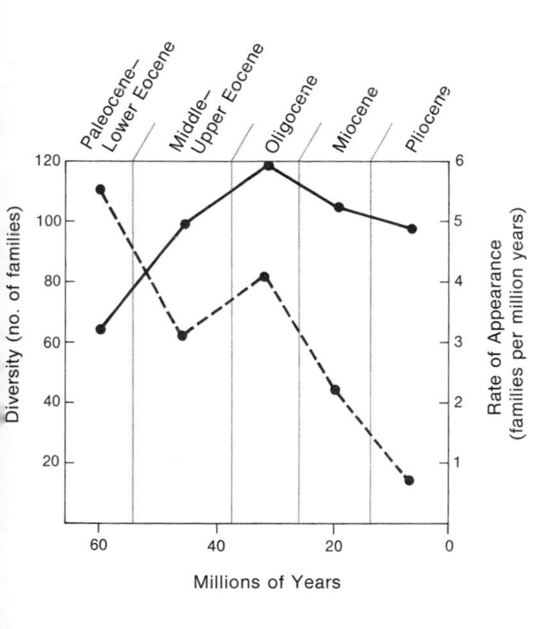

FIGURE 10-9.
Rates of appearance (broken line) of families of terrestrial placental mammals and their standing diversities (solid line) during the Cenozoic Era. Mammals have appeared at rates as high as nearly six families per million years, or one order of magnitude greater than the highest bivalve rate. [After Stanley, 1973.]

restrial placental mammals. Within each group diversity varies through time; the rates of change are quite different for these two groups, with mammals being far more volatile.

Taxonomic group rates include both anagenetic and cladogenetic events. While they may measure evolutionary activity, they usually do not indicate the amount of evolutionary "advance," i.e., increased structural complexity or adaptive potential, or even the amount of change that occurs during a consistent evolutionary trend. For those purposes, character rates are clearly best. However, group rates within carefully selected lineages, such as the direct line of equids leading from *Hyracotherium* (the "dawn horse") to *Equus* (the living horse), must approximate anagenetic rates along such trends. There are eight more or less successive genera of equids during the 60 million years involved in the horse line, giving an average rate of about 0.13 genera per million years (Simpson, 1953).

One way of obtaining a rough estimate of cladogenetic vs. anagenetic activity in any group is to compare the group rates of taxa in one category with those of taxa in the next higher category. In Figure 10-10, generic rates are compared with familial rates in trilobites. During the early Cambrian a radiation created many genera and a rather large number of families, one for about every five genera. However in middle Cambrian to lower Ordovician time the rate of appearance of genera nearly doubled, while families continued to appear at about

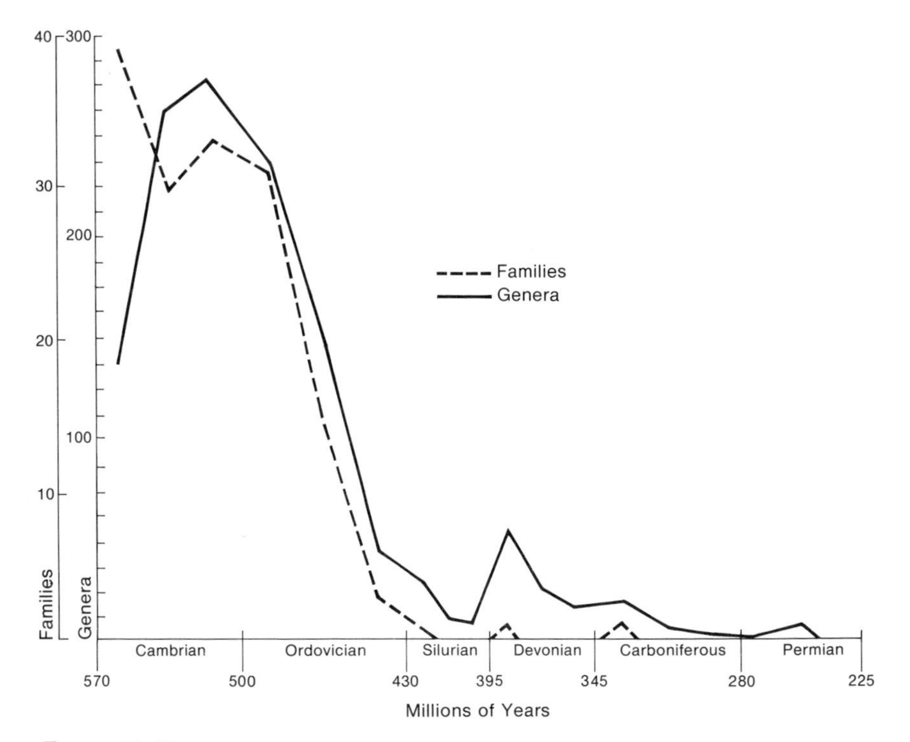

FIGURE 10-10
Contrasting rates of appearance of families and genera of trilobites, an extinct group of arthropods. Note that the proportion of families to genera is highest in the early Cambrian, falls during the upper Cambrian, and is quite low in Silurian to Permian time. The early genera tended to be rather distinctive and are often classed in separate families. Later genera, however, tended to be less novel and so to be placed in proportionately fewer families, suggesting a lower level of anagenetic change for each cladogenetic divergence.

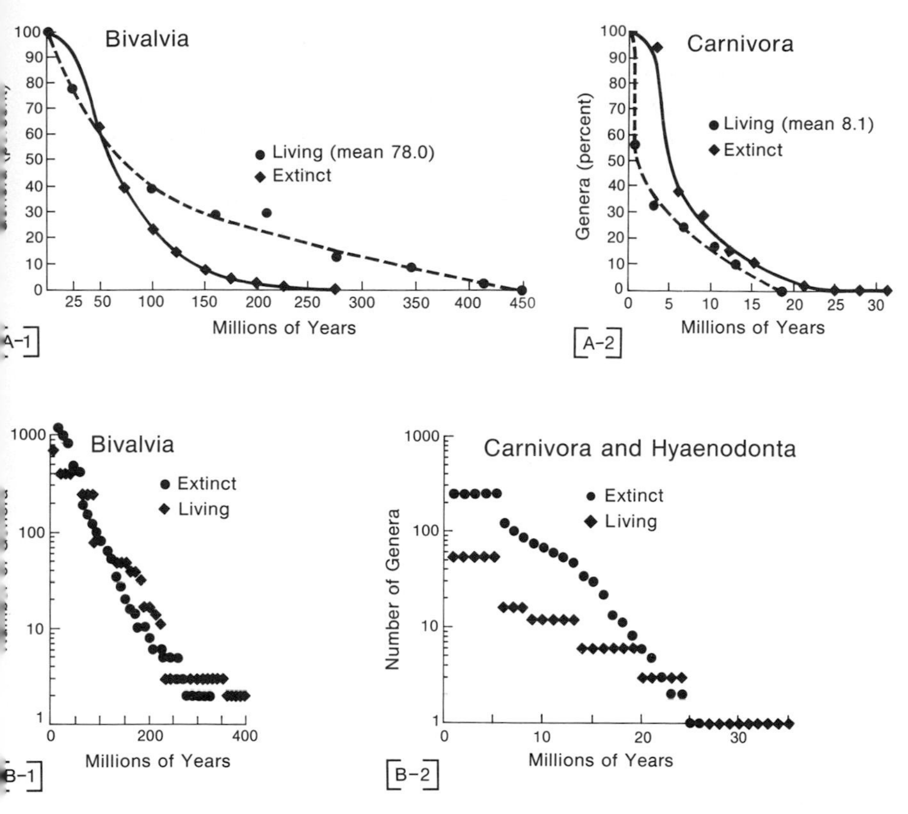

FIGURE 10-11.
A. Duration frequencies of genera of Bivalvia (Mollusca) and of terrestrial Carnivora (Mammalia) with arithmetic coordinates. [After Simpson, 1958.] **B.** Duration frequencies of genera of Bivalvia (Mollusca) and of Carnivora and Hyaenodonta (Mammalia) plotted on a logarithmic ordinate and therefore somewhat analogous to the conventional survivorship curves employed by demographers for living populations. [After Van Valen, 1973.]

the same rate. The elaboration of new genera did not lead very often to novel families, but to mere variations on an established family theme. This trend was continually enhanced; during the Devonian, although nearly a hundred new genera appear, only one new family is described. Little or no real evolutionary advance was taking place.

Still another approach to ascertain rates of evolution is to examine the frequency distribution of the lengths of time during which taxa of some group existed. Figure 10-11 shows the frequency distribution of the durations of genera of bivalves and land carnivores in percent; the former generally endured much longer, although the patterns are similar. When such duration patterns are plotted with a logarithmic ordinate, they become comparable to the survivorship curves of population biology. Sometimes these curves plot as a fairly straight line (Figure 10-11B; Van Valen, 1973). Simpson (1944, 1953) has noted that in some groups there are many taxa that survive longer than would be expected on the basis of such a survivorship distribution. These unexpectedly long-lived taxa have evolved only very slowly; they include famous "living fossils," such as the brachiopod *Lingula*, which has survived over 400 million years without much morphological change, and the reptile *Sphenodon*, which resembles Upper Triassic forms living over 190 million years ago. Even more impressive survivors are found among blue-green algae. Many modern-looking fossils appear about 900 million years ago; some of them form whole assemblages with modern aspect (Schopf and Black, 1971).

Simpson recognized three types of evolutionary rates. *Horotelic* rates are those that have the average distribution of duration frequencies within groups, as exemplified by the curves in Figure 10-12; both rapidly and slowly evolving lineages may be horotelic. *Bradytelic* rates are exceedingly slow, and are manifest when slowly evolving lineages exceed the number expected on the basis of horotelic distributions; these lineages survive long beyond expectations. *Tachytelic* rates are those more rapid than fast horotelic rates. Simpson suggests that tachytelic rates are associated with the first invasions of new adaptive zones, and represent evolutionary accelerations along phyletic lines. (A sudden rash of cladogenesis would not qualify as tachytely unless it involved rapid anagenetic changes.) Horotely represents the usual pace of evolution within an adaptive zone, including faster and slower episodes as evolutionary opportunities wax and wane. Bradytely is achieved by lineages whose adaptive zones happen to endure indefinitely. This is more likely for flexibly adapted forms.

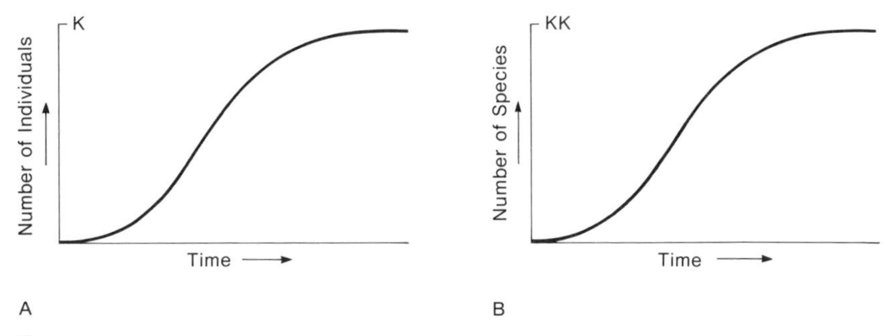

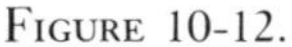

FIGURE 10-12.
A. The equilibrium model of population density. The equilibrium population size, K, is a function of the carrying capacity of the environment for this species; the rate at which a population approaches K is a function of its reproductive potential, r. **B.** The equilibrium model of species diversity. The equilibrium diversity, KK, is a function of the capacity of the environment to accommodate species; the rate at which species numbers in a given environment approach KK is a function of rr, the diversification potential, which depends on rates of cladogenesis, immigration, and local extinction.

REGULATION OF DIVERSITY

Past changes in the composition of the world's biota are well known. In the sea, trilobites and brachiopods have given way to snails and clams as the dominant skeletonized forms. On land, amphibians were largely replaced by dinosaurs and these in turn by mammals. We shall trace this progression in Chapter 13. These changes were not always gradual and uneventful, but frequently were accompanied by major waves of extinctions as groups declined or disappeared, and by waves of diversifications as still other groups expanded. Indeed, the species diversity in the marine realm alone, at any one time, has varied through at least one order of magnitude during the Phanerozoic. Before examining the probable causes of extinctions and diversifications, it is appropriate to examine the factors that may regulate diversity.

The total number of species that can coexist in the world at any given time must be limited. There are only so many resources, and they can only be partitioned among so many species, each of which must maintain a sufficiently large population to persist. We understand many of the factors that limit the carrying capacity of the environment for populations of individual species; they are trophic resources (chiefly food for animals, and nutrients and sunlight for plants) and living space. The severity of the operation of these factors depends upon population density. As density increases the resource demand increases, and the rate of population increase falls until an equilibrium value is approximated; this value is called the *carrying capacity*, symbolized by K (Figure 10-12A). Factors that act in this way are termed *density-dependent* factors.

Because each species requires a certain amount of trophic resources and habitat space to maintain a viable population structure, there also must be some carrying capacity for the number of species, which we symbolize as KK. If a region or an adaptive zone is relatively empty of species, we can imagine immigration and diversification occurring until the region reaches its approximate species capacity (Figure 10-12B). If the region is so isolated that immigration is unlikely, then diversification must operate chiefly alone. The question arises whether the biospace (the habitable world as a whole), or at least portions of it, has ever actually filled to its approximate capacity. It is clear that all parts of the biospace have not always been filled. There were few terrestrial species in the early Paleozoic, for example, and probably few marine species in the late Precambrian.

There is some evidence that evolution does operate to fill biospace in time. At present, the number of species living together in natural associations or communities varies from place to place in seemingly logical patterns. In different environmental regimes that seem otherwise similar, species diversity is highest where there are the most habitat opportunities, that is, where the environment is most heterogeneous spatially. In the sea, for example, more species live along rocky shores, which are extremely varied in habitat, than live in association with the more monotonous sandy beaches in the same regions. There is abundant evidence that the terrestrial biota of islands increases in species diversity with increases in the diversity of available habitats (MacArthur and Wilson, 1967).

On the other hand, when regions of similar spatial heterogeneity are compared, those with the temporally more stable trophic resource supplies contain more species (Valentine, 1971; see also Margalef, 1968). Marine examples include communities living in muddy bottoms in shallow tropical waters, where productivity is reasonably stable the year round, contrasted with muddy-bottom communities in high latitudes, where productivity is extremely seasonal; the former are relatively rich in species, the latter are poor. Even in the deep sea, which entirely lacks photosynthetically based primary productivity and has relatively low amounts of trophic resources, species diversity is very high for benthic communities on comparable substrates (Hessler and Sanders, 1967; Sanders, 1968). The trophic supplies are highly stable and rather evenly distributed (Hessler and Jumars, 1974). Thus it appears that trophic resource stability and spatial heterogeneity are important regulators of species diversity within communities.

The explanation for diversity regulation by trophic resource regimes is that in environments where trophic resources fluctuate over significantly long periods of the life cycle of members of a given population, the population requires a larger share of energy than does a comparable population living where trophic resources fluctuate less. Populations in trophically unstable environments tend to be more broadly adapted on the average, eating a wider variety of food and living in a wider range of habitats than those in trophically stable environments. Presumably, this relative flexibility is achieved by selection as an adaptation to inclement seasons. Food shortages are more easily dealt with by generalized feeders; specialists rely upon only a few food items and thus, if these happen to become scarce, may suffer severe mortality or even extinction. Habitat specializations, commonly favored by selection where resources are stable, may become unfavorable where resources are uncertain, for they restrict population sizes and distribution patterns in situations where large size and ubiquity are advantageous. It is not possible, the argument goes, to pack very many species with such broad ecospaces into a given portion of biospace. In stable regions, however, specialization is permitted or favored to insure some unique trophic or habitat resource for each species. Specialized species have small ecospaces and many of them can be packed into a given portion of biospace. (For a more extended discussion of these propositions see Valentine, 1973b.)

Spatial heterogeneity is due in many cases to the organisms themselves, which greatly elaborate upon the inorganic physical framework of an environment to provide numerous additional habitats. For example, in terrestrial ecosystems the diversity level of animal life is affected greatly by the density, both horizontal and vertical, of the plant cover. In deserts and steppes, where woody plants are widely separated from each other, two kinds of habitats exist. In the open habitats the nature of the animal life is governed largely by the physical environment: the nature of the soil or rock, the amount of sun or shade, and the diurnal and seasonal fluctuations of temperature. Around the shrubs special habitats exist, created by the protection which they afford, in which annual plants, insects, small reptiles, and other forms of life can maintain themselves even though they cannot survive in the intervening open spaces.

A contrasting ecosystem is that of the tropical rain forest. In areas such as the Amazon and Congo Basins and the lowlands of Malaya, the ground is completely covered with both plant life and the decaying remains of leaves and other plant material, so that the diversity of habitats depends almost entirely upon the nature of the plant cover. Vertical layering is most conspicuous. One set of environments is subterranean, occupied by worms, nematodes, and many kinds of microorganisms, plus burrowing vertebrates and insects. The next layer is the forest floor, to which little light penetrates, so that many of the smaller plants are saprophytes or parasites, and the animals are either predators or feeders on decaying matter. Above this there may be two or three layers of arboreal habitats, formed by trees of different sizes. Each of these layers has a characteristic amount of light and air currents, and a distinctive assemblage of plants, including both the crowns of trees and the epiphytes that rest on their branches. These provide separate habitats that may be variously partitioned by the distinctive fauna of a particular layer. The uppermost layer, represented by the crowns of the highest trees, is in some ways the most favorable, since the greatest amount of light is present and there is the greatest opportunity for movement by active animals, such as birds, lepidoptera, and other insects. Many of the tropical animals are highly specialized in food or habitat, and in other components of adaptation, such as reproduction.

In the sea the coral reef association is a highly diverse complex of communities that owes its richness in great part to the biogenic reef structure, which breaks the shallow-water zone into a number of distinctive environments. Each of the numerous habitats is occupied by a vertically layered series of partially unique species assemblages. These habitats range from exposed, wave-swept buttresses and channels along the reef front to quiet lagoon-margin pools and deep lagoon floors. Many of the animals form commensal or parasitic relationships and, like tropical forest animals, many are highly specialized. Diversity on reefs greatly exceeds that in comparable shallow-water environments where reefs are not developed. However, it is clear that biological heterogeneity as in tropical forests and reefs does not provide a final explanation of the diversity pattern. The problem is rather to explain the causes that permit such heterogeneity to arise in these places.

The degree of regionation or provincialization of the biota is important in establishing global diversity levels (Valentine, 1973b, c, d). For example, the mammalian regions are separated by barriers to immigration, chiefly topographic (such as oceans) and climatic (Figure 10-13; Sclater and Sclater, 1899; Simpson, 1968). Each region contains many endemic lineages. If the barriers were removed the numbers of lineages would decline markedly. In the sea the poikilothermic ("cold-blooded") shelf animals are strongly limited by climate and by topographic barriers (land or deep waters). Today the longest coastlines, and therefore the longest stretches of continental shelves, trend north–

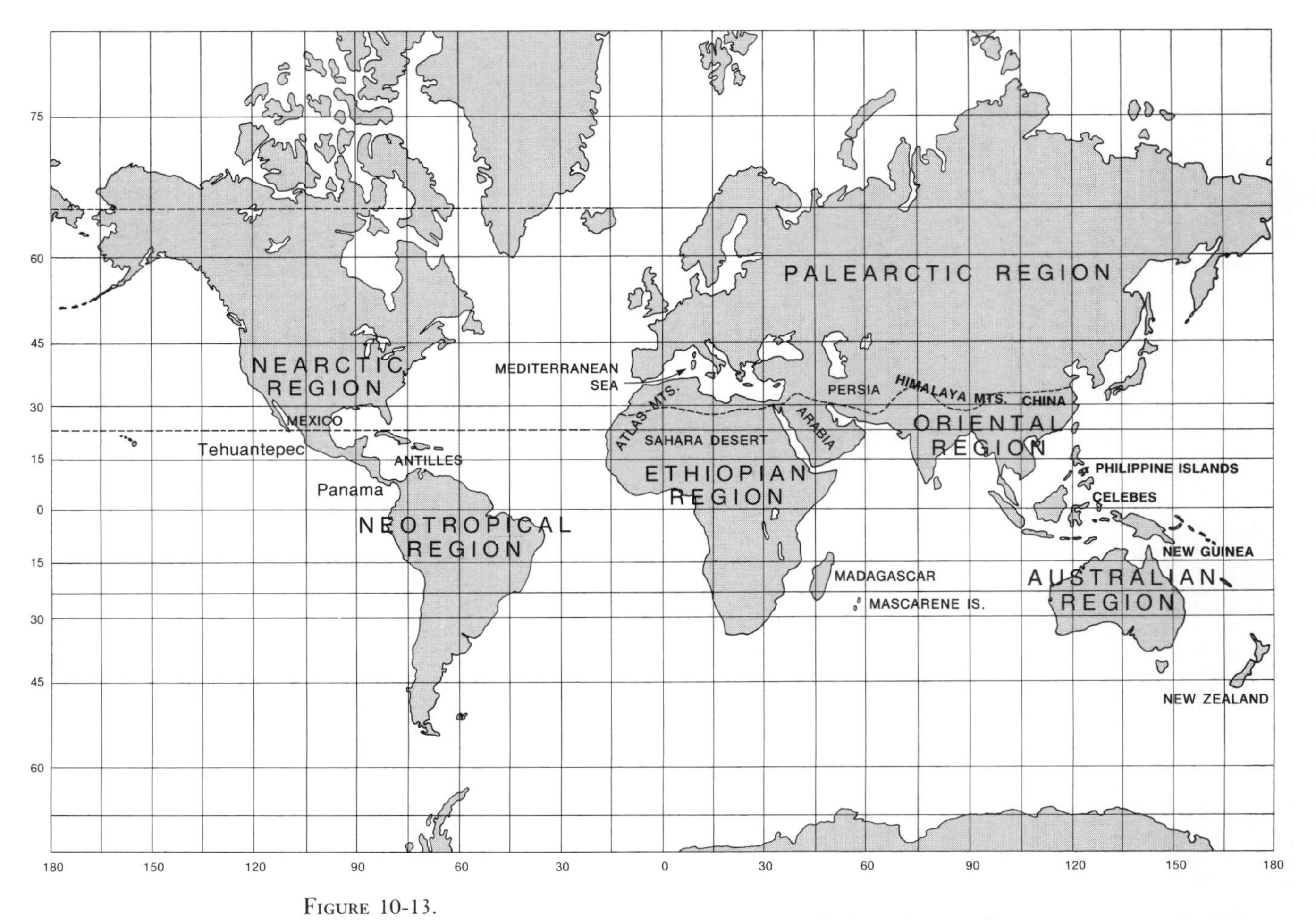

FIGURE 10-13.
The six main regions of the present geographical distribution of mammals.

south and are separated by continents and by deep sea. Environmental discontinuities occur at intervals along the shelves, owing to geographic changes in marine climate that depend upon ocean circulation patterns and coastal topography. Large numbers of species find that the environmental changes across such discontinuities are limiting and end their ranges near such places, which form provincial boundaries. Chains of separate provinces are therefore found along each north–south continental coast (Figure 10-14). Over 30 marine provinces exist at present, each containing a large endemic element. If the numbers of provinces were to decrease, the number of species in the marine biota would decrease accordingly.

Many other factors are known to affect local diversity levels. Predators reduce prey populations to relatively low levels, and therefore the prey do not consume as large a share of the space and food resources as they would otherwise (Paine, 1966). These resources are thus freed for utilization by additional species. Competitors may wrest parts of the resources from populations that

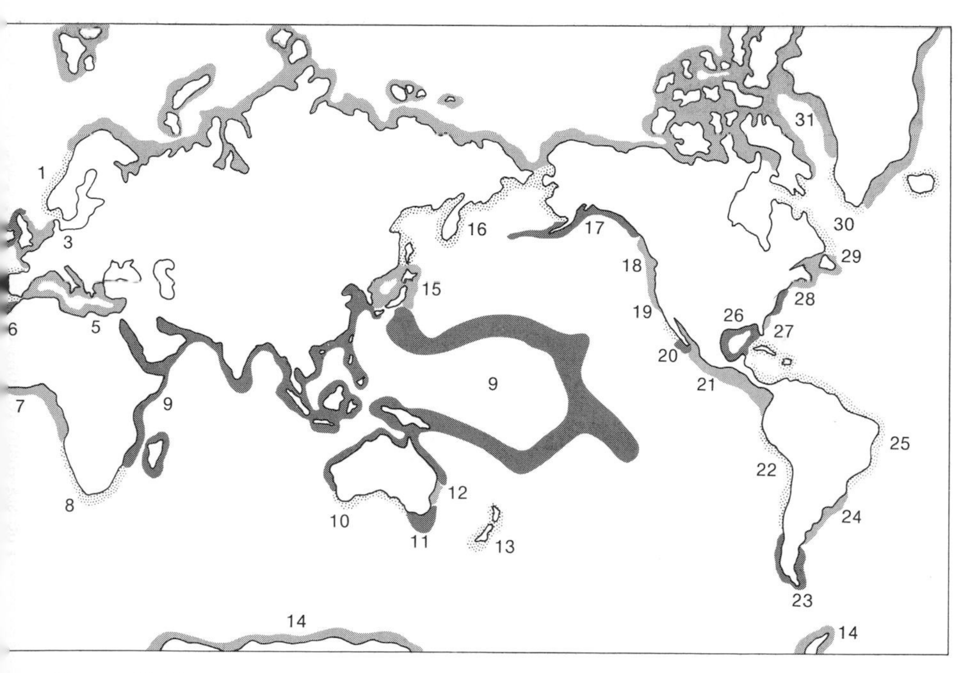

FIGURE 10-14.
The 31 main biotic provinces of shallow-water marine invertebrates in the ocean. (1) Norwegian. (2) Caledonian. (3) Celtic. (4) Lusitanian. (5) Mediterranean. (6) Mauritanian. (7) Guinean. (8) South African. (9) Indo-Pacific. (10) South Australian. (11) Maugean. (12) Peronian. (13) Zealandian. (14) Antarctic. (15) Japonic. (16) Bering. (17) Aleutian. (18) Oregonian. (19) Californian. (20) Surian. (21) Panamanian. (22) Peruvian. (23) Magellanic. (24) Patagonian. (25) Caribbean. (26) Gulf. (27) Carolinian. (28) Virginian. (29) Nova Scotian. (30) Labradorian. (31) Arctic. [After Valentine, 1973b.]

would otherwise utilize them, and thus competition can serve to raise diversity (Connell, 1961; Miller, 1967). Physical disturbances, from forest fires and storms at sea to the pounding of drift logs against intertidal rocks, can free habitat space and food, permitting additional species to appear within the disturbed communities (for a marine example, see Dayton, 1971). The elucidation of such ecological mechanisms is most illuminating because it provides insight into the many sorts of intricate relationships established by natural selection. However, it seems unlikely that they regulate the diversity capacity, for in general they are not diversity-dependent factors. The concept of species capacity implies an equilibrium condition for any given environmental state, towards which evolution tends. As an environment fills with species, the rate of cladogenesis should slow and evolutionary activity in general tail off, until we can imagine a virtual cessation at equilibrium. That an equilibrium has ever been actually established seems unlikely, for as we have seen, more or less continual environmental changes occur.

PATTERNS OF EXTINCTION

The general results of environmental changes are predictable from evolutionary theory; as the quality of biospace changes, species must evolve new adaptations or become extinct. The fossil record suggests that more or less continuous extinctions and pseudoextinctions are the rule, but that extinction rates vary greatly through time (Newell, 1956; Simpson, 1944; and see Figures 10-8, 10-9). Extinction rates for families of shallow-water marine invertebrates during Phanerozoic time are shown in Figure 10-15.

Recall that when extinct families or genera of higher animal groups are treated as if they formed a cohort, that is, as if they had all originated at the same time, the distribution of their duration frequencies sometimes plots close to a straight line on a logarithmic time scale (Figure 10-11B; Van Valen, 1973). If the extinct taxa had indeed formed a cohort, such a plot would indicate that the probability of extinction was the same at any time for any of the taxa. That is, as time passed, the probability of extinction of the surviving taxa would neither increase nor decrease. However, all genera and families of animal groups do *not* originate at the same time, so that the plot indicates instead that when extinctions do occur, young and old taxa are equally vulnerable. Lineages do not learn in time to survive better. This indication is in perfect accord with the synthetic theory of evolution by natural selection. Species are adapted to their ambient environments, not to some future possibilities. When the environment does change, it deteriorates for most extant taxa regardless of their previous longevities. To be sure, some ambient environments fluctuate and thus require a flexible adaptive strategy of their inhabitants, but this is a different matter from lineages acquiring special abilities to contend with future environmental regimes. The slopes of the extinction curves do vary from group to group, and this indicates that the probability of extinction is higher in some groups than in others.

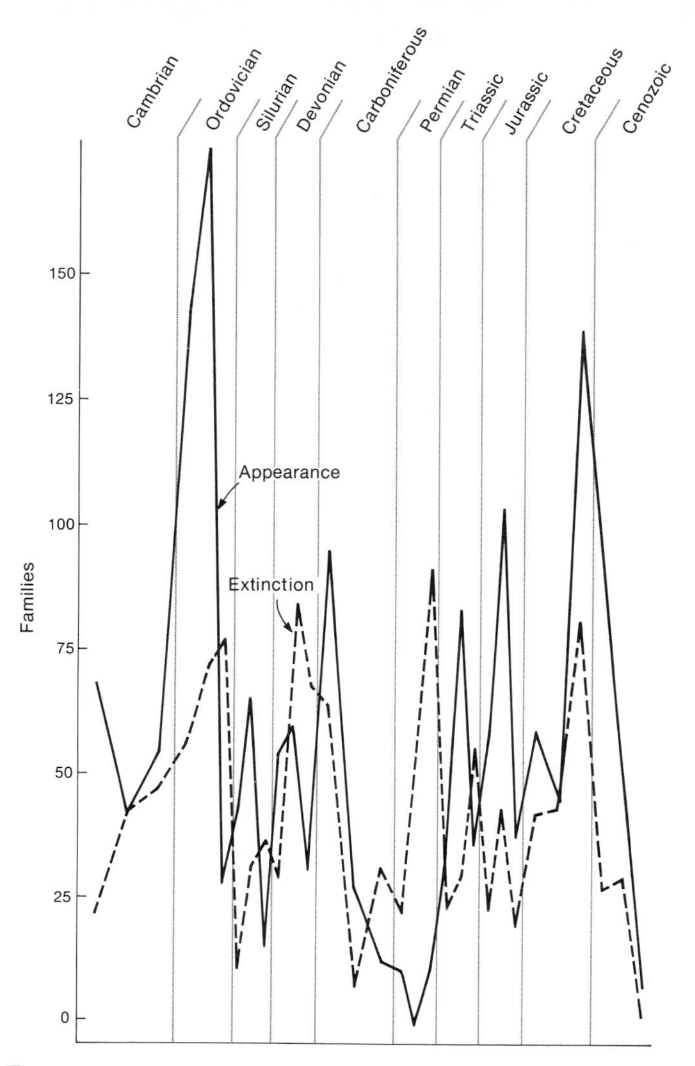

FIGURE 10-15.
First appearances (solid line) and last appearances (broken line) in the fossil record of families of well-skeletonized shelf invertebrates during the Phanerozoic. The curve is plotted on three points per period. Important times of family diversification occur in the early Ordovician, early Carboniferous, and Triassic, Jurassic, and Cretaceous periods. Important waves of family extinctions occur in the Ordovician, Devonian, late Permian, and late Cretaceous periods. [After Valentine, 1969.]

At one time it was theorized that lineages, like individuals, had life cycles, and that they proceeded from youthful through mature to geronitic stages, terminating by extinction. Lineages that were about due for extinction could be identified by certain characteristics of "racial senescence" or "phylogerontism." Among these characteristics were increases in body size, appearance of spines and similar elaborate skeletal features, and the loss of teeth or other structures. Each of these features has now been adequately explained as an adaptation that improves fitness in some mode of life; no compelling biological reason for phylogerontism was ever put forward, and no evidence for this notion has appeared. Indeed, that extinctions may tend to occur independently of the age of taxa suggests at once that phylogerontism is incorrect, and that the whole analogy of individual life cycles with those of higher taxa is mistaken. The fossil record clearly shows that small as well as large, plain as well as fancy, and toothed as well as toothless lineages are extinguished by changing conditions, not by any innate aging process. It is true that highly specialized forms are less able to cope with many environmental changes than are flexibly adapted ones. Specialists commonly possess elaborate or otherwise unusual morphologies for their groups, and this may have led to the curious theories of racial senescence.

It is convenient to distinguish two types of extinction that intergrade. One results from a lowering of the species capacity in part or all of the biosphere; the other, not related to diversity regulation, results simply from environmental changes (either in biotic or physical factors) that prove deleterious to the adaptations of some lineages. The former is a diversity-dependent, the latter a diversity-independent type of extinction.

Diversity-independent extinctions seem to occur more or less continually. Many diversity-independent factors are climatic; as temperatures rise or fall or rainfall increases or decreases, extinctions occur. Such extinctions ordinarily free resources employed by the extinct lineage, so that these resources may then support new species. Immigrant populations that happen to be preadapted to the new environmental conditions may have little competition for these resources. The eventual achievement of the status of independent species by these immigrant populations, allopatric with respect to their source stocks, would be common. Thus, diversity-independent extinctions and cladogenesis may operate hand in hand, continually turning over some fraction of the species in most habitats (MacArthur and Wilson, 1967). During such turnovers a group may significantly change its relative dominance in the biota, according to whether or not it is favored, relative to other groups, by the directions of environmental change.

Diversity-dependent extinctions may occur locally or globally, and evidently have sometimes occurred on such a scale as to affect greatly the level of diversity of the entire planetary biota. Drops in species capacity that lead to extinction cannot be made up by diversification until the capacity rises again. The lowering of species capacity must usually involve changes in trophic resource regimes or in the spatial heterogeneity of the environment. From the previous discussion of environmental change, it is clear that these factors may be subjected to either episodic or continual changes.

Variability in trophic resource regimes increases as seasonality increases. This can occur when continents move into higher latitudes or when the effects of continentality are increased. For example, in shallow marine waters nutrient supplies can become seasonal or otherwise irregular owing to perturbations in the water column (upwelling, downwelling, seasonal change in water types) that are associated with continentality (Valentine, 1971). Increases in continentality are expected when sea levels are low and expose more area of continental platform; when continents surround relatively small oceans; or when continents coalesce to form larger continents (Valentine and Moores, 1970, 1972). In the terrestrial realm, local increases in seasonality can result from lowered sea levels or continental assembly, especially in the interiors as they become more remote from the ocean. All these changes, among many others, tend to lower the species capacity of the affected environments, which would lead to extinctions.

Spatial heterogeneity also varies through time. On a global scale such variation results from two interrelated events: the fragmentation or assembly of continents and the increase or decrease in latitudinal thermal gradients and therefore of climatic variety. Both these phenomena result ultimately from global tectonic processes. The climatic gradient, for example, depends in part upon the poleward transport of heat, which can be altered by changes in the positions of continents and such physiographic features as mountain ranges. When the climatic gradient is high, such as today, most species cannot range far latitudinally, since the temperature changes are great. Therefore, a large number of latitudinal climatic provinces are formed, each with a complement of climatically limited species. However, if the climatic gradient were reduced because of continental drift, provinciality would be reduced and extinctions would occur (Valentine, 1968). For example, if the gradient were reduced because of warming poles, then low-latitude zones of warm water would expand and warm-water species could extend their ranges poleward. By contrast, high-latitude, cool-water species would have nowhere to go and would become extinct as the climatic regime to which they are adapted disappeared.

The collision of continents leads to diversity-dependent extinctions in the marine environment because pairs of shallow-water provinces formerly separated by topographic barriers become coalesced. This brings two (or more) biotas that had previously been more or less isolated into association, and the species capacity is filled by a single, new biota. This may cause extinction of the equivalent of the entire endemic fraction of one of the biotas. The general cause of these extinctions is the reduction of spatial heterogeneity at the provincial level, but the proximal cause would ordinarily be competition, frequently associated with mortality rises due to the introduction of new predators, parasites, or amensals from one provincial biota into the other. An example of such extinctions is afforded by the collision of North America and Europe during the Silurian–Devonian, when two formerly separate biotas coalesced. This resulted in the elimination of many North American lineages and their replacement by European stocks (Boucot, 1968).

An additional major cause of extinctions in the marine realm is associated with changes in sea level, which can significantly reduce the habitable areas in whole shelf provinces. In the terrestrial realm continental assembly can bring separate biotas into association to cause extinctions. Drops in sea level, on the other hand, may actually raise terrestrial diversity as a whole by adding to the area and to the heterogeneity of that environment. The events leading to major diversity-dependent extinctions nearly always involve changes in the environmental regime, which prove deleterious to some lineages and result in many diversity-independent extinctions as well.

The temporal variation in rates of extinction observed in the fossil record of, say, families, can be ascribed to temporal changes in the rates of environmental change. What, then, is responsible for the characteristic differences in extinction rates between different higher taxa, such as bivalves and mammals? One suggestion is that competition is primarily responsible, that the groups with the higher turnover rates are those in which species compete most vigorously among themselves (Stanley, 1973). Thus, mammals exhibit more intense interspecific competition than bivalves. This correlation is plausible since the competition pressures in times of other stress might prove decisive. Indeed, it is plausible to expand this suggestion to a general case, that species whose populations are under heavy ecological pressure, from whatever source, are especially vulnerable to extinction, other things being equal. Quite often such forms have the more active, vigorous modes of life. They frequently have high metabolic rates (see McAlester, 1970) and require a relatively large energy base.

Rates of extinction of species are naturally higher than those of genera and families, but they are impossible or difficult to estimate from the fossil record. The average terrestrial mammal species probably persists for only 50,000 years or so, while the average marine invertebrate may persist perhaps 1,000,000 years. Why do not more of them persist longer, evolving appropriate adaptations as novel environments appear? The answer seems to be that evolution has no foresight, and acts continually to optimize adaptation, prorated over competing selection pressures. High extinction rates are partly a measure of the success of evolution in achieving a high degree of adaptation to older conditions, and of the effectiveness of the conservative, stabilizing forces that make for the coadaptation of gene pools. It is hardly surprising that many species, trapped in a web of competition and predation, fail to meet new environmental challenges. Invading immigrants sometimes have distinct advantages in this respect; they may have preadaptations to the new physical conditions, and they may be free at first from the pressures of predation, parasitism, and competition that eventually will develop and which can tax their abilities to maintain viable population structures.

PATTERNS OF DIVERSIFICATION

Chapters 6, 7, and 8 dealt with speciation and with the origin of higher taxa, which can raise diversity levels. The fossil record documents the history of diversification, indicates that diversification has operated at different rates at different times, and provides some clues as to the causes of particular historic diversification episodes. Most of these seem to be connected with one of three situations: with replacement of diversity-independent extinctions, which is nearly a continuous process; with rises in species capacities, with or without previous extinctions, which provide opportunity for speciation; and with the invasion and exploitation of new environments or of old environments in radically new ways.

These situations are all found in the history of marine families of skeletonized invertebrate phyla, whose diversification is plotted in Figure 10-15 (p. 343). The early peak in Figure 10-15, during the Ordovician, is ascribed chiefly to the exploitation of newly evolved modes of life, chiefly of suspension-feeding and predation in epifaunal communities. Such diversifications do not necessarily involve much environmental change, although in this case it is likely that the species capacity was rising, perhaps due to an increase in trophic resource stability. The smaller peaks of diversification during the Silurian, Devonian, and Carboniferous tend to follow extinction peaks, and may involve the replacement of extinct lineages by radiation of the survivors into vacated biospace.

The Permo–Triassic extinctions occurred during a time of little diversification, which together with other evidence suggests that it was due largely to diversity-dependent factors. The diversification peaks that followed appear to be due to increases in species capacities (owing to the breakup of the Permo–Triassic supercontinent), although they may be somewhat complicated by replacements of lineages that were extinguished by diversity-independent causes. The last diversification peak, in particular, appears to have resulted from a rise in species capacity due to provincialization—the creation of new endemic biotas in many regions. This resulted partly from a rising latitudinal climatic gradient, which provided opportunities for new species to inhabit climatically novel regions, from which most older species were excluded.

New climatic provinces, containing many novel endemic lineages, appeared as the climate became differentiated. Indeed, whole new suites of such provinces appeared as the American continents separated from Europe and Africa during the Cretaceous (Kurtén, 1967; Valentine, 1969). Both the terrestrial biotas and the shallow marine biotas of the daughter continents became more widely separated as drift continued. Gene flow between populations on different daughter continents, terrestrial and marine alike, must have declined as their separation increased. Allopatric speciation thus ensued. The phylogenetic trees of the marine taxa involved in such diversifications show a multiple branching of species as the lineages invading the new ocean became differentiated and then subdivided further by isolation within new provinces. The detailed patterns depend upon the precise sequences of immigration, climatic differentiation, and geographic separation.

Diversity changes within higher taxa are complicated. In the early history of groups, diversity commonly rises as the new group exploits the available biospace. Subsequently, diversity changes represent a balance sheet of the average success of some lineages in diversifying and the average failure of other lineages to endure. It appears that some ground plans are usually superior under given conditions, and, as lineages of their inferior contemporaries become extinct, are more likely than others to fill the vacant biospace. Thus, through numerous fluctuations in environmental factors of all kinds, and through numerous extinctions and speciations, one higher taxon may gradually assume dominance while another declines, perhaps even to extinction. The decline of the eurypterids, a group of large swimming arthropods that flourished during the mid-Paleozoic, correlates with the expansion of the fishes, suggesting an evolutionary replacement of the latter in the biospace of the former.

Environmental change creates a variety of outcomes. For a large fraction of species change spells extinction. This occurs at some "background" level, except when the change involves some key events—extensive fragmentation or assembly of the continents, the acquisition of some critical geographic pattern that markedly changes the world climatic pattern, changes in the oceanic current pattern, or has some similar pervasive effect. Usually such major events involve a large diversity-dependent effect. During or shortly after key events there are peaks or waves of extinction or diversification. All species eventually suffer extinction or "pseudoextinction." When surplus diversity capacity becomes available, the vacant biospace is filled by immigration or by diversification of local lineages. Local diversification is suggested by the species swarms of stable old lakes, where flocks of closely related, often congeneric species have arisen, partitioning the environment on a rather fine scale (see Brooks, 1950; Kozhov, 1963; Fryer and Iles, 1969). Such situations may be interpreted as a response to an environment that was undersaturated in species but from which most of the appropriate immigrants were barred; species capacities are reached by radiation of a few stocks.

11

Cosmic Evolution And the Origin of Life

The problem of the origin of new life forms has been called the "mystery of mysteries." Surely an even deeper mystery is the origin of the very first living things. There is a rich variety of nonscientific ideas on this problem. Most, if not all cultures have created tales of the origins of life. Many of these involve some kind of omnipotent Being believed to be responsible not merely for the first spark of life but also for the variety of life forms.

The notion of a Creator is nonscientific principally because there is no type of evidence that cannot be explained as the work of a Creator. Falsification of the creation hypothesis, or creationism, is impossible, therefore the whole idea lies beyond the purview of science. However, it is certainly possible to test, at least in principle, hypotheses that life originated as a result of some particular natural events. It is a perfectly legitimate scientific occupation to seek to identify processes and conditions for the origin of life through natural causes.

An increasing number of scientists from a variety of disciplines have become interested in the problem of the origins of life, or biogenesis. Research has been spurred by developments in space exploration, which has brought the question of life on other worlds into the realm of practical interest. In fact, recent scientific activity has resulted in a large array of naturalistic hypotheses of the origins of life, perhaps rivaling those of creationism in number and variety. However, since the naturalistic explanations are amenable to scientific testing, several theories have already been discarded, and the number of working hypotheses will certainly continue to narrow as more information is developed. Eventually we may have a very good idea of the probable history of the origins of life. Although we are far from a final solution today, there are no objections to the assumption that life did indeed arise through natural causes and no insuperable difficulties to testing the idea.

There is no evidence that life is presently arising from nonliving matter. On the contrary, attempts to test the idea of natural "spontaneous generation" of life on today's earth have all failed. Although recorded experiments date from the seventeenth century, the early work was not conclusive. Pasteur's work in the late nineteenth century convinced the scientific community generally that life was arising only from life. One reason why life today is not arising from nonliving matter is that organic materials are quickly taken up as food by the pervasive existing microorganisms, preventing any lengthy processes of nonbiotic evolution. Moreover, the chemical conditions required for the spontaneous generation of life are probably quite different from those prevailing today.

Most current work seeks to determine the conditions under which life could have originated on earth in the past. There is also a possibility that life originated elsewhere, and in some manner found its way to our planet. If this should be so, conditions for the origin of life need not ever have been realized during earth history, and the problem becomes quite different. It is also possible that life has originated independently on several worlds, whether or not it originated on earth. There is thus interest in the "general" conditions of the origin of life as well as in the particular conditions that may have existed on earth. Since we know little or nothing of environmental conditions on any planet outside our solar system, and since no life beyond earth is now known, let us proceed by first examining the current evidence and hypotheses for the origin of life on earth. We shall then consider problems of the possible existence, quality, and quantity of life elsewhere.

The evolution of living organisms consists of changes in gene frequency and organization. Genetic material must have three fundamental properties: it must contain information for the construction and operation of an organism; it must be able to replicate this information and ability in descendants; and it must be able to change if evolution is to occur. As we have seen in Chapters 2 and 3, nucleic acid molecules possess these properties in living cells: information is contained in the genetic code; developmental and metabolic effects are achieved through proteins with structural and enzymatic functions; and replication occurs during meiosis or mitosis. Replicable (heritable) variations in the genetic information are furnished by mutation processes.

It has been suggested that a minimal organism could be simply a naked gene, with the ability to replicate and to "feed" on material in the environment, although other possible minimal organisms are also conceivable. Nucleic acids are not good prospects for naked genes, for they operate only with the assistance of proteins. Indeed, if nucleic acids form the blueprints of life, then proteins are the tools, workers, and building materials. Many proteins are required in order to synthesize a protein, or to replicate nucleic acids. A minimal organism with nucleic acid genes, organized along the lines of living cells, would require about 45 proteins (Morowitz, 1966). Perhaps a minimum primitive genetic system employing nucleic acid genes could be assembled from 30 molecular species (Miller and Orgel, 1974). It is too much to expect that any such constellation of complex interacting molecules appeared by chance to initiate life at one step. We must look for a process that would gradually create such materials and organization from simple beginnings. The original materials must have been produced abiotically. Could this have happened on earth?

THE PRIMITIVE EARTH

It is a tenable hypothesis that the cosmos as we know it originated in a gigantic explosion, which blew matter from a primordial fireball, scattering outwards the material of the universe in an ever-expanding volume. Clouds of dust and gases, chiefly helium and hydrogen, spread through space, attenuated in some regions and dense in others. Stars can be formed from the condensation of such clouds; our own sun probably formed from a rotating mass of dust and gas that was flattened by the spin and which collapsed upon a central region due to the force of gravity. As the gaseous disk condensed inwards, the spin of the central region must have increased in order to conserve momentum. At some point the spin could have become so great that material would be ejected owing to centrifugal force. However, as such a point was approached, and the centrifugal force increased, matter in the outer portions of the condensing disk would tend to be more attracted to local dense patches than to the central region. Thus, at some distances from the proto-sun, matter would be condensed into separate bodies. These would be the planets, which lie in the plane of the sun's rotation. The solar satellites, including the earth and presumably the other planets and asteroids and the earth's moon as well, appear to date from about 4.6 billion years ago. (For a readable account of hypotheses of planetary origins see McLaughlin, 1965.)

The composition of a planet formed in this way should correspond with the composition of the gas and dust in the region from which it condensed, unless the original materials were fractionated during planetary formation. In the universe at large lighter elements vastly predominate, and even in the larger planets of our solar system helium and hydrogen are exceedingly abundant. Earth, however, is quite deficient in many lighter elements. It is a relatively small planet, and presumably its gravitational field is inadequate to retain them, considering its thermal environment. The moon is too small to retain any atmosphere at all.

Elements other than just hydrogen and helium are also relatively deficient in the present earth's atmosphere, and some of these are surprisingly heavy, e.g., neon and argon, with atomic weights of 20 and 40, respectively. These two noble gases do not form compounds with other elements, and thus at the earth's surface they must exist in the atmosphere (or hydrosphere) rather than being locked into solid compounds. Their near absence indicates that some event during the earth's formation (probably heating) caused the escape of quite heavy volatile materials, up to atomic weights of at least 40. Elements such as nitrogen, and compounds such as methane, ammonia, water, and volatile carbon compounds, which are retained in the earthly atmosphere at present, must have been lost as well, for they are lighter than argon. The present atmosphere and oceans do not represent primordial volatile materials, but must have been expelled from the earth's interior after this escape event.

The timing of the appearance of the present atmosphere and hydrosphere in their most primitive states is not certain. Probably the early loss of the two noble gases occurred during condensation of the earth itself, which must have been accompanied by extensive heating. Expulsion of volatiles from the interior of the earth is continuing today, as witnessed by volcanic activity. Nevertheless, it is likely that a considerable portion of the primitive atmosphere and oceans accumulated during the earliest stages following formation of the planet, when heat from radioactive sources raised the temperature of the earth's interior over that at present. This heat source is greatly reduced today, owing to the decay of a sizable fraction of the original planetary supply of radioisotopes.

If we only knew the general composition of volatiles that escaped from the early planetary interior, we could estimate the composition of the early atmosphere and oceans. Unfortunately, our knowledge is very limited. Looking at present volcanic emanations is not too much help, partly because today's volcanic volatiles are largely recycled from the surface, and partly because conditions in the earth's interior must have changed greatly during geological time. We must seek indirect evidence, such as the nature of early geological deposits. These are exceedingly scanty before the first traces of life are found, but they provide useful data that when joined with evidence from other sources permit some generalizations. Table 11-1 depicts the timing of some of the main events in earth history. The age of the earth is estimated partly by determining the time at which lead isotope products of the uranium decay series must have begun accumulating.

The earliest earth rocks now known are about 3.8 billion years old (Moorbath, O'Nions, and Pankhurst, 1973). Few localities with rocks of this great age are known. However, there is evidence that the ages of some rocks have been underestimated due to difficulties in radiometric dating, and subsequent work may reveal other very old rocks (Goldich and Hedge, 1974); if so, this will be an exciting development. Interestingly, some of the earliest rocks include sediments and have compositions that suggest a continental derivation. Presumably a hydrosphere and atmosphere were present. It has been suggested (Cloud, 1974) that life may have followed shortly after the appearance of chemical weathering and subaqueous deposition. The earliest appearance of these processes is still unknown, but appears to antedate 3.8 billion years. The actual earliest appearance of life so far detected is in rocks that are about 3.3 billion years old. There is as yet a very long and poorly known period of time—well over a billion years from the origin of the planet—during which biogenesis occurred.

The sedimentary rocks that remain from 3.8 to about 1.9 billion years ago, which appear to have been deposited chiefly in shallow epicontinental seas, generally suggest an anoxic atmosphere; the minerals are mostly in low states of oxidation relative to today (although more highly oxidized layers do appear locally).

TABLE 11-1. Major events in the history of the earth. [After Cloud, 1974.]

Events in biosphere	Time (billions of years)	Events in planetary environment		
Permian–Triassic extinctions	0.22		(Plate tectonics)	Free oxygen in atmosphere
First well-mineralized skeletons	0.57			
First body fossils	0.70			
First metazoans?	1.0			
Possible early eukaryotes	1.3			
		Cratonal sediment, chiefly oxidized	(Plate tectonics probable)	
Possible early eukaryotes	1.91 2.21	Most banded iron formations		
(Prokaryotes becoming diverse)			(Global tectonics unlike present)	Chiefly anoxic atmosphere, probably reducing
		Cratonal sediment, chiefly unoxidized		
Oldest dated fossils (prokaryotic autotrophs)	3.31	Rocks chiefly granitic and gneissic, sediments extensive		
	3.81	Oldest dated rocks		
		Record not known		
	4.61	Origin of Earth		

The most famous examples are the banded iron formations, which contain vast amounts of ferrous iron. This indicates that free oxygen was not available in the atmosphere; otherwise the iron would be in a ferric state (Cloud, 1968).

That the primitive atmosphere lacked free oxygen is reasonable from other standpoints. Free oxygen is available in quantity from only two sources. One is photodissociation of water vapor, which occurs high in the atmosphere. Hydrogen, freed from water molecules, leaks out into space, while the heavier oxygen is retained. The extent of this process in the past, and the rate at which photodissociated oxygen could accumulate, is much debated. Oxygen is also produced from water via photosynthesis. It is commonly believed that the bulk of free oxygen was produced by plants. Before the advent of life, and indeed until plant biomass became large, this source of oxygen was absent or unimportant.

One final indication that oxygen was not ordinarily present in the early atmosphere, and therefore was not dissolved in early oceans, is that organic compounds are stable in the presence of hydrogen (that is, under reducing conditions) but are quickly destroyed in the presence of oxygen (Haldane, 1933). The compounds required as precursors or building blocks for living systems must have originated and accumulated in a reducing environment. Even though we are fairly sure that the early atmosphere was reducing, we are still not at all sure of its composition, and it is very hard to find evidence to bear upon this question.

PREBIOTIC ORGANIC COMPOUNDS

The abiotic synthesis of organic materials is reasonably commonplace in nature. These abiotic materials have relatively simple, small molecules. Complex organic macromolecules, such as proteins or nucleic acids, are not known to be forming abiotically today. Indeed, even the much simpler amino acids, sugars, and nucleosides that form the building blocks of most large organic molecules are not known to be forming in nature at present except in living systems. However, some of these have long been synthesized artificially from simple organic compounds, which in turn can result from inorganic reactions. Nearly all these building-block molecules have now been successfully synthesized in laboratories under conditions that plausibly approximate those of a primitive earth; such syntheses are called *prebiotic.*

The key experiment in the prebiotic synthesis of organic compounds was performed by Miller in 1953. At that time it was believed that the primitive atmosphere had probably been composed chiefly of water vapor, methane, hydrogen, and ammonia. Such a gas mixture was circulated through a closed apparatus by steam from boiling water, and subjected to an electrical spark discharge between tungsten electrodes (Figure 11-1). The gas mixture was then condensed and added to the boiling water for recirculation; any nonvolatiles that appeared would accumulate in the water. This apparatus was permitted to run for a week. The results were spectacular. Several amino acids were synthesized (alanine, glycine, glutamic acid, and aspartic acid) and a number of other organic compounds appeared; about 15 percent of the carbon in the gas mixture became incorporated into these compounds.

Since this original experiment, a large array of organic compounds has been obtained by numerous workers in a wide range of presumably prebiotic conditions. Sometimes organic molecules are formed in large yields. The most interesting products of prebiotic syntheses include purines, pyrimidines, sugars, and 18 of the 20 amino acids occurring naturally in proteins. Some of these building-block molecules can be used in turn to form polymers under prebiotic conditions. Polypeptide chains have been obtained prebiotically, although nucleosides have proven to be very difficult to synthesize under plausible prebiotic conditions, and nucleotides have not yet been obtained; the origin of

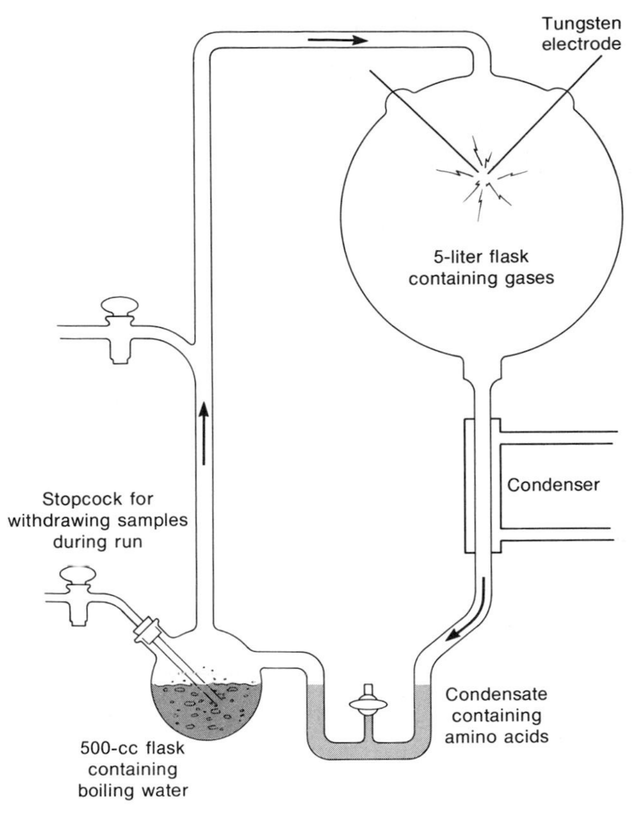

FIGURE 11-1.
Diagram of the apparatus used by Miller in experiments of early prebiotic synthesis. A gas mixture of methane, hydrogen, and ammonia is circulated through the apparatus, together with water vapor, by boiling water in a 500-cc flask. The mixture is subjected to an electrical discharge in a 5-liter flask. The gas is then condensed and the condensate trapped in tubing at the bottom of the apparatus, where nonvolatile constituents accumulate. When the experiment was performed, several amino acids appeared, along with a number of other organic compounds.

these substances remains an important problem in prebiotic chemistry (Miller and Orgel, 1974).

In geological time the amount of organic material that might accumulate from prebiotic syntheses is quite large. By the rates at which electrical discharges occur in the atmosphere today, enough organic material to form a layer three feet deep over the entire surface of the earth could accumulate in only 100,000 years, assuming the efficiency of natural syntheses approximated that of laboratory conditions (Orgel, 1973). Such a high efficiency is very unlikely, but it is clear that abundant amounts of organic material could have been synthesized in a geologically short time. Furthermore, other sources of energy, especially ultraviolet radiation, may well have contributed to prebiotic syntheses. In a primitive anoxic atmosphere ultraviolet radiation at the earth's surface would have been much higher than it is today, for it is largely screened out now by ozone in the upper atmosphere. Still other sources of energy may have been present, such as heat from volcanoes and hot springs. Orgel (1973) has estimated that the prebiotic oceans and lakes may have contained as much as a gram of organic material per liter—about a third the concentration of an average-strength chicken bouillon.

There is therefore no particular difficulty in accounting for a supply of prebiotic organic compounds, some of which are rather complex, some of which are monomers of macromolecules, and some of which are even polymerized macromolecules. One of the surprises arising from experiments in prebiotic synthesis is the ease with which the very compounds most critical as precursors or building blocks of the more important biological molecules can be formed. It is a striking fact that of all the thousands upon thousands of organic compounds that we know, these very ones should be most easily obtainable in plausible prebiotic syntheses. Indeed, we have learned that such organic compounds are by no means restricted to the earth. Relatively simple organic compounds have been detected in interstellar dust clouds by radioastronomical techniques. The first three to be detected are important intermediates in prebiotic syntheses: formaldehyde, which can give rise to sugars; hydrogen cyanide, from which amino acids and adenine can be derived; and cyanoacetylene, from which pyrimidine bases can be formed (Orgel, 1973). Furthermore, specimens of one class of stony meteorite, the carbonaceous chondrites, contain a number of organic compounds, including such amino acids as glycine, valine, and alanine. We do not know whether these compounds were formed via syntheses similar to prebiotic reactions on earth, but their existence indicates that some chemicals that are products of life may in fact occur widely throughout the universe as a result of abiotic reactions.

CONCENTRATION OF PREBIOTIC COMPOUNDS

Given that prebiotic syntheses could create a rich assortment of the compounds required for early life, we are still left with the problem of how these materials were properly concentrated and then organized so as to acquire the properties of living matter. This question is particularly important since polymers, such as proteins and polynucleotides, form spontaneously only from highly concentrated solutions. A number of possible concentration sites have been suggested.

Evaporation in shallow pools or arms of the sea or in lakes may result in markedly increasing the concentration of dissolved substances, leading eventually to their precipitation; salt deposits are forming in this manner at present. In waters rich in prebiotic organic substances—the prebiotic soup—concentration may have occurred by evaporation and even by precipitation of layers of organic substances onto the substrate. When heated, dry amino acid mixtures may polymerize to form somewhat heterogeneous, large polypeptides called protenoids (see Fox and Dose, 1972). It has been argued that periods of concentration of organic material (including amino acids) by evaporation may have alternated with dry periods of thermal synthesis of polypeptides. A difficulty with this model is that the polymerizations occur at temperatures well above the boiling point of water. Temperatures of this sort are uncommon in nature at the earth's surface, although they may occur locally in association with volcanic activity. Some investigators have therefore postulated that volcanic heat played a critical role in the origins of life. However, polypeptides quickly hydrolyze in warm solutions. A very special sequence of events must thus be postulated to concentrate materials, form polymers by thermal synthesis, preserve them, and make them available for further prebiotic developments.

Freezing is another possible concentration mechanism. It operates today in concentrating sea salts, for as freezing water is locked up in ice the remaining solution becomes increasingly more concentrated. Organic materials in a prebiotic soup could reach saturation in this manner and subsequently be precipitated from solution. In a very cold environment, where freezing is common, large pools of cold water rich in organic matter might be common. The low temperatures would retard the rate of decomposition of organic molecules, and perhaps prebiotic syntheses could go forward in such a site.

Another possible concentration mechanism for prebiotic materials is the formation of colloidal particles of organic materials; rather than dissolving, the particles separate from water as minute droplets in suspension. Russian workers have been particularly interested in colloidal droplets as possible sites of prebiotic reactions. A. I. Oparin (1953) has called organic colloidal droplets *coacervates*. Such droplets can dissolve or absorb many organic compounds; their interiors provide nonaqueous environments wherein important prebiotic syntheses can be imagined. Unfortunately, data are still lacking on the sorts of prebiosynthetic pathways that might be possible within coacervates.

Mineral particles may have played important roles in biogenesis. Minerals are naturally occurring substances of definite internal structures and chemical compositions that are constant or that vary within set ranges. They may have been involved in the origins of life in several ways. Most minerals occur as crystals, but natural crystal lattices are structurally imperfect, for they contain dislocations and other mechanical imperfections and may include ions that do not "fit" the ideal mineral structure.

Some ions that substitute for each other within crystal structures carry different electrical charges, which can give rise to excess positive or negative charges on mineral surfaces. A charge imbalance can also arise when some sites within a crystal are not occupied by any ions at all. Because of the net charges, and of surface irregularities due to dislocations, many crystals are rather efficient catalysts. Additionally, they readily adsorb a wide variety of organic materials, including polymers, onto their surfaces. Indeed, certain clay minerals adsorb polymers more easily than monomers, so that if polymers were being created from monomers in solution, they might be preferentially adsorbed onto a mineral surface and preserved from immediate decomposition, and thus concentrated. Peptide chains have been formed experimentally in this manner on the common clay mineral montmorillonite (Paecht-Horowitz, Berger, and Katchalsky, 1970). Layered silicate minerals, which form a high proportion of marine and lacustrine substrates, could have provided sites of concentration of complex molecules. Minerals with particular surface configurations and charge patterns may even have provided templates for certain classes of polymers. Although we cannot as yet evaluate the relative historical importance of various concentration mechanisms to the origin of life, a number of processes can be postulated which would have effected concentration. In principle, the problem of concentration of prebiotic materials does not pose insuperable difficulties to a hypothesis of natural biogenesis.

THE ORIGIN OF NATURAL SELECTION

The next questions seem more interesting and are more difficult. The only plausible way to generate a complex gene–enzyme system is to select for it; random "chance" processes are not effective unless favorable materials or trends are preserved. To illustrate this point, consider the following example adapted from Cairns-Smith (1971).

The workings of chance are sometimes illustrated by imagining monkeys seated at typewriters randomly producing sequences of letters and symbols. Given a monkey with an infinite life span and patience and a typewriter of infinite durability, any particular sequence should eventually be produced—that of *Hamlet*, or the Bible, or a textbook on evolution. However, the time required to produce with a reasonable likelihood any of these works, complete from beginning to end, exceeds by billions upon billions of times the known duration of the universe. Even works much shorter than *Hamlet* would require an impossibly long time to be produced by a monkey. Take for instance the first sentence of *Origin of Species:*

> When on board HMS Beagle, as naturalist, I was much struck with certain facts in the distribution of the inhabitants of South America, and in the geological relations of the present to the past inhabitants of that continent.

Given a 30-symbol typewriter with a key randomly struck every second, this sentence would be typed out (disregarding spaces and capitals) in something over 10^{180} years.

Let us make the job somewhat easier on the monkey. We shall require only that he produce each word in the sentence. After he produces the first word, "when," we shall select it from his typescript, and he need not produce it again; the next word we require is "on." After each word is successfully produced the monkey may proceed to the next word. The words are thus produced by a random process but preserved by selection. Under these rules the monkey would produce the entire sentence in only about 18,159,335,148 years—an improvement of 170 orders of magnitude, although still several times longer than the age of the earth. If we require only that the monkey produce the first letter, "w," which is then selected from the typescript, and then the following letter, "h," which is again selected, and so on, we find that the sentence will be completed after only 1 hour 33 minutes and 30 seconds. Again, the constituent units (in this case letters) are produced randomly but preserved by selection.

The process of selection is the key to speeding up the production of the sentence from 10^{180} years to about an hour and a half. If the whole sentence is produced by only a random process, it takes more or less forever. If selection can act upon randomly produced words, it takes much less time, and if selection can act upon randomly produced individual letters, it takes very little time. When randomness is completely eliminated and the entire sentence is selected by Darwin, speed of production is limited only by typing skill.

For practical purposes, the probability of creating any particular average protein from a prebiotic supply of amino acids by random processes is as remote as the probability of a monkey typing our sentence; it would require more time than is available even at the highest imaginable rate of protein synthesis. Selection must have been involved. Now, natural selection implies that the selectee—molecule or organism—is fitter than alternatives. In other words, for natural selection to be effective the selectee must be more successful in leaving offspring with qualities that promote the fitness. The selectee must be able not only to reproduce itself but to replicate or quasireplicate the desirable properties.

Since both proteins and nucleic acids (plus other simpler substances) are required to develop and reproduce organisms living today, an obvious question is which of these substances arose first. No clear answer is available. To date, protein-like materials have proven easiest to synthesize prebiotically; perhaps they arose first. This implies that there was some sort of replicating genetic material associated with early proteins upon which selection could act. This replicating material could have been a protein itself. We cannot say that this is impossible, but no self-replicating proteins are known. Nucleic acids seem difficult to form, but they do replicate; perhaps the earliest gene was a simple nucleic acid. The trouble with nucleic acids is that they don't do anything themselves. They seem to have no interesting qualities for selection to explore—except perhaps replication efficiency—until they become associated with protein synthesis. It is the proteins that can perform the actual operations involved in the metabolic processes of life and proto-life.

Thus there is a paradox. Both nucleic acids and proteins are required to function before selection can act at present, and yet the origin of this association is too improbable to have occurred without selection. There can be no question that the elegant and precise molecular mechanisms associated with transcription and, especially, translation (Chapter 2) have evolved from much simpler precursor mechanisms. But all living organisms employ the same intricate transcription mechanics, and the identical genetic code (Table 2-1). There are no remaining traces of any primitive precursor genetic systems; we therefore have no direct evidence as to how natural selection originated.

Failing any direct evidence, we can at least attempt to construct models in order to test the hypothesis that evolution of the genetic system is plausible or at least possible. One model of the primitive genetic system postulates that the first genes were minerals, minute crystals associated with proteins as concentration sites and to some extent as templates (Cairns-Smith, 1971). Certain minerals, e.g., clays or other layered silicates, were in very large supply as products of early rock weathering. They might have favored the formation of classes of polymers for which they provided good templates. Selection would favor a polymer whose activities preserved or even increased the supply of its mineral template, perhaps by helping to weather a parent mineral to produce the clay "gene." The association of two (and eventually many) different polymers in mutually helpful aggregates would be a second step; some polymers might have aided in mineral formation, while others provided adhesion for the polymer colony, and still others cleaved prebiotic organic compounds to provide building block materials for the colony, and so forth.

The polymers in any such early colonies would have been proteins, but eventually a nucleic acid was added to a colony, presumably for some property other than protein-coding or even self-replication. Perhaps its original function was mechanical, acting as a fiber or in some other structural capacity. This would be a very early example of preadaptation. The ability of the nucleic acid to code for proteins that were themselves useful to the colony eventually became manifest. Once the nucleic acid was contributing to colony fitness, any process that would replicate it *in situ* would be heavily favored. Thus, a nucleic acid (probably a simple RNA at first) would replace minerals as genetic material. Presumably, the nucleic acid gradually took over the task of coding for all the proteins employed in the colony, including those associated with its own replication. Eventually DNA evolved (presumably coded by RNA originally) and an essentially modern nucleic acid—protein system of gene—enzyme was present. At some point membranes were developed around the colonies and those membrane-bound units became cells.

Cairns-Smith's scenario is highly speculative. We do not know that life followed such a protoadaptive pathway from prebiotic materials to living cells. We have presented this particular speculation because it illustrates one way in which the origin of selection in a genetic system can be accounted for. Other scenarios may be proposed, but all are similarly speculative at this time.

THE NATURE OF PROTOCELLS

Genetic systems that are transitional between the most primitive systems and the present elaborate ones have been suggested (Woese, 1967). Perhaps some simple early nucleic acid genes evolved into tRNA (Chapter 2). The early code may have been much less specific than today's, with a codon specifying a class of a few similar amino acids. Translation would have been less accurate and somewhat ambiguous, and the proteins resulting from translation would not all be precisely alike but would belong to classes with similar amino acid sequences. Assuming a fairly high correlation between composition and function, proto-proteins within a given class would have nearly all functioned in somewhat similar fashions, but would not have achieved the high functional specificity of modern proteins. The evolution of the genetic code into its present form may have been influenced by some original relationship between primitive nucleic acid molecules and amino acids, perhaps a stereochemical one, which does not exist in modern systems. On the other hand, it may have arisen by chance insofar as its particular order is concerned. In either event the genetic code is now "frozen," protected against further change, since any mutational change in the code would affect all the proteins in the individual mutant; the chance of such a change producing a viable organism would be effectively zero.

There were probably fewer codons in the early genetic system, and indeed at some stage the code may have specified an amino acid (or class thereof) by using only the first two of the nucleotide triplets of the codon. It can be seen from Table 2-1 (p. 28) that in this even either phenylalanine or leucine would be specified by UU—, the third position containing any nucleotide base; several other of the first pairs of nucleotides would specify more than one amino acid even today. Part of the reason why this stage is suspected during the evolution of the code is that it is chiefly in the third position that the code is degenerate. If this notion is correct, then selection led in time to increasingly specific functions, subdividing and refining nucleic acid specification by the incorporation of the third nucleotide into the recognition process of the codon. This sort of refinement need not have involved lethal mutations, since it must be assumed that the generalized proteins of the time were already functioning with either of the amino acids previously specified by the first two nucleotides of the codon.

The evolution of prokaryotic and unicellular eukaryotic forms is considered in the next chapter; all these forms share a common genetic code, of course. Precursor forms with primitive nucleic acid genes and primitive codes must have existed, and they were undoubtedly living beings—we shall call them protocells. Can we infer anything about them? Certainly their properties must have varied through time, and there may have been quite diverse arrays of protocells at times.

We can infer that protocells were much less capable of internal regulation, and poor in homeostatic mechanisms when compared with prokaryotes. They must also have been exceptionally generalized in function. Their rather nonspecific proteins must have reacted on a variety of substrates, and protocells probably utilized wide varieties of food items. Presumably they and their simpler ancestors derived materials and energy from abiotically synthesized organic matter—a continuation of the prebiotic soup from which they had sprung. Thus, the first organisms were heterotrophs. Although reproducing by protocell division with replication of their primitive nucleic acid genes, daughter cells would not be closely similar to each other because the primitive translation process was ambiguous. Rather, daughter and sister protocells would bear generalized "family" resemblances. Mutation rates may have been much higher on the primitive earth with its high surface levels of ultraviolet radiation. Mutations would have been much more likely to produce a viable protocell than a viable prokaryote, since the genetic codes were imprecise (Stebbins, 1969). Insofar as their possession of great intrapopulation variability is concerned, protocells would have resembled sexually reproducing eukaryotes more than prokaryotic clones, although of course they would not have mendelized.

Selection upon such a variety of phenotypes would have evolutionary consequences only when the underlying genotypes were different with respect to the functions under selection. Thus, selection would favor both fit gene products and those gene mechanisms that tended to reproduce the fit gene products more faithfully. As translation was improved, selection for more specialized functions became possible and the protocells probably became more specialized by habitat (which would lower the variety of conditions with which a given genotype would have to cope) and by energy source. At some point one or more lineages developed the ability to utilize solar energy, reducing the requirements for organic materials, and eventually leading to the complex processes of photosynthesis that freed plants from reliance on organic compounds.

Photosynthesis doubtless evolved in small steps, beginning with some simple precursor. There may even have been a switch in the biochemistry of photosynthesis. Granick has suggested that magnetite crystals containing sulfur impurities might have acted as mineral catalysts in an early form of photosynthesis (Stebbins, 1969). They might have employed ultraviolet light to dissociate water molecules; the energy available from this process could then be used for synthesis of organic molecules. We have no idea just when in the history of the protocell the use of radiant energy to promote biochemical reactions first developed. Since the prebiotic soup should have been used up as protocells flourished, some writers describe the evolution of photosynthesis as a race against time. Others suggest that photosynthesis may have developed very early indeed, so the energy sources for life were never especially at hazard (Stebbins, 1969).

The prebiotic soup must have given sustenance to lineages of heterotrophs; these would have been reduced to feeding on biotic detritus, or alternatively would have advanced to feeding on photosynthetic protocells, when the soup course was finished. Alas, the primitive heterotrophs (and primitive autotrophs, for that matter) are all extinct. Modern organisms are traceable only to a common ancestral lineage with a modern transcription system and genetic code. This ancestor must have appeared well after the prebiotic soup had been consumed and photosynthesis had evolved. Therefore, the ancestor was itself an autotroph, or else photosynthesis evolved all over again, a most unlikely possibility. The only protocells that did not become extinct without issue are those in the direct ancestral pathway of the last common ancestor of modern organisms, and they have been transformed beyond all hope of reconstruction. All

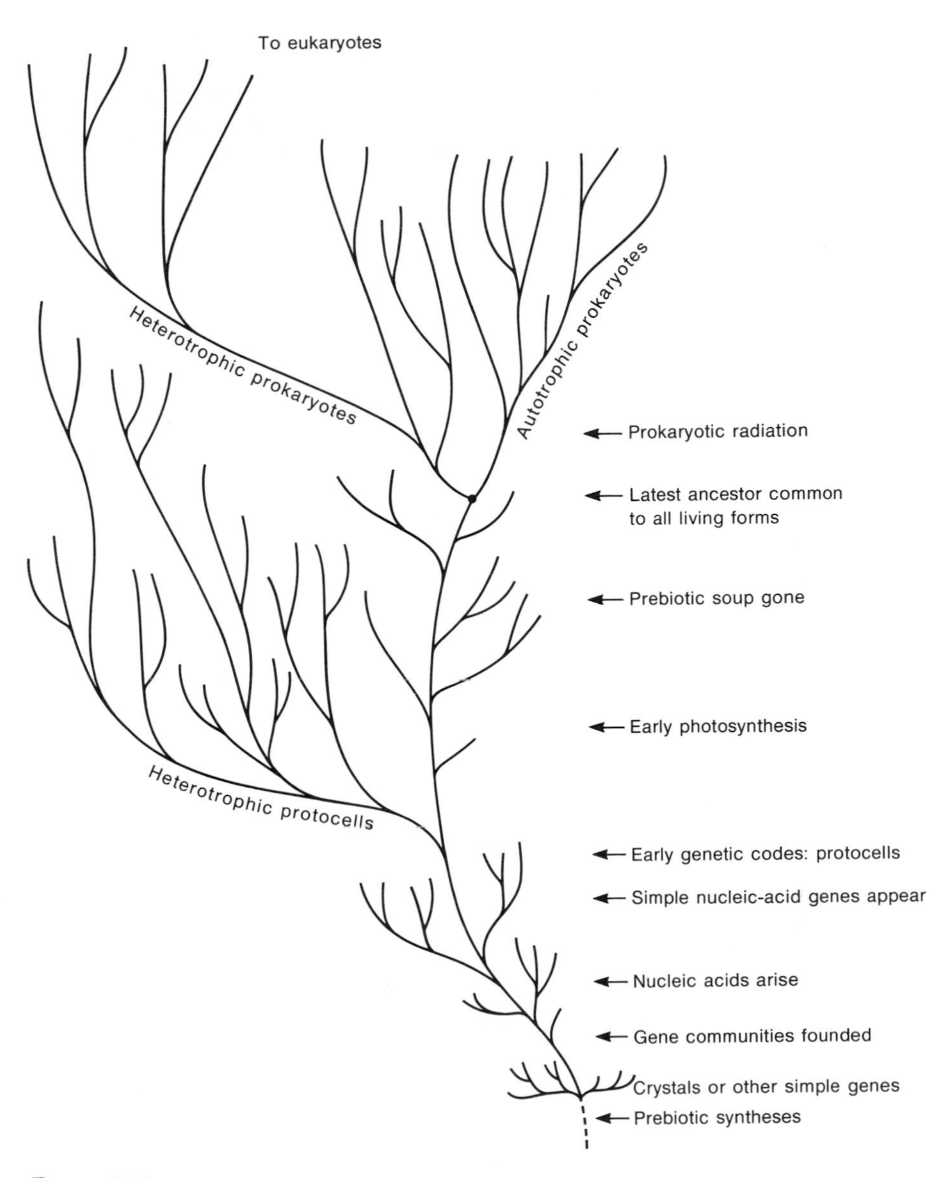

FIGURE 11-2.
Some of the important events during the rise of living systems and their evolution into protocells, represented on a phylogenetic tree. The trunk consists at first of simple gene and gene–protein associations, which can be regarded as heterotrophic, and which evolved into heterotrophic protocells. Photosynthesis developed among some of these lineages (represented on the right), which thus became autotrophic. Within the autotrophic protocells the genetic code was refined and translation developed until it attained its modern form. One of the more modernistic autotrophic protocell species became the common ancestor of all living organisms, diversifying into heterotrophic as well as autotrophic prokaryotes and leading to eukaryote descendants. [After Cairns-Smith, 1971.]

modern heterotrophs, including the entire animal kingdom, are therefore descended from an autotrophic ancestor. Some of the inferred key events during the origin of life and the development of protocells, and the general relations of protocells to prokaryotes and eukaryotes, are depicted in Figure 11-2.

EXTRATERRESTRIAL LIFE

That life could have originated on earth as a product of prebiotic chemical processes is, as we have seen, strongly suggested by experimental results and calculations, and by geological evidence on primitive earth conditions. The simple organic compounds that are most essential as precursors of more complex substances turn out to be among the most common products of plausible prebiotic syntheses. The larger molecules that developed from these intermediates evidently became associated in chemical systems (that included simpler substances as well) that eventually acquired the properties of life. These living systems displayed properties (emergent properties) that could not have been predicted from a knowledge of the prebiotic systems alone—a common phenomenon when higher levels of organization are formed out of lower.

This all seems natural enough. We suppose that the prebiotic processes utilized the materials at hand to produce complex systems and eventually life. Suppose that we could turn the clock back to prebiotic times and then permit primitive earth processes to take their course again; would life again develop, and would it be the same? We do not really know, but we can make reasonable guesses. From the experiments on prebiotic syntheses we would expect that amino acids, proteins, sugars, and many other important classes of molecules would appear, and we would have to suppose that life itself would have at least an average chance of developing. However, the chances that precisely the same substances would appear again are vanishingly small. The chances of developing even one of the same proteins by random processes are essentially nil considering the time available. It is, however, conceivable that selection, while improving the functioning of small, primitive, generalized proteins, would discover a given efficient protein more than once; the number of efficient proteins must be much smaller than the number of inefficient ones (Cairns-Smith, 1971). We cannot specify the probabilities of this occurrence. Nevertheless, it seems extremely unlikely that whole suites of similar proteins would appear in separate biogenetic experiments. The second time around, life would have a different biochemistry, although it might still function with enzymes and genes according to similar principles. We might repeat this experiment many times; we would expect life to be a common development, but never to be identical biochemically. Whether nucleic acid would be likely to appear is not at all certain; if it did not, alternate efficient genes would have to be discovered.

Assuming that some successful prebiotic experiments took place, and supposing that they ran for a few billion years, there is still little chance that any resulting life forms would closely resemble the present ones. The pathway that has led from the earliest living organism to even the most primitive organisms now alive has been long and crooked, with many tricky stretches where key innovations must have occurred. Photosynthesis had to evolve, for example, to provide an adequate energy source for future evolution. Translation is another case in point. Some mechanism that does the sort of job that translation accomplishes must develop in order for heredity to be accurate. Many other key events must occur during the evolution of a eukaryotic sort of cell (Chapter 12). It is by no means a foregone conclusion that any given key innovation will eventually develop, even over billions of years. It is easy to imagine life remaining at the protocell level in many of our experiments.

Nevertheless, one evolutionary trend might appear in some experiments: the trend towards increasing complexity in newly evolved types. It has certainly been a striking trend on earth. It occurs as modifications are required in an existing, co-adapted form–function complex. Prevailing adaptations would be destroyed if the organism were stripped down for redesign; instead, the organism is modified chiefly by elaboration of existing features or invention of novelties while maintaining functional integrity. This tends to produce ever more complex and elaborate organisms. Furthermore, modifications that increase the independence of organisms from their environments should be favored, so that as adaptation proceeds under changing conditions, homeostatic improvements may occur. Selection should take any opportunity that arises to improve the lineage in general ways, as well as in the special ways that are adaptive to new conditions. Increased homeostasis involves increased self-regulation, which, in keeping with our previous point, should be achieved chiefly through increased complexity. The trend to complexity is preserved by the information-storage system in DNA (Stebbins, 1969).

Thus, if multicellular organisms do appear, they may well evidence an increasing complexity through time. However, we could hardly expect men to appear in any of our experiments, any more than we could expect a given species of butterfly. It is likely that complex animals have as many as 100,000 gene loci (see Chapter 3); numerous alleles can and do exist for many loci. The possible gene combinations for 100,000 genes with 10 alleles each is $10^{100,000}$. The number of *possible* structural genes is far, far higher, as we have seen. The number of ways in which a suite of 100,000 genes can be organized via a regulatory network is also extremely great, though impossible to calculate in a general way. Even for a single lineage the potential pathways for evolution seem enormous in number. When we realize that for practical purposes the assortment of genes and their recombination owing to segregation and crossing-over create different outcomes every time, then the results of evolutionary processes can be seen to be unpredictable in detail by their very nature.

We may predict that evolution will nearly always lead towards adaptation, but we cannot predict the adaptive pathways that a lineage will happen to follow. The possibilities, while not literally limitless, are many orders of magnitude greater than the number of stars or planets. Even in experiments with complex animals we would not expect to obtain phylum-level ground plans that much resembled, say, arthropods or chordates. Their presence and dominance in many adaptive zones has depended upon the appearance of long chains of morphological solutions to problems in adaptation. Each link in these chains has probabilistic components.

In summary, no one can calculate how often life would arise in separate experiments, nor how often it would achieve a complex form. It is unlikely to be very often. We can, however, be fairly certain that it would not resemble present life very closely on the molecular level, and that the life forms would have somewhat distinctive architectures, even if the physical history of the planet were the same each time. Each species today is unique, and each species in alternate worlds would be unique also. Monarch butterflies would never again appear; neither would man. Intelligence of the sort that can create technological societies is unique to man out of many millions of species in earth history. It is obviously not possible to calculate the chances that it would ever rise again, but they are very small indeed (Dobzhansky, 1972).

Granting that the possibility of obtaining a man-like creature is vanishingly small even given an astronomical number of attempts, and that the universe must be characterized, in Simpson's phrase, by the nonprevalence of humanoids, there is still some small possibility that another intelligent species has arisen, one that is capable of achieving a technological civilization.

It is an interesting contrast that, while evolutionists have generally been pessimistic about the chances of encountering intelligent extraterrestrial life, other scientists have been more enthusiastic. Efforts have been mounted to communicate with extraterrestrial civilizations, and a burgeoning literature has developed, stimulated in part by the development of space exploration. The study of extraterrestrial life, exobiology, is a curious field of science, since its subject matter has never been observed and may not exist (see Simpson, 1964). Nevertheless, it holds great intrinsic interest for many, and exobiological questions can be framed that in principle are amenable to scientific test.

Discussions of the possible presence and extent of advanced technical civilizations beyond the earth are often based on a formulation by Frank Drake, which evaluates the number of such civilizations (N) in our galaxy in the following way:

$$N = R f_p n_e f_l f_i f_c L,$$

where R is the average rate of formation of stars in the galaxy, f_p the fraction of stars with planets, and n_e the average number of planets (per star with planets) with environments suitable for the development of life. Thus, the first three factors in Drake's expression estimate the extent of locales that are suitable for biogenesis. Astronomical data exist to permit the estimation of R. It is probable that the average rate of star formation has been about 10 stars per year. It is lower than this at present, but may have been higher when the galaxy was young (Shklovskii and Sagan, 1966).

There are essentially no observational data on the number of stars with planets, but astronomers generally agree that it is a very large number. There are some theoretical grounds to suggest that all stars on the main sequence of stellar evolution develop planets as a matter of course. As the number of stars in our galaxy is on the order of 10^{11}, this implies a vast number of planets. How many of these possible planets are capable of supporting life? Estimates vary widely and complications abound. For example, since the origin of the galaxy, the amount of heavy elements should have increased significantly owing to nuclear reactions in the interior of stars. The first generation of solar systems formed early in galactic history may have been deficient in heavy elements relative to our system. This should very much lower the chances of there being appropriate locales for the development of life in such systems, at least of life in any form of which we have an inkling. We can sum up estimates of the number of planets with environments suitable for biogenesis as usually lying between 10^{-2} (Shapley, 1958) and 10^{10} (Shklovskii and Sagan, 1966) per galaxy, a phenomenal range.

For the next factor (f_l), the fraction of suitable planetary locales on which life actually developed, we have only the earth experience to guide us; at this writing we are not sure whether Mars has life, or whether it is suitable even if it doesn't. Since life appears to have originated within a relatively short time on earth, once suitable conditions developed, then we are inclined towards a value not much smaller than one for f_l. However, the degree of confidence in this guess is low; the value of f_l could be very small.

The last three parameters of Drake's expression are concerned with the number of advanced technical civilizations in the galaxy, given that there are planets on which life has arisen: f_i is the fraction of planets with life on which beings with intelligence and manipulative ability arise; f_c is the fraction of these planets on which advanced technical civilizations appear; and L is the mean lifetime of such a civilization. Certainly all stars should not provide equal opportunities for the development of intelligent life. Large stars remain on the main sequence for less time than small stars; the planets of the larger ones that stay less than 10^8 years can probably be eliminated as suitable for advanced civilizations; there just is not enough time. On the other hand, small stars may not be luminous enough to support advanced life. Suitably luminous and long-lived stars on the main sequence make up about 10 percent of all stars. Nevertheless, the chances of intelligence arising are likely to be so small that the number of suitable planets would be insufficient to expect any other technical civilization ever being developed. If this is so, the exercise ends at this point.

These speculations are interesting up to a point, but we are a long way from being able to subject them to convincing tests. Perhaps the reader would like to make his own evaluation of the likely number of advanced technical civilizations that are "out there" somewhere. Drake's expression may serve as a good starting point. Miller and Orgel (1974) have produced a form in which readers may insert their own guesses as to the values of the various parameters involved and total up their estimates, which are as likely to be correct as the estimates of exobiologists.

In a somewhat different vein, there has been speculation that not only does extraterrestrial life abound, but that life on earth has in fact originated elsewhere. An early hypothesis of this sort was proposed by Svante Arrhenius, who in 1908 suggested that simple, minute life forms were carried high into their home atmospheres and that some escaped into space, perhaps by electrostatic ejection. Drifting through space under the influence of radiation pressure, these life forms might in time innoculate planet after planet, each planet in turn serving as a new source, and thus they might come to pervade parts of the galaxy. These hypothetical life forms were called *panspermia*. It seems possible that successful escape from or colonization of planets could be effected by panspermia. However, it is now known that radiation in space is so great that organisms small enough to be plausible panspermia could not be appropriately shielded and would be killed long before accomplishing an interstellar voyage (Shklovskii and Sagan, 1966). Other natural sources of extraterrestrial life might include meteorites, as suggested by Kelvin. The possibility of such relatively large objects arriving upon earth from beyond the solar system is so small that it has probably never happened.

If microorganisms cannot get to earth from other solar systems by natural means, perhaps they have been sent here (Crick and Orgel, 1973). They could have been placed in spaceships that were appropriately shielded and directed toward likely planets. *Directed panspermia* may have been motivated by a desire to seed planets to insure that the galaxy contained inhabited worlds other than the source world. Or microorganisms may not have been sent, but brought by advanced space travellers. Gold (1960) speculates that modern earthly life has evolved from microorganisms left behind as pollution by untidy visitors—the so-called picnic (or garbage) hypothesis. Ball (1973) suggests that the reason we have not been contacted by highly developed extraterrestrial civilizations, which he supposes to be present, is that they are deliberately avoiding us. Perhaps to them we are a zoo, or even a laboratory. Presumably terrestrial life could have been seeded or could have originated naturally in either of these cases. Since we are quite sure that our present terrestrial life did evolve from microorganisms on earth, since it all has a similar biochemistry and there is fossil evidence of the general course of evolutionary events (Table 11-1), it is at least certain that advanced stages of extraterrestrial life, if they were ever delivered to earth, have not survived or remained.

12

Evolution of Prokaryotes And Unicellular Eukaryotes

The outcome of evolution is the hierarchical organization of life described in general terms in Chapter 8. The purpose of this and the following chapters is to show how the processes that have been recognized and analyzed at the subspecific and specific levels of evolution, and discussed in Chapters 2 through 7, can account for the origin of this entire hierarchy.

THE KINGDOMS OF ORGANISMS

The classical division of the world of life into two kingdoms, animals and plants, has such a low informative and instructional value as to be almost useless for our purpose. At the level of unicellular organisms the distinction between animals and plants is based upon a single characteristic, the presence or absence of photosynthesis, and so is highly artificial. Much more instructive are the various modern schemes that recognize several kingdoms, and use several different characteristics for distinguishing them. That of Whittaker (1969), which recognizes five kingdoms (Figure 12-1), is most satisfactory. These kingdoms represent three grades of advancement: prokaryote (Monera), unicellular eukaryote (Protista), and multicellular or coenocytic eukaryote. The first two grades are each represented by a single kingdom, but at the multicellular level three kingdoms are recognized: plants, animals, and fungi. The reasons for doing this are discussed later in this chapter.

First, however, the historical reason for revising the biologist's conception of kingdoms must be reviewed. It results directly from the greatly increased knowledge of microorganisms acquired by biologists during the past century.

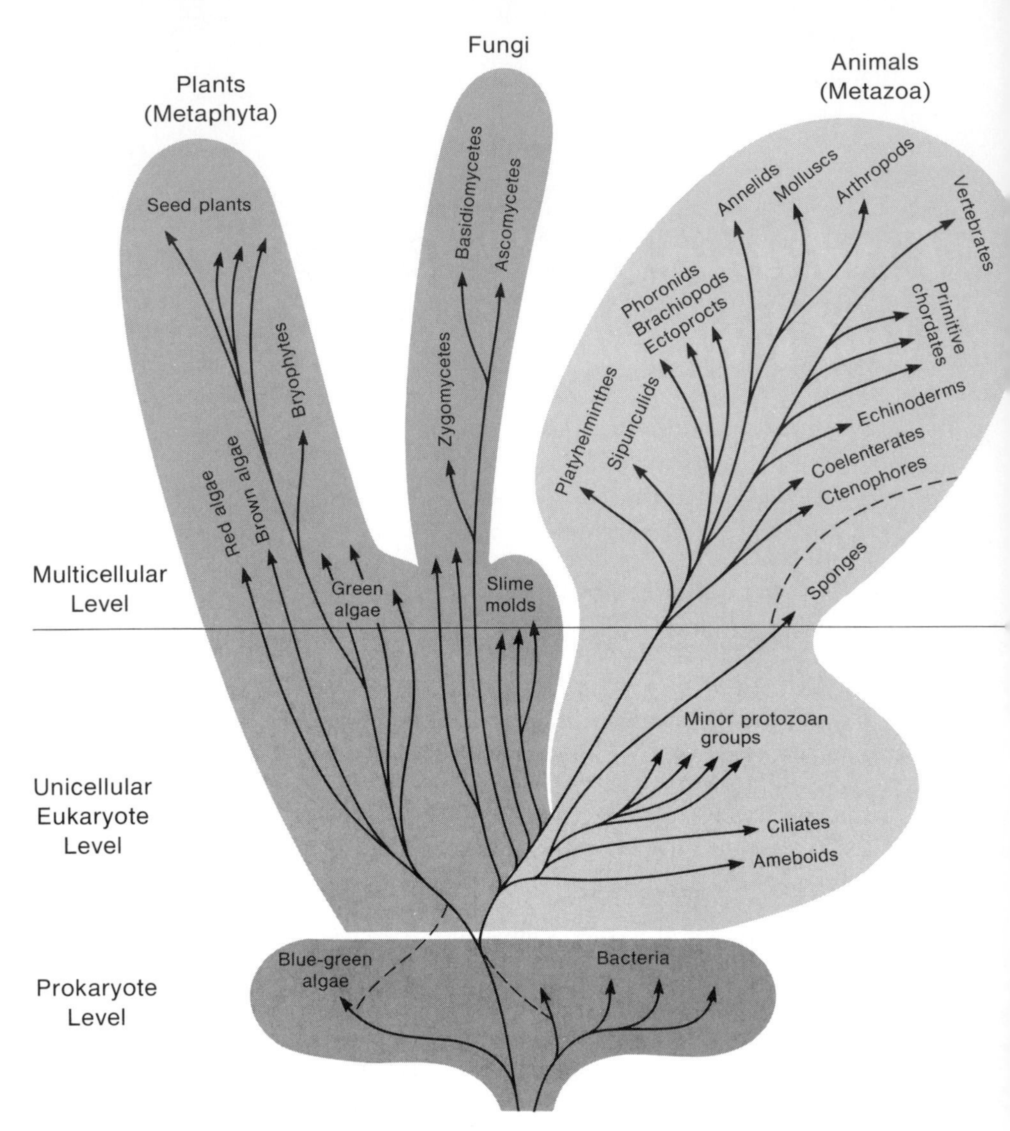

FIGURE 12-1.
Phylogenetic chart of living organisms, based upon the concept of five kingdoms. Three major grades of advancement are recognized: prokaryote, unicellular eukaryote, and multicellular. Adaptive radiation or cladogenesis has occurred at each level. At the prokaryote level, two phyla are recognized, bacteria and blue-green algae. At the unicellular eukaryote level, extensive cladogenesis has led to the formation of many different classes and orders, but no distinctive phyla. From some of these classes several phylogenetically distinct groups of multicellular organisms have arisen, only five of which have advanced to the grade that includes tissue differentiation and elaboration of distinctive form at the visible or macroorganism level. Three of these are basically photosynthetic: red algae, brown algae, and the green algal–archegoniate–seed plant line. Because they resemble each other in being autotrophic (with progressive expansion of surface area) and, in the case of most aquatic forms, having similar reproductive propagules, these three groups are placed in the plant kingdom. One line, the fungi,

From about 1870 to 1940 the relationships between different groups of unicellular eukaryotes or protista were intensively investigated and the modern discipline of protozoology was founded. These investigations confused rather than clarified the problem of the relationships between different kinds of microorganisms. In spite of the presence of chlorophyll and the ability to perform photosynthesis, many newly discovered unicellular organisms were recognized by zoologists as animals and classified as Protozoa because of their motility, their ability to ingest food particles, and their resemblance to other colorless Protozoa. These same organisms, of which the best known example is *Euglena*, were classified by botanists as plants, because of their photosynthetic ability. Each systematist could present logical arguments to support his claim. In this way a sort of "no man's land" of organisms arose, containing species and genera that were classified in one way, with one set of terms, by zoologists, and in an entirely different way by botanists. This confusion has persisted to the present day, as is evident when one compares modern textbooks of zoology, such as that of Villee, Walker, and Barnes (1973), with a corresponding textbook of botany, such as that of Raven and Curtis (1970), or at a higher level of sophistication, the *Traité de Zoologie* (Grassé, 1952) with the first volume of the *Traité de Botanique* (Chadefaud, 1960).

During the same period the position and relationships of the bacteria became equally confused as a result of two conflicting tendencies. Since they do not ingest food and are very different from Protozoa, bacteria have never been recognized by zoologists as animals. They are, however, sufficiently plant-like to have attracted the attention of botanists, who have placed them in the plant kingdom as a separate division or phylum in most textbooks and general treatises. At the same time, a study of bacteria has become an essential part of medical science, and a large school of medical bacteriologists have devised their own methods of investigating bacterial relationships. In general, they have had little communication with botanists, even with mycologists, interested in bacteria. The resulting cross currents and lack of communication have led to a confusing situation (see Ainsworth and Sneath, 1962).

The problem was clarified and progress was made toward its solution by two important developments in the middle of this century: the rise of molecular biology and the invention of the electron microscope. Modern research has revealed a previously unsuspected cleft between two major groups of organisms, one that cuts right across the preexisting classification of animal and plant kingdoms. With respect to their biochemistry, genetics, and cellular and nuclear fine structures, organisms can be divided into two sharply distinct groups, *eukaryotes* and *prokaryotes* (Stanier, Doudoroff, and Adelberg, 1970). The prokaryotes

---

consists of heterotrophs that absorb rather than ingest food, and thus share with plants expansion of surface and a sessile mode of life. The third line includes the multicellular animals or Metazoa, which are heterotrophic ingestors, most of which remain compact in bodily form and develop internal rather than external membranes for absorption. [Based on Whittaker, 1969.]

include the bacteria and blue-green algae; all other cellular organisms are eukaryotes. Table 12-1 summarizes the principal differences between these two groups. On the basis of biochemical differences, McLaughlin and Dayhoff (1970) have concluded that the evolutionary distance between prokaryotes and eukaryotes is twice that between animals, plants, and fungi.

The distinctive characteristics of prokaryotes are the absence of important structural features and reproductive cycles, such as mitosis, meiosis, and syngamy, which are present in most or all eukaryotes. That this absence in prokaryotes is a primary condition rather than a secondary loss is evident from the Precambrian fossil record, which shows that prokaryotic cells appeared

TABLE 12-1. Principal differences between prokaryotes and eukaryotes.

	Prokaryotes	Eukaryotes
Nucleus		
Nuclear membrane	No	Yes
Mitotic spindle	No	Yes (no)*
Meiotic reduction of chromosome number	No	Yes (no)
Recombination by union of entire gametes	No	Yes (no)
Chromosomes		
Number of different nonhomologous chromosomes in basic genome	1	2–600+
Histones and acidic proteins complexed with DNA	No	Yes (no)
Regular condensation and relaxation of chromosomes during mitotic cycle	No	Yes (no)
Cytoplasm		
Mitochondria	No	Yes
Chloroplasts	No	Yes or no
Metabolic enzymes in cellular membrane	Yes	No
Cilia or flagella with 9-plus-2 fiber arrangement	No	Yes or no
Endoplasmic reticulum	No	Yes
Vacuoles	No	Yes (no)
Golgi apparatus	No	Yes or no
Phagocytosis, pinocytosis	No	Yes or no
Amoeboid movement, cytoplasmic streaming	No	Yes or no
Molecular synthesis		
Nature of cytoplasmic ribosomes	70S	80S
Nature of organellar ribosomes	No	70S

*Information in parentheses applies to a small group of mesokaryotic forms described later in this chapter.

long before eukaryotes (Barghoorn, 1971; Schopf, 1975). Consequently, evolutionists are now faced with a major problem: How did the first major ascent in evolutionary grade, the origin of eukaryotes from prokaryotes, take place? This problem is discussed later in this chapter.

The separation of prokaryotes as a kingdom distinct from eukaryotes does not solve the problem of the distinction between plants and animals. Whittaker's solution is to recognize at the level of kingdoms a second advance in grade, that from unicellular to multicellular or coenocytic eukaryotes. This procedure localizes all the previous artificial separations between autotrophic and heterotrophic organisms either at the level of unicellular eukaryotes (Protista) or of prokaryotes (Monera), i.e., bacteria vs. blue-green algae. The difference at these levels between autotrophy and heterotrophy is no more significant than other cytological, physiological, and biochemical differences, and in most instances is useful chiefly for distinguishing between orders, families, or even genera within a family. The logic of recognizing Protista as a single and separate kingdom becomes clear when the distribution of autotrophy vs. heterotrophy among them is compared with that of other characteristics, such as the chemistry of the cell wall, the position and fine structure of flagellae, and the organization of the mitotic apparatus. For instance, with respect to these other characteristics the facultatively autotrophic *Euglena* resembles various heterotrophic flagellates and is completely different from the autotrophic flagellate *Chlamydomonas*. Calling *Euglena* a plant and placing it with *Chlamydomonas* rather than with the flagellates it resembles (in characteristics other than autotrophy) is a highly artificial procedure. Another course adopted by some botanists is to classify *Chlamydomonas* with plants because it resembles them in having a cellulose wall, and exclude *Euglena* because the chemical structure of its cell wall is completely different. This procedure creates other problems. Red and brown algae, which are plant-like in many respects, including photosynthetic ability, nevertheless have walls that do not contain cellulose or that differ considerably from those of typical green plants in other respects.

Unicellular eukaryotes, therefore, are here recognized as a single, probably monophyletic kingdom, the Protista, and kingdoms that are parallel branches or clades of major dimensions are recognized only at the grade of multicellular or coenocytic organisms. This procedure raises two additional problems: where to draw the boundaries between Protista and the kingdoms of multicellular organisms, and how many kingdoms to recognize at the multicellular or coenocytic grade. The first problem does not arise with respect to animals, since the gap between Protista and the animal kingdom, which according to the present treatment is synonymous with the Metazoa, is so great that the origin of Metazoa is still obscure (Wells, 1968). On the other hand, the different phyla of algae and fungi are connected to various groups of unicellular forms through a series of intermediates, which may be colonial cell aggregates, small coenocytes, or have low degrees of cellular differentiation. One might justly ask, therefore, why reject the artificial distinction between autotrophs and heterotrophs at the level of Protista, while erecting more or less artificial boundaries between unicellular and multicellular autotrophs?

Our first answer is that the world of life is not divided into a few sharply separated kingdoms, which can be easily delimited. If plants are to be separated from ciliate Protista, which are surely much more like animals than plants, artificial boundaries must be established somewhere. Second, the advantage of recognizing kingdoms at the somewhat indistinct boundary between the unicellular and the multicellular or coenocytic grade is admittedly heuristic. It calls attention to the great importance of this advance in grade for the evolution of life, and invites studies directed at learning how this advance came about.

Once the policy is adopted of recognizing several parallel kingdoms at the multicellular grade, the question arises: How many should be recognized? This question is not easy to answer. If one adopts the rule that each kingdom should be monophyletic, including only a single advance to the multicellular grade, two difficulties arise. First, the number of these advances is relatively large; an estimate of 16 is given later in this chapter. Second, their actual number is largely a matter of inference based upon comparisons between modern forms, since multicellular microorganisms lack a significant fossil record. Clearly the phylogenetic approach, because of its uncertainties, is a weak base upon which to found categories as broad as kingdoms.

The approach adopted here separates three kingdoms of multicellular or coenocytic organisms on the basis of fundamental differences in their mode of nutrition. Plants are autotrophic, except for a few saprophytes or parasites among seed plants, which are obviously of secondary origin. Fungi are heterotrophic and, like bacteria, absorb their food in a predigested form. Animals are heterotrophic ingesters. All three of these kingdoms are probably polyphyletic. Among animals, sponges have probably had a different origin from the remaining phyla (Chapter 13; and see Wells, 1968). Among fungi, the water molds, or oömycetes, probably arose from heterotrophic flagellates independently of terrestrial fungi (Whittaker, 1969). The plant kingdom is probably polyphyletic even if, following Margulis (1970), the brown and red algae are excluded from it and only green algae plus land plants are included. There are good reasons for believing that the coenocytic green algae were derived from Protista independently of the cellular forms (Preston, 1968; Fott, 1974). Even the filamentous green algae now appear to have had at least three separate origins from unicellular forms: the Conjugales, Ulotrichales, and *Klebsormidium* group, which are the only green algae that form typical cross walls from a cell plate (Pickett-Heaps and Marchant, 1972). Once polyphylesis is admitted, the addition of the brown and red algae to the plant kingdom as separate and independent phyla becomes reasonable because of the numerous parallelisms between their morphological and ecological evolution and that of green algae. In our view, kingdoms of multicellular organisms are aggregates of phyla that, considered together, have a heuristic value because they represent similar ways of achieving advances in grade.

The principal heuristic value inherent in this system is as follows. First, the origin of multicellular organisms having differentiation or division of labor among the parts of their bodies involved two major ascents in grade. The first was the evolution of the eukaryotic cell. Without the division of labor among the organelles and membranes evolved in this kind of cell, cellular differentiation such as exists in animals and plants could not evolve. Some of the essential characteristics of the eukaryotic cell evolved only once. Many writers have pointed out the similarity between all eukaryotic cells with respect to: cilia and flagella, when present; ribosomes and other protein-synthesizing machinery of nuclear origin (Table 12-1); and, with few exceptions, the mitotic apparatus and molecular organization of the chromatin of their chromosomes.

The second advance in grade, the origin of multicellular or multinucleate differentiated organisms, took place many times in various evolutionary lines, and followed the first advance only after tens or hundreds of millions of years had elapsed and extensive adaptive radiation had taken place at the unicellular level (see Table 11-1, p. 353). The initial evolution of the eukaryotic type of cellular organization was a necessary, but not sufficient, condition for the origin of multicellularity and differentiation.

A second kind of informational content inherent in this system is the recognition of the three principal types of nutrition: autotrophy, heterotrophy with absorption of soluble food, and heterotrophy with ingestion of solid particles and internal digestion.

The earliest true cells were probably heterotrophic, and surrounded by firm cell walls (Barghoorn, 1971). They excreted enzymes into the surrounding medium and absorbed soluble food that had been externally digested. This form of nutrition is predominant in prokaryotes, since only the relatively advanced and, for prokaryotes, structurally complex blue-green algae plus a few groups of photosynthetic and sulfur bacteria have evolved autotrophy. On the other hand, all three forms of nutrition exist in unicellular eukaryotes or Protista. Not infrequently two different methods, such as autotrophy or ingestion, as well as ingestion or absorption, can be carried out by the same cell. At the unicellular level differentiation with respect to these three basic methods of nutrition does not require extensive differentiation with respect to structure, physiology, or gene content.

The situation is completely different in multicellular organisms. Both autotrophy and heterotrophy with absorption require extensive contact between cell surfaces and the external medium, whether it be water, air, or the cellular environment of a parasite's host. In a multicellular organism this can be achieved only by extending its external surface area more or less proportionately with its size. Large autotrophs and absorbers are, by their nature, sedentary, so that compactness of bodily form is not an advantage to them. Furthermore, because of their sedentary nature they can attain wide distributions and can colonize new habitats only with the aid of propagules: cells or multicellular structures especially adapted for dispersal. This explains the origin of spores, motile gametes, and, in the more advanced forms, seeds. Many sedentary ingesters have evolved similar propagules. The situation in heterotrophic ingesters is completely different. Since they find their food by active motion, they must remain relatively compact and symmetrical, no matter what their size. As size increases, each organ increases the surface area of its internal membranes or cellular layers relative to its size (Rensch, 1947). If they are predators, efficiency of nutrition usually depends upon activity and keenness of the senses, hence the successive evolution of more efficient sense organs and locomotor apparatus. If they are small and nonpredatory, their compactness makes them the logical prey of predators. Increasing efficiency of sense organs and locomotor apparatus is adaptive for most multicellular ingesters, i.e., animals, the chief exceptions being those that acquire, secondarily, armored coverings or noxious, unpalatable characteristics. At the same time, the sense organs and locomotor apparatus that are vital for nutrition and escape can be modified with relative ease to serve as aids in sexual union. Hence, the distinction between adaptations for survival and for reproduction is far less marked in animals than in plants.

At the multicellular level the distinction between plants and animals is sharp and clear. Moreover, fungi are almost as distinct from multicellular plants as are animals. They are, like bacteria, heterotrophic absorbers, so that with respect to their ecology and many of their biochemical properties the smaller fungi are more like bacteria than higher plants. In addition, the chitinous cell walls of most fungi are completely different from those of plants, and certain genetic characteristics, such as the heterokaryotic condition in Ascomycetes and the dikaryotic condition in Basidiomycetes, are also distinctive. The recognition of fungi as a kingdom separate from animals and plants, and of mycology as a discipline separate from botany, has both a phylogenetic basis and a justification in terms of its informational content.

Some groups of organisms do not fit into a system of five kingdoms based upon three grades. The best known of these are the slime molds, both cellular (Acrasiales) and plasmodial (Myxomycetes), and the Volvocales. The slime molds, during their stages of growth and metabolism, retain the properties of heterotrophic, ingesting unicellular Protista. Nevertheless, they have evolved the ability to form propagules, differentiated cells that enable them to reproduce in a manner essentially like that of fungi, by means of spores. The most highly evolved Volvocales, particularly the genus *Volvox*, have many properties of colonial autotrophic Protista, but their mature colonies have a polarized differentiation, and carry out a process of "gastrulation" not unlike that of animals (Raven and Curtis, 1970).

Two ways of handling these anomalous groups seem reasonable. One way would be to stretch somewhat the definition of Protista, and treat the anomalous groups as unusual members of that kingdom. The other would be to erect little "principalities" for them, and thus to exclude them from all of the five kingdoms. The latter procedure has greater heuristic value, since it calls attention to the fact that multicellularity and differentiation have evolved many times, in response to immediate adaptive advantages in particular environments, and that only some of these multicellular organisms have undergone extensive further evolution.

ADAPTIVE RADIATION OF PROKARYOTES

The earliest known cellular organisms, bacteria of the African strata known as Fig Tree Chert, are about 3.3 billion years old (Barghoorn, 1971). The first appearance of unicellular eukaryotes was "earlier than 1,300 million years ago and possibly earlier than 1,700 million years ago" (Schopf *et al.*, 1973). Consequently, for more than 1,500 million years the only living organisms on the earth were prokaryotes (see Table 11-1, p. 353). Since the oldest fossils that could be regarded as multicellular or multinucleate organisms are not more than 700 million years old (see below), the period when evolution was exclusively at the prokaryote level was twice as long as that encompassing the entire evolution of multicellular eukaryotes. Since the evolution of prokaryotes is largely biochemical rather than structural, it remains, to a large extent, a closed book. Nevertheless, on the basis of reasonably reliable information about the nature of the earth during the era of prokaryote evolution, some inferences about the nature of prokaryote evolution can be made.

The first important point is that when life first arose the oxygen content of the atmosphere was much lower than in more recent times, so that aerobic metabolism, if it existed at all, had to be at a much lower rate than in modern aerobes (Chapter 11; Cloud, 1968a). The greater part of the oxygen in the modern atmosphere was and is produced by the activity of photosynthesizing autotrophs. The origin of a truly oxidizing atmosphere took place, probably, not much more than a billion years ago. The greater part of prokaryote evolution, therefore, was based upon anaerobic or weakly aerobic metabolism.

The most significant single event of this evolution was the origin of photosynthesis. It took place relatively soon after the origin of cellular life. The fossil record of blue-green algae is instructive in this connection. Their modern forms have molecules of chlorophyll that are identical with those of higher plants, and even the carotenoid pigment associated with their chlorophyll resembles that found in eukaryotes (Raven, 1970). Although some of them can carry out anaerobic metabolism, most of them are aerobic, and perform photosynthesis with H_2O as the hydrogen donor. Hence, by the time blue-green algae appeared, photosynthesis similar to that of higher plants either existed or was represented by a similar process. The oldest fossils of blue-green algae are calcareous deposits known as stromatolites, similar to the mineral layers that surround colonies of marine forms in some parts of the modern world, particularly Australia (Figure 12-2). The oldest stromatolites are in the Buluwaya formation of South Africa, 2.7 billion years old (Barghoorn, 1971). They are 500 million years younger than the oldest known bacteria, but 1.4 billion years older than the earliest eukaryotes.

The complex, sophisticated kind of photosynthesis carried out by blue-green algae must have been preceded by a whole series of more simple, primitive methods of autotrophy. The only methods known to us are those of three groups of photosynthetic bacteria: the green bacteria *(Chlorobium, Pelodictyon)*,

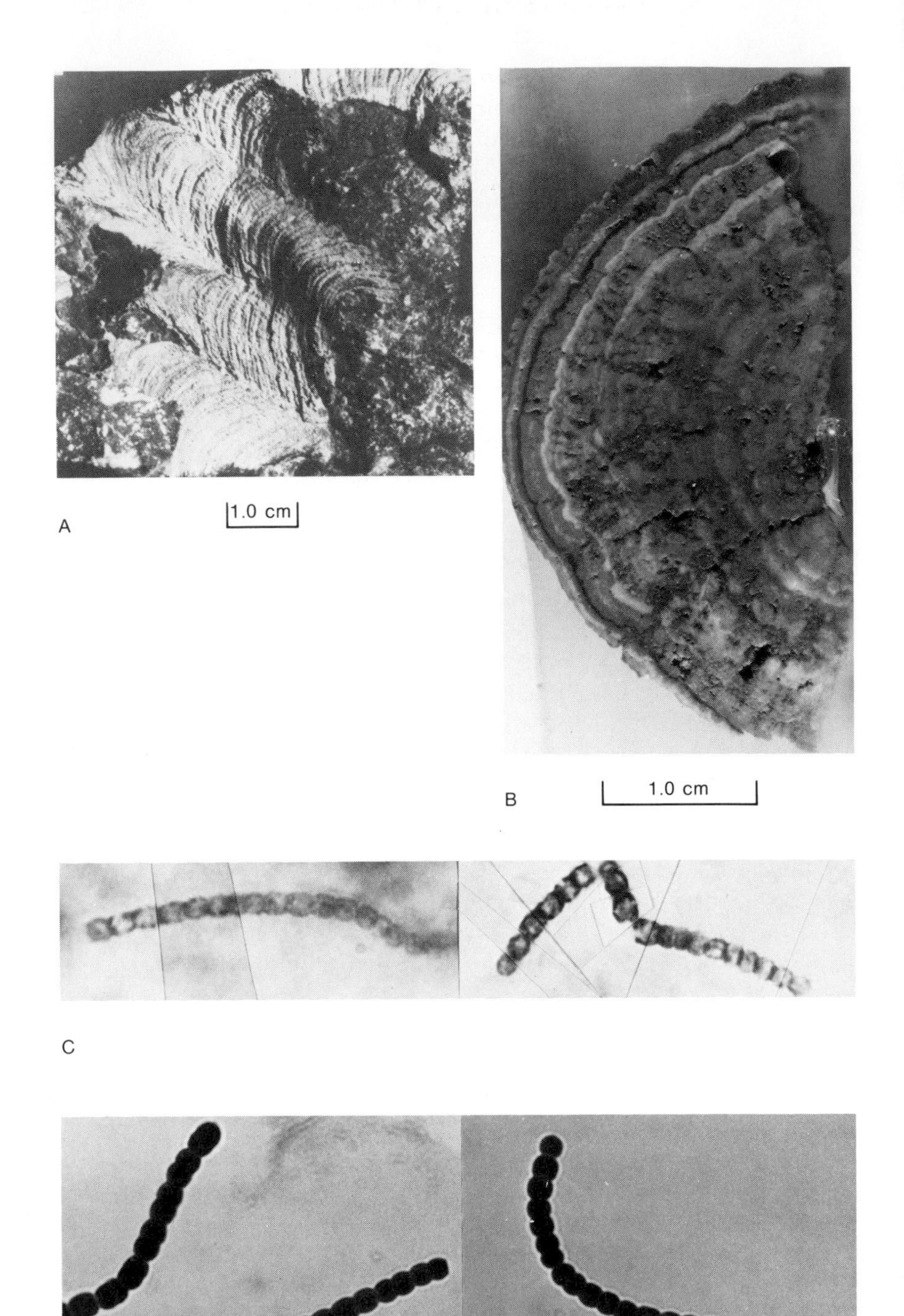

A
1.0 cm
B
1.0 cm
C
D
0
10
20
μm

purple sulfur bacteria *(Thiospirillum, Lamprocystis),* and purple nonsulfur bacteria *(Rhodospirillum, Rhodopseudomonas, Rhodomicrobium)* (Stanier, Doudoroff, and Adelberg, 1970). All these forms have chlorophyll molecules that resemble those of other photosynthesizers with respect to the tetrapyrrole porphyrin ring surrounding a central atom of magnesium, but differ with respect to certain side chains. None of the forms has a strictly aerobic kind of photosynthesis characteristic of most blue-green algae and all higher plants. The green bacteria and purple sulfur bacteria are strictly anaerobic, and require sulfide, thiosulfate, or perhaps hydrogen as an electron donor. The purple nonsulfur bacteria carry out photosynthesis anaerobically, but use organic compounds as the principal electron donors, and can become aerobic and heterotrophic in the presence of oxygen and low illumination. Slow, inefficient methods of photosynthesis like that found in these bacteria must have preceded aerobic photosynthesis of the modern type. Although fossil evidence may never be adequate to solve the problem, the assumption that forms similar to photosynthetic bacteria evolved prior to blue-green algae is quite reasonable.

The total range of adaptive radiations carried out by bacteria before the evolution of eukaryotes may never be known (Sneath, 1974). The ecology of the extinct species cannot be inferred, and one cannot expect to find among contemporary bacteria representatives of all groups that must have existed. If the relative diversity of modern forms is any kind of a guide, one must conclude that this radiation was much more restricted in its earlier stages, when oxygen was limited and only anaerobic forms could exist, than at later periods, when the principal radiants must have been aerobic. The contemporary anaerobic free-living bacteria are, besides the photosynthetic bacteria, principally of three groups (Stanier, Doudoroff, and Adelberg, 1970). The methane bacteria inhabit the bottoms of ponds or quiet streams, where they obtain energy by the anaerobic oxidation of molecular hydrogen. The *Clostridium* group is characterized by the ability to form highly resistant spores, and live in moist soil. Finally, there is a large series of relatively little-known anaerobic bacteria living in the mud at the bottom of the sea, chiefly at depths of 1,000 to 2,000 meters (Kriss, 1963). In addition, some pathogenic bacteria, including members of the *Clostridium* group, as well as some of the bacteria that inhabit the mouth and digestive tract of animals, are obligate anaerobes. These latter groups, obviously, are of relatively recent origin. In addition, many groups of bacteria, such as the lactic-acid bacteria, are facultative anaerobes, which are able to utilize

---

FIGURE 12-2 (opposite).
Precambrian fossils of blue-green algae and their modern counterparts. **A.** Stromatolite, formed by a colonial alga, from the Bulawaya Formation of Rhodesia, about 3,200 million years old. **B.** Recent stromatolite from Lake Stephanie, Ethiopia, about 5,000 years old. **C.** Filaments of *Nostoc* from the Bitter Springs Chert, Australia, about 900 million years old. **D.** Living filaments of the blue-green alga *Nostoc.* [Photos courtesy of J. W. Schopf.]

oxygen for respiration in an aerobic atmosphere but which, in its absence, can derive energy from anaerobic metabolism.

The extent to which aerobic metabolism enabled bacteria to undergo additional adaptive radiation even before the evolution of eukaryotes may be estimated by the wide range of ecological niches in the contemporary world inhabited by aerobic prokaryotes. The blue-green algae have already been mentioned; their range of habitats extends from hot springs, at temperatures of 73° to 75°C, to the icy wastes of Antarctica. They are also much more tolerant of polluted water than are eukaryotic photosynthesizers. Derivatives of the blue-green algae that have reverted to heterotrophy are the filamentous gliders, such as *Beggiatoa* and *Thiothrix,* which live in sulfur springs. Other strictly aerobic free-living bacteria are the nitrogen-fixing soil bacteria (*Azotobacter, Beijerinckia*), the colonial actinomycetes, the highly salt-tolerant *Halobacterium,* and the free-living members of the group of spirochaetes. These organisms are remarkable for the long flagella-like structures contained in their cell wall, which enable them to carry out their characteristically rapid spiral motility. Judging from the diversity of these modern forms, the advent of oxygen in the earth's atmosphere greatly expanded the possibilities for adaptive radiation and evolution at the prokaryote level.

The advent of an oxygen-rich atmosphere required many readjustments of cellular metabolism. One of them must have consisted of ways by which cells could overcome the toxic effect of oxygen on obligate anaerobes. Of the many evolutionary strategies for acquiring this adaptation, one was particularly striking: the origin of bioluminescence in bacteria, and presumably in their protistan descendants. McElroy and Seliger (1962) have produced a good argument to support their hypothesis that the biochemical reactions responsible for bioluminescence or fluorescence evolved from the vestiges of mechanisms that in the earliest aerobic organisms reduced molecular oxygen directly and quickly, thus preventing it from exerting toxic effects, which oxygen has when it accumulates in the cells that are normally anaerobic. The cells that produce luminescence produce a compound, luciferin aldehyde, which reacts with the potentially toxic peroxide produced by free oxygen to form the corresponding organic acid and water. The organism responsible for the phosphorescent glow so often seen in seawater, the dinoflagellate *Noctiluca,* may have acquired its property via an evolutionary descent from a luminous prokaryote.

EVOLUTION IN VIRUSES

For many years viruses were not regarded as organisms at all, so that their evolution need not have been considered in connection with that of cellular organisms. Now, however, molecular biologists generally accept the hypothesis that viruses have been derived from detached portions of cellular organisms (Joklik, 1974). They contain typical nucleic acids, either DNA or RNA or both (Chapter 2), and with the aid of enzymes provided by the host synthesize proteins, just like cellular organisms.

Three further points established by Joklik (1974) in his review of virus evolution are significant to evolution in general. First, groups of viruses differ so radically from each other with respect to both the cross affinities of their nucleic acids and the immunological relationships of their proteins that they almost certainly have been derived separately from cellular organisms. Second, the members of the same viral group, with similar nucleic acids and proteins, include parasites of very different organisms. Among the closely knit group of Reoviridae, for instance, some attack mammals, others birds, while still others have plants and/or insects as hosts. In most of these groups with heterogeneous host affinities, insects appear to be the common denominators, so that their role as vectors may have been highly significant in the origin of the group.

Third, some groups of viruses show clear evidence of evolutionary lineages. This is evident partly from their diversity of form (Figures 12-3 and 12-4).

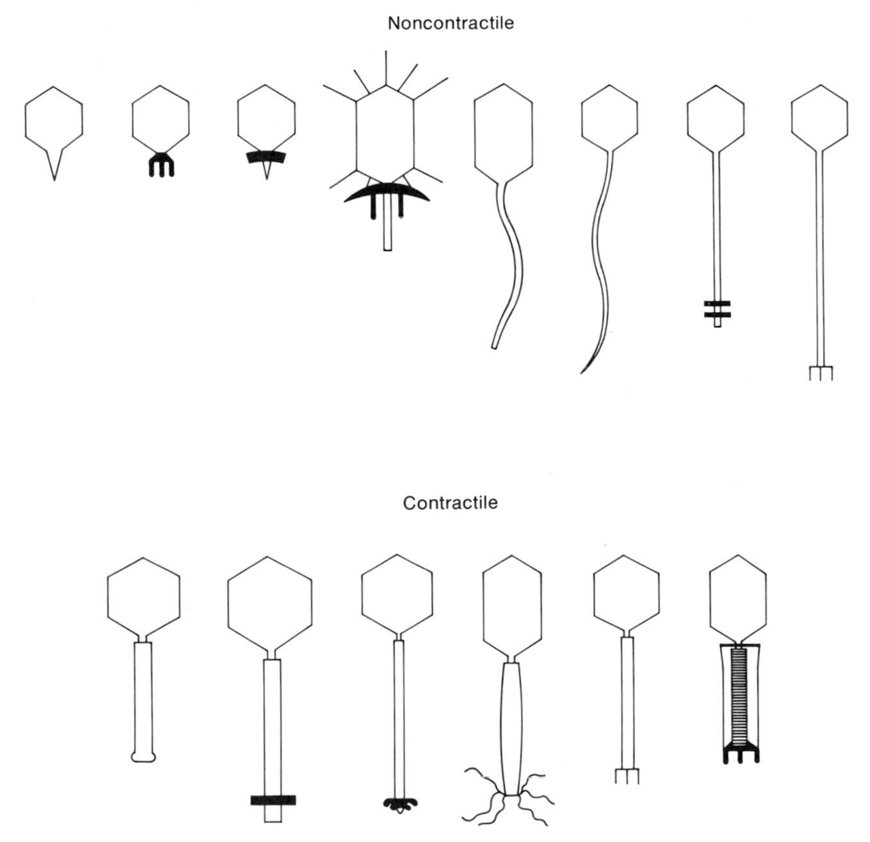

FIGURE 12-3.
Comparative morphology of bacteriophage viruses, showing various degrees of complexity. Among the simplest are the noncontractile viruses having short wedge-shaped tails, such as bacteriophage *T3* of *Escherichia coli* (top left). The most complex is the contractile phage *T4* (lower right; see Figure 12-4 for an electron micrograph). Among viruses having intermediate complexity, one of the best known is lambda (top row, third from right). [After Bradley, 1971.]

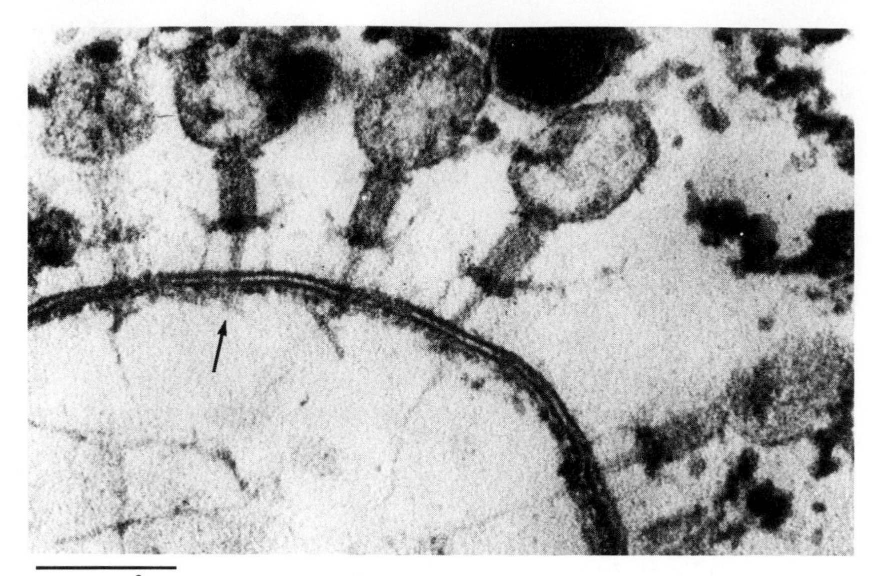

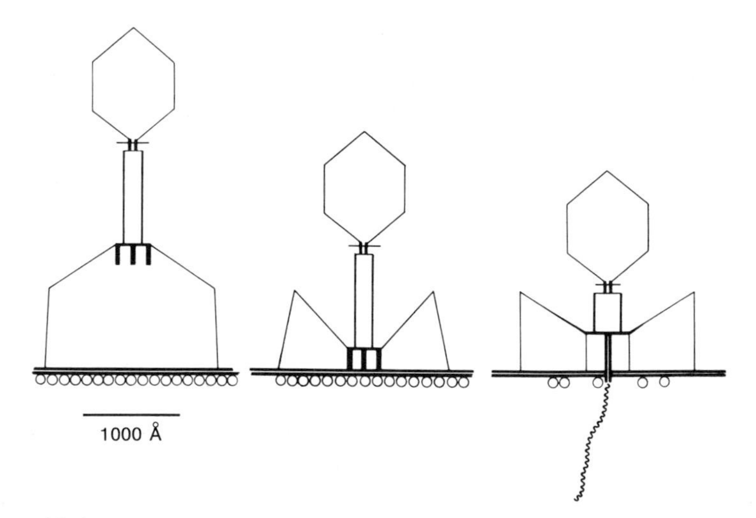

FIGURE 12-4.
Above. Virions of bacteriophage *T4* adsorbed on a cell wall of the bacterium *Escherichia coli* and in the process of injecting their DNA into the cell. Electron micrograph of a thin section magnified 186,000 times. *Below.* Schematic representation of the process illustrated in the photograph above. [Courtesy of T. F. Anderson.]

Particularly important in this connection are the capsid-forming bacteriophages. The most highly evolved of these, such as the well-known *T4* phage, have elaborate tails and tail fibers that act like a hypodermic syringe in injecting DNA into the host cell (see Figure 12-4). In different phages of this group a whole series of intermediate kinds of tail mechanisms have been recognized. These viruses have clearly undergone extensive evolution of the conventional sort after originating as detached portions of cells.

THE ORIGIN OF EUKARYOTES

The first advance in grade after the origin of cellular life was that of the eukaryotic cell. Its origin can be regarded as the most significant single advance of organic evolution, since until cells had acquired the complexity and the degree of division of labor among membranes and organelles that are associated with the eukaryotic condition, multicellular organisms, having differentiated cells and tissues, could not evolve.

There is general agreement among biologists that the basic organization of eukaryotic cells evolved only once. This is because eukaryotic cells possess certain complicated organelles that are strikingly similar to each other in their supramolecular structure. The most conspicuous of these are the mitochondria, the plastids of autotrophic eukaryotes, and flagella or cilia having the 9 + 2 organization of fibrils. Among eukaryotes lacking these organelles, only the red algae can be postulated with reasonable probability as never having had such flagella. A second similarity is the mitotic apparatus in nearly all eukaryotes, including red algae. Biochemical characteristics common to eukaryotes are the Embden–Meyerhof pathway of glucose metabolism, the universal presence of Krebs-cycle oxidations, and particularly the protein-synthesizing machinery. The ribosomes of eukaryotes are larger than those of prokaryotes, and their activity is insensitive to chloramphenicol, which inhibits protein synthesis in prokaryotes, while it is sensitive to cyclohexamide, to which prokaryotic cells are resistant.

Not unexpectedly, many biologists have speculated about the origin of the eukaryotic cell, a cardinal event of evolution. The most complete hypothesis is that of Margulis (1970), who has marshalled a great wealth of morphological, biochemical, and paleontological facts that, she believes, support the hypothesis that the cellular organelles of eukaryotes, even including flagella, centrioles, and spindle fibers, are derived from symbiotic prokaryotes that originally existed in a host cell and later became modified in various ways (Figure 12-5A). This hypothesis has been supported and somewhat amplified by Raven (1970), as well as by Schnepf and Brown (1971) and Taylor (1974), but has been severely attacked by others, particularly Alsopp (1969), Raff and Mahler (1972), and Bogorad (1975).

Raff and Mahler postulate that mitochondria originated from invaginations of the inner cellular membrane, which first formed respiratory vesicles bound by membranes. These vesicles became converted into mitochondria by the

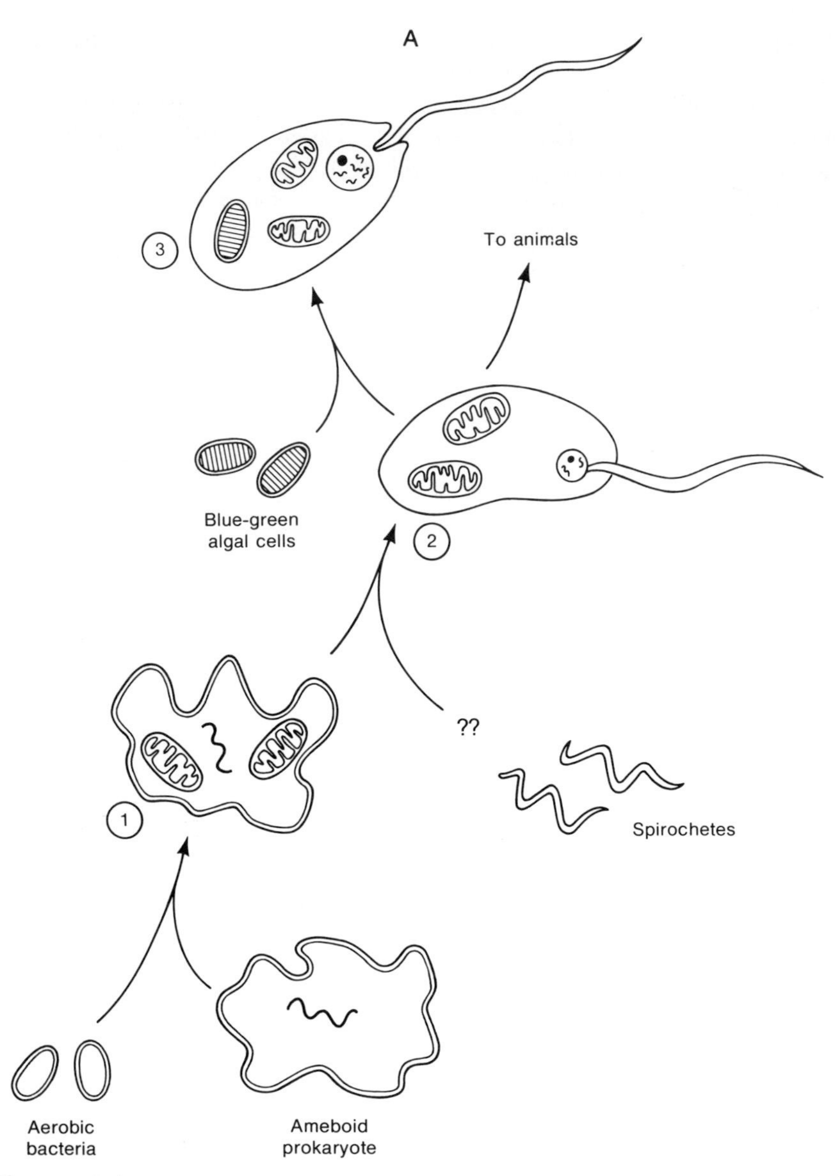

FIGURE 12-5.
The origin of the eukaryotic cell.

A. The *symbiosis theory* of the origin of the eukaryotic cell. *A-1.* A prokaryotic cell having a flexible cell wall, capable of ingesting bacterial cells. *A-2.* An intermediate cell, in which ingested bacteria are evolving into mitochondria. According to the theory, spirochetes became ingested and converted into the nuclear fibrils, centrioles, and flagella of a primitive unicellular eukaryote (shown immediately above the spirochetes). *A-3.* Blue-green algal cells became incorporated into the colorless unicellular eukaryote to form a simple green flagellate (top). Animals were derived from more highly evolved descendants of a colorless unicellular eukaryote, or flagellated protozoan. [After Marguilis, 1970.]

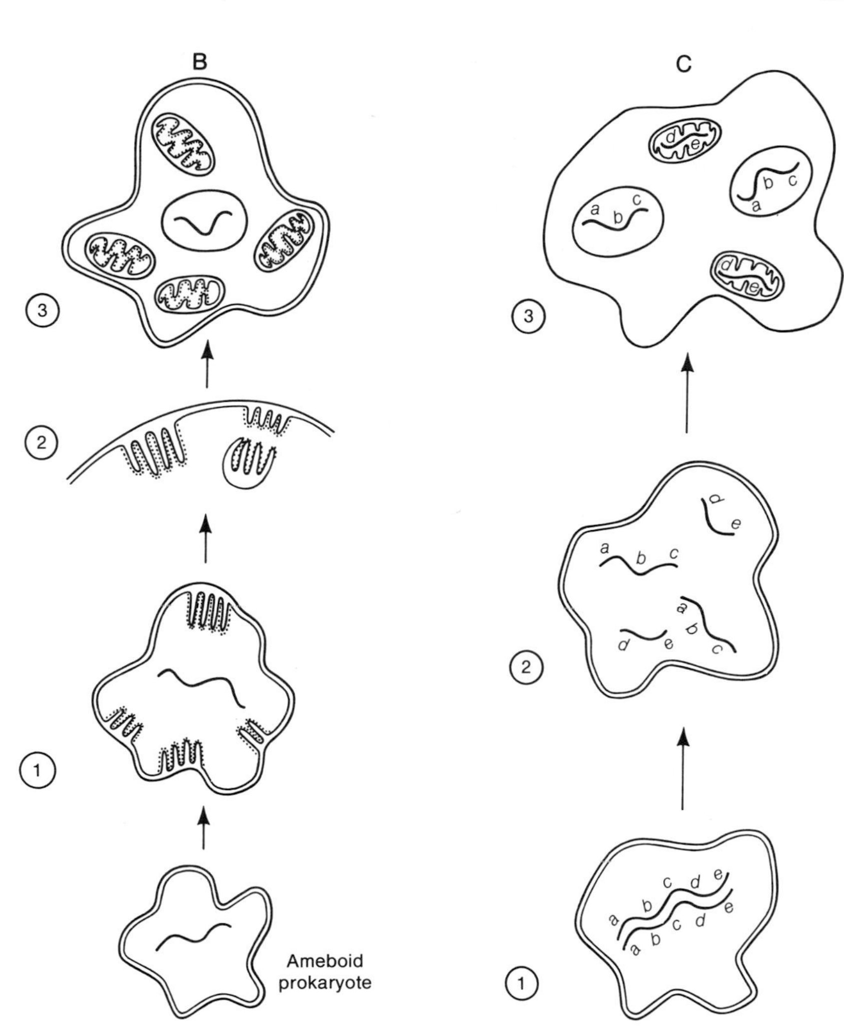

B. The origin of mitochondria according to the *invagination theory*. *B-1.* A prokaryotic cell similar to that shown in part *A*. Extensive invaginations are evolved, derived from the inner cell membrane. *B-2.* Enzymes associated with glycolysis and respiration become associated with these invaginations. *B-3.* The invaginated membranes lose contact with the cell membrane and become converted into mitochondria. [In part after Raff and Mahler, 1972.]

C. Bogorad's *cluster clone theory*. *C-1.* A prokaryote having a single duplicated strand of DNA on which are five genes (indicated by letters). *C-2.* A similar cell in which the DNA strand or chromosome has broken into two parts containing different genes. *C-3.* Retention of some of these genes in a nucleus, which has acquired a nuclear membrane, and conversion of other DNA segments into mitochondria through the acquisition of separate membranes. Neither the method nor the source of these membranes is explained by Bogorad. [After Bogorad, 1975.]

addition of a protein-synthesizing DNA plasmid, derived from another prokaryote (Figure 12-5B). Raff and Mahler say nothing about the origin of plastids, but reject outright the origin of flagella and the mitotic apparatus from symbiont microorganisms. They believe that the difference in modern eukaryotes between nuclear-directed and mitochondria-directed protein synthesis and other metabolic activities is due to divergent selective pressures that have acted since the acquisition of mitochondria.

Bogorad offers an alternative hypothesis for the origin of the eukaryote nucleus and the DNA contained in plastids and mitochondria (Figure 12-5C). Emphasizing the well-known fact that some of the proteins of which these organelles consist are coded by nuclear genes, while others are coded by genes located in the organelles themselves, he has suggested the "cluster–clone" hypothesis. He believes that all of the DNA found in eukaryote cells originally existed in a nuclear area, perhaps surrounded by a common nuclear membrane. Later, clusters of genes derived from the proto-nucleus became separated and surrounded by their own membranes, thus forming the plastids and mitochondria (Figure 12-5C).

At first sight, the hypothesis of symbiosis and the suggested alternatives appear to be diametrically opposed. Actually, they are not so contradictory when one examines the qualifications made in each theory to harmonize with the facts. An intermediate theory, containing elements of both extreme hypotheses, may be acceptable. The biochemical similarities between plastids, mitochondria, and prokaryotic cells cannot be denied. One striking similarity is between the enzyme superoxide dismutase of chicken liver mitochondria and the corresponding enzymes in *Escherichia coli:* both are very different from the dismutases of nuclear origin found in bovine erythrocytes (Steinman and Hill, 1973). Another similarity, which concerns the origin of chloroplasts, is the extensive homology between sequences of oligonucleotides found in the chloroplast DNA of the red alga *Porphyridium* and corresponding sequences recognized in prokaryotes, such as *E. coli, Bacillus subtilis,* and the blue-green alga *Anacystis nidulans.* Nevertheless, the similar sequences could have been introduced into the evolving prokaryotic cell in the form of plasmids, as Raff and Mahler have suggested. Whichever theory is accepted, evolutionary modification of both nuclear and organellular DNA during early eukaryote evolution must be postulated.

The strongest case for the symbiosis theory concerns the origin of plastids. The chlorophylls of eukaryotes are very similar to those of the prokaryotic blue-green algae; in addition, certain accessory pigments, particularly the phycobilins of red algae, are also similar (Klein and Cronquist, 1967). Furthermore, the cells of blue-green algae and the chloroplasts of red algae contain organelles known as phycobilisomes, which are so similar in their fine structure that they are almost certainly homologous (Stanier, 1974). There are numerous other examples of unquestioned symbiosis between heterotrophic eukaryotes and blue-green algae. Two genera of unicellular flagellates, *Glaucocystis* and *Cyanophora*, which have typical eukaryotic nuclei, contain photosynthesizing organelles (or perhaps symbionts) that in some respects are intermediate between blue-green algae cells and plastids (Schnepf and Brown, 1971). Furthermore, the chloroplast DNA of another autotrophic flagellate, *Euglena,* has been shown by hybridization techniques to be very similar to that of blue-green algae (Pigott and Carr, 1972).

There is every reason to believe that some or all autotrophic eukaryotes have acquired their photosynthetic apparatus by evolution from blue-green algae. If one rejects the symbiosis hypothesis and assumes that plastids have originated by invagination, one must assume a direct connection via cell lineages from blue-green algae to autotrophic flagellates. This connection is very hard to accept, chiefly because of the great differences in the chemistry and the ultramicroscopic structure of cell walls between blue-green algae, the most primitive autotrophic eukaryotes, and typical green plants. The rigid walls of blue-green algae, like those of bacteria, have a thick network consisting of murein, which includes, covalently bonded together, N-acetylmuramic acid and a sugar, N-acetylglucoseamine. This complex is in no way related to the cellulose of higher plants, or to the mucilages based on galactose that form the cell walls of red algae. Furthermore, autotrophic eukaryotes described in the next section, such as dinoflagellates and *Euglena,* which have chromosomes and a mitotic apparatus intermediate between prokaryotes and eukaryotes, do not have rigid cell walls at all, but thin, somewhat gelatinous pellicles. One can logically postulate that organisms having thin flexible walls evolved from the lower grade of rigid-walled autotrophic prokaryotes and later gave rise to typical plants, algae as well as land plants.

Since in autotrophic organisms flexible cell walls are less adaptive than rigid ones, the loss of the rigid cell wall during the evolution of an autotrophic eukaryote would be very hard to explain. If, however, the first eukaryotic cell was heterotrophic and thin-walled, as postulated by the symbiosis hypothesis, the origin of the primitive autotrophic flagellates by ingestion of small blue-green algae and their conversion to plastids would be a plausible phenomenon, and one analogous to many well-documented evolutionary sequences. The acquisition of rigid cell walls having a different structure would certainly be adaptive in a number of ecological situations.

On the other hand, the internal membranes of eukaryotic cells, particularly the nuclear membrane and the endoplasmic reticulum, are best explained as invaginations. Since the prokaryote chromosomes are attached to the nuclear membrane, at least when they divide (Stanier, Doudoroff, and Adelberg, 1970), an invagination of this part of the cellular membrane would automatically place the chromosomes inside the nucleus. The connections and molecular similarity between flagella and centrioles could be explained by assuming that in the earliest eukaryotes the nuclear membrane was still attached to the outer cell membranes at the time when both the mitotic apparatus and flagella were evolving simultaneously. The DNA of the basal body of the flagellum could then be explained by the addition of a plasmid, as suggested by Raff and Mahler.

The most plausible explanation of the complex series of changes required for the origin of the eukaryotic cell is therefore that both symbiosis and invagination were involved. The most disputed problem, that of the origin of mitochondria, may become settled when more biochemical data become available. As Taylor (1974) has maintained, the origin of flagella and the mitotic apparatus in symbiosis is highly improbable. Indeed, at the present state of knowledge no plausible hypothesis exists about the origin of flagella and the mitotic apparatus.

Both the invagination and symbiosis hypotheses can provide equally well a clue to the ecological conditions that prevailed during the prokaryote–eukaryote transition, and the type of organism that carried it out. Either of these methods would occur most easily in a relatively large cell having a thin wall and flexible cell membrane, i.e., an amoeboid prokaryote. The only organisms of this kind now alive are the mycoplasmas. It is highly unlikely these organisms are the ancestors of eukaryotes, since they inhabit chiefly the blood plasma of vertebrates, but wall-less forms of conventional bacteria, known as the L-phase, have been induced by a number of cultural techniques. Some of them breed true, apparently as a result of appropriate mutations (Smith, 1971). When derived from bacteria having the less specialized gram-positive type of cell wall, these forms are adapted to media having an osmotic strength of NaCl as high as 0.25 to 1.1 M. One might imagine that free-living gram-positive marine bacteria, living in the ooze at the bottom of seashore pools, would find a medium of the right osmotic strength, and a habitat rich in available organic matter. If a prokaryote should begin occupation of this habitat, the ability to ingest the organic particles, derived from decaying marine organisms, would have a high adaptive value, so that the selective pressure for losing the murein mesh and acquiring the amoeboid form of nutrition would be strong. At the same time, a strong selective pressure would exist for transferring the enzymes responsible for glycolysis and respiration, which in bacteria are embedded in the external cellular membrane, to an internal position as associates of separate organelles. This could be done efficiently either by invagination or by ingestion of smaller bacteria. Subsequently, selection would favor increased activity of enzymes located in the internal organelles and inactivation of those remaining on the external cellular membrane, so that the latter would become devoted solely to such functions as selective permeability and extensibility in connection with the formation of pseudopodia and the ingestion of food.

Since the marine environment postulated would have a low oxygen content, particularly if the atmosphere above it contained less oxygen than the contemporary atmosphere (Fischer, 1965), these first amoeboid prokaryotes would be anaerobes. However, once they had acquired the division of labor between external and organelle membranes postulated in the last paragraph, they would be ready for further evolution toward more efficient metabolism by cellular activity. For this new phase of evolution, aerobic metabolism, motility by means of flagella, and autotrophy, either separately or combined with each other, would have a high adaptive value. If the amoeboid prokaryotes already had mitochondria or proto-mitochondria, the number of these organelles per cell, and hence the rate of aerobic metabolism, could be increased by their differential reproduction relative to cell division. Flagella could be acquired by ingesting various other prokaryotes and by digesting all of their parts except the DNA of their chromosomes, which could be broken up into plasmids that could either be rejected or incorporated into the nucleus depending upon the adaptive value of the genes contained in them. As already mentioned, autotrophy would be acquired by ingestion of small unicellular blue-green algae and their modification into plastid symbionts.

The suggestion of Sonea (1972), that bacterial plasmids derived from ingested cells contributed greatly to building up the chromosomal complements of eukaryotic cells, is one of the most plausible explanations yet offered for this fundamental part of the evolution of the eukaryotic nucleus. The transitional forms may well have evolved endonuclease enzymes specially adapted to the conversion of bacterial DNA's into such plasmids. The existence in viruses of endonucleases having this kind of specificity has recently been demonstrated by several workers (Middleton, Edgell, and Hutchison, 1972; Johnson, Lee, and Sinsheimer, 1973).

The partial endosymbiosis hypothesis, in the ecological version just proposed, is supported further by the reclassification of amoeboid Protozoa or Sarcodina proposed by Jahn, Bovee, and Griffith (1974). This greatly alters the phylogenetic position of the genus *Pelomyxa*, which Margulis singled out for particular attention because its cells contain large numbers of inclusions of various sorts (Figure 12-6). In earlier classifications of the Sarcodina this genus was regarded as relatively specialized, but according to Jahn, Bovee, and Griffith, it may be primitive.

The present version of the symbiosis hypothesis has a direct bearing upon transspecific evolution in general for the following reasons. First, it postulates that the event that triggered off this major advance in grade was one of many adaptive radiations carried out by anaerobic prokaryotes, and one that was distinctive because of the highly specialized and unusual environment the radiant entered. Second, it postulates that the one major way of food-getting that does not exist in free-living prokaryotes—ingestion of solid particles—was the primary method in eukaryotes. This is in accord with the fact that ingestion is more common among contemporary heterotrophic protists than is the bacterial method of external digestion and absorption. Finally, it postulates that the evolution of autotrophic organisms did not follow a direct line from blue-green algae to eukaryotic protists and algae, but involved the intermediate stage of heterotrophy followed by symbiosis.

These ideas are all completely compatible with the hypothesis of Margulis as well as with that of Raff and Mahler, at least concerning the mitochondria, the only organelles the latter authors considered. It is incompatible with the cluster–clone hypothesis of Bogorad, which postulates a direct continuity from blue-green algae to photosynthetic eukaryotes.

THE ORIGIN OF MITOSIS AND SEX

After the eukaryotes originated they evolved for at least 600 million years, and probably longer at the unicellular level (Schopf, 1975; and see Table 11-1, p. 353). This evolution included the invasion of a large number of habitats

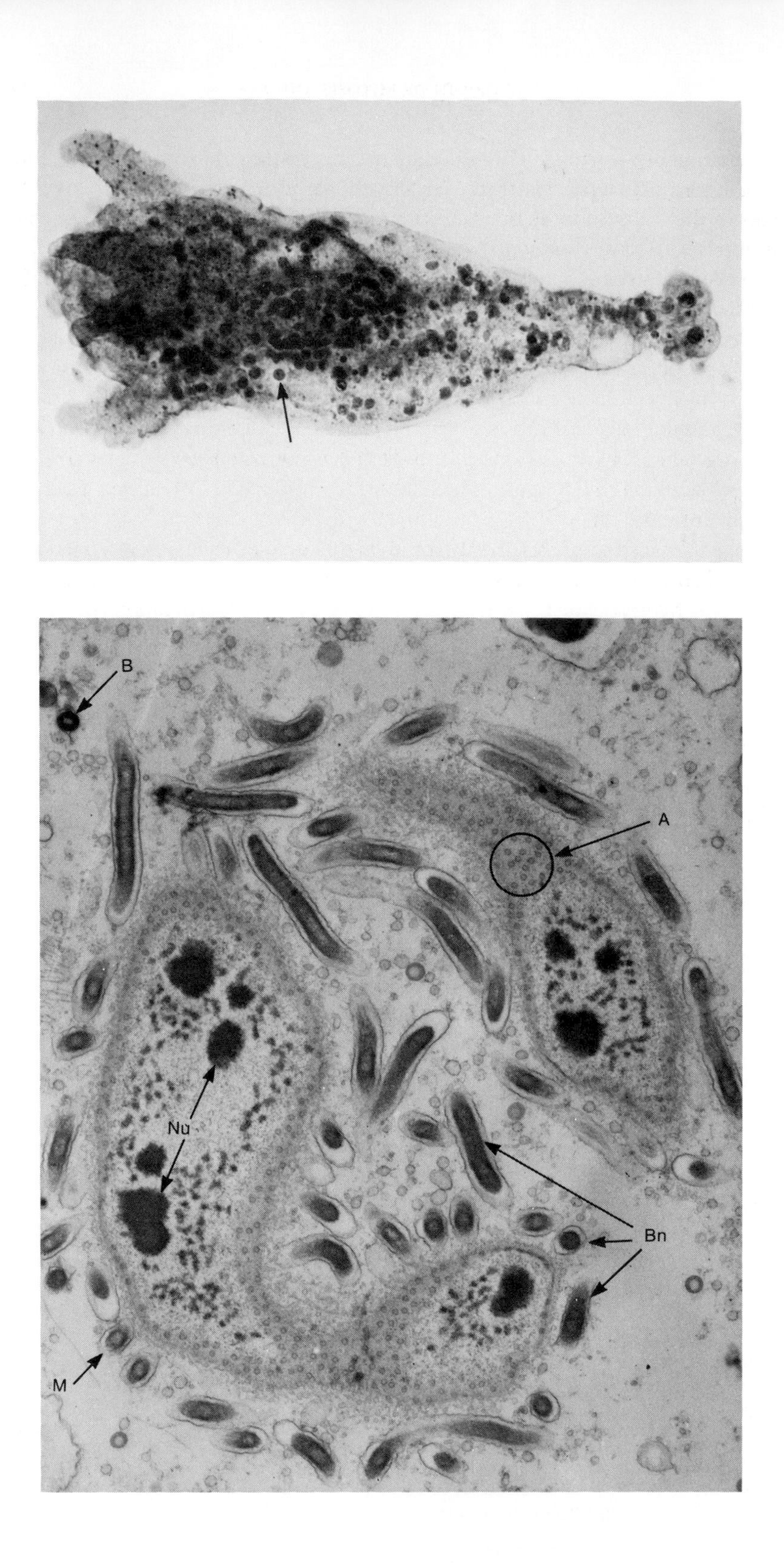
B
A
Nu
Bn
M

(Corliss, 1974; Tappan, 1974), although present data do not permit definite conclusions as to what this adaptive radiation was like. Nevertheless, the perfection of the mitotic cycle and the origin of meiosis and the sexual cycle most certainly took place during this period.

Analyses of cell division in some groups of flagellates give clues to the origin of the mitotic cycle, including chromosome splitting, condensation of chromosomes during prophase, relaxation at the end of telophase, and the formation of the mitotic spindle. Of greatest interest are the dinoflagellates, which have a kind of nucleus designated as mesokaryotic (Kubai and Ris, 1969; Dodge, 1971; Soyer, 1972). Their nuclear membrane contains pores, and they have nucleoli like eukaryotes. The dinoflagellates contain several chromosomes rather than a single "genophore" as in typical prokaryotes. Their chromosomal fibers, however, are densely packed together in a very different way from those of eukaryote chromosomes. Furthermore, the metabolic nuclei, between mitoses, contain chromatin that is nearly or quite as condensed as that of metaphase and anaphase chromosomes; the cycle of condensation and elongation so characteristic of other eukaryote chromosomes is absent (Soyer and Haapala, 1974; Haapala and Soyer, 1974). They contain acidic proteins, but little or no basic protein or histone (Rizzo and Nooden, 1972). During cell division, typical mitotic spindles are not formed, and the chromosomes remain within the nuclear membrane. The separation of the daughter chromosomes apparently results from elongation of this membrane, and so is somewhat similar to the separation of chromosomes or genophores in a dividing bacterial cell. The cytoplasmic organelles of dinoflagellates are typical of eukaryotes. The situation in *Euglena* is somewhat similar (Leedale, 1967). On the basis of these observations, as well as a less anomalous but still aberrant situation in a parasitic protozoan, *Syndinium*, Ris and Kubai (1974) distinguish three successive stages in the origin of typical eukaryotic mitosis. Whether or not this scheme has general validity will not be clear until more transitional examples have been studied by means of modern electron-microscopic techniques.

With respect to the origin of sexual reproduction, two challenging questions present themselves. First, in what kinds of organisms did sex first arise? And second, what was the adaptive advantage that caused sexual reproduction to become predominant in higher organisms?

Until very recently no clear answer could be given to the first question. A survey of modern unicellular organisms (Grell, 1967) concluded that those having the least-specialized cell structure—Chrysomonadina, Cryptomona-

---

FIGURE 12-6 (opposite).
Above: The giant amoeba *Pelomyxa palustria,* showing the large number of inclusions (arrow) contained in the cell. × 51. *Below:* Tangential view of a nucleus of this cell with surrounding cytoplasmic structures. × 9,775. *A,* annuli of nuclear membrane; *Nu,* nucleoli; *Bn,* perinuclear bacteria; *B,* other bacteria in the cytoplasm; *M,* delicate cytoplasmic microtubules. [Photographs courtesy of E. W. Daniels, Argonne National Laboratory.]

dina, Euglenoidea, Dinoflagellata, and Amoebina—are not known to possess sex. Recently, however, earlier, doubtful records of a sexual cycle have been confirmed for the dinoflagellate genus *Crypthecodinium* (Gyrodinium) by clear-cut morphological, cytological, and genetic evidence (Beam and Himes, 1974; Tuttle and Loeblich, 1974; Himes and Beam, 1975). Since this organism is mesokaryotic, its sexuality suggests strongly that sexual cycles evolved contemporaneously with the perfection of the mitotic cycle. Moreover, the sequence of meiotic divisions in this genus is not fully developed. Most probably, flagellate and ameboid protists with a fully evolved mitotic cycle but not known to possess sex either have an undetected sexual cycle or are asexual derivatives of sexual ancestors. The presence of fully developed sexual cycles in more specialized protists, such as *Chlamydomonas* and its relatives *Actinosphaerium* and *Actinophrys,* polymastigoid flagellates, Foraminifera, Sporozoa, and ciliates, is in harmony with this conclusion. Paleontological data (Schopf, 1972) suggest that both mitosis and meiosis were present in microorganisms, probably algae, preserved in the Bitter Springs Formation of Australia, about 900 million years old.

The adaptive value of the sexual cycle has recently been debated, and the conventional statement, that it favors new adaptive combinations of genes, has been challenged (Maynard Smith, 1971; Williams, 1975). Williams points out that a well-adapted diploid genotype, in order to produce offspring that are equally well adapted to its particular environment, must pay a "50 percent cost of meiosis." This is because each gamete produced by meiosis contains only 50 percent of the parental alleles, including, presumably, those that compose the organism's adaptive gene combination. They are normally replaced in the zygote by genes derived from another individual, which may or may not be equally well adapted to the environment of the parent being considered. Obviously, if an organism occupies a constant, homogeneous environment, natural selection will favor those offspring that are genetically most like their parents. Since such offspring are always produced by asexual reproduction, the initial steps of evolution toward sexuality will lead to offspring with lowered reproductive fitness relative to the asexual offspring, and so will be eliminated by selection. The argument that sexuality can evolve because of its long-term advantages, particularly in a changing environment, is regarded as inadmissible by Williams, since natural selection is not predictive but operates from generation to generation.

According to Williams, the solution to this dilemma lies in the hypothesis that the first sexually reproducing organisms possessed an alternation between sexuality and asexuality, each process serving a somewhat different function. Asexuality served to maintain the population in a particular habitat, while sexuality made possible the exploration of new and slightly different habitats, thus increasing the ecological and geographic amplitude of the species. Examples of organisms having this duality are hydroid corals and molluscs among animals, and perennial plants, particularly those that, like the strawberry, reproduce efficiently by such vegetative structures as stolons, rhizomes, and bulblets. Efficient exploration of the environment calls for cross-fertilization at random and high fecundity. This produces a great excess of gametes and zygotes with slightly different genotypes. If these can move about actively or passively, each habitat similar to the parental one can be colonized by a large number of genetically different zygotes or young offspring. Natural selection can then weed out those that are poorly adapted and favor those that are particularly well suited to the new environment. If the latter become capable of asexual reproduction, they can "capture" the new habitat and continue the cycle.

The early phylogenetic history of multicellular organisms supports this hypothesis. Most of the phyla and classes of Metazoa, except for terrestrial vertebrates, arthropods, and molluscs, include some organisms that have efficient asexual reproduction, or at least the capacity to produce a great excess of gametes during their lifetime. Sponges (Porifera), colonial coelenterates, pelecypod molluscs, most algae, bryophytes, most spore-bearing vascular plants, and the great majority of seed plants are all good examples. Unfortunately, the exceptions—animals having either shorter lives or lower fecundity or both, as well as annual flowering plants—include nearly all of the multicellular organisms that have been studied intensively by geneticists. Hence, according to Williams, our knowledge of population genetics is biased toward those organisms with relatively low fecundity. Recent information about the gene pools of the more fecund organisms, based upon allozyme studies, should help to remove this bias.

Williams' hypothesis could explain why sexual cycles are rare or absent in free-living protists, such as *Euglena* and the majority of flagellates that make up the nannoplankton of the oceans. Once an organism has become adapted to the planktonic mode of life, exploration of new habitats is no longer necessary, and natural selection favors maximum production of genotypes similar to the successful parent and obtained by asexual reproduction.

Combining Williams' hypothesis and the one previously suggested for the origin of eukaryotes themselves, the following hypothesis may explain the very first origins of sexuality in primitive eukaryotes. These ameboid cells were asexually able to reproduce genetically similar ameboids, and thus could saturate the habitat of a particular tide pool. But without sex their future success depended either upon considerable phenotypic flexibility or the ability of passively dispersed propagules to find a new habitat sufficiently similar to the old one to make new colonization possible. Colonization might be aided by the evolution of active movements propelled by flagella. Actively motile propagules could also acquire the ability to attract each other by chemical means and so produce zygotes. If the latter then should settle down in a new pool and give rise by meiosis to four ameboid progeny, the chances that one of them could successfully colonize this new habitat by asexually produced progeny would be greatly increased. The hypothesis suggests that the first sexually reproducing eukaryote protists may have resembled the modern cellular slime molds.

It must be emphasized that the "cost of meiosis" argument applies only to diploid organisms. Protists, such as *Chlamydomonas*, as well as many green algae, lower fungi, and probably the earliest sexual organisms, such as *Crypthecodinium*, are strictly haploid or monoploid during their asexual or vegetative reproduction; the only diploid cell is the zygote. Such organisms can evolve mechanisms for genetic exploration of new habitats without paying any cost at all. Consequently, the really critical question to ask is: What was the initial advantage of diploidy, and why is it almost the only condition present among all phyla of Metazoa?

The most plausible answer to this question is that the first diploid organisms possessed marked heterosis or hybrid vigor. The advantage of delaying meiosis until many cell generations after fertilization (so that the somatic cells became diploid) was a particular rather than a general one. It existed only in the progeny of two parents having good combining ability. Diploidy helped to stabilize this highly adaptive heterotic condition. If diploidy and heterosis then became regularly associated in the population, reversions to haploidy would have been less fit and would have been eliminated. Later, the diploid condition made possible the enrichment of the gene pool through mutations, particularly those that gave rise to recessive alleles that were eliminated with difficulty or were retained indefinitely because of their contribution to superior heterozygotes. A high correlation exists between the diploid state and the existence of complex developmental patterns (Stebbins, 1960; Raper and Flexer, 1970). This is most easily recognized by comparing parallel lines of evolution, one of them diploid (or predominantly so) and the other haploid. Examples are comparisons of the larger diploid brown algae with the haploid green algae; dikaryotic basidiomycete fungi with the monokaryotic ascomycetes; and the diploid vascular archegoniates ("Pteridophytes") with the haploid gametophytes of nonvascular ones ("Bryophytes"). These comparisons justify the conclusion of Raper and Flexer that the superiority of diploidy lies in the fact that two similar genomes, intimately associated, can exert subtle control upon minute steps of complex sequences of development and differentiation.

Another question raised by Williams (1975) is, "Why have most higher animals, which have acquired reduced fecundity and no longer perpetuate themselves by means of genetic exploration of the environment, nevertheless retained sexual reproduction, and are able to withstand the cost of meiosis?" He answers this question in part by noting that once sexuality has evolved into a highly complex and harmoniously integrated sequence of events, the pathway to successful asexuality passes through several intermediate stages that will usually be less adaptive. This adaptive "valley" is difficult to cross. A second answer for land vertebrates as well as many insects is that maternal care can provide a buffer that greatly reduces the significance of small genetic differences between sibling offspring. In birds and mammals the entire "cost of meiosis" is probably paid by parents in the form of feeding and otherwise taking care of their offspring.

THE ORIGINS OF MULTICELLULAR ORGANISMS

Even a cursory review of the simpler multicellular organisms indicates that the grade of multicellularity was achieved many times independently, and probably for rather different reasons in different groups. Recent intensive studies—particularly of cell-wall structure and chemistry; cytoplasmic organelles, including cilia and flagella; the number of nuclei per cell or coenocyte; the chromosomal or sexual cycle; and the plastic pigments in autotrophic forms—have tended to increase our estimate of the number of separate origins. The following list is a minimal number (17) of simple multicellular kinds of organisms that we believe have arisen independently from unicellular Protista.

Autotrophic. Red algae; brown algae; filamentous golden algae; Vaucheriales; siphonalean green algae; *Acetabularia* and other Dasychadales; Volvocales; Conjugales; Ulotrichales and other uninucleate green algae; phragmoplast-forming green algae (*Klebsormidium* line).

Heterotrophic and assimilating. Oomycetes; Chytridiomycetes; Zygomycetes.

Heterotrophic and ingesting. Plasmodial slime molds (Myxomycetes); cellular slime molds (Acrasiales); metazoan phyla; and sponges (Porifera).

Ideally, one would like to explain each of these separate origins as the outcome of a particular kind of adaptive radiation at the level of unicellular eukaryotes. Unfortunately, this is impossible, owing to the absence of fossils of the delicate organisms the earliest members of each line must have been, and the absence from the contemporary biota, probably through extinction, of most transitional forms. The origin of Metazoa is discussed in Chapter 13. In the case of the earliest plant groups a reasonable hypothesis can be formed based upon two facts. First, the common denominator of differentiation with respect to the vegetative parts is polarity. All groups that evolved beyond diversification within a single order acquired filaments or other cellular aggregates in which one or more cells are adapted to anchoring the plant to a substrate and others are differentiated into the tip of a filament or the margin of a flat, sheetlike structure. Second, the simpler aquatic algae having polarity reproduce asexually by means of motile cells that are themselves polarized and resemble unicellular protists in all cytological details. When these cells settle down to form a filament, their front end, bearing flagella, becomes attached to the substrate, and the first cross wall of the future filament is formed at right angles to the polarized axis of the motile cell. Hence, the polarization of the filament, which eventually in higher plants became transferred to the polarization root-shoot, did not arise *de novo*, but was derived directly from that of the motile unicellular ancestor.

On the basis of these facts a plausible hypothesis for the origin of the multicellular filament has been suggested (Stebbins, 1974b). Unicellular autotrophs could colonize turbulent or flowing water only if they became attached to a firm substrate, to avoid being swept away. The flagella of many unicellular forms are viscid and adhesive, and so would be the natural organelles for attaching the cell to the substrate. Natural selection under these conditions would favor cells having the most viscid flagella, and mutations that replaced typical flagella with more elongate, branching structures similar to the rhizoids of algal cells would have a particularly high adaptive value. Once this happened, reproduction would depend upon the ability to produce, by mitosis, cells capable of differentiating into motile zoospores, which would have a very high selective value. On the other hand, the most successful attached filaments would be those in which a considerable amount of photosynthetic surface was produced before zoospores were initiated. This could be acquired by cellular elongation before the onset of the first cell division, or by mitotic divisions that produced cylindrical, photosynthesizing filament cells before producing cells capable of differentiating into zoospores. Ever since the nineteenth-century observations of Julius Sachs, botanists have known that when a plant cell becomes relatively long and narrow, the mitotic spindle becomes oriented parallel to its long axis and the first cross wall is formed at right angles to this axis. Hence, selection for increased photosynthetic surface would result directly in the evolution of a polarized filament. Further evolution forming branched filaments, flat plates, or three-dimensional structures would depend upon processes that as yet are not understood.

There is no reason to believe that the origin of multicellular organisms, any more than the origin of the eukaryotic cell, depended upon processes other than natural selection of genetic combinations fitting the earliest multicellular plants and animals to particular environmental niches. The most general cause favoring the later success of multicellular organisms was the adaptive value of increased size (Bonner, 1965b, 1974).

13

The Evolutionary History of Metazoa

METAZOAN GRADES AND MAJOR BODY PLANS

Metazoa may be defined as animals that share a common multicellular ancestry with coelenterates and flatworms. Although there is a good fossil record of the major groups that have well-mineralized skeletons, the origins and earliest evolution of the metazoan phyla cannot be documented from fossil evidence. Nevertheless, from the fossils that we do have, from comparative studies among living organisms, and from interpretations of the functional significance of primitive characters, it is possible to produce models of metazoan origins and diversifications.

As noted in Chapter 8, there are several types of fossil remains, and each has a different probability of preservation. We shall place them in three classes here: trace fossils (trails, burrows, tracks); soft-bodied fossils (chiefly impressions or carbon films); and rigidly skeletonized fossils, which are usually impregnated with minerals (chiefly carbonates or phosphates). Although soft-bodied animals frequently possess hydrostatic skeletons, fossil skeletons are of mineralized or protein-tanned structures. Both soft-bodied and rigidly skeletonized fossils are called *body fossils*. So far as the record is now understood, trace fossils appear earliest, then soft-bodied forms, and finally mineralized skeletons.

Traces left by the activities of organisms are difficult to discriminate from sedimentary structures left by many inorganic processes. Although there are occasional claims of very old animal fossils, these can usually be clearly ascribed to inorganic causes (Cloud, 1968b, 1973); a few remain doubtful. The earliest traces that appear likely to be valid organic traces were recently discovered in Zambia (Clemmay, 1976). They are probably near 1,000 million years old.

They are horizontal burrows, up to about 10 mm in diameter, that have been back-filled with sediment, presumably by detritus-feeders that ingested the sediment itself. The next earliest known burrows may be less than 700 million years old. The long barren interval between these fossils is puzzling. Sediment deposits from this interval, confidently interpreted as shallow marine deposits, have been searched for fossils without success. The burrows that appear sometime after 700 million years ago (their ages are not closely determined as yet) include horizontal and vertical burrows and sea-bottom trails (Glaessner, 1969; Banks, 1970). The abundance and diversity of traces increase progressively in younger rocks until, during the Cambrian Period (beginning 570 million years ago), they reach a richness in shallow water that they maintain for hundreds of millions of years (Crimes, 1974).

Among the earliest authenticated records of soft-bodied invertebrate fossils are those of the Ediacaran fauna, named for an unusually large assemblage from the Ediacara formation, South Australia. This fauna seems to have lived between at least 680 and 580 million years ago and is found on several continents that are at present widely dispersed: Europe, Africa, North America, and Australia (Glaessner, 1971). The fauna is principally composed of medusae, pennatuloids, and other forms that appear also to have been coelenterates, and a small variety of worms of uncertain affinities, but perhaps including annelid-like forms.

The next younger assemblage of soft-bodied fossils of importance is from the Burgess Shale, British Columbia, Canada. It is of Middle Cambrian age and postdates the appearance of mineralized skeletons by a few tens of millions of years. For over 65 years this fauna has been known to exist and is now undergoing a much needed reappraisal. The new results are most impressive (Conway Morris, in press). About a quarter of the fauna represents groups that have already appeared by mid-Cambrian time, many with mineralized skeletons. These include coelenterates, molluscs, trilobite arthropods, brachiopods, echinoderms, and hemichordates. Another quarter is represented by non-trilobite arthropods, including sorts that are otherwise unknown; sponges constitute another quarter. Of the remainder, many specimens represent the earliest undisputed records of living phyla: Annelida (which may be represented in the Ediacara fauna), Priapuloida, Nemertea, and Chordata. Finally, a few specimens belong to groups that are distinct from living phyla; they represent extinct phyla that do not seem ancestral to any living animals. There are at least three of these; only one has as yet been adequately described (Conway Morris, 1976), and is interpreted as a lophophorate coelomate.

The contrast between this soft-bodied assemblage, teeming with invertebrate phyla, and the depauperate late Precambrian fauna thus far known, is striking. There are only a few other occurrences that provide us with important glimpses of some of the soft-bodied associations of later times. An important one is in the lower Carboniferous of Illinois, which has yielded a form that is not readily assignable to any living phylum (Richardson and Johnson, 1970).

The earliest mineralized skeletons appear slightly earlier than the generally accepted base of the Cambrian, below the Tommotian Stage of the Russian platform (Matthews and Missarzhevsky, 1975). The fossils are minute and of uncertain affinities. Additional skeletons appear in the Tommotian, including small coiled shells that are probably molluscs, and small conical elements that may be associated with the tentacles of primitive suspension feeders. In the next stage many more phyla appear, so that within the first two Cambrian stages, seven skeletonized phyla are known. The last skeletonized phylum to appear, the Ectoprocta, does so in the earliest Ordovician. Figure 13-1 summarizes our knowledge of the appearance in the fossil record of those phyla that have been formally described.

During the Cambrian and Ordovician, radiations among the durably skeletonized phyla led to the establishment of numerous classes and orders, so that the number of taxa in those categories was considerably greater than at present for these same phyla (Figure 13-8). There is no reason to believe that the phyla for which records are poor—the soft-bodied groups—were not represented also by large numbers of classes and orders at that time.

To summarize, the earliest animal fossils (traces) may date from one billion years. Body fossils appear later and include coelenterates, possibly annelids, and a number of enigmatic forms. Then shortly after 570 million years ago, durably skeletonized forms that include several phyla of coelomate metazoans appeared in quantity. The next younger soft-bodied assemblage known is also rich in advanced metazoan types. By the early Ordovician, all living phyla that have durable skeletons are present; a large number of phyla that have no hard parts at all, including some extinct ones, also have been discovered. It appears that among marine invertebrates diversity of anatomical architecture as represented by phyla, and major modifications of these basic body plans as represented by classes and orders, reached higher levels in the early Paleozoic than ever again.

EVOLUTION OF THE TISSUE GRADE

The earliest metazoans are unknown; the earliest yet described had already attained a tissue grade of organization. The adaptive and phylogenetic pathways that led to this grade have been inferred from functional and morphological studies of living organisms, which can be interpreted in a number of ways. As a result, there are a number of different historical models to explain both the evolution of metazoan grades and the radiations of body plans within grades.

It is widely accepted that metazoans evolved from protozoans, though there is a dispute as to the ancestral type. The first significant advance towards metazoans must have been the development of a multicellular organism. As we have seen (Chapter 12), the step to multicellularity occurred several times. The ancestral type usually suggested for the metazoans is a flagellate protozoan (Hyman, 1940; Kerkut, 1960), which developed a colonial habit not unlike living *Volvox,* leading eventually to obligate multicellularity. Hadzi (1963) suggests that the protozoan ancestors were ciliates and that multicellular metazoans arose from the division of a multinucleate syncytium. Among the difficul-

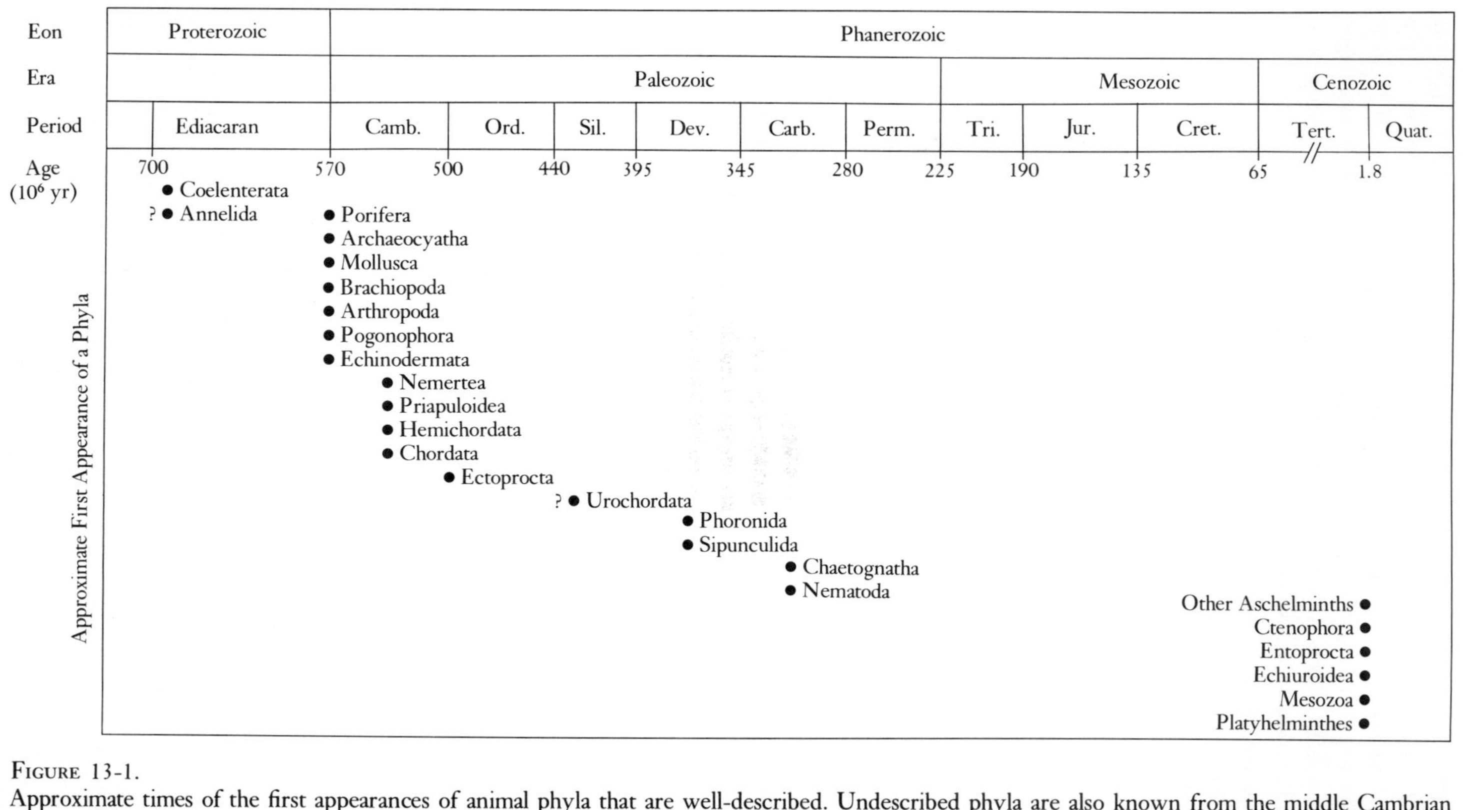

FIGURE 13-1.
Approximate times of the first appearances of animal phyla that are well-described. Undescribed phyla are also known from the middle Cambrian period (at least three) and the Pennsylvanian period in the Carboniferous (one).

ties faced by this hypothesis is the highly specialized nuclear behavior of living ciliates, which is unlike that of metazoan cells.

The advantages of multicellularity per se flow chiefly from the sizes, shapes, and repetitions of cellular machinery that are achieved through a modular construction. Among them are increased homeostasis, longevity, and reproductive potential. Like the primitive eukaryotes (Chapter 12), the metazoans must have evolved in the sea, where there would be numerous ecological opportunities for these primitive animals. For example, important adaptive zones in the sea contain detritus- or suspension-feeders, which utilize particles in the water column by sweeping them from the water or by receiving a rain of descending particles on the sea floor. In shallow water particles are likely to be somewhat patchily distributed, so that a large surface area for particle capture is advantageous. One way that primitive organisms can increase their areas is for their replicating cells to remain in contact. Uneven distribution of food items can then be corrected by transferring ingested materials or their degraded products across cell walls; in this way each cell has access to a more stable food supply, enhancing homeostasis.

The adaptation of the colony could be enhanced if growth patterns and morphogenetic movements were specified within the genome of the colony to control its size and shape. To accomplish this, the regulatory repertory would have to enlarge from control of individual cell functions to the integration of many genetically identical cells. Evolution of multicellular control could proceed through the intercellular exchange of substances, such as hormones or larger molecules originally employed in internal cell regulation but which become inductors. Once some genetic multicellular integration was achieved, variations in the shapes of colonies could have arisen through recombination and mutation, and natural selection would establish an array of shapes, each adapted to a distinctive range of habitats. For example, a dome-like or hummocky shape can be advantageous in benthic colonies because it increases the surface area of the colony. Furthermore, water currents flowing over irregular surfaces become turbulent, so that a hummocky colony living in fairly quiet water is washed by a disproportionately larger volume of water than a flat sheet of cells; of course, increased current flow increases the probability of food. Cylindrical or cup shapes also permit colonies to obtain food particles from a larger segment of the water column. Strands of cells extending above the colony proper serve a similar purpose.

Establishment of even simple shapes, while making for more efficient exploitation of resources by a colony, often results in an imbalance in food-uptake at different locations within a colony. In domical shapes, for example, the cells towards the center tend to receive more food than peripheral ones. Natural selection would promote differentiation to increase the feeding efficiency of the central cells, those chiefly responsible for nourishing the colony. Reproduction might well be suppressed in such cells but selected in well-nourished cells at neighboring localities, while peripheral cells would be differentiated for strength in support of the colony. Patterns of differential selective pressures on cells would vary among colonies of different shapes. Suspension-feeding colonies would be modified in shape to promote efficient current patterns for feeding. Other forms, especially those feeding on benthic detritus, would be required to creep along in order to exploit new areas of the substrate. A rim or some pattern of ventral cells could be specialized for locomotion, while other cells specialized in ingestion of food. In some lineages feeding cells came to be located in partial invaginations in order to increase cell number and localize digestive capability. Floating or swimming modes of life are often postulated for primitive multicellular colonies, with food items trapped by secretions or swept into pockets or locations specialized for accepting them.

During the evolution of intercellular communication, regulation, and differentiation, a point is reached where the cells no longer form a colony, but an integrated individual organism. We can imagine an array of simple organisms. The early forms at this grade are all extinct. Although reconstructions are speculative, they indicate the selective advantages of small steps along an adaptive pathway from unicellular to truly multicellular organisms. Perhaps the most interesting hypotheses concerning the earliest metazoans are those by Haeckel (1874, and elsewhere) and by Metschnikoff (1883, and elsewhere). Haeckel postulated that metazoans arose through a hollow spherical layer of cells, the *blastaea.* The posterior cells of this sphere are inferred to have become specialized for feeding and to have invaginated, thus producing a two-layered *gastraea.* The interior cells became endoderm and the hollow of the invagination became a primitive gut. The evolution of a jellyfish-like coelenterate from such a primitive form can easily be imagined. Metschnikoff, however, believed that the primitive metazoan was solid, and that digestion (which is intracellular in lower metazoans) occurred in cells that filled the interior by ingression from the external cell layer. The planula larvae of most coelenterates is in fact a living analog of such a form. Invagination to produce an endoderm and a primitive gut space is generally rare in metazoan ontogeny (Hyman, 1940). To the extent that such ontogenies indicate phylogenetic pathways, the evidence favors Metschnikoff. However, if the very early ontogenetic stages evolved simply as effective developmental steps rather than reflecting phylogenetic development, then we are left without direct evidence of the precise phylogenetic and adaptive pathways during earliest metazoan evolution.

There are three living multicellular animal groups with especially primitive features: the Porifera (sponges), the Coelenterata (jellyfish, corals), and the Ctenophora (comb jellies). It is worthwhile to speculate briefly upon plausible pathways leading from simple multicellular forms to these groups.

The multicellular sponges display cellular differentiation, although they lack nervous systems or well-defined organs. Some are solitary, while others are considered "colonial" because they have numerous openings or osculae and can be considered as collections of individuals. However, there seems to be no differentiation of function whatsoever among individuals within a "colony." The body walls are composed of outer and inner cell layers separated by a gelatinous substance containing amoebocytes. The cell layers are not homologous with the ectoderm and endoderm of other animals; sponge development is quite unique. Most sponges secrete mineralized skeletal spicules, which are first found as fossils in the Lower Cambrian.

Since cells of the inner layer closely resemble certain flagellate protozoans, it is commonly suggested that sponges have evolved from such ancestors. Since their unique development suggests that they evolved independently of other animal groups, they are often classified as Parazoa, separate from the Metazoa. It is quite possible that they evolved well after the Metazoa, which appear in the fossil record at least 100 million years earlier. Sponges illustrate that animals are perfectly viable at the simpler multicellular grades of organization.

Coelenterates possess two cell layers, ectoderm and endoderm—the *diploblastic* condition (Figure 13-2). Space between the layers is filled with a fibrous, gelatinous substance, the *mesoglea,* the jelly of jellyfish. Cell differentiation from the cell layers to form tissues is well advanced; gonads are present, and a relatively simple nerve net is developed. A number of evolutionary routes could lead to the coelenterate condition. A conical, detritus-feeding, multicellular animal with either internal or invaginated digestive tissues, which we have previously considered, has essentially reached a diploblastic state. However, such an animal would be a detritus feeder or a grazer and somewhat vagile, and would tend to be polarized anteroposteriorly and develop bilateral symmetry. Nevertheless, a coelenterate ancestor of this general sort has been postulated (Jägersten, 1955).

A diploblastic state could also develop via the evolution of a cup-shaped suspension feeder, the cells located centrally within the cup bẹcoming specialized for ingestion and then becoming internalized to promote digestive efficiency. A sessile benthic habitat commonly correlates with a radial symmetry. There is evidence from comparative coelenterate embryology and anatomy, however, that the more primitive form is a medusa, a pelagic animal (Rees, 1957). Hand (1959, 1963) has argued that the coelenterates arose from a pelagic planula-like organism and that their radial symmetry is adaptive to pelagic life. The mesoglea in medusae is highly functional, serving to antagonize muscles and to provide support, shape, and flotation for the body (Hand, 1959; Chapman, 1966). Rees (1966) has suggested a simple coelenterate ancestor that has four hollow tentacles and a tetraradiate stomach (Figure 13-2). He postulates a bentho-pelagic habitat for this form. One can derive such an animal from any of the more generalized multicellular organisms discussed above. Hadzi (1963, and references therein) has sought to derive coelenterates from bilateral flatworms, but this notion has not stood up well to critical examination (see Hand, 1959; Rees, 1966).

Ctenophora or comb jellies are another diploblastic group that are sometimes treated as a subphylum of Coelenterata (Komai, 1963) or as a separate phylum (Hyman, 1940). However they may be classified, their origin appears to be tied up with that of the coelenterates, with which they share numerous common features. Clearly, the precise adaptive pathway and the precise succession of direct ancestral forms of the coelenterates is not known. However, our

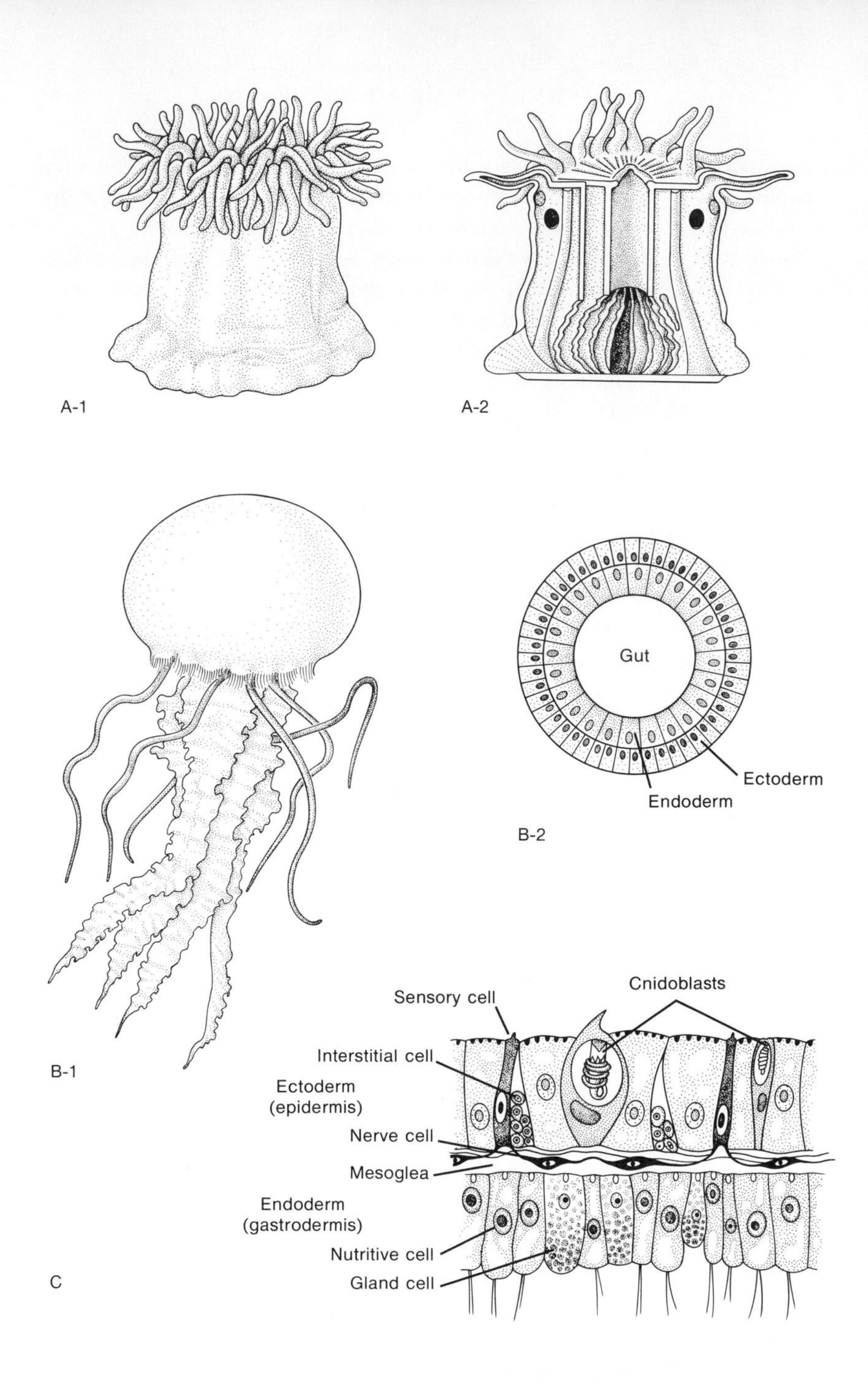

FIGURE 13-2.
Generalized architecture of some coelenterates. **A.** (1) A sea anemone (class Anthozoa); (2) vertical cross-section of a sea anemone showing tentacles surrounding mouth, gastric cavity, and an internal mesentary with associated muscles and gonads. **B.** (1) A jellyfish (class Scyphozoa); (2) horizontal cross-section of a simple jellyfish showing two-layered body wall. **C.** Close-up of a segment of a coelenterate body wall with detail of layers and cell types.

problem lies not in imagining a reasonable adaptive route, but in determining which among a wide variety of plausible possibilities is the historical pathway. In the absence of a fossil record we may never be able to decide this question definitively, although information on molecular evolution and increased understanding of the early environmental framework may limit our speculations. At any rate, the evolutionary processes reviewed in previous chapters are clearly adequate to develop the diploblastic condition.

One trend among the coelenterates is particularly notable as an illustration of the evolutionary tendency toward complexity. This is the development and elaboration of colony formation among individual multicellular organisms. It has occurred a number of times in Coelenterata as well as in other metazoan phyla. The general advantages of coloniality for coelenterates are similar to those that we outlined for primitive multicellular organisms. Noncolonial aggregations of individuals are commonly advantageous, for clumping can lead to useful surface topographies of the aggregates that are advantageous to the individuals, especially in suspension feeding (Barnes and Powell, 1950) and in reproduction. One disadvantage of aggregation is that it frequently results in overcrowding. The integration of aggregates into colonies can mitigate this problem by controlling the population density; the most advantageous spacing distances and patterns between individuals can be evolved, once these features are under genetic control. Furthermore, coordination among colonial individuals, which are physically attached by tissues, can be achieved through metabolites or common nerve paths. Such coordination can lead to improved efficiency in obtaining food, in eliminating metabolic wastes, or in warning of predatory attack or of the onset of other threatening conditions (Horridge, 1957; Mackie, 1963). The commonness of coloniality can easily be explained by such advantages.

Once colonial organization appears, a further advantage may be gained through the evolution of specialization. Some individuals are so placed within a colony that they encounter food, competitors, or predators more frequently than other cells, and selection for distinctive functions follows. In extreme cases, individuals specializing in feeding, reproduction, defense, or other functions lose nearly all but one major function, employed to the advantage of the colony. This development implies an elaboration of the machinery of genetic regulation. Colonial individuals originate by budding and have identical genomes, yet they become extremely differentiated during ontogeny. In the coelenterate class Scyphozoa, which includes such pelagic animals as the Portuguese man-of-war and the by-the-wind-sailor, differentiation of individuals is so extreme as to produce superorganisms of individuals (Mackie, 1963). One individual forms the float, others serve as feeding organs, digestive organs, or reproductive organs. The same sort of adaptive path that has led from single cells to multicellularity has led a diploblastic line from solitary individuals to a colony that is virtually at a higher grade of organization. However, this particular trend towards complexity ends in the scyphozoans; higher Metazoa arose along other adaptive pathways, although in response to some of the same selective pressures.

EVOLUTION OF THE TRIPLOBLASTIC GRADE

Instead of proceeding from compound superorganisms, higher Metazoa evolved by attaining a truly new major grade of organization at the level of the individual organism. Among the simplest living organisms at this new grade are flatworms, the phylum Platyhelminthes (Figure 13-3). In the place of the gelatinous mesoglea layer, a meshwork of cells is developed to form a tissue, the *mesenchyme.* This is considered to represent a third tissue layer, or mesoderm, developed between ectoderm and endoderm in all higher Metazoa; the triple layered structure is *triploblastic.*

Flatworms are small elongate animals that employ one of three chief methods of locomotion (Barrington, 1967). First, smaller species move by the activity of cilia packed on the flattened ventral surface. At larger body sizes the increase in mass is disproportionate to the increase in area and therefore the cilia become too few to provide adequate propulsion. Instead, locomotion is enhanced by the body wall musculature, which can provide great force. The second method is for these muscles to "inch" the worms along. A third method is for these muscles

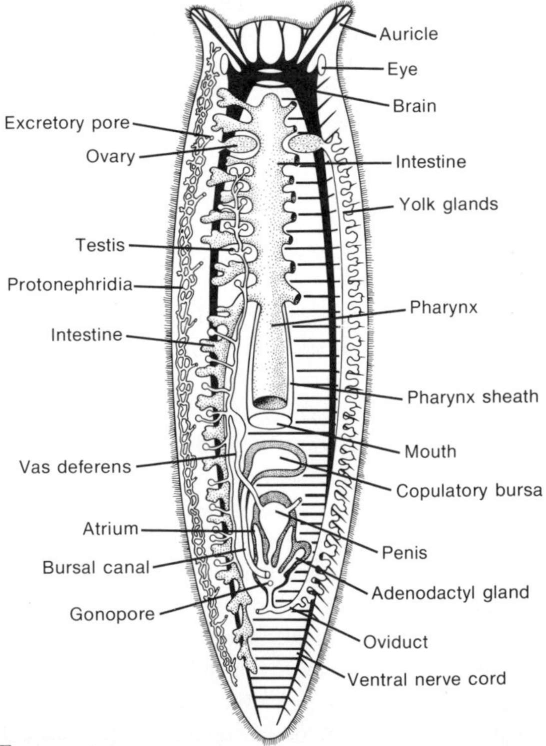

FIGURE 13-3.
Architecture of a typical flatworm. Note the serial repetition of intestinal diverticulae and consequent placement of series of gonads between them; nervous and excretory systems are also seriated.

to cause waves (pedal waves) of alternate swellings and hollows in the body. The swellings adhere to the substrate, and the waves are passed backwards by coordinated muscular contractions. New waves form as the anterior portion of the body extends forward beyond the previous leading wave. The elongate, bilaterally symmetrical flatworm is well adapted to directional creeping on reasonably solid substrates. The relatively solid mesenchyme provides the internal rigidity required for locomotion by pedal waves. Flatworms cannot burrow well, although they can penetrate loosely packed sediments, such as sands, making their way within particle interstices.

The only simple living organism that much resembles the imagined flatworm ancestor is the coelenterate planula larva. It has been suggested that flatworms developed from early multicellular organisms resembling planulae. These ancestors may be imagined as metazoan larvae that underwent neoteny, thereby dropping the adult stages out of their life cycles (diploblastic forms, perhaps) and remaining on the substrate as worms.

A benthic diploblastic adult has also been postulated as the flatworm ancestor (Lang, 1884); some living coelenterates creep, and some ctenophorans, e.g., *Coeloplana,* are highly convergent with flatworms in general body form. However, the ctenophoran and platyhelminth ground plans are very distinct, and this pathway is unlikely. Hadzi (1963) and some previous workers have proposed a phylogeny in which flatworms developed from ciliate syncytia, which in turn gave rise to the coelenterates. Some difficulties with this hypothesis were mentioned earlier. At any rate, the rise of the triploblastic grade does not present any conceptual difficulties, although again we cannot trace a precise phylogenetic route.

The development of the flatworm ground plan involved still other elaborations. Although the flattened form does not require specialized breathing apparatus (since oxygen exchange can occur over the general body surface), the elongate shape and triploblastic construction give rise to problems of nourishment and waste elimination. As body size increases, the interior cells become increasingly remote from food sources in the absence of a circulatory system. The endoderm-lined digestive tract opens ventro-medially and runs both posteriorly and anteriorly to service the length of the organism. In some turbellarian flatworms the endodermal lining of the gut is greatly increased in area by a series of laterally paired diverticulae, which extend the digestive spaces into the lateral body regions (see Figure 13-3). Excretion is provided by protonephridia, and reproduction is provided by special tissues that are concentrated into gonads. These organs are placed in the spaces between gut diverticulae, one on each side in series. This arrangement is of obvious advantage in serving an elongate organism (Clark, 1964).

The *pseudometamerous* body plan of flatworms, with a bilaterally symmetrical serial arrangement of internal organs, is found in other phyla, such as Nemertea. These are larger triploblastic worms that possess a few circulatory vessels beneath the body-wall musculature but still lack special respiratory organs. Many are far less flattened than the platyhelminths and some lineages have thick cores of mesenchyme. The thickened mesenchyme is associated with a type of locomotion that foreshadows developments in later metazoans: peristaltic locomotion. Annular waves of contractions, passing entirely around the body, run consecutively backward along the body length. The dense mesenchymal tissue is not easily compressible, and when squeezed in by say a constricting crevice, it bulges out to each side, so that it forms a hydrostatic skeleton to antagonize the body-wall muscles. The worm flows forward as the waves pass backwards; this process is aided by ciliary activity on the body surface. Nemertines are thus able to burrow, for the peristaltic system permits the pushing aside of enclosing sediment. Nemertine burrowing is probably restricted by the relatively limited hydraulic efficiency of tissues, and by the lack of respiratory and advanced circulatory systems.

EVOLUTION OF THE COELOM

A number of other triploblastic lower metazoan phyla are known, each presenting a fascinating puzzle in adaptation. For our purposes, however, it is best to examine the next evolutionary advance in grade, which freed the invertebrates from many ecological constraints and permitted their diversification into all the extant higher metazoans. This was accomplished by the development of the coelom, an internal, fluid-filled space surrounded by mesoderm and lined with a special tissue, the peritoneum. Unlike the gut, which represents a space that is continuous with the external environment, the coelom is entirely internal. It does communicate with the exterior through special ducts that transmit wastes (renal ducts) or reproductive products (gonoducts).

The origin of the coelomic cavity is uncertain; common theories are that it arose first as gonoducts, which were later expanded, or that it arose as outpockets of the gut that migrated into the mesoderm. Clark (1964) has argued that the primitive function of the coelom is as a hydrostatic skeleton (see also Chapman, 1958), and that the coelom arose during the evolution of peristaltic locomotion. Since there are a number of different invertebrate phyla, the coelomates may be polyphyletic. (For discussions of this problem see Hyman, 1959; Clark, 1964; Barrington, 1967; and several papers in Dougherty, 1963.)

Four or five basically different coelomate body plans can be recognized among living organisms, but we do not know which, if any, is the most primitive. Although the early coelomates have left fossil traces, these are not distinctive enough to permit the erection of phylogenetic trees. The simplest coelomate architecture consists of a hollow-bodied worm (Figure 13-4A,B). Among living phyla, Sipunculida may have had a longitudinally unregionated coelom from the beginning. Other worms with annelidan affinities have little segmentation, but this is probably a secondary condition. Unsegmented worms are commonly detritus feeders, living in shallow burrows or nestling in borings or crevices and gathering food by ingestion of sediments or by trapping detritus on mucous nets. These worms are relatively sedentary. Introverts are commonly developed for burrowing and feeding.

A second coelomic architecture is found in forms that have two longitudinal coelomic compartments separated by a septum. This situation occurs among the lophophorate phyla Phoronida, Brachiopoda, and Ectoprocta. The phoronids are vermiform and may be closest to the common ancestral type of lophophorate (Figure 13-4C,D). These worms are sessile suspension feeders. Most form a burrow, from which they extend their lophophores above the substrate to feed. The lophophore consists of a set of tentacles surrounding the mouth. Within each tentacle is a coelomic lumen; these spaces are continuous within the tentacular crown, but are separated from the trunk coelom by a septum. The tentacles are manipulated by intrinsic muscles that are antagonized by the tentacular coelom. The cylindrical trunk with its coelomic cavity can antagonize body-wall muscles in burrowing. The hydrostatic systems of the trunk and lophophore function somewhat independently.

The deuterostome invertebrate phyla, the group from which the chordates have arisen, possess a coelomic plan similar to that of the lophophorates. These phyla include Echinodermata, Hemichordata, Urochordata, and some minor groups. The coelom in this alliance is primitively divided into three longitudinal compartments. Presumably this plan arose because three separate hydrostatic functions evolved. One compartment was probably used in burrowing, and another to manipulate tentacles or some similar structure for feeding. It has been suggested that the third compartment may have originally been associated with locomotion, perhaps to move up and down a burrow (Clark, 1964). Whatever their primitive function, the coelomic architectures of the deuterostomes are much modified during ontogeny.

The division of the coelom into two or three regions is termed *oligomerous,* a name that applies to both the lophophorate and deuterostome phyla. Rudiments of a third coelomic space may be present in some members of lophophorate phyla. Therefore, these phyla are sometimes considered to be descended from a stock with three coelomic regions and to share a common coelomate ancestry with the deuterostomes.

A much higher degree of coelomic partitioning, a *metamerous* condition, is found among Annelida (Figure 13-4E,F). The body is divided into numerous longitudinal segments, each of which may contain a complement of organs, usually including a pair of nephridia, a pair of coelomoducts, a pair of nerve ganglia, a pair of lateral blood vessels, and frequently a pair of parapodia. In forms that are most vagile the segments are nearly all separated by transverse septa that prevent the transfer of coelomic fluid from one segment to the next, and insure that each segment deforms somewhat independently. Segments are tied together in groups of three or more by bundles of muscle fibers in the body wall. The internal fluid pressures resulting from muscular contraction during peristalsis are thus localized within a few segments, and succeeding peristaltic waves attain a large measure of mechanical independence, which makes for a relatively high locomotory efficiency.

In annelid lineages that have become largely sedentary, such as suspension-feeding burrowers (*Arenicola*) and tube-dwellers (serpulids), many of the septa are reduced or obsolete (Clark, 1964). These animals pursue a way of

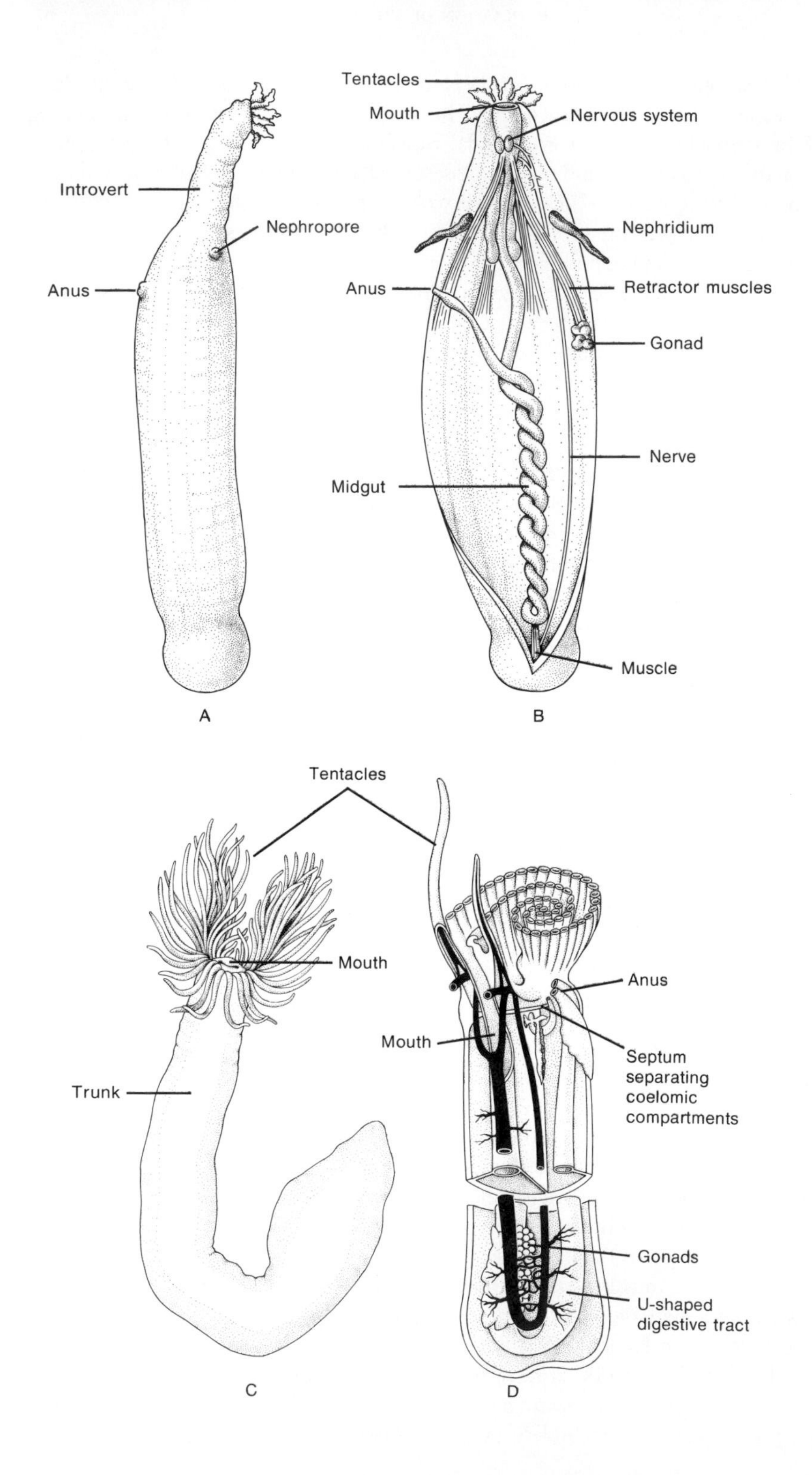
Tentacles
Mouth
Nervous system
Introvert
Nephropore
Nephridium
Anus
Anus
Retractor muscles
Gonad
Nerve
Midgut
Muscle
A
B
Tentacles
Mouth
Anus
Mouth
Septum separating coelomic compartments
Trunk
Gonads
U-shaped digestive tract
C
D

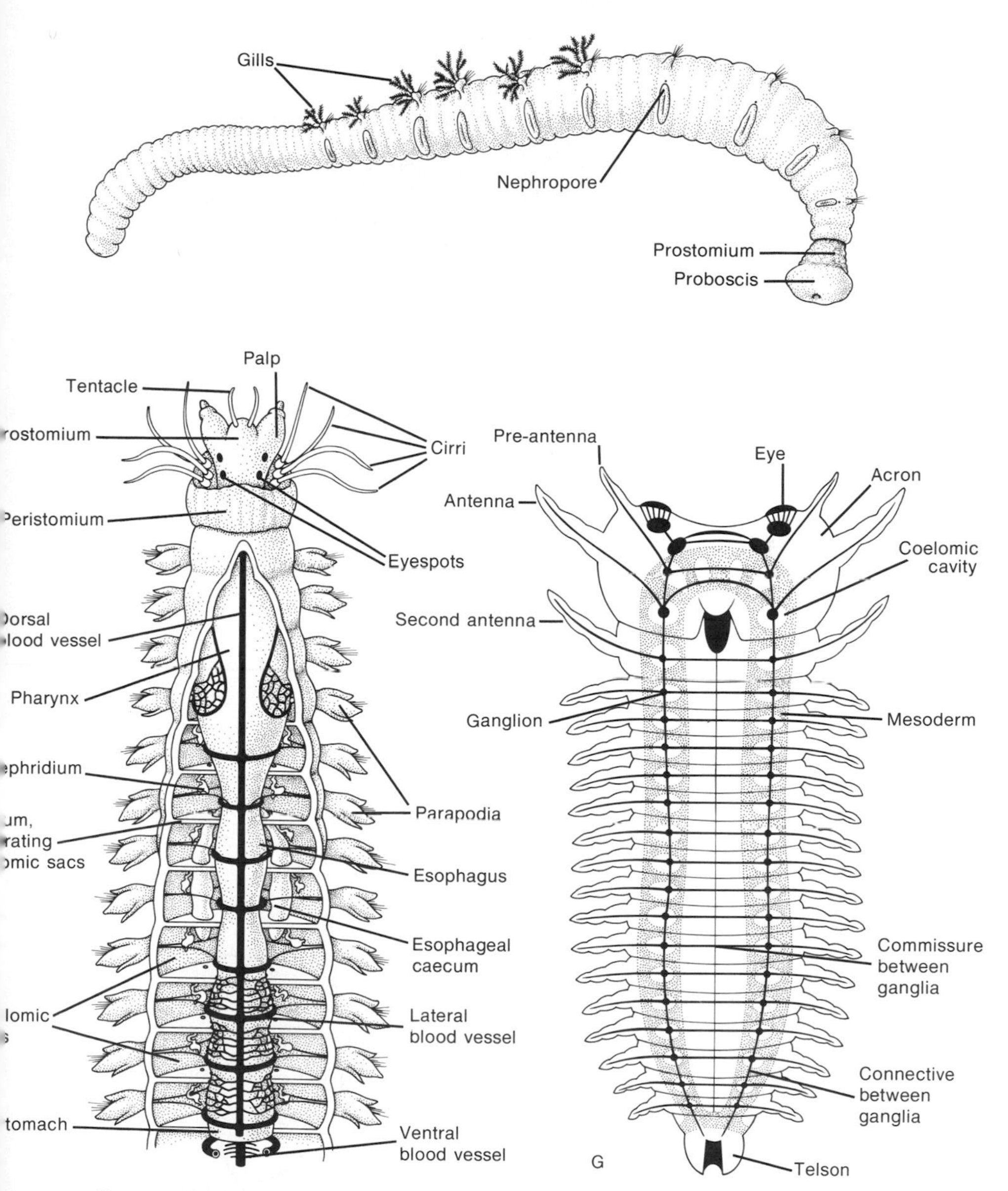

FIGURE 13-4.
Some coelomate architectures. **A** and **B.** Sipunculids *(Sipunculus),* which entirely lack coelomic regionation (amerous architecture). **C** and **D.** Phoronids *(Phoronis),* which have a coelom divided into tentacular and trunk portions by a septum (oligomerous architecture). **E** and **F.** Annelids (*Arenicola* and *Nereis,* respectively), which have a highly segmented coelom and regularized repetition of organs (metamerous architecture). **G.** A generalized arthropod, showing metamerous architecture; intersegmental septae are absent.

Figure continued next page

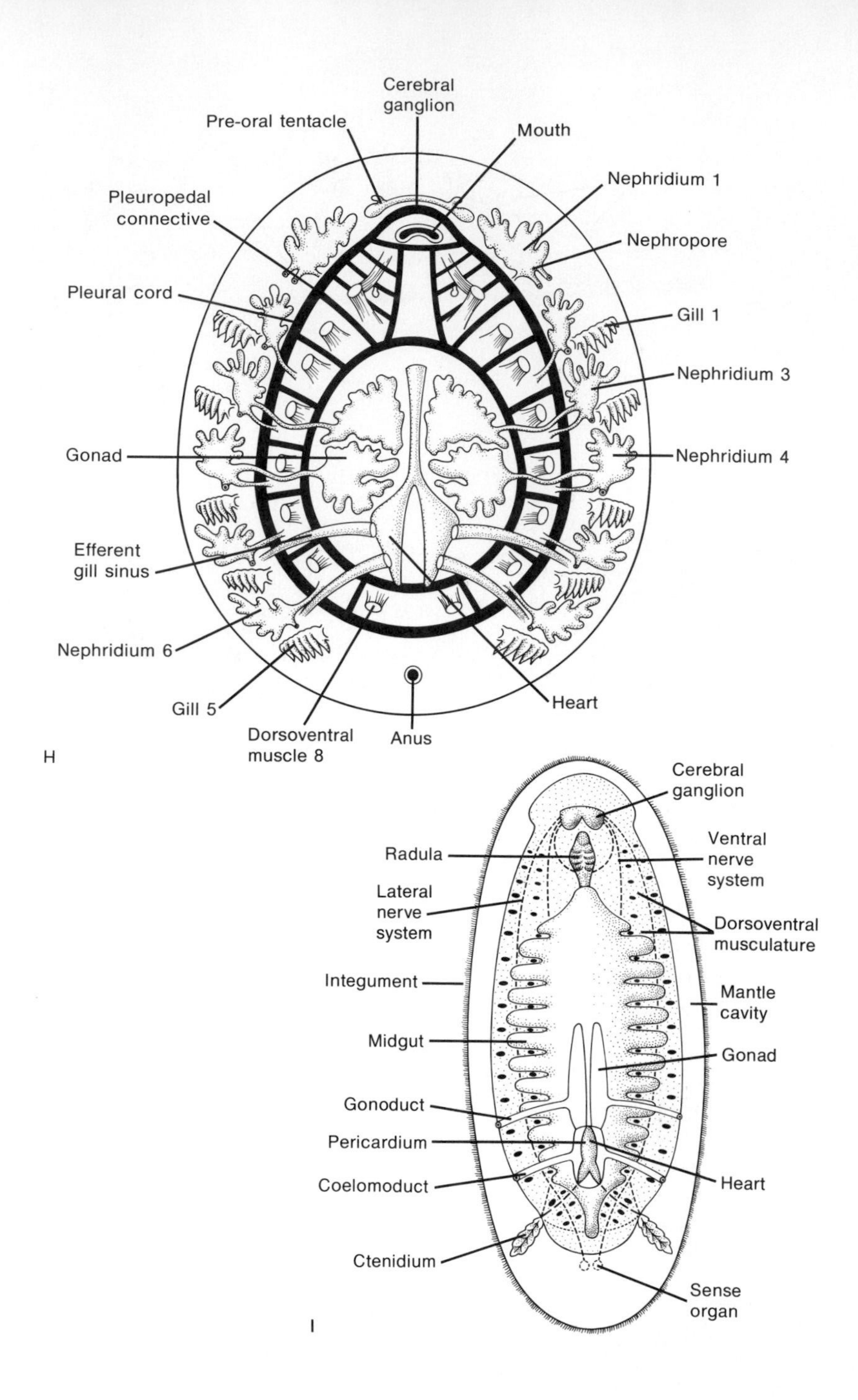

FIGURE 13-4 (continued)
H. A primitive mollusc (*Neopilina,* a monoplacophoran) showing serial repetition of some organs; there is however no longitudinal coelomic regionation or segmentation. **I.** A possible ancestral mollusc, with an architecture reminiscent of a flatworm.

life reminiscent of the phoronids; their coelomic architecture is simplified although the regularized repetition of organ systems and musculature is retained, indicating that complete segmentation was the primitive condition. Annelid ground plans seem to have evolved primitively for efficiency in more or less continuous burrowing, a habit that is useful to infaunal detritus feeders and predators. Arthropods, which have descended from proto-annelidan stocks, have developed jointed exoskeletons and do not require a hydrostatic skeleton. Accordingly, there is no septation between their body segments (Figure 13-4G).

A final type of coelomic architecture is displayed by the molluscs. Their coelom is neither regionated nor segmented; yet the primitive molluscs (Monoplacophora) display extensive serial repetition of organ systems (Figure 13-4H,I), including gills (five pair), nephridia (six pair), pedal musculature (eight pair each of two types), gonads (two pair), and a number of other organs (Lemche and Wingstrand, 1959). This sort of pseudometamerism recalls the flatworm condition. The monoplacophorans have rather extensive coelomic spaces. There are no data on monoplacophoran locomotion; presumably they move by pedal waves, like large flatworms and gastropods. If so, their coelom is not involved. A thorough functional account of the monoplacophorans is greatly needed.

Each of the distinctive coelomic plans can be explained as an adaptation to a distinctive mode of life. Each group has played an important role in the evolution of the biosphere, except perhaps the sipunculids. Lophophorates dominated many marine communities during the Paleozoic, and molluscs dominate these communities today. Arthropods are the most diverse terrestrial phylum today, while deuterostomes, which gave rise to the chordates and hence to man, have always been important marine forms. Even though we cannot yet certify a phylogeny among the early coelomate types, we can erect an adaptive model to explain many features of their anatomy and of the early fossil record.

EVOLUTION OF EARLY METAZOAN RADIATIONS

From the adaptive pathways likely followed in the development of metazoan ground plans, and from the data of the fossil record, one can erect a model of the early diversification of the metazoans. The horizontal burrows dated about 1,000 million years indicate the presence of animals with some sort of hydrostatic skeletons. These could be diploblastic but are more likely triploblastic animals with skeletons that are either mesenchymal (tissue), pseudocoelomic (body spaces with fluid but lacking a peritoneal lining), or coelomic (fluid spaces surrounded by a peritoneum). If so, much metazoan evolution had preceded them. Clemmay (1976) suggests that the early burrowers may have become extinct, since traces do not appear again for 300 million years or so. Of course, metazoans at the flatworm grade might have persisted, to give rise eventually to higher grades again. This problem must await further evidence.

From the appearance of the traces near 700 million years ago there is a record of continuous metazoan presence with increasing complexity of type and expansion of the ecospace (the inhabited functional regions of the biosphere). Although it is mechanically possible that the burrows of Ediacaran time were all made by non-coelomates, it is much more likely that primitive coelomate worms are represented, including detritus feeders and perhaps predators and suspension feeders.

Data from cytochrome sequencing (Chapter 9) tend to support this judgment. According to work by Brown and his colleagues (1972), annelids and deuterostomes separated about 750 million years ago, molluscs and deuterostomes about 720 million years ago. Further work may adjust these dates. It is not certain that the difference between them is real. The dates do suggest that the coelomates were present and that at least three of the major coelomate ground plans had diverged during the late Precambrian. A tentative schedule of phylogenetic events leading to living coelomate phyla is depicted in Figure 13-5. The phylogeny shown is frankly speculative. We reason that the earliest coelomates were not yet segmented or regionated but were seriated, resembling flatworms with coeloms, and that from these forms: (1) the *metamerous annelids* diverged through coelomic segmentation, which promoted improved detritus feeding habits; (2) *oligomerous lineages* diverged as suspension feeders (lophophorates) living within the sediments of the seabed, and perhaps as sessile particle trappers (deuterostomes); (3) the *pseudometamerous molluscan* lineages diverged by development of a dorsal visceral mass, freeing the foot for creeping; and (4) *amerous sipunculids* diverged by suppression of seriation, developing an introvert for feeding. Whatever the actual pathways of descent may be, it is clear that the development of the coelom paved the way for new grades of organization. One immediate effect was to permit efficient burrowing into the sediments of the sea floor—an *infaunal* mode of life—opening up a region of biospace that today is densely populated by a large variety of animals. The diversification of coelomic architecture in the late pre-Cambrian is implied by the diversity of coelomic plans in Cambrian faunas.

With the widespread occurrence of durable skeletons in the Cambrian, the fossil record becomes much better. In all phyla the early skeletonized forms (for which modes of life have been worked out) were essentially *epifaunal,* that is, they lived upon rather than within the seafloor (Valentine, 1973a, b). Although they must have descended from ancestors with one or another of the coelomic architectures described (Figure 13-5), they were no longer vermiform, but had developed novel body plans.

Peristaltic locomotion is very inefficient on a surface, when only a short arc of a worm's cross section would actually contact the substrate. The energy required to maintain body shape and generate annular peristaltic waves would not be repaid by locomotory progress (Clark, 1964). One solution, followed by arthropods, was to develop a chitinous jointed exoskeleton with intrinsic musculature to operate appendages for locomotion. Arthropodization may have been the first adaptive pathway leading to a novel epifaunal coelomate.

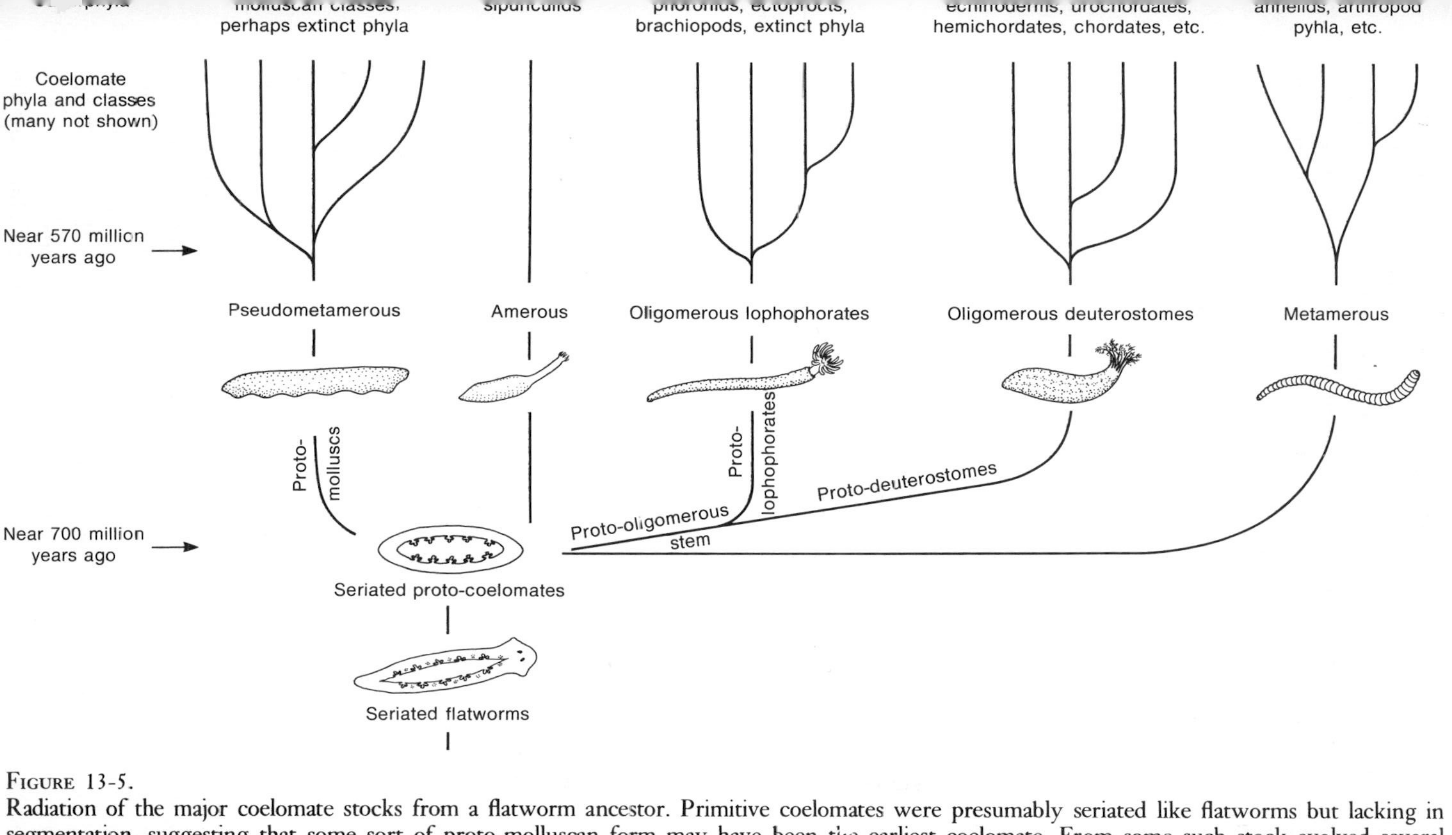

FIGURE 13-5.
Radiation of the major coelomate stocks from a flatworm ancestor. Primitive coelomates were presumably seriated like flatworms but lacking in segmentation, suggesting that some sort of proto-molluscan form may have been the earliest coelomate. From some such stock evolved several distinctive ground plans, including the metamerous one of annelids, the bi- or tri-regionated oligomerous ones of lophophorates and deuterostomes, and the unsegmented, unseriated one of sipunculids. The radiation of phyla and classes within the major lineages is highly diagrammatic. [After Valentine, 1973a.]

Another solution, favored by unsegmented lineages, was to employ pedal waves, thereby consolidating body-wall muscles and moving the viscera into a dorsal position to free the locomotory foot. Body shape and muscular purchase is afforded by a consolidated body-wall musculature and a rigid exoskeletal cap. A seriated vermiform ancestor would result in a primitive monoplacophoran, creeping on or at shallow depths within the substrate.

Infaunal worms that were nearly or quite sessile seem also to have given rise to epifaunal descendants (Hyman, 1959). Lophophorate oligomerous worms similar to the phoronids probably developed into solitary epibenthic forms, such as the brachiopods, along several different pathways (Valentine, 1975), and also into the minute colonial epibenthic phylum Ectoprocta (Farmer, Valentine, and Cowen, 1973). Although the brachiopod skeleton seems to have developed in part for protection, it developed chiefly to enhance suspension feeding, first by framing and supporting the lophophore and later by bringing the feeding currents completely inside the mantle cavity, providing for increased homeostasis.

It is not necessary here to pursue hypotheses as to the rise of each epifaunal lineage. The central point is that the appearance of skeletons nearly 580 million years ago signals the rise of relatively diverse epifaunal communities, and suggests that a number of living phyla had their origin at about that time. As Cloud (1949) has pointed out, the ground plans of many of the durably skeletonized phyla do not make sense in the absence of the skeleton. The anatomical modifications associated with the shifting ecospaces of these lineages were coadapted with skeletonization. The brachiopods, for example, have a soft-part anatomy that is precisely adapted to reside within a bivalved exoskeleton, with pedicle, visceral, and lophophoral spaces and a complicated musculature associated with valve functions. Without a skeleton, a brachiopod would have to be redesigned. The simplest such design would closely resemble a short species of phoronid, which may closely resemble the brachiopod ancestor. We can be reasonably certain that the appearance of the first primitive brachiopod skeleton in the fossil record is very close to the time of evolution of the association of features that constitute the brachiopod ground plan.

A similar line of reasoning holds for many other phyla and classes, although not for groups like corals and bryozoans, in which mineralized skeletons form no part of their anatomical ground plans. These two colonial groups do require rigid skeletons in order to occupy all of the habitats and to perform the array of ecological functions they have achieved, but their ground plans could have been evolved long before their skeletons appear in the record. Indeed, their skeletons first appear well after the bulk of invertebrate groups; perhaps they originated near the early Cambrian but did not become rigidly skeletonized until later. The arthropods present another special case. A rigid skeleton forms an integral part of their ground plan, but it is commonly formed only of organic material that does not preserve well. Aside from a few questionable impressions from Ediacaran time, we first see arthropods when a mineralized carapace was evolved by trilobites, very likely for shallow burrowing or grubbing to form superficial depressions or pits; possibly this was a feeding behavior.

It seems, then, that quite a number of invertebrate phyla originated near the beginning of the Cambrian. It is not known just what events localized the important radiations at this time. A significant environmental change must have occurred. One possibility is that oxygen levels, which had been rising as a result of growing photosynthetic activities (Chapter 12), reached a critical level that permitted the appearance of metazoans in the late Precambrian, and then a higher critical level that permitted the appearance of mineralized skeletons near the lower Cambrian boundary (Cloud, 1968b; Rhoads and Morse, 1971). Another suggestion is that plate-tectonic activity resulted in a geographic configuration that permitted a reduction in seasonality, with a consequent increase in the stability of trophic resources. This might have permitted the rise of diverse epifaunal communities (Valentine, 1973b). In any case, as the epifaunal invasion proceeded there were radiations of each major coelomate type into a number of lineages, many of which have skeletons. The results of these radiations are depicted in Figure 13–5. The novel body plans of many phyla with mineralized skeletons, and probably of soft-bodied phyla, must have evolved relatively rapidly, and this must have been due in large part to changes in the regulatory portion of the genome. Three aspects of regulatory change make this plausible. First, old regulatory functions may continue to operate as new ones evolve, thus maintaining adaptation of the lineage during important evolutionary trends. Second, repatterning of the regulatory apparatus may achieve new architectural relations by employing components that are already well developed, thereby altering the relationships among structural genes, batteries of genes, and sets of gene batteries through a hierarchical apparatus. Third, the morphological changes that arise from spatial or temporal additions or repatterning of gene batteries commonly involve the ground plans of the organisms, and the resulting morphological distances between ancestor and descendant are therefore considered to be large. This is especially true when such changes are compared to the morphological results of structural gene evolution, which tends to develop only general morphological variations. The results of regulatory evolution appear to be greater and to occur more rapidly because they can achieve novel biological architectures (Valentine and Campbell, 1975), and may affect numerous genes simultaneously.

EVOLUTION OF THE CHORDATES

For our purposes it is best to restrict the phylum Chordata to the subphyla Cephalochordata and Vertebrata. The most primitive living chordates are probably members of the subphylum Cephalochordata, typified by amphioxus (Figure 13-6D). Amphioxus has a solid dorsal supportive structure, the notochord, which is regarded as a precursor to the vertebral column. The notochord is neither cartilaginous nor bony, but consists of a sheathed series of vacuolated cells that clearly function as a supporting structure. It is present in vertebrate (including mammalian) embryos, and persists into the adult forms of some fishes, such as lampreys and even sturgeons. Despite its notochord, amphioxus is a rather sessile filter feeder. It burrows in sand and ingests water through the mouth, expelling it through gill slits—the feeding method inferred for primitive fishes.

A similar feeding system is found in two invertebrate deuterostome phyla: Hemichordata (enteropneusts, pterobranchs), which lack a notochord, and Urochordata (tunicates), which have tadpole-like larvae that possess notochords in their tails (Figure 13-6C). The gill slits of these invertebrates are very similar to those of the primitive vertebrates, and most workers have concluded that they are all inherited from a common ancestor.

The earliest known vertebrate fossils are scraps of bone of early Ordovician age that in all probability represent a group of jawless fishes, the ostracoderms (Bockelie and Fortey, 1976). These primitive fishes have a well-developed dermal armor of bone. They do not have paired fins like more advanced fishes, and have very small mouths. Presumably they were filter feeders like amphioxus, ingesting water through the mouth and expelling it through gill slits (Stahl, 1974). The ostracoderms became quite diverse and evidently gave rise to jawed fishes with paired fins, which appear in Devonian rocks. Thereafter the ostracoderms tended to become more restricted, and most groups died out. Modern lampreys may represent an ostracoderm lineage that has retained many primitive features—they are jawless and lack paired fins, for example—while losing their bony skeletons. It is likely that in lampreys and some other fishes cartilage is an embryonic adaptation retained in the adult stage and is therefore a neotenous feature in such forms (Stahl, 1974).

A plausible deuterostome phylogeny (Figure 13-7) begins with infaunal oligomerous worms that feed by means of a tentacular system resembling a lophophore. From this stock epifaunal forms evolved, probably in the late Precambrian or early Cambrian. These included Echinodermata, a phylum in which the tentacular system was elaborated into an extensive coelomic water-vascular system. This group developed a unique skeletal ground plan based upon numerous small plates or ossicles, and diversified into a large number of classes and orders during the early Paleozoic. Other early epifaunal deuterostome lineages remained without mineralized skeletons, and from these sprang the hemichordates. They developed gill slits, presumably to enhance the feeding system. The slits provided an outlet for feeding currents, which could be directed into the mouth by cilia associated with the tentacular system. Perhaps the tunicates evolved from a hemichordate stock; they have become completely sessile as adults and seem little more than filtering baskets.

The chordates seem to have arisen from an early deuterostome invertebrate. Since the tunicate larvae possess notochords and gill slits and somewhat resemble our expectations for a primitive chordate, it has been suggested that such a larva might have evolved into a chordate by developing the ability to reproduce in the larval stage—a classic case of neoteny, if true (Chapter 10). It is also possible that adult members of an early deuterostome worm lineage became swimmers, and that the notochord was elaborated to enhance this function. The urochordates could be an offshoot of this hypothetical stock,

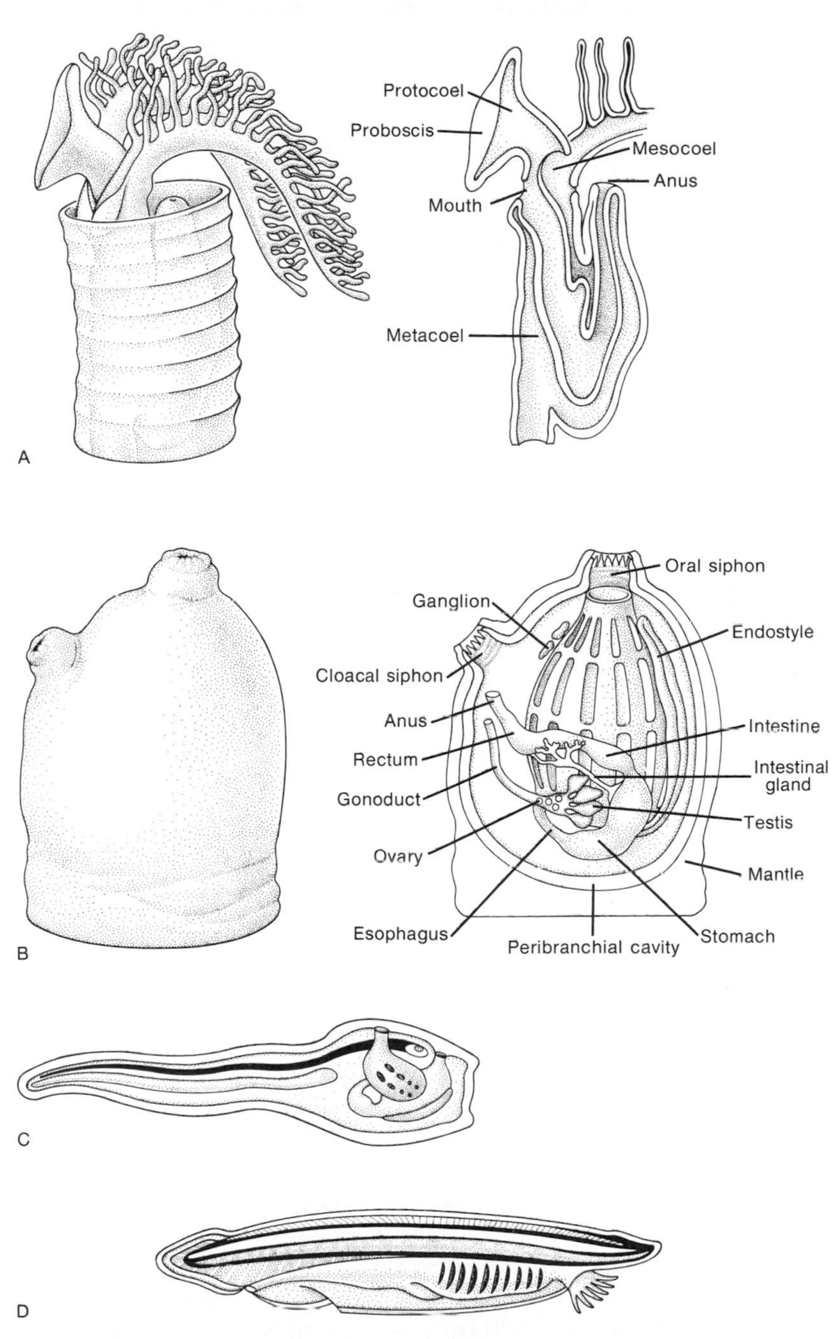

FIGURE 13-6.
Some deuterostome architectures. **A.** Hemichordate: a pterobranch and a cross-section through a pterobranch. The three coelomic compartments typical of deuterostomes (protocoel, mesocoel, and metacoel) are indicated. **B.** Urochordate: a tunicate and an anatomical scheme of a tunicate. **C.** Planktonic larva of a tunicate, resembling a tadpole, with notochord. **D.** Cephalochordate: an amphioxus.

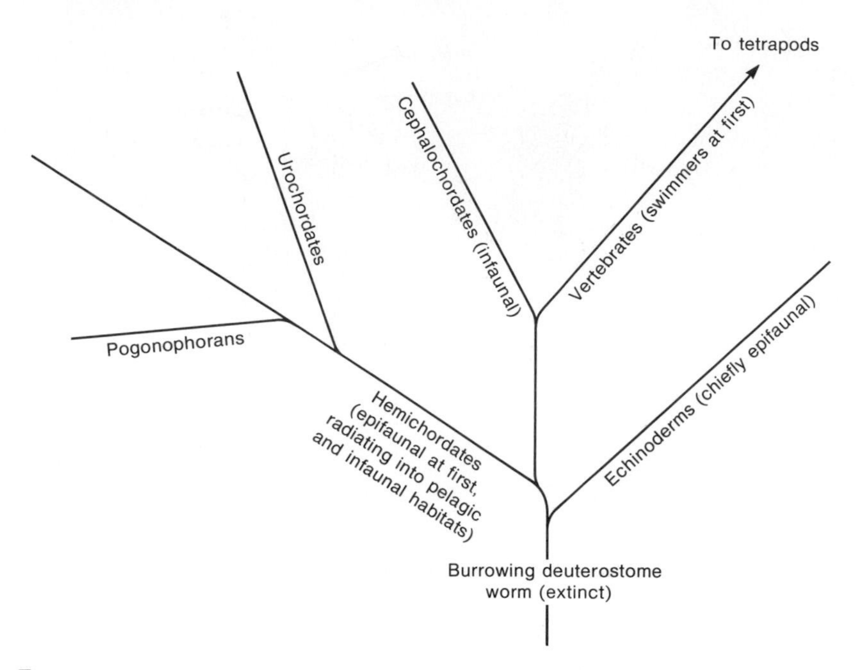

FIGURE 13-7.
A possible phylogenetic tree of Deuterostomia. Assuming primitive burrowing worms were the ancestors, radiation into the epifauna in late Precambrian or early Cambrian times produced one epibenthic group with well-mineralized skeletons (the echinoderms), another epibenthic group with a nonmineralized organic skeleton (the hemichordates), and perhaps a swimming group (proto-cephalochordates). The latter gave rise to cephalochordates and fishes, leading of course to the land vertebrates and man.

and amphioxus a more direct descendant. Still another phylogenetic hypothesis, argued most forcefully by Jeffries (1967, 1968), derives the vertebrates from carpoids, bilaterally symmetrical organisms with strong calcareous platy skeletons that are usually assigned to Echinodermata. Carpoids appear in the earliest Middle Cambrian and disappear in the Middle Devonian. Jeffries views them as a transitional group between echinoderms and chordates, with their more advanced members resembling huge armored tunicate larvae. However, the earliest known chordate is soft-bodied and is also of Middle Cambrian age.

Vertebrate bone is certainly a unique skeletal material, but many of the skeletonized invertebrate phyla also have unique shell structures. For epifaunal invertebrates and for the chordates, the appearance of mineralized skeletons seems to have heralded a new level of success, or at least diversification. Skeletonization was usually associated with the establishment of a new anatomical architecture that was coadapted with the skeleton to function in a novel environment or novel mode. Many coelomates became skeletonized once the epifaunal ground plans evolved, and commonly the skeleton proved a key to much further elaboration and diversification. The most general reason for the success of mineralized skeletons seems to be that they employ much less energy and are simpler to maintain than skeletal systems based on fluids.

Thus, for functions that require support and rigidity, mineralized skeletons are superior. Functions that require a variable skeletal system, such as burrowing, are performed efficiently by hydraulic systems; fluid skeletons are often employed for such purposes. Mineralized skeletons can be put to many uses other than support, such as protection and thermoregulation, and so are often subjected to a variety of positive selective pressures. It is therefore not surprising that once a skeleton has appeared within a lineage it should be extensively developed in a short time, and that skeletonized lineages should be potentially versatile and frequently capitalize upon environmental opportunities to become highly diverse. Skeletonization is frequently a significant component of the adaptation that permits invasion and exploitation of a new region of biospace (Chapter 8).

PHANEROZOIC MARINE DIVERSIFICATION AND EXTINCTION

Proceeding through the record of Phanerozoic time, it becomes increasingly easier not only to reconstruct phylogenies, but also to judge the temporal changes in the environmental regime and to follow the trends of structural changes in ecological units, such as biotic communities and provinces. From such information we can estimate the levels of diversity in each taxonomic category at any given time. Figure 13-8 shows the recorded Phanerozoic levels of diversity of durably skeletonized marine invertebrates in different taxonomic categories. These phyla do not exhibit much change, although the record of soft-bodied animals indicates that total phylum diversity has decreased. The recorded diversity of lower categories quickly rose during the Cambrian and Ordovician, reached a plateau during the middle Paleozoic, and then declined in the late Paleozoic to a low near the close of the era, about 225 million years ago. The diversity curves are rather generalized and do not show short-term variations. Figure 13-9 depicts the numbers of appearances and disappearances of these same taxa during the Phanerozoic. There are numbers of major waves of extinctions, each commonly followed closely by a wave of diversification. The Cambrian extinction wave is even accompanied by a high rate of diversification. Obviously, an excess of extinction over diversification lowers the diversity level, just as the opposite combination raises it. The great wave of extinction recorded in the late Permian is neither accompanied nor closely followed by significant recorded diversification, so that early Triassic diversity is low (Rhodes, 1967).

The patterns of appearances and disappearances that the record reveals have been charted independently by a number of investigators, using somewhat different taxa for different purposes (for example, Newell, 1956, 1967; Harland *et al.*, 1967; Valentine, 1969; Flessa and Imbrie, 1973; Pitrat, 1974). The results agree very well. Considering the biases in the record (see p. 323ff.), however,

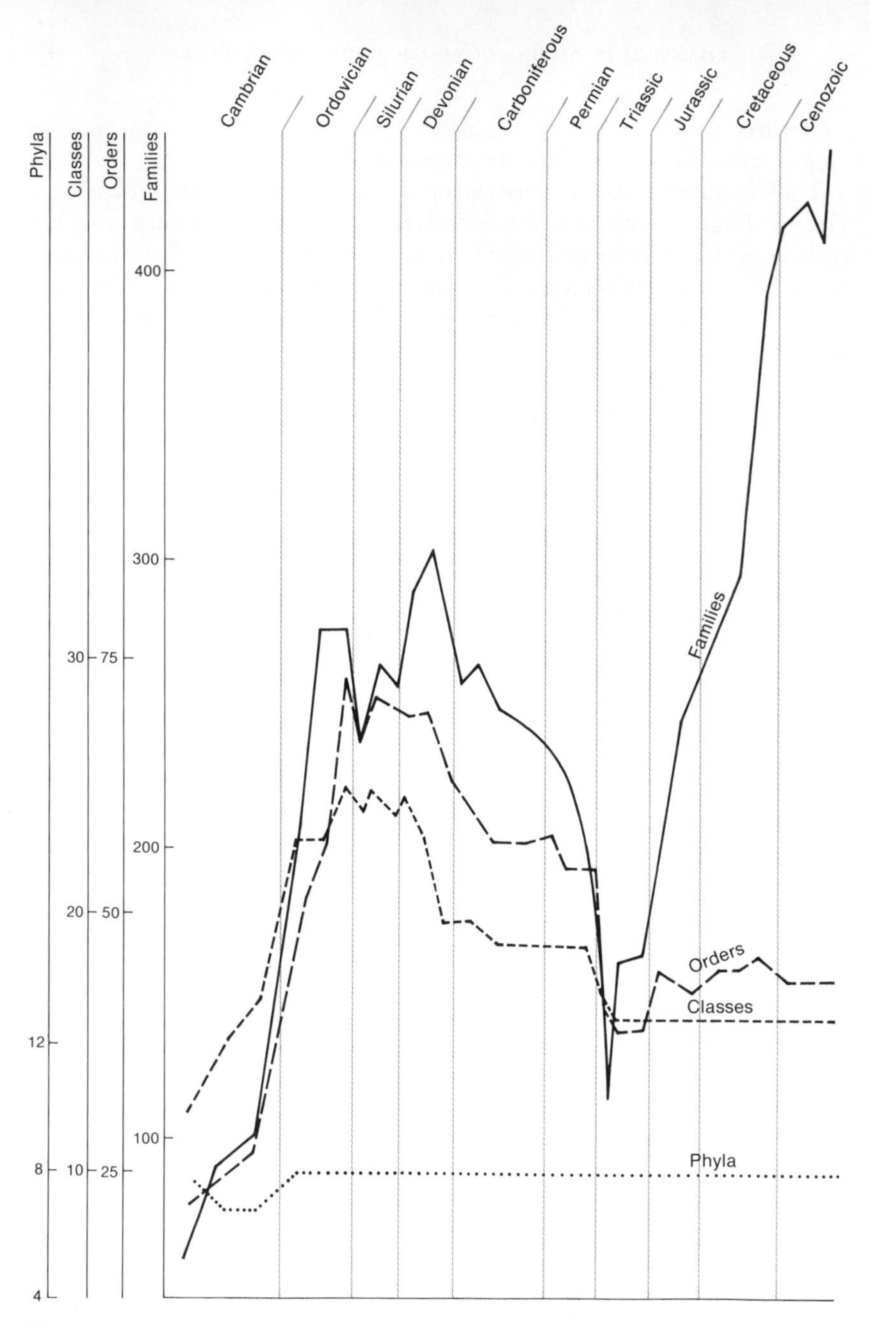

FIGURE 13-8.
Diversification of different taxonomic categories of marine animals during the Phanerozoic. The taxa used for the graph are all well-skeletonized benthic invertebrates of shallow seas, the taxa for which the fossil record is most complete. Nine phyla are represented. Note that the vertical scale (number of taxa) is different for each category. The diversity of the skeletonized phyla remains about the same throughout the Phanerozoic periods; classes and orders reach early peaks in the mid-Paleozoic, and decline to low levels near the Permian–Triassic boundary. Families also peak and decline at these times, but then recover to reach their highest diversity in more recent times. [After Valentine, 1969.]

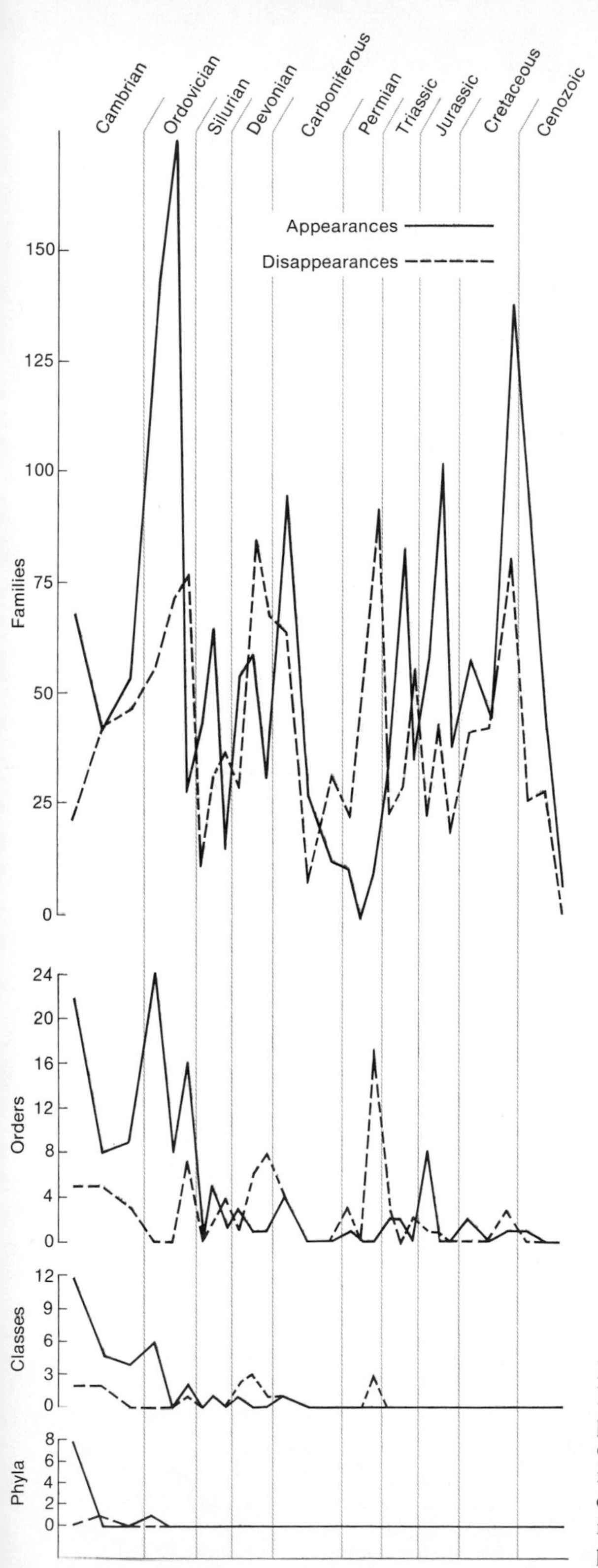

FIGURE 13-9.
Rates of appearances and disappearances of different taxonomic categories of well-skeletonized shelf invertebrates throughout the Phanerozoic periods. Most of the activity in higher taxa occurs during earlier periods. [After Valentine, 1969.]

it is not certain that the preserved fossils reflect the diversity of their times, even for well-skeletonized animals. Since most of the lineages that disappear do not reappear, so far as we can tell, most of the extinctions are probably real; they are not pseudo-extinctions. Even so, many actual extinctions must have occurred somewhat later than the disappearances suggest. Furthermore, the first appearances of taxa must have lagged to varying degrees beyond the times of their evolution, although for most taxa the appearances must indicate at least a growing abundance and success. The general patterns of diversification and extinction are probably real enough, but not precisely recorded.

However, the recorded levels of diversity must be considerably biased. If the fossil record varies in its completeness from one time to another, as it obviously does, then the record of diversity levels will also be biased. It has been suggested that nearly all changes in diversity of fossil faunas, including

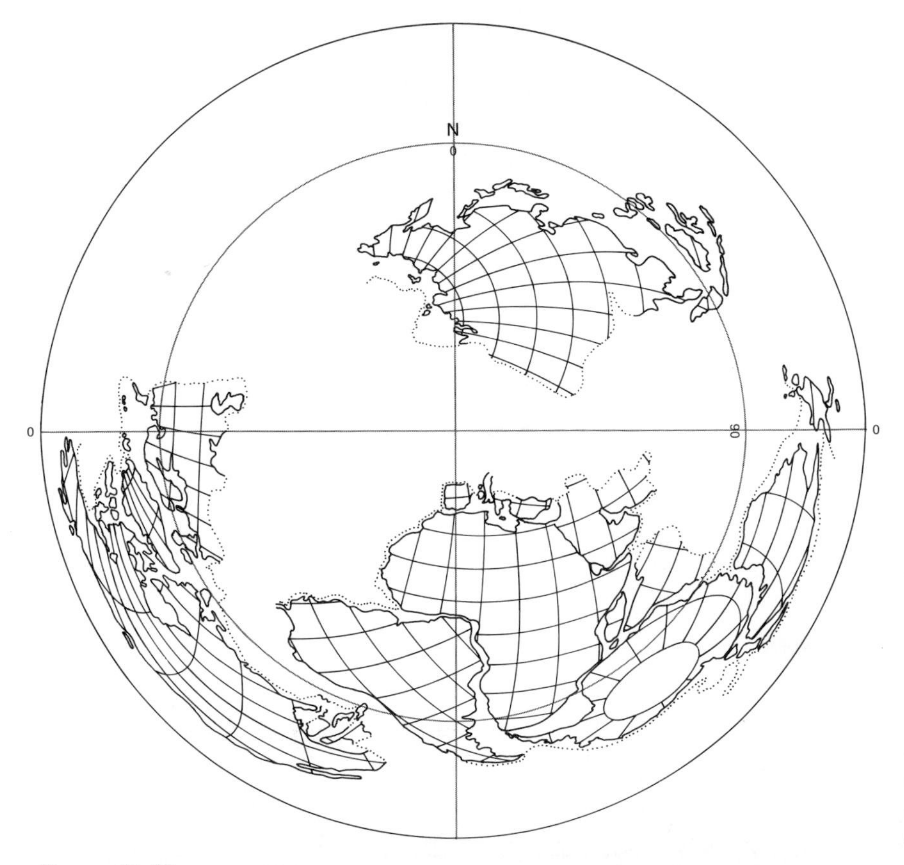

FIGURE 13-10.
World geography about 380 million years ago in the early Devonian, as inferred from paleomagnetic measurements on continental rocks. Most land is in the Southern Hemisphere, with South America lying on the South Pole and Asia alone in the Northern Hemisphere. Modern continental outlines have been retained to facilitate reading; longitudes are somewhat arbitrary. [After Briden, Drewry, and Smith, 1974.]

the Permian–Triassic low point, are mere artifacts of the incompleteness of the record (Raup, 1973). A closer look at Permian–Triassic history is warranted, especially as the mass extinctions of that time are well-documented.

Among the major environmental changes during the late Paleozoic was the gradual assembly of the continental masses into the supercontinent Pangaea (Figures 13-10, 13-11). As a result, the species capacity of the marine realm must have been greatly lowered, for several reasons. First, the number of marine provinces was lowered as continents coalesced, which brought into association biotas that had formerly been separated by topographic (chiefly deep-sea) or climatic barriers. Permian provinciality was not great; there were usually only four or five provinces or less. The Triassic seems to have begun with two provinces that coalesced into a single province later, though definitive studies are lacking. It can be conservatively estimated that a reduction of five provinces

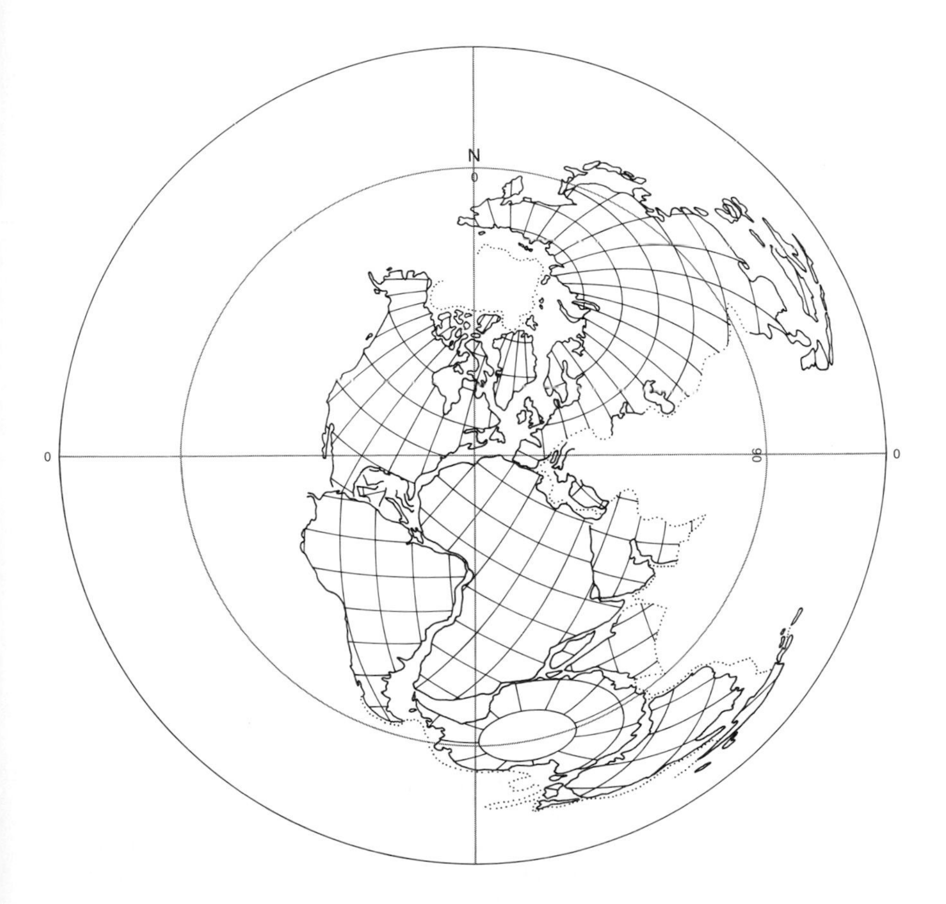

FIGURE 13-11.
World geography about 250 million years ago, during the Permian. The single continent is called Pangaea. The ocean is over 17,000 miles across near the equator. [After Briden, Drewry, and Smith, 1974.]

to two would have entailed the extinction of well over 50 percent of species; the exact percentage would depend upon the pattern and extent of endemism.

Second, the increase in continental mass would raise the "continentality," the extent of perturbation of atmospheric and oceanic conditions, to create more environmental fluctuations, more seasonality, and a consequent increase in the average range of environmental factors in an average place. Fluctuations in trophic resources would have increased in many regions, and as this is presumably a diversity-dependent factor, it would result in an additional drop in the species capacity. The more specialized forms would be at a disadvantage and would tend to form a disproportionately large proportion of extinct lineages.

Third, at about the time Pangaea was finally assembled, a major marine regression occurred, limiting the extent of shallow marine habitats; in theory, this would have increased continentality even further. Again, the species capacity must have fallen. And fourth, the decrease in shelf area accompanying the regression was considerable (Ronov, 1968) and would have led to a decreased capacity for species in the shelf benthos.

From all these causes, which follow from changes in environmental structure and regime due to processes associated with well-established historical events, large numbers of lineages would have become extinct (Valentine and Moores, 1972). The proximal causes of extinction must have varied greatly among the lineages. Pressure from competition or predation by introduced lineages or those lineages best preadapted to the novel conditions, as well as simple habitat or trophic resource failure, operate in a variety of ways to produce extinctions. New environmental configurations must have resulted in extinctions as well as novel adaptations.

So long as the configuration of Pangaea remained stable, marine diversity would remain low. However, during the middle Triassic period and later, shallow seas spread over large portions of the continental platform. This should have ameliorated the "continental" conditions, and indeed recorded diversity did rise, though this may be partly due to the better fossil record left from the more widespread seas. In the Jurassic and Cretaceous periods Pangaea was strongly fragmented into a number of continents that gradually dispersed (Figure 13-11). In shallow water diversity rose steeply as allopatric populations were created by the widening oceanic barriers, and cladogenesis became widespread. As the continents continued to disperse, they assumed a configuration that gave rise to steep latitudinal thermal gradients (Valentine and Moores, 1972; Valentine, 1973b), permitting the formation of chains of marine provinces along north–south trending continental shelves. These provincial chains were repeated on each side of large continents and on each side of oceans, so that the total species capacity increased enormously. Additionally, average environmental stability was doubtless increased by breaking the supercontinent into smaller masses, again increasing species capacity. All this would lead to an increase in species diversity by as much as an order of magnitude, which would easily account for the increased diversity found in the late Mesozoic and Cenozoic fossil record (Valentine, 1973d).

With the rise of the theory of plate tectonics we have achieved the ability to reconstruct in broad outline many important changes in the environmental framework of the past. It has therefore become possible to compare broad features of the biotic and environmental successions. For the Permian–Triassic events and for later times, such a comparison suggests that major extinctions and diversifications represent evolutionary adjustments of the biota to the changing environmental configurations. The lowered sea level and restricted epicontinental seas, responsible in part for the poor geological record of the marine late Permian and early Triassic, are also responsible in part for the extinctions, so that the correlation of these factors can be explained. The later provincialization, which served to greatly enhance diversification, also restricted the range of many species and therefore their likelihood of becoming fossilized. Even so, the fossil record indicates increasing marine diversity during the late Mesozoic and Cenozoic. The broad trends in marine diversification suggested by the fossil record are quite plausible, and are tentatively accepted. It is also plausible that the major Paleozoic trends, revealed in the fossil record, reflect actual changes among the Paleozoic faunas; there is no special reason why this record might be suspect.

If the major trends of diversification are tentatively accepted, then Phanerozoic biotic history may be treated as consisting of two main phases (Valentine, 1973c). The first begins with the late Precambrian appearance of metazoans and the Cambro-Ordovician diversification, and ends with the great late Paleozoic decline in diversity. It was a time of relatively few continents, relatively little climatic zonation in the sea, and low provinciality. The earliest Paleozoic animals appear to have been chiefly detritus feeders and other generalized forms, and to have been of relatively low diversity. Perhaps that was a time of high trophic resource instability, due to continental aggregation or to a special continental configuration; we are not yet certain of late Precambrian and Cambrian geography. It is possible that the invention of the coelom was associated with the onset of unstable trophic regimes that increased the advantages of infaunal detritus feeding. The rise of mid-Paleozoic diversity may be ascribed to increasing provinciality and to increasing stabilization of trophic resources, which permitted specialization; both factors can be associated with separating land masses. Certainly, numerous specialized suspension feeders and predators evolved at this time. This entire phase was terminated by the creation of Pangaea.

The second phase began as Pangaea fragmented, and it continues today. It is a time of the development of many continents, of high latitudinal thermal gradients in the sea, and of high provinciality (there are over 30 marine provinces at present). As we have seen, it is characterized by great increases in diversification at lower taxonomic levels.

Both phases were occasionally punctuated by waves of extinction and diversification (Figure 13-9), and many of the extinctions correlate well with key plate-tectonic events, such as continental collisions (Valentine and Moores, 1970, 1972; Flessa and Imbrie, 1973). There are also many major transgressions and regressions, at least some of which correlate in time with plate-tectonic events. Environmental changes brought about by sea-level changes commonly require extinctions or provide opportunities for diversification (Moore, 1954; Newell, 1956). Paleontologists have not yet worked out the detailed reconstructions of each wave of extinction and diversification. It appears that episodic environmental changes associated with geological processes are adequate to explain them, but we must reserve final judgment until earth and life histories are better known.

The effects of these environmental changes on the taxonomic hierarchy seem to vary among categories. Diversification during the early Phanerozoic from generalized ancestral lineages evidently occurred in a relatively empty biospace during a rising species capacity. This combination of circumstances has never been repeated in the sea. The production of large numbers of quite different ground plans and their major variants occurred relatively rapidly. Although a much larger number of classes and orders arose than exist today (Figure 13-8), they probably did not include many species. When severe waves of extinctions were experienced, the chances of extinction of even classes and orders must have been relatively high. The chances would have been enhanced if the taxon happened to be poorly preadapted to the new environmental regimes. When new diversifications occurred, the vacated biospace would have been repopulated by suitably modified descendants of surviving lineages; a new class or order would not have evolved from some generalized ancestral stock to replace the loss. The newly modified lineages are usually accorded lower taxonomic rank than the extinct lineages, since they are merely variations on preexisting lower taxa. Thus, as wave after wave of extinctions occurred, the number of higher-level taxa tended to decline relative to the lower-level taxa that evolved to fill vacated biospace. Although today there are probably more species than ever before in the oceans—possibly eight or 10 times the average during the Phanerozoic—there are evidently many fewer classes and orders than existed 400 to 500 million years ago (Valentine, 1969).

Some functional modes have been developed time and again in a wide variety of higher-level taxa. Epibenthic suspension feeding is a case in point (Valentine, 1973b). A shallow subtidal rock today may support annelids, cirripedes, bivalves, gastropods, brachiopods, ectoprocts, tunicates, sponges, and even phoronids and crinoids, all feeding on suspended material. There is some partitioning of food among such an assemblage, but then there is some partitioning even between congeneric species, and suspension feeding tends to result in rather similar food intakes in different taxa (Jørgensen, 1966). The ground plans of many of the phyla and classes represented evolved in order to exploit entirely different modes of life besides suspension feeding. Nevertheless, although these very distinct species employ very different principles of food acquisition, their functional convergence is truly great. Lineages of these phyla have evolved down twisting pathways of opportunity, radiating again and again into newly opened biospaces to acquire a wide diversity of novel adaptations, including the adaptations of sessile suspension feeders.

This diversity of adaptation, combined with the large numbers of species of most phyla, classes, and orders today, serves to make the higher-level taxa more immune to extinction. Comparing extinctions of equal severity, fewer higher-level taxa are eliminated in later periods. The last extinction of a skeletonized marine class occurred about 225 million years ago, and the last of an order about 65 million years ago. Peaks of equal height in extinction and diversification curves, such as those for families in Figure 13-9, usually indicate that the earlier wave was more severe with respect to both the percentage of total species affected and eliminations or appearances of distinctive adaptive types.

THE TERRESTRIAL INVASIONS

The development of a skeletonized swimming form, the fish, was an extremely successful event that led to several waves of subsequent radiation. A fish-like mode of life is obviously an apt way to cope with the problems of life in the sea and in terrestrial aquatic environments. It is not certain whether the first fishes arose in fresh or in salt water. Nevertheless, selection created a variety of fish forms in both fresh-water and marine environments, certainly in response to opportunities provided by the mode of life and probably to challenges afforded by environmental changes as well. As selection explored the adaptive potentials of fishes, two morphological innovations proved to have special consequences. These are the development of jaws (from gill arches) and paired fins. Jaws expanded the food sources of fishes, which must have soon become important predators. Paired fins improved swimming balance and orientation, which would promote efficient predation. Jawed fishes quickly diversified and eventually replaced the jawless ostracoderms, both in marine and fresh waters, so that by Middle Devonian time early ray-finned and lobe-finned fishes were well established. The ray-finned forms eventually gave rise to a vast majority of modern fishes. Lobe-fins were represented by lungfishes, coelacanths, and a group called rhipidistians (Figure 13-12). The rhipidistians became extinct near the end of the Permian, but they have an important place in evolutionary history because the amphibians evolved from early members of this group.

In some rhipidistian lineages the endoskeleton was increasingly ossified, legs developed from fin supports, and air breathing (not at all a unique development among fish) appeared. These features were coadapted with many fish-like features to form components of what must have been a highly adaptive mode of life. It is not certain that legs evolved in response to selection for terrestrial locomotion; they may well have originally served for locomotion in very shallow water, perhaps for quick plunges at the small fish that constituted the chief diet of early rhipidistians and amphibians. The earliest known amphibians, the ichthyostegids, had paddle-like tails and sensory systems associated with a lateral line, indicating aquatic habits (Figure 13-13). Their axial skeletons were essen-

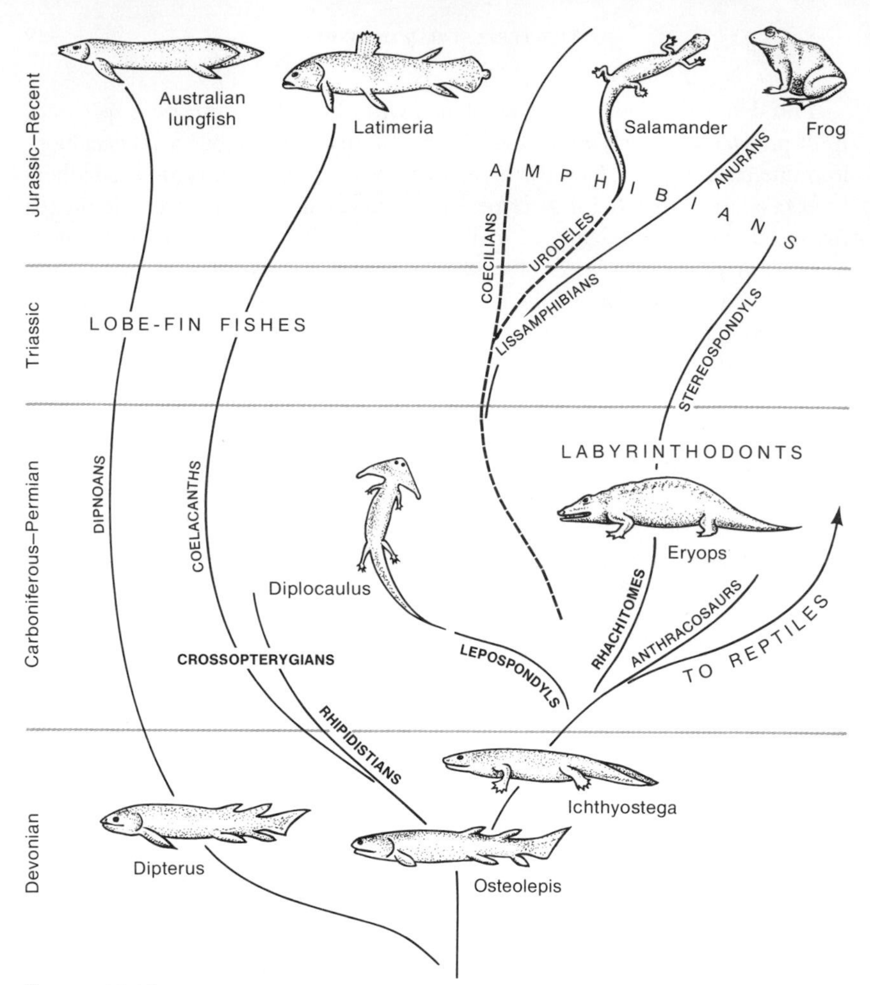

FIGURE 13-12.
Generalized phylogeny of lobe-finned fishes and amphibians. Devonian lobe-fins are classed in three groups: (1) dipnoans, represented today by lungfishes; (2) coelacanths, represented today by the "living fossil" *Latimeria* from deep water in the western Indian Ocean; and (3) rhipidistians, extinct as fishes although all land vertebrates are among their descendants. The tetrapods developed from primitive rhipidistians through such early amphibians as the ichthyostegids, radiating into disparate groups including large extinct amphibians, familiar living amphibians, and primitive reptiles. [After Colbert, 1969.]

tially rhipidistian, but they had bony girdles to support both hindlimbs and forelimbs.

Thus, the ichthyostegids departed from a typical fish-like structure. The fishes have evolved numerous bizarre specializations. We would probably regard the ichthyostegids as merely another interesting evolutionary excursion of the fishes were it not for the fact that they gave rise to the tetrapods, which conquered terrestrial environments. Although many amphibian lineages remained aquatic (some were legless, perhaps neotenic), some groups became adapted to prolonged adult life on land.

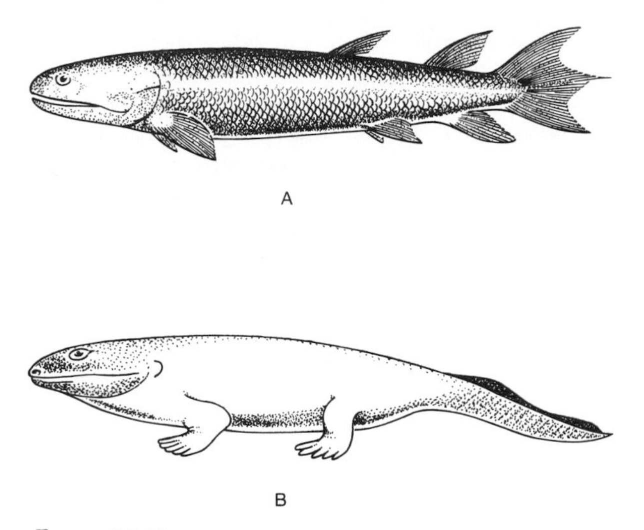

FIGURE 13-13.
The fish–amphibian transition. **A.** A progressive lobe-finned rhipidistian fish *(Eusthenopteron)* of the Devonian; animals of this sort were ancestral to the amphibians. **B.** Early ichthyostegid amphibian of the Devonian, probably highly aquatic, descended from a *Eusthenopteron*-like stock.

As the fossil record of terrestrial habitats is generally poor, the course of the biotic invasions of the land is known only in broad outline. Plants must have been established in fresh water from at least the early Paleozoic, and invertebrates no doubt followed closely. Fish may have originated in fresh water, and if not, were represented there at least by Silurian time. During the middle Paleozoic or earlier, life moved beyond the edges of fresh-water lakes, ponds, and swamps to become established on land. The earliest known vascular (land) plants are Silurian, and probably lived in damp regions near fresh-water bodies. Arthropods must have evolved terrestrial habits soon after, perhaps living at first in subterranean cracks (Ghilarov, 1959) and feeding on the plant detritus then enriching the swampside soils; soon they were feeding on the plants themselves. By Devonian time the green belts were spreading widely into lowland terrains and trees appeared; also at this time flying insects developed. Some rhipidistians followed the burgeoning land biota into terrestrial environments, developing into amphibian tetrapods during the Devonian. Probably their terrestrial foods were chiefly arthropods.

The pressures that cause lineages to evolve in novel habitats are numerous. Populations suffer much internal competition between individuals for food and space, and commonly there is much competition from populations of other species as well. Furthermore, populations are usually heavily preyed upon, and may be subjected to periodic or regular inclement physical conditions. Given the opportunity, selection will act so as to minimize these pressures, even if this requires pioneering in an unusual environment. Scientists have frequently attempted to identify particular pressures as the most important in given cases. For early amphibians, for example, one theory suggests that desiccation of ponds led the ancestral tetrapods to migrate overland to larger pools, and thus to acquire gradually terrestrial adaptations, including limbs (Romer, 1958). Another theory suggests that overpopulation may have caused rhipidistian–amphibian stocks to evolve some independence from the aquatic habitat and to exploit terrestrial food sources for parts of their life cycles (Inger, 1957). We are not really sure what the pressures actually were; the complexity of ecological interactions is so great that even if the situation had been studied at the time, an accurate assessment of the contributions of all relevant factors might not have been forthcoming. Clearly, preadaptations commonly permit the successful invasion of a new habitat, especially when the creation of novel modes of life are involved. Partly for this reason, it is tempting to favor the idea that limbs were originally an aquatic adaptation.

The primitive amphibians became increasingly adapted to terrestrial life; improvements were achieved by different lineages at different times. The body could have been raised from the ground as a locomotory improvement and skin layers thickened to impede desiccation. Stahl (1974) gives an account of the functional modifications that occurred in early amphibians. The most diverse amphibian group to appear during the Carboniferous was the labyrinthodonts, which include the primitive ichthyostegids as a stem group and which diversified into a number of distinctive phyletic lineages (Figure 13-12). As has so commonly happened, it was one of the earlier lineages (the anthracosaurs) that eventually gave rise to the next major advance in grade.

REPTILES, DINOSAURS, AND MAMMALS

Reptiles are known from the late Carboniferous, but may have descended from an amphibian stock that diverged from other anthracosaurs in the early Carboniferous. They possess eggs that can develop out of water, and skins that prevent undesirable water loss to the environment. In principle, these features free the reptile from necessary dependence upon life in water at any stage of the life cycle. The detailed adaptive pathways that led to the reptiles are not yet understood. Important steps along that pathway seem to have included the development and elaboration of a membrane around the embryo, presumably for respiration at first, the subsequent acquisition of a horny protective covering around the egg to prevent desiccation, and finally the mineralization of the egg shell. Physiological and structural changes required of the embryo itself include the evolution of physiological systems to produce safe metabolites that are relatively nontoxic and can be stored in an enclosed egg. The reptilian egg may have evolved to prevent desiccation in impermanent ponds or drying swamp shallows, or to protect eggs from the numerous aquatic predators, which was accomplished by depositing them on land.

Many of the early reptiles, the captorhinomorphs, lived chiefly or entirely in water (Romer, 1959), but some seem to have inhabited forests fringing large swamps (Carroll, 1969). They soon diversified, and from their basal stock arose distinctive lineages, including the pelycosaurs (Figure 13-14). These forms spread into moist terrestrial lowlands, which they eventually shared with advanced captorhinomorphs and other reptiles. Clearly, their integuments were proof against excessive desiccation. Numerous modifications of their skeletal morphology occurred as they radiated into a variety of habitats, developing modes of life that included feeders on plants, on fish, and on each other. Probably by early Permian time a group of active predators had split off from the pelycosaurs, in turn diversifying into herbivores as well as carnivores of

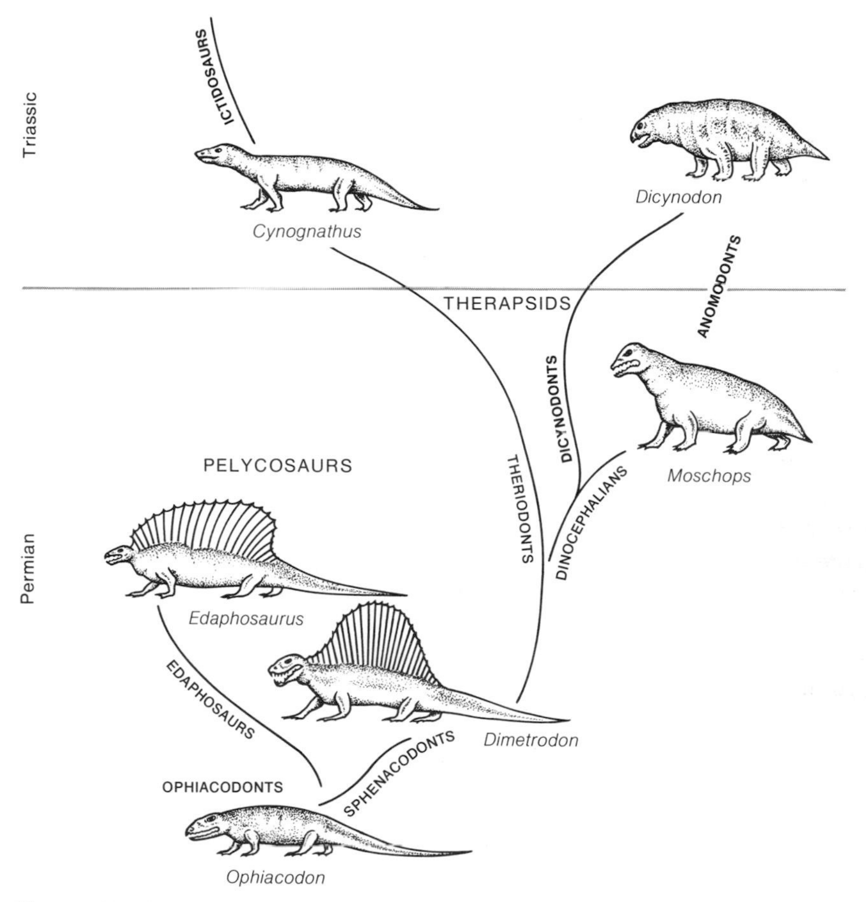

FIGURE 13-14.
Generalized phylogeny of pelycosaurs and the therapsid reptiles, from which mammals evolved. Primitive captorhinomorph reptiles (such as *Ophiacodon*) gave rise to pelycosaurs, of which one branch (characterized by *Dimetrodon,* a predator) produced the therapsids. The therapsids diversified, with one branch (characterized by *Cynognathus,* a predator) giving rise to mammals, possibly through a little-known group called ictidosaurs. [After Colbert, 1969.]

several types and coming to dominate a wide range of middle and upper Permian terrestrial habitats. These were the therapsids (Figure 13-14). It is from predatory therapsids that mammals evolved. Developments in axial skeletons, limbs, and skulls in several therapsid groups in the Triassic produced features that were to be emphasized and continued in mammals (Romer, 1959; Stahl, 1974). It is possible to reconstruct, from clues presented by therapsid skeletons, features of soft-part anatomy that are now possessed by mammals, though it is not really certain which of these were acquired by therapsids and which by their early mammalian descendants. Such features include the development of a diaphragm for breathing, hair, improved agility, and the facial skin and muscles that permitted suckling of young; even warm-bloodedness may have appeared in therapsids.

Because so many of these distinctively mammalian characteristics are difficult or impossible to infer from skeletal remains, and since they were doubtless acquired at different times, paleontologists have had to adopt arbitrary skeletal definitions of the Mammalia (Figure 13-15). Suggested requirements for mammalian status include jaws that contain a joint between dentary and squamosal bones, and teeth (especially molars) that are more complex than those of reptiles and that are not continually replaced during life. This definition is especially practical, since teeth and jaws are common fossil items. Based on such fragmentary fossils, we can construct a phylogenetic model for late therapsids and early mammals that depicts mammals as radiating from a single founding lineage (Hopson and Crompton, 1969). It is not certain that this is in fact the

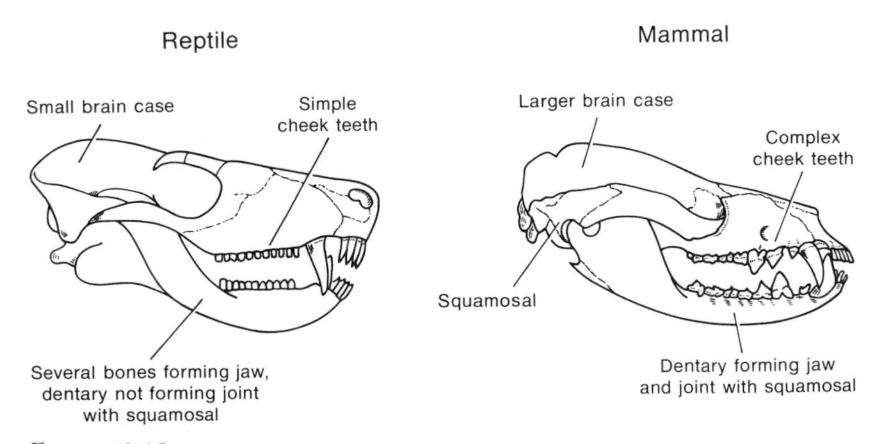

FIGURE 13-15.
Comparison of reptilian and mammalian skull features. The mammals have a larger brain case, differentiated dentition, and a single lower jaw bone, the dentary, which forms a joint with the squamosal bone at the base of the skull. In mammal-like reptiles there are additional jaw bones, one of which, the articular, joins with the squamosal. Two joints are present in ictidosaurs.

historical case. The egg-laying monotremes (platypus and echidna) may represent a lineage descended independently of other mammals.

Mammals diversified and modernized at first only gradually; in some lineages the trend toward a longer fetal period eventually led to the appearance of the placenta, probably during the early Cretaceous and possibly in Asia. By the close of the Cretaceous, placentals were undergoing the diversification that produced the primates (Van Valen and Sloan, 1965); this diversification extended into the early Tertiary and gave rise to modern mammalian groups. Throughout their long Mesozoic history, which embraces fully two thirds of their existence, mammals were completely overshadowed in abundance, diversity, and size by reptiles, particularly by the dinosaurs. Dinosaurs probably arose from a lineage that diverged from the captorhinomorphs, only a few million years before the mammals appeared. The results of their spectacular radiations are well known, particularly the much depicted large brontosaurs, stegosaurs, ceratopsians, and the great predators. Birds also developed from a dinosaur stock. As mammals have now become far more successful than reptiles, the questions arise why reptiles were so successful in the Mesozoic, and why they have declined to their present status. There are now plausible, though unproven, answers.

These answers have emerged from studies, by Ostrum and others, that have reevaluated such features as posture and thermoregulation in reptiles and mammals. Bakker (1971) has proposed a model in which dinosaurs and early mammals each inherited a sprawling posture from their (separate) reptilian ancestors. By late Triassic time dinosaur lineages had achieved erect posture and probably became quite active, though mammals were still sprawling or semi-erect. Several of the larger dinosaur groups, including the famous predator *Tyrannosaurus*, are envisioned as having fast locomotory gaits. This indicates high body temperatures and suggests homoiothermy. If dinosaurs did attain this physiological grade, it would be easier to explain their numerous adaptive radiations, for we know that the level of activity provided by homoiothermy opens wide arrays of life modes similar to those later developed among larger mammals. It would also account for the apparent ease with which dinosaurs prevented mammals from encroaching upon their biospace; mammal homoiothermy would not have conferred any special competitive advantage. Indeed, visions of charging tyrannosaurs have led some paleontologists to be grateful that dinosaurs have become extinct.

Although some dinosaurs were rather small, down to about the size of a rooster, they were larger than contemporary mammals; even their hatchlings were larger than most of their adult mammalian contemporaries. Large animals have surface–volume ratios favorable to heat retention; and since dinosaurs lacked obvious insulating features, such as feathers or fur, they might have been unable to maintain body heat with small sizes. Early mammals, though small, were furry and could thus achieve homoiothermy despite their small sizes. According to this model, then, dinosaurs and mammals each achieved homoiothermy and high activity levels; the former radiated into biospace appropriate for large animals, the latter for small ones. It was not until dinosaurs declined that mammals radiated into the vacated habitats and took up many of the abandoned modes of life. Although the notion that dinosaurs were warm-blooded has been attacked, it has also been stoutly defended, and remains an attractive hypothesis. (For an interesting exchange between detractors and proponents, see the articles in *Evolution,* 28:491–504, and references therein.)

What extinguished the dinosaurs? There is no lack of hypotheses, but none has been adequately tested. Since dinosaurs descended from reptiles, they traditionally have been considered as ectotherms. Cold-blooded terrestrial animals are restricted in habits and habitats by thermoregulatory problems. Climatic changes to cool weather, to hot weather, or to highly seasonal climates, could in theory easily exceed dinosaur ectothermal potentials, and have often been suggested as a major factor in their extinction. Even if the dinosaurs were homoiotherms, their hatchlings would have been too small to be able to cope with cool temperatures (Stahl, 1974). Climatic change thus remains a likely cause of dinosaur extinction, partly because it would have a wide variety of secondary effects that could increase dinosaur mortality and contribute to breakdowns in their population structure (for example, food sources could be affected).

PRIMATES

Although the placental mammals originated while dinosaurs still existed, the Mesozoic lineages were chiefly small, generalized insectivore-like forms. However, they began to differentiate somewhat near the end of the Cretaceous and underwent an explosive diversification that established most modern orders during the Paleocene, filling many places in terrestrial ecosystems that had been vacated by extinct dinosaurs. As the post-Pangaean continents became increasingly isolated owing to plate-tectonic processes, distinctive mammalian assemblages developed on different continents, although there were some subsequent interchanges (as across the Bering Strait region and the Panamanian Isthmus).

The earliest primates appeared among the late Cretaceous orders of placental mammals, differing probably only a little from the insectivores from which they arose. Simons (1972) gives a careful resumé of fossil primates. Cretaceous primates are represented only by teeth, so that little is known of their habitats, and Paleocene deposits have yielded few postcranial remains. In the Eocene rather modern, lemur-like prosimians appeared. These Eocene primates had shorter faces and larger eyes than their ancestors, and probably had arboreal habitats. As rodents were differentiating at this time with much apparent success on the ground, the adoption of arboreal habits by primates may have been due in part to their exclusion from forest floors.

Monkeys arose in the late Eocene or early Oligocene. The living Old World (catarrhine) and New World (platyrrhine) monkeys seem to have been separated for a long time, and it is sometimes suggested that they arose independently from non-monkey ancestors, thus representing a case of parallel evolution. This is possible, but their common ancestor would have to have been well removed from the prosimian stocks that have living descendants; all monkeys and apes have proven to be rather similar when compared by techniques of molecular biochemistry, and rather different from living prosimians (Goodman, 1976; see Chapter 9).

Apes are catarrhines; they are not known for certain until the Oligocene (see Chapter 14). The earliest known ape (*Aegyptopithecus*) had a small brain, a tail, and hands and feet that indicate an arboreal, quadrupedal life. By Miocene time several species assigned to the genus *Dryopithecus* had appeared. Modern chimpanzees, orangutans, and gorillas probably descended from this assemblage. Man seems to have descended from the catarrhines also (Chapter 14).

There is nothing very unusual in the evolutionary processes that gave rise to the hominid lineage, although the evolution of the human brain has certainly had spectacular consequences—but so did the evolution of the reptilian egg, the development of the chordate skeleton, the epifaunal radiation of the invertebrates, and the invention of the coelom, to choose only a few of many landmark events in animal evolution. In each case the innovations evolved through selection and solved an immediate adaptive problem that opened the way into vast regions of biospace.

Evolutionary advances have been steadily accompanied by extinctions, which are clearly an integral part of the evolutionary patterns. They seem an inevitable consequence of adaptation to the ambient environment in a changing biosphere. Peak extinction levels probably correspond with reductions in the capacity of the environment to accommodate species. Today, diversity appears to be near a record high. Evolutionists therefore have a rich biosphere to study, furnished with a wide variety of organisms, communities, and provinces. One wonders what the future of this biosphere may be, and whether the potentials are as great as previously. We shall discuss such questions after considering the special properties of one last species, our own.

14

Evolution of Mankind

Man, *Homo sapiens,* is only one of several million species now living. Yet science devotes more attention to the study of this single species than to all others put together. This is as it should be: we are humans, and anything pertinent to understanding man is important and interesting. It may even be argued that all biology, and even all science, should be anthropocentric, i.e., relevant to man. We study the biological world, inanimate nature, the planet earth, and the remotest galaxies in part at least in order to understand ourselves and our place in the cosmos, to comprehend who we are, where we came from, and where we may be going.

The ancient Greek philosophers assumed that the world was serenely fixed and unchanging. Although changes were observed happening everywhere, these changes were declared illusions. Plato believed that individual human beings were more or less distorted copies of a sublimely beautiful, immutable Idea of Man. Wisdom grew from the ability to discern this ideal in its imperfect manifestations in the women and men we actually see. Plato's world view was static. The evolutionary world view first appeared in Judeo-Christian thought.

In the Middle Ages people believed that the world was created for the sake of man. In the Renaissance Copernicus, Galileo, Kepler, Newton, and others discovered the immensity of the universe, and man appeared to recede in importance in the cosmos. It seemed hardly credible that so vast a universe existed for so relatively miniscule an object as man. In the seventeenth century Descartes made a gallant attempt to restore man to center stage. He believed that the universe is a machine, and that living bodies are also machines. Descartes likened the body of a living man to a watch in good order, and a dead body to the same watch "when it is broken and when the principle of its movement ceases to act." All living bodies can be understood in terms of the physical principles that operate in nonliving nature. Thus Descartes was, as a biologist, a mechanist rather than a vitalist. Yet he drew a sharp distinction between humans and animals. He regarded animals as mechanical automata, lacking consciousness and souls. Humans, on the other hand, had souls, and this distinction alone set man above the rest of nature.

The Cartesian view proved unconvincing. No cogent explanation of how a nonmechanical soul could govern a mechanical body was forthcoming. In fact, it was argued that if the soul had even the slightest influence on material processes, it would therefore be just another mechanical force. Most philosophers in the eighteenth century decided that man was a machine-like animal, albeit a more complicated one. When Linnaeus put forward a classification of living beings (1735-1766) he unhesitatingly placed man, together with the apes, in the zoological order of primates (Anthropomorpha). To Linnaeus this did not mean that mankind and apes were related by descent from a common ancestor. In Linnaeus's time every existing species was regarded as separately created, and in about the same condition in which we observe it now. Nevertheless, in the eighteenth and early nineteenth centuries, biologists such as Buffon, Tyson, Camper, and Cuvier could not fail to be impressed by implications of finding more and more basic similarities between human and animal bodies. And among animals, apes and monkeys resemble us most.

In his great work *Origin of Species* (1859) Darwin refrained from dealing with human evolution. Only near the end of the book did he write that with evolutionary studies "Much light will be thrown on the origin of man and his history." Darwin's forbearance was in vain. His "light" dazzled those to whom darkness was more comfortable. Vehement attacks on Darwin and his theory were launched before as well as after *The Descent of Man* (1871), which dealt specifically with the evolutionary origins of mankind. The idea that man could somehow be related to all living things seemed to conflict with established religious dogma. To have apes and monkeys as our closest living relatives seemed to deprive the human estate of its vaunted dignity. Although enlightened philosophers and theologians saw that these allegations were groundless, the urgent task that biologists had to face was to prove that the idea of common descent was valid.

Evidence had to be assembled to demonstrate beyond doubt that the living world is the outcome of evolutionary development. Evidence of human evolution had to be particularly strong, because where man was concerned the conflict with traditional beliefs was most acute. The mutually sustaining efforts of biologists, paleontologists, and anthropologists have made the theory of the evolutionary descent (or rather ascent) of man impregnable. Whoever takes the effort to familiarize himself with the relevant evidence cannot doubt that humans have evolved from nonhuman ancestors. And yet some antievolutionists persist. Some of them are simply ignorant of the evidence, while others have so prejudged the question that no evidence is meaningful to them.

STRUCTURAL SIMILARITIES AND DIFFERENCES

More than two millennia ago Aristotle classified man among the species of warm-blooded animals. He observed that "Of all animals man has the largest brain in proportion to his size; and it is larger in men than in women." He also noted mankind's uniqueness when he wrote, "Man is a political animal." In 1555 Belon compared a human skeleton with that of a bird, and identified the homologous bones. In 1699 Tyson dissected an infant chimpanzee and found 48 traits in common with man and only 34 traits in common with monkeys. In 1863, i.e., after the publication of *Origin of Species* but before *The Descent of Man,* T. H. Huxley produced his greatest work, *Man's Place in Nature.* A thorough analysis of all available data led him to conclude that ". . . man differs less from the chimpanzee or the orang, than these do even from the monkeys, and that the difference between the brains of the chimpanzee and of man is almost insignificant, when compared with that between the chimpanzee brain and that of a lemur." He invites his readers to ". . . find in the lowly stock whence man has sprung the best evidence of the splendor of his capacities; and will discern in his progress through the Past, a reasonable ground of faith in his attainment of a nobler Future."

The discovery of remarkable similarities in body structure between man and animals need not hide the existence of important differences. Man's normal posture while walking is upright. To be sure, he is not the only bipedal animal, not even the only bipedal mammal. Kangaroos and some rodents move by leaping on their hind legs; many others occasionally stand up in order to see their surroundings better. But only man can comfortably stand and walk with his trunk erect. The transition from the quadrupedal locomotion of our remote ancestors to our bipedal stance ushered in numerous structural changes. The bones of the pelvis and the musculature connecting the pelvis with the legs and the spine underwent changes in shape and strength (Figure 14-1). To support the weight of the abdominal viscera, the pelvis became bowl-like. The vertebrate column has developed an S-shaped curve. The conformation in man of the bones of the pelvis and the sacrum is so characteristic that whenever these skeletal parts are preserved in fossil relatives and ancestors of our species, it is possible to infer with a high degree of assurance whether the respective animals did or did not walk erect. Bipedal, instead of quadrupedal locomotion also led to changes in the human foot. To support the weight of the body the tarsal and heel bones became strong and closely articulated. The toes, however, became relatively short; the great toe is not opposable to the other digits, as it is in most apes and monkeys.

The arms of apes are longer than their legs, while in man the opposite is true (Figure 14-2). Relieved of use in walking, the anterior extremities in man became shorter and less powerfully muscular than the legs; they are specialized for precision manipulation of various objects. The ability to use and make tools for all kinds of purposes initiated the invention and perfection of a variety of technologies. Man is a tool-using and tool-making animal (Oakley, 1961). Some pessimists believe that for man to call his species *Homo sapiens* (man the

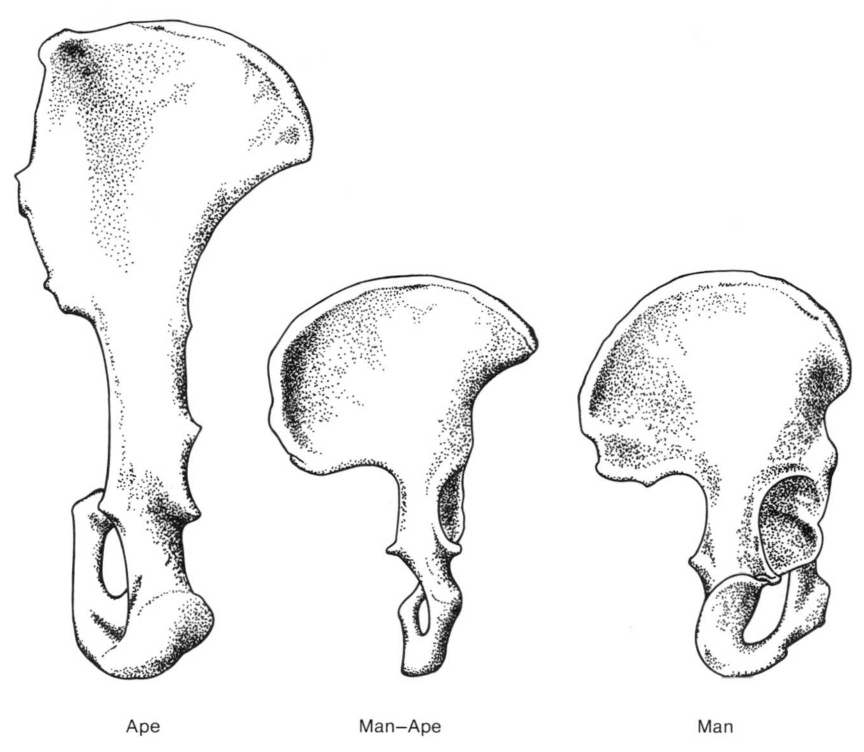

FIGURE 14-1.
The pelvises of apes, man-apes *(Australopithecus)*, and man, reflecting differences between quadrupeds and bipeds. The upper part of the human pelvis is wider and shorter than that of apes. The lower part of the man-ape pelvis resembles that of apes, whereas the upper part resembles that of man. [After Washburn, 1960.]

wise) is unwarranted arrogance, but it cannot be gainsaid that he is *Homo faber* (man the maker). The development of technologies contributed powerfully toward the biological success of the species. Man is unquestionably the most generally competent and dominant outcome of the evolutionary process. This statement should not be construed to mean that a breakdown and extinction of the human species is unthinkable—many successful species have ended by dying out. But no species before man has had any directing influence over its evolution. Hence, if mankind were to become extinct, it would be the first instance of evolutionary suicide of a biological species.

The biological, evolutionary, ascendance of man is due not to bodily prowess but to the preeminence of his brain. The human brain is large and has a remarkably expanded frontal lobe. To be sure, the human brain is not the largest in the animal kingdom. Elephants and whales have still larger brains, but they are relatively smaller in proportion to body size. Table 14-1 gives some data on cranial capacities in higher primates. Although adult gorilla males are about two and a half times heavier than human males, their brains weigh less than half the human brain.

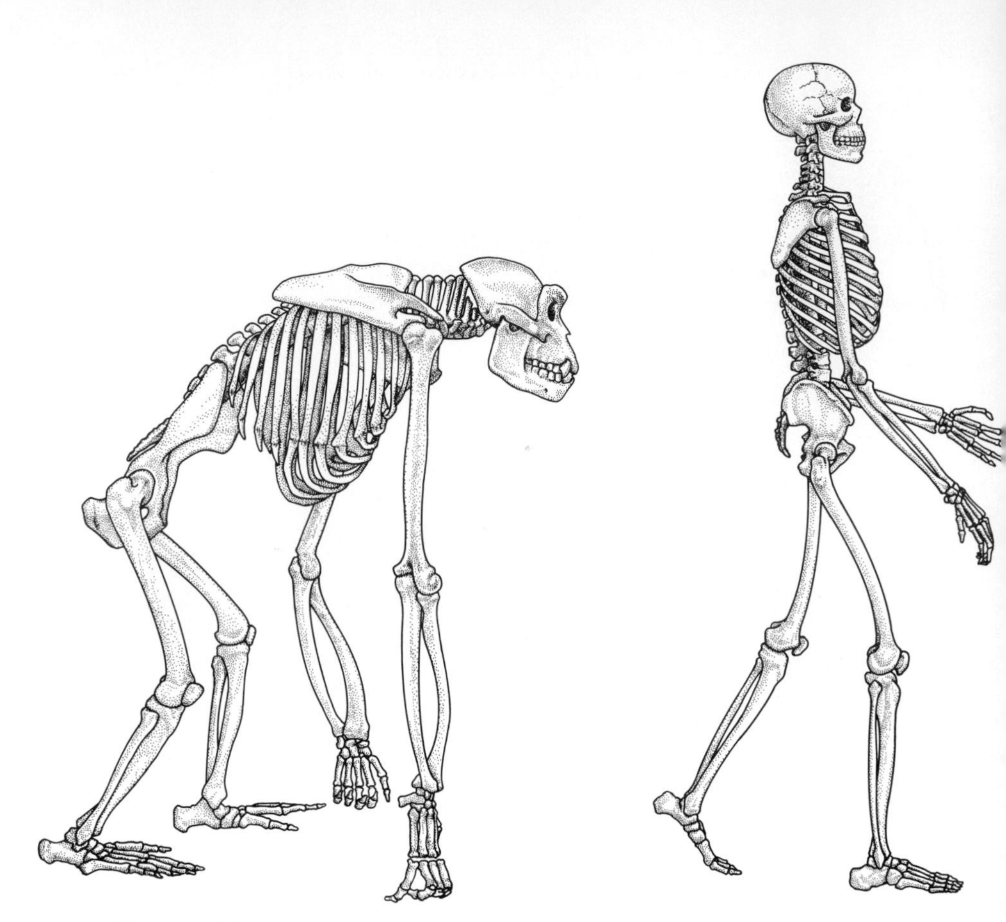

FIGURE 14-2.
Skeletons of ape and man. In apes the arms are longer than the legs; in man the opposite is true. The long, straight pelvis of the ape provides support for quadrupedal locomotion; the short, broad pelvis of man curves backward, carrying spine and torso in bipedal position. [After Washburn, 1960.]

TABLE 14-1. Means and approximate ranges of cranial capacities in higher primates. [Data from Tobias, 1971; Campbell, 1972.]

	Mean (cm^3)	Range (cm^3)
Hylobates lar (gibbon)	103	82–125
Pan (chimpanzee)	383	282–500
Gorilla (gorilla)	505	340–752
Pongo (orangutan)	405	275–540
Australopithecus africanus	588	435–815
Homo erectus	950	755–1225
Homo sapiens	1330	1000–2000

Even though the human species is the most successful outcome of the evolutionary process, it is not free of weaknesses and flaws. This is by no means strange. Natural selection is the guiding agent of evolution, but it is not an all-seeing and all-wise pilot. It adapts, as best it can, a living species to the environments prevailing in a given place at a given time, but it cannot know the future. Hence, biological species frequently combine excellence in some of their parts with astonishing imperfections in others. The human species has both.

Although the enlargement of the brain in man has made our species the pinnacle of creation, it has also caused problems in restructuring some anatomical and physiological traits. The pelvis of a mammal must have a large enough opening through which the infant can pass at birth. Although the female pelvis is somewhat wider than in the male, the globular head of the infant with its voluminous brain is barely able to squeeze through. Childbirth in man is agonizingly painful, which is a biological absurdity. The performance of a function so essential for the perpetuation of the species should, theoretically, be as pleasurable to the individual as copulation is, and for the same reason. The hardship of childbirth has imposed a limit on further enlargement of the head and consequently of the brain. It has also caused human infants to be born at a less-advanced stage of development than other primates. Soon after birth, infants of apes and monkeys are able to hold firmly onto their mother's body. Human infants are utterly helpless at birth; they have to be carried and succored by the parents for considerable time. To be sure, the prolonged childhood and delayed sexual maturity in man has been turned to good use. It gives time for enculturation and training, essential for a human individual to become a functional member of his society. This is an example of preadaptation (scc Chapter 4).

The restructuring of the human skull has been considerable. To balance the head on top of a vertically held spinal column, the opening of the brain capsule (foramen magnum) and the neck joint had to be moved forward to the middle of the bottom of the skull. The face became shortened, the palate small and U-shaped, the dental arch rounded, and the teeth decreased in size. This is particularly true of the canine teeth, which in other primates are large enough to be used for defense and offense; human canines do not project perceptibly above the rest of the dentition. Instead of using his teeth, man defends himself with his hands, which carry the weapons he invents. Whether or not the degeneration of the teeth, which makes the services of dentists so necessary, can be considered evolutionary degradation is an open question.

CHEMICAL SIMILARITIES AND DIFFERENCES

All living bodies are astonishingly similar in chemical composition. Not only do they contain atoms of the same elements, but often in similar proportions. The same classes of compounds, particularly nucleic acids and proteins, are found everywhere. Proteins are composed of the same 20 amino acids, and the amino acids, with rare exceptions, are represented only by left optical isomers. Energy carriers (such as adenosine triphosphate, ATP) and enzymes with identical functions (such as cytochrome *c*) are present in most diverse organisms. To an evolutionist, these chemical similarities are meaningful; they affirm that man is kin to all that lives.

More novel and more interesting are the differences between related compounds in man and other animals. Though some enzymes may play similar physiological functions in man and other organisms, they are often more or less distinct in their chemical composition. The achievements of molecular biology have made possible the detection and quantification of such distinctions. Hemoglobins are oxygen-transporting pigments in the blood of all vertebrate and some invertebrate animals. In adult humans the most abundant form of hemoglobin is A, each molecule of which consists of two alpha and two beta chains, composed respectively of 141 and 146 amino acids and an iron-containing heme group (see Chapter 9).

The sequences in which the amino acids are arranged in human alpha and beta hemoglobin chains have been worked out by Ingram and many other investigators (see Figure 9-17, p 295). Many mutational variants have been discovered, most of them carried in very few or even in single individuals. However, some variants, such as hemoglobin S, are widespread in populations of large geographic regions, where they confer on their carriers a relative immunity to endemic diseases, especially to malarial fevers. A large majority of mutant hemoglobins differ from the prevalent, or "normal," hemoglobin A by substitution of single amino acids in either the alpha or beta chains. Comparison of the amino acid sequences in the hemoglobins of man and other animals is fascinating. The hemoglobin A of the chimpanzee is identical with the human one. The gorilla alpha chain differs from the human one in a single amino acid substitution: glutamic acid in man and aspartic acid in the gorilla at position 23 in the chain. Between man and the gorilla there is also one amino acid difference in the beta chain. The differences between human and gorilla hemoglobins are not much greater than variants in human individuals. One is tempted to ask: could a man live with gorilla hemoglobin, and vice versa? So far no answer to this question is forthcoming.

The chimpanzee and gorilla were judged to be man's closest relatives on the basis of their external and anatomical similarities, as well as their chromosome numbers (46 in man, 48 in the chimpanzee and gorilla; see Chapter 3). The amino acid sequences in their hemoglobins confirm this judgement. Zuckerkandl, Buettner-Janusch, and their co-workers found that among primates morphologically less similar to man than the chimpanzee and gorilla, the hemoglobins are also less similar. The alpha chain of the baboon differs from human alpha in several amino acids, and the beta chains show even more differences. A comparison of man and *Lemur fulvus* shows six substitutions in the alpha chain and 23 substitutions in the beta chain. Studies by many workers disclose the following amino acid substitutions in the hemoglobin chains of various animals compared to man (these figures show so-called mutational distances, i.e., the minimum numbers of nucleotide substitutions in the DNA of the genes needed to yield the amino acid changes observed; see Chapter 9):

Alpha Chains		*Beta Chains*	
Gorilla	1	Gorilla	1
Macaque	5	Macaque	10
Mouse	19	Mouse	31
Sheep	26	Sheep	33
Pig	20	Pig	28
Horse	22	Horse	30
Rabbit	28	Rabbit	16
Chicken	45	Kangaroo	54
Carp	93		
Lamprey	113		

The similarities as well as the differences in the above list are remarkable. As the animals examined become more and more obviously different from man, their hemoglobins become increasingly different. A single mutation could conceivably transform the alpha or beta chains of the gorilla into the respective chains of man; the coincidence of the several mutations needed to transform a human alpha chain into that of a mouse, or even of a macaque, has a probability indistinguishable from zero. It took the accumulation of many separate mutations over millions of years to build up the differences observed. And yet, even the hemoglobins of organisms as different as man and chickens, or man and fish, have preserved similarities in amino acid arrangements that are far too great to be explained by chance. The similarities are marks of common ancestry.

Several similarities in amino acid sequence have also been observed in comparisons of other proteins of man and various organisms. The work of Margoliash, Fitch, Jukes, and their colleagues on cytochrome *c*, myoglobins, fibrinopeptides, and other proteins discloses increasing divergence of the amino acid sequences in more and more different organisms (Chapter 9).

MAN'S EARLY ANCESTORS AND RELATIVES

When Darwin put forward his conclusion that the human species evolved from ape-like ancestors, the evidence of fossils was almost wholly missing. Of course, in 1856 the skull of the Neanderthal Man had been unearthed in Germany, but the interpretation of this find was at once mired in controversy. Some authorities deemed the Neanderthal to be a pathological specimen of modern man, while others saw it as a "missing link" between *Homo sapiens* and his ape-like ancestors. The brain size of the Neanderthal was equal to or even slightly above the average for modern man, but the shape of the brain capsule, the presence of heavy brow ridges, and a strong and chinless mandible made him distinctive. Numerous Neanderthal skulls and skeletons have since been found in a territory stretching from France and Spain to Palestine and central Asia. Some authors regarded him as a separate species, which became extinct and was replaced by *Homo sapiens.* At present, the consensus is that the Neanderthals were a race, a subspecies, of man, *Homo sapiens neanderthalensis,* adapted to the severe climatic conditions of the last Glacial Period (called Würm in Europe, and Wisconsin glaciation in America). It is quite probable that some Neanderthal genes have survived and are carried in living men.

Although prehuman fossil remains are rare relative to those of many other animals, the interest in human ancestry led many investigators to search for remains all over the Old World (Pilbeam, 1972). Numerous discoveries have been made during the last half century, and it seems that the rate of discovery is accelerating. The literature on fossil man is enormous, but riddled with contention and confusion. This state of affairs has two causes; one is the fragmentariness of the remains, requiring some guesswork in reconstruction, the other is an unnecessary proliferation of identifying names. The former can be helped by making new and more complete finds, the latter by greater restraint on the part of the discoverers, who tend to describe every scrap of fossil bone as a new species or even genus (Campbell, 1974). Some contend that this is innocuous—what, after all, is in a name? In fact, however, the proliferation of names is seriously confusing. Suppose that at some period of the past two or several hominid or prehominid species lived simultaneously. Only one of the species could be our ancestor; the rest must have become extinct without issue. On the other hand, a species that at one time level was divided in any number of races or subspecies could, at subsequent time levels, be transformed into and remain a single species, whether it subdivided into races or not. Theoretically, it could also split into two or several species. In the account that follows we give what seems to us the simplest and most probable version of the evolutionary development of man. The reader is warned that other researchers have somewhat different interpretations.

As early as middle Miocene times, 15 to 25 million years ago, there lived in Africa, Europe, and Asia several species of apes, constituting the subfamily Dryopithecinae, which were probably the common ancestors of the present-day apes (Pongidae) and humans (Hominidae). In late Miocene and early Pliocene, eight to 14 million years ago, there lived in India *Ramapithecus,* and in central Africa an identical or very similar form (needlessly given a separate generic name, *Kenyapithecus*). Unfortunately, only some fragments of jaws and teeth have been found, but these bones seem to be intermediate between those of the earlier Dryopithecinae and the later Hominidae. If *Ramapithecus* is indeed a hominid (and not everyone agrees that it is), the hominid and the pongid stems diverged some 15 to 20 million years ago (Simons, 1972).

The southern man-apes, australopithecines, first found in South Africa in 1924, are now almost unanimously accepted as links between the ape-like and human stages of our ancestry. They lived from at least five million to perhaps less than a million years ago (the dates are not as certain as we would like to have them). Most of the known remains have been found in southern and east-central Africa, with some very fragmentary and doubtful finds in Indonesia. There existed at least two species—a larger and less man-like *Australopithecus robustus (= Zinjanthropus boisei)*, and a smaller, more gracile *Australopithecus africanus*. The former apparently died out without issue, while the latter is our likely ancestor—although at least one anthropologist contended that we are dealing not with two species, but with males and females of a single species!

The body structure of the australopithecines is a fascinating mixture of human, ape, and intermediate traits. The shape of their pelvis and leg bones leaves no doubt that these creatures walked erect (Figure 14-1). Their dentition was also more human-like than ape-like. And yet their brains were small: 450–600 cm^3, about the size of a gorilla brain. Of course, the gorilla is a much larger animal, so that the *Australopithecus* brain was not small in relation to the body size. We would probably not have recognized their faces as human at all. Most significant is the fact that at least some of the australopithecines were making and using very primitive stone tools ("Oldowan industry"). Although van Lawick-Goodall (1971) has observed that chimpanzees in the wild on occasion also make some tools (such as sticks for the extraction of termites from their nests), the performance of the australopithecines is on a different level. The remains of *Homo habilis,* found by Leakey and studied in detail by Tobias, are associated with stone tools (hence the name *habilis*). The bones of *Homo habilis* are in many ways so similar to those of *Australopithecus africanus* that placing the former species in the genus *Homo* together with modern man is questionable. This is less consequential than it may seem, however; while a species, at least in sexual outbreeding organisms, is a biologically nonarbitrary entity, a genus is a matter of convenience. *Homo,* or *Australopithecus, habilis* represents a significant step from animality to humanity (Figure 14-3).

Homo (or *Pithecanthropus*) *erectus* is another link in the history of man's evolutionary emergence. Remains of *Homo erectus* were discovered in Java in 1891, more than a quarter of a century before the australopithecines, by a Dutch physician E. Dubois (Figure 14-4). The estimated cranial capacity of this creature was 850–900 cm^3; this is about midway between chimpanzees and modern man, distinctly greater than in the australopithecines. The conformation of the femur bone clearly shows that the creature walked erect. Remains of races of *Homo erectus* were subsequently encountered in several places in Asia, Africa, and Europe. The best known is Peking Man, *Homo erectus pekinensis;* the remains of at least 40 individuals have been discovered. Peking Man lived 500,000–800,000 years ago in northern China, and had an average cranial capacity of about 1000 cm^3 (which is at the lower limit for modern man). Not only did he make stone tools, but he was the earliest known user of fire.

At present, the evolutionary sequence *Australopithecus africanus* → *Australopithecus* (or *Homo*) *habilis* → *Homo erectus* → *Homo sapiens* seems to be the most plausible one. However, the recent work of Arambourg, Howell, Richard Leakey, Patterson, and their colleagues in southern Ethiopia (Omo) and northern Kenya (Lake Rudolf) has raised some questions that cannot yet be

FIGURE 14-3.
Mandible (top) and left parietal bone (bottom) from *Homo habilis.* [Courtesy of Dr. P. V. Tobias.]

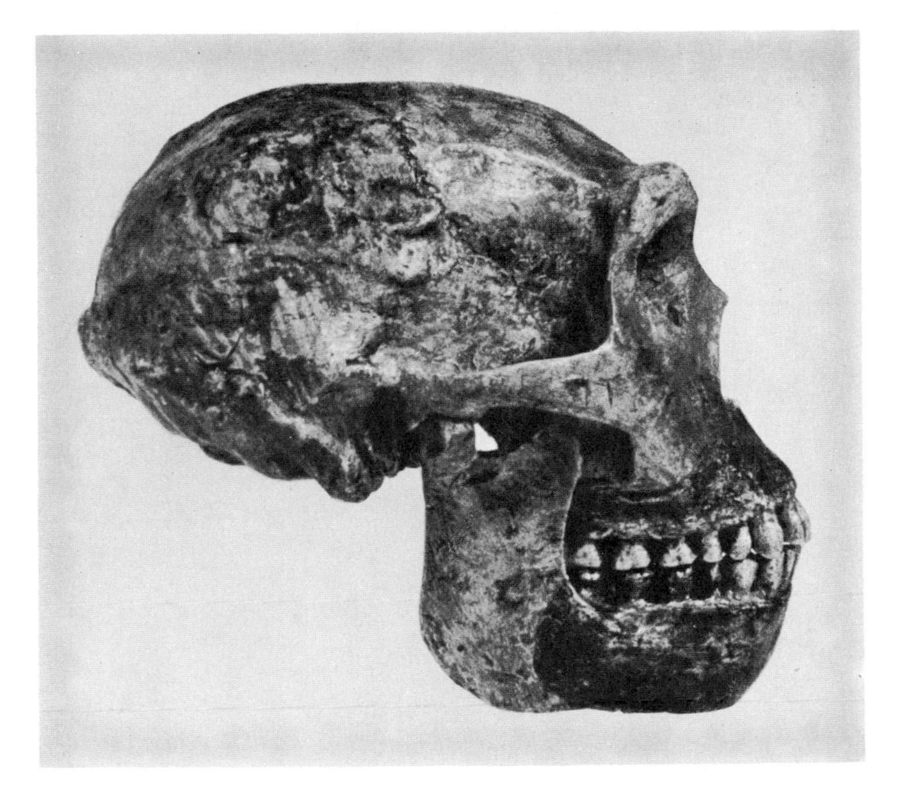

FIGURE 14-4.
Homo erectus, or Java Man, whose remains were unearthed in 1891 by Eugène Dubois. Characteristics of *H. erectus* include the smallness and flatness of the cranium, the heavy browridge, and both the sharp bend and the ridge for muscle attachment at the rear of the skull. The robustness of the jaw adds to the skull's primitive appearance. In most respects (size being one exception) the teeth of *H. erectus* resemble those of modern man.

answered to everyone's satisfaction. Some fragmentary remains of australopithecines have been found in strata of great antiquity—two to five million years old. Both *Australopithecus robustus* and *A. africanus* lineages seem to be present. On the other hand, some australopithecine remains are much younger, less than one million years. Most unexpected, however, is Leakey's discovery at Lake Rudolf of a more advanced hominid, allegedly a species of the genus *Homo,* dated about 2.6 to 2.9 million years. If this claim is validated, it would seem to follow that *Homo* and *Australopithecus* lived simultaneously and in about the same geographic region. Our ideas of the evolutionary sequence of hominids would then have to be revised.

The transition from *Homo erectus* to *Homo sapiens* also needs clarification. Several fragmentary remains (Swanscombe in England, Steinheim in Germany), which have been dubbed "pre-sapiens," have been dated at the last interglacial period (Riss–Würm). The Neanderthals who inhabited Europe during the last (Würm) period of the Ice Age seem to resemble modern *H. sapiens* rather less than the older pre-sapiens did, despite the fact their brain was as large as that of modern man. Some 35–40 thousand years ago the Neanderthals were abruptly replaced in Europe by a full-fledged *H. sapiens:* Cro-Magnon people, whose skeletal parts are indistinguishable from those of modern man.

Homo sapiens (or pre-sapiens) developed from *H. erectus* perhaps during the penultimate (Riss) glaciation or even earlier, probably in tropical and subtropical climates. It subsequently spread to Europe and during the last stage of the Ice Age (Würm) formed the race *neanderthalensis,* adapted to the severe climatic conditions of that time. This race was finally replaced by a less rugged but culturally more advanced *H. sapiens* people. Thereafter, the success of human populations depended on the possession of superior technologies rather than greater bodily strength. That *neanderthalensis* was a subspecies of *H. sapiens* rather than a separate species has been substantiated by the discovery of intermediates in caves of Mount Carmel in Palestine (McCown and others). Some researchers have interpreted these intermediates as "hybrids," although this interpretation is not a necessary one if by "hybrids" one means the immediate progeny of parents who belong to different subspecies (or species). A more plausible view is that Palestine was then a territory with a population intermediate between *H. sapiens sapiens* and *H. s. neanderthalensis,* just as at present we find territories inhabited by populations transitional between white and Negroid races. The Neanderthals no longer exist at present. It is possible that they were mostly killed off by the invading *H. sapiens*—the earliest suspected case of genocide. It is just as plausible that they interbred with the invaders, and we may carry some of their genes.

CULTURE—THE HUMAN DOMAIN

P. Teilhard de Chardin and G. G. Simpson are great scientists with radically different philosophical backgrounds and convictions. Yet they agree that mankind is an extraordinary outcome of the evolutionary process. Teilhard (1959) wrote: "Man, as science is able to reconstruct him today, is an animal like others. . . . Yet, to judge by the biological results of his advent, is he not in reality something altogether different?" And Simpson (1964) wrote that man "is another species of animal, but not just another animal. He is unique in peculiar and extraordinarily significant ways." What are these significant ways? They pertain not so much to physical characteristics as to mental capacities.

Biological classification always was and still is based very largely on comparative morphology, the study of the homology of body parts. This is all that is directly available for examination in fossil remains of our ancestors and relatives. Behavior is not recorded in fossil forms; it can only be inferred occasionally from the environmental conditions in which a fossil is found, or from the presence of artifacts.

On morphological grounds, mankind is classified as the sole surviving species of the family Hominidae; nobody would erect a separate zoological order, let alone a class or a phylum, for this single species. Yet mankind's paramount distinctive attribute is culture. Culture is a store of information and behavior patterns, transmitted by instruction and learning, by example and imitation. The central role in the transmission of culture belongs not to genes but to human symbol systems. According to Langer (1966), "The trait that sets human mentality apart from every other is its preoccupation with symbols . . . ; all human activity is based on the appreciation and use of symbols. Language, religion, mathematics, all learning, all science and superstition, even right and wrong, are products of symbolic expression rather than direct experience." A symbol is an object or act the meaning of which is not self-evident, but rather is socially agreed upon, or, according to Leslie White, "bestowed upon it by those who use it." Human languages are symbolic, in contrast to so-called animal "languages," which consist of signs rather than symbols. The meaning of a sign is apparent, and does not require social agreement. "A sign is anything that announces the existence or the imminence of some event, the presence of a thing or a person, or a change in a state of affairs" (Langer, 1966). Communication by signs is widespread; animals as well as men use signs, but only humans create symbols.

Because the transmission of culture (sometimes called "cultural heredity") is radically different from the transfer of genes from parents to offspring, culture has been called "superorganic." This label can easily be misleading. Only possessors of human genotypes can acquire, transmit, or creatively change or innovate culture. Let us make this crystal clear: human genotypes are indispensable for culture, but they do not decide which one of the many existing variant cultures will be acquired. There are no genes for American, European, Chinese, Hindu, or Hottentot cultures; a human child can assimilate any culture in which she or he is brought up. Likewise, human genes are indispensable for learning and using language, but they do not decide which particular language will be learned.

A novel biological system can arise by evolution only if the genetic raw materials for its construction are available for natural selection. At least some of the genetic building blocks from which the new system is to be made up must be older than the system itself. For this reason, it is a challenging problem for evolutionists to discover whether or not rudiments of symbolic communication and other unique human qualities may be found on the animal level. Two types of research may be expected to be informative. Information may come from observations on the behavior of animals, especially those most closely related to man. One may also try to induce some man-like behaviors in animals in captivity. Both approaches have been successfully utilized.

Important observations on wild chimpanzees have been made by van Lawick-Goodall (1971) and her collaborators, on gorillas by Schaller (1964), and on baboons and other primates by Devore (1965) and Hamburg (1971). Tool-*making* is very rare, even the making of tools for use in the immediate future, but instances of tool-*using* have been observed in a variety of animals. For example, chimpanzees are known to "fish" for termites. A twig is pushed down into a termite nest, withdrawn, and the termites clinging to it are scooped with the lips. A leafy twig cannot be so utilized, and the chimpanzee transforms it into an effective tool by stripping the leaves off. Sticks are often used to enlarge holes or openings where food can be found, or to investigate objects the animal is afraid to touch. Bunches of leaves are used to get water when the chimpanzee cannot reach water with its lips, and to wipe off dirt from its body. Chimpanzees have been seen using stones to open hard-shelled fruits. This tool-using and tool-making may develop into a kind of rudimentary "culture," since the juveniles attentively watch and imitate the behavior of their elders.

Several investigators have attempted to induce learned symbolic behavior in higher primates by making them acquire rudiments of human symbolic language. Chimpanzees and some other higher primates have large inborn "vocabularies" of calls, each of which is readily understood by other individuals. Nevertheless, these animals have proven to be remarkably refractory in experiments that attempt to teach them human speech. Hayes and Hayes (1954) had little success, despite much effort, teaching a chimpanzee child to speak. Gardner and Gardner (1969) accomplished more with another chimpanzee by teaching it a sign language, consisting of gestures originally developed for communication between deaf-mutes. Premack (1971) was even more successful. He utilized as symbols a variety of figures cut out from colored plastics. A chimpanzee first learned symbols that stood for objects, then symbols for "yes" and "no," for "similar" and "different," and for colors and shapes. With this "vocabulary" of symbols, the ape was able to understand and use not only words but also what amounted to simple sentences.

The beginning of language can not be dated, even approximately. The fact that without exception all human populations have language suggests (but does not prove) great antiquity for this form of communication. There is no doubt that communication by language, and the genes that made it possible, were promoted by natural selection. The cognitive structures on which both language and tool-making depend are the foundations of the control of the environment by culture. While all organisms adapt to their environments by changing their genes, man alone adapts mainly, though not exclusively, by creating the environments that suit his genes. In so doing, man has often abused the environment in which he lives, and the abuse becomes more and more serious as the number of people increases. Yet it can be affirmed without hesitation that it was *genetically based* culture that raised man to the role of "lord of creation."

MIND, SELF-AWARENESS, DEATH-AWARENESS

Without doubt, the human mind sets our species apart from nonhuman animals. Unfortunately, what we call the mind is notoriously refractory to scientific study. It is something that everyone apprehends by introspection, not by objective examination or measurement by some scientific means. Except in early infancy or during sleep or narcosis, every human being is conscious of himself as something apart from the environment and other people. In point of fact, self-awareness is the most immediate and incontrovertible of all realities. For Descartes, the only self-evident truth, one that was positively beyond doubt, was his own existence: *Cogito ergo sum* ("I think, therefore I am"). How do we know, however, that our human neighbors have similar minds and self-awareness? We cannot possibly know because we cannot enter into their streams of consciousness and experience directly their feelings or perceptions. The question is not an idle philosophical puzzle. Consider, for example, that between five and 10 percent of males (and about one percent of females) in the United States have genes that make them red–green color blind. Do they see red and green as I see these colors, or do they see blue and grey? Only by analogy can I infer that other humans have self-awareness—they usually act as I do in situations in which I know that my self-awareness is involved (Dobzhansky, 1967).

The great physiologist Sherrington (1953) wrote: "A radical distinction has therefore arisen between life and mind. The former is an affair of chemistry and physics; the latter escapes chemistry and physics." This sounds like the talk of a vitalist, which Sherrington was not. He stated only the truth that mind is scientifically an unknown, though existentially the basic entity. Manuals of psychology and neurophysiology often do not even mention the slippery words "mind" or "self-awareness." An evolutionist, however, is obligated to deal with them, since the emergence of mankind is incomprehensible without them.

We infer the existence of self-awareness, or mind, in people other than ourselves only by analogy with our own introspective experiences. Therefore, when it comes to the question of whether or not some rudiments of self-awareness may be present among other animals, conclusive evidence is unobtainable. No wonder that competent scientists are far from unanimous in their judgements. Some are willing to ascribe the beginnings of mind to some mammals (apes, monkeys, dogs), or even to all animals with developed nervous systems. Other scientists make mind an exclusively human possession. For example, Teilhard de Chardin (1959), in a now-famous statement, wrote: "Admittedly the animal knows. But it cannot know that it knows—this is quite certain." We need not take a stand here on this controversy. Human self-awareness obviously differs greatly from any rudiments of mind that may be present in nonhuman animals. The magnitude of the difference makes it a difference in kind, and not one of degree. Owing primarily to this difference, mankind became an extraordinary and unique product of biological evolution.

One form of behavior that is universal in man and wholly absent in animals, and which may prove to be trustworthy, though indirect evidence of self-awareness, is ceremonial burial of the dead. Burial rites are very diverse, taking the form of interment, cremation, exposure to carrion-eating birds, mummification, etc. Rarely, if at all, are cadavers treated as rubbish. Yet this is exactly what happens in animals other than man. Animals are generally indifferent to the death of conspecific individuals, with the important exception of the attitude of mothers toward young progeny. Colonial insects, such as ants and bees, throw cadavers out of the nest together with other debris; termites eat dead members of the colony as they do any other substances having food value. Ceremonial burial is not a recent invention in human culture; Neanderthals left many specially prepared and "decorated" burial sites.

Ceremonial burial is evidence of self-awareness because it represents an awareness of death. There is no indication that individuals of any species other than man know that they will inevitably die. In man, any individual past childhood who is not a low-grade mental defective is aware of the inevitability of death. Foreseeing the remote future and planning for future contingencies require capabilities that we know exist only in the human mind. Self-awareness and death-awareness are probably causally related and appeared together in evolution. They appeared because they enhanced the adaptedness of their possessors. The adaptive role of self-awareness is sufficiently obvious, no matter how elusive self-awareness may be. It is an integral part of the complex of adaptations that include the use of symbols, language, and hence acquisition and transmission of culture.

The adaptive function of death-awareness is not as clear. What conceivable advantage could our remote ancestors at the dawn of humanity derive from knowing that they would inevitably die? Naturally, this knowledge would become useful at a rather advanced stage of cultural development, when parents began to make provision for the welfare of their offspring after their own demise. Many animals to which there is no reason to ascribe death-awareness achieve the same end by means of genetically programmed instinctual behavior. Thus, many insects build and provision nests for offspring they will never see. However, it is improbable that death-awareness first functioned for the benefit of progeny. It is most probable that death-awareness arose originally not because it was adaptively useful by itself, but because it was a by-product of self-awareness, which was adaptive. There is no single gene for self-awareness and death-awareness; rather, a complex of many genes is responsible for these characters.

ETHICS AND VALUES

Most people are capable of distinguishing between good and evil, and most feel some compulsion to promote what they value as good. Quite often they do not succeed, either because of circumstances or because they succumb to the temptation of what they ordinarily regard as evil. Nevertheless, all human societies about which we know anything have had moral codes, compliance with which is enforced by persuasion or sanctions. There are two interesting sources of ethics and values—cultural and biological. The ethical standards of every individual are imparted, mainly in childhood and youth, by other members of the society for which the individual is being prepared. Ethics are acquired, not biologically inherited. Yet, according to Waddington's (1960) happy phrases, humans are genetically determined "ethicizing beings" and, especially in childhood and youth, "authority acceptors."

Every member of a human society must become familiar with the ethical and value systems of his society. Failure to do so makes him a misfit or an outcast; it jeopardizes his and his progeny's success and survival. Therefore, natural selection has exerted pressure to insure that every member of the human species comes into the world with a genetic endowment making him an "ethicizing being." This selection pressure has been going on for perhaps a million or more years (say, 40,000 generations). Yet there are many cultures, and they differ, sometimes greatly, in their value systems. This would make a specific genetically predetermined ethic counterproductive. What is inherited is a potential, the ability to "ethicize." Its expression is governed by the culture into which the individual is born. Or, to put it differently, the evolutionary development that made the ascent of man possible has not endowed mankind with any particular system of ethics; it has, however, made human beings capable of learning various kinds of ethics, values, and morals.

Two kinds of questions present themselves in this connection. First, since our ancestors were nonhuman animals, have we inherited from them some genetically conditioned instincts or drives that may agree, or clash, with our culturally implanted ethics? Second, do animal species other than man possess some rudimentary capacity for ethics? Both the optimistic belief that man is inherently good and the pessimistic belief that he is inherently depraved antedate scientific biology and anthropology. Great religions, particularly Christianity, have incorporated both beliefs; in Christianity the doctrine of original sin coexists with the doctrine of divine grace.

Much energy has been wasted on disputes over whether aggression and violence or kindness and cooperation are inborn genetic drives, instincts, or "imperatives" in the human species. Ethologists, students of animal behavior, have discovered a variety of aggressive behaviors in many animals: threatening displays, dominance rivalries, and territorial defenses. However, aggression against individuals of the same species is usually ritualized; the threat of violence from a stronger individual is answered by a gesture of submission, which defuses the aggression. The trouble with man, according to Lorenz (1966), is that he has developed increasingly vicious means of aggression—from stones to hydrogen bombs—without possessing ritualized behavior to appease aggressors.

Let us suppose, for the sake of argument, that all or some members of the human species do have genetically conditioned tendencies toward aggressive and domineering behavior. It still would not follow that these tendencies could not be brought under control. One can be brought up to be gentle or violent, peaceful or aggressive. Anthropologists have ample evidence that different cultures demand different modes of behavior, and that these demands are usually complied with. A culture may encourage extroversion or introversion, combativeness or meekness, swagger or modesty, profligacy or parsimony.

Competent biologists disagree whether animals other than man possess capacities for ethics. Some ascribe rudiments of moral sense, for example, to dogs, which occasionally behave as though they have guilt feelings and bad conscience. Others believe that ethics and values are exclusively in the human domain. Without trying to settle this problem, one may nevertheless say without contradiction that certain kinds of behavior found in animals would be ethical or altruistic, and others unethical and egotistic, *if these behaviors were exhibited by men* (see Chapter 4). For example, the behavior of workers and soldiers among ants and termites impresses us as a model of altruism, or unselfish devotion to the common good. By contrast, in some quasi-social animals and birds the exclusion of subordinate males from feeding grounds by dominant individuals strikes us as cruel because the excluded individuals are effectively condemned to die. Examples of this sort can be multiplied indefinitely. At the same time, one must agree with Simpson (1964) that "it is nonsensical to speak of ethics in connection with any animal other than man. . . . There is really no point in discussing ethics, indeed one might say that the concept of ethics is meaningless, unless the following conditions exist: (a) There are alternative modes of action; (b) man is capable of judging the alternatives in ethical terms; and (c) he is free to choose what he judges to be ethically good. Beyond that, it bears repeating that the evolutionary functioning of ethics depends on man's capacity, unique at least in degree, of predicting the results of his action."

Some human behavior patterns that are charged with ethical evaluations can reasonably be supposed to have been shaped in evolution under control of natural selection. Not surprisingly, these behavior patterns are exactly the ones that most resemble the behaviors of nonhumans. The human family is perhaps the oldest social institution, having also the most evident biological function. Motherhood has always been esteemed a virtue, even in societies in which women were subjugated and treated little better than slaves. Children have always been cherished and loved; parents often and willingly suffer discomfort, self-abnegation, and self-sacrifice for the benefit of their own children, less often for those of their relatives, and still less for unrelated children. Desire to have children is regarded as "natural" and good. Failure to take care of children once they are born and to provide for their sustenance exposes the parents to condemnation; it is regarded as a sign of moral turpitude. Infanticide, which has been practiced in many societies, is generally considered hateful and horrifying, although in some situations it was virtually indispensable for self-preservation of the remaining family and of the tribe.

All these attitudes and evaluations favor reproductive efficiency and hence are consistent with the demands of natural selection (see Chapter 4). This can hardly be said of many other ethics and values that are recognized in most, if not all, human societies as valid. For example, it is wrong to steal, swindle, rob, waylay, or murder other people, especially members of one's own group or society and, by extension, any human being. This is wrong even if so doing is profitable, the misdeed is undetected, and no vengeance or retribution is to be feared. On the contrary, honesty, generosity, and veracity are praiseworthy, especially if they bring hardships to persons who practice them. Human life, that of a stranger no less than that of a relative, is sacred, with the significant exception of war. Life is to be preserved at all costs (including that of incurably ill persons whose existence may be sheer misery). At the summit of ethics, we have the commandments of universal love (including one's enemies), service to others, and resistance to evil.

At the risk of oversimplification, we may distinguish two kinds of ethics: family and group (or species) ethics (Campbell, 1972; Dobzhansky, 1973b). *Family ethics* are shared by man with the "quasi-ethics" of at least some animals; in animals as well as in men, many family ethics are genetically conditioned dispositions (although in man they may be overcome by an exercise of will). Family ethics can be envisaged as products of natural selection, which established the genetic bases of these ethics in our ancestors as well as in other animal species. *Group ethics* are products not of biological but of cultural evolution. They confer no advantage and may be disadvantageous to individuals who practice them, although they are indispensable to the maintenance of human societies. Natural selection has not made man inherently evil (as is so readily assumed by believers in original sin or proponents of territorial and other "imperatives"). Whatever proclivities to selfishness and hedonism man may have, he also has a genetically established educability that permits him to counteract these proclivities by means of culturally derived group ethics. Natural selection for educability and plasticity of behavior, rather than for genetically fixed egoism or altruism, has been the dominant directive factor in human evolution.

DOES MANKIND CONTINUE TO EVOLVE GENETICALLY?

Four or five million years is a rather short time on the evolutionary scale. Yet within this interval there evolved a radically novel and unique form of life—the human species, mankind. It is the only species that possesses culture. Like in any other biological species, every human generation inherits from its parents, and transmits to its descendants, a genetic endowment coded in the genes. But unlike any other species, every human generation inherits and also transmits a body of knowledge, customs, and beliefs that are not coded in the genes. They are acquired, by every individual for himself, through imitation and learning. As pointed out above, the existence of culture is dependent on the possession of human genotypes. Nevertheless, the mode of the transmission of culture is quite unlike that of biological heredity.

Man adapts to and controls his environments, by means of culture. As we have seen in previous chapters, adaptive evolution of species other than mankind occurs by substitution of genes that are favorable in new environments for those that were fitting in old environments. Birds, bats, and insects are fliers descended from ancestors who did not fly. To become fliers they had to evolve genetically for millions of years. Man has become the most powerful flier of all, by constructing flying machines, not by reconstructing his genotype. Adaptations of the species by culture can be vastly more rapid than adaptation by genetic alterations. Changed genes can be transmitted only to children and other direct descendants; cultural changes can be passed regardless of genetic relationships. Gene substitution by selection usually takes many generations; cultural changes can, in principle, be made within a generation. The parents and grandparents of the millions of workers who now operate complex machines were peasants or farmers, tied to the soil they cultivated with the aid of relatively simple tools.

The question that suggests itself is whether mankind has continued to evolve genetically after it started to evolve culturally? A viewpoint that appeals to many social and to some natural scientists is that the biological evolution of mankind is a thing of the past. It became superfluous when the evolution of culture began. A species endowed with cultural adaptability no longer needs the relatively clumsy and slow apparatus of genetic change. Marxist philosophers add a special twist to the above argument: our species became human when it started to "work." Henceforth, this species became subject to sociological, rather than biological laws. It is admitted that culture and cultural evolution can occur only in carriers of human genes. Since every person is such a carrier, genetic variables need not, so it is argued, be considered any further. This is only a half-truth, however, and a mischievous one because it sounds so plausible.

For several million years the evolution of our prehuman and early human ancestors exhibited a trend of gradual but rather steady increase of the brain volume. This trend culminated perhaps some 200,000 years ago in *Homo sapiens* and its Neanderthal race, whereupon no further increase of the skull capacity has come to pass. As pointed out above, this may have been due to a limit imposed by the difficulties of childbirth. Dogmatic assertions of some authors, to the effect that the mental capacities of members of the human species have also remained unchanged, or have increased or decreased, must not be taken on faith. The simple fact is that there is no way to measure these capacities in humans who lived centuries ago, not to speak of millennia. Buddha, Ikhnaton, Socrates, or Aristotle may have been equal or superior in wisdom to any living person, but it would be naive to think that the average intellectual capacities of their contemporaries were equal to these intellectual giants, and that since their time the average human intellect has decreased.

For perhaps as long as two million years cultural changes were increasingly preponderant over genetic ones. Yet this has not made all humans genetically identical. Genetic variations underlie differences in physical features, health and disease, learning abilities, and in other behavioral and personality traits. Genetic variability is a necessary but not sufficient condition for the occurrence of evolutionary changes. Suppose that people are genetically variable, but the variants are adaptively neutral; if the carriers of all genes have on the average the same chances to survive, reproduce, and bring up the same numbers of children to maturity, no evolutionary alterations will take place. (More precisely, only random fluctuations of gene frequencies could occur under these conditions.) What is the situation in human populations? Some genetic variants seem to be adaptively neutral. For example, the details of the facial features, though doubtless genetically variable, seem neither to enhance nor hinder significantly the biological, reproductive, success of their possessors.

However, numerous genetic variants in human populations are far from selectively neutral. McKusick (1975) has catalogued 2,336 genetic variants causing diseases and malformations in man, ranging from mild defects and weaknesses to complete lethals. Of these, the mode of inheritance has been well established for about one thousand variants, while the rest are still uncertain in some respects. Accumulation or diminution of frequencies of favorable or unfavorable genes in human populations cannot fail to have consequences for public health and the general welfare of a human society. The genetic and cultural evolutions of mankind are not independent but interdependent. They are tied together in a system of feedback relationships.

For those who doubt that mankind is still evolving biologically, it may be useful to ask the question: If we wished to stop the evolution of a species, how could this be achieved? The mutation process constantly supplies the genetic raw materials from which evolutionary changes can be constructed. We can try to minimize the mutation rates, but there is no known way to arrest mutation completely. Moreover, to stop or at least to slow down evolution one would have to do away with the process of natural selection. To this end, *every* child born (really, every fertilized egg) would have to be maintained to maturity (alternatively, survival would have to be entirely at random with respect to the individual's genotype). Every person would have to mate and become the parent of the same number of children at exactly the same age. Obviously, nothing like this has ever happened in mankind, nor is it likely.

EUPHENICS AND EUGENICS

No species before man could select its evolutionary destiny. Armed with knowledge, man can do so. He can steer evolution in the direction he regards good and desirable. Or he may elect to drift on the evolutionary current oblivious of consequences. One thing he probably cannot do is to have evolution stand still.

We have seen in Chapter 4 that human populations, like those of any sexually reproducing species, carry heavy burdens of genetic defects. These burdens grow heavier as time goes on. Progress in the biological and medical sciences permits more and more hereditary diseases to be "cured". Of course, only the symptoms that are the manifestations of the defective genes are corrected, not the genes themselves. For example, a child with phenylketonuria can grow up and develop reasonably normally on a diet free of phenylalanine. Timely surgery lets a child with pyloric stenosis survive and enjoy good health.

Euphenics is the science of management by environmental manipulation of the manifestations of genetic endowments. The symptoms of many genetic diseases that are now incurable will surely be assuaged by progress in euphenics. Whether euphenics will be sufficient to cure all hereditary diseases in the long run is a different question. The genetic endowment itself may require rehabilitation or improvement. This is the realm of *eugenics*. Negative and positive eugenics may be distinguished. The former aims to relieve human populations of genetic diseases and defects; the latter is a more ambitious project—improvement of the genetic endowment of mankind. Of course, the distinction is not clear-cut. An elimination or decrease in the incidence of genetic infirmities constitutes an improvement of the gene pool of the species. One may wish, however, to go much beyond this relatively modest goal and aim for the creation of novel genetic endowments, superior to any existing ones. Some people go so far as to contemplate making a "new man."

Several eugenic procedures could be put in operation today with the biological technology now available. More could be made available by intensified basic and applied research. Still others seem fanciful daydreams, but they may some day be realities. A fundamental caveat must always be kept in mind in evaluating eugenic measures: the practicability of any and all methods depends as much on their social acceptability as on biological efficacy.

Genetic counseling, which is becoming increasingly widespread, has eugenic implications (Lipkin and Rowley, 1974). Prospective parents are provided with the best available information about the genetic nature of a given trait or illness, and about the chances of its transmission to their offspring. Suppose the probability is even that the offspring of prospective parents will be afflicted with a serious genetic defect or disease, or will be "normal," i.e., free of that disease. So advised, the prospective parents may abstain from conception, or may take their chances on a "normal" child. If they choose the latter course, they will at least have been forewarned of what may be in store for their child and for themselves. The recently developed technique for amniocentesis gives genetic counseling a wider applicability and a more secure basis. Amniocentesis involves withdrawal of a sample of the amniotic fluid surrounding the fetus inside the mother's womb. The fluid is tested for the presence or absence of certain enzymes, and the chromosome complements of the cells floating in the fluid are examined for traits diagnostic of certain genetic abnormalities. The prospective mother can then be informed whether or not the fetus she carries has a certain genetic defect. She may then elect to ward off the affliction by abortion, or may let the child be born with the defect.

Any improvement in the gene pool that can be expected from genetic counseling will be too slow and too limited in the opinion of some authorities. The great geneticist H. J. Muller devoted much of his life to elaborating his proposal for so-called germinal choice or prenatal adoption (1967). His eugenic program was to collect the semen of genetically superior men, preserve it in a deep-frozen condition in "sperm banks," and eventually use it to artificially inseminate as many women as would want it, as long as the supply would last. The biological technology needed to put Muller's program in execution is available. To this extent, his project is not unrealistic. The spermatozoa of various mammals, including man, can be preserved in a deep-frozen condition without loss of fertilizing ability. There is no certainty that the preservation of the sperm in a deep-frozen state for long periods will not induce genetic changes of an undesirable sort. Muller evidently felt that the risk is worth taking. The difficulty with his project is social rather than biological. Although artificial insemination of women with the sperm of unknown donors is now practiced fairly widely, whenever husbands are sterile, nevertheless Muller's program conflicts with established mores. Even those who do not object to the elitist philosophy implicit in this program usually prefer to raise their own biological children. Either Muller was ahead of his time, or human emotions are not as easily managed as he assumed.

Recent advances in the biological sciences have furthered our understanding of the evolutionary forces that act upon the human species. Some techniques for managing the human gene pool are already available, and can be improved further. Other techniques, much more powerful than existing ones, could possibly be developed. Yet it is incautious, in our opinion, to take for granted that these latter techniques will necessarily be discovered, and to speak and write as if they were already at hand. They may conceivably be invented tomorrow, but on the other hand they may elude our grasp for a long time, even forever.

Parents superior in intelligence, artistic ability, or other qualities often produce average or below-average children. The opposite is also true: outstanding individuals frequently come from average or below-average parents. Insofar as these situations have genetic bases, they stem from genetic recombination at the meiotic stages of development. We do not inherit all the genes of our parents, only one half of their genes, and brothers and sisters inherit different assortments of the parental genes. Moreover, a gene does not exert its effects independently of other genes—genes interact in the process of development. It is the particular constellation of genes that constitutes the genetic foundation of one's personality. Children are really never copies of their parents.

A technique has been devised that would insure that a child would be a true genetic copy of one of its parents. Every somatic cell carries the same full complement of the chromosomes and genes. Gurdon and others, working with toads and frogs, succeeded in transplanting the nucleus of a somatic cell into an ovum whose nucleus had been removed. The ovum was then induced to develop without fertilization. All cells in the resulting embryo had nuclei descended from the implanted one. This technique is known as *cloning*. No cloning experiments have been made with human ova, but there is no reason why such operations could not some day be successful. One could obtain several individuals genetically identical to the donor of the nuclei and as similar to each other as monozygotic twins. Such individuals would be members of a clone. Instead of sperm and ova banks, one would then have to institute banks of tissue cultures, perpetuating the cells of selected donors.

The possibilities that would be opened by cloning humans are staggering. One could produce as many individuals as desired with genes exactly like those of some genius or hero. One could dispense with the ordinary method of reproduction, and produce a species consisting only of men or only of women. The idea of cloning humans is audacious enough. Bolder visionaries promise still greater wonders, such as chemical alteration of deliberately chosen genes in the chromosomes of sex cells or body cells, in predetermined and beneficial directions. Still further, one can think of synthesizing new kinds of genes with desirable properties, and inserting them into the chromosomes to replace some of the existing but less advantageous genes, or else adding new genes to the genetic equipment. New genes could, perhaps, make a "new man."

These bold projects of genetic engineering have some support, if only a faint one, in actual discoveries. Genetic transformation of strains of pneumococci, bacteria responsible for some varieties of pneumonia, was observed several decades ago. Avery showed that the "transforming principle" that brings about the directed genetic changes in these bacteria is DNA (see Chapter 2). Transduction involves transfer of a section of so-called temperate bacteriophages (bacterial viruses). One is tempted to speculate that some day a nonpathogenic virus may be discovered that could transfer genetic materials, and leave them implanted in human chromosomes. Artificial synthesis of a gene was achieved by H. G. Khorana; the gene acts like the one in yeast coding for the enzyme alanine transferase. J. Beckwith and his collaborators isolated a gene, or group of genes constituting an operon, that enables the bacteria to metabolize the milk sugar lactose. Of course, these discoveries are far from being usable in genetic engineering.

Knowledge is said to make men free. Yet knowledge often brings problems and responsibilities that do not trouble the ignorant. The discovery of evolution is a case in point. Can mankind drift on the evolutionary current oblivious of the direction in which it is being carried? Or should it determine the direction in which it wishes to move? Some scientists would start sperm banks and utilize their "deposits" immediately; cloning and genetic engineering have to wait only until at least a crude technique is devised. These scientists are sure they know what is good for mankind, and have no fear that mistakes could result in irreversible harm instead of improvement. Others reject application of any genetic technology to man as "unnatural," and condemn all research to devise such technology as ethically doubtful, if not unethical.

Man is biologically unique. The human species has acquired its unique genetic qualities in the natural process of biological evolution. One of these unique qualities is that man can and does modify and transform nature. To some people, including oddly enough some biologists, this makes anything man touches "unnatural." Thus, it has been argued that natural selection is no longer operating in mankind because mankind is pampered by culture. But it was natural selection that made man develop culture and eventually made him unable to live without it. It is inconceivable that man would wish to reverse this process, and anyway he would find it very difficult to do so if he wished. One should certainly be wary of any eugenic measures, and should evaluate them both from the standpoint of their biological effectiveness and their psychological, social, and ethical acceptability. And yet, experimentation aimed at developing and perfecting techniques of genetic selection and genetic engineering should receive every encouragement. Wherever possible, such experimentation should use nonhuman materials; but wherever necessary human materials should not be excluded a priori, but should be used with reasonable and effective precautions.

15

The Future of Evolution

Prophesying the future by extrapolation from the past is a hazardous occupation. This hazard is particularly great with respect to organic evolution because of the complexity of the factors involved. Nevertheless, the account presented in the preceding chapters is as sound a basis as any that scientists might offer at present for predicting the future of living beings on this planet.

For the foreseeable future, organic evolution will be bound up intimately with the future of mankind. Earlier chapters of this book have emphasized the importance of changing environments as catalysts of evolutionary change. In particular, the extinction of species, appearance of new predators or sources of food, and changes in the abundance of other species are important evolutionary stimuli for most populations of animals and plants. During the past two thousand years such changes have been produced by the activity of mankind on a vastly greater scale than at any previous time in the earth's history.

Many species have become extinct, and others, associated with mankind as weeds or pests, have become much more common and widespread. Furthermore, the gene pools of many species are becoming reorganized in particular ways that fit them better to the new conditions that mankind has created.

At present, habitats unaltered by man are rare throughout the entire earth, and in a century or less will exist only if they are intentionally preserved. Many larger mammals living when *Homo sapiens* first evolved are either extinct, or represented only by their domesticated descendants, or, like the bison of Eurasia and North America and elephants of Asia and Africa, are being preserved in a quasi-domesticated state. Few if any of the remaining species of mammals can evolve independently of man's influence. The same is true to a lesser degree of other vertebrates, as well as the larger forms of plant life, particularly the trees of the forests. The fossil record of insects indicates that, even before mankind appeared, the evolution of this class had already become confined to the elaboration of existing themes, represented by Tertiary genera (Riek, 1970). The same is true of marine invertebrate life as well as algal life. Organic evolution independent of man's influence will not occur on a large scale in the foreseeable future.

A possible exception might be the extinction of *Homo sapiens* and his replacement by some other related species. Yet this is highly improbable for the following reasons. In the past species of large mammals became extinct because of unfavorable competition with or predation from other species of mammals, or because of drastic changes in the rest of their environment. Neither of these causes operates against mankind today. Other species of mammals exist only because men wish to preserve them, or at least tolerate them. Environmental factors have been neutralized by man's ability to control effectively his environment. As was pointed out in the last century by A. R. Wallace, man's distinctive form of adaptation is his ability to transform the environment to suit his needs (see Chapter 14). If a new ice age should arise, similar to that which transformed the Northern Hemisphere during the Pleistocene epoch, northeastern North America and northern Europe would become outposts of culture, as Alaska, Greenland, and Antarctica now are, and cultural centers nearer the Equator might assume dominant positions. Other less drastic and slower changes, such as vulcanism, mountain-building, and the advance of continental seas, would have similar effects in redistributing rather than eliminating mankind and human culture.

Two other possibilities remain: "species suicide," as a result of a global war, and evolution toward some distinct species of superman. The former prospect has begun to recede in recent years and one can only hope that the trend will continue. Further evolution of the hominid line toward a new species, genetically distinct and reproductively isolated from *Homo sapiens*, is highly unlikely. The reasons for this are: the worldwide distribution of human populations; the ease of communication, which makes virtually impossible the establishment of isolates protected from gene flow by space or any other factor; and the leveling effect upon human anatomy of adaptation by cultural transformation of the environment. Of course, natural selection will continue to operate indefinitely on our species. With respect to visible anatomical characteristics, however, it will continue to take primarily the form of normalizing selection, as it has during the past 25,000 to 50,000 years. Selection pressure for changes in the gene pool will affect chiefly those genes that control invisible characteristics, such as immunity to disease and psychological reactions to the shocks imposed by modern civilization.

One might ask, Doesn't this convert human evolution into a completely unique set of processes, the outcome of which, for this reason alone, cannot be predicted? Of course a unique course of evolution might be expected because of man's unique way of interacting with his environment. But recent observations have shown that long-continued morphological or anatomical stability of a species is not nearly as unusual as it was once thought to be (see Chapter 10). Termites are the oldest organisms to have evolved, in part, by virtue of some ability to transform a localized habitat into one suited to their needs. In most insects of the termite order, reproductive individuals leave the termitary, the borings in wood that their species has created, for only short periods of time. This way of life is associated with a remarkable evolution of the digestive system through symbiosis with protozoa and other organisms, and with the ability to differentiate drastic phenotypic modifications in the form of castes. Nevertheless, with respect to many anatomical features, such as wing structure and body segmentation, the reproductive castes of termites have retained a remarkable stability during more than a hundred million years (Emerson, 1955). In man the possibility of phenotypic caste differentiation does not exist, except through such cultural symbols as clothing. Anatomical stability of a single species is the most probable future for *Homo sapiens*.

THE EVOLUTIONARY FUTURE OF MANKIND

In the future, organic and cultural evolution will remain inextricably bound together, as they have been during the past few thousand years. Moreover, the dominance of cultural over organic evolution as an agent of change for both populations and their environments is bound to increase rather than diminish. This new dominance of human cultural evolution represents a change in the earth's history as drastic as, if not more than, any that has taken place on the earth since the appearance of cellular life.

For an evolutionist looking ahead as well as behind himself, the most crucial question of modern times is, What is evolution going to be like under the dominance of human culture? We have already stated that we cannot answer this question by simple extrapolation from past events of organic evolution, as we could if mankind were just another biological species. Valid extrapolations depend upon estimates of how extensive and significant the differences are between mankind and all other organisms that have inhabited the earth. Mankind has three aspects, which are often in conflict with each other, sometimes violently so. These aspects have existed ever since the time when tool-making, teaching and learning, and language became basic characteristics of the hominid line. They have become increasingly significant, with accelerating speed, during the last few thousand years, and have exploded into dominance as a result of the industrial revolution and its consequences for communication, organization, and planning. The three aspects are: man as a biological animal; man as an intelligent and social being, aware of himself and his responsibility to society; and man as a transformer of his environment.

There is a conflict between man the biological animal and man the social being. In Chapter 14 the close anatomical and biochemical resemblance of man to the higher primates was pointed out. The rise of culture has been so rapid that the genetic evolution of man's capacity for mental or emotional adjustment has been unable to keep pace with the revolutionary alterations of man's environment. The rise of human intelligence, and mankind's distinctive way of coping with his environment, was probably accomplished during a stage when the dominance and aggressiveness of single individuals, which often led to lethal conflicts, had a high adaptive value. This adaptive value may be inferred by extrapolation from recent investigations of the cultural and genetic structure of certain aboriginal modern societies. In their way of life these aboriginal societies resemble more closely than any others the conditions that existed during the Pleistocene epoch. Neel (1970) has reported that among the Xavante Indians and similar tribes of the Matto Grosso and neighboring regions of Brazil, the effect of an unusually intelligent and aggressive male upon the gene pool of his community is much greater than was previously believed. This is because (1) there are lethal rivalries for the position of chief in a village; (2) the successful chief has a much larger number of wives and children than other males; and (3) his sons have a better chance of acquiring wives and children than other males, because they grow up in favored positions. One can easily infer that if a series of human communities were competing with each other for occupancy of a region in which intelligence and organizing ability had a high adaptive value, a community having a social structure like the Xavante's could adapt more quickly to such conditions than could one in which every male, regardless of his mental ability, would have an almost equal chance of transmitting his genes to the next generation, or even one in which the combined intelligence of a single couple was related to their reproductive success.

In many modern societies occupying habitats similar to those of our preagricultural ancestors, aggression that includes forcible exclusion of other males from mating and preeminent leadership in fighting can lead to a great excess of the aggressors' genes in subsequent generations. Washburn (1966) has pointed out that aggression of one kind or another is general throughout primate societies, and is accentuated by crowding, as with the rhesus monkeys of India. During all but the last 2,000 years of the approximately 75,000 years in which *Homo sapiens* has existed in his modern form, expanding populations have invaded new territories. As a result, aggressiveness accompanied by higher intelligence and a more efficient culture undoubtedly led to the spread of the aggressors' genes at the expense of those possessed by less aggressive and less intelligent populations. The modern world is saturated with human populations, and at a time when dwindling sources of food and energy are a major concern populations expand at their own peril. Aggressiveness is likely to encounter opposing force, leading to wholesale destruction and damage to both the aggressors and their intended victims; in any case, superior cultural traits are more likely to spread through peaceful cultural diffusion than through aggressive conquest. Under these conditions, gene combinations that promote aggression, along with aggressive cultural traditions, have a negative adaptive value, and in the future could have even more harmful effects. They must be controlled at all costs.

Control of aggression by reduction of the frequency of genes that promote it has not taken place because human populations have so recently saturated the earth that the negative adaptive values of such genes have had no time to exert their effects. Dangerous aggressive traits, which probably have some genetic basis, are still with us. The survival of mankind depends upon its ability to counteract aggressive traits by cultural means. The great flexibility of human social behavior, correctly emphasized by Wilson (1975), justifies the optimistic conclusion that such cultural changes can be made.

Another serious conflict is between man as a transformer of his environment, on the one hand, and man as a member of a community that must exist in harmony with the environment. The source of this conflict is not hard to identify. With modern tools and technology man can easily do things that appear to relieve the hardships, suffering, and grief of his daily life for the present and immediate future; but estimates of the ultimate consequences of these actions are hard to make. Even when man knows that the consequences of his actions may be disastrous in the far future, the majority of human beings are willing to risk this danger for the sake of immediate satisfaction.

The staggering problems facing modern society can be understood and solved more easily if they are regarded in the light of these conflicts rather than as the results of the malevolence or callousness of personalities or special interests. These conflicts cannot be resolved by turning back the clock of technological progress in the hope that social progress can be induced to catch up. Their solution requires prolonged efforts to develop new balances between technological advances and more intelligent control of their deleterious effects.

Cultural evolution, like organic evolution, is a sequence of events integrated with each other in a highly complex fashion. In neither case can sequences be reversed so that populations revert to conditions that prevailed in the remote past, however much we might wish to do so. The only possibility is conscious guidance of evolution, of all forms of life as well as our own species.

ARTIFICIAL CONTROL OF ORGANIC EVOLUTION

Artificial control over the evolution of those species of animals and plants that are essential to mankind, i.e., domestic animals and food plants, has been practiced for more than 10,000 years, and is now well advanced. Although domestic animals differ only quantitatively from their wild ancestors, some cultivated plants are novelties. The most striking example is maize or Indian corn. Until it was analyzed genetically, *Zea mays* was regarded by botanists as forming a monotypic genus distinct from its nearest wild relative, *Euchlaena mexicana* or teosinte. This is because the corn cob, with its surrounding husks and enormously long styles or silks that receive the pollen, is as distinctly novel a structure as any that exists among wild species of flowering plants. Although opinions differ as to the precise ancestry of maize, botanists believe that the ancestor of corn was a wild grass, either teosinte or something similar to it, which did not bear ears like those of modern corn. Fossil corn cobs provide us with a record of the gradual evolution of this novel structure under human guidance during the past 7,000 years. Bread wheat (*Triticum aestivum*) is likewise a man-created species; its hexaploid chromosome number is due to the incorporation of the D genome from *Aegilops squarrosa.* Man's ability to breed new races of domestic animals that are highly distinctive in appearance and behavior was recognized by Darwin as strong evidence in favor of evolution by selection. May it not also be possible to breed men and women having superior physical, mental, and emotional characteristics?

Recent efforts and prospects in this direction were reviewed in Chapter 14. Two kinds of proposals for "genetic engineering"—artificial production of identical genotypes by cloning, and the artificial implantation of new and supposedly valuable genes into human DNA—are ways of directly guiding human evolution that will receive further consideration in this chapter.

The prospect of cloning higher animals is both highly promising and, in its potential applications to man, terrifying. For the animal breeder, particularly breeders of large animals like cattle, it would be of enormous value. A single high-yielding individual of good quality could be converted into a persistent variety, like the varieties of fruit trees and many ornamental flowers. The application of cloning to humans raises the specter envisaged by such authors as Aldous Huxley in his *Brave New World:* genetic castes of workers comparable to those of social insects.

Proposals for improving the human species by cloning individuals believed to be superior in qualities other than the capacity for continued hard work are even more dangerous. Superiority in most characteristics regarded as desirable, such as ego strength, sociability vs. aloofness, tolerance, self-control, and sense of responsibility, is determined as much or more by cultural factors as by differences in heredity. In fact, a comparison of monozygotic and dizygotic twins with respect to these characteristics concluded that heritabilities for many such differences are not significant (Dworkin *et al.,* 1976). Cultural differences affect these traits more than genetic differences. Consequently, cloned progeny of individuals selected as superior would retain this superiority only under precisely defined regimes of upbringing. Cloning would insure superiority of socially desirable traits only if it were accompanied by regimented upbringing. It would be effective only in a society so restrictive upon individual freedoms as to be intolerable.

An ambitious proposal for positive eugenics, involving a highly sophisticated form of "genetic engineering," is the artificial implantation of supposedly better genes into the nuclei of human zygotes. This proposal invites the same objections as human cloning. It is hardly possible in the foreseeable future that specific alleles of the human genotype, except for normal alleles at loci for genetic diseases, will be identified as undoubtedly good for mankind. Moreover, the most successful societies contain a great variety of genotypes, from which training and education produce an even greater variety of phenotypes. The conscious guidance of human biological or organic evolution toward a presumably better condition, even if it should become technically possible, would be undesirable.

It is impossible at present because genes that might contribute toward better social adjustment have not been identified, and are highly unlikely to be recognized in the foreseeable future. It will always be undesirable because there is not just one, but many "good" kinds of human beings. Among the finest qualities of mankind are social flexibility and an infinite diversity of individuals. Human society cannot afford to risk losing or degrading either of these qualities in the vain search for a utopian ideal.

THE GUIDANCE OF CULTURAL EVOLUTION

The only alternative to genetic improvement of mankind is cultural improvement. Because of the flexibility and diversity of human nature, the continued improvement of society by progressive cultural change in many directions is an attainable goal. The key to progress is education, particularly that which molds the character of the infant and the growing child. The enormous complexity of the problem, together with some suggestions on how it might be solved, has been discussed in detail by Hamilton (1974). Leadership can be provided by psychologists, sociologists, anthropologists, political scientists, philosophers, and others who are primarily interested in human behavior, as well as by evolutionary biologists. The latter can provide a perspective that will help social leaders define the problems facing mankind and search for the most practical solutions.

One tool for determining what kinds of cultural change are practical and how they could be accomplished is provided by analogies between cultural and organic evolution. Such analogies must emphasize both similarities and differences between the two kinds of change. In earlier chapters of this book the effectiveness of natural selection was shown to be based upon the richness and diversity of gene pools in populations. Human populations are equally rich and diverse with respect to cultural traits. The improvement of society by cultural change must rely upon the richness and diversity of this cultural pool, analogous to the gene pools of animals and plants. The cultural pool consists of inventions, technical skills, constitutions, laws and customs, religious attributes, systems of education, and aesthetic achievements such as literature, music, painting, and sculpture. Differences between the cultural pools of nations or smaller communities correspond to differences in the gene pools of biological species. Just as mutations increase variation in gene pools, so new ideas and inventions can increase cultural pools. Individual traits of gene pools are transferred from one population to another by migration, crossing between individuals, and genetic recombination; cultural traits are transferred by cultural diffusion, which often includes migration. As the phenotypic effects of genes that pass from one population into another having a different adaptive norm are altered by acquiring different modifiers, so ideas and inventions, diffusing from one culture to another, acquire different properties when adjusted to suit the mores of the society they enter.

Cultural evolution can be characterized as the transformation of cultural pools in response to alterations of the cultural environment. Its nearest analog in organic evolution is alteration in gene pools in response to changes in the biotic environment, i.e., coevolution of predators and prey, flowers and pollinators, parasites and their hosts, etc. The analogy begins to break down when the forces that determine change and direction are considered. Natural selection based upon differential reproduction has no counterpart in cultural evolution. To a certain extent, cultural pools can be altered by conscious selection or rejection of new inventions or customs. On the other hand, the conscious planning and foresight that can guide cultural evolution into new channels has no counterpart in organic evolution.

Certain factors, however, tend to reduce this difference. Often, human plans are either not realized or produce unexpected side effects. The leaders of the industrial revolution, when they crowded workers into densely populated slums and established popular newspapers in order to disseminate establishment ideas, did not foresee the possibility that crowding and mass communication could be turned against them in the form of unionization of labor. Likewise, the founders of the first Marxist state, the USSR, failed to realize that the incentives offered to diligent workers and cooperative scientists and engineers could form the basis for a new elite, and so divert the cultural pool away from the Marxian ideal of an egalitarian society. Intellectual leaders of the newly independent states of South Asia have held the highest possible ideals and ambitions in planning for a prosperous society, only to find them received with cynicism, the legacy of age-old class distinctions, poverty, and misery (Myrdal, 1967).

Undoubtedly, cultural evolution can produce change far more rapidly than organic evolution. This fact is amply evident from the revolutionary changes produced during the past two centuries. Two factors are largely responsible for this great speed. First, cultural diffusion is far more rapid than gene flow, since it can affect an entire population in much less time than the course of a single generation. Second, by producing profound changes in the environment in a short time it stimulates greatly its own continued progress. The chain reactions that are released through interactions between culturally evolving societies and the altered environments they generate are far more rapid than those resulting from interactions between two kinds of animal or plant species during organic evolution.

With respect to natural barriers of isolation, the analogy between organic and cultural evolution breaks down almost completely, and the two kinds of evolution move in opposite directions. As pointed out in Chapters 6 and 7, reproductive isolation resulting from the erection of one or more kinds of barriers to gene exchange is an almost inevitable result of continued evolutionary divergence. On the other hand, one of the principal effects of cultural evolution in recent years has been to increase enormously the rate of cultural diffusion, breaking down cultural barriers that previously had existed for centuries or even millennia. Fads such as Coca Cola and rock and roll have spread rapidly throughout the world. Other fads have likewise spread so far and rapidly that in some respects different subcultures within the same community may differ more widely from each other than cultures in widely separated continents. Many American university professors have realized that they are more at home in a faculty gathering in Oxford, Paris, Barcelona, Padua, Tübingen, Helsinki, Jerusalem, Canberra, Santiago, or Novosibirsk than at a meeting of Rotarians or labor unionists in their own community.

One question that might be posed is: Does this breakdown of cultural barriers carry the danger that eventually all mankind will share a common monotony of traits and institutions, and is the richness of local cultural pools seriously threatened? Some Westerners are understandably impatient with the barriers to cultural exchange that nations such as China and Burma have erected, as well as with the attempts of some North American blacks to set up a "culture within a culture." We are not ready to condemn these moves as inimical to society; they may point the way to devices for retaining the distinctness and richness of certain cultural pools. At the same time, contacts between nations having different ideologies promote diffusion of important humanitarian elements. Such diffusion can counteract the factors that tend to split the world into economically prosperous and advanced nations on the one hand, and poor and backward nations on the other.

Analogies to cultural evolution can also be found in the more extended phases of evolution, such as organic evolution at the transspecific level. Both kinds of evolution are basically opportunistic, and do not involve inevitable progress from lower to higher states. As one cannot arrange all vertebrates, or all insects, from lower to higher, so one cannot assign levels of advancement to all human cultures, as some nineteenth-century anthropologists tried to do (Morgan, 1877). The legacy of such "evolutionism" is the reluctance of many twentieth-century anthropologists to recognize evolution as a property of individual cultures.

In one respect, cultural evolution is completely different from organic evolution. Ethical systems cannot be a property of animals since animals are incapable of value judgments. On the other hand, codes of ethics are inherent in and necessary for human societies, although these codes differ from one society to another, and none is perfect.

TOWARD A CULTURAL AND ECOLOGICAL EQUILIBRIUM

Two interpretations of the world situation in the latter part of the twentieth century form the basis of our philosophy. First, the extreme urgency of solving immediately some of the world's major problems has receded. Since the Cuban missile crisis of 1962 the desire of the superpowers to avoid a major nuclear conflict, or even a minor one that might escalate, has become increasingly evident (see Pruitt and Snyder, 1969). The "population time bomb," which a few years ago was described with terrifying predictions and drastic recommendations (Ehrlich, 1970), seems to have a longer fuse than was imagined, and the will to extinguish it or to reduce its force is growing (Freedman and Berelson, 1974; Westoff, 1974; Demeny, 1974). With respect to the impending crises of population and energy resources, some computer-based predictions

(Meadows *et al.*, 1972) have lost part of their force (Cole *et al.*, 1973). The latter authors do not deny that very serious problems exist and that societies need to alter drastically their ways of thinking and planning. Nevertheless, they suggest that mankind has the time necessary to plan carefully without making decisions that might alleviate the crisis but generate even greater problems for future generations.

Second, all the major crises that we face are interrelated: the threat of nuclear conflict, the widening gap between rich and poor nations, the twin problems of overpopulation and exhaustion of food supplies, the exhaustion of energy resources, and the irremediable pollution of the earth's biosphere (Passmore, 1974). They are the outcome, at least in part, of conflicts between the three aspects of man: the biological animal, the member of a purposeful society, and the transformer of his environment. They also stem from a philosophy of exploitation, expansion, and aggression without which mankind could not have evolved modern civilization, but which must now be altered into a philosophy of stability, control, and commonality if human culture is to survive.

In summary, evolutionary guidelines for solving the problems related to the future of evolution can be stated briefly. First, the urgency of the situation requires that cultural rather than genetic solutions be sought. Second, the basis for improvement must be the flexibility and diversity of cultural traits, comparable to the genetic flexibility and diversity that exist in populations of animals. Third, the drastic alterations in the human environment that have taken place in very recent times require equally drastic reorganization of human standards and goals. Exploitation, expansion, and aggression must be replaced by conservation, population control, and tolerance. Finally, societies cannot advance toward improved conditions without considering, in addition to standards that are technological, ethical standards that are strictly human in nature.

16

Philosophical Issues

The study of biological evolution begets more, as well as more urgent, philosophical questions than any other branch of the natural sciences. Problems such as the origin and future of man and man's relationships with the rest of nature are the subject of evolutionary biology but also raise important philosophical questions. A question may be regarded as philosophical if it cannot be approached by the methods characteristic of the empirical sciences but is nevertheless amenable to rational investigation. (The methodological trademarks that distinguish the empirical sciences from other forms of knowledge will be identified below.)

Various kinds of philosophical questions arise from the study of evolutionary biology: epistemological questions, dealing with the sources and grounds of knowledge in evolutionary biology; metaphysical questions, concerned with the nature and ultimate significance of man and the universe; ethical questions, concerned with right and wrong human conduct; and questions of aesthetics, concerned with the nature of beauty and the criteria for artistic judgment.

Chapters 14 and 15 dealt more or less incidentally with the ethical implications of man "engineering" his own biological future and man's management of nature, questions that also have important sociological and political implications. A systematic treatment of the ethical issues arising in evolutionary biology would take us far afield and will not be attempted here. This chapter is primarily concerned with epistemological questions, such as the goals of science and its characteristic methodology, the relationships between biology and the physical sciences, the role of teleological explanations in evolutionary biology, and the logical structure of the theory of natural selection. The notion of biological progress, a metaphysical question, will also be discussed. Evolution appears to be obviously progressive, at least in some sense, and evolutionary biologists often speak of evolutionary progress. The first issue to be discussed is the nature of the scientific method and its application to the study of evolution. Some scientists as well as philosophers erroneously assume that the scientific method used in evolutionary biology is different from that used in the physical sciences.

EMPIRICAL SCIENCE

The Greek philosopher and scientist Aristotle said that man naturally desires to know. Man seeks understanding of natural phenomena and of himself because of sheer intellectual curiosity and also because such understanding gives him the means to manipulate and utilize the environment for his benefit, because it improves his strategic position in the world. Man's manipulation of the environment can be traced to the first tools constructed by *Australopithecus,* and much later to the discovery of fire and other achievements by *Homo erectus.*

All human cultures, including primitive ones, develop explanations for natural phenomena. At least since the discovery of agriculture, mankind has understood certain causal relationships between rain, sunlight, the seasons, and the life cycles of plants and animals. Yet the birth of empirical science (or simply science) is commonly traced to the sixteenth and seventeenth centuries. What are the traits that distinguish science from the common-sense knowledge of primitive cultures and from nonscientific forms of knowledge?

One distinguishing characteristic is that science seeks the systematic organization of knowledge. Science formulates statements about observable phenomena. Common sense also provides knowledge about natural phenomena, and often is correct. Common sense tells one that children resemble their parents and that good seeds produce good crops. Common sense, however, shows little interest in systematically establishing connections between phenomena that do not appear to be obviously related. By contrast, science is concerned with formulating general laws and theories that manifest patterns of relations between very different kinds of phenomena. Science develops by discovering new relationships, and particularly by integrating statements, laws, and theories, which previously seemed to be unrelated, into more comprehensive laws and theories.

Another distinguishing characteristic of science is that it strives to explain why observed events do in fact occur. Although knowledge acquired in the course of ordinary experience is frequently accurate, it seldom provides explanations of why phenomena occur as they do. Practical experience tells us that children resemble one parent in some traits and the other parent in other traits, or that manure increases crop yield. But it does not provide explanations for these phenomena. Science, on the other hand, seeks to formulate explanations for natural phenomena by identifying the conditions that account for their occurrence.

Seeking the systematic organization of knowledge and trying to explain why events are as observed are characteristics that distinguish science from common-sense knowledge. However, these characteristics are also shared by other forms of knowledge, such as mathematics and philosophy. The characteristic that distinguishes the empirical sciences from other systematic forms of knowledge is that scientific explanations must be subject to the possibility of *empirical falsification.* Falsifiability is indeed the criterion of demarcation that sets science apart from other forms of knowledge (Popper, 1934). A scientific hypothesis (or theory) must be empirically testable. A hypothesis is tested by ascertaining whether or not precise predictions derived as logical consequences from the hypothesis agree with the state of affairs found in the empirical world. A hypothesis that is not subject to the possibility of rejection by observation and experiment cannot be regarded as scientific.

Science may be defined as the systematic organization of knowledge about the universe on the basis of explanatory principles subject to the possibility of empirical falsification. Another definition is the following: "Science is an exploration of the material universe that seeks natural, orderly relationships among observed phenomena and that is self-testing" (Simpson, 1964, p. 91). Many other definitions have been proposed, but seeking a "perfect" definition is a futile endeavor. Science is a complex enterprise that cannot be adequately defined in a compact statement. Thus, rather than attempt to formulate an adequate definition, we shall further examine the methodology of science.

THE SCIENTIFIC METHOD

The idea that induction is the method of science is a common misconception, which can be traced to the English statesman and essayist Francis Bacon (1561–1626). Bacon had an important and influential role in shaping modern science by his criticism of the prevailing metaphysical speculations of medieval scholastic philosophers. In the nineteenth century the most ardent and articulate proponent of inductivism was John Stuart Mill (1806–1873), an English philosopher and economist.

Induction was proposed as a method of achieving *objectivity* while avoiding subjective preconceptions, and of obtaining *empirical* rather than abstract or metaphysical knowledge. In its extreme form, inductivism holds that the scientist should observe any phenomena that he encounters in his experience, and record them without any preconceptions as to what to observe or what the truth about them might be—truths of universal validity are expected eventually to emerge. The methodology proposed may be exemplified as follows. A scientist measuring and recording everything that confronts him observes a tree with leaves. A second tree, and a third, and many others, are all observed to have leaves. Eventually, he formulates a universal statement, "All trees have leaves."

The inductive method fails to account for the actual process of science.

First of all, no scientist works without any preconceived plan as to what kind of phenomena to observe. Scientists choose for study objects or events that, in their opinion, are likely to provide answers to questions that interest them. Otherwise, as Darwin (1903) wrote, "one might as well go into a gravelpit and count the pebbles and describe the colours." A scientist whose goal was to record carefully every event observed in all waking moments of his life would not contribute much to the advance of science; more likely than not, he would be considered mad by his colleagues.

Moreover, induction fails to arrive at universal truths. No matter how many singular statements may be accumulated, no universal statement can be logically derived from such an accumulation of observations. Even if all trees so far observed have leaves, or all swans observed are white, it remains a logical possibility that the next tree will not have leaves, or the next swan will not be white. The step from numerous singular statements to a universal one involves logical amplification. The universal statement has greater logical content —it says more—than the sum of all singular statements.

Another serious logical difficulty with induction as a scientific method is that scientific hypotheses and theories are formulated in abstract terms that do not occur at all in the description of empirical events. Mendel observed in the progenies of hybrid plants that alternative traits could be segregated into certain ratios. Repeated observations of these ratios could never have led inductively to the formulation of his hypothesis that "factors" (genes) exist in the sex cells and are arranged according to certain rules. These "factors" were not observed, and thus could not be included in observational statements. The most interesting and fruitful scientific hypotheses are not simple generalizations. Instead, scientific hypotheses are creations of the mind, imaginative suggestions as to what might be true.

Induction fails in all three counts pointed out: as a method that insures objectivity and avoids preconceptions, as a logical method to reach universal truths, and as a description of the process followed by scientists in the formulation of hypotheses.*

A more nearly correct concept of the methods of science is provided by the so-called *hypothetico-deductive* model. The explicit formulation of this model may be traced to William Whewell (1794–1866) and William Stanley Jevons (1835–1882) in England, and to Charles S. Peirce (1839–1914) in the United States. The most precise characterization of the scientific method has been expounded by Karl R. Popper (1934). Scientists, of course, practiced the hypothetico-deductive method long before it was adequately defined by philosophers. Eminent practitioners of this method include Blaise Pascal (1623–1662) and Isaac Newton (1642–1727) in the seventeenth century, and, among nineteenth-century biologists, Claude Bernard (1813–1878) and Louis Pasteur (1822–1895) in France, Charles Darwin (1809–1882) in England, and Gregor Mendel (1822–1884) in Austria. These and other suc-

*For more extensive criticisms of inductivism see, for example, Popper (1934), Kuhn (1962), Hempel (1966), and the very readable account of Medawar (1969).

cessful scientists practiced the hypothetico-deductive method even if some of them claimed to be inductivists in order to conform to the claims of contemporary philosophers.

Science is a complex enterprise that essentially consists of two interdependent episodes, one imaginative or creative, the other critical. To have an idea, advance a hypothesis, or suggest what might be true is a creative exercise. But scientific conjectures or hypotheses must also be subject to critical examination and empirical testing. Scientific thinking may be characterized as a process of invention or discovery followed by validation or confirmation. One process concerns the acquisition of knowledge, the other concerns the justification of knowledge.

Scientists like other people *acquire* knowledge in all sorts of ways: from conversation with other people, from reading books and newspapers, from inductive generalizations, and even from dreams and mistaken observations. Newton is said to have been inspired by a falling apple. Kekulé had been unsuccessfully attempting to devise a model for the molecular structure of benzene. One evening he was dozing in front of the fire. The flames appeared to Kekulé as snake-like arrays of atoms. Suddenly one snake appeared to bite its own tail and then whirled mockingly in front of him. The circular appearance of the image inspired in him the model of benzene as an hexagonal ring. Darwin proposed his hypothesis of the origin of coral reefs before he had ever seen a coral reef. The model to explain the evolutionary diversification of species came to Darwin while riding in his coach and observing the countryside. "I can remember the very spot in the road . . . when to my joy the solution came to me. . . . The solution, as I believe, is that the modified offspring . . . tend to become adapted to many and highly diversified places in the economy of nature" (Darwin, 1958).

Hypotheses and other imaginative conjectures are the initial stage of scientific inquiry. It is the imaginative conjecture of what might be true that provides the incentive to seek the truth and a clue as to where we might find it (Medawar, 1967). Hypotheses guide observation and experiment because they suggest what to observe. The empirical work of scientists is guided by hypotheses, whether explicitly formulated or simply in the form of vague conjectures or hunches about what the truth might be. But imaginative conjecture and empirical observation are mutually interdependent processes. Observations made to test a hypothesis are often the inspiring source of new conjectures or hypotheses.

Although the conception of an idea is the starting point of scientific inquiry, this process is not the subject of investigation of logic or epistemology. The complex conscious and unconscious events underlying the creative mind are properly the interest of empirical psychology. The creative process is obviously not unique to scientists. Philosophers as well as poets, novelists, and other artists are also creative; they too advance models of experience and they also generalize by induction. What distinguishes science from other forms of knowledge is the process by which this knowledge is justified or validated.

THE CRITERION OF DEMARCATION

Testing a hypothesis (or theory) involves at least four different activities. First, the hypothesis must be examined for internal consistency. A hypothesis that is self-contradictory or not logically well-formed in some other way should be rejected. Second, the logical structure of the hypothesis must be examined to ascertain whether it has explanatory value, i.e., whether it makes the observed phenomena intelligible in some sense, whether it provides an understanding of why the phenomena do in fact occur as observed. A hypothesis that is purely tautological should be rejected because it has no explanatory value. Third, the hypothesis must be examined for its consistency with hypotheses and theories commonly accepted in the particular field of science, or to see whether it represents any advance with respect to well-established alternative hypotheses. Lack of consistency with other theories is not always ground for rejection of a hypothesis, although it will often be. Finally, the hypothesis must be tested empirically.

A hypothesis (or theory) is tested empirically by ascertaining whether or not predictions about the world of experience derived as logical consequences from the hypothesis agree with what is actually observed. The critical element that distinguishes the empirical sciences from other forms of knowledge is the requirement that scientific hypotheses be empirically falsifiable. Scientific hypotheses cannot be consistent with all possible states of affairs in the empirical world. A hypothesis is scientific only if it is consistent with some but not with other possible states of affairs not yet observed in the world, so that it may be subject to the possibility of falsification by observation. The predictions derived from a scientific hypothesis must be sufficiently precise that they limit the range of possible observations with which they are compatible. If the results of an empirical test agree with the predictions derived from a hypothesis, the hypothesis is said to be provisionally corroborated; otherwise it is falsified.

The requirement that a scientific hypothesis be falsifiable has been appropriately called the *criterion of demarcation* of the empirical sciences because it sets apart the empirical sciences from other forms of knowledge (Popper, 1934). A hypothesis that is not subject, at least in principle, to the possibility of empirical falsification does not belong in the realm of science.

The requirement that scientific hypotheses be falsifiable rather than simply verifiable may seem surprising at first. It might seem that the goal of science is to establish the "truth" of hypotheses rather than attempt to falsify them. It is not so. There is an asymmetry between falsifiability and verifiability of universal statements that derives from the logical nature of such statements. A universal statement can be shown to be false if it is found inconsistent with even one singular statement, i.e., a statement about a particular event. But, as was pointed out in the discussion of induction, a universal statement can never be proven true by virtue of the truth of particular statements, no matter how numerous these may be.

Consider a hypothesis, H_1, from which a certain number of consequences,

$C_1, C_2, \ldots, C_n$, are logically derived. Assume that it is found that C_1, $C_2, \ldots, C_n$ are true. It does not necessarily follow that H_1 is true. Consider the argument: If H_1 is true, then C_1 must also be true; it is the case that C_1 is true. Therefore H_1 is true. This erroneous kind of inference is called by logicians the *fallacy of affirming the consequent.* The conclusion is invalid even if both premises are true. It may in fact be the case that there is some other hypothesis, H_2, from which the same consequences or predictions can be derived. Then $C_1, C_2, \ldots, C_n$ might still be true because H_2 is true, even if H_1 is false.

The proper form of logical inference for conditional statements is what logicians call the *modus tollens* (*modus* = mode; *tollens* = to take away, to reject). It may be represented by the following argument. If H_1 is true, then C_1 must also be true; but evidence shows that C_1 is not true; therefore H_1 is false. This is a correct form of inference; if both premises are true, the conclusion necessarily follows. It is possible to show the falsity of a universal statement, but it is never possible to demonstrate conclusively the truth of a universal statement concerning the empirical world.

The asymmetry between verification and falsification is recognized in the statistical methodology of testing hypotheses. The hypothesis subject to test, the *null hypothesis,* may be rejected if the observations are inconsistent with it. If the observations are consistent with the predictions derived from the hypothesis, the proper conclusion is that the test has failed to falsify the null hypothesis, not that its truth has been established. The requirement that scientific hypotheses be falsifiable has another parallel in statistical inference, namely in the requirement that the power of the test be greater than zero. Statisticians recognize two kinds of errors: a Type I error, the probability of rejecting the null hypothesis when it is true, usually represented as α; and a Type II error, the probability of not rejecting the hypothesis when it is false, symbolized as β. Scientists pay considerable attention to Type I errors, and thus choose α levels sufficiently low. It is unfortunate that many scientists pay little attention to Type II errors. Yet the power of the test depends on the probability, $1 - \beta$, of rejecting the null hypothesis when it is wrong. Thus, small levels for both α and β are desirable. Although for any given test the magnitudes of α and β are inversely related, the value of β may be reduced by increasing the sample size or the number of replications in a test.

Just as the power of a statistical test must be greater than zero, so it is more generally required of tests of scientific hypotheses that they have a positive probability of resulting in the rejection of the hypothesis if it is false. A scientific hypothesis divides all particular statements of fact into two nonempty subclasses. First, we have the class of all statements with which it is inconsistent, the class of the "potential falsifiers" of the hypothesis. Second, there is the class of all statements that the hypothesis does not contradict, the class of "permitted" statements. A hypothesis is scientific only if the class of its potential falsifiers is not empty, because it makes assertions only about its potential falsifiers—it asserts that they are false. "Not for nothing do we call the laws of nature 'laws': the more they prohibit the more they say" (Popper, 1934).

The empirical or information content of a hypothesis is measured by the class of its potential falsifiers. The larger this class, the greater the information content of the hypothesis. A hypothesis asserts that its potential falsifiers are false; if any of them is true, the hypothesis is proven false. A hypothesis or theory consistent with all possible states of affairs in the natural world lacks empirical content, and thus does not belong in the realm of science.

Scientific hypotheses can only be accepted provisionally, since their truth can never be conclusively established. This does not mean that we have the same degree of confidence in all hypotheses that have not yet been falsified. A hypothesis that has passed many empirical tests may be said to be "corroborated." The degree of corroboration is not simply a matter of the number of tests, but rather their severity. Severe tests are precisely those that are very likely to have outcomes incompatible with the hypothesis if the hypothesis is false. The more precise the predictions being tested, the more severe the test. A so-called critical or crucial test is an experiment for which competing hypotheses predict alternative, mutually exclusive outcomes. A critical test thus will corroborate one hypothesis and falsify the others.

The larger the variety of severe tests withstood by a hypothesis, the greater its degree of corroboration. Hypotheses or theories may thus become established beyond reasonable doubt. The hypothesis of evolution, that new organisms come about by descent with modification from dissimilar ancestors, is an example of a hypothesis corroborated beyond reasonable doubt. This is what is claimed by biologists who state that evolution is a fact rather than a theory or hypothesis. In ordinary usage, the terms "hypothesis" and "theory" sometimes imply a lack of sufficient corroboration. The evolutionary origin of organisms is compatible with virtually all known facts of biology, and has passed a wide variety of severe tests.

The *modus tollens* is a logically conclusive method—if a necessary consequence of a premise is false, then the premise must also be false. Nevertheless, the process of falsification is subject to human error. It is possible, for example, that an observation or experiment contradicting a hypothesis may have been erroneously performed or erroneously interpreted. Thus, it is usually required, particularly in the case of important or well-corroborated hypotheses, that the falsifying observation be repeatable.

The *modus tollens* may lead to erroneous conclusions if the prediction tested is not a necessary logical consequence from the hypothesis. The connection between a hypothesis and specific predictions derived from it is often not a simple matter. The logical validity of an inference may depend not only on the hypothesis being tested, but also on other hypotheses, whether explicitly stated or not, as well as on assumptions concerning the particular conditions under which the deduced inferences obtain (boundary conditions). If a particular prediction is falsified, it follows that the hypothesis tested as well as other hypotheses necessarily implied and the boundary conditions cannot all jointly be true. The possibility exists that one of the subsidiary hypotheses or some assumed condition may be false. A proper test of a hypothesis thus tests the validity of all the hypotheses and conditions involved.

MENDEL AND THE SCIENTIFIC METHOD

Gregor Mendel's classic paper, "Experiments in Plant Hybridization," is an eminent example of the use of the scientific method in biology. First published in 1866, this paper established single-handedly the basic principles and fundamental theory of heredity, from which a whole new branch of science, genetics, would develop. The paper is also remarkable because of Mendel's explicit and lucid awareness of the requirements of the scientific method. Mendel formulated hypotheses, examined their consistency with previous results, then submitted the hypotheses to severe empirical tests and suggested additional tests that might be performed.

Mendel's genius is evident in his recognition of the conditions required to formulate and test a theory of inheritance: different traits should be considered individually, alternative states of the traits should differ in clear-cut ways, and ancestry of the plants should be precisely known (which, in turn, requires that only true-breeding lines be used in the experiment, and that the origin of the pollen, and not only the eggs, be controlled). Mendel's hypotheses were formulated in probabilistic terms; accordingly, he obtained large samples and subjected them to statistical analysis.

Mendel studied the transmission of seven different traits in the garden pea, *Pisum sativum,* including the color of the seed (yellow versus green), the configuration of the seed (round versus wrinkled), and the height of the plant (tall versus dwarf). The results of Mendel's experiments are too well known to need detailed presentation here, but it is worth pointing out the various stages of his methodology. His first series of experiments was with plants that differ in a single trait. The regularities observed led to certain generalizations having the form of law-like statements: only one of the two traits (the *dominant* trait) appears in the F_1 progenies; after self-fertilization, three-fourths of the F_2 progenies exhibit the dominant trait, and one-fourth exhibit the other (*recessive*) trait; the F_2 plants exhibiting the recessive trait breed true in the F_3 and following generations, but the plants exhibiting the dominant trait are of two kinds, one-third breed true, the other two-thirds are hybrids. Mendel tested these generalizations by repeating his experiments for each of the seven characters. These generalizations were summarized in a law, later called the Principle of Segregation: hybrid plants produce seeds that are one-half hybrid, one-fourth pure breeding for the dominant trait, and one-fourth pure breeding for the recessive trait.

Mendel tested the hypothesis of segregation by deriving and verifying additional predictions. For example, he predicted that after n generations of self-fertilization the ratio of true-breeding to hybrid plants in the progeny of a hybrid should be $2^n - 1$ to 1. He explicitly stated that this prediction would obtain only if the following condition obtained, that all plants have "an average quality of fertility . . . in all generations."

The study of the offspring of crosses between plants differing in two traits (e.g., round and yellow seeds in one parent, wrinkled and green seeds in the other parent) led him to formulate a second law-like statement, later called the Principle of Independent Assortment: "The principle applies that [in] the offspring of the hybrids in which several essentially different characters are combined . . . the relation of each pair of different characters in hybrid union is independent of the other differences in the two original parental stocks." He corroborated this principle by examining progenies of plants differing in various combinations of two traits, as well as in plants differing in three and four traits. He correctly predicted and corroborated experimentally that in the progenies of plants hybrid for n characters there will be 3^n different classes of plants.

Textbooks give credit to Mendel for having formulated the principles of segregation and independent assortment, often called Mendel's First and Second Law of Inheritance. The formulation and experimental testing of these two laws take up only approximately the first half of Mendel's paper. Midway through the paper Mendel advanced what he properly called a hypothesis or theory to account for his previous results and for the two laws. In the second half of the paper predictions are derived from the theory and tested.

Mendel's theory of inheritance contains the following elements: (1) for each character in any plant, whether hybrid or not, there is a pair of hereditary "factors"; (2) these two factors are inherited one from each parent; (3) the two factors of each pair segregate during the formation of the sex cells, so that each sex cell receives only one factor; (4) each sex cell receives one or the other factor of a pair with a probability of one-half; (5) alternative factors for different characters associate at random in the formation of the sex cells. Mendel's well-deserved eminence as a scientist rests particularly on the formulation of this theory of heredity. Mendel was also quite aware of the logical status of his proposal, namely that it was a hypothesis and therefore required experimental corroboration. In the paragraph following the formulation of his theory, Mendel wrote that "this hypothesis would fully suffice to account for the development of the hybrids in the separate generations," i.e., the hypothesis is consistent with his previous observations. But, quite appropriately, he recognized that further experimental tests were called for: "In order to bring these assumptions to an experimental proof the following experiments were designed." The experiments are two series of back-crosses that confirm segregation and independent assortment in the egg cells, and then in the pollen cells.

The theory was further tested with experimental crosses using a different plant, *Phaseolus.* The ratios observed in the progenies of hybrids were the same as in *Pisum* and thus consistent with the theory. With respect to two traits, flower and seed color, Mendel observed in the F_2 progenies "a whole series of colors . . . from purple to pale violet and white." Mendel correctly conjectured that "these enigmatic results . . . might be explained . . . if we might assume that the color of the flowers and seeds of *Ph* [*aseolus*] *multiflorus* is a combination of two or more entirely independent colors, which individually act like any other character in the plant." After showing how this hypothesis accounts for the gamut of colors, Mendel added: "It must, nevertheless, not be forgotten that the explanation here attempted is based on a mere hypothesis, only supported by the very imperfect result of the experiment just described. It would, however, be well worthwhile to follow up the development of color in hybrids by similar experiments." To the very end, Mendel was fully aware of the demand of the scientific method.

Some authors have suggested that Mendel must have had a fairly clear conception of what he expected to find even before he began the first experiments reported in his paper. The experiments appear too well designed for things to be otherwise. Indeed, it might be the case that Mendel had performed some previous experiments, or that he came upon the idea of the binary determination of traits from considerations of the existence of two sexes, or in some other way. Whether or not Mendel had a preconception of what the results of his experiments would be is irrelevant to the integrity of his scientific accomplishments or his masterly understanding of the requirements of the scientific method. Mendel's paper remains a model of the scientific method.

DARWIN AND THE SCIENTIFIC METHOD

Few scientists in the nineteenth century or at any earlier time equal Mendel's clear delineation of the scientific method he was pursuing. In the English-speaking countries, scientists advanced hypotheses and tested them in their work, but often claimed in their writings to be following the orthodoxy of inductionism proclaimed by philosophers as the method of good science. Darwin is a remarkable example of this discrepancy.

In his *Autobiography* Darwin says that he proceeded "on true Baconian principles and without any theory collected facts on a wholesale scale" (1958, p. 119). The opening paragraph of *Origin of Species* conveys the same impression:

> When on board H. M. S. *Beagle,* as naturalist, I was much struck with certain facts in the distribution of the inhabitants of South America, and in the geological relations of the present to the past inhabitants of that continent. These facts seemed to me to throw some light on the origin of species—that mystery of mysteries, as it has been called by one of our greatest philosophers. On my return home, it occurred to me, in 1837, that something might perhaps be made out on this question by *patiently accumulating and reflecting on all sorts of facts which could possibly have any bearing on it.* After five years' work I allowed myself to speculate on the subject, and drew up some short notes; these I enlarged in 1844 into a sketch of the conclusions, which then seemed to me probable: from that period to the present day I have steadily pursued the same object. [Emphasis added]

Darwin claims in many other writings to have followed the inductivist canons. The facts are very different from these claims, however. Darwin's notebooks and private correspondence show that he entertained the hypothesis of the evolutionary transmutation of species shortly after returning from the voyage of the *Beagle*, and that the hypothesis of natural selection occurred to him in 1838—several years before he claims to have allowed himself for the first time "to speculate on the subject." Between the return of the *Beagle* on October 2, 1836, and publication of *Origin of Species* (and, indeed, until the end of his life) Darwin relentlessly pursued empirical evidence to corroborate the evolutionary origin of organisms, and to test his theory of natural selection.

Why this disparity between what Darwin was doing and what he claimed? There are at least two reasons. First, in the temper of the times "hypothesis" was a term often reserved for metaphysical speculations without empirical substance. This is the reason why Newton, the greatest theorist of all scientists, also claimed, *Hypotheses non fingo* ("I fabricate no hypotheses"). Darwin expressed distaste and even contempt for empirically untestable hypotheses. He wrote of Herbert Spencer: "His deductive manner of treating any subject is wholly opposed to my frame of mind. His conclusions never convince me . . . . His fundamental generalizations (which have been compared in importance by some persons with Newton's Laws!) which I daresay may be very valuable under a philosophical point of view, are of such a nature that they do not seem to me to be of any strictly scientific use. They partake more of the nature of definitions than of laws of nature. They do not aid me in predicting what will happen in any particular case" (1958, p. 109).

There is another reason, a tactical one, why Darwin claimed to proceed according to inductive canons. He did not want to be accused of subjective bias in the evaluation of empirical evidence. Darwin's true colors are shown in a letter to a young scientist written in 1863: "I would suggest to you the advantage, at present, of being very sparing in introducing theory in your papers (I formerly erred much in Geology in that way); *let theory guide your observations*, but till your reputation is well established, be sparing of publishing theory. It makes persons doubt your observations" (F. Darwin, 1903, 2:323; see also Hull, 1973). Nowadays scientists, young or not, often report their work so as to make their hypotheses appear as conclusions derived from the evidence at hand, rather than as preconceptions tested by empirical observations.

Darwin rejected the inductivist claim that observations should not be guided by hypotheses. The statement quoted earlier, "A man might as well go into a gravel-pit and count the pebbles and describe the colours," is followed by this telling remark: "How odd it is that anyone should not see that all observation must be for or against some view if it is to be of any service!" (F. Darwin, 1903, 1:195). He acknowledged the heuristic role of hypotheses, which guide empirical research by telling us what is worth observing, what evidence to seek. He confesses: "I cannot avoid forming one [hypothesis] on every subject" (1958, p. 141).

Darwin was an excellent practitioner of the hypothetico-deductive method of science, as modern students of Darwin have abundantly shown (De Beer, 1964; Mayr, 1964; Ghiselin, 1969; Hull, 1973). Darwin advanced hypotheses in multiple fields, including geology, plant morphology and physiology, psychology, and evolution, and subjected his hypotheses to empirical test. "The line of argument often pursued throughout my theory is to establish a point as a probability by induction and to apply it as a hypothesis to other parts and see whether it will solve them" (Darwin, 1960). Popper (1934) has not only made clear that falsifiability is the criterion of demarcation of the empirical sciences from other forms of knowledge, but also that falsification of seemingly true hypotheses contributes to the advance of science. Darwin recognized this: "False facts are highly injurious to the progress of science, for they often endure long; but false views, if supported by some evidence, do little harm, for every one takes a salutary pleasure in proving their falseness; and when this is done, one path towards error is closed and the road to truth is often at the same time opened" (Darwin, 1871, 2nd ed., p. 606).

Some philosophers of science have claimed that evolutionary biology is a historical science that does not need to satisfy the requirements of the hypothetico-deductive method. The evolution of organisms, it is argued, is a historical process that depends on unique and unpredictable events, and thus is not subject to the formulation of testable hypotheses and theories. Such claims emanate from a monumental misunderstanding. There are two kinds of questions in the study of biological evolution (Dobzhansky, 1951, pp. 11–12). One concerns history: the study of phylogeny, the unraveling and description of the actual course of evolution on earth that has led to the present state of the biological world. The scientific disciplines contributing to the study of phylogeny include systematics, paleontology, biogeography, comparative anatomy, comparative embryology, and comparative biochemistry. The second kind of question concerns the elucidation of the mechanisms or processes that bring about evolutionary change. These questions deal with causal, rather than historical, relationships. Population genetics, population ecology, paleobiology, and many other branches of biology are the relevant disciplines.

There can be little doubt that the causal study of evolution proceeds by the formulation and empirical testing of hypotheses, according to the same hypothetico-deductive methodology characteristic of the physicochemical sciences and other empirical disciplines concerned with causal processes. This book contains numerous examples of hypotheses, and their tests, advanced to account for the various processes underlying biological evolution (see also Ayala, 1975b). But even the study of evolutionary history is based on the formulation of empirically testable hypotheses. Consider a simple example. For many years specialists proposed that the evolutionary lineages leading to man separated from the lineage leading to the great apes (chimpanzee, gorilla, orangutan) before the lineages of the great apes separated from each other. Some recent authors have suggested instead that man, chimpanzees, and gorillas are more closely related to each other than the chimpanzee and the gorilla are to the other apes. A wealth of empirical predictions can be derived logically from these competing hypotheses. One prediction concerns the degree of similarity between enzymes and other proteins. It is known that the rate of amino acid substitutions is approximately constant when averaged over many proteins and long periods of time (Chapter 9). If the first hypothesis is correct, the average amount of protein differentiation should be greater between man and the African apes than among these and orangutans. On the other hand, if the second hypothesis is correct, man and chimpanzees should have greater protein similarity than either one has with orangutans. These alternative predictions provide a critical empirical test of the hypotheses. The available data favor the second hypothesis. Man, chimpanzees, and gorillas appear to be phylogenetically more closely related to each other than any one of them is related to orangutans (Goodman, 1976).

Certain biological disciplines relevant to the study of evolution are largely descriptive and classificatory. Description and classification are necessary activities in all branches of science, but play a greater role in certain biological disciplines, such as systematics and biogeography, than in other disciplines, such as population genetics. Nevertheless, even systematics and biogeography occasionally use the hypothetico-deductive method and formulate empirically testable hypotheses.

MECHANISM AND VITALISM

The relationships between the biological and physical sciences, and between organisms and inorganic matter, are of considerable interest to the philosophy of science. Many papers, books, and symposia have been devoted to these relationships in recent years (e.g., Koestler and Smythies, 1969; Ayala and Dobzhansky, 1974). The issue at stake is sometimes called "the problem of reductionism" or "the question of reduction." Few philosophical issues have been more actively debated in recent years, particularly among scientists, than the question of reduction. The debate, however, involves several different issues, not always properly distinguished. Issues about the relationship between the biological and physical sciences fall into at least three domains, which may be called "ontological," "methodological," and "epistemological." We shall identify the issues in each domain, and then consider each domain in turn.

Reductionistic questions arise, first, in what may be called the ontological, the structural, or the constitutive domain. The issue here is whether or not physicochemical entities and processes underlie all living phenomena. Are organisms constituted of the same components as those making up inorganic matter? Or do organisms consist of other entities besides molecules, atoms, and ultimately subatomic particles? Other questions are related to these. Are organisms nothing else than aggregations of atoms and molecules? Do organisms exhibit properties other than those of their constituent atoms and molecules?

Second, there are reductionist questions that might be called methodological, procedural, or strategical. These questions concern the strategy of research and the acquisition of knowledge, the approaches to be followed in the investigation of living beings. The general question is whether biological problems should always be investigated by studying the underlying (ultimately, physical) processes, or whether they should not also be studied at higher levels of organization, such as the cell, the organism, the population, and the community.

The third type of reductionistic question concerns issues that may be called epistemological, theoretical, or explanatory. The fundamental issue here is whether or not the theories and laws of biology can be derived from the laws and theories of physics and chemistry. Epistemological reductionism is concerned with the question whether biology may not be a separate science, but simply a special case of physics and chemistry.

Distinguishing the various kinds of questions being asked in the debate over reductionism is the first step towards solving the issues. Much argument and confusion has resulted from a failure to identify the type of question being argued in particular instances. It is not untypical, for example, to see a reductionist concerned primarily with the ontological question accusing a self-proclaimed anti-reductionist of being a vitalist, while the latter may be an epistemological anti-reductionist but also in fact an ontological reductionist.

In the ontological domain reductionism versus anti-reductionism, in its extreme form, resolves into mechanism versus vitalism. The mechanist position is that organisms are ultimately made up of the same atoms that make up inorganic matter, and nothing more. Vitalists argue that organisms are made up not only of material components (atoms, molecules, and aggregations of them) but also of some nonmaterial entity, variously called entelechy, vital force, *élan vital,* radial energy, and the like. Aristotle (384–322 B.C.), the great Greek philosopher who was also the best biologist of his time, is sometimes said to be the first systematic proponent of vitalism. The modern controversy over mechanism versus vitalism dates from the seventeenth century, when René Descartes (1596–1650) proposed that animals are nothing else than complex machines. Early in the twentieth century vitalism was defended by such philosophers as Henri Bergson (1859–1941), and by some biologists, notably Hans Driesch (1867–1941). At present vitalism has no distinguished proponents among biologists, and few if any among philosophers.

Vitalism has been excluded from science because it does not meet the requirements of a scientific hypothesis. Vitalism is not a hypothesis subject to the possibility of empirical falsification and therefore leads to no fruitful observations or experiments. Moreover, all available evidence indicates that organisms and life processes can be explained without recourse to any substantive nonmaterial entity.

Ontological reductionism claims that organisms are exhaustively composed of nonliving parts. No substance or other residue remains after all atoms making up an organism are taken into account. Ontological reductionism also implies that the laws of physics and chemistry fully apply to all biological processes at the level of atoms and molecules.

Ontological reductionism does not necessarily claim, however, that organisms are nothing but atoms and molecules. The idea that because something consists of "something else" it is nothing but this "something else" is an erroneous inference, called by philosophers the "nothing but" fallacy. Organisms consist exhaustively of atoms and molecules, but it does not follow that they are nothing but heaps of atoms and molecules. A steam engine may consist only of iron and other materials, but it is something other than iron and the other components. Similarly, an electronic computer is not only a pile of semiconductors, wires, plastic, and other materials. Organisms are made up of atoms and molecules, but they are highly complex patterns, and patterns of patterns, of these atoms and molecules. Living processes are highly complex, highly special, and highly improbable patterns of physical and chemical processes.

A much debated reductionist question that belongs in the ontological domain is whether organisms exhibit *emergent properties,* or whether their properties are simply those of their physical components. For example, are the functional properties of the kidney simply the properties of the chemical constituents of that organ? It must be pointed out, first, that the question of emergent properties is not exclusive to biology, but applies to all complex systems. The general formulation of this question is whether the properties of a particular object are simply the properties of its component parts, organized in certain ways.

Whether complex systems exhibit emergent properties is largely a spurious issue that can be solved as a matter of definition. Consider the following question. Are the properties of common salt, sodium chloride, simply the properties of sodium and chlorine when they are associated according to the formula NaCl? If among the properties of sodium and chlorine we include their association into table salt and the properties of the latter, the answer is yes. In general, if among the properties of an object we include the properties that the object has when associated with other objects, it follows that the properties of complex systems, including organisms, are also the properties of their component parts. However, this is simply a definitional maneuver that contributes little to understanding the relationships between complex systems and their parts.

In common practice the properties of an object do not include the properties of systems of which the object may form a part. There is a good reason for this. No matter how exhaustively an object is studied in isolation, there is usually no way to ascertain all the properties that it may have in association with any other object. Among the properties of hydrogen we do not usually include the properties of water, ethyl alcohol, proteins, and human beings. Nor do we include among the properties of iron those of the steam engine.

The question of emergent properties may also be formulated in a somewhat different manner. Can the properties of complex systems be *inferred* from knowledge of the properties that their component parts have in isolation? For example, can the properties of benzene be predicted from knowledge about oxygen, hydrogen, and carbon? Or, at a much higher level of complexity, can the behavior of a cheetah chasing a deer be predicted from knowledge about the atoms and molecules making up these animals? Formulated in this manner, the issue of emergent properties is an epistemological question, not an ontological one. It asks whether the laws and theories accounting for the behavior of complex systems can be derived as logical consequences of the laws and theories that explain the behavior of their component parts. (Epistemological questions of reductionism are discussed later in this chapter, p. 491ff.)

REDUCTIONISM VERSUS COMPOSITIONISM

One outstanding characteristic of living beings is the complexity of organization connoted by the name "organism." A hierarchy of levels of complexity runs from atoms, through molecules, macromolecules, organelles, cells, tissues, multicellular organisms, populations, and communities. Some biological disciplines focus on one or a few of these levels of complexity of organization: cytology is the study of cells, histology the study of tissues, and ecology the study of populations and communities. Yet biological disciplines are identified more by the kinds of questions asked, and the kinds of answers sought, than by the level of organization investigated.

Methodological reductionism claims that living phenomena are best studied at lower levels of complexity, ultimately at the level of atoms and molecules. For example, genetics should seek to understand heredity ultimately in terms of the behavior and structure of DNA, RNA, enzymes, and other macromolecules, rather than in terms of whole organisms, the level at which the Mendelian laws of inheritance are formulated. Methodological reductionism has its counterpart in what may be called methodological compositionism (Simpson, 1964), which claims that to understand organisms we must first explain their organization—not only how organisms and groups of organisms are organized, but also what functions the organization serves. Accordingly, organisms and groups of organisms should be studied as wholes, as well as in their component parts.

Methodological reductionism in its extreme form claims that biological research should be conducted only at the level of physicochemical component parts and processes. Research at other levels, it is argued, is not worth pursuing since biological phenomena can ultimately be understood only at the molecular and atomic levels. Methodological compositionism in its extreme form makes the opposite claim, namely, that biological research is worth pursuing only at the level of whole organisms, populations, and communities. Research at lower levels of organization may be good physics or good chemistry, but it has no biological significance.

It is unlikely that any thoughtful scientist would advocate the extreme form of either compositionism or reductionism. Extreme methodological reductionism would imply the unreasonable claim that genetic investigations should not have been undertaken until the discovery of DNA as the hereditary material, or that a moratorium should be declared in ecology until we can investigate the physicochemical processes underlying ecological interactions. Similarly, extreme methodological compositionism would imply that understanding the structure of DNA or the enzymatic processes involved in replication is of no significance to the study of heredity, or that the investigation of physicochemical reactions in the transmission of nerve impulses is of no importance to the understanding of animal behavior.

The moderate version of methodological reductionism emphasizes the success of the analytical method in science, and the obvious fact that the understanding of living processes at any level of organization is much advanced by knowledge of the underlying processes. Moderate methodological reductionists claim that the best strategy of research is to investigate any given biological phenomenon at increasingly lower levels of organization as this becomes possible, and ultimately at the level of atoms and molecules.

These claims of moderate methodological reductionists are legitimate. Reductionist analysis is of great heuristic value, i. e., it serves to discover and to stimulate investigation; much can be learned about a phenomenon through the investigation of its component elements or processes. In biology the most impressive achievements of the last few decades have been those of molecular biology. But there is little justification for any exclusionist claim that research should always proceed by investigation of increasingly lower levels of integration. The only criterion of validity of a research strategy is its success. Compositionist as well as reductionist approaches, synthetic as well as analytic methods of investigation, are justified if they further our understanding of a phenomenon, if they increase knowledge. Reductionist and compositionist approaches to the study of a biological problem are complementary; often the best strategy of research is an alternation between analysis and synthesis.

Investigation of a biological phenomenon at higher levels of complexity often contributes to the understanding of the phenomenon itself. Compositionist investigations are also heuristic. It is doubtful that the structure and functions of DNA would have been known as readily as they were if there had been no previous knowledge of Mendelian genetics. The problem of the specificity of the immune response of antibodies proved refractory to a satisfactory solution as long as the structure alone of antibodies and antigens was taken into consideration. The natural selection theory of antibody function emerged only when antibodies were considered in their organismic milieu. Although the idea of clonal selection was logically inadequate and quite vague at first, it had an enormous heuristic value in helping to understand how the specificity of antibodies comes about (Edelman, 1974).

THE REDUCTION OF THEORIES

When philosophers of science speak of reductionism, they are generally referring neither to ontological nor to methodological issues, but to epistemological reductionism. In biology the central question of epistemological (theoretical, explanatory) reductionism is whether the laws and theories of biology can be shown to be special cases of the laws and theories of the physical sciences.

Science seeks to discover patterns of relations among vast kinds of phenomena in such a way that a number of principles explain a large number of propositions concerning those phenomena. Science advances by developing gradually more comprehensive theories, i. e., by showing that theories and laws that had hitherto appeared as unrelated can in fact be integrated in a single theory of greater generality. For example, Mendel's theory of heredity can explain diverse observations about many kinds of organisms, such as the proportions in which traits are transmitted from parents to offspring, why progenies exhibit some traits inherited from one parent and some from the other parent, and why the offspring may exhibit traits not present in their parents. The discovery that the behavior of chromosomes during meiosis is connected with the Mendelian principles made possible the explanation of many additional observations concerning heredity; for example, why certain traits are inherited independently from each other, while other traits are more often transmitted together. Further discoveries made possible the development of a unified theory of inheritance of great generality, which explains many diverse observations, including the distinctness of individuals, the adaptive nature of organisms and their traits, and the discreteness of species.

The connection among theories has sometimes been established by showing that the tenets of a theory or branch of science can be explained by the tenets of another theory or branch of science of greater generality. The less general theory (or branch of science), called the secondary theory, is then said to have been reduced to the more general or primary theory. Epistemological reduction of one branch of science to another takes place when the theories or experimental laws of a branch of science are shown to be special cases of the theories and laws formulated in some other branch of science. The integration of diverse scientific theories and laws into more comprehensive ones simplifies science and extends the explanatory power of scientific principles, and thus conforms to the goals of science.

The reduction of a theory or even a whole branch of science to another has occurred repeatedly in this history of science (Nagel, 1961; Popper, 1974). One of the most impressive examples is the reduction of thermodynamics to statistical mechanics, made possible by the discovery that the temperature of a gas reflects the mean kinetic energy of its molecules. Several branches of physics and astronomy have been to a large extent unified by their reduction to a few theories of great generality, such as quantum mechanics and relativity. A large sector of chemistry was reduced to physics after it was shown that the valence of an element bears a simple relation to the number of electrons in the outer orbit of the atom. Parts of genetics were to some extent reduced to chemistry after discovery of the structure and mode of replication and action of the hereditary material, DNA.

The impressive successes of these reductions have led some authors to claim that the ideal of science is to reduce all the natural sciences, including biology, to a comprehensive physical theory that would provide a common set of principles of maximum generality capable of explaining all observations about natural phenomena.

Nagel (1961) has formulated the two conditions that are necessary and sufficient to effect the reduction of one theory or branch of science to another. These are the condition of derivability and the condition of connectability. The *condition of derivability* states simply that in order to reduce a branch of science to another it is necessary to show that the laws and theories of the secondary science can be derived as logical consequences from the laws and theories of the primary science.

The *condition of connectability* is based on the rule that no terms can appear in the conclusion of a demonstrative argument that do not appear in the premises. The reduction of one science (theory) to another takes the form of a deductive argument in which one of the premises is the primary science (theory), and the conclusion the secondary. Generally, however, the experimental laws and theories of a branch of science contain distinctive terms that do not appear in other branches of science. For the deduction to be logically valid there must be another premise that establishes the connection between the terms of the primary and secondary sciences. For example, the reduction of thermodynamics to statistical mechanics required the definition of "temperature" in terms of "kinetic energy." The reduction of the theories or experimental laws of genetics to those of chemistry requires that such terms as "gene" and "chromosome" be defined in terms of "hydrogen bond," "nucleotide," "deoxyribonucleic acid," "histone protein," and the like.

Whenever the conditions of connectability and derivability are satisfied, the epistemological reduction of a theory to another becomes logically feasible. If all the experimental laws and theories of one branch of science can be reduced to those of another, the former science is said to have been completely reduced to the latter.

It bears repetition that epistemological reduction is not a question of whether the *properties* of certain kinds of objects, such as organisms, result from the properties of other kinds of objects, such as inorganic compounds. Scientific laws and theories consist of propositions about the natural world. The reduction of one science to another is a matter of deriving one set of *propositions* from another. It is therefore a legitimate epistemological question to ask whether the *statements* concerning the properties of organisms, but not the properties themselves, can be logically deduced from statements concerning the properties of their physical components. The question of epistemological reduction can only be settled by empirical investigation of the logical consequences of propositions, and not by discussions about the "nature" of things or their properties.

It follows from the previous comments that questions of epistemological reduction can only be properly answered by referring to the actual state of development of the scientific disciplines involved. Certain parts of chemistry were reduced to physics after the modern theory of atomic structure was advanced half a century ago. That reduction could not have been accomplished before such a development. If the reduction of one science to another is not possible at the present stage of development of the two disciplines, claims that such a reduction will be accomplished in the future carry little weight, since such claims are not based on developments from existing theories but are merely expressions of hope.

There are some extreme positions about the question of epistemological reductionism that can be easily discounted. Some vitalists claim that biology is in principle irreducible to the physical sciences because living phenomena are the manifestation of nonmaterial principles, such as vital forces, entelechies, and so on. Epistemological anti-reductionism is thus predicated on ontological anti-reductionism. Vitalism is not an empirical hypothesis because it does not lend itself to the possibility of empirical falsification.

At the other end of the spectrum is the claim that the epistemological reduction of biology to the physical sciences is not only possible but the most important task of biologists at present. The impressive successes of molecular biology during recent decades have moved some people to claim that the only worthy and truly scientific biological investigations are those leading to the explanation of biological phenomena in terms of their underlying physicochemical components and processes. *Nevertheless, epistemological reduction of biology to the physical sciences is not possible at present.* In the current stage of scientific development a great many biological terms, such as "organ," "species," "consciousness," "mating propensity," "fitness," "competition," and "predator," cannot be defined adequately in physicochemical terms. Nor is there any class of statements and hypotheses in physics and chemistry from which every biological law can be derived logically. Therefore, neither the condition of connectability nor the condition of derivability—the two necessary conditions for epistemological reduction—can be satisfied.

A moderate reductionist position is probably not uncommon among biologists. Although the reduction of biology to chemistry cannot be effected at present, it is claimed that it is possible in principle and therefore a goal to be actively sought; actual reduction of biology to the physical sciences is made contingent upon further progress in the biological or physical sciences or both. This moderate form of epistemological reductionism is often based on convictions about ontological reductionism. It is generally accepted by biologists that living beings are exhaustively made up of physical components. It does not follow, however, that organisms are *nothing but* physical systems. Ontological reductionism does not entail epistemological reductionism. From the fact that organisms are exhaustively composed of atoms and molecules, it does not follow that the behavior of organisms can be exhaustively explained by the laws advanced to explain the behavior of atoms and molecules.

The claim that the reduction of biology to chemistry will eventually be possible is contingent upon unspecified, and at present *unspecifiable,* scientific advances. It is, therefore, a position that cannot be convincingly argued rationally. Moreover, there are reasons to believe that complete reduction of biology to physics will never be possible. Popper has shown that no major case of epistemological reduction (including such model cases as the reduction of thermodynamics to statistical mechanics) "has ever been *completely* successful: there is almost always an unresolved residue left by even the most successful attempts at reduction" (Popper, 1974, p. 260; see also Hull, 1974). It does not follow, however, that scientists should not attempt to reduce biological theories to those of the physical sciences, whenever such an undertaking seems likely to be successful. On the contrary, epistemological reductions are very successful forms of scientific explanation. A great deal is learned from epistemological reductions even when they are unsuccessful or incomplete, because much understanding is gained by the partial success, and valuable problems arise from the partial failure (Popper, 1974). The reduction of Mendelian genetics to molecular genetics has been far from completely successful (Hull, 1974). Yet there can be little doubt that much has been learned from what has been accomplished up to the present.

Some authors have claimed that even if some parts of biology can be reduced to physics and chemistry now or in the future, a complete reduction is impossible in principle because biological disciplines employ patterns of explanation that do not occur in the physical sciences (Simpson, 1964; Ayala, 1968b). Historical explanations are sometimes mentioned as distinctive of the biological and social sciences. They are not, however, because they also occur in some physical sciences, such as astronomy and geology. Although historical explanations do play a role in evolutionary theory, the theory is primarily concerned with causal explanations of evolutionary processes. Teleological explanations, as expounded below, are appropriate in biology but appear to be neither necessary nor appropriate in the explanation of physical phenomena.

DARWIN'S CONCEPTUAL REVOLUTION

The publication in 1859 of Darwin's *Origin of Species* opened a new era in the intellectual history of mankind. The discoveries of Copernicus, Kepler, Galileo, and Newton had gradually led to a conception of the universe as a system of matter in motion governed by natural laws. The earth was found to be not the center of the universe but a small planet rotating around an average star; the universe appeared as immense in space and in time; the motions of the planets around the sun could be explained by simple laws—the same laws that accounted for the motion of objects in our planet. These and many other discoveries greatly expanded human knowledge. But the conceptual revolution that went on through the seventeenth and eighteenth centuries was the realization that the universe obeys immanent laws that can account for natural phenomena. The workings of the universe were brought into the realm of science—explanation through natural laws. Physical phenomena could be reliably predicted whenever the causes were adequately known. Darwin completed the Copernican revolution by drawing out for biology the ultimate conclusions of the notion of nature as a lawful system of matter in motion. The adaptations and diversity of organisms, the origin of novel and highly organized forms, even the origin of man himself could now be explained by an orderly process of change governed by natural laws.

Before Darwin, the origin of organisms and their marvelous adaptations was most frequently attributed to the design of an omniscient Creator. God had created the birds and bees, the fish and corals, the trees in the forest, and best of all Man. God had given man eyes so that he might see, and had provided fish with gills to breathe in water. Philosophers and theologians often argued that the functional design of organisms evinced the existence of an all-wise Creator. In the thirteenth century St. Thomas Aquinas had used the argument-from-design as his "fifth way" to demonstrate the existence of God. In the nineteenth century the English theologian William Paley argued in his *Natural Theology* (1802) that the functional design of the human eye provided conclusive evidence of an all-wise Creator. It would be absurd to suppose, wrote Paley, that the human eye by mere chance "should have consisted, first, of a series of transparent lenses . . . ; secondly of a black cloth or canvas spread out behind these lenses so as to receive the image formed by pencils of light transmitted through them, and placed at the precise geometrical distance at which, and at which alone, a distinct image could be formed . . . ; thirdly of a large nerve communicating between this membrane and the brain." The eight Bridgewater Treatises, written by eminent scientists and philosophers, were published between 1833 and 1840 to set forth "the Power, Wisdom, and Goodness of God as manifested in the Creation." The mechanisms and vital endowments of man's hand gave, according to Sir Charles Bell, author of one Bridgewater Treatise, incontrovertible evidence that the hand had been designed by the same omniscient Power that had created the world.

The apparent strength of the argument-from-design to demonstrate the existence of a Creator is obvious. Wherever there is function or design we look for its author. A knife *is made* to cut and a clock *is made* to tell time; their functional designs are contrived by a knifemaker and a watchmaker. The structures, organs, and behaviors of living beings are directly organized to serve certain functions. Thus the functional design, or teleology, of organisms and their features would seem to argue for the existence of a designer. Darwin showed that the directive organization of living beings could be explained as the result of a natural process—natural selection. There was no need to resort to a Creator or other external agent. The origin and adaptation of organisms were thus brought into the realm of science.

Darwin recognized that organisms are teleologically organized. Organisms are adapted to certain ways of life and their parts adapted to perform certain functions. Fish are adapted to live in water, kidneys are designed to regulate the composition of blood, the hand of man is made for grasping. Darwin accepted the facts of adaptation and then provided a natural explanation for those facts. He brought the teleological aspect of living beings into the realm of science: he provided ample evidence for the occurrence of evolution, and for this he is appropriately given credit. More revolutionary, however, was the fact that he extended the Copernican revolution to the world of living things. The origin and adaptive nature of organisms could now be explained, like the phenomena of the inanimate world, as the result of natural laws manifested in natural processes. Darwin's theory was opposed in certain religious circles, not so much because he proposed the evolutionary origin of living things (this had been proposed many times before, even by Christian theologians), but because he excluded God from the explanation of the adaptation of organisms. The Roman

Catholic church's opposition to Galileo in the seventeenth century was similarly motivated.

The central argument of the theory of natural selection was eloquently summarized by Darwin in *Origin of Species:*

> As more individuals are produced than can possibly survive, there must in every case be a struggle for existence, either one individual with another of the same species, or with the individuals of distinct species, or with the physical conditions of life. . . . Can it, then, be thought improbable, seeing that variations useful to man have undoubtedly occurred, that other variations useful in some way to each being in the great and complex battle of life, should sometimes occur in the course of thousands of generations? If such do occur, can we doubt (remembering that more individuals are born than can possibly survive) that individuals having any advantage, however slight, over others, would have the best chance of surviving and of procreating their kind? On the other hand, we may feel sure that any variation in the least degree injurious would be rigidly destroyed. This preservation of favorable variation and the rejection of injurious variations, I call Natural Selection.

Darwin's argument addresses the problem of explaining the adaptive nature of organisms. Darwin argues that adaptive variations ("variations useful in some way to each being") must occasionally appear, and that these are likely to increase the reproductive chances of their carriers. Over the generations favorable variations will be preserved, injurious ones will be eliminated. In one of the few places where he uses the term "adaptation" or its cognates in *Origin of Species,* Darwin adds: "I can see no limit to this power [natural selection] in slowly and beautifully *adapting* each form to the most complex relations of life." Natural selection was proposed by Darwin primarily to account for the adaptive organization of living beings; it is a process that promotes or maintains adaptation. Evolutionary change through time (anagenesis) and evolutionary diversification (cladogenesis) are not directly promoted by natural selection, but they often ensue as by-products of natural selection fostering adaptation. "One problem of great importance . . . is the tendency in organic beings to diverge in character as they become modified. . . . The solution [to this problem] is that the modified offspring of all dominant and increasing forms tend to become *adapted* to many and highly diversified places in the economy of nature" (Darwin, 1958).

TELEOLOGICAL EXPLANATIONS

Teleology (from the Greek *telos* = end) is "the use of design, purpose, or utility as an explanation of any natural phenomenon" (*Webster's Third New International Dictionary,* 1966). An object or a behavior is said to be teleological or telic when it gives evidence of design or appears to be directed toward certain ends. The behavior of human beings is often teleological. A person who buys an airplane ticket, reads a book, or cultivates the earth is trying to achieve a certain end: getting to a given city, acquiring knowledge, or getting food. Objects and machines made by people also are usually teleological: a knife is made for cutting, a clock is made for telling time, a thermostat is made to regulate temperature. Features of organisms are teleological as well: a bird's wings are *for* flying, eyes are for seeing, kidneys are constituted for regulating the composition of the blood. The features of organisms that may be said to be teleological are those that can be identified as adaptations, whether they are structures like a wing or a hand, or organs like a kidney, or behaviors like the courtship displays of a peacock. Adaptations are features of organisms that have come about by natural selection because they serve certain functions and thus increase the reproductive success of their carriers.

Inanimate objects and processes (other than those created by men) are not teleological because they are not directed toward specific ends, they do not exist to serve certain purposes. The configuration of a sodium chloride molecule depends on the structure of sodium and chlorine, but it makes no sense to say that that structure is made up so as to serve a certain end. The shape of a mountain is the result of certain geological processes, but it did not come about so as to serve a certain end. The motion of the earth around the sun results from the laws of gravity, but it does not exist in order to satisfy certain ends or goals. We may use sodium chloride as food, a mountain for skiing, and take advantage of the seasons, but the use that we make of these objects or phenomena is not the reason why they came into existence or why they have certain configurations. On the other hand, a knife and a car exist and have particular configurations precisely in order to serve the ends of cutting and transportation. Similarly, the wings of birds came about precisely because they permitted flying, which was reproductively advantageous. The mating display of peacocks came about because it increased the chances of mating and thus of leaving progeny.

The previous comments point out the essential characteristics of telic phenomena, i. e., phenomena whose existence and configuration can be explained teleologically. We may now propose the following definition. *Teleological explanations account for the existence of a certain feature in a system by demonstrating the feature's contribution to a specific property or state of the system.* Teleological explanations require that the feature or behavior contribute to the existence or maintenance of a certain state or property of the system. Moreover, and this is the essential component of the concept, the contribution *must be the reason why the feature or behavior exists at all.*

The configuration of a molecule of sodium chloride contributes to its property of tasting salty and therefore to its use as food, not vice versa; the potential use of sodium chloride as food is not the reason why it has a particular molecular configuration or tastes salty. The motion of the earth around the sun is the reason why seasons exist; the existence of the seasons is not the reason why the earth moves about the sun. On the other hand, the sharpness of a knife can be explained teleologically because the knife has been created precisely to serve the purpose of cutting. Motorcars and their particular configurations exist because they serve transportation, and thus can be explained teleologically. (Not all features of a car contribute to efficient transportation—some features are added for aesthetic or other reasons. But as long as a feature is added because it exhibits certain properties—like appeal to the aesthetic preferences of potential customers—it may be explained teleologically. Nevertheless, there may be features in a car, a knife, or any other man-made object that need not be explained teleologically. That knives have handles may be explained teleologically, but the fact that a particular handle is made of pine rather than oak might simply be due to the availability of material. Similarly, not all features of organisms have teleological explanations.)

Many features and behaviors of organisms meet the requirements of teleological explanation. The hand of man, the wings of birds, the structure and behavior of kidneys, the mating displays of peacocks are examples already given. In general, as pointed out above, those features and behaviors that are considered adaptations are explained teleologically. This is simply because adaptations are features that come about by natural selection. Among alternative genetic variants that may arise by mutation or recombination, the ones that become established in a population are those that contribute more to the reproductive success of their carriers. The effects on reproductive success are usually mediated by some function or property. Wings and hands acquired their present configuration through long-term accumulation of genetic variants adaptive to their carriers. An alternative feature may be due to a single gene mutation, e.g., the presence of normal hemoglobin rather than hemoglobin S in humans. One amino acid substitution in the beta chain in humans results in hemoglobin molecules less efficient for oxygen transport. The general occurrence in human populations of normal rather than S hemoglobins is explained teleologically by the contribution of hemoglobin to effective oxygen transport and thus to reproductive success. The difference between peppered-gray and melanic moths is due to one or only a few genes. The replacement of gray moths by melanics in polluted regions is explained teleologically by the fact that melanism decreases the probability of predation in such regions. The predominance of peppered forms in nonpolluted regions is similarly explained.

Not all features of organisms need to be explained teleologically, since not all come about as a direct result of natural selection. Some features may become established by genetic drift, by chance association with adaptive traits, or in general by processes other than natural selection. Proponents of the neutrality theory of protein evolution argue that many alternative protein variants are adaptively equivalent. Most evolutionists would admit that at least in certain cases the selective differences between alternative protein variants must be virtually nil, particularly when population size is very small. The presence in a population of one amino acid sequence rather than another adaptively equivalent to the first would not then be explained teleologically. Needless to say, in such cases there would be amino acid sequences that would not be adaptive. The presence of an adaptive protein rather than a nonadaptive one would be explained teleologically; but the presence of one protein rather than another among those adaptively equivalent would not require a teleological explanation.

NATURAL AND ARTIFICIAL TELEOLOGY

In the previous section some man-made objects and adaptive traits of organisms served as examples of teleological phenomena. We may now distinguish several kinds of teleological phenomena (Ayala, 1968b, 1970). Actions or objects are *purposeful* when the end-state or goal is consciously intended by an agent. Thus, a man mowing his lawn is acting teleologically in the purposeful sense; a lion hunting deer and a bird building a nest have at least the appearance of purposeful behavior. Objects resulting from purposeful behavior exhibit *artificial* (or *external*) teleology. A knife, a table, a car, and a thermostat are examples of systems exhibiting artificial teleology: their teleological features were consciously intended by some agent.

Systems with teleological features that are not due to the purposeful action of an agent but result from some natural process exhibit *natural* (or *internal*) teleology. The wings of birds have a natural teleology; they serve an end, flying, but their configuration is not due to the conscious design of any agent. We may distinguish two kinds of natural teleology: *determinate* or necessary, and *indeterminate* or nonspecific. Determinate natural teleology exists when a specific end-state is reached in spite of environmental fluctuations. The development of an egg into a chicken, or of a human zygote into a human being, are examples of determinate natural teleological processes. The regulation of body temperature in a mammal is another example. In general, the homeostatic processes of organisms are instances of determinate natural teleology. Two types of homeostasis are usually distinguished—physiological and developmental—although intermediate conditions exist. Physiological homeostatic reactions enable organisms to maintain certain physiological steady states in spite of environmental shocks. The regulation of the concentration of salt in blood by the kidneys, or the hypertrophy of muscle owing to strenuous use, are examples of physiological homeostasis. Developmental homeostasis refers to the regulation of the different paths that an organism may follow in the progression from fertilized egg to adult. The process can be influenced by the environment in various ways, but the characteristics of the adult individual, at least within a certain range, are largely predetermined in the zygote.

Indeterminate or nonspecific teleology occurs when the end-state served is not specifically predetermined, but rather is the result of selection of one from among several available alternatives. For teleology to exist, the selection of one alternative over another must be deterministic and not purely stochastic. But what alternatives happen to be present may depend on environmental and/

or historical circumstances and thus the specific end-state is not generally predictable. Indeterminate teleology results from a mixture of stochastic (at least from the point of view of the teleological system) and deterministic events. The adaptations of organisms are teleological in this indeterminate sense. The wings of birds require teleological explanations: the genetic constitutions responsible for their configuration came about because wings serve to fly and flying contributes to the reproductive success of birds. But there was nothing in the constitution of the remote ancestors of birds that would necessitate the appearance of wings in their descendants. Wings came about as the consequence of a long sequence of events, where at each stage the most advantageous alternative was selected among those that happened to be available; but what alternatives were available at any one time depended at least in part on chance events. In spite of the role played by stochastic events in the phylogenetic history of birds, it would be mistaken to say that wings are not teleological features. Again, there are differences between the teleology of an organism's adaptations and the nonteleological potential uses of natural inanimate objects. A mountain may have features appropriate for skiing, but those features did not come about so as to provide skiing slopes. On the other hand, the wings of birds came about precisely because they serve flying. The explanatory reason for the existence of wings and their configuration is the end they serve—flying—which in turn contributes to the reproductive success of birds. If wings did not serve an adaptive function they would have never come about or would gradually disappear over the generations.

The indeterminate character of the outcome of natural selection over time is due to a variety of nondeterministic factors. The outcome of natural selection depends, first, on what alternative genetic variants happen to be available at any one time. This in turn depends on the stochastic processes of mutation and recombination, and also on the past history of any given population. (What new genes may arise by mutation and what new genetic constitutions may arise by recombination depend on what genes happen to be present—which depends on previous history.) The outcome of natural selection depends also on the conditions of the physical and biotic environment. Which alternatives among available genetic variants may be favored by selection depends on the particular set of environmental conditions to which a population is exposed.

Some evolutionists have rejected teleological explanations because they have failed to recognize the various meanings that the term "teleology" may have (Pittendrigh, 1958; Mayr, 1965, 1974b; Williams, 1966; Ghiselin, 1974). These biologists are correct in excluding certain forms of teleology from evolutionary explanations, but they err when they claim that teleological explanations should be excluded altogether from evolutionary theory. In fact, they themselves often use teleological explanations in their works, but fail to recognize them as such, or prefer to call them by some other name, such as "teleonomic." Teleological explanations, as explained above, are appropriate in evolutionary theory, and are recognized by most evolutionary biologists and philosophers of science who have thoughtfully considered the question (Beckner, 1959; Nagel, 1961; Simpson, 1964; Dobzhansky, 1970; Ayala, 1968b, 1970; Wimsatt, 1972; Hull, 1974). Which kinds of teleological explanations are appropriate and which ones are inappropriate with respect to various biological questions may be briefly specified.

Mayr (1965) has pointed out that teleological explanations have been applied to two different sets of biological phenomena. "On the one hand is the production and perfection throughout the history of the animal and plant kingdoms of ever-new and ever-improved DNA programs of information. On the other hand is the testing of these programs and their decoding throughout the lifetime of each individual. There is a fundamental difference between end-directed behavioral activities or developmental processes of an individual or system, which are controlled by a program, and the steady improvement of the genetically coded programs. This genetic improvement is evolutionary adaptation controlled by natural selection." The "decoding" and "testing" of genetic programs of information are the issues considered, respectively, by developmental biology and functional biology. The historical and causal processes by which genetic programs of information come about are the concern of evolutionary biology. Grene (1974) uses the term "instrumental" for the teleology of organs that act in a functional way, such as the hand for grasping; "developmental" for the teleology of such processes as the maturation of a limb; and "historical" for the process (natural selection) producing teleologically organized systems.

Organs and features such as the eye and the hand have determinate (and internal) natural teleology. These organs serve determinate ends (seeing or grasping) but have come about by natural processes that did not involve the conscious design of any agent. Physiological homeostatic reactions and embryological development are processes that also have determinate natural teleology. These processes lead to end-states (from egg to chicken) or maintain properties (body temperature in a mammal) that are on the whole determinate. Thus, Mayr's "decoding" of DNA programs of information and Grene's "instrumental" and "developmental" teleology, when applied to organisms, are cases of determinate natural teleology (Mayr prefers to use the term "teleonomy" for this type of teleology). Human tools (such as a knife), machines (such as a car), and servomechanisms (such as a thermostat) also have determinate teleology, but of the artificial kind, since they have been consciously designed.

The process of natural selection is teleological but only in the sense of indeterminate natural teleology. It is not consciously intended by any agent, nor is it directed towards specific or predetermined end-states. Yet the process is far from random or completely indeterminate (Chapter 4). Among the genetic alternatives available at any one time, natural selection favors those that increase the reproductive success of their carriers in the particular environmental circumstances in which the organisms live. Reproductive success is, of course, mediated by some adaptive function, say flying, that is determined by the genetic variants that are favored by natural selection.

Some authors exclude teleological explanations from evolutionary biology because they believe that teleology exists only when a specific goal is purposefully sought. This is not so. Terms other than "teleology" could be used for natural (or internal) teleology, but this might in the end add more confusion than clarity. Philosophers as well as scientists use the term "teleological" in the broader sense, to include explanations that account for the existence of an object in terms of the end-state or goal that they serve.

The process of evolution by natural selection is not teleological in the purposeful sense. Thomas Aquinas and the natural theologians of the nineteenth century erroneously claimed that the directive organization of living beings evinces the existence of a Designer. The adaptations of organisms can be explained as the result of natural processes without recourse to consciously intended end-products. There is purposeful activity in the world, at least in man; but the existence and particular structures of organisms, including man, need not be explained as the result of purposeful behavior.

Lamarck (1809) erroneously thought that evolutionary change necessarily proceeded along determined paths from simpler to more complex organisms. Similarly, the evolutionary philosophies of Bergson (1907), Teilhard de Chardin (1959), and such theories as *nomogenesis* (Berg, 1926), *aristogenesis* (Osborn, 1934), *orthogenesis*, and the like are erroneous because they all claim that evolutionary change necessarily proceeds along determined paths. These theories mistakenly take embryological development as the model of evolutionary change, regarding the teleology of evolution as determinate. Although there are teleologically determinate processes in the living world, like embryological development and physiological homeostasis, the evolutionary origin of living beings is teleological only in the indeterminate sense. Natural selection does not in any way direct evolution toward any particular organisms or toward any particular properties.

Teleological explanations are fully compatible with causal explanations (Nagel, 1961; Ayala, 1970). It is possible, at least in principle, to give a causal account of the various physical and chemical processes in the development of an egg into a chicken, or of the physicochemical, neural, and muscular interactions involved in the functioning of the eye. It is also possible in principle to describe the causal processes by which one genetic variant becomes eventually established in a population. But these causal explanations do not make it unnecessary to provide teleological explanations where appropriate. Both teleological and causal explanations are called for in such cases.

One question biologists ask about features of organisms is "What for?" That is, "What is the function or role of a particular structure or process?" The answer to this question must be formulated teleologically. A causal account of the operation of the eye is satisfactory as far as it goes, but it does not tell all that is relevant about the eye, namely that it serves to see. Moreover, evolutionary biologists are interested in the question why one particular genetic alternative rather than others came to be established in a population. This question also calls for teleological explanations of the type: "Eyes came into existence because they serve to see, and seeing increases reproductive success of certain organisms in particular circumstances." In fact, eyes came about in several independent evolutionary lineages: cephalopods, arthropods, vertebrates.

There are two questions that must be addressed by a teleological account of evolutionary events. First, there is the question of how a genetic variant contributes to reproductive success; a teleological account states that an existing genetic constitution (say, the allele coding for a normal hemoglobin beta chain) enhances reproductive fitness better than alternative constitutions. Then there is the question of how the specific genetic constitution of an organism enhances its reproductive success; a teleological explanation states that a certain genetic constitution serves a particular function (for example, the molecular composition of hemoglobin has a role in oxygen transport).

Both questions call for teleological hypotheses that can be empirically tested. It sometimes happens, however, that information is available on one or the other question but not for both. In population genetics the fitness effects of alternative genetic constitutions can often be measured while the mediating adaptive function responsible for the fitness differences may be difficult to identify. We know, for example, that in *Drosophila pseudoobscura* different inversion polymorphisms are favored by natural selection at different times of the year (Chapter 4) but we are largely ignorant of the physiological processes involved. In a historical account of evolutionary sequences the problem is occasionally reversed: the function served by an organ or structure may be easily identified, but it may be difficult to ascertain why the development of that feature enhanced reproductive success and thus was favored by natural selection. One example is the large brain of man, which makes possible culture and other important human attributes. We may advance hypotheses about the reproductive advantages of increased brain size in the evolution of man, but these hypotheses are notoriously difficult to test empirically.

Teleological explanations in evolutionary biology have great heuristic value. They are also occasionally very facile, precisely because they may be difficult to test empirically. Every effort should be made to formulate teleological explanations in a fashion that makes them readily subject to empirical testing. When appropriate empirical tests cannot be formulated, evolutionary biologists should use teleological explanations only with the greatest restraint (see Williams, 1966).

THE THEORY OF NATURAL SELECTION

According to the theory of evolution expounded in this book, natural selection is the process responsible for the adaptations of organisms, and also the main process by which evolutionary change comes about. The principle of natural selection, together with some subsidiary and generally well-corroborated hypotheses (such as the Mendelian theory of inheritance), can explain a large number of phenomena in the living world, such as the diversity of organisms, their gradual change through time, and their remarkable adaptations to their environments. The theory of evolution is indeed the single most encompassing biological theory.

Two criticisms of the theory of evolution by natural selection have been raised. One is that the theory can explain all conceivable states of affairs in the living world and therefore is not subject to the possibility of empirical falsification. The other criticism is more fundamental. It claims that the principle of natural selection is circular; it is able to explain all conceivable evolutionary outcomes because it lacks empirical content. This alleged circularity would explain why the theory of natural selection is not empirically falsifiable.

These criticisms are mistaken. We shall show that the theory of natural selection is not circular and that it can be properly tested empirically.

Claims of circularity may imply circularity of either definition or argumentation. Circularity of definition occurs when the terms to be defined appear in the definition. Circularity of argumentation occurs when the conclusion is logically included in one of the premises. Circularity of definition is not a serious criticism in empirical science. It can be resolved by replacing the circular definition by a noncircular one. Natural selection is often defined as "differential reproduction of alternative genetic variants," which is not a circular definition. Definitions of natural selection need not be circular.

Recently, some philosophers have leveled the more serious charge against the theory of natural selection that it is purely tautological (Himmelfarb, 1962; Smart, 1963; Manser, 1965; Flew, 1967; Barker, 1969; Grene, 1974). According to these critics, arguments for natural selection proceed approximately as follows. One premise states that alternative genetic variants that confer higher fitness to their carriers will increase in frequency over the generations at the expense of genetic variants with lower fitness. The second premise is empirical, establishing that a particular genetic variant confers high fitness in a particular situation. The conclusion drawn is that the particular genetic variant (observed to have higher fitness) will gradually increase in frequency. The catch, it is claimed, is the empirical premise. Since evolutionists measure fitness by observing which genetic variants leave greater numbers of progeny, the conclusion of the argument is not a conclusion at all: the conclusion has been used to establish the second premise of the argument. Grene (1974, p. 86) writes: "What have we? Once more, tautology: well, after all, what survives survives. . . . When the theory [of natural selection] is summed up in a formula for measuring differential gene ratios, you have a theorem universally applicable because empty, totally comprehensive because it expresses a simple identity."

These criticisms are mistaken. First, the critics erroneously equate fitness with changes in gene frequencies. Although fitness differences are likely to lead to changes in gene frequencies, not all gene frequency changes are due to fitness differences. Gene frequencies change by natural selection, but also by drift, mutation, and migration. Whether a particular evolutionary change is due to fitness differences, i.e., to natural selection, can be tested empirically. Second, the critics err because they fail to consider a critical premise in the argument. Natural selection, as Darwin saw it, is postulated to explain *adaptation*—why organisms exhibit features that are end-directed. Evolutionary change is simply a consequence of natural selection promoting the adaptation of organisms to their environments.

The theory of evolution by natural selection advances arguments of the following general form. Among alternative genetic variants, some result in features that are useful to their carriers as adaptations to the environment. Individuals possessing useful adaptations are likely to leave, on the average, greater numbers of progeny than individuals lacking them (or having less useful adaptations). Therefore useful adaptations become established in populations.

To explain a particular adaptation, a valid selectionist argument has to show (1) that natural selection is involved at all; and (2) that natural selection favors the particular adaptation. These two claims can be tested empirically. Those who claim that the theory of natural selection is circular erroneously claim two identities: that changes in gene frequencies are the same as differences in fitness, and that fitness is the same as adaptation. These identities do not always obtain. Whether natural selection is involved in a particular genetic change, and whether natural selection favors a particular adaptation are questions to be resolved empirically. As pointed out in the previous section, there are two kinds of problems encountered in explanations of evolution by natural selection. One is to determine whether natural selection is involved in a certain genetic change; the second is to identify the particular adaptation involved in the genetic change.

Many examples of adaptations have been given in previous sections of this chapter and elsewhere in the book. Adaptation is, nevertheless, a concept difficult to define (Bock and Wahlert, 1965; Williams, 1966; Dobzhansky, 1968; Ayala, 1969). Adaptations can be recognized in individuals—whether physiological, morphological, or behavioral—as well as at the level of the population. Some operational ways of measuring the adaptation of populations have been considered by Dobzhansky (1968), Ayala (1969b), and others. Williams (1966) has proposed that a useful criterion for identifying individual adaptations is whether an analogy can be established between some human artifact and the feature presumed to be an adaptation. A mammalian oviduct may be seen as a mechanism for conveying an early embryo to the uterus; the uterus may be seen as designed for the protection and nourishment of the embryo. Ayala (1968b, 1970) has suggested utility as a criterion for identifying adaptations. A feature of an organism is regarded as an adaptation if it has utility for the organism and if such utility explains the presence of the feature.

Adaptation and fitness are in any case different concepts. Fitness is simply a measure of the reproductive efficiency of one genetic constitution relative to alternative ones in the same population. Fitness does not always go hand-in-hand with adaptation. One example is provided by the *t* alleles in the house mouse. Homozygotes for these alleles are lethal or sterile. The *t* alleles distort segregation in heterozygous males in such a way that a majority (up to 95 percent) of the spermatozoa carry the mutant rather than the normal allele. A *t* allele introduced in a population by either mutation or migration will increase in frequency due to natural selection. Yet this increase reduces the adaptation of the individual carriers and the population. Other examples of genetic variants with high fitness that decrease adaptation can be given (Ayala, 1969).

The theory of evolution proposes natural selection as the process that accounts for the structural and functional adaptation of organisms to their environments. The observed adaptations of organisms were the facts that Darwin set himself to explain as the result of natural processes. The connection between fitness, i.e., systematic differences in reproductive efficiency that result in natural selection, and adaptation needs to be demonstrated for each particular case of adaptation. This is done by means of empirically falsifiable hypotheses claiming that carriers of a given adaptation have greater reproductive efficiency than organisms lacking that adaptation.

It is surprising that critics of the theory of natural selection would argue that evolutionists measure fitness simply by observing changes in gene frequencies from one generation to the next in a given population. Changes in gene frequencies may be due to any of the variety of causes discussed in other chapters of this book. For example, genetic drift may result in substantial genetic evolution over generations, particularly in small populations (Chapter 4). Also, the neutrality theory of molecular evolution postulates that much protein evolution is not due to natural selection (Chapter 9). In natural selection arguments, the claim that a given genetic change is due to fitness differences is a hypothesis empirically falsifiable by appropriate observations and experiments. Many examples have been given in earlier chapters of this book; two will be briefly summarized here. Certain chromosomal arrangements are observed to change in frequency throughout the year in natural populations of *Drosophila pseudoobscura* (Dobzhansky, 1971). One possible explanation is that these changes are due to seasonal changes in the fitness of the chromosomal arrangements. This hypothesis is subject to empirical tests in the laboratory (Wright and Dobzhansky, 1946), as well as in nature, e.g., by observing whether the seasonal changes are repeated from one year to the next (Dobzhansky and Ayala, 1973). Populations of *Biston betularia* and other moths have experienced considerable evolution throughout the last hundred years. In some localities melanic forms have totally replaced the typical grayish moths. A great variety of empirical observations and tests have corroborated the hypothesis that the evolution of melanic forms is due to fitness changes associated with industrial pollution of the environment. The particular adaptation involved has also been identified as the cryptic coloration of melanic forms that are more likely to escape predation than light gray moths when resting on blackened trees.

THE CONCEPT OF PROGRESS

The earliest organisms living on earth were no more complex than some bacteria and blue-green algae. Three billion years later their descendants include the orchids, the bees, the dolphins, and man. Thus, biologists sometimes speak of "evolutionary progress" to refer to certain evolutionary trends, such as advances in complexity of organization, increased homeostasis, or the development of social organization. The evolutionary literature abounds in such terms as "lower (or higher) organisms," "more (or less) advanced," and the like. Yet evolutionary lineages may be "progressive" with respect to one or a few attributes but not with respect to others. Or they may not be "progressive" at all by any reasonable definition; bacteria and blue-green algae today are not very different from their ancestors of two or three billion years earlier. Moreover, many evolutionary lineages have become extinct.

What do we mean by "biological progress"? In what sense, if any, can we say that evolution is progressive? The term "progress" may be clarified by comparing it with other related terms used in biological discourse. These terms are "change," "evolution," and "direction."

"Change" means alteration, whether in the location, the state, or the nature of a thing. Progress implies change, but the opposite is not true—not all changes are progressive. Molecules of oxygen and nitrogen in the air of a room change locations continuously; such change would not generally be regarded as progressive. The mutation of a gene from a functional allele to a nonfunctional one is a change of state, but definitely not a progressive one.

"Evolution" can also be distinguished from "progress," although both terms imply that *sustained* change has occurred. Evolutionary change is not necessarily progressive. The evolution of a species may lead to its extinction, a change that is not progressive, at least not for that species.

"Direction" and "progress" are also distinguishable. The concept of "direction" implies that a series of changes have occurred that can be arranged in a linear sequence, with respect to some property or feature, such that elements in the later part of the sequence are more different from early elements of the sequence than from intermediate elements. Directional change may be "uniform" or "net" (see below), depending on whether each succeeding member is invariably more different from the first than each preceding member is, or whether directional change occurs only on the average.

In discussions of evolution "directionality" is sometimes equated with "irreversibility": the process of evolution is said to have a direction because it is irreversible. Evolutionary changes are irreversible (see Chapter 1), except perhaps in some trivial sense, as when a previously mutated gene mutates back to its former allelic state. Direction, however, implies more than irreversibility. Consider a new pack of cards with each suit arranged from ace to ten, knave, queen, king, and with the suits arranged in the sequence spades, clubs, hearts, diamonds. If we shuffle the cards thoroughly, the order of the cards changes, and the change cannot be reversed by shuffling. We may shuffle again and again until the cards are totally worn out without ever restoring the original sequence. The change of order of the cards is irreversible but not directional.

Irreversible and directional changes occur in the inorganic as well as the living world. The second law of thermodynamics describes sequential changes that are irreversible but also directional, and indeed uniformly directional. Within a closed system, entropy always increases; that is, a closed system passes continuously from less to more probable states. The concept of direction applies to what in paleontology are called "evolutionary trends." A trend occurs in a phylogenetic sequence when a feature persistently changes through time in the members of the sequence. Trends are commonly seen in fossil sequences that are sufficiently long to be called "sustained" (Simpson, 1953).

Consider the trend toward a gradual reduction of the number of dermal bones in the skull roof in the evolutionary sequence from fish to man, or the trend toward increased molarization of the last premolars in the phylogeny of the Equidae from *Hyracotherium* in early Eocene to *Haplohippus* in early Oligocene. These trends represent directional change, but it is not obvious that they should be labeled progressive. To label them progressive we would need to agree that the directional change had been in some sense for the better. To consider a directional sequence progressive, we need to add an evaluation, namely that the condition in the later members of the sequence represents a *betterment* or improvement. Directionality in a sequence may be accepted without any such evaluation being added. Progress implies directional change, but the opposite is not true.

The concept of progress contains two elements: one descriptive—that directional change has taken place; the other axiological (= evaluative)—that the change represents a betterment (Goudge, 1961). The notion of progress requires that a value judgment be made of what is better and what is worse. However, the axiological standard of reference need not be a moral one; not all forms of progress are moral. To recognize progress, an evaluation of better versus worse is required, but not necessarily one of right versus wrong, or of good versus evil. "Better" may simply mean more efficient, or more abundant, or more complex, without any reference to moral values or standards.

Progress, then, may be defined as *systematic change in a feature belonging to all members of a sequence in such a way that posterior members of the sequence exhibit an improvement of that feature.* More simply, progress may be defined as directional change toward the better. The antonym of progress is "retrogression," or directional change for the worse. The two elements of the definition, namely directional change and improvement according to some standard, are jointly necessary and sufficient for the occurrence of progress. Directional change (and progress) may be observed in sequences that are spatially rather than temporally ordered. *Clines* are examples of directional change recognized along a spatial dimension. In evolutionary discourse, however, temporal (historical) sequences are of greatest interest.

Various kinds of progress can be distinguished by attending to either one of the two essential elements of the definition. We shall later refer to different types of evolutionary progress based on different standards. Here we shall further clarify the concept of progress by making two distinctions that relate to the descriptive element of the definition, namely the requirement of directional change. These distinctions also apply, therefore, to the concept of direction.

If we attend to the *continuity* of the direction of change, we can distinguish two kinds of progress: uniform and net. *Uniform progress* takes place whenever every later member of a sequence is better than every earlier member of the sequence according to a certain standard. Let m_i represent the members of a sequence, temporally ordered from 1 to n, and let p_i measure the state of the feature under evaluation. There is uniform progress if it is the case for every m_i and m_j that $p_j > p_i$ for every j greater than i. *Net progress* does not require that every member of a sequence be better than all previous members of the sequence and worse than all its successors; it requires only that later members of the sequence be better *on the average* than earlier members. Net progress allows for temporary fluctuations of value. If the members of a sequence, m_i, are linearly arranged over time, net progress occurs whenever the regression (in the sense used in mathematical statistics) of p on time is significantly positive.

Some authors have argued that progress has not occurred in evolution because, no matter what standard is chosen, fluctuations of value can be found in every evolutionary lineage. This criticism is valid against uniform, but not net evolutionary progress. Also, neither uniform nor net progress require that progress continue forever, or that any specified goal be achieved. The *rate* of progress may decrease with time; progress means only a gradual improvement in the members of a sequence. It is possible that a progressive sequence may tend asymptotically toward a finite goal, that is, continuously approach but never reach the goal.

The distinction between uniform and net progress is similar but not identical to the distinction between uniform and perpetual progress made by Broad (1925) and Goudge (1961). Simpson (1949, 1953) implicitly makes a distinction between uniform and net progress in his enlightening discussion of evolutionary progress. He uses terms like "universal," "invariable," "constant," and "continuous" for what we have called uniform progress (although he also uses these terms with other meanings).

Other types of progress (and directional change) can be distinguished. With respect to the *scope* of the sequence considered, progress can be either general or particular. *General progress* is that which occurs in all historical sequences of a given domain of reality, and from the beginning of the sequences until their end. *Particular progress* is that which occurs in one or several, but not all historical sequences, or progress that takes place during part but not all of the duration of the sequence or sequences.

In biological evolution general progress would be any kind of progress that can be predicated of the evolution of all life from its origin to the present. If a type of progress is predicated of only one or several lines of evolutionary descent, it is a particular kind of progress. Progress that embraces only a limited period of the existence of life is also a particular kind of progress. Some writers have denied that evolution is progressive because not all evolutionary lineages exhibit advance. Some evolutionary lineages, like those leading to certain parasitic forms, are retrogressive by certain standards; and many lineages have become extinct without issue. These considerations may be valid criticism against the idea of general progress, but not necessarily against that of particular forms of progress.

THE EXPANSION OF LIFE

We have established that the concept of "progress" involves an axiological element. Discussions of evolutionary progress require that a choice be made of the standard by which organisms and evolutionary events are to be evaluated. A decision must also be made as to what direction of change represents improvement. These decisions are in part subjective, but they should not be arbitrary; biological knowledge should guide them. A standard is valid if it enables one to say enlightening things about the evolution of life. The choice of appropriate standards depends on how much relevant biological information is available, and whether the evaluation can be made.

There is no standard according to which *uniform* progress can be said to have occurred in the evolution of life. Changes of direction, slackening, or reversals have occurred in all evolutionary lineages, no matter what feature is considered (Simpson, 1949, 1953). The question, then, is whether *net* progress has occurred in the evolution of life, and in what sense.

The next question is whether there is any criterion by which net progress can be said to be a *general* feature of evolution, or whether identifiable progress applies only to particular lineages or during particular periods. One conceivable standard of progress is the increase in the amount of genetic information stored in organisms. DNA contains the information that, in interaction with the environment, directs the development and behavior of organisms. Net general progress can be said to have occurred if organisms living at a later time have, on the average, greater content of genetic information than their ancestors. One difficulty, insuperable at least for the present, is that there is no way in which the genetic information contained in the whole DNA of an organism can be measured. The amount of information is not simply related to the amount of DNA, since we know that many DNA sequences are repetitive. There are ways to measure, at least approximately, the "complexity" of DNA in a given organism, i.e. the total length of different DNA sequences (see Chapter 3). But a large fraction of the DNA does not encode information in the form of codon triplets, and much DNA may have nonsense messages.

Another possible criterion of general progress is the expansion of life. According to Simpson (1949) there is in evolution "a tendency for life to expand, to fill in all the available spaces in the livable environments, including those created by the process of that expansion itself." In principle, the expansion of life can be measured by at least four different though related criteria: (1) the number of *kinds* of organisms, i.e., the number of species; (2) the number of individuals; (3) the bulk of living matter (biomass); and (4) the total rate of energy flow. Increases in the number of individuals or their total bulk may not be an unmixed blessing, as is the case now for mankind, but they can be a measure of biological progress. Net progress might be a general feature of the evolution of life by any one of these four standards of progress.

Living organisms have a tendency to multiply exponentially ad infinitum without intrinsic constraints. This is simply a consequence of the process of biological reproduction: each organism is capable of producing, on the average, more than one progeny throughout its lifetime. However, the tendency of life to expand encounters extrinsic constraints of various sorts. Once a certain species has come to exist, its expansion is limited at least in two ways. First, the supply of resources accessible to organisms is limited. Second, favorable conditions for multiplication do not always occur. Predators and competitors, together with climatic conditions, are the main factors interfering with the multiplication of organisms. Drastic and gradual changes in the weather, as well as geological events, at times lead to vast decreases in the size of some populations and even of the whole of life. Because of these constraints the tendency of life to expand has not always succeeded. Nevertheless, it appears certain that life has, on the average, expanded throughout most of the evolutionary process.

The number of extant biological species is almost certainly greater than two million, and may be as large as six million. Although it is difficult to estimate the number of plant species that existed in the past (since well-preserved plant fossils are rare), the number of animal species can be roughly estimated. Approximately 150,000 animal species live in the seas today, probably a larger number than the total number of animal species that existed in the Cambrian (600 million years ago), when no animal or plant species lived on land. The number of land animal species is probably at a maximum now, even if we exclude the insects. Insects make up about three-quarters of all animal species, and about half of all species, including plants. Insects did not appear until the Carboniferous, some 350 million years ago. More species of insects exist now than at most, probably all, times in the past. On the whole, it appears that the number of living species is probably greater in recent times than it ever was before, and that at least on the average a gradual increase in the number of species has characterized the evolution of life (chapter 13).

The expansion of the number of species operates as a positive feedback process. The greater the number of species, the greater the number of environments that are created for new species to exploit. Once plants came into existence it was possible for animals to exist; and animals sustain large numbers of species of other animals, as well as parasites and symbionts.

The number of individuals living on earth today is not known with any reasonable approximation, even if we exclude microorganisms. The median number of individuals for insect species is estimated to be around 2×10^8, but some species may consist of more than 10^{16} individuals—and there are more than one million insect species! The number of individuals of *Euphausia superba,* the small krill eaten by some whales, may be greater than 10^{20}. There can be little doubt that the number of individual animals and plants and their total biomass are greater now than they were in the Cambrian. Very likely they are also greater than they have been at most times since the beginning of life. Even if we include microorganisms, it is probable that the number of living individuals has increased, on the average, throughout the evolution of life. The total bulk of living matter has also probably increased, on the average, since larger organisms have generally appeared later in time.

The rate of energy flow has probably increased in the living world faster than the total biomass. One effect of living things is to retard the dissipation of energy that covers the earth's surface. Green plants store radiant energy from the sun that would otherwise be converted into heat. Animals, although they dissipate energy individually, since they have higher rates of catabolism than of anabolism, provide a new path for the flow of energy; their interactions with plants increase the total rate of energy flow through living matter.

PARTICULAR FORMS OF EVOLUTIONARY PROGRESS

As stated above, the concept of progress is based on an evaluation of better versus worse relative to some criterion. Particular forms of evolutionary progress, which obtain only in certain evolutionary sequences and usually for only a limited span of time, may be identified using a variety of criteria. Simpson (1949) has examined several criteria, including dominance, invasion of new environments, replacement, improvement in adaptation, adaptability and possibility of further progress, increased specialization, control over the environment, increased structural complexity, increase in general energy or level of vital processes, and increase in the range and variety of adjustments to the environment. For each of these criteria, Simpson has shown in what evolutionary sequences and for how long progress has taken place. Stebbins (1969) has proposed a law of "conservation of organization" that accounts for evolutionary progress as a small bias towards increased complexity of organization. Other criteria of progress have been examined by Huxley (1942, 1953), Rensch (1947), Williams (1966), and Ayala (1974b).

No single criterion is a priori the best. Any criterion that furthers understanding of the evolutionary process is valid. Different criteria illuminate different features of evolution. As an example, let us consider the ability of organisms to obtain and process information about the environment (Ayala, 1974b). This ability is of evolutionary interest because it contributes to the biological success of organisms, particularly animals. It is particularly relevant to the evolution of man, since, among the differences that mark off man from all other animals perhaps the most fundamental is man's greatly developed ability to perceive the environment, and to react flexibly to it. Whereas all organisms become genetically adapted to their environments, man is also able to create environments that suit his genetic constitution.

Increased ability to gather and process information about the environment is sometimes expressed as evolution towards "independence from the environment." This latter expression is misleading. No organism can be truly independent of the environment. Allegedly, the evolutionary sequence fish ⟶ amphibian ⟶ reptile is an example of evolution towards independence from the environment. Reptiles, birds, and mammals are indeed independent of *water* as a living medium, but their lives depend on the conditions of the land.

They have not become independent of the environment, but rather have exchanged dependence on one environment for dependence on another.

"Control over the environment" has been linked to the ability to gather and use information about the state of the environment. However, in the strictest sense, control over the environment is exercised only by the human species. All organisms *interact* with the environment, but not all can *control* it. Nest-building among birds, the construction of beehives, or the building of dams by beavers do not represent control over the environment except in a trivial sense. The ability to control the environment started with the australopithecines, the first group of organisms that can be called human. They are considered to be men precisely because they were able to produce devices such as rudimentary pebble and bone tools to manipulate the environment. The ability to obtain and process information about the conditions of the environment enables organisms to avoid unsuitable environments and to seek suitable ones. It has developed in many organisms because it is a useful adaptation.

All organisms interact selectively with the environment. The cell membrane of a bacterium permits certain molecules but not others to enter the cell; selective molecular exchange occurs also in the inorganic world. But this can hardly be called a form of information processing. Certain bacteria when placed on an agar plate move about in a zig-zag pattern that is almost certainly random. The most rudimentary ability to gather and process information about the environment can be found in certain single-celled eukaryotes. A *Paramecium* swims in a sinusoid path, ingesting the bacteria that it encounters as it swims. Whenever it meets unfavorable conditions, like unsuitable acidity or salinity in the water, the *Paramecium* checks its advance, turns, and starts in a new direction. Its reaction is purely negative. The *Paramecium* apparently does not *seek* food or a favorable environment, but simply avoids unsuitable conditions.

A somewhat greater ability to process information about the environment occurs in the single-celled *Euglena*. This organism has a light-sensitive spot by which it can orient itself in the direction from which the light originates. *Euglena's* motions not only avoid unsuitable environments but actively seek suitable ones. An amoeba represents further progress in the same direction; it reacts to light by moving away from it, and also actively pursues food particles.

An increase in the ability to gather and process information about the environment is not a general characteristic of the evolution of life. Progress in this respect occurred in certain evolutionary lines but not in others. Today's bacteria are not more progressive by this criterion than their ancestors of one billion years ago. In many evolutionary sequences some very limited progress took place in the early stages, without further progress through the rest of their history. By this criterion, animals are in general more advanced than plants; vertebrates are more advanced than invertebrates; mammals are more advanced than reptiles, which are more advanced than fish. The most advanced organism by this criterion is man.

The ability to obtain and process information about the environment has progressed little in the plant kingdom. Plants generally react to light and to gravity. Geotropism is positive in the root, but negative in the stem. Plants also grow toward light; the parts of some plants, such as the sunflower, follow the daily course of the sun. Plants are also hydrotropic, that is, the roots tend to grow toward water. The response to gravity, water, and light is basically due to differential growth rates; a greater elongation of cells takes place on one side of the stem or root than on the other side. Some plants react also to tactile stimuli; their tendrils twine around whatever they touch. *Mimosa* and carnivorous plants, like the Venus flytrap (*Dionaea*), have leaves that close rapidly when touched.

In multicellular animals the ability to obtain and process information about the environment is mediated by the nervous system. All major groups of animals, except the sponges, have nervous systems. The simplest nervous system among living animals occurs in coelenterates, which include hydra, corals, and jellyfishes. Each tentacle of a jellyfish reacts individually, and only if it is directly stimulated; there is no coordination of the information received by different parts of the animal. Moreover, jellyfishes are unable to learn from experience. A limited form of coordinated behavior occurs in echinoderms, which include starfishes and sea urchins. Whereas coelenterates possess only an undifferentiated nerve net, echinoderms possess a nerve ring and radial nerve cords in addition to a nerve net. When the appropriate stimulus is encountered, a starfish reacts with direct and unified actions of the whole body.

The most primitive form of a brain occurs in such organisms as planarian flatworms, which also have numerous sensory cells and eyes without lenses. The information gathered by these sensory cells and organs is processed and coordinated by the central nervous system and the rudimentary brain. A planarian worm is capable of some variability of responses and some simple learning —similar stimuli do not always produce similar responses. In the ability to gather and process information about the environment, planarian flatworms have progressed further than starfishes, and starfishes have progressed farther than sea anemones and other coelenterates. But none of these organisms has progressed very far by this criterion. Among invertebrates, the most progressive groups of organisms are arthropods, but vertebrates have progressed much further than any invertebrates.

Arthropods, which include the insects, have complex forms of behavior. Precise visual, chemical, and acoustic signals are obtained and processed by many arthropods, particularly in their search for food and their selection of mates. Vertebrates are generally able to obtain and process much more complicated signals and to produce a much greater variety of responses than the arthropods. The vertebrate brain has an enormous number of neurons with an extremely complex arrangement. Among vertebrates, progress in the ability to deal with environmental information is correlated with development of the neopallium. The neopallium is involved in the association and coordination of all kinds of impulses from all receptors and brain centers. It first appeared in reptiles; in mammals it has expanded to become the cerebral cortex, which covers most of the cerebral hemispheres. The larger brain of vertebrates permits them also to have a large number of neurons involved in information storage or memory.

The ability to perceive the environment, and to integrate, coordinate, and react flexibly to what is perceived, has developed in its most advanced state in man. This incomparable advancement is perhaps the most fundamental characteristic that marks off *Homo sapiens* from all other animals. Symbolic language, complex social organization, control over the environment, the ability to envisage future states and work toward them, values and ethics are all developments made possible by man's greatly developed capacity to obtain and organize information about the state of the environment.

The ability to obtain information about the environment and react to it is useful to organisms as an adaptation. It is an acceptable criterion of progress because it illuminates certain features of the evolution of life. However, it is not necessarily better or worse than other criteria of progress. Other criteria may help to discern other features of evolution, and thus be worth examining. Needless to say, organisms are more or less progressive depending on what criterion of progress is used. By certain criteria, flowering plants are more progressive than many animals.

Literature Cited

Ainsworth, G. C., and P. H. A. Sneath, eds. 1962. *Microbial Classification. 12th Symposium of the Society for General Microbiology*. Cambridge University Press.

Alexander, M. L. 1949. Note on genetic variability in natural populations of *Drosophila*. Univ. of Texas Publ. No. 4920, pp. 63–69.

———. 1952. Gene variability in the *americana–texana–novamexicana* complex of the *virilis* group of *Drosophila*. Univ. of Texas Publ. No. 5204, pp. 73–105.

Allard, R. W., and J. Adams. 1969. The role of intergenotypic interactions in plant breeding. *Proc. XII. Intern. Congr. Gen.,* 3:349–370.

Allard, R. W., and A. L. Kahler. 1972. Patterns of molecular variation in plant populations, *Proc. 6th Berkeley Symp. on Math. Stat. Prob.,* vol. 5, pp. 237–254.

Alsopp, A. 1969. Phylogenetic relationships of the procaryotes and the origin of the eucaryotic cell. *New Phytol.,* 68:591–612.

Amadon, D. 1950. The Hawaiian honeycreepers (Aves, Drepaniidae). *Bull. Amer. Mus. Nat. Hist.,* 95:151–262.

Anderson, D. T. 1973. *Embryology and Phylogeny in Annelids and Arthropods*. Pergamon Press, Oxford, England.

Anderson, E. 1949. *Introgressive Hybridization.* Wiley, New York.

Anxolabehere, D., and G. Periquet. 1972. Variation de la valeur sélective de l'hétérozygote en fonction de fréquences alléliques chez *Drosophila melanogaster. C. R. Acad. Sci. Paris,* 275:2755–2757.

Arnheim, N., and R. Steller. 1970. Multiple genes for lysozyme in birds. *Arch. Biochem. Biophys.,* 141:656–661.

Avery, O. T., C. M. MacLeod, and M. McCarty. 1944. Studies on the chemical nature of the substance inducing transformation of pneumococcal types. Induction of transformation by a deoxyribonucleic acid fraction isolated from pneumococcus Type III. *J. Exp. Med.,* 79:137–158.

Avise, J. C., and F. J. Ayala. 1975. Genetic differentiation in speciose versus depauperate phylads: Evidence from the California minnows. *Evolution,* 29:411–426.

Avise, J. C., and G. G. Kitto. 1973. Phosphoglucose isomerase gene duplication in the bony fishes: An evolutionary history. *Biochem. Gen.,* 8:113–132.

Axelrod, D. I. 1939. A Miocene flora from the western border of the Mohave Desert. *Carnegie Inst. Wash. Publ.* 516, pp. 1–129.

Ayala, F. J. 1965a. Relative fitness of populations of *Drosophila serrata* and *Drosophila birchii. Genetics,* 51:527–544.

———. 1965b. Evolution of fitness in experimental populations of *Drosophila serrata. Science,* 150:903–905.

———. 1966. Evolution of fitness. I. Improvement in the productivity and size of irradiated populations of *Drosophila serrata* and *Drosophila birchii. Genetics,* 53:833–895.

———. 1967. Evolution of fitness. III. Improvement of fitness in irradiated populations of *Drosophila serrata. Proc. Nat. Acad. Sci. U.S.A.,* 58:1919–1923.

———. 1968a. Genotype, environment, and population numbers. *Science,* 162: 1453–1459.

———. 1968b. Biology as an autonomous science. *Amer. Sci.,* 56:207–221.

———. 1969a. Evolution of fitness. V. Rate of evolution in irradiated populations of *Drosophila. Proc. Nat. Acad. Sci. U.S.A.,* 63(3):790–793.

———. 1969b. An evolutionary dilemma: Fitness of genotypes versus fitness of populations. *Canad. J. Cytol. Gen.,* 11:439–456.

———. 1970. Teleological explanations in evolutionary biology. *Phil. of Sci.,* 37:1–15.

———. 1974a. Biological evolution: Natural selection or random walk? *Amer. Sci.,* 62:692–701.

———. 1974b. The concept of biological progress. In *Studies in the Philosophy of Biology,* F. J. Ayala and T. Dobzhansky, eds. Macmillan, London.

———. 1975a. Genetic differentiation during the speciation process. *Evol. Biol.,* 8:1–78.

———. 1975b. Scientific hypotheses, natural selection, and the neutrality theory of protein polymorphism. In *The Role of Natural Selection in Human Evolution,* F. M. Salzano, ed. North Holland, Amsterdam.

———, ed. 1976. *Molecular Evolution.* Sinauer, Sunderland, Mass.

Ayala, F. J., and C. A. Campbell. 1974. Frequency-dependent selection. *Ann. Rev. Ecol. Syst.,* 5:115–138.

Ayala, F. J., and T. Dobzhansky, eds. 1974. *Studies in the Philosophy of Biology*. Macmillan, London; and University of California Press.

Ayala, F. J., D. Hedgecock, G. S. Zumwalt, and J. W. Valentine. 1973. Genetic variation in *Tridacna maxima,* an ecological analog of some unsuccessful evolutionary lineages. *Evolution,* 27:177–191.

Ayala, F. J., J. R. Powell, and Th. Dobzhansky. 1971. Polymorphisms in continental and island populations of *Drosophila willistoni. Proc. Nat. Acad. Sci. U.S.A.,* 68:2480–2483.

Ayala, F. J., M. L. Tracey, L. G. Barr, J. F. McDonald, and S. Pérez-Salas. 1974a. Genetic variation in natural populations of five *Drosophila* species and the hypothesis of the selective neutrality of protein polymorphisms. *Genetics,* 77:343–384.

Ayala, F. J., M. L. Tracey, L. G. Barr, and J. G. Ehrenfeld. 1974b. Genetic and reproductive differentiation of the subspecies *Drosophila equinoxialis caribbensis. Evolution,* 28:24–41.

Ayala, F. J., M. L. Tracey, D. Hedgecock, and R. C. Richmond. 1974c. Genetic differentiation during the speciation process in *Drosophila. Evolution,* 28:576–592.

Ayala, F. J., and J. W. Valentine. 1974. Genetic variability in a cosmopolitan deep-water ophiuran, *Ophiomusium lymani. Marine Biol.,* 27:51–57.

Babcock, E. B., and G. L. Stebbins. 1938. *The American Species of Crepis. Carnegie Inst. Wash. Publ.* No. 504.

Bachmann, K., O. B. Goin, and C. J. Goin. 1974. Nuclear DNA amounts in vertebrates. In "Evolution of Genetic Systems," H. H. Smith, ed., *Brookhaven Symp. Biol.* No. 23, pp. 419–450.

Baker, H. G. 1959. Reproductive methods as factors in speciation in flowering plants. *Cold Spring Harbor Symp. Quant. Biol.,* 24:177–191.

Bakker, R. T. 1971. Dinosaur physiology and the origin of mammals. *Evolution,* 25:636–658.

Ball, J. A. 1973. The zoo hypothesis. *Icarus,* 19:347–349.

Banks, N. L. 1970. Trace fossils from the late Precambrian and Lower Cambrian of Finmark, Norway. In "Trace Fossils," T. P. Crimes and J. C. Harper, eds., *Geol. J.* (spec. issue), 3:19–34.

Barghoorn, E. 1971. The oldest fossils. *Sci. Amer.,* 224(5):30–42.

Barker, A. D. 1969. An approach to the theory of natural selection. *Philosophy,* 44:274.

Barnes, H., and H. T. Powell. 1950. The development, general morphology and subsequent elimination of barnacle populations, *Balanus crenatus* and *B. balanoides,* after a heavy initial settlement. *J. Animal Ecol.,* 19:175–179.

Barrington, E. J. W. 1967. *Invertebrate Structure and Function.* Houghton Mifflin, Boston.

Beadle, G. W., and E. L. Tatum. 1941. Genetic control of biochemical reactions in *Neurospora. Proc. Nat. Acad. Sci. U.S.A.,* 27:499–506.

Beam, C. A., and M. Himes. 1974. Evidence for sexual fusion and recombination in the Dinoflagellate *Crypthecodinium (Gyrodinium) cohnii. Nature,* 250:435–436.

Becak, W. 1969. Gene action and polymorphism in polyploid species of amphibians. *Genetics* (suppl.), 61:183–190.

Beckner, M. 1959. *The Biological Way of Thought.* Columbia University Press.

Benado, M. B., F. J. Ayala, and M. M. Green. 1976. Evolution of experimental "mutator" populations of *Drosophila melanogaster. Genetics,* 82:43–52.

Berg, E. S. 1926. *Nomogenesis or Evolution Determined by Law.* London; reissued 1969, M.I.T. Press.

Bergson, H. 1907. *L'Évolution Créatrice.* [1911] Creative Evolution, New York.

Bernstein, S. C., L. J. Throckmorton, and J. L. Hubby. 1973. Still more genetic variability in natural populations. *Proc. Nat. Acad. Sci. U.S.A.,* 70:3928–3931.

Berry, W. B. N. 1968. *Growth of a Prehistoric Time Scale.* Freeman, San Francisco.

Bidney, D. 1953. *Theoretical Anthropology.* Columbia University Press.

Blair, A. B. 1941. Variation, isolating mechanisms, and hybridization in certain toads. *Genetics,* 26:398–417.

Bock, W. J. 1965. The role of adaptive mechanisms in the origin of higher levels of organization. *Syst. Zool.,* 14:272–287.

———. 1970. Microevolutionary sequences as a fundamental concept in macroevolutionary models. *Evolution,* 24:704–722.

Bock, W. J., and G. von Wahlert. 1965. Adaptation and the form–function complex. *Evolution,* 19:269–299.

Bockelie, T., and R. A. Fortey. 1976. An early Ordovician vertebrate. *Nature,* 260: 36–38.

Boesiger, E. 1962. Sur le degré d'hétérozygotie des populations naturelles de *Drosophila melanogaster* et son maintien par la sélection sexuelle. *Bull. Biol. France Belgique,* 96:3–122.

———. 1974. Evolutionary theories after Lamarck and Darwin. In *Studies in the Philosophy of Biology,* F. J. Ayala and Th. Dobzhansky, eds. Macmillan, London; and University of California Press.

Bogorad, L. 1975. Evolution of organelles and eukaryotic genomes. *Science,* **188**: 891–898.

Bohm, D. 1969. Some comments on Maynard Smith's contributions. In *Towards a Theoretical Biology,* vol. 2, C. H. Waddington, ed., pp. 98–105. Edinburgh University Press.

Bonnell, M. L., and R. K. Selander. 1974. Elephant seals: Genetic variation and near extinction. *Science,* **184**:908–909.

Bonner, J. T. 1965a. *The Molecular Biology of Development.* Oxford University Press, New York.

———. 1965b. *Size and Cycle, An Essay on the Structure of Biology.* Princeton University Press.

———. 1974. *On Development: The Biology of Form.* Harvard University Press.

Boucot, A. J. 1968. Origins of the Silurian fauna. *Geol. Soc. Amer., Prog. with Abstracts,* 1968 Meetings, pp. 33–34.

Bowman, R. I. 1961. Morphological differentiation and adaptation in the Galapagos finches. *Univ. Calif. Publ. Zool.,* **58**:1–302.

Bradley, D. E. 1971. A comparative study of the structure and biological properties of bacteriophages. In *Comparative Virology,* K. Maramorosch and E. Kurstak, eds., Academic, New York, pp. 208–253.

Bradshaw, A. D. 1971. Plant evolution in extreme environments. In *Ecological Genetics and Evolution,* R. Creed, ed. Blackwell, Oxford.

Brewbaker, J. L. 1964. *Agricultural Genetics.* Prentice–Hall, Englewood Cliffs, New Jersey.

Briden, J. C., G. E. Drewry, and A. G. Smith. 1974. Phanerozoic equal-area world maps. *J. Geol.,* **82**:555–574.

Britten, R. J., and E. H. Davidson. 1969. Gene regulation for higher cells: A theory. *Science,* **165**:349–357.

———. 1971. Repetitive and non-repetitive DNA sequences and a speculation on the origins of evolutionary novelty. *Quart. Rev. Biol.,* **46**:111–138.

Britten, R. J., and D. E. Kohne. 1968. Repeated sequences in DNA. *Science,* **161**: 529–540.

Broad, C. D. 1972. *The Mind and Its Place in Nature.* Kegan Paul, London.

Brooks, J. C. 1950. Speciation in ancient lakes. *Quart. Rev. Biol.,* **25**:131–176.

Brown, A. W. A. 1967. Genetics of insecticide resistance in insect vectors. In *Genetics of Insect Vectors of Disease,* J. W. Wright and R. Pal, eds., pp. 505–552. Elsevier, Amsterdam.

Brown, D. D., and J. B. Gurdon. 1964. Absence of ribosomal RNA synthesis in the anucleolate mutant of *Xenopus laevis. Proc. Nat. Acad. Sci. U.S.A.,* **51**:139–146.

Brown, R. H., M. Richardson, D. Boulter, J. A. M. Ramshaw, and R. P. S. Jeffries. 1972. The amino acid sequence of cytochrome *c* from *Helix aspersa* Müller (garden snail). *Biochem. J.,* **128**:971–974.

Brues, A. M. 1964. The cost of natural selection vs. the cost of not evolving. *Evolution,* **18**:379–383.

Bullman, O. M. B. 1970. *Treatise on Invertebrate Paleontology. Part V, Graptolithina,* 2nd ed. Geological Society of America and University of Kansas.

Bush, G. L. 1975. Modes of animal speciation. *Ann. Rev. Ecol. and Syst.,* 6:339–361.

Cairns, J. 1963. The chromosome of *Escherichia coli. Cold Spring Harbor Symp. Quant. Biol.,* 28:43–46.

Cairns-Smith, A. G. 1971. *The Life Puzzle.* Oliver and Boyd, Edinburgh.

Campbell, B. G. 1974. *Human Evolution,* 2nd ed. Aldine, Chicago.

Campbell, D. T. 1972. On the genetics of altruism and the counter-hedonic components of human culture. *J. Soc. Issues,* 28:21–37.

Carlquist, S. 1970. *Hawaii. A Natural History.* American Museum of Natural History, New York.

Carlson, P. S., H. H. Smith, and R. D. Dearing. 1972. Parasexual interspecific plant hybridization. *Proc. Nat. Acad. Sci. U.S.A.,* 69:2292–2294.

Carroll, R. L. 1969. Problems of the origin of reptiles. *Biol. Rev.,* 44: 393–432.

Carson, H. L. 1964. Population size and genetic load in irradiated populations of *Drosophila melanogaster. Genetics,* 49:521–528.

———. 1973. Reorganization of the gene pool during speciation. In "Genetic Structure of Populations," N. E. Morton, ed., *Pop. Gen. Monog.* III, University Press Hawaii, pp. 274–280.

———. 1975. The genetics of speciation at the diploid level. *Amer. Nat.,* 109:83–92.

Carson, H. L., D. E. Hardy, H. T. Spieth, and W. S. Stone. 1970. The evolutionary biology of the Hawaiian Drosophilidae. In *Essays in Evolution and Genetics in Honor of Th. Dobzhansky,* M. K. Hecht and W. C. Steere, eds., pp. 437–543. Appleton–Century–Crofts, New York.

Carson, H. L., and K. Y. Kaneshiro. 1976. *Drosophila* of Hawaii: Systematics and ecological genetics. *Ann. Rev. Ecol. Syst.,* 7:311–345.

Cavalli-Sforza, L. L. 1973. Some current problems of human genetics. *Amer. J. Human Gen.,* 25:82–104.

Cavalli-Sforza, L. L., and W. F. Bodmer. 1971. *The Genetics of Human Populations.* Freeman, San Francisco.

Chadefaud, M. 1960. *Traité de Botanique,* vol. 1. Masson, Paris.

Champion, A. B., E. M. Prager, D. Wachter, and A. C. Wilson. 1974. Microcomplement fixation. In *Biochemical and Immunological Taxonomy of Animals,* C. A. Wright, ed., Academic Press, London, pp. 397–416.

Chapman, G. 1958. The hydrostatic skeleton in the invertebrates. *Biol. Rev.,* 33: 338–371.

———. 1966. The structure and functions of the mesogloea. In "The Cnidaria and Their Evolution," W. J. Rees, ed. *Symp. Zool. Soc. London,* 16:147–168.

Chargaff, E. 1951. Structure and function of nucleic acids as cell constituents. *Fed. Proc.,* 10:654–659.

Chetverikov, S. S. 1926. On certain aspects of the evolutionary process from the standpoint of genetics. *Zhurnal Exp. Biol.,* 1:3–54 (Russian); English translation in *Proc. Amer. Phil. Soc.,* 105:167–195 (1959).

Cisne, J. L. 1974. Trilobites and the origin of the arthropods. *Science,* 186:13–18.

Clark, R. B. 1964. *Dynamics in Metazoan Evolution.* Clarendon Press, Oxford.

Clausen, J. 1951. *Stages in the Evolution of Plant Species.* Cornell University Press.

Clausen, J., D. D. Keck, and W. W. Hiesey. 1940. Experimental studies on the nature of species. *Carnegie Inst. Washington Publ.* No. 520, pp. 1–452.

Cleland, R. E. 1964. The evolutionary history of the North American evening primroses of the *"biennis"* group. *Proc. Amer. Phil. Soc.,* 108:88–98.

Clemmay, H. 1976. World's oldest animal traces. *Nature,* 261:576–578.

Cloud, P. 1949. Some problems and patterns of evolution exemplified by fossil invertebrates. *Evolution,* 2:322–350.

———. 1968a. Atmospheric and hydrospheric evolution on the primitive earth. *Science,* 160:729–736.

———. 1968b. Pre-metazoan evolution and the origin of the Metazoa. In *Evolution and Environment,* E. T. Drake, ed., Yale University Press, pp. 1–72.

———. 1973. Pseudofossils: A plea for caution. *Geology,* 1:123–127.

———. 1974. Evolution of ecosystems. *Amer. Sci.* 62:54–66.

Cocks, L. R. M. 1971. Facies relationships in the European Lower Silurian. *Mém. Bur. Rech. Géol. Min.* No. 73, pp. 223–227.

Colbert, E. H. 1969. *Evolution of the Vertebrates,* 2nd ed. Wiley, New York.

Cole, H. S. D., C. Freeman, M. Jahoda, and K. L. R. Pavitt. 1973. *Thinking About the Future: A Critique of the Limits to Growth.* Chatto and Windus, London.

Coluzzi, M. 1970. Sibling species in *Anopheles* and their importance in malariology. *Misc. Publ. Entom. Soc. Amer.,* 7:63–72.

Comings, D. E., and T. A. Okada. 1970. Whole-mount electron microscopy of meiotic chromosomes and the synaptonemal complex. *Chromosoma,* 30:269–286.

Connell, J. H. 1961. The influence of interspecific competition and other factors on the distribution of the barnacle *Chthamalus stellatus. Ecology,* 42:710–723.

Conway Morris, S. 1976. A new Cambrian Lophophorate from the Burgess Shale of British Columbia. *Palaeontology,* 19:199–222.

———. Forthcoming. The Burgess Shale. In *Encyclopedia of Paleontology,* R. W. Fairbridge and D. Jablonski, eds. Dowden, Hutchison, and Ross, Stroudsburg, Pa.

Cope, E. D. 1896. *The Primary Factors of Organic Evolution.* Open Court, Chicago.

Cory, L., and J. J. Manion. 1955. Ecology and hybridization in the genus *Bufo* in the Michigan–Indiana region. *Evolution,* 9:42–51.

Corliss, J. O. 1974. Time for evolutionary biologists to take more interest in protozoan phylogenetics? *Taxon,* 23:497–522.

Correll, D. 1950. *Native Orchids of North America North of Mexico.* Chronica Botanica, Waltham, Mass.

Craig, G. B., and W. A. Hickey. 1967. Genetics of *Aedes aegypti.* In *Genetics of Insect Vectors of Disease,* J. W. Wright and R. Pal, eds., pp. 67–131. Elsevier, Amsterdam.

Crick, F. H. C., and L. E. Orgel. 1973. Directed panspermia. *Icarus,* 19:341–346.

Crimes, T. P. 1974. Colonisation of the early ocean floor. *Nature,* 248:328–330.

Crow, J. F. 1957. Genetics of insect resistance to chemicals. *Amer. Rev. Entom.,* 2:227–246.

Crow, J. F., and M. Kimura. 1970. *An Introduction to Population Genetics Theory.* Harper and Row, New York.

Crumpacker, D. W. 1967. Genetic loads in maize (*Zea mays* L.) and other cross-fertilized plants and animals. *Evol. Biol.,* 1:306–424.

Darevsky, I. S. 1966. Natural parthenogenesis in a polymorphic group of Caucasian Rock Lizards related to *Lacerta saxicola* E. Eversmann. *J. Ohio Herpet. Soc.,* 5:115–152.

Darlington, C. D. 1940. Taxonomic species and genetic systems. In *The New Systematics,* J. Huxley, ed., Clarendon Press, Oxford, pp. 137–160.

Darwin, C. 1859. *On the Origin of Species by Means of Natural Selection* (6th ed., 1972). Murray, London.

———. 1871. *The Descent of Man and Selection in Relation to Sex* (2nd ed., 1889). Murray, London.

———. 1958. *The Autobiography of Charles Darwin, 1809–1882,* Nora Barlow, ed. Collins, London.

———. 1960. *Darwin's notebooks on transmutation of species,* G. De Beer, ed. *Bull. Brit. Mus.,* 2:23–200.

Darwin, F. 1903. *More Letters of Charles Darwin,* 2 vols. Murray, London.

Davidson, E. H. 1968. *Gene Activity in Early Development.* Academic Press, New York.

Davidson, E. H., and R. J. Britten. 1973. Organization, transcription, and regulation in the animal genome. *Quart. Rev. Biol.,* 48:565–613.

Davidson, G., H. E. Patterson, N. Coluzzi, G. F. Mason, and D. W. Micks. 1967. The *Anopheles gambiae* complex. In *Genetics of Insect Vectors of Disease,* J. W. Wright and R. Pal, eds. Elsevier, Amsterdam.

Davidson, J. N., I. Leslie, R. M. S. Smellie, and R. Y. Thomson. 1950. Chemical changes in the developing chick embryo related to the deoxyribonucleic acid content of the nucleus. *Biochem. J.,* 46:xl.

Dawood, M. M., and M. W. Strickberger. 1969. The effects of larval interaction on viability of *Drosophila melanogaster.* III. Effects of biotic residues. *Genetics,* 63: 213–220.

Dayhoff, M. O. 1969. Computer analysis of protein evolution. *Sci. Amer.,* 221(1): 86–95.

———. 1972. *Atlas of Protein Sequences.* National Biomedical Research Foundation, Washington, D.C.

Dayton, P. K. 1971. Competition, disturbance, and community organization: The provision and subsequent utilization of space in a rocky intertidal community. *Ecol. Monogr.,* 41:351–389.

De Beer, G. 1964. *Charles Darwin, A Scientific Biography.* Doubleday, Garden City, New York.

Demarly, Y. 1963. Génétique des tétraploides et amélioration des plantes. *Ann. Amélior. Plant.,* 13:307–400.

De Maupertius, P. L. M. 1746. *The Earthly Venus,* translated by S. B. Boas (reprinted 1966). Johnson, New York.

Demeny, P. 1974. The populations of the underdeveloped countries. *Sci. Amer.,* 231(3):148–159.

Demerec, M. 1937. Frequency of spontaneous mutations in certain stocks of *Drosophila melanogaster. Genetics,* 22:469–478.

Demerec, M., and U. Fano. 1945. Bacteriophage resistant mutants in *Escherichia coli. Genetics,* 30:119–136.

Denis, H., and J. Brachet. 1969a. Gene expression in interspecific hybrids. I. DNA synthesis in the lethal cross *Arbacia lixula* ♂ × *Paracentrotus lividus* ♀. *Proc. Nat. Acad. Sci. U.S.A.,* 62:194–201.

———. 1969b. Gene expression in interspecific hybrids. II. RNA synthesis in the lethal cross *Arbacia lixula* ♂ × *Paracentrotus lividus* ♀. *Proc. Nat. Acad. Sci. U.S.A.,* 62:438–445.

———. 1970. Expression du génome chez les hybrides interspécifiques. Fidélité de la transcription dans la croisement létal *Arbacia lixula* ♂ × *Paracentrotus lividus* ♀. *Eur. J. Biochem.,* 13:86–93.

De Vore, I., ed. 1965. *Primate Behavior: Field Studies on Monkeys and Apes.* Holt, Rinehart and Winston, Boston.

De Wet, J. M. J., and J. R. Harlan. 1966. Morphology of the compilospecies *Bothriochloa intermedia. Amer. J. Bot.,* 53:94–98.

Dewey, J. F., and J. M. Bird. 1970. Mountain belts and the new global tectonics. *J. Geophys. Res.,* 75:2625–2647.

Dickerson, G. 1955. Genetic slippage in response to selection for multiple objectives. *Cold Spring Harb. Symp. Quant. Biol.,* 20:213–214.

Dobzhansky, Th. 1937. *Genetics and the Origin of Species* (lst ed.); 2d ed., 1941; 3d ed., 1951. Columbia University Press.

———. 1955. A review of some fundamental concepts and problems of population genetics. *Cold Spring Harb. Symp. Quant. Biol.,* 20: 1–15.

———. 1967. *The Biology of Ultimate Concern.* New American Library, New York.

———. 1968. On some fundamental concepts of Darwinian biology. In *Evolutionary Biology,* vol. II, Th. Dobzhansky, M. K. Hecht, W. C. Steere, eds., pp. 1–34. Appleton–Century–Crofts, New York.

———. 1970. *Genetics of the Evolutionary Process.* Columbia University Press.

———. 1971. Evolutionary oscillations in *Drosophila pseudoobscura.* In *Ecological Genetics and Evolution,* R. Creed, ed. Blackwell, Oxford.

———. 1972. Darwinian evolution and the problem of extraterrestrial life. *Persp. Biol. Med.,* 15:157–175.

———. 1973a. Nothing in biology makes sense except in the light of evolution. *Amer. Biol. Teacher,* 35:125–129.

———. 1973b. Ethics and values in biological and cultural evolution. *Zygon,* 8:261–281.

Dobzhansky, Th., and F. J. Ayala. 1973. Temporal frequency changes of enzyme and chromosomal polymorphisms in natural populations of *Drosophila. Proc. Nat. Acad. Sci. U.S.A.,* 70:680–683.

Dobzhansky, Th., H. Levene, B. Spassky, and N. Spassky. 1959. Release of genetic variability through recombination. III. *Drosophila prosaltans. Genetics,* 44:75–92.

Dobzhansky, Th., and B. Spassky, 1947. Evolutionary changes in laboratory cultures of *D. pseudoobscura. Evolution,* 1:191–216.

———. 1953. Genetics of natural populations. XXI. Concealed variability in two sympatric species of *Drosophila. Genetics,* 38:471–484.

———. 1963. Genetics of natural populations. XXXIV. Adaptive norm, genetic load and genetic elite in *D. pseudoobscura. Genetics,* 48:1467–1485.

———. 1967. An experiment on migration and simultaneous selection for several traits in *D. pseudoobscura. Genetics,* 55:723–734.

Dobzhansky, Th., B. Spassky, and N. Spassky. 1952. A comparative study of mutation rates in two ecologically diverse species of *Drosophila. Genetics,* 37:650–664.

———. 1954. Rates of spontaneous mutation in the second chromosomes of the sibling species, *Drosophila pseudoobscura* and *Drosophila persimilis. Genetics,* 39:899–907.

Dodge, J. D. 1971. A dinoflagellate with both a mesocaryotic and a eucaryotic nucleus. I. Fine structure of the nuclei. *Protoplasma,* 73:145–157.

Dodson, C. H. 1967. Relationships between pollinators and orchid flowers. *Atlas Simp. Biota Amazonia,* 5:1–72.

Doolittle, R. F. 1970. Evolution of fibrinogen molecules. *Thrombosis et Diatheiss Haemorrhagica* (suppl.), 39:25–42.

Doty, M. S. 1957. Rocky intertidal surfaces. In "Treatise on Marine Ecology and Paleoecology," Vol. 1, J. W. Hedgpeth, ed. *Geol. Soc. Amer. Mem.* No. 67, pp. 535–585.

Doty, P., J. Marmur, J. Eigner, and C. Schildkraut. 1960. Strand separation and specific recombination in deoxyribonucleic acids: physical chemical studies. *Proc. Nat. Acad. Sci. U.S.A.,* 46:461–476.

Dougherty, E. C., *et al.,* eds. 1963. *The Lower Metazoa, Comparative Biology and Phylogeny.* University of California Press.

Dubinin, N. P. 1966. *Evolution of Populations and Radiation.* Atomizdat, Moscow.

Dubinin, N.P., D. D. Romashov, M. A. Heptner, and Z. A. Demidova. 1937. Aberrant polymorphism in *Drosophila fasciata* Meig. *Biol. Zhurnal,* 6:311–354.

Durham, J. W. 1967. The incompleteness of our knowledge of the fossil record. *J. Paleon,* 41:559–565.

Dworkin, R. H., B. W. Burke, B. A. Maher, and I. I. Gottesman. 1976. A longitudinal study of the genetics of personality. *J. Personality and Social Psych.,* 34:510–518.

Ehrlich, P. R. 1970. *The Population Bomb.* Ballantine Books.

Eicher, D. L. 1968. *Geologic Time.* Prentice–Hall, Englewood Cliffs, New Jersey.

Eiseley, L. 1959. Charles Darwin, Edward Blyth, and the theory of natural selection. *Proc. Amer. Phil. Soc.,* 103:94–158.

Eldredge, N. 1971. The allopatric model and phylogeny in Paleozoic invertebrates. *Evolution,* 25:156–167.

Eldredge, N., and S. J. Gould. 1972. Punctuated equilibria: An alternative to phyletic gradualism. In *Models in Paleobiology,* T. J. M. Schopf, ed., Freeman, Cooper, San Francisco, pp. 82–115.

Emerson, A. E. 1955. Geographical origins and dispersion of termite genera. *Fieldiana Zool.,* 37:465–521.

Ephrussi, B. 1972. *Hybridization of Somatic Cells.* Princeton University Press.

Falconer, D. S. 1964. *Introduction to Quantitative Genetics.* Oliver and Boyd, London.

Fankhauser, G. 1945. The effects of change in chromosome number on amphibian development. *Quart. Rev. Biol.,* 20:20–78.

Farmer, J. D., J. W. Valentine, and R. Cowen. 1973. Adaptive strategies leading to the ectoproct ground plan. *Syst. Zool.,* 22:233–239.

Farris, J. S. 1972. Estimating phylogenetic trees from distance matrices. *Amer. Nat.,* 106:645–668.

Ferone, R., M. O'Shea, and M. Yoell. 1970. Altered dihydrofolate reductase associated with drug-resistance transfer between rodent plasmodia. *Science,* 167:1263–1264.

Fischer, A. G., B. C. Heezen, R. E. Boyce, D. Bukry, R. G. Douglas, R. E. Garrison, S. A. Kling, V. Krasheninnikov, A. P. Lisitzin, and A. C. Pimm. 1970. Geological history of the western North Pacific. *Science,* 168:1210–1214.

Fisher, R. A. 1930. *The Genetical Theory of Natural Selection.* Clarendon Press, Oxford.

Fitch, W. M. 1966. The relation between frequencies of amino acids and ordered trinucleotides. *J. Mol. Biol.,* 16:1–27.

———. 1973. Aspects of molecular evolution. *Ann. Rev. Genetics,* 7:343–380.

———. 1975. Evolutionary rates in proteins and the cost of natural selection: Implications for neutral proteins. In *The Role of Natural Selection in Human Evolution,* F. M. Salzano, ed., pp. 43–56. Elsevier, New York.

———. 1976. Molecular evolutionary clocks. In *Molecular Evolution,* F. J. Ayala, ed., pp. 160–178. Sinauer, Sunderland, Mass.

Fitch, W. M., and E. Margoliash. 1967. Construction of phylogenetic trees. *Science,* 155:279–284.

———. 1970. The usefulness of amino acid and nucleotide sequences in evolutionary studies. *Evol. Biol.,* 4:67–109.

Flessa, K. W., and J. Imbrie. 1973. Evolutionary pulsations: Evidence from Phanerozoic diversity patterns. In *Continental Drift, Sea Floor Spreading and Plate Tectonics,* D. H. Tarling and S. K. Runcorn, eds., Academic Press, London.

Flew, A. 1967. *Evolutionary Ethics.* Macmillan, London.

Ford, E. B. 1971. *Ecological Genetics,* 3rd ed. Chapman and Hall, London.

Fott, B. 1974. The phylogeny of eucaryotic algae. *Taxon,* 23:446–461.

Fowler, J. A. 1964. The *Rana pipiens* problem, a proposed solution. *Amer. Nat.,* 98:213–219.

Fox, S. W., and K. Dose. 1972. *Molecular Evolution and the Origin of Life.* Freeman, San Francisco.

Fraenkel-Conrat, H., and B. Singer. 1957. Virus reconstitution: Combination of protein and nucleic acid from different strains. *Biochim. Biophys. Acta,* 24:540–548.

Freedman, R., and B. Berelson. 1974. The human population. *Sci. Amer.,* 231(3): 30–39.

Fromm, E. 1964. *The Heart of Man.* Harper and Row, New York.

Fryer, G., and T. D. Iles. 1969. Alternate routes to evolutionary success as exhibited by African cichlid fish of the genus *Tilapia* and the species flocks of the Great Lakes. *Evolution,* 23:359–369.

Galau, G. A., R. J. Britten, and E. H. Davidson. 1974. A measurement of the sequence complexity of polysomal messenger RNA in sea urchin embryos. *Cell,* 2:9–20.

Galau, G. A., W. H. Klein, M. M. Davis, B. J. Wold, R. J. Britten, and E. H. Davidson. 1976. Structural gene sets active in embryos and adult tissues of the sea urchin. *Cell,* 7:487–505.

Ganong, W. F. 1901. The cardinal principles of morphology. *Bot. Gaz.,* 31:426–434.

Gardner, R. A., and B. T. Gardner. 1969. Teaching sign language to a chimpanzee. *Science,* 165:664–672.

Garn, S. M. 1961. *Human Races.* Thomas, Springfield.

Gershenson, S. 1945. Evolutionary studies on the distribution and dynamics of melanism in the hamster (*Cricetus cricetus* L.). II. Seasonal and annual changes in the frequency of black hamsters. *Genetics,* 30:233–251.

Ghilarov, M. S. 1959. Adaptations of insects of soil dwelling. *Proc. Int. Congr. Zool.,* 15:354–357.

Ghiselin, M. T. 1969. *The Triumph of the Darwinian Method.* University of California Press.

———. 1974. *The Economy of Nature and the Evolution of Sex.* University of California Press.

Gierer, A., and G. Schramm. 1956. Infectivity of ribonucleic acid from tobacco mosaic virus. *Nature,* 177:702–703.

Glaessner, M. F. 1969. Trace fossils from the Precambrian and basal Cambrian. *Lethaia,* 2:369–393.

———. 1971. Geographic distribution and time range of the Ediacara Precambrian fauna. *Bull. Geol. Soc. Amer.,* 82:509–514.

Gold, T. 1960. Cosmic garbage. *Air Force and Space Digest,* May, p. 65.

Goldberg, R. B., G. A. Galau, R. J. Britten, and E. H. Davidson. 1973. Sequence content of sea urchin embryo messenger RNA. *Proc. Nat. Acad. Sci. U.S.A.,* 70:3516–3520.

Goldich, S. S., and C. E. Hedge. 1974. 3,800-Myr granitic gneiss in southwestern Minnesota. *Nature,* 252:467–468.
Goldschmidt, R. B. 1940. *The Material Basis of Evolution.* Yale University Press.
———. 1948. Ecotype, ecospecies and macroevolution. *Experientia,* 4:465–472.
———. 1952. Evolution as viewed by one geneticist. *Amer. Sci.,* 40:84–98.
Goodman, M. 1976. Protein sequences in phylogeny. In *Molecular Evolution,* F. J. Ayala, ed., Sinauer, Sunderland, Mass., pp. 141–159.
Goodman, M., G. W. Moore, and G. Matsuda. 1975. Darwinian evolution in the genealogy of hemoglobin. *Nature,* 253:603–608.
Gottesman, I. I., and L. L. Heston. 1972. Human behavioral adaptations: Speculations on their genesis. In *Genetics, Environment and Behavior,* L. Ehrman, G. S. Omenn, and E. Caspari, eds., Academic Press, New York.
Gottlieb, L. D. 1973. Enzyme differentiation in *Clarkia franciscana, C. rubicunda* and *C. amoena. Evolution,* 27:205–214.
Goudge, T. A. 1961. *The Ascent of Life.* University Press, Toronto.
Gould, S. J. 1973. Positive allometry of antlers in the "Irish Elk," *Megaloceros giganteus. Nature,* 244:375–376.
Grant, K. A., and V. Grant. 1964. Mechanical isolation of *Salvia apiana* and *Salvia mellifera. Evolution,* 18:196–212.
Grant, V. 1949. Pollination systems as isolation mechanisms in flowering plants. *Evolution,* 3:82–97.
———. 1963. *The Origin of Adaptations.* Columbia University Press.
———. 1971. *Plant Speciation.* Columbia University Press.
Grant, V., and K. A. Grant. 1965. *Flower Pollination in the Phlox Family.* Columbia University Press.
Grassé, P. 1952. *Traite dé Zoologie, I. Phylogénie, Protozoaires: Generalités, Flagellés.* Masson, Paris.
———. 1973. *L'evolution du vivant.* Albin Michel, Paris.
Green, M. M. 1970. The genetics of a mutator gene in *Drosophila melanogaster. Mutation Res.,* 10:353–363.
———. 1973. Some observations and comments on mutable and mutator genes in *Drosophila. Genetics* (suppl.), 73:187–194.
Grell, K. 1967. Sexual reproduction in Protozoa. In *Research in Protozoa,* T. T. Chen, ed., Pergamon Press, Oxford, pp. 147–214.
Grene, M. 1969. Notes on Maynard Smith's "Status of neo-Darwinism." In *Towards a Theoretical Biology,* vol. 2, C. H. Waddington, ed., pp. 97–98.
———. 1974. *The Understanding of Nature. Essays in the Philosophy of Biology.* Reidel, Boston.
Gustafsson, A. 1946–47. *Apomixis in Higher Plants.* Lunds Universitets Arsskrift.

Haapala, O. K., and M. O. Soyer. 1974. Size of circular chromatids and amounts of haploid DNA in the dinoflagellates *Gyrodinium cohnii* and *Prorocentrum micans. Hereditas,* 76:83–90.
Hadzi, J. 1963. *The Evolution of the Metazoa.* Pergamon Press, Oxford.
Haeckel, E. 1866. *Generelle Morphologie der Organismen,* vol. 2. Georg Reimer, Berlin.
———. 1874. The gastraea theory, the phylogenetic classification of the animal kingdom, and the homology of the germ lamellae. *Quart. J. Microsc. Sci.,* n.s., 14:142–165.
Haldane, J. B. S. 1932. *The Causes of Evolution.* Harper, New York.

Haldane, J. B. S. 1933. *Science and Human Life.* Harper, New York.

———. 1957. The cost of natural selection. *J. Gen.,* 55:511–524.

Hallam, A., and S. J. Gould. 1975. The evolution of British and American Middle and Upper Jurassic *Gryphaea:* a biometric study. *Proc. Roy. Soc. London (B),* 189:511–542.

Hamburg, D. H. 1971. Aggressive behavior in chimpanzees and baboons in natural habitats. *J. Psychiat. Res.,* 8:385–398.

Hamilton, H. J. 1974. *The Evolution of Societal Systems. Publication Center for Futures Research,* University of Southern California.

Hamilton, W. D. 1964. The genetic theory of social behaviour. *J. Theoret. Biol.,* 1:1–16, 17–52.

Hamrick, J. L., and R. W. Allard. 1972. Microgeographical variation in allozyme frequencies in *Avena barbata. Proc. Nat. Acad. Sci. U.S.A.,* 69:2100–2104.

Hand, C. 1959. On the origin and phylogeny of the coelenterates. *Syst. Zool.,* 8:191–202.

———. 1963. The early worm: A planula. In *The Lower Metazoa, Comparative Biology and Phylogeny,* E. C. Dougherty *et al.,* eds., University of California Press, pp. 33–39.

Harberd, D. 1961. Observations on population structure and longevity of *Festuca rubra* L. *New Phytol.,* 60:184–206.

Harland, W. B., *et al.,* eds. 1967. *The Fossil Record, A Symposium with Documentation.* Geological Society, London.

Harris, H. 1966. Enzyme polymorphisms in man. *Proc. Roy. Soc. Lond. (B),* 164: 298–310.

Harris, H., and D. A. Hopkinson. 1972. Average heterozygosity in man. *J. Human Gen.,* 36:9–20.

Hayes, R. J., and C. Hayes. 1954. The cultural capacity of chimpanzee. *Human Biology,* 26:288–303.

Hedges, R. W. 1972. The pattern of evolutionary change in bacteria. *Heredity,* 28:39–48.

Hempel, C. G. 1966. *Philosophy of Natural Science.* Prentice–Hall, Englewood Cliffs, New Jersey.

Hennig, W. 1950. *Grundzüge einer Theorie der Phylogenetischen Systematik.* Deutscher Zentralverlag, Berlin; English edition, *Phylogenetic Systematics,* 1966, University of Illinois Press.

Hershey, A. D., and M. Chase. 1952. Independent functions of viral protein and nucleic acid in growth of bacteriophage. *J. Gen. Physiol.,* 36:39–56.

Hessler, R. R., and P. A. Jumars. 1974. Abyssal community analysis from replicate box cores in the central North Pacific. *Deep-Sea Res.,* 21:185–209.

Hessler, R. R., and H. L. Sanders. 1967. Faunal diversity in the deep sea. *Deep-Sea Res.,* 14:65–79.

Himes, M., and C. A. Beam. 1975. Genetic analysis in the dinoflagellate *Crypthecodinium (Gyrodinium) cohnii:* Evidence for unusual meiosis. *Proc. Nat. Acad. Sci. U.S.A.,* 72:4546–4549.

Himmelfarb, G. 1962. *Darwin and the Darwinian Revolution.* Doubleday, New York.

Hinegardner, R. 1976. Evolution of genome size. In *Molecular Evolution,* F. J. Ayala, ed., Sinauer, Sunderland, Mass., pp. 179–199.

Holliday, R., and J. E. Pugh. 1975. DNA modification mechanism and gene activity during development. *Science,* 187:226–232.

Hood, L., J. H. Campbell, and S. C. R. Elgin. 1975. The organization, expression, and evolution of antibody genes and other multigene families. *Ann. Rev. Gen.,* 9:305–353.

Hopson, J. A., and A. W. Crompton. 1969. Origin of mammals. *Evol. Biol.,* 3:15–72.

Horridge, G. A. 1957. The coordination of the protective retraction of coral polyps. *Phil. Trans. Roy. Soc. London (B),* 240:495–529.

Howells, W. W. 1966. *Homo erectus. Sci. Amer.,* 215(11):46–53.

Hoyer, B. H., B. J. McCarthy, and E. T. Bolton. 1964. A molecular approach in the systematics of higher organisms. *Science,* 144:959–967.

Hoyer, B. H., and R. B. Roberts. 1967. Studies on nucleic acid interactions using DNA–agar. In *Molecular Genetics,* pt. II, H. J. Taylor, ed., Academic Press, New York, pp. 425–479.

Hoyer, B. H., N. W. van de Velde, M. Goodman, and R. B. Roberts. 1972. Examination of hominid evolution by DNA sequence homology. *J. Human Evol.,* 1:645–649.

Hull, D. 1970. Contemporary systematic philosophies. *Amer. Rev. Ecol. Syst.,* 1:19–54.

———. 1973. *Darwin and His Critics.* Harvard University Press.

———. 1974. *Philosophy of Biological Science.* Prentice–Hall, Englewood Cliffs, New Jersey.

Hungate, R. E. 1966. *The Rumen and its Microbes.* Academic Press, New York.

Hutchinson, G. E. 1957. Concluding remarks. *Cold Spring Harbor Symp. Quant. Biol.,* 22:415–427.

Huxley, J. S. 1939. *Man in the Modern World.* Harper, New York.

———. 1942. *Evolution: The Modern Synthesis.* Harper, New York.

———. 1953. *Evolution in Action.* Harper, New York.

———. 1957. The three types of evolutionary process. *Nature,* 180:454–455.

Hyman, L. H. 1940. *The Invertebrates: Protozoa through Ctenophora.* McGraw–Hill, New York.

———. 1959. *The Invertebrates: Smaller Coelomate Groups.* McGraw–Hill, New York.

Inger, R. F. 1957. Ecological aspects of the origins of the tetrapods. *Evolution,* 11:373–376.

———. 1958. Comments on the definition of genera. *Evolution,* 12:370–384.

Ives, P. T. 1950. The importance of mutation rates in evolution. *Evolution,* 4:236–252.

Jacob, F., and J. Monod. 1961. Genetic regulatory mechanisms in the synthesis of proteins. *J. Mol. Biol.,* 2:318–356.

Jägersten, G. 1955. On the early phylogeny of the Metazoa: The bilaterogastrula theory. *Zool. Bidr. Uppsala,* 30:321–354.

Jahn, T. L., E. C. Bovee, and D. L. Griffith. 1974. Taxonomy and evolution of the Sarcodina: A reclassification. *Taxon,* 23:483–496.

Jeffries, R. P. S. 1967. Some fossil chordates with echinoderm affinities. *Symp. Zool. Soc. London,* 20:163–208.

———. 1968. The subphylum Calcichordata (Jeffries, 1967), primitive fossil chordates with echinoderm affinities. *Bull. Brit. Mus. Geol.,* 16:243–339.

Johannsen, W. 1909. *Elemente der exakten Erblichkeitslehre.* Gustav Fischer, Jena.

Johnson, F. M., C. G. Kanapi, R. H. Richardson, M. R. Wheeler, and W. S. Stone. 1966. An analysis of polymorphisms among isozyme loci in dark and light *Drosophila ananassae* strains from American and Western Samoa. *Proc. Nat. Acad. Sci. U.S.A.,* 56:119–125.

Johnson, M. S. 1971. Adaptive lactate dehydrogenase variation in the crested blenny, *Anoplorchus. Heredity,* 27:205–226.

Johnson, P. H., A. S. Lee, and R. L. Sinsheimer. 1973. Production of specific fragments of φX174 replicative form DNA by a restriction enzyme from *Haemophilus parainfluenzae,* endonuclease HP. *J. Virol.,* 11:596–599.

Joklik, W. K. 1974. Evolution in viruses. In "Evolution in the Bacterial World," *24th Symp. Soc. Gen. Microbiol.,* M. J. Carlile and J. J. Skehel, eds., Cambridge University Press, pp. 293–320.

Jørgensen, C. B. 1966. *Biology of Suspension Feeding.* Pergamon Press, Oxford.

Jukes, T. H., and R. Holmquist. 1972. Evolutionary clock: Nonconstancy of rate in different species. *Science,* 177:530–532.

Kamemoto, H., and K. Shindo. 1964. Meiosis in interspecific and intergeneric hybrids of *Vanda. Bot. Gaz.,* 125:132–138.

Karpechenko, G. D. 1927. Polyploid hybrids of *Raphanus sativus* L. × *Brassica oleracea* L. *Zeit. ind. Abst. Vererbungslehre,* 48:1–58.

Kauffman, S. 1971. Gene regulation networks: A theory for their global structures and behaviors. In *Current Topics in Developmental Biology,* A. A. Moscona and A. Monroy, eds., Academic Press, New York, pp. 145–182.

Kedes, L. H., and M. L. Brinstiel. 1971. Reiteration and clustering of DNA sequences complementary to histone messenger RNA. *Nature New Biol.,* 230:165–169.

Kennedy, W. P. 1967. Epidemiologic aspects of the problem of congenital malformations. *Birth Defects,* 3(2):1–18.

Kerkut, G. A. 1960. *The Implications of Evolution.* Pergamon Press, Oxford.

Kerr, W. E., and Z. V. da Silveira. 1972. Karyotypic evolution of bees and corresponding taxonomic implications. *Evolution,* 26:197–202.

Kessler, K. F. 1880. On the law of mutual help. (In Russian) *Bull. St. Petersburg Soc. Naturalist,* 11:124–136.

Kessler, S. 1969. The genetics of *Drosophila* mating behavior. II. The genetic architecture of mating speed in *Drosophila pseudoobscura. Genetics,* 64:421–433.

Kimura, M., and T. Ohta. 1971. Protein polymorphism as a phase of molecular evolution. *Nature,* 229:467–469.

———. 1972. Population genetics, molecular biometry, and evolution. In *Proc. Sixth Berkeley Symp. Math. Stat. Prob.,* vol. 5, pp. 43–68.

King, J. L., and T. H. Jukes. 1969. Non-Darwinian evolution. *Science,* 164:788–798.

King, M. C., and A. C. Wilson. 1975. Evolution at two levels: Molecular similarities and biological differences between humans and chimpanzees. *Science,* 188:107–116.

Kitzmiller, J. B., G. Frizzi, and R. H. Baker. 1967. Evolution and speciation within the *maculipennis* complex of the genus *Anopheles.* In *Genetics of Insect Vectors of Disease,* J. W. Wright and R. Pal, eds., Elsevier, Amsterdam.

Klein, R. M., and A. Cronquist. 1967. A consideration of the evolutionary and taxonomic significance of some biochemical, micromorphological, and physiological characters in the thallophytes. *Quart. Rev. Biol.,* 42:105–296.

Koehn, R. K., and D. J. Rasmussen. 1967. Polymorphic and monomorphic serum esterase heterogeneity in catastomid fish populations. *Biochem. Gen.,* 1:131–144.

Koestler, A., and J. R. Smythies. 1969. *Beyond Reductionism.* Hutchinson, London.

Kohne, D. E., J. A. Chiscon, and B. H. Hoyer. 1972. Evolution of primate DNA sequences. *J. Human Evol.* 1:627–644.

Kojima, K. 1971. Is there a constant fitness value for a given genotype? *Evolution,* 25:281–285.

Komai, T. 1963. A note on the phylogeny of the Ctenophora. In *The Lower Metazoa, Comparative Biology and Phylogeny,* E. C. Dougherty *et al.,* eds., University of California Press, pp. 181–188.

Koopman, K. F. 1950. Natural selection for reproductive isolation between *Drosophila pseudoobscura* and *Drosophila persimilis. Evolution,* 4:135–148.

Kozhov, M. 1963. Lake Baikal and its life. *Monographiae Biologicae* 11.

Kretchimer, N. 1972. Lactose and lactase. *Sci. Amer.* 227(4):70–78.

Kriss, A. E. 1963. *Marine Microbiology (Deep Sea).* Wiley, New York.

Kropotkin, E. 1902. *Mutual Aid. A Factor in Evolution.* Doubleday, New York.

Kubia, D. F., and H. Ris. 1969. Division of the dinoflagellate *Gyrodinium cohnii* (Schiller): A new type of cellular reproduction. *J. Cell Biol.,* 40:508–528.

Kuhn, T. S. 1962. *The Structure of Scientific Revolutions.* University of Chicago Press.

Kullenberg, B., and G. Bergström. 1973. The pollination of *Ophrys* orchids. In *Nobel Symposium 25: Chemistry in Botanical Classification,* pp. 253–258.

Kurtén, B. 1967. Continental drift and the palaeogeography of reptiles and mammals. *Commentationes Biologica, Soc. Scient. Fennica,* 31:1–8.

Lack, D. 1947. *Darwin's Finches.* University Press, Cambridge.

Laird, C. D., and B. J. McCarthy. 1968. Magnitude of interspecific nucleotide sequence variability in *Drosophila. Genetics,* 60:303–322.

Laird, C. D., B. L. McConaughy, and B. J. McCarthy. 1969. Rate of fixation of nucleotide substitutions in evolution. *Nature,* 224:149–154.

Lamarck, J. B. 1809. *Zoological Philosophy,* Translated by H. Elliot. Reprinted 1963, Hafner, New York.

Lane, C. D., G. Marbaix, and J. B. Gurdon. 1971. Rabbit haemoglobin synthesis in frog cells: The translation of reticulocyte 9S RNA in frog oocytes. *J. Mol. Biol.,* 61:73–91.

Lang, A. 1884. Die Polycladen des Golfes von Neapel. *Fauna Flora Golfes Neapel,* Monogr. 11.

Langer, S. R. 1966. The lord of creation. In *The Borzoi College Reader,* C. Muscatine and M. Griffith, eds. Knopf, New York.

Langley, C. H., and W. M. Fitch. 1974. An examination of the constancy of the rate of molecular evolution. *J. Mol. Evol.,* 3:161–177.

Laporte, L. F. 1968. *Ancient Environments.* Prentice–Hall, Englewood Cliffs, New Jersey.

Lederberg, J., and E. M. Lederberg. 1952. Replica plating and indirect selection of bacterial mutants. *J. Bacteriol.,* 63:399–406.

Leedale, G. F. 1967. *Euglenoid Flagellates.* Prentice–Hall, Englewood Cliffs, New Jersey.

Lemche, H., and K. G. Wingstrand. 1959. The anatomy of *Neopilina galathaea* Lemche, 1957 (Mollusca: Tryblidiacea). *Galathea Report* (Copenhagen), 3:1–71.

Lenz, L., and D. E. Wimber. 1959. Hybridization and inheritance in orchids. In *The Orchids,* C. Withner, ed., Ronald Press, New York.

Lerner, I. M. 1958. *The Genetic Basis of Selection.* Wiley, New York.

———. 1968. *Heredity, Evolution, and Society.* Freeman, San Francisco.

Levene, H. 1953. Genetic equilibrium when more than one ecological niche is available. *Amer. Nat.,* 87:331–333.

Levene, H., O. Pavlovsky, and Th. Dobzhansky. 1954. Interaction of the adaptive values in polymorphic experimental populations of *Drosophila pseudoobscura*. *Evolution,* 8:335–349.

Levin, D. A. 1970. The exploitation of pollinators by species and hybrids of *Phlox*. *Evolution,* 24:367–377.

———. 1975. Somatic cell hybridization: Application in plant systematics. *Taxon,* 24:261–270.

Levins, R. 1968. *Evolution in Changing Environments.* Princeton University Press.

Levy, B., and B. J. McCarthy. 1975. Messenger RNA complexity in *Drosophila melanogaster. Biochemistry,* 14:2440–2446.

Levy, M., and D. A. Levin. 1971. The origin of novel flavonoids in *Phlox* allotetraploids (glycosidating enzymes/gene repressions). *Proc. Nat. Acad. Sci. U.S.A.,* 68: 1627–1630.

Lewis, H. 1966. Speciation in flowering plants. *Science,* 152:167–172.

Lewis, H., and W. L. Bloom. 1972. The loss of a species through breakdown of a chromosomal barrier. *Symp. Biol. Hungarica,* 12:61–64.

Lewontin, R. C. 1968. Evolution. In *International Encyclopedia of the Social Sciences,* vol. 5, pp. 202–209. Macmillan, New York.

———. 1972. The apportionment of human diversity. *Evol. Biol.,* 6:381–398.

———. 1974. *The Genetic Basis of Evolutionary Change.* Columbia University Press.

Lewontin, R. C., and J. L. Hubby. 1966. A molecular approach to the study of genic heterozygosity in natural populations. II. Amount of variation and degree of heterozygosity in natural populations of *Drosophila pseudoobscura*. *Genetics,* 54:595–609.

Li, C. C. 1955. *Population Genetics.* University of Chicago Press.

Liapunova, E. A., and N. N. Vorontsov. 1970. Chromosomes and some issues of the evolution of the ground squirrel genus *Citellus* (Rodentia: Sciuridae). *Experientia,* 26:1033–1038.

Lima-de-Faria, A. 1956. The role of the kinetochore in chromosome organization. *Hereditas,* 42:85–160.

Lin, C. C., B. Chiarelli, L. E. M. De Boer, and M. M. Cohen. 1973. A comparison of the fluorescent karyotypes of the chimpanzee (*Pan troglodytes*) and man. *J. Human Evol.,* 2:311–321.

Lipkin, M., and P. T. Rowley, eds. 1974. *Genetic Responsibility.* Plenum, New York.

Littlejohn, M. J. 1965. Premating isolation in the *Hyla ewingi* complex (Anura, Hylidae). *Evolution,* 19:234–243.

Loeblich, A. R., Jr., and H. Tappan. 1964. Foraminiferida. In *Treatise on Invertebrate Paleontology, Part C,* R. C. Moore, ed. Protista, Lawrence, Kansas.

Lorenz, K. 1966. *On Aggression.* Harcourt, Brace, and World, New York.

Lull, R. S. 1922. *Organic Evolution.* Macmillan, New York.

Luria, S. E., and M. Delbrück. 1943. Mutations of bacteria from virus sensitivity to virus resistance. *Genetics,* 28:491–511.

MacArthur, R. H., and E. O. Wilson. 1967. *The Theory of Island Biogeography.* Princeton University Press.

MacGillavray, H. J. 1968. Modes of evolution mainly among marine invertebrates. *Bijdragen tot de Dierkunde,* 38:69–74.

Mackie, G. O. 1963. Siphonophores, bud colonies, and superorganisms. In *The Lower Metazoa, Comparative Biology and Phylogeny,* E. C. Dougherty *et al.,* eds., University of California Press, pp. 329–337.

Mandel, M. 1969. New approaches to bacterial taxonomy: Perspective and prospects. *Ann. Rev. Microbiol.,* 23:239–274.
Manser, A. R. 1965. The concept of evolution. *Philosophy,* 40:18–34.
Manwell, C., and C. M. A. Baker. 1970. *Molecular Biology and the Origin of Species.* University of Washington Press.
Maren, T. H. 1967. Carbonic anhydrase: Chemistry, physiology, and inhibition. *Physiol. Rev.,* 47:595–781.
Margalef, R. 1968. *Perspectives in Ecological Theory.* University of Chicago Press.
Margoliash, E. 1972. The molecular variations of cytochromes *c* as a function of the evolution of species. *The Harvey Lectures,* 66:177–247.
Margulis, L. 1970. *Origin of Eukaryotic Cells.* Yale University Press.
Marinković, D. 1967a. Genetic loads affecting fecundity in natural populations of *D. pseudoobscura. Genetics,* 56:61–71.
———. 1967b. Genetic loads affecting fertility in natural populations of *Drosophila pseudoobscura. Genetics,* 57:701–709.
Marinković, D., and F. J. Ayala. 1975a. Fitness of allozyme variants in *Drosophila pseudoobscura.* I. Selection at the *Pgm-1* and *Me-2* loci. *Genetics,* 79:85–95.
———. 1975b. Fitness of allozyme variants in *Drosophila pseudoobscura.* II. Selection at the *Est-5, Odh,* and *Mdh-2* loci. *Genetical Research,* 24:137–149.
Marmur, J., R. Rownd, and C. L. Schildkraut. 1963. Denaturation and renaturation of deoxyribonucleic acid. *Prog. Nucl. Acid Res.,* 1:231–300.
Massaro, E. J., and C. L. Markert. 1968. Isozyme patterns of salmonid fishes: Evidence for multiple cistrons for lactate dehydrogenase polypeptides. *J. Exp. Zool.,* 168:223–238.
Matthews, S. C., and V. V. Missarzhevsky. 1975. Small shelly fossils of late Precambrian and early Cambrian age: A review of recent work. *Jour. Geol. Soc. London,* 131:289–304.
Maynard Smith, J. 1970. Genetic polymorphism in a varied environment. *Amer. Nat.,* 104:230–234.
Maynard Smith, J., and K. C. Sondhi. 1960. The genetics of a pattern. *Genetics,* 45:1039–1050.
Mayr, E. 1931. Birds collected during the Whitney South Sea Expedition. XII. Notes on *Halcyon chloris* and some of its subspecies. *Amer. Mus. Novitates,* 469:1–10.
———. 1942. *Systematics and the Origin of Species.* Columbia University Press.
———. 1960. The emergence of evolutionary novelties. In *Evolution after Darwin,* vol. 1, S. Tax, ed., University of Chicago Press, pp. 349–380.
———. 1963. *Animal Species and Evolution.* Harvard University Press.
———. 1964. Introduction. In *On the Origin of Species,* C. Darwin, Harvard University Press.
———. 1965. Cause and effect in biology. In *Cause and Effect,* D. Lerner, ed., Free Press, New York, pp. 33–50.
———. 1969. *Principles of Systematic Zoology.* McGraw–Hill, New York.
———. 1970. *Populations, Species and Evolution.* Harvard University Press.
———. 1972. Sexual selection and natural selection. In *Sexual Selection and the Descent of Man,* B. G. Campbell, ed., Aldine, Chicago. pp. 87–104.
———. 1974a. Cladistic analysis or cladistic classification? *Z. Zool. Syst. Evolutionsforsch.,* 12:94–128.
———. 1974b. Teleological and teleonomic, a new analysis. In *Boston Studies in the Philosophy of Science,* XIV, R. S. Cohen and M. W. Wartofsky, eds., pp. 91–117. Reidel, Boston.

McAlester, A. L. 1970. Animal extinctions, oxygen consumption, and atmospheric history. *J. Paleon.,* 44:405–409.

McCarthy, B. J., and R. B. Church. 1970. The specificity of molecular hybridization reactions. *Ann. Rev. Biochem.,* 39:131–150.

McCarthy, B. J., and M. N. Farquhar. 1974. The rate of change of DNA in evolution. In "Evolution of Genetic Systems," *Brookhaven Symp. Biol.* No. 23, pp. 1–41.

McCracken, R. D. 1971. Lactase deficiency: An example of dietary evolution. *Current Anthro.,* 12:479–517.

McDonald, J. F., and F. J. Ayala. 1974. Genetic response to environmental heterogeneity. *Nature,* 250:572–574.

McElroy, W. D., and H. H. Seliger. 1962. Origin and evolution of bioluminescence. In *Horizons in Biochemistry,* M. Kasha and B. Pullman, eds., Academic Press, New York, pp. 91–101.

McKusick, V. A. 1975. *Mendelian Inheritance in Man,* 4th ed. Johns Hopkins Press, Baltimore.

McLaughlin, D. B. 1965. The origin of the earth. In *Stratigraphy and Life History,* M. Kay and E. H. Colbert, eds., Wiley, New York, pp. 669–698.

McLaughlin, P. J., and M. O. Dayhoff. 1970. Eukaryotes versus prokaryotes: An estimate of evolutionary distance. *Science,* 168:1469–1471.

McLean, J. H. 1969. Marine shells of southern California. *Los Angeles Co. Mus. Nat. Hist. Sci. Ser.* No. 24, Zool. 11.

McMillan, C. 1964. Ecotypic differentiation within four North American prairie grasses. I. Morphological variation within transplanted community fractions. *Amer. J. Bot.,* 51:1119–1128.

Meadows, D. H., D. L. Meadows, J. Randers, and W. W. Behrens III. 1972. *The Limits to Growth.* Universe Books, New York.

Medawar, P. B. 1967. *The Art of the Soluble.* Methuen, London.

———. 1969. *Induction and Intuition in Scientific Thought.* American Philosophical Society, Philadelphia.

Meselson, M., and F. W. Stahl. 1958. The replication of DNA in *Escherichia coli. Proc. Nat. Acad. Sci. U.S.A.,* 44:671–682.

Metshnikoff, E. 1883. Researches on the intracellular digestion of invertebrates. *Quart. J. Microsc. Sci.,* 24:89–111.

Middleton, J. H., M. H. Edgell, and C. A. Hutchison. 1972. Specific fragments of ϕX174 deoxyribonucleic acid produced by a restriction enzyme from *Haemophilus aegyptius,* endonuclease Z. *J. Virol.,* 10:42–50.

Milani, R. 1967. The genetics of *Musca domestica* and of other muscoid flies. In *Genetics of Insect Vectors of Disease,* J. W. Wright and R. Pal, eds., Elsevier, Amsterdam, pp. 315–369.

Milkman, R. 1973. Electrophoretic variation in *E. coli* from natural sources. *Science,* 182:1024–1026.

Miller, R. S. 1967. Pattern and process in competition. In *Advances in Ecological Research,* 4:1–74.

Miller, S. J. 1953. A production of amino acids under possible primitive earth conditions. *Science,* 117:528.

Miller, S. J. and L. E. Orgel. 1974. *The Origins of Life on the Earth.* Prentice-Hall, Englewood Cliffs, New Jersey.

Mirov, N. T. 1967. *The Genus Pinus.* Ronald, New York.

Mirsky, A. E., and H. Ris. 1949. Variable and constant components of chromosomes. *Nature,* 163:666–667.

Mirsky, A. E., and H. Ris. 1951. The deoxyribonucleic acid content of animal cells and its evolutionary significance. *J. Gen. Physiol.*, 34:451–462.
Mivart, St. George. 1871. *Genesis of Species.* Macmillan, London.
Moorbath, S., R. K. O'Nions, and R. J. Pankhurst. 1973. Early Archaean age for the Isua iron formation, West Greenland. *Nature,* 245:138–139.
Moore, B. P. 1969. Biochemical studies in termites. In *Biology of Termites,* K. Krishna and F. M. Weesner, eds., Academic Press, New York, pp. 407–432.
Moore, R. C. 1954. Evolution of late Paleozoic invertebrates in response to major oscillations of shallow seas. *Bull. Mus. Comp. Zool. Harvard Coll.,* 112:259–286.
Morgan, L. H. 1877. *Ancient Society.* Holt, New York.
Morgan, T. H. 1932. *The Scientific Basis of Evolution.* Norton, New York.
Morowitz, H. J. 1966. The minimum size of cells. In *Principles of Biomolecular Organisation,* G. E. W. Wolstenholme and M. O'Connor, eds., Churchill, London, pp. 446–462.
Mourão, C. A., F. J. Ayala, and W. W. Anderson. 1972. Darwinian fitness and adaptedness in experimental populations of *Drosophila willistoni. Genetica,* 43:552–574.
Mukai, T. 1964. The genetic structure of natural populations of *D. melanogaster.* I. Spontaneous mutation rate of polygenes controlling viability. *Genetics,* 50:1–19.
———. 1969. The genetic structure of natural populations of *Drosophila melanogaster.* VII. Synergistic interaction of spontaneous mutant polygenes controlling viability. *Genetics,* 61:749–761.
———. 1970. Spontaneous mutation rates of isozyme genes in *D. melanogaster. Drosophila Info. Serv.,* 45:99.
Muller, H. J. 1950. Our load of mutations. *Amer. J. Human Gen.,* 2:111–176.
———. 1967. What genetic course will man steer? In *International Congress of Human Genetics,* J. F. Crow and J. V. Neel, eds., Johns Hopkins Press, Baltimore.
Muller, H. J., and W. D. Kaplan. 1966. The dosage compensation of *Drosophila* and mammals as showing the accuracy of the normal types. *Gen. Res.,* 8:41–59.
Müller, P. 1972. Der neotropische Artenreichtum als biogeographisches Problem. *Zool. Medelingen,* 47:88–110.
Murray, J. 1972. *Genetic Diversity and Natural Selection.* Hafner, New York.
Myrdal, G. 1967. *Asian Drama: An Inquiry into the Poverty of Nations.* Twentieth Century Fund, New York.

Nagel, E. 1961. *The Structure of Science.* Hartcourt, Brace, and World, New York.
Neel, J. V. 1970. Lessons from a "primitive" people. *Science,* 170:815–822.
Nei, M. 1972. Genetic distance between populations. *Amer. Nat.,* 106:283–291.
———. 1975. *Molecular Population Genetics and Evolution.* American Elsevier, New York.
Newell, N. D. 1956. Catastrophism and the fossil record. *Evolution,* 10:97–101.
———. 1967. Revolutions in the history of life. *Geol. Soc. Amer. Spec. Papers* No. 89, pp. 63–91.
Neyfach, A. A. 1971. Steps of realization of genetic information in development. *Current Topics Devel. Biol.,* 6:45–77.
Nobs, M. A. 1963. Experimental studies on species relationships in *Ceanothus. Carn. Inst. Wash. Publ.* No. 623, pp. 1–94.

Oakley, K. P. 1961. *Man the Tool-maker.* British Museum of Natural History, London.
Ohno, S. 1969. The preferential activation of maternally derived alleles in development of interspecific hybrids. In *Heterospecific Genome Interaction,* V. Defendi, ed., pp. 137–150. Wistar Institute Press, Philadelphia.

Ohno, S. 1970. *Evolution by Gene Duplication.* Springer, New York.

Oparin, A. I. 1953. *The Origin of Life.* Dover, New York.

Orgel, L. E. 1973. *The Origins of Life: Molecules and Natural Selection.* Wiley, New York.

Ornduff, R., B. Bohm, and N. A. M. Saleh. 1973. Flavonoids of artificial interspecific hybrids of *Lasthenia. Biochem. Syst.,* 1:147–151.

Osborn, H. F. 1934. Aristogenesis, the creative principle in the origin of species. *Amer. Nat.,* 68:193–235.

Paecht–Horowitz, M., J. Berger, and A. Katchalsky. 1970. Prebiotic synthesis of polypeptides by heterogeneous polycondensation of amino acid adenylates. *Nature,* 228:636.

Paine, R. T. 1966. Food web complexity and species diversity. *Amer. Nat.,* 100: 65–75.

Passmore, J. 1974. *Man's Responsibility for Nature: Ecological Problems and Western Traditions.* Duckworth, London.

Pentzos–Daponte, A., E. Boesiger, and A. Kanellis. 1967. Fréquences de gènes mutants dans plusieurs populations naturelles de *Drosophila subobscura* de Grèce. *Ann. Fac. Sci. Univ. Aristot. Thessaloniki.,* 10:133–152.

Petit, C., and L. Ehrman. 1969. Sexual selection in *Drosophilia. Evol. Biol,* 3:177–223.

Pickett–Heaps, J. D., and H. J. Marchant. 1972. The phylogeny of the green algae: A new proposal. *Cytobios,* 6:255–264.

Pigott, G. H., and N. G. Carr. 1972. Homology between nucleic acids of blue-green algae and chloroplasts of *Euglena gracilis. Science,* 175:1259–1261.

Pilbeam, D. 1972. *The Ascent of Man.* Macmillan, New York.

Pitrat, C. W. 1974. Vertebrates and the Permo-Triassic extinction. *Paleogeogr., Paleoclimat., Paleoec.,* 14:249–264.

Pittendrigh, C. S. 1958. Adaptation, natural selection, and behavior. In *Behavior and Evolution,* A. Roe and G. G. Simpson, eds., pp. 390–416. Yale University Press.

Popp, R. A. 1969. Studies on the mouse hemoglobin loci. X. Linkage of duplicate genes at the α-chain locus, *Hba. J. Heredity,* 60:131–133.

Popper, K. R. 1934. *Logik der Forschung.* Vienna. English edition, *The Logic of Scientific Discovery,* 1959, Hutchinson, London.

———. 1974. Scientific reduction and the essential incompleteness of all science. In *Studies in the Philosophy of Biology,* F. J. Ayala and T. Dobzhansky, eds., Macmillan, London.

Powell, J. R. 1971. Genetic polymorphisms in varied environments. *Science,* 174: 1035–1036.

Powell, J. R., H. Levene, and Th. Dobzhansky. 1973. Chromosomal polymorphism in *Drosophila pseudoobscura* used for diagnosis of geographic origin. *Evolution,* 26:553–559.

Prakash, S., R. C. Lewontin, and J. L. Hubby. 1969. A molecular approach to the study of genic heterozygosity in natural populations. IV. Patterns of genic variation in central, marginal and isolated populations of *Drosophila pseudoobscura. Genetics,* 61:841–858.

Premack, D. 1971. Language in chimpanzee? *Science,* 172:808–822.

Preston, N. 1968. Plants without cellulose. *Sci. Amer.,* 218(6):102–108.

Pruitt, D. G., and R. C. Snyder, eds. 1969. *Theory and Research on the Causes of War.* Prentice–Hall, Englewood Cliffs, New Jersey.

Punnett, R. C. 1930. Genetics, mathematics, and natural selection. *Nature,* 126: 595–597.

Raff, R. A., and H. R. Mahler. 1972. The non-symbiotic origin of mitochondria. *Science,* 177:575–582.

Raper, J. R., and A. S. Flexer. 1970. The road to diploidy with emphasis on a detour. *Symp. Soc. Gen. Microbiol, Prokaryotic and Eukaryotic Cells,* pp. 401–432.

Raup, D. 1973. Taxonomic diversity during the Phanerozoic. *Science,* 177:1065–1071.

Raven, P. H. 1970. A multiple origin for plastids and mitochondria. *Science,* 169: 641–646.

Raven, P., and H. Curtis. 1970. *Biology of Plants.* Worth, New York.

Rees, H. 1974. DNA in higher plants. In "Evolution of Genetic Systems," *Brookhaven Symp. Biol.* No. 23, pp. 394–418.

Rees, H., and M. H. Hazarika. 1969. Chromosome evolution in *Lathyrus. Chrom. Today,* 2:158–165.

Rees, W. J. 1957. Evolutionary trends in the classification of the capitate hydroids. *Bull. Br. Mus. (Nat. Hist.) Zool.,* 4:1–60.

———. 1966. The evolution of the hydrozoa. In "The Cnidaria and their Evolution," W. J. Rees, ed. *Symp. Zool. Soc. London,* 16:199–221.

Reichlin, M., M. Hay, and L. Levine. 1964. Antibodies to human A_1 hemoglobin and their reaction with A_2, S, C, and H hemoglobins. *Immunochemistry,* 1:21–30.

Rensch, B. 1947. *Evolution Above the Species Level.* Columbia University Press.

———. 1954. *Neuere Probleme der Abstammungslehre.* Enke, Stuttgart.

Rhoads, D. C., and J. W. Morse. 1971. Evolutionary and ecologic significance of oxygen-deficient marine basins. *Lethaia,* 4:413–428.

Rhodes, F. H. T. 1967. Permo–Triassic extinction. In *The Fossil Record,* W. B. Harland *et al.,* eds., pp. 57–76. Geological Society of London.

Richardson, E. S., Jr., and R. G. Johnson. 1970. The Mazon Creek faunas. *N. Amer. Paleont. Convention, Chicago, 1969, Proc.,* I:1222–1235.

Riek, E. F. 1970. Fossil history. In *The Insects of Australia,* Division of Entomology, Commonwealth Scientific and Industrial Research Organization (CSIRO), Canberra, pp. 168–186.

Ris, H., and D. F. Kubai. 1974. An unusual mitotic mechanism in the parasitic protozoan *Syndinium* sp. *J. Cell. Biol.,* 60:702–720.

Ritossa, F. M., K. C. Atwood, and S. Spiegelman. 1966. On the redundancy of DNA complementary to amino acid transfer RNA and its absence from the nucleolar organizer region of *Drosophila melanogaster. Genetics,* 54:663–676.

Ritossa, F. M., and S. Spiegelman. 1965. Localization of DNA complementary to ribosomal RNA in the nucleolus organizer region of *Drosophila melanogaster. Proc. Nat. Acad. Sci. U.S.A.,* 53:737–745.

Rizzo, P. J., and L. D. Nooden. 1972. Chromosomal proteins in the dinoflagellate alga *Gyrodinium cohnii. Science,* 176:796–797.

Roberts, D. F. 1953. Body weight, race, and climate. *Amer. J. Phys. Anthro.,* 11: 533–558.

Robertson, F. W. 1959. Studies in quantitative inheritance. XII. Cell size and number in relation to genetic and environmental variation of body size in Drosophila. *Genetics,* 44:869–896.

Robson, G. C., and O. W. Richards. 1936. *The Variations of Animals in Nature.* Longmans Green, London.

Romer, A. S. 1958. Tetrapod limbs and early tetrapod life. *Evolution,* 12:365–369.

———. 1959. *The Vertebrate Story* (4th ed. of *Man and the Vertebrates*). University of Chicago Press.

Ronov, A. B. 1968. Probable change in the composition of seawater during the course of geological time. *Sedimentology,* 10:25–43.

Sanders, H. L. 1968. Marine benthic diversity: A comparative study. *Amer. Nat.,* 102:243–282.

Sanger, F., and E. O. P. Thompson. 1953. The amino acid sequence in the glycyl chain of insulin. *Biochem. J.,* 53:353–374.

Sankaranarayanan, K. 1964. Genetic loads in irradiated experimental populations of *Drosophila melanogaster. Genetics,* 50:131–150.

———. 1966. Some components of the genetic loads in irradiated experimental populations of *Drosophila melanogaster. Genetics,* 54:121–130.

Sarich, V. M., and A. C. Wilson. 1966. Quantitative immunochemistry and the evolution of primate albumins: Microcomplement fixation. *Science,* 154:1563–1566.

———. 1967. Immunological time scale for hominid evolution. *Science,* 158:1200–1203.

Schaller, G. B. 1964. *The Year of the Gorilla.* University of Chicago Press.

Schildkraut, C., J. Marmur, and P. Doty. 1961. The formation of hybrid DNA molecules and their use in studies of DNA homologies. *J. Mol. Biol.,* 3:595–617.

Schildkraut, C., L. Wierzchowski, J. Marmur, D. M. Green, and P. Doty. 1962. A study of the base sequence homology among the T series of bacteriophages. *Virology,* 18:43–55.

Schindewolf, O. H. 1936. *Paläontologie, Entwicklungslehre und Genetik. Kritik und Synthese.* Borntraeger, Berlin.

———. 1950. *Grundfragen der Paläontologie.* Schweizerbart, Stuttgart.

Schnepf, E., and R. M. Brown, Jr. 1971. On relationships between endosymbiosis and the origin of plastids and mitochondria. In *Origin and Continuity of Cell Organelles,* J. Reinert and H. Ursprung, eds., Springer, New York, pp. 299–322.

Schopf, J. W. 1970. Precambrian micro-organisms and evolutionary events prior to the origin of vascular plants. *Biol. Rev.,* 45:319–352.

———. 1972. Evolutionary significance of the Bitter Springs (Late Precambrian) microflora. *Proc. XXIV Int. Geol. Congress, Sect. 1, Precambrian Geol.,* Montreal, pp. 68–77.

———. 1974. Paleobiology of the Precambrian: The age of blue-green algae. *Evol. Biol.,* 7:1–43.

———. 1975. The age of microscopic life. *Endeavour,* 122:51–58.

Schopf, J. W., and J. M. Blacic. 1971. New microorganisms from the Bitter Springs Formation (Late Precambrian) of the north-central Amadeus Basin, Australia. *J. Paleontology,* 45:925–960.

Schopf, J. W., B. N. Haugh, R. E. Molnar, and D. F. Satterthwait. 1973. On the development of metaphytes and metazoans. *J. Paleon.,* 47:1–9.

Schopf, T. J. M., and L. S. Murphy. 1973. Protein polymorphism of the hybridizing seastars *Asterias forbesi* and *Asterias vulgaris* and implications for their evolution. *Biol. Bull.,* 145:589–597.

Schull, W. J., and J. V. Neel. 1965. *The Effects of Inbreeding on Japanese Children.* Harper and Row, New York.

Schultz, R. J. 1969. Hybridization, unisexuality, and polyploidy in the teleost *Poeciliopsis* (Poeciliidae) and other vertebrates. *Amer. Nat.,* 103:605–619.
Sclater, W. L., and P. L. Sclater. 1899. *The Geography of Mammals.* Kegan, Paul, Trench, Trübner, London.
Selander, R. K. 1976. Genic variation in natural populations. In *Molecular Evolution,* F. J. Ayala, ed., Sinauer, Sunderland, Mass., pp. 21–45.
Selander, R. K., and D. W. Kaufman. 1973. Self-fertilization and genetic population structure in a colonizing land snail. *Proc. Nat. Acad. Sci. U.S.A.,* 70:1186–1190.
Selander, R. K., S. Y. Yang, and W. G. Hunt. 1969. Polymorphism in esterases and hemoglobin in wild populations of the house mouse (*Mus musculus*). *Studies in Genetics,* V., University Texas Publ. No. 6918, pp. 271–338.
Selander, R. K., M. H. Smith, S. Y. Yang, W. E. Johnson, and J. B. Gentry. 1971. Biochemical polymorphisms and systematics in the genus *Peromyscus.* Univ. Texas Publ. No. 7103, pp. 49–90.
Selander, R. K., S. Y. Yang, R. C. Lewontin, and W. E. Johnson. 1970. Genetic variation in the horseshoe crab (*Limulus polyphemus*), a phylogenetic "relic." *Evolution,* 24:402–414.
Shapley, H. 1958. *Of Stars and Men.* Beacon Press, Boston.
Sherrington, C. 1953. *Man on His Nature.* Doubleday-Anchor, Garden City.
Shields, G. F., and N. A. Straus. 1975. DNA–DNA hybridization studies of birds. *Evolution,* 29:159–166.
Shklovskii, I. S., and C. Sagan. 1966. *Intelligent Life in the Universe.* Holden–Day, San Francisco.
Shull, A. F. 1936. *Evolution.* McGraw–Hill, New York.
Sibley, C. G. 1954. Hybridization in the red-eyed towhees of Mexico. *Evolution,* 8:252–290.
Simons, E. L. 1972. *Primate Evolution.* Macmillan, New York.
Simpson, G. G. 1944. *Tempo and Mode in Evolution.* Columbia University Press.
———. 1949. *The Meaning of Evolution.* Yale University Press.
———. 1953. *The Major Features of Evolution.* Columbia University Press.
———. 1961. *Principles of Animal Taxonomy.* Columbia University Press.
———. 1964. *This View of Life.* Harcourt, Brace, and World, New York.
———. 1968. *Evolution and Geography.* Oregon State System of Higher Education, Eugene.
Simpson, G. G., and W. S. Beck. 1965. *Life,* 2nd ed. Harcourt, Brace, and World, New York.
Smart, J. J. C. 1963. *Philosophy and Scientific Realism.* Humanities Press, New York.
Smith, E. B. 1970. Pollen competition and relatedness in *Haplopappus* section *Isopappus* (Compositae). *Amer. J. Bot.,* 57:874–880.
Smith, P. F. 1971. *The Biology of Mycoplasmas.* Academic Press, New York.
Sneath, P. H. A. 1974. Phylogeny of microorganisms. In "Evolution in the Microbial World," *24th Symp. Soc. Gen. Microbiol.,* M. J. Carlile and J. J. Skehel, eds., Cambridge University Press, pp. 1–39.
Sneath, P. H. A., and R. R. Sokal. 1973. *Numerical Taxonomy.* Freeman, San Francisco.
Sonea, S. 1972. Bacterial plasmids instrumental in the origin of eukaryotes? *Rev. Can. Biol.,* 31:61–63.
Soyer, M. O. 1972. Les ultrastructures nucléaires de la Noctiluque (Dinoflagellé libre) au cours de sporogenèse. *Chromosoma* (Berl.), 39:419.

Soyer, M. O., and O. K. Haapala. 1974. Structural changes of dinoflagellate chromosomes by pronase and ribonuclease. *Chromosoma* (Berl.), 47:179–192.

Spassky, B., R. C. Richmond, S. Pérez–Salas, O. Pavlovsky, C. A. Mourão, A. S. Hunter, H. Hoenigsberg, T. Dobzhansky, and F. J. Ayala. 1971. Geography of the sibling species related to *Drosophila willistoni,* and the semispecies of the *Drosophila paulistorum* complex. *Evolution,* 25:129–143.

Spencer, W. P. 1957. Genetic studies on *Drosophila mulleri.* I. Genetic analysis of a population. Univ. Texas Publ. No. 5721, pp. 186–205.

Spieth, P. T. 1975. Population genetics of allozyme variation in *Neurospora intermedia. Genetics,* 80:785–805.

Stahl, B. J. 1974. *Vertebrate History: Problems in Evolution.* McGraw–Hill, New York.

Stanier, R. Y. 1974. The origins of photosynthesis in eukaryotes. In "Evolution in the Microbial World," *24th Symp. Soc. Gen. Microbiol.* Cambridge University Press, pp. 219–240.

Stanier, R. Y., M. Doudoroff and E. A. Adelberg. 1970. *The Microbial World.* Prentice-Hall, Englewood Cliffs, New Jersey.

Stanley, J. M., 1973. Effects of competition on rates of evolution, with special reference to bivalve mollusks and mammals. *Syst. Zool.,* 22:486–506.

Stanley, W. M., and E. G. Valens. 1961. *Viruses and the Nature of Life.* Dutton, New York.

Stebbins, G. L. 1950. *Variation and Evolution in Plants.* Columbia University Press.

———. 1958. The inviability, sterility, and weakness of interspecific hybrids. *Adv. in Gen.,* 9:147–215.

———. 1959. The role of hybridization in evolution. *Proc. Amer. Phil. Soc.,* 103: 231–251.

———. 1960. The comparative evolution of genetic systems. In *Evolution After Darwin,* vol. 1, S. Tax, ed., University of Chicago Press, pp. 197–226.

———. 1966. *Processes of Organic Evolution.* Prentice–Hall, Englewood Cliffs, New Jersey.

———. 1969. *The Basis of Progressive Evolution.* University of North Carolina Press.

———. 1970. Variation and evolution in plants: Progress during the past twenty years. In *Essays in Honor of Th. Dobzhansky,* M. K. Hecht and W. C. Steere, eds., Plenum, New York, pp. 173–208.

———. 1971. *Chromosomal Evolution in Higher Plants.* E. Arnold, London.

———. 1974a. *Flowering Plants: Evolution Above the Species Level.* Harvard University Press.

———. 1974b. Adaptive radiation and the origin of form in the earliest multinuclear organisms. *Syst. Zool.,* 22:478–485.

Stebbins, G. L., and L. Ferlan. 1956. Population variability, hybridization, and introgression in some species of *Ophrys. Evolution,* 10:32–46.

Steinman, H. M., and R. L. Hill. 1973. Sequence homologies among bacterial and mitochondrial superoxide dismutases. *Proc. Nat. Acad. Sci. U.S.A.,* 70:3725.

Stephens, S. G. 1950. The internal mechanisms of speciation in *Gossypium. Bot. Rev.,* 16:115–149.

Stern, C. 1973. *Principles of Human Genetics,* 3rd ed. Freeman, San Francisco.

Stevenson, A. C. 1959. The load of hereditary defect in human populations. *Radiation Res.* (suppl.), 1:306–325.

———. 1961. Frequency of congenital and hereditary disease. *Brit. Med. Bull.,* 17:254–259.

Straw, R. M. 1956. Floral isolation in *Penstemon. Amer. Nat.,* 90:47–53.

Sturtevant, A. H., and E. Novitsky. 1941. The homologies of the chromosome elements in the genus *Drosophila*. *Genetics*, 26:517–541.

Stutz, H. C., and L. K. Thomas. 1964. Hybridization and introgression between *Cowania* and *Purshia*. *Evolution*, 18:183–195.

Suomalainen, E., and A. Saura. 1973. Genetic polymorphism and evolution in parthenogenetic animals. I. Polyploid Curculionidae. *Genetics*, 74:489–508.

Sved, J. A. 1968. Possible rates of gene substitution in evolution. *Amer. Nat.*, 102: 283–292.

———. 1971. An estimate of heterosis in *Drosophila melanogaster*. *Gen. Res.*, 18: 97–105.

Sved, J. A., and F. J. Ayala. 1970. A population cage test for heterosis in *Drosophila pseudoobscura*. *Genetics*, 66:97–113.

Sved, J. A., T. E. Reed, and W. F. Bodmer. 1967. The number of balanced polymorphisms that can be maintained in a natural population. *Genetics*, 55:469–481.

Tappan, H. 1974. Protistan phylogeny: Multiple working hypothesis. *Taxon*, 23: 271–276.

Tashian, R. E., M. Goodman, R. E. Ferrell, and R. J. Tanis. 1976. Evolution of carbonic anhydrase in primates and other mammals. In *Molecular Anthropology*, M. Goodman and R. E. Tashian, eds., Plenum, New York.

Tashian, R. E., R. J. Tanis, R. E. Ferrell, S. K. Stroup, and M. Goodman. 1972. Differential rates of evolution in the carbonic anhydrase isozymes in catarrhine primates. *J. Mol. Evol.*, 1:545–552.

Taylor, F. R. J. 1974. Implications and extensions of the serial endosymbiosis theory of the origin of eukaryotes. *Taxon*, 23:229–258.

Taylor, J. H. 1969. Replication and organization of chromosomes. *Proc. XII Int. Cong. Gen.*, 3:177–189.

Taylor, J. H., P. S. Woods, and W. L. Hughes. 1957. The organization and duplication of chromosomes as revealed by autoradiographic studies using tritium-labelled thymidine. *Proc. Nat. Acad. Sci. U.S.A.*, 43:122–128.

Teigs, O. W., and S. M. Manton. 1958. The evolution of the Arthropods. *Biol. Rev.*, 33:255–337.

Teilhard de Chardin, P. 1959. *The Phenomenon of Man*. Harper, New York.

Temin, R. G. 1966. Homozygous viability and fertility loads in *Drosophila melanogaster*. *Genetics*, 53:27–46.

Terzaghi, E., Y. Okada, G. Streisinger, J. Emrich, M. Inouye, and A. Tsugita. 1966. Change of a sequence of amino acids in phage T4 lysozyme by acridine-induced mutations. *Proc. Nat. Acad. Sci. U.S.A.*, 56:500–507.

Tinbergen, N. 1968. On war and peace in animals and man. *Science*, 160:1411–1418.

Tobari, Y. N., and K. Kojima. 1972. A study of spontaneous mutation rates at ten loci detectable by starch gel electrophoresis in *Drosophila melanogaster*. *Genetics*, 70:397–403.

Tobias, P. V. 1971. *The Brain in Hominid Evolution*. Columbia University Press.

Tracey, M. L., and F. J. Ayala. 1974. Genetic load in natural populations: Is it compatible with the hypothesis that many polymorphisms are maintained by natural selection? *Genetics*, 77:569–589.

Treffers, H. P., V. Spinelli, and N. O. Belser. 1954. A factor (or mutator gene) influencing mutation rates in *Escherichia coli*. *Proc. Nat. Acad. Sci. U.S.A.*, 40:1064–1071.

Trimble, B. K., and J. H. Doughty. 1974. The amount of hereditary disease in human populations. *Ann. Human Gen.,* **38**:199–209.
Trivers, R. L. 1972. Parental investment and sexual selection. In *Sexual Selection and the Descent of Man,* B. G. Campbell, ed., Aldine, Chicago, pp. 136–179.
Tucker, J. M. 1952. Evolution of the California oak *Quercus alvordiana. Evolution,* **6**:162–180.
Turesson, G. 1922. The genotypical response of the plant species to its habitat. *Hereditas,* **3**:211–350.
Tuttle, R. C., and A. R. Loeblich III. 1974. Genetic recombination in the dinoflagellate *Crypthecodinium* (*Gyrodinium*) *cohnii. Science,* **185**:1061–1062.

Uzzell, T. 1970. Meiotic mechanisms of naturally occurring unisexual vertebrates. *Amer. Nat.,* **104**:433–445.

Valentine, D. H., and S. R. J. Woodell. 1963. Studies in British Primulas. X. Seed incompatibility in intraspecific and interspecific crosses at diploid and tetraploid levels. *New Phytol.,* **62**:125–143.
Valentine, J. W. 1968. Climatic regulation of species diversification and extinction. *Geol. Soc. Amer. Bull.,* **79**:273–276.
———. 1969. Patterns of taxonomic and ecological structure of the shelf benthos during Phanerozoic time. *Palaeontology,* **12**:684–709.
———. 1970. How many marine invertebrate fossil species? A new approximation. *J. Paleon.,* **44**:410–415.
———. 1971. Resource supply and species diversity patterns. *Lethaia,* **4**:51–61.
———. 1973a. Coelomate superphyla. *Syst. Zool.,* **22**:97–102.
———. 1973b. *Evolutionary Paleoecology of the Marine Biosphere.* Prentice–Hall, Englewood Cliffs, New Jersey.
———. 1973c. Plates and provinciality, a theoretical history of environmental discontinuities. In "Organisms and Continents Through Time," N. F. Hughes, ed., *Spec. Papers in Palaeontology* No. 12, and *Syst. Assoc. Publ.,* **9**:79–92.
———. 1973d. Phanerozoic taxonomic diversity: A test of alternate models. *Science,* **180**:1078–1079.
———. 1975. Adaptive strategy and the origin of grades and ground-plans. *Amer. Zool.,* **15**:391–404.
Valentine, J. W., and C. A. Campbell. 1975. Genetic regulation and the fossil record. *Amer. Sci.,* **63**:673–680.
Valentine, J. W., and E. M. Moores. 1970. Plate-tectonic regulation of faunal diversity and sea level: A model. *Nature,* **228**: 657–659.
———. 1972. Global tectonics and the fossil record. *J. Geol.,* **80**:167–184.
Van der Pijl, L., and C. Dodson. 1966. *Orchid Flowers, Their Pollination and Evolution.* University of Miami Press.
Van Lawick-Goodall, J. 1971. *In the Shadow of Man.* Houghton Mifflin, Boston.
Van Valen, L. 1973. A new evolutionary law. *Evol. Theory,* **1**:1–30.
Van Valen, L., and R. E. Sloan. 1965. The earliest primates. *Science,* **150**:743–745.
Vaurie, C. 1950. Notes on some Asiatic nuthatches and creepers. *Amer. Mus. Novit.* No. 1472, pp. 1–39.
Vendrely, R., and C. Vendrely. 1948. La teneur du noyau cellulaire en acide désoxyribonucléique à travers les organes, les individus, et les espèces animales. Techniques et premiers résultats. *Experientia,* **4**:434–436.

Vendrely, R., and C. Vendrely. 1949. La teneur du noyau cellulaire en acide désoxyribonucléique à travers les organes, les individus, et les espèces animales. Étude particuliere des Mamiferes. *Experientia,* 5:327–329.

Villee, C. A., W. F. Walker, Jr., and R. D. Barnes. 1973. *General Zoology,* 4th ed. Saunders, Philadelphia.

Vine, F. J. 1969. Sea-floor spreading—new evidence. *J. Geol. Educ.,* 17:6–16.

Vogel, S. 1972. Pollination von *Orchis papilionacea* L. in den Schwarmbahnen von *Eucera tuberculata* F. *Jahresb. Nat. Vereins Wuppertal,* 25:67–74.

Waddington, C. H. 1960. *The Ethical Animal.* Allen and Unwin, London.

———. 1969. Paradigm for an evolutionary process. In *Towards a Theoretical Biology.* vol. 2, C. H. Waddington, ed., Edinburgh University Press, pp. 106–123.

Wall, J. R., and T. L. York. 1957. Inheritance of seedling cotyledon position in *Phaseolus* species. *J. Hered.,* 48:71–74.

Wallace, B. 1958. The average effect of radiation-induced mutations on viability in *Drosophila melanogaster. Evolution,* 12:532–556.

Wallace, B., and C. Madden. 1953. The frequencies of sub- and supervitals in experimental populations of *Drosophila melanogaster. Genetics,* 38:456–470.

Walter, H. 1971. *Ecology of Tropical and Subtropical Vegetation.* Translated by D. Mueller-Dombois (from the original German edition, 1964). Oliver and Boyd, Edinburgh.

Washburn, S. L. 1966. Conflict in primate society. In *Conflict in Society,* A. de Reuck and J. Knight, eds., Churchill, London, pp. 3–15.

———. 1960. Tools and human evolution. *Sci. Amer.,* 203(9):62–75.

Watson, J. D., and F. H. Crick. 1953a. A structure for deoxyribose nucleic acid. *Nature,* 171:737.

———. 1953b. Genetical implications of the structure of DNA. *Nature,* 171:964.

Webster, T. P., W. P. Hall, and E. E. Williams. 1972. Fission in the evolution of a lizard karyotype. *Science,* 177:611–613.

Wells, M. J. 1968. *Lower Animals.* McGraw–Hill, New York.

Westoff, C. F. 1974. The populations of the developed countries. *Sci. Amer.,* 231(3): 108–120.

Westoll, T. S. 1949. On the evolution of the Dipnoi. In *Genetics, Paleontology, and Evolution,* G. L. Jepson, E. Mayr, and G. G. Simpson, eds., Princeton University Press. pp. 121–184.

Wexelsen, H. 1965. Studies in tetraploid red clover. Fertility and seed yield. Inbreeding and heterosis effects. *Sci. Rep. Agric. Coll. Norway,* 44:1–23.

White, M. J. D. 1973. *Animal Cytology and Evolution,* 3rd ed. Cambridge University Press.

Whiteley, A. H., and H. R. Whiteley. 1972. The replication and expression of maternal and paternal genomes in a blocked echinoid hybrid. *Devel. Biol.,* 29:183–198.

Whiteley, H. R., and A. H. Whiteley. 1975. Changing populations of reiterated DNA transcripts during early echinoderm development. In *Current Topics in Developmental Biology,* A. A. Moscona and A. Mamay, eds., Academic Press, New York.

Whitt, G. S., W. F. Childers, and P. L. Cho. 1973. Allelic expression at enzyme loci in an intertribal hybrid sunfish. *J. Hered.,* 64:55–61.

Whittaker, R. H. 1969. New concepts of kingdoms of organisms. *Science,* 163:150–160.

Wiedmann, J. 1969. The heteromorphs and ammonoid extinction. *Biol. Rev.,* 44:563–602.

Williams, B. J. 1973. *Evolution and Human Origins.* Harper and Row, New York.

Williams, C. B. 1960. The range and pattern of insect abundance. *Amer. Nat.,* 94:137–151.

Williams, G. C. 1966. *Adaptation and Natural Selection.* Princeton University Press.

———. 1975. *Sex and Evolution.* Princeton University Press.

Williams, N. H., and C. H. Dodson. 1972. Selective attraction of male Euglossine bees to orchid floral fragrances and its importance in long-distance pollen flow. *Evolution,* 26:84–95.

Wilson, A. C., C. R. Maxson, and V. M. Sarich. 1974. Two types of molecular evolution: Evidence from studies of interspecific hybridization. *Proc. Nat. Acad. Sci. U.S.A.,* 71:2843–2847.

Wilson, A. C., and E. M. Prager. 1974. Antigenic comparison of animal lysozymes. In *Lysozyme,* E. F. Ossermen, R. E. Canfield, and S. Beychock, eds., pp. 127–141. Academic Press, New York.

Wilson, E. O. 1963. Pheromones. *Sci. Amer.,* 208(5):100–114.

———. 1971. *The Insect Societies.* Belknap, Cambridge.

———. 1975. *Sociobiology, the New Synthesis.* Belknap, Cambridge.

Wilson, E. O., and W. H. Bossert. 1971. *A Primer of Population Biology.* Sinauer, Stamford, Conn.

Wimsatt, W. C. 1972. Teleology and the logical structure of function statements. *Stud. Hist. Phil. Sci.,* 3:1–80.

Woese, C. 1967. *The Genetic Code: The Molecular Basis for Genetic Expression.* Harper and Row, New York.

Wooding, G. L., and R. F. Doolittle. 1972. Primate fibrinopeptides: Evolutionary significance. *J. Human Evol.,* 1:553–563.

Woodworth, C. M., E. R. Leng, and R. W. Jugenheimer. 1952. Fifty generations of selection for protein and oil in corn. *Agron. J.,* 44:60–66.

Wright, S. 1931. Evolution in Mendelian populations. *Genetics,* 16:97–159.

———. 1932. The roles of mutation, inbreeding, crossbreeding, and selection in evolution. *Proc. VI Int. Cong. Gen.,* 1:356–366.

———. 1942. Statistical genetics and evolution. *Bull. Amer. Math. Soc.,* 48:223–246.

———. 1955. Classification of factors of evolution. *Cold Spring Harbor Symp. Quant. Biol.,* 20:16–24.

———. 1969. *Evolution and the Genetics of Populations,* vol. 2. University of Chicago Press.

Wright, S., and Th. Dobzhansky. 1946. Genetics of natural populations. XII. Experimental reproduction of some changes caused by natural selection in certain populations of *Drosophila pseudoobscura. Genetics,* 31:125–150.

Wynn-Edwards, V. C. 1962. *Animal Dispersion in Relation to Social Behavior.* Oliver and Boyd, Edinburgh.

Ziegler, A. M. 1966. The Silurian brachiopod *Eocoelia hemisphaerica* (J. de C. Sowerby) and related species. *Palaeontology,* 9:523–543.

Zuckerkandl, E. 1965. The evolution of hemoglobin. *Sci. Amer.,* 212(5):110–118.

Zuckerkandl, E., and L. Pauling. 1965. Evolutionary divergence and convergence in proteins. In *Evolving Genes and Proteins,* V. Bryson and H. J. Vogel, eds., Academic Press, New York, pp. 97–166.

Name Index

Adams, J., 116
Adamson, M., 263
Adelberg, E. A., 371, 379, 387
Ainsworth, G. C., 232, 371
Albertus Magnus, 263
Alexander, M. L., 37
Allard, R. W., 116, 131–132, 192, 204
Allison, A. C., 111
Alsopp, A., 383
Amadon, D., 248
Anaximander, 9
Anderson, D. T., 327
Anderson, E., 219, 225
Anderson, T. F., 382
Anderson, W. W., 45
Antonovics, J., 122
Anxolabehere, D., 116
Aquinas, Thomas, 9, 263, 496, 503
Arambourg, C., 447
Aristotle, 95, 263, 440, 458, 475, 488
Arnheim, N., 303
Arrhenius, S., 368
Atwood, K. C., 83
Auerbach, C., 59
Avery, T. O., 20
Avise, J. C., 80, 286–287
Axelrod, D. I., 224
Ayala, F. J., 32–35, 45–46, 55, 66–67, 70–71, 95, 116–117, 161–162, 180, 192–193, 198, 201, 213, 286–288, 290, 310, 486–487, 495, 500, 502–503, 506–507, 513
Babcock, E. B., 230–231
Bacon, Francis, 476
Bachmann, K., 76–77
Baker, C. M. A., 37
Baker, H. G., 203–204
Bakker, R. T., 435
Ball, J. A., 368
Barghoorn, E., 377
Barker, A. D., 505
Barnes, H., 405
Barnes, R. D., 371
Barrington, E. J. W., 406, 408
Bateson, W., 15–16, 58, 166
Baur, E., 188
Beam, C. A., 392
Becak, W., 228
Beck, W. S., 267
Beckner, M., 502
Beckwith, J., 462
Bell, Sir Charles, 476
Belon, 440
Belser, N. O., 70
Benado, M. B., 71
Berelson, B., 472
Berg, E. S., 503
Berger, J., 358
Bergson, Henri, 488, 503
Bergström, G., 208
Bernard, Claude, 14, 477
Bernstein, S. C., 55
Berry, W. B. N., 315
Bird, J. M., 321

Birnstiel, M. L., 85
Black, J. M., 336
Blair, A. B., 173
Bloom, W. L., 217, 219
Blumenbach, J. F., 139
Blyth, Edward, 96
Bock, W. J., 248–249, 251, 254, 506
Bockelie, T., 418
Bodmer, W. F., 110, 140, 160, 162, 164
Boesiger, E., 10, 37, 129
Bogorad, L., 383, 385–386, 389
Bolton, E. T., 276
Bonnell, M. L., 53
Bonner, J., 29, 255, 396
Bovee, E. C., 389
Bowman, R. I., 186
Boyd, W., 140
Brachet, J., 211
Bradshaw, A. D., 122, 133
Brewbaker, J. L., 38
Briden, J. C., 424–425
Britten, R. J., 29, 83, 85–87, 255–258
Broad, C. D., 510
Brooks, J. C., 348
Brown, A. W. A., 72, 121
Brown, D. D., 83
Brown, R. H., 414
Brown, R. M. Jr., 383, 387
Brues, A. M., 164
Buddha, 458
Buettner-Janusch, J., 444
Buffon, Georges, 128, 148, 439
Bullman, O. M. B., 237
Bush, G. L., 198, 201–202

Cairns, J., 23
Cairns-Smith, A. G., 358, 360, 363–364
Campbell, B. G., 446
Campbell, C. A., 116, 255, 417
Campbell, D. T., 457
Carlquist, S., 273
Carlson, P. S., 212
Carr, N. G., 387
Carroll, R. L., 433
Carson, H. L., 71, 187, 190, 198, 200–202, 217, 273, 275–276
Carus, Lucretius, 96
Cavalli-Sforza, L. L., 110, 140, 160, 162
Cerami, A., 63
Chadefaud, M., 371
Chambers, Robert, 96
Champion, A. B., 291
Chapman, G., 403, 408
Chargaff, E., 21
Chase, M., 20
Chetverikov, S. S., 16, 37, 129
Chiscon, P. A., 283
Church, R. B., 279
Cisne, J. L., 327
Clark, R. B., 407–409, 414
Clausen, J., 148, 177, 188, 195–196, 204, 217–218
Cleland, R. E., 204, 206
Clemmay, H., 397, 413
Cloud, P., 352, 377, 397, 416–417
Cocks, L. R. M., 328
Colbert, E. H., 5, 430, 433
Cole, H. S. D., 472
Coluzzi, M., 172
Comings, D. E., 92
Connell, J. H., 342
Cope, E. D., 244
Copernicus, 438, 495
Corliss, J. O., 391
Correll, D., 207
Cory, L., 173
Cowen, R., 416
Craig, G. B., 37
Crick, F. H., 21, 23, 368
Crimes, T. P., 398
Crompton, A. W., 434
Cronquist, A., 386
Crow, J. F., 32, 72, 156, 163–164
Crumpacker, D. W., 37
Curtis, H., 371, 376
Cuvier, Georges, 263, 439

Daniels, E. C., 390
Darevsky, I. S., 230
Darlington, C. D., 217
Darwin, Charles, 4, 7, 10–18, 29–32, 39, 96–100, 109, 118, 121, 128–129, 165–166, 186, 242, 263, 359, 439, 445, 469, 477–478, 484–486, 495–497, 507
Darwin, F., 485
Davidson, E. H., 29, 83, 85–87, 255–258
Davidson, G., 172
Davidson, J. N., 72
Dawood, M. M., 116
Dayhoff, M. O., 294, 299, 372
Dayton, P. K., 342
De Beer, G., 485
Delbrück, M., 17, 65
Demarly, Y., 225
Demeny, P., 472
Demerec, M., 65, 70
Denis, H., 211
Descartes, R., 438–439, 453, 488

DeVore, I., 452
De Vries, H., 16, 58–59, 129, 166, 188, 204–205
De Wet, J. M. J., 226
Dewey, J. F., 321
Dickerson, G., 39
Dobzhansky, Th., 2, 5, 17, 19, 23, 34–44, 66, 69–70, 135, 138, 173–174, 200, 213–216, 255, 272, 366, 453, 457, 486–487, 502, 506–507
Dodge, J. D., 391
Dodson, C. H., 176, 207–209
Doolittle, R. F., 303
Dose, K., 357
Doty, M. S., 247
Doty, P., 276
Doudoroff, M., 371, 379, 387
Dougherty, E. C., 408
Doughty, J. H., 107
Drake, F., 366–367
Drewry, G. E., 424–425
Driesch, Hans, 488
Dubinin, N. P., 37, 69–70
Dubois, Eugène, 447, 449
Dufour, L., 174
Dunn, L. C., 37, 124
Durham, J. W., 324

Edelman, G., 491
Edgell, M. H., 389
Ehrlich, P. R., 472
Ehrman, L., 119–120, 174
Eicher, D. L., 316–317
Eiseley, L., 96
Eldredge, N., 326
Emerson, A. G., 466
Empedocles, 9, 96
Ephrussi, B., 212

Falconer, D. S., 38
Fankhauser, G., 229
Fano, U., 65
Farmer, J. D., 416
Farquhar, M. N., 281
Farris, J. S., 286
Ferlan, L., 209, 221
Ferone, R., 232
Fischer, A. G., 319, 388
Fisher, R. A., 16, 32, 129
Fitch, W. M., 79, 268, 282, 298–301, 303, 310–312
Flessa, K. W., 421, 427
Flew, A., 505
Flexer, A. S., 394
Ford, E. B., 3, 35, 115, 122
Fortey, R. A., 418
Fowler, J. A., 216
Fox, S. W., 357
Fraenkel-Conrat, H., 20
Freedman, R., 372
Fryer, G., 348

Galau, G. A., 29, 87
Galileo, 438, 495, 497
Ganong, W. F., 4
Gardner, B. T., 452
Gardner, R. A., 452
Garn, S. M., 145–146
Garstang, W., 260
Georghiou, G., 121
Gershenson, S., 2
Ghiselin, M. T., 485, 501
Gierer, A., 20
Glaessner, M. F., 398
Goin, C. J., 76–77
Goin, O. B., 76–77
Goldberg, R. B., 83
Goldich, S. S., 352
Goldschmidt, R. B., 59, 166, 199, 242
Goodman, M., 304, 310, 487
Gottesman, I. I., 152–153, 469
Gottlieb, L. D., 217
Goudge, T. A., 509–510
Gould, S. J., 244–245, 326–327
Grant, K. A., 176, 209
Grant, V., 5, 176–177, 184, 196, 198–199, 202, 204, 209–210, 219–221, 223, 226, 229
Grassé, P., 129, 371
Gray, Asa, 12
Green, M. M., 70–71
Gregory, 122
Grell, K., 391
Grene, M., 502, 505
Griffith, D. L., 389
Gurdon, J. B., 27, 83, 461
Gustafsson, A., 203

Haapala, O. K., 391
Hackett, 172
Hadzi, J., 399, 403, 407
Haeckel, E., 242–243, 402
Haldane, J. B. S., 16–17, 129, 163–164, 213, 354
Hall, W. P., 93
Hallam, A., 326–327
Hamburg, D. H., 452
Hamilton, H. J., 470

Hamilton, W. D., 127
Hamrick, J. L., 131–132
Hand, C., 403
Harberd, D., 203
Harlan, J. R., 226
Harland, W. B., 421
Harris, H., 52–53, 140, 192
Hay, M., 290
Hayes, C., 452
Hayes, R. J., 452
Hazarika, M. H., 75
Hedge, C. E., 352
Hedges, R. W., 232
Hempel, C. G., 477
Hennig, W., 238, 241
Hershey, A. D., 20
Hessler, R. R., 338
Heston, L. L., 152–153
Hickey, W. A., 37
Hiesey, W. W., 148, 195
Hill, R. L., 386
Himes, M., 392
Himmelfarb, G., 505
Hinegardner, R., 73, 77
Holliday, R., 255
Holmquist, R., 310
Hood, L., 85
Hooker, Joseph, 12
Hopkinson, D. A., 53, 140
Hopson, J. A., 434
Horridge, G. A., 405
Howells, W. W., 447
Hoyer, B. H., 276–279, 282–283
Hubby, J. L., 47, 52, 62, 156
Hull, D., 485, 494–495, 502
Hungate, R. E., 2, 4
Hutchinson, G. E., 11, 247
Hutchison, C. A., 389
Huxley, Aldous, 469
Huxley, J. S., 17, 236, 513
Huxley, T. H., 440
Hyman, L. H., 399, 402–403, 408, 416

Iles, T. D., 348
Imbrie, J., 421, 427
Inger, R. F., 270, 432
Ingram, V., 444

Jacob, F., 29, 255
Jägersten, G., 403
Jahn, T. L., 389
Jeffries, R. P. S., 420
Jenkin, Fleeming, 15, 18, 98, 100
Jevons, William Stanley, 477
Johannsen, W., 16, 29, 98, 129
Johnson, F. M., 52
Johnson, M. S., 161
Johnson, P. H., 389
Johnson, R. G., 398
Joklik, W. K., 380–381
Jordan, K., 166
Jørgensen, C. B., 428
Jukes, T. H., 156, 305, 310
Jumars, P. A., 338

Kahler, A. L., 204
Kamemoto, H., 209
Kanellis, A., 37
Kaneshiro, K., 275
Kaplan, W. D., 35
Karpechenko, G. D., 178
Kastritsis, C., 174
Katchalsky, A., 358
Kauffman, S., 255
Kaufman, D. W., 53
Keck, D. D., 148, 195
Kedes, L. H., 85
Kekulé, August, 478
Kelvin, 368
Kennedy, W. P., 38
Kepler, Johannes, 438, 495
Kerkut, G. A., 399
Kerr, W. E., 229
Kessler, K. F., 98
Kessler, S., 181
Kettlewell, H. B. D., 122
Khorana, H. G., 32
Kimura, M., 156, 159, 163–164, 305, 307, 309–310
King, J. L., 156, 305
King, M. C., 194, 260
Kitto, G. G., 80
Kitzmiller, J. B., 172
Klein, R. M., 386
Kleinschmidt, 166
Koehn, R. K., 161
Koestler, A., 487
Kohne, D. E., 86, 283
Kojima, K., 68, 116
Komai, T., 403
Koopman, K. F., 181
Kozhov, M., 348
Kretchmer, N., 152
Kriss, A. E., 379
Kropotkin, Peter, 98
Kubai, D. F., 391
Kuhn, T. S., 477
Kullenberg, B., 208
Kurtén, B., 347

Lack, D., 186
Laird, C. D., 277–281

Lamarck, Jean-Baptiste, 9–11, 29, 96, 128–130, 165, 263, 503
Lamprecht, 199
Lane, C. D., 27
Lang, A., 407
Langer, S. R., 451
Langley, C. H., 310–312
Lansteiner, Karl, 139
Laporte, L. F., 320
Leakey, L., 447
Leakey, Richard, 447, 449
Lederberg, E. M., 17, 65
Lederberg, J., 17, 65
Lee, A. S., 389
Leedale, G. F., 391
Lemche, H., 413
Lenz, L., 209
Lerner, I. M., 38, 97
Levene, H., 116–117, 135, 173
Levin, D. A., 209, 212
Levine, L., 290
Levins, R., 117, 156
Levy, B., 87
Levy, M., 212
Lewis, H., 198, 217, 219
Lewontin, R. C., 35, 37, 39, 41, 47, 50, 52, 62, 115, 121, 130, 140, 144, 156, 162–164, 192
Li, C. C., 32
Liapunova, E. A., 94
Lima-de-Faria, A., 92
Lin, C. C., 94
Linnaeus, C., 9, 12, 128, 139, 165, 182, 263, 439
Lipkin, M., 460
Littlejohn, M. J., 174, 182
Loeblich, A. R., Jr., 327
Loeblich, A. R., III, 392
Lorenz, K., 98, 455
Lotsy, J. P., 188, 190
Ludwig, W., 156
Lull, R. S., 244
Luria, S. E., 17, 65
Lyell, Sir Charles, 12

MacArthur, R. H., 338, 344
MacGillavray, H. J., 326
Mackie, G. O., 405
MacLeod, C. M., 20
Madden, C., 41
Mahler, H. R., 383–384, 386–387, 389
Malthus, T. R., 97
Manning, J., 63
Manser, A. R., 505
Mandel, M., 232
Manton, S. M., 327
Manwell, C., 37
Marbaix, G., 27
Marchant, H. J., 374
Margoliash, E., 62, 79, 268, 300–301
Margulis, L., 374, 383–384, 389
Maren, T. H., 302
Margalef, R., 338
Marinković, D., 44, 162
Marion, J. J., 173
Market, C. L., 228
Marmur, J., 25
Massaro, E. J., 228
Matsuda, G., 311
Matthew, Patrick, 96
Matthews, S. C., 398
Maupertuis, 9
Maxson, C. R., 260
Maynard Smith, J., 39, 117, 392
Mayr, E., 5, 7, 17, 118, 137, 147, 166, 184–185, 187, 196, 198, 202, 228, 240–241, 246, 270, 485, 501–502
McAlester, A. L., 346
McCarthy, B. J., 87, 276–281
McCarty, M., 20
McConaughty, B. L., 281
McCracken, R. D., 150
McDonald, J. F., 117, 161
McElroy, W. D., 380
McFadden, E. S., 226
McKusick, V. A., 38, 53, 63, 71, 459
McLaughlin, D. B., 351
McLaughlin, P. J., 372
McLean, J. H., 252
McMillan, C., 205
McNeilly, T. S., 133
Meadows, D. H., 472
Medawar, P. B., 477–478
Mendel, Gregor, 14–15, 99, 100, 242, 477, 482–484
Meselson, M., 23
Metschnikoff, E., 402
Middleton, J. H., 389
Miescher, F., 20
Milani, R., 37
Milkman, R., 72
Mill, John Stuart, 476
Miller, S. J., 350, 354–355, 367
Miller, R. S., 342
Mirov, N. T., 172
Mirshky, A. E., 20, 72
Missarzhevsky, V. V., 398
Mivart, St. George, 242
Monod, J., 29, 255
Moorbath, S., 352
Moore, B. P., 4

Moore, G. W., 310
Moore, J. A., 177
Moore, R. C., 428
Moores, E. M., 345, 426–427
Morgan, T. H., 15–16, 34, 59, 188, 472
Morowitz, H. J., 350
Morris, C., 398
Morse, J. W., 417
Mourão, C. A., 45
Mukai, T., 68–69, 71, 154
Muller, H. J., 35, 59, 155, 460–461
Müntzing, A., 226
Murphy, L. S., 53
Murray, J., 198, 201
Myrdal, G., 471

Nagel, E., 492, 502–503
Neel, J. V., 107, 467
Nei, M., 156, 285
Newell, N. D., 342, 421, 428
Newton, Isaac, 438, 477–478, 485, 495
Nilsson-Ehle, H., 16
Nirenberg, M., 27
Nobs, M. C., 149, 189
Novitski, E., 93

Oakley, K. P., 440
Ohno, S., 77, 81–82, 212
Ohta, T., 156, 159, 305, 307, 309–310
Okada, T. A., 92
O'Nions, R. K., 352
Oparin, A. I., 357
Orgel, L. E., 350, 355–356, 367–368
Ornduff, R., 212
Osborn, H. F., 503

Paecht-Horowitz, M., 358
Paine, R. T., 341
Paley, W., 496
Pankhurst, R. J., 352
Pascal, Blaise, 477
Passmore, J., 473
Pasteur, Louis, 14, 350, 477
Patterson, B., 447
Pauling, L., 80
Peirce, Charles S., 477
Pentzos-Daponte, A., 37
Periquet, G., 116
Petit, C., 119
Pickett-Heaps, J. D., 374
Pigott, G. H., 387
Pilbeam, D., 446
Pitrat, C. W., 421
Pittendrigh, C. S., 501
Plato, 263, 438
Plinius, 165
Popp, R. A., 80
Popper, K. R., 476–480, 486, 492, 494–495
Powell, J. R., 117, 135, 161, 405
Prager, E. M., 293
Prakash, S., 62
Premack, D., 452
Pruitt, D. G., 472
Pugh, J. E., 255
Punnett, R. F., 16

Raff, R. A., 383, 385–387, 389
Raper, J. R., 394
Rasmussen, D. J., 161
Raup, D., 425
Raven, P. H., 371, 376–377, 383
Reed, T. E., 164
Rees, H., 75–76
Rees, W. J., 403
Reichlin, M., 290
Rensch, B., 147, 166, 236, 376, 513
Rhoads, D. C., 417
Richards, O. W., 15
Richardson, E. S., Jr., 398
Riek, E. F., 465
Ris, H., 20, 72, 391
Ritossa, F. M., 83
Rizzo, P. J., 391
Roberts, D. F., 151
Roberts, R. B., 278–279
Robertson, F. W., 39
Robertson, W. R., 92
Robson, G. C., 15
Romer, A. S., 432–434
Rowley, P. T., 460

Sachs, J., 396
Sagan, C., 366–368
Sanders, H. L., 338
Sanger, F., 294
Sankaranarayanan, K., 71
Sarich, V. M., 260, 290–292
Saura, A., 72
Schaller, G. B., 452
Schildkraut, C., 276
Schindewulf, O. H., 242
Schnepf, E., 383, 387
Schopf, J. W., 318, 336, 377–378, 389, 392
Schopf, T. J. M., 53
Schramm, G., 20
Schull, W. J., 107
Schultz, R. J., 230

Sclater, P. L., 339
Sclater, W. L., 339
Sears, E. R., 226
Selander, R. K., 53, 56, 161, 192
Seliger, H. H., 380
Semenov-Tian-Shansky, 166
Sherrington, C., 453
Shields, G. F., 281
Shindo, K., 209
Shklovskii, I. S., 366–368
Shull, A. F., 15
Sibley, C. G., 220
Silveira, Z. V., 229
Simons, E. L., 436, 446
Simpson, G. G., 3, 5, 7–8, 17, 198, 235–236, 245–246, 251, 265, 267, 269–270, 327, 332, 334–336, 339, 342, 366, 450, 456, 476, 490, 495, 502, 509–511, 513
Singer, B., 20
Sinsheimer, R. L., 389
Sloan, R. E., 435
Smart, J. J. C., 505
Smith, A. G., 424–425
Smith, E. B., 176
Smith, P. F., 388
Smith, W., 12
Smythies, J. R., 487
Sneath, P. H. A., 232, 237, 270, 286, 371
Snyder, R. C., 472
Socrates, 458
Sokal, R. R., 237, 270, 286
Sondhi, K. C., 39
Sonea, S., 389
Sonneborn, T. M., 177
Soyer, M. O., 391
Spallanzani, L., 9
Spassky, B., 39, 41–43, 66, 69–70, 289
Spassky, N., 69–70
Spencer, H., 485
Spencer, W. P., 37
Spiegelman, S., 83
Spieth, P. T., 72
Spinelli, V., 70
Stahl, B. J., 418, 432, 434, 436
Stahl, F. W., 23
Stanier, R. Y., 371, 379, 386–387
Stanley, S. M., 332–333, 346
Stebbins, G. L., 17, 115, 172, 177, 195, 200, 202–203, 205, 207, 209, 211–214, 216–217, 220–221, 223–231, 362, 365, 394–395, 513
Steinman, H. M., 386
Steller, R., 303
Stephens, S. G., 179
Stern, C., 142–143
Stevenson, A. C., 38, 107
Straus, N. A., 281–282
Straw, R. M., 209
Strickberger, M. W., 116
Sturtevant, A. H., 93
Stutz, H. C., 219
Suomalainen, E., 72
Suzuki, D. T., 64
Sved, J. A., 45, 164

Tappan, H., 327, 391
Tashian, R. E., 302, 310
Taylor, F. R. J., 383, 388
Taylor, J. H., 23
Teigs, O. W., 327
Teilhard de Chardin, P., 450, 453, 503
Temin, R. G., 44
Terzaghi, E. Y., 60–61
Thomas, L. K., 219
Thompson, E. O. P., 294
Tobari, Y. N., 68
Tobias, P. V., 447–448
Tracey, M. L., 45–46
Treffers, H. P., 70
Trivers, R. L., 127
Trimble, B. K., 107
Tucker, J. M., 196, 219, 223
Turesson, G., 147, 195
Tuttle, R. C., 392

Uzzell, T., 230

Valentine, D. H., 212
Valentine, J. W., 55, 247, 255, 324, 338–339, 341, 343–345, 347, 414–417, 421–423, 426–428
Van der Pigl, L., 207–208
Van Lawick-Goodall, J., 447, 452
Van Valen, L., 332, 335–336, 342
Vaurie, C., 182
Vendrely, C., 72
Vendrely, R., 72
Villee, C. A., 371
Vogel, S., 209
Von Wahlert, G., 506
Vorontsov, N. N., 94

Waagen, W., 58
Waddington, C. H., 455
Wagner, M., 166, 198
Walker, W. F., Jr., 371
Wall, J. R., 200

Wallace, A. R., 12, 97, 180, 465
Wallace, B., 154
Walter, H., 2
Washburn, S. L., 441–442, 467
Watson, J. D., 21, 23
Webster, T. P., 93
Weismann, A., 30, 129
Wells, M. J., 374
Westoff, C. F., 472
Westoll, T. S., 330
Wexelsen, H., 225
Whewell, W., 477
White, L., 451
White, M. J. D., 92–93, 115, 131, 201, 228–230, 271
Whiteley, A. H., 211–212
Whiteley, H. R., 211–212
Whitt, G. S., 212
Whittaker, R. H., 369–370, 373–374
Wiedmann, J., 327
Williams, B. J., 148
Williams, E. E., 93
Williams, G. C., 392–394, 501, 504, 506, 513
Williams, N. H., 209
Wilson, A. C., 194, 260, 290–293
Wilson, E. O., 126, 338, 344, 468
Wimber, D. E., 209
Wimsatt, W. C., 502
Wingstrand, K. G., 413
Woese, C., 360
Woodell, S. R. J., 212
Wooding, G. L., 303
Woodworth, C. M., 38
Wright, S., 8, 16–17, 116, 126, 129, 157, 159, 167, 245, 247, 507
Wynn-Edwards, V. C., 127

York, T. L., 200

Ziegler, A. M., 328–329
Zuckerkandl, E., 80, 444

Subject Index

Abortion, 107, 460
Acacia, 200
Acetabularia, 395
Acetic-saline-giemsa staining method, 77
Achillea, 207
Achondroplastic dwarfism, 103f.
Acmaea, 251f.
Acquired characteristics, inheritance of, 10, 15ff., 29–30
Acquisition of knowledge, 478, 487
Acrasiales, 376, 395
Acrididae, 290
Actinosphaerium, 392
Actinophrys, 392
Adaptation, 29, 32, 64f., 71, 95f., 117, 120ff., 146, 454, 495ff., 502ff., 506f., 514
  and mutation (*see* Mutation *and* Neutral alleles)
  and variation, 497
  as a teleological feature, 501ff.
  in individuals, 492ff., 503, 506
  of eye and hair color, 149
  of geographical variation, 147
  of populations, 506
  of race differences, 159f.
  to insect pollinators, 204
Adaptedness, of racial traits, 146 (*see also* Adaptation)
Adaptive
  evolution, 457
  function, 501
  modification along the lines of least resistance, 4
  pathway, 247, 254
  peaks and valleys, 167f., 170, 180, 245, 248
  radiation, 8, 186, 370
  type, 245ff., 259
  value, 101, 115, 121, 167f. (*see also* Fitness *and* Selective value)
  zone, 246f., 249, 254, 336
Aedes aegypti, 37
Aegilops squarrosa, 469
Aegyptopithecus, 437
Agamic complex, 230f.
Aggression, 455, 467
Agonistic display, 118
Agropyron, 213
Agrostis tenuis, 133
Albinism, 104f.
Albumin, 290ff.
Alkylating agents, 61f.
Allen's rule, 147, 150
Alleles, 31, 142
  frequencies, 50f.
  modifying fitness, 46
  neutral, 154ff., 164, 305f., 308, 459
  substitutions, 308
  variation (*see* Genetic variation)
Allometry, 245
Allopolyploidy, 178, 180, 188, 226, 229 (*see also* Polyploidy)

Allozymes, 18, 192, 194 (*see also* Enzymes)
  polymorphism, 136, 161
Alpha hemoglobin. *See* Hemoglobin.
Altitudinal variation, 133, 149
Altruism, 126f., 456
Ameria, 415
Amerous architecture, 411, 414
Amino acids, 25, 27f., 153 (*see also* Protein)
  differences, 309f.
  replacements, 55, 290, 294
  substitutions, 50, 62, 290
Amino acid sequences, 47, 50, 60, 153, 193, 260, 266, 284, 297, 499
  α chain of hemoglobin, 295
  β chain of hemoglobin, 295
  chymotrypsinogen A, 82
  cytochrome *c*, 296ff., 308, 311f.
  trypsinogen, 82
Ammonites, 327
Amniocentesis, 460
Amoeba dubias, 73, 514
Amphibians, 56, 73, 75, 325, 337, 429ff.
Amphioxus, 417ff.
Amphiploidy, 212, 214, 221
Amphiuma means, 74
Anagenesis, 236ff., 241, 246, 252, 254, 324, 334, 336, 497
Analogy, 263, 270, 305
Analytic method, 491
Aneuploidy, 58 (*see also* Chromosome)
Aniridia, 65
Annelida, 73, 327, 398f., 408f., 411, 413f.
Annidation, 156
Anolis, 93
Anopheles, 161, 172, 183
Anoplorchus, 161
Anthoxanthum, 133
Anthozoa, 404
Antibiotics, 122
Antibodies, 290ff. (*see also* Antigens; Blood groups; *and* Immunology)
Anticodon, 25, 84
Anti-evolutionists, 439
Antigens, 290f. (*see also* Antibodies; Blood groups; *and* Immunology)
Antireductionism, 488
Antirrhinum, 188, 225
Apogamy, 72, 130
Apoidea, 228
Apomictic complexes, 230
Arabidopsis thaliana, 74
Arbacia lixula, 211
Archetype, 260
Arenicola, 409, 411
Argument from design, 496
Aristogenesis, 503
Armerica, 204
Artemia, 228, 230
Arthropods, 266, 326f., 334, 348, 398, 411, 413f., 416, 431, 515
Artificial
  selection, 38f., 47, 97, 121, 125, 130f. (*see also* Selection)
  teleology, 500, 502 (*see also* Teleology)
Artiodactyles, 303
Asclepias, 210
Ascomycetes, 376, 394
Asexual reproduction, 132, 170
Astrapia, 185
Asterias, 53
Asthenosphere, 321
Atmosphere, 351ff., 356
Auditory signals, 174, 184
Australopithecus, 441ff., 514
Authority acceptors, 455
Autogenetic theories, 129
Autopolyploidy, 224ff. (*see also* Polyploidy)
Autotrophs, 362ff., 375, 388
  flagellates, 387
  protocell, 363
Auxotroph mutants, 59, 64
Avena, 30, 131f., 147, 228
Axiological standards, 509
Azotobacter, 380

Baboon, 278, 292f., 302 (*see also* Primates)
  antiserum, 292
  DNA, 279
Baconian principles, 484
Bacteria, 23, 68f., 74, 370f., 376f., 387, 514
  anaerobic, 379
  clones of, 130
  conjugation of, 232
  L-phase of, 388
  plasmids, 389
Bacteriophages, 23, 25, 383, 462 (*see also* Viruses)
  T2, 69
  T3, 381
  T4, 382
Balancer chromosomes, 39ff., 45 (*see also* Chromosome)
Balanced lethal system, 109, 206
Balance theory of population structure, 31, 34ff., 156
Balancing selection, 35, 46, 107f., 111ff., 115f., 156 (*see also* Natural selection)

Balanced polymorphism, 109, 120, 155f., 164
Banded iron formation, 353
Basidiomycetes, 376, 394
Beagle, H.M.S., 485
Beggiatoa, 380
Beijerinckia, 380
Benthic invertebrates, 422
Benzene, 478
Bergmann's rule, 147, 150f.
Beta hemoglobin. *See* Hemoglobin.
Biogenesis, 349, 352, 357, 364ff.
Biogeography, 263
Biological classification, 165, 451
Bioluminescence, evolution of, 380
Biomass, 512f.
Biospace, 247, 251f., 326, 337f., 342, 347f., 421, 428, 435
Biosphere, 414
Biotic provinces, 341
Bipedal locomotion, 440f.
Bisexuality, 170
Biston betularia, 123, 507
Bitter Springs Chert, 378
Bivalents, 178
Bivalvia, 332f., 335
Blackflies, 271
Blastaea, 402
Blood groups, 37, 47, 63, 139ff., 144f., 160f.
  ABO, 38
  association with disease, 160
  classical, 142
  differences, 143, 161
  genes, 140f.
  gene frequencies, 162
  incompatibilities, 160
  polymorphisms, 38
  rhesus, 142
Blood theory of heredity, 98f.
Blue-green algae, 370, 375, 377f., 380, 386f.
Bobbed mutants, 84
Body weight and annual temperature, 151
Body size and antler size, 244
Bombus, 209
Bombykol, 175
Bony fishes, evolutionary history, 80f.
Bos taurus, 85
Bothriochloa, 229
Boundary conditions, 481
Boundary zones, 137
Brachiopods, 327ff., 331, 336f., 398, 409, 416
Bradytelic evolution, 336
Brassica oleracea, 178, 226
Bridgewater Treatises, 496
Britten–Davidson model of gene regulation, 257
Brown algae, 374, 394f.
Bryophytes, 393f.
Bryozoans, 416
Bufo, 75f., 173, 182
Bulawaya Formation, 378
Bullfrog, 298f.
Bumble bee, 176
Burgess Shale, 398
Burial rites, 454
Butterfly, 72, 92, 182

Callithrix jacchus, 293
Cambrian, 400, 414, 420, 512
Campanula, 266
Candida, 300f.
Canis, 169, 267, 270
Captorhinomorphs, 433, 435
Carbon-14, 317
Carbonic anhydrase, 302f.
Carboniferous, 330, 343, 512
Cardinalis cardinalis, 282
Carnivora, 169f., 335
Carnivorous plants, 515
Carpodacus purpureus, 282
Carrying capacity, 337
Cartesian view, 439
Caryophyllaceae, 207
Castanea, 223f.
Castes in India, 139
Catalpa, 223
Catastomus clarkii, 161
Catastrophic speciation, 217 (*see also* Speciation)
Categories of classification, 242, 331f., 334
Catharus guttatus, 281f.
Catholic church, 497
Cattleya, 208f.
Causal explanations, 495, 503
Causes of evolution, 129, 486
Ceanothus, 172
Cell
  differentiation, 29
  organelles, 74
Cenozoic, 316, 333
Central nervous system, 515
Centrarchidae, 212
Centromere, 89, 92
Cephalochordata, 417ff.
Ceratodus, 330
Cercocebus galeritus, 291
Cercopithecus, 291, 293
Cerebral cortex and hemispheres, 515

Ceremonial burial, 454
Chance, 7, 64, 156, 199, 305
Change and progress, 508ff. (*see also* Progress)
Chaos chaos, 73
Character displacement, 249
Chelicerata, 327
Chemical evolution, 9
Chemical stimuli, 174
Chemotherapeutic agents, 122
Chicken, 39, 278f., 299ff., 303, 311
Chimpanzee, 94, 194, 260, 278ff., 302f., 437, 442 (*see also* Primates)
Chironomus, 88, 272
Chlamydomonas, 66, 69, 373, 392f.
Chlorobium, 377
Chlorophyll, 37f., 377
Chloroplasts, 25, 387
Chondrodystrophy. *See* Achondroplastic dwarfism.
Chondrostei, 80f.
Chordata, 254f., 409, 413, 417, 420
Christianity, 455, 496
Chromatin, 391
Chromatography, 294
Chromosome
  aberration, 75ff., 92ff., 112ff.
  acrocentric, 90, 93f.
  arrangement, 40, 90, 112ff., 136, 161f., 217, 271ff., 276, 289, 502
  change, 30
  deletion, 57, 75ff., 90, 294
  duplication, 58, 75ff., 83, 90ff., 268 (*see also* Gene duplication)
  evolution, 75ff.
  fission, 92ff.
  frequency, 114, 140
  fusion, 92ff.
  homozygotes, 45
  human, 94
  inversion, 40, 87ff., 112ff., 136, 161f., 190, 216f., 271ff., 276, 289, 507
  lethal, 42, 45f.
  mutation, 57
  number, 21, 103, 190f., 214
  phylogeny, 271
  polymorphism, 120, 133, 161, 190
  repair, 70
  repatterning, 195, 216f.
  segregation, 206
  size, 75
  sterility, 178, 211
  translocation, 58, 87, 94, 109, 115, 198f., 205f., 216f.
  variants, 108, 147
  volume, 75
Chymotrypsin, 81f., 294
Chytridiomycetes, 395
Ciliates, 399, 401, 407
Circularity
  of argumentation, 505
  of definition, 505
Citellus, 94
Civetone, 175
Clades, 236
Cladistic classification, 238ff., 270
Cladogenesis, 236, 239, 241, 246, 252, 324, 334, 336, 342, 344, 370, 497
Clarkia, 78, 198f., 217ff., 227
Classical theory of population structure, 34ff., 155
Classification, 165, 233f., 237, 240f., 270
  biological, 165, 451
  categories, 242, 331f., 334
  cladistic, 238ff., 270
  phenetic, 237f., 270
  race, 144f.
Climatic
  barrier, 137
  gradient, 345
Clines, 133, 509
Clock of evolution, 308f., 312f.
Clones, 130ff., 362, 461f., 469
Clonal selection, 491
Clostridium, 379
Cluster-clone theory, 385f.
Coacervates, 357
Coadaptation, 129, 242, 246, 346
Co-dominance, 29
Codon, 23ff., 47, 83f., 298, 327, 360, 511
Coefficient of dominance, 102
Coelacanths, 429f.
Coelenterata, 73, 254f., 393, 398ff., 515
Coelom, 399, 408ff., 427
Coelophana, 407
Cognitive structure, 452
Colisella, 251f.
Colobus polykomos, 291
Comb jellies, 402f.
Commandment of universal love, 457
Communication, 451f.
Communities, 337ff., 347
Comparative anatomy, 14, 263, 270, 303, 381, 451
Comparative embryology, 263
Competition, 31, 126, 341, 346, 405, 426, 431
Compilospecies, 226

Complement fixation, 290
Complexity
of DNA, 85f., 511
of ecosystems, 2
of organization, 508
of viruses, 381
Compositae, 207, 217, 266
Compositionism, 491
Concealed
genetic load, 106
variation, 39, 44, 47
Conchopoma, 330
Congenital defects, 38
Conjugales, 374, 395
Connectability, condition of, 492ff.
Connection between theories, 492
Conscious evolution, 468, 471
Consciousness, 439
Constancy of evolutionary rates, 308ff.
Continentality, 322f., 344f., 348, 426
Control over the environment, 514
Convergent evolution, 265f., 270, 305, 326f.
Cooperation, 127, 455
Copernican revolution, 495f.
Coral, 339, 402, 416
Co-repressor, 255
Corroboration, degree of, 481
Coryodoras aeneus, 74
Courtship, 118, 173f.
Cowania, 219f.
Crayfish, 266
Creationism, 346
Crepis, 181
Cretaceous, 319, 343
Cri du chat syndrome, 77
Criterion
of demarcation, 476, 479, 486
of progress, 516
of utility, 506
Cro-Magnon Man, 450
Crossing-over, 80, 89f.
Cruciferae, 207
Crypthecodinium, 392f.
Cryptic
coloration, 507
structural hybridity, 216
Ctenodus, 330
Ctenophora, 402f., 407
Culture, 139, 451ff., 470f.
barriers, 472
evolution of, 9, 457ff.
Cyanophora, 386
Cychnoches chlorochilon, 207
Cynognathus, 433
Cyprinidae, 286
Cytochrome *c,* 62, 65, 296ff., 308ff., 414, 444

Dactylis, 38, 227
Danaus plexippus, 234
Darwinian fitness. *See* Adaptation; Fitness; *and* Selective value.
Darwin's finches, 186f., 248f.
Dasycercus, 267
Dasychadales, 395
Dasyurus, 267
DDT, 8, 121
Death-awareness, 453f.
Deep-sea
organisms, 53ff.
ridge, 319
trench, 319f.
Deficiency. *See* Chromosome deletion.
Deletion. *See* Chromosome deletion.
Delphinium, 221
Delta hemoglobin. *See* Hemoglobin.
Deme, 30
Demonstrative argument, 493
Dendraster excentricus, 211
Dendrogram, 270, 284, 286, 299, 302
Density-dependent factor, 337
Derivability, condition of, 492ff.
Descent of Man, The, 439f.
Deterministic processes, 157, 164
Detritus feeders, 401ff., 408, 413f., 427
Deuterostomes, 260, 409ff.
Developmental
homeostasis, 106, 500
sterility, 179, 211ff.
Devonian, 330, 343, 424, 430f.
lobe-finned fish, 430
Dichanthelium, 229
Dimer, 49f., 62
Dimetrodon, 433
Dinoflagellates, 387, 391f.
Dinosaurs, 240, 254, 337, 435f.
Dioecy, 170
Dionaea, 515
Diploblasts, 403, 405, 407, 413
Diplococcus pneumoniae, 20
Diploidy, advantage of, 394
Diplopteraster multipes, 55
Dipnoi, 330, 430
Dipnorhynchus, 330
Dipterus, 330
Direction
and progress, 508ff.
concept of, 509

Disappearance of taxonomic categories, 423
Discontinuity of organic diversity, 168f., 492
Dissociation. *See* Chromosome fusion.
Divergence, 166, 302, 311
  invertebrates and vertebrates, 311
  mammals and birds, 302, 311
  marsupials and placental mammals, 302
Diversification, 336, 346ff., 421, 428f.
Diversity, 95, 116, 168, 338f., 341ff., 422ff., 495
DNA, 20ff.
  chloroplast, 387
  competitor, 277
  content per cell, 72ff.
  content of chromosomes, 74
  content of mitochondria, 74
  denatured, 276
  deoxyribose sugar, 22
  duplexes, 280, 283
  Drosophila, 277ff.
  hybridization, 280ff.
  nucleotide additions, 60, 294
  phylogenies, 281
  programs of information, 502
  repetitive, 29, 78, 85f., 257
  replication, 23f., 57, 61
  satellite, 86
  sequence, 59f., 78, 80, 271, 277ff.
  viruses, 73
Domesticated species, 121, 139, 151, 153, 169
Dominance, 29, 37ff., 102f., 108, 482
Dominance behavior, 455
Down's syndrome, 103
Drepanididae, 248, 286
Drift. *See* Genetic drift.
Drosophila
  ananassae, 52, 93f.
  atrimentum, 275
  bifasciata, 174
  birchii, 34, 66ff.
  equinoxialis, 49f., 183, 287ff.
  flavomontana, 290
  funebris, 277ff.
  grimsawi, 275
  Hawaiian, 186, 192
  hawaiiensis, 275
  hydei, 37
  imaii, 174
  insularis, 183, 288f.
  melanogaster, 37ff., 59, 68–87, 93ff., 154, 174, 277ff.
  miranda, 190, 272
  mulleri, 37
  nebulosa, 287f.
  novamexicana, 37
  paulistorum, 183ff., 287ff.
  pavlovskiana, 183, 288f.
  persimilis, 41, 43, 69, 173, 177ff., 181, 183f., 190, 198, 200, 213, 272
  primaeva, 273, 275
  prosaltans, 41ff.
  pseudoobscura, 41–52, 62, 66, 69, 88, 93, 106, 112–119, 133–147, 156, 161, 173–184, 190, 198, 200, 213, 272, 504, 507
  repleta, 273
  serrata, 32–34
  sibling species, 183
  simulans, 278–281
  subobscura, 37, 93
  tropicalis, 183, 288f.
  villosipedis, 275
  virilis, 56, 190f.
  willistoni, 41ff., 52ff., 69f., 93f., 161, 183f., 192ff., 287ff.
  world fauna, 186
Dryopithecus, 437, 446
Dugesia, 228
Duplication, 58, 75f., 90, 266 (*see also* Chromosome duplication *and* Gene duplication)

East Pacific Rise, 319
Echinodermata, 73, 259, 398, 409, 418, 420, 515
Ecogeographic factors, 196
Ecogeographical rules, 147
Ecological
  isolation, 171, 173, 209 (*see also* Isolating mechanisms)
  niche, 3, 156, 168, 186
Ecospace, 247ff., 251, 338, 414, 416
Ecotype, 147ff.
Ectoderm, 403, 406
Ectoprocta, 399, 409, 416
Ediacaran, 398, 414, 416
Education, key to progress, 470
Effective population size, 157, 159, 306ff. (*see also* Population size)
Effector molecules, 25
Egg production, 38f.
Egotism, 126
Ehrharta, 225
Élan vital, 488
Electrophoresis, 47, 50, 52, 56, 62, 68, 133, 140, 192ff., 260, 284ff., 290

Elodea canadensis, 203
Elymus, 203f., 213, 225
Embryological development, 502f.
Emergent properties, 489
Empirical
content of a hypothesis, 481
falsification, 476, 479, 505, 507
observation, 478
prediction, 486
psychology, 478
research, 485
science, 475
testing, 478, 485ff.
world, 480
Enculturation, 443
End-directed features, 506 (*see also* Teleology)
Endoderm, 403, 406
Endoplasmic reticulum, 387
Endosymbiosis hypothesis, 389
Entelechy, 488, 494
Enteropneusts, 418
Entropy, 508
Environment, ability to perceive, 513ff.
Environmental
change, 66
hyperspace, 248
influence, 129
manipulation, 460
Enzymes, 18, 47, 133, 140, 144, 192 (*see also* Alleles; Gene; *and* Polymorphism)
Eocoelia, 328f., 332
Eocene, 319
Ephedra fragilis, 74
Epiceratodus, 330
Epifaunal, 414
Epistatic interaction, 29, 64, 188, 213
Epistemological reduction, 474, 488ff.
Epistemology, 478
Epoxide, 61
Equilibrium model of population density, 336
Equisetum, 228
Equus, 334
Erica, 266
Escherichia coli, 20, 23, 27, 59, 64, 69f., 72, 84f., 232, 381f., 386
Ethical
behavior, 456
being, 455
questions, 474
standards, 455
Ethics, 454ff.
Ethological isolation, 118, 171, 173, 177, 180f., 184f., 192f., 195, 203, 287 (*see also* Isolating mechanisms)
Ethology, 263
Ethyl methanesulfonate, 61
Euchlaena mexicana, 468
Eugenics, 459f., 463
Euglena, 371, 373, 387, 393, 514
Euglossine bees, 209
Eukaryote, 8, 83, 255, 361ff., 369, 371ff., 386ff., 401, 514
origin of, 384ff.
Eukaryotic regulatory systems, 258
Euphausia superba, 512
Euphenics, 459f.
Euphorbias, 266
Eurypterids, 348
Eusthenopteron, 431
Evolutionary
clock, 308f., 312f.
taxonomy, 270
world view, 438
Exobiology, 366, 368
Expansion of life, 511f.
Experimental law, 492
Extinction, 252, 342ff., 421, 424, 426ff., 437
of the human species, 441
rates, 342
Extraterrestrial life, 364, 366, 368

Facilitation, 116
Facultative methods of reproduction, 132, 231
Falciparum malaria, 110f.
Fallacy of affirming the consequence, 480
Falsifiability, 476, 479f. (*see also* Empirical falsification; Empirical testing; *and* Verifiability)
Falsification of universal statements, 480
Family ethics, 457
Fecundity, 39, 44, 47
Felis, 169, 267
Fertility, 63, 84, 188
Festuca, 38, 133, 203
Fetal hemoglobin, 79
Fibrinogen, 302
Fibrinopeptides, 302, 311f., 445
Ficus, 209, 221
Filter feeder, 418
Fish–amphibian transition, 431
Fitness, 32, 35, 39, 45ff., 66, 98–126, 153, 155, 163f., 168, 170, 174, 180, 242, 258, 305f., 505ff. (*see also* Adaptation *and* Selective value)
Flagella, 383, 388

Flagellate protozoan, 399, 403
Flatworm ancestor, 406, 415
Fleurantia, 330
Flexibility of human nature, 468ff.
Floral characters, 210
Forelimb of vertebrates, 264
Flowering plants, 210, 273
Fossil record, 263, 269, 399, 408, 414, 426
Founder effect, 157, 188
Frame-shift mutations, 60f., 65f.
France, 37
Frequency-dependent selection, 115ff., 156
Frieleia halli, 55
Functional design, 496 (*see also* Adaptation)
Fundamental Theorem of Natural
  Selection, 32
Future
  of evolution, 464
  of mankind, 464

Galago, 278f., 281, 283
Galápagos Islands, 12f., 186f.
Galeopsis tetrahit, 226
Gamete, 55, 170
Gametic
  pool, 157
  isolation, 171, 176 (*see also* Isolating
    mechanisms)
Gamma hemoglobin. *See* Hemoglobin.
Gel electrophoresis. *See* Electrophoresis.
Gene (*see also* Alleles)
  arrangement, 112, 134ff., 145 (*see also*
    Chromosome arrangements)
  coadaptation (*q.v.*)
  combinations, 3, 167f.
  diffusion, 160, 162
  duplication, 77ff., 92, 268, 302f. (*see also*
    Chromosome duplication)
  flow, 132, 160, 198 (*see also* Migration)
  frequency gradients, 145
  implantation, 469
  integrator, 256ff.
  lethal, 40, 63f., 69f.
  multiple copies, 78, 83
  phylogenies, 268f.
  pool, 4, 6, 30f., 35, 99f., 102, 104, 108,
    117, 126, 129, 132, 154, 157, 460
  regulation, 29, 84, 86, 256
  regulatory, 29, 56, 194, 256, 258, 260,
    401, 417
  responsible for isolating mechanisms, 194
  structural, 24f., 28ff., 47, 55f., 60, 83ff.,
    194, 255ff., 284
Generalizations, 482
Genesis, book of, 9
Genetic
  abnormalities, 460
  burden, 105ff., 153, 155, 181
  code, 23f., 27f., 50, 59, 262, 298,
    359, 361ff.
  counseling, 460
  death, 154, 163f.
  defects, 107, 460
  differentiation, 139f., 160, 163, 188ff.,
    284f., 288, 294
  diseases, 108, 460
  distance, 180, 192ff., 285, 290
  drift, 31, 108, 126, 157ff., 193, 246,
    305ff., 499, 507
  engineering, 462f., 469
  identity, 193f., 285
  information, 21, 23f., 262, 276, 511
  load, 111, 105f., 108, 163
  polymorphisms, 36, 117, 129
  programs of information, 502f.
  recombination, 4ff., 40, 44f., 88, 167, 170,
    202, 230, 232, 499, 501
  regulation, 255, 405
  revolution, 200
  structure of populations (*see* Balance theory
    of population structure *and* Classical
    theory of population structure)
  technology, 462
  uniformity, 117
  variation, 3f., 31ff., 46ff., 55, 70, 72, 98f.,
    105ff., 121, 135, 156, 188, 458,
    499, 501ff.
Genic
  disharmony, 211ff.
  sterility, 179, 211
Genome, 401, 417
Genome size. *See* DNA, content per cell.
Genotype, 28ff., 47, 98, 137, 255, 362
Gentians, 266
Geographic
  diversity, 128
  gradients, 137
  isolation, 12, 138
  race, 134, 146, 166
  speciation, 166, 201
  species formation, 166
  theory of speciation, 196
  variation, 161
Geological
  processes, 498
  time, 308, 315f.

Geology, 495
Geospizinae, 187
Geotaxis, 39
Geotropism, 515
Gibbon, 279, 281, 283, 291f., 442
Giemsa staining method, 77
Gila, 209, 221, 286f.
Glaucocystis, 386
Glaucomys, 267, 270
Gloger's rule, 147
Golden whistler, 137f.
Gonyaulax polyedra, 73
Gorilla, 291ff., 437, 442 (*see also* Primates)
Gossypium, 179, 216, 227
Grade, 236, 259, 435
Greek philosophers, 438
Green monkey, 281, 283, 293 (*see also* Primates)
Grillus domesticus, 74
Ground plans, 237, 242, 258f., 366, 413, 416f., 428
Group selection, 125ff. (*see also* Natural selection)
Gryphaea, 327

Habitat-correlated variations, 161
Habitat isolation, 171ff. (*see also* Isolating mechanisms)
Haldane's dilemma, 163f.
"Haldane's Law," 213
Halobacterium, 380
Haplohippus, 509
Haploidy, 58
Hardy–Weinberg Equilibrium, 99f., 105, 157, 224
Hedgehog, 278f.
Heme–heme interactions, 79
Hemichordata, 398, 409, 418ff.
Hemignathus lucidus, 249, 251
Hemoglobin, 78f., 111, 268, 444
  A, 62f., 79, 109, 111, 293, 444
  A2, 79f.
  alpha, 62, 79f., 266ff., 295, 303, 309ff., 444f.
  B, 65
  beta, 62f., 79f., 111, 266ff., 295, 303, 444f., 504
  delta, 79f., 266ff., 303
  deoxygenated, 63
  gamma, 79f., 266ff.
  genes, 79
  S, 63, 444, 499
Hemolytic disease, 142, 160
Heredity, 4, 18, 30ff, 57, 64, 57, 124, 451, 457
  blood theory, 98f.
Hermaphrodite animals, 75, 131, 170, 173
Hesperoleucus, 286f.
Heterodimer, 50, 81
Heterogametic sex, 52
Heterokaryotype, 112f., 116
Heterosis, 109, 115, 164, 205, 394
Heterotic selection, 111f., 116, 155, 163 (*see also* Natural Selection)
Heterotrophs, 360ff., 375
Heterozygosity, 51ff. (*see also* Genetic variation)
Heterozygous carriers, 111
Heuristic role of hypotheses, 485
Hieracium umbellatum, 207
Histones, 84, 391
Historical explanations, 495
Holocarpha, 196, 204
Holosteans, 80f.
Homeomorph, 327
Homeostasis, 116, 120, 508
  developmental, 106, 500
Homeostatic reaction, 502
Homo
  erectus, 442, 446ff., 475
  faber, 441
  habilis, 447f.
  sapiens, 69, 234, 291, 293, 438ff., 516
Homogeneous environments, 117
Homokaryotypes, 113f.
Homologous
  chromosomes, 87
  features, 270
  genes, 268
  polypeptides, 293
  proteins, 294, 298f.
  sequences, 279
Homology, 263ff., 276ff., 293ff., 303
Homosequential species, 192
Homozygosity, frequency of, 51ff. (*see also* Genetic variation)
Honeybee queen substance, 175
Honeycreeper, 186
Honeysuckle, 266
Horotelic rate, 336
Horse, 300f.
Housefly, 121
House sparrow, 281f.
Human
  DNA, 279, 282, 469
  evolution, 439

Human (*continued*)
  language, 451
  populations, 36, 112, 139, 152
  races, 135, 138f., 144, 163
  societies, 455, 457, 459
  species *(See Homo)*
Huntington's chorea, 68
Hyaenodonta, 335
Hybrid, 179, 212
  breakdown, 171, 177ff., 210, 213 (*see also* Isolating mechanisms)
  cell cultures, 212
  DNA, 276ff. (*see also* DNA)
  inviability, 109, 171, 177, 210ff (*see also* Isolating mechanisms)
  lethality, 181
  polyploids, 214
  speciation, 220
  sterility, 109, 171, 177, 180, 210 (*see also* Isolating mechanisms)
  swarms, 196, 220, 223
  vigor, 109
Hybridization, 157
  artificial, 130
  as an evolutionary catalyst, 219
  reaction, 276
Hydrogen bond, 21f., 61, 293
Hydrosphere, 351f.
Hyla, 75, 182
Hylobates lar, 291, 442 (*see also* Primates)
Hymenoptera, 207
Hypotheses
  heuristic role, 485
Hypothetico-deductive method, 477ff.
Hyracotherium, 7, 334, 509

Ice Age, 450
Ichneumonid wasp, 207
Ichthyostegids, 429ff.
Ictidosaurs, 433f.
Immanent laws, 495
Immune response, 491
Immunological distance, 290ff.
Immunological techniques, 260, 290ff., 303
Inborn genetic drive, 455
Inbreeding, 106
Incest, 106
Incipient species, 182, 287 (*see also* Species *and* Speciation)
Indeterminate teleology. *See* Teleology.
Inducible system, 255
Inductive method, 476ff.
Industrial melanism, 3, 72, 122, 146
Infanticide, 456
Infaunal mode of life, 414
Information content of a hypothesis, 481
Information processing, 514
Informational macromolecules, 263, 313
Insects, 273, 339, 431, 512
  colonial, 454
  species, 72
Insecticide, 107, 121f.
Insectivore, 436
Insemination reaction, 176
Instinct, 455
Insuline, 294, 312
Integrator genes, 256ff.
Interglacial period (Riss–Würm), 450
Interspecific
  copulation, 174
  hybrid, 178, 188
Introgression, 180, 219f., 223, 226
Introspection, 453
Invagination hypothesis, 385, 388
Inversion (*see also* Chromosome, inversion)
  heterozygote, 87ff., 112, 115
  polymorphism, 115, 200, 504
IQ, 101
Iris, 219
Irish elk, 244
Irradiated populations, 67f.
Irreversibility, 8, 508
Isolating mechanisms, 31, 129, 171ff., 194ff. (*see also* Reproductive isolation)
  development of, 180
  developmental sterility, 179, 211ff.
  evolutionary origin of, 170
  hybrid breakdown, 171, 177ff., 210
  hybrid inviability, 109, 171, 177, 210, 212
  hybrid sterility, 109, 171, 177, 180, 210
  mechanical, 171, 174
  postmating, 171, 181–182, 210
  premating, 171ff.
  seasonal, 171ff.
  segregational sterility, 211ff., 222
  sexual, 173, 185
  temporal, 171

Java Man, 449
Judeo-Christian thought, 438
Junco, 281f.
Jurassic, 80f., 319, 343

K, 337
KK, 337

Kangaroo, 299ff.
Karyotype, 103
Kenyapithecus, 446
Kin selection, 125ff. (*see also* Natural selection)
Klebsormidium, 374, 395

Lactose intolerance, 150ff.
Lambda phage (λ), 25, 381
Lamprey, 78, 418
Lamprocystis, 379
Language, 454
Latimeria, 430
Lathyrus, 75
Lavinia, 286f.
Law-like statements, 482f.
Law of conservation of organization, 513
Laws of biology, 491
Laws of nature, 485
Layia, 204, 218
Lemur, 279, 436, 444 (*see also* Primates)
Lepomis, 212
Lepidosiren, 74, 330
Lepore hemoglobin, 80
Leptolepiform, 80f.
Lethal
  chromosome, 42, 45f.
  gene, 40
  mutation, 63f., 69f.
Liliaceae, 207
Lingula, 241, 336
Lithosphere, 320f.
Lock-and-key theory, 174
Lolium, 38
Lonicera, 266
Lophophorates, 409, 413ff.
Loxops, 248f., 251
Lungfish, 73f., 330f., 429f.
Luxuriance, 177
Lycopodium, 228
Lysandra atlantica, 92
Lysozymes, 60, 292f.
  baboon, 292
  chimpanzee, 292
  gorilla, 292
  man, 292
  orangutan, 292

Macaca, 291, 293 (*see also* Primates)
Macromutation, 59
Madiinae, 217
Magnoliaceae, 228
Malaria, 111, 172, 183
Mammal, 56, 73f., 236, 240f., 260, 302, 311, 325, 332ff., 340, 346, 433ff.
Man, 38, 52, 62, 65, 69, 85, 194, 279ff., 291ff., 300ff., 310, 451
  ancestors of, 445
  ascendance of, 441
  biological animal, 466
  biologically unique, 462
  evolution of, 458
  intelligent and social being, 466
  lord of creation, 452
  place in nature, 440
  political animal, 440
  transformer of his environment, 466, 468
Marmota, 267
Marsupials, 265, 267
  anteater, 266
  Australian, 266f.
Mating
  advantage, 119f.
  advantage of minority males, 120
  call of tree frogs, 182
  display, 498
  ritual, 173
Mechanical isolation, 171, 174 (*see also* Isolating mechanisms)
Mechanistic doctrine, 439, 487f.
Medicago sativa, 225
Megaloceros, 244f.
Megaspore, 206
Meiosis, 88, 91, 124, 170, 177f.
  cost of, 392, 394
  divisions of, 88f.
  prophase of, 91
Melanoplus differentialis, 74
Melospiza melodia, 282
Mendelian
  genetics, 47, 192
  laws of ineritance, 490
  population, 30f., 132, 125, 137, 171, 195
  principles, 492
  recombination, 131f.
  segregation, 130, 167
Mental
  capacities, 450
  disabilities, 38
  disorders, 38
  retardation, 64
Mesenchyme, 406ff., 413
Mesocoel, 419
Mesoderm, 406, 408
Mesoglea, 403, 406
Mesokaryotic, 391f.
Mesozoic, 259, 316

Messenger RNA (mRNA), 25ff., 60f., 83f., 211, 255, 282
and polyribosome complexes, 84
Metabolism, aerobic, 388
Metacentric chromosome, 90, 93f.
Metacoel, 419
Metameria, 415
Metamerous architecture, 409, 411ff.
Metaphysical
knowledge, 476
questions, 474
speculation, 485
Metazoa, 256, 373, 393, 414
Methodological reductionism, 487, 490f.
Methodology of science, 475ff.
Metronomic clock, 308 (*see also* Molecular clock of evolution)
Metrosideros, 200
Microcomplement fixation, 291f.
Microgeographic race, 132ff., 146
Micropterus, 212
Middle Ages, 438
Migration, 31, 72, 126, 157f., 249, 251, 276
Milk diet, 150, 152
Mimosa, 515
Minnow fishes, 286f.
Miocene, 446
Mirounga angustirostris, 53
Missense mutation, 62, 66 (*see also* Mutation)
Mitochondria, 25, 298, 383ff.
Mitotic apparatus, origin of, 396ff.
MN blood group, 38 (*see also* Blood group)
Modus tollens, 480f.
Molecular biology, 444, 491
Molecular clock of evolution, 308f., 312f.
Molecular genetics, 17, 30, 36, 47
Molluscs, 73, 259, 333, 335, 392, 395, 413ff.
Mongoloidism. *See* Down's syndrome.
Monoecy, 37, 170
Monophyletism, 234ff., 239–240, 253
Monoplacophora, 413, 416
Monosomy, 58
Monotremes, 435
Morals. *See* Ethics.
Morphological
evolution, 326, 329f.
variation, 36f., 47, 59, 68
Morphospecies, 325, 330
Mosaic evolution, 331
Mouse, 69, 86, 124, 267, 278f., 309
Multicellularity, 30, 66, 370, 394
Mus, 69, 161, 267
Musca domestica, 37, 121
Musk deer, 175
Mustard gas, 59, 61
Mustelidae, 169
Mutation, 3ff., 15f., 31, 34, 57, 64ff., 70, 100, 102, 107f., 112, 129ff., 153, 158ff., 201f., 242, 258, 305, 307, 499, 501
and selection, 102
classification of, 57
distance, 299, 445
effects, 62
equilibrium, 100
pressure, 34, 156
rate, 59, 68ff., 103ff., 154, 158, 306, 308, 362
theory, 59
Mutator gene, 70f.
Mycoplasm, 388
Mylopharodon, 286f.
Myoglobin, 78, 269, 312, 445
Myrmecobius, 266f.
Myrmecophaga, 266f.
Myxoderma sacculatum ectenes, 55
Myxomycetes, 376, 395

Nagasaki, 107
Nannoplankton, 393
Natural laws, 495f.
Natural phenomena, 495
Natural selection, 4f., 11, 14ff., 31ff., 65ff., 84, 95–203 *(passim),* 305ff., 393, 443, 452, 455ff., 462, 485, 496ff. (*see also* Selection *and* Selective advantage)
critics of, 507
differential, 126
directional, 4, 108, 120ff.
disruptive, 117
diversifying, 4, 116f., 156, 163
is teleological, 502
theory of, 504ff.
theory of antibody function, 491
Natural
science, 474
teleology, 500, 503
theology, 496, 503
Neanderthal Man, 445f., 450, 454, 458
Nearchaster aciculosus, 55
Necturus maculosus, 74
Nemertea, 398, 407f.
Neo-Darwinism, 129
Neo-Lamarckism, 129
Neopallium, 515
Neopilina, 411
Neoteny, 260f., 407, 418
Nereis, 411
Neurospora, 59, 69, 72, 300f.

Neutral alleles, 154ff., 164, 305f., 308, 459
Neutrality theory of evolution, 130, 303ff., 499, 507
New World monkeys, 292 (*see also* Primates)
Nicotiana, 212, 221
Nitrocellulose membrane filter, 276f., 281
Nitrogen base, 21f., 61f.
Noctiluca, 380
Nomogenesis, 503
Non-Darwinian evolution. *See* Neutrality theory of evolution.
Nonsense mutation, 66
Norm of reaction, 29
Normalizing selection, 4, 96, 102, 105ff., 117ff., 154f. (*see also* Natural selection)
Nostoc, 378
Notch phenotype, 77
"Nothing but" fallacy, 488
Notemigonus, 286f.
Notoacmaea, 251
Notorycotes, 267
Nucleic acid. *See* DNA *and* RNA.
Nucleolar organizer, 83
Nucleolus, 83
Nucleotide, 21ff. (*see also* DNA and RNA)
  sequence, 47, 57, 60, 64, 83, 153, 276, 284, 293
  substitution, 60, 63, 65, 79f., 287, 295, 298, 301, 304, 312
  triplets, 153
Null hypothesis, 480
Nullosomic, 58
Numerical taxonomy, 237f., 241, 270f.

Ocean basins, 320, 322f.
Oceanic islands, 126, 161, 186, 188
Ocelli, 39
Oenothera, 15, 59, 92, 109, 204ff., 221
Oldowan industry, 447
Oligomerous architecture, 409, 411ff.
Omo, 447
Ontological reductionism, 488ff.
Oomycetes, 374, 395
Operator gene, 25, 29, 255
Operon, 29, 462
Ophacodon, 433
Ophioglossum, 92, 228
Ophiomusium lymani, 53, 55
Ophrys, 221
Orangutan, 282, 291ff., 302, 437, 442 (*see also* Primates)
Orchid, 176, 207ff., 221
Orchis, 209
Ordovician, 343
Origin of life, 9, 363
Origin of Species, 14, 31, 358, 439f., 484f., 495, 497
Original sin, 457
Orthodon, 286f.
Orthogenesis, 17, 242ff., 503
Orthologous gene, 268, 303
Orthologous protein, 303
Ostracoderms, 418, 429
Otiorrhynchus scaber, 72
Overdominance, 109 (*see also* Heterosis)

Pachycephala pectoralis, 137f.
Pacific ridge, 319
Paedomorphosis, 260
Paeonia, 213
Paleo-ecology, 324
Paleontological
  evidence, 269
  record, 308
  time, 312
Paleozoic, 316, 422
Pan, 94, 291ff., 442 (*see also* Chimpanzee)
Pangaea, 322, 425ff., 436
Pangenesis, theory of, 14
Panmictic population, 30, 99, 306
Panneutralism, 156ff (*see also* Neutrality theory of evolution)
Panspermia, 368
Panthera, 169
Papio, 291, 293 (*see also* Primates)
Paracentric inversion, 58, 88f. (*see also* Chromosome inversion)
Paracentrotus lividus, 211
Parallel evolution, 265ff., 270, 305, 326
Paralogous
  gene, 268f., 303
  protein, 303
Paramecium, 177, 514
Parapatric speciation, 198, 201f. (*see also* Speciation)
Parascaris equorum, 92
Parasitism, 31, 77, 346
Parental care, 127
Parthenogenesis, 72, 75, 125, 130ff., 170
Particulate heredity, 18
Partula, 201
Passer domesticus, 281f.
Passerella melodia, 137
Passerine birds, 186
Pathogenic bacteria, 122
Pedal waves, 407, 413, 416
Pedicularis, 209

Peking Man, 447
Pelecypod, 393
Pelodictyon, 377
Pelomyxa, 389f.
Pelvis,
  apes, 441
  man, 441
  man-ape, 441
Pelycosaurs, 433
Penguin, 299ff.
Pennsylvanian, 400
Penstemom, 209, 221
Pentlandia, 330
Peptide bond, 27, 83 (*see also* Amino acids *and* Protein)
Pericentric inversion, 58, 89f., 94 (*see also* Chromosome inversion)
Peristaltic locomotion, 408f., 414
Peritoneum, 408, 413
Permian, 334, 343, 425
Permian–Triassic boundary, 422
Peromyscus polionotus, 161
Pesticides, resistance to, 39, 72, 121f., 147
Petaurus, 267
Pgi gene duplication, 80f.
Phalaris, 225
Phaneropleuron, 330
Phanerozoic, 316, 333, 343, 422f.
Phasolomys, 267
Phaseolus, 200, 483
Phenetic classification, 237f., 270
Phenotype, 28ff., 47, 98, 242, 252, 362
Phenotypic variation, 98, 129
Phenylketonuria, 64, 104f., 460
Pheromones, 174f.
Philosophy of science, 474ff., 486f.
Phleum pratense, 38
Phlox, 209, 266
Phoronids, 409, 411, 413, 416
Phoronis, 411
Phosphate group, 22
Photodissociation, 353
Photosynthesis, 353, 362f., 365, 370, 377, 379, 417
Phototaxis, 39
Phyla, 370
  ages of, 315
Phylloxera, 130
Phylogeny, 234ff., 251, 262f., 271–304
  first tree, 243
  amphibians, 430
  carp and mammals, 310
  Deuterostomia, 420
  Drosophila, 272, 275, 288
  genes, 303
  globins, 79, 303f., 310
  lobe-finned fishes, 430
  man, 292
  pelycosaurs, 433
  primate species, 283, 292, 302
  the five kingdoms, 370
  therapsid reptiles, 433
Physical disabilities, 38
Physics, 492
Physiological homeostasis, 500
Phylogerontism, 343f.
Picture-winged *Drosphila,* 273, 276
Piltdown Man, 331
Pinus, 74, 172f., 205
Pipilo, 219
Pisum sativum, 482
Pithecanthropus erectus, 447
PKU, 64
Placental mammals, 76, 266f., 435f.
Planarian flatworms, 515
Plantago maritima, 205
Planula larva, 407
Plasmids, 389
Plasmodium falciparum, 110f.
Plastids, 386f.
Platanus, 223f.
Plate-tectonics, 322, 417, 427
Platyhelminthes, 406f.
Pleiotropism, 29
Pliocene, 446
Poa, 229f.
Pogonichthys, 286f.
Pogonophora, 325
Poikilothermic organisms, 339
Poisson distribution, 285, 290
Polar bodies, 88
Polarity differentiation, 395
Polemoniaceae, 209
Pollinator specificity, 209
Pollinium, 206
Polyhaploids, 226
Polymastigoid flagellates, 292
Polymers, 21
Polymorphism, 35, 100, 107ff., 128, 143, 146, 153ff. (*see also* Genetic variation)
  as adaptation, 146
  balanced, 164
  blood group, 140
  species, 325
Polynucleotide chains, 21, 23
Polypeptides, 25ff., 47
Polyploidy, 58, 75ff., 204, 212, 224ff.
  in animals, 228

and apomoxis, 229
in chrysanthemums, 224
complexes, 227
in cotton, 224
conservatism of, 224
in wheat, 224
Polyphyletism, 234ff., 408
Polyribosome, 84
Polytene chromosomes, 75ff., 87f., 112, 190, 271ff., 275
Polytypic species, 128, 138
Polytypism, 128f., 146, 153
Pongo, 291ff., 442, 446
Population, 30f., 128 (*see also* Mendelian population)
cage, 113, 121
genetics, 99
interaction with environment, 2ff.
size, 33
structure, 34
time bomb, 472
Porifera, 395, 402
Porphyry, 263
Porthetria dispar, 168
Postadaptation, 246
Postmating isolation, 171, 180f., 210 (*see also* Isolating mechanisms)
Potential falsifier, 480f.
Potentilla, 148f., 207, 229
Preadaptation, 125, 246, 249, 254, 346, 360, 426, 432, 443
Prebiotic synthesis, 354, 357, 359ff.
Precambrian, 378, 414, 420
Predation, 31, 127, 341, 346f., 405, 413, 426f., 431
Prediction, 481
Predictive theory of evolution, 130
Preferential pairing, 214, 216, 226
Premating isolation, 171ff. (*see also* Isolating mechanisms)
Presbytis entellus, 291
Priapuloida, 398
Primary science, 492f.
Primates, 303, 435f.
African, 291f.
Asiatic, 292
baboon, 278, 292f., 302
capuchin, 279, 283
catarrhine, 436f.
chimpanzee, 94, 194, 260, 278ff., 302f., 437, 442
cranial capacity in higher p., 442
DNA, 278, 283
galago, 278f., 281, 283
gibbon, 279, 281, 283, 291f., 442
gorilla, 291ff., 437, 442
green monkey, 281, 283, 293
macaque, 291, 293
orangutan, 282, 291ff., 302, 437, 442
rhesus, 278f., 283, 293, 302–303
Primitive
character complex, 330
coelomates, 414f.
culture, 475
mollusc, 411
unicellular eukaryote, 384
vascular plants, 73
Primrose, 109, 266
Primula, 204, 212, 266
Principle of independent assortment, 483
Principle of segregation, 482
Probability of fixation, 306f., 310
Progress, 121, 236, 475, 507ff.
axiological standards of, 509
definition, 509
in the medical sciences, 459
Prokaryote, 230, 232, 255, 361ff.
Prokaryote–eukaryote transition, 388
Promoter sites, 25
Prosimians, 436f.
Protein, 47
change, 313
content, 38f.
differentiation, 487
evolution, 306, 308f.
functional properties of, 161
phylogenies, 299
primary structure, 284, 293f., 299, 301
quarternary structure, 293
secondary structure, 284, 293
sequencer, 294
sequencing in the study of phylogeny, 313
synthesis, 83f.
tertiary structure, 284, 293
variation, 50, 53, 499 (*see also* Genetic variation)
Protenoids, 357
Protista, 371, 373f., 376, 395
Protocell, 361f., 364f.
Proto-cephalochordates, 420
Protocoel, 419
Proto-molluscan, 415
Protoporphyrin ring, 78
Protopterus, 330
Prototrophic organisms, 59
Protozoa, 73, 371, 389
Province, 341, 345, 347, 425ff.
Provincial boundaries, 341

Prunus avium, 38
Pseudocoelom, 413
Pseudocopulation, 176, 207
Pseudodiversification, 332
Pseudonestor, 249, 251
Pseudoextinction, 332, 342, 348, 424
Pseudometameria, 415
Pseudometamerous architecture, 407, 413f.
Pseudo-wild strain, 60f.
Psilopsida, 73
Psilotum, 228
Psittirostra, 248f., 251
Pteraster jordani, 55
Pteriodophytes, 394
Pterobranchs, 418
Ptychocheilus, 286f.
Pure lines, 130ff.
Pure race, 132, 139
Purine, 21f., 60ff.
Purposeful behavior, 95, 500, 503
Purshia, 219f.
Pygmies, 150
Pyrimidine, 21f., 60f.

Quadrupedal locomotion, 440, 442
Quantitative inheritance, 16
Quantum mechanics, 492
Quantum speciation, 197f., 201f., 217 (*see also* Speciation)
Quasinormal chromosome, 41ff.
Quaternary, 319
Quercus, 196, 219, 223
Quinacrine fluorescent staining, 94
Quinacrine mustard, 77

Rabbit, 281, 300ff., 309f.
Race, 128, 135ff., 162ff., 184
  boundaries, 137, 145
  classification, 144f.
  differences, 144ff., 153, 160, 162
  formation, 139
  human, 135, 138ff., 163, 450
  labels, 145
  senescence, 343
  studies, 144
  traits, 153
  types, 139
  variation, 139, 144f., 163
Radial energy, 488, 513
Radiation, 71, 241, 247, 249, 252, 259, 326, 348, 399, 413ff., 429
  coelomate, 415
  of body plan, 399
  of phyla and classes, 415
Radioactive decay, 309
Radish–cabbage hybrid, 178f.
Ramapithecus, 446
Rana, 177, 216, 311
  Random drift. *See* Genetic drift.
Random mating, 99, 119, 155
  equilibrium, 53
  population, 51, 306
Random process, 157
Random sampling, 47, 50, 56, 305
Range of reaction, 29
Ranunculaceae, 207
Raphanobrassica, 178f.
Raphanus sativus, 178, 226
Rapid speciation, 188 (*see also* Speciation)
Rate of
  adaptation, 34, 70
  allelic substitution, 306ff.
  amino acid replacement, 305, 309f., 486
  appearance of families, 333f.
  appearance of genera, 334
  cladogenesis, 336
  development, 39, 44, 47
  energy flow, 513
  evolutionary change, 305f.
  evolution, 67, 71f., 286, 305ff., 327f., 331f., 335
  gene substitution, 164
  immigration, 336
  local extinction, 336
  molecular evolution, 313
  mutation, 68ff., 158
  neutral mutation, 306
  nucleotide change, 281ff.
  nucleotide substitution, 305, 309, 312
  progress, 510
  protein evolution, 312
  substitution of selective alleles, 308
Rattlesnake, 299ff.
Receptor gene, 256ff.
Recessive, 29
  defect, 106
  gene, 41, 44
  lethal, 70
  mutant, 37
  trait, 482
Reciprocal translocation. *See* Chromosome translocation.
Recombinant chromosome, 44
Recombination. *See* Genetic recombination.
Recombinational speciation, 180, 221f.
Red algae, 374, 386f., 395
Reductionism
  epistemological, 474, 488f., 490ff.

methodological, 287, 490f.
of Mendelian genetics to molecular genetics, 495
of one science to another, 491ff.
of thermodynamics to statistical mechanics, 492
ontological, 488ff.
Redundancy of the genetic codes, 23, 47, 284
Regionated oligomer, 415
Regulation of diversity, 337
Regulatory genes, 29, 56, 194, 256, 258, 260, 401, 417
Relative fitness. *See* Fitness.
Relativity, 492
Religion, 439, 455
Renaissance, 438
Repetitive DNA. *See* DNA, repetitive.
Replication of DNA. *See* DNA, replication.
Repressible system, 255
Repressor, 255
Reproductive
community, 30f., 131ff. (*see also* Mendelian population)
efficiency, 39, 124f.
fitness, 118, 504 (*see also* Fitness)
isolation, 7, 137, 170, 176–196, 201ff., 232, 249 (*see also* Isolating mechanisms)
modalities, 132
structures of flowering plants, 266
success, 498f., 501, 504 (*see also* Fitness)
Reptile, 56, 73f., 240f., 299, 336, 339, 430ff.
Reptilian skull, 434
Resistance to pesticides, 39, 72, 121f., 147
Reticulate evolution, 204
Retrogression, 509
Rhesus monkey, 278f., 283, 293, 302f. (*see also* Primates)
alleles, 160
blood group gene, 142
DNA, 278f.
Rhipidistians, 429ff.
Rhizome, 392
Rhodomicrobium, 379
Rhodopseudomonas, 379
Rhodospirillum, 379
Ribonucleotide, 25
Ribose, 21
Ribosome, 25f., 83f., 383
Ribosomal protein, 83
Ribosomal RNA (rRNA), 25
Richardsonius, 286f.
Riss glaciation, 450
RNA, 20f., 24, 30, 211f., 360
activator, 256
polymerase enzyme, 25
ribosomal, 25, 84
sequence, 83
virus, 73
"Robertsonian" change, 92 (*see also* Chromosome fission *and* Chromosome fusion)
Rosaceae, 207, 228
Rubus, 229f.
Rumina decollata, 53
Rye grass, 38

Saccharomyces, 299ff.
Sagenodus, 330
Saguinus oedipus, 293
Saimiri sciureus, 293
Salmonella typhimurium, 69
Salvia, 176
Sampling errors, 157, 305 (*see also* Genetic drift)
Saprophytic way of life, 77
Sarcodina, 389
Satellite DNA, 86
Scaphiopus holbrookii, 74
Scholastic philosophers, 476
Science, definition of, 476
Scientific hypothesis, 476ff., 480f.
Scientific method, 475ff., 484
Scrophulariaceae, 209, 266
Scyphozoa, 404f.
Sea level, 318, 323, 345, 426ff.
Seasonal
changes, 114, 134, 507
isolation, 172f. (*see also* Isolating mechanisms)
Sea urchin, 74, 85ff.
Secondary science, 492f.
Second law of thermodynamics, 505
Segmentation, 408f., 413, 415
Segregational sterility, 211ff., 222 (*see also* Isolating mechanisms)
Selection (*see also* Natural selection)
against recessives, 101
coefficient, 101, 104, 115, 158f., 307f.
and dominance, 102
rate, 121
Selective advantage, 112, 124, 164
Selective value, 100f. (*see also* Adaptation *and* Fitness)
Self-awareness, 453f.
Self-fertilization, 131f., 170, 173, 204, 482
Selfish behavior, 127

Self-pollinating species, 131
Semiconservative model of DNA replication, 23
Semilethal chromosome, 41ff.
Semispecies, 182, 184f., 192f., 196, 198, 205, 223, 287f.
  definition, 184
  of *Drosophila paulistorum,* 185, 288
  in plants, 221
Senecio virgaurea, 207
Sensor genes, 256–261
Septum, 411
Sequinator, 294
Sequoia, 2, 166
Serial
  homology, 266, 268
  repetition, 406
Seriated
  flatworms, 415
  proto-coelomates, 415
Serpulids, 409
Sex-linked genes, 31, 37, 50ff.
Sex pheromones, 175
Sex ratio, 124f.
Sexual
  behavior, 289
  caste, 127
  cycle, adaptive value, 392
  dimorphism, 118
  isolation, 173, 185 (*see also* Isolation mechanisms)
  maturity, 443
  nonhermaphroditic species, 132
  outbreeding species, 132, 138
  recombination, 129, 232
  reproduction, 99, 131f., 170
  selection, 118f., 156, 163
Sexuality, 132
Shallow-water marine invertebrates, 341
Siamang, 291f. (*see also* Primates)
Sibling species, 165, 182ff., 190, 193f., 196, 204, 216, 260, 287, 325
  of mosquitoes, 172
  of *D. willistoni,* 183, 289
Sickle-cell anemia, 63, 109f.
Sign (vs. symbol), 451
Silurian, 334
Sipunculids, 408, 411, 413
Sipunculus, 411
Sitta, 182
Skin
  cancer, 148
  color, 139, 148f.
  pigmentation, 144
Slow loris, 279 (*see also* Primates)
Snail, 73, 337
Snapdragon, 188
Social
  Darwinism, 98
  hierarchy, 127
  insects, 126f.
  institution, 456
  organization, 508
  science, 495
Soft-bodied fossil, 397
Solenobia, 228, 230
Solidago sempervirens, 207
Somatic
  cell, 72
  hybrid vigor, 177
Song sparrow, 137
Sophophora, 93
Soul, 439
Spacer sequences of DNA, 83
Spalax, 202
Spatial heterogeneity, 337f., 344f.
Specialized form, 247, 253
Speciation, 170, 190ff., 288, 347f.
  differences between animals and plants, 202
  in plants, 204
  on oceanic islands, 186
  parapatric, 198, 201f.
  quantum, 197f., 201f., 217
  stasipatric, 201
  sympatric, 166, 198, 202
Species, 31, 107, 166, 168, 170, 180, 184, 194
  definition, 171, 230
  allopatric, 176
  capacity, 337, 342, 344, 346ff., 425., 428
  continental, 186, 188
  differences in allozymes, 192
  genetic differences, 160, 163, 188ff., 285, 294
  hybrid, 179f.
  origin, 165
  phylogeny, 268
  problem, 166
  suicide, 465
Spermatozoa, 72, 89
Sperm bank, 461f.
Sphenodon, 336
Spizella arborea, 282
Sponges, 73, 374, 393, 398, 402f., 515
Spontaneous generation, 350
Sporozoa, 392
Sprekelia formosissima, 74
Squamosal bone, 434
Stabilizing selection, 117 (*see also* Natural selection)

Stasigenesis, 236, 241
Stasipatric speciation, 201
Statistical mechanics, 493f.
Statistical methodology of testing hypotheses, 480
Steinheim, 450
Sterile
  caste, 127
  female, 126
  individual, 154
Sterility, 63, 181, 218
  in artificial hybrids, 218
  developmental, 179, 211ff. (*see also* Isolating mechanisms)
  of male hybrids, 192, 287
Stipa, 225
Stochastic clock. *See* Molecular clock of evolution.
Stochastic process, 65, 157, 164, 501 (*see also* Neutrality theory of evolution)
Stream of consciousness, 453
Streptomycin resistance and sensitivity, 66
Stromatolite, 377ff.
Strongylocentrotus purpuratus, 85f., 211
Strophosomus capitatus, 72
Structural genes, 24f., 28ff., 47, 55f., 60, 83, 85, 87, 194, 255ff., 284
Struggle for life, 98
Subduction, 320, 322
Submetacentric chromosome, 90, 94
Subsidiary hypothesis, 481
Subspecies, 128, 134, 137, 145, 165, 184, 193f., 198, 200, 223, 287
"Substitutional" genetic load, 163f. (*see also* Genetic load)
Subvital chromosomes, 41ff.
Supergenes, 115, 200, 205, 217
Superorganic (culture), 451
Supervital chromosomes, 41, 43, 106
Suppression, genetic, 64
Suppression of seriation, 414
Surplus individuals, 127
Survivorship curves, 335
Suspension feeders, 401ff., 414, 427f.
Swanscombe, 450
Sympatric speciation, 166, 198, 202 (*see also* Speciation)
Sympatry, 251
Symbol, 451, 454
Symbiosis hypothesis, 387ff.
Sympetalae, 266
Symphalangus syndactylus, 291
Syndinium, 391
Synergism, 116
Syngamy, 170
Synthetic
  chemicals, 122
  theory of evolution, 5, 17, 129
Systematic
  category, 165
  organization of knowledge, 476
  mutation, 59

Tachytelic rate, 336
t alleles, 124, 506
Talpa, 267
Taraxacum, 229
Tasmanian wolf, 267, 270
Taste-blindness polymorphism, 108
Tautology, 479
Tautomeric shift, 61f.
Taxa, 237ff., 241, 245ff., 260, 331, 334, 342, 344, 348, 421, 428
Taxonomic categories, 234, 247, 399, 428
  appearance of, 343, 423
Taxonomic hierarchy, 233f.
Teleological explanation, 474, 495ff.
  definition, 498
Teleological feature, 500ff.
Teleology, 95, 495ff.
Teleonomy, 501
Teleost fish, 74, 76, 80f.
Telic phenomena, 498
Telocentric chromosome, 92
Temperature-sensitive lethals, 64
Temperature stability of DNA, 280ff.
Temporal isolation, 171 (*see also* Isolating mechanisms)
Terminator codon, 23, 27, 60, 64
Territorial defense, 455
Territorial imperative, 457
Terrestrial mammal, 292, 330
Tetracentron, 228
Tetramer, 50, 62
Tetraodon fluviatilis, 74
Tetraploidy, 58, 179
Tetrapods, 430
Tetrasomy, 58
Thelytoky, 230
Thermal stability, 280ff.
Thermodynamics, 493f.
Thiospirillum, 379
Thiothrix, 380
Threat display, 455
Thymine, 21f., 25, 62
Thylacinus, 267, 270
Time-phyletic rate, 332
Tissue grade, 399
Tmesipteris, 228
Toad, 74

Tool-making and tool-using, 452
Trace fossils, 324, 397
Transcription, 24ff., 28, 84, 262, 359, 362
Transduction, 232
Transient polymorphism, 108 (*see also* Polymorphism)
Transfer RNA (tRNA), 25ff., 64, 84, 153, 360
  alanine, 26
  aminoacyl, 27
  genes, 84
Transformation, 20, 232
Transforming principle, 462
Transition, 60f., 65
Translation, 24ff., 262, 359, 361, 363, 365
Translocation. *See* Chromosome translocation.
Transposition, 58, 90
Transspecific evolution, 5ff., 228, 233
Transversion, 60f., 65, 70
Tribolium, 120
Trifolium, 38, 225
Trillium, 217
Trilobites, 327, 334, 337, 398, 416
Triplet. *See* Codon.
Triploblastic organisms, 406–408, 413
Triploidy, 58
Trisomy, 58
Triticum aestivum, 226, 468
Trochilidae, 207
Trochodendron, 228
Truncation selection, 164 (*see also* Natural selection)
Trypsin, 81, 294
Trypsinogen, 81, 83
Tunicates, 418
Turbellarians, 407
Turbidimetry, 290
Turtle, 299ff.
Type I and Type II errors, 480
Typological mode of thought, 139
Tyrannosaurus, 435

Ulotrichales, 374, 395
Ultraviolet radiation, 61
Umbelliferae, 207
Unicellular
  organism, 68, 170
  eukaryote, 370
Unified theory of inheritance, 492
Unirama, 327
Unit of information, 23
Universality of the genetic code, 28
Universal statements, 477, 480
Universal truth, 477
Uracil, 21, 25, 28, 61
Urochordata, 260, 409, 418f.
Urodela, 74
Uronemus, 330
Urostyla caudata, 73
Urisidae, 169
Use and disuse, 30
Useful variation, 32, 102, 107

Values, 454, 456
Value systems of society, 455
Vanda, 209
Variety, 166
Vaucheriales, 395
Venus flytrap, 515
Verifiability, 479
Vertebrates, 56, 73, 339, 417, 420, 515
  brain, 515
  serum, 291
  species, 53
Viability, 39ff., 44, 47, 63f., 70, 84
  of homozygotes, 41, 45
Vicia faba, 23
Viola, 188f.
Viruses, 20f., 68, 73
  evolution of, 380f.
Vitamin D., 148
Vitamin K, 122
Vitalism, 96, 439, 487f., 493f.
Viverridae, 169
Volvocales, 376, 395
Volvox, 376, 399

Warfarin, 122
Watusi, 150
Wheat, 62, 65, 298f.
Wild-type allele, 34f.
Winteraceae, 228
Wisconsin glaciation, 446
World geography
  250 million years ago, 425
  380 million years ago, 424
Würm glaciation, 446, 450

Xavante indians, 467
X chromosome, 31, 50, 69f., 77f., 93, 124
Xenopus, 83ff.
X-rays, 34, 59, 66
Xylocopa, 207

Y chromosome, 84, 124

Zea mays, 38, 69, 106, 468
Zinjanthropus boisei, 447
Zonotrichia albicollis, 282
Zygomycetes, 395